THE UFAW HANDBOOK
ON THE CARE AND MANAGEMENT
OF LABORATORY ANIMALS

The UFAW Handbook on the Care and Management of Laboratory Animals

Edited by UFAW

Foreword by

C. W. Hume,
O.B.E., M.C., B.Sc.

FIFTH EDITION

CHURCHILL LIVINGSTONE
EDINBURGH LONDON AND NEW YORK
1976

CHURCHILL LIVINGSTONE
Medical Division of Longman Group Limited

Distributed in the United States of America by
Longman Inc., 19 West 44th Street, New York,
N.Y. 10036 and by associated companies,
branches and representatives throughout
the world.

First Edition   1947
Second Edition   1957
   Reprinted   1959
Third Edition   1967
Fourth Edition   1972
Fifth Edition   1976

ISBN 0 443 01404 3

Library of Congress Cataloging in Publication Data
Universities Federation for Animal Welfare.
   The UFAW handbook on the care and management of
laboratory animals.

   Includes bibliographies and index.
   1. Laboratory animals. I. Title.
SF406.U54 1976   636.08'85   76-8407

Printed in Great Britain by
T. & A. Constable Ltd., Edinburgh

Foreword

By Major C. W. Hume, o.b.e., m.c., b.sc.
President and Founder of UFAW

UFAW was founded in 1926 as the University of London Animal Welfare Society. Its work expanded and in order to allow of a wider membership the Universities Federation for Animal Welfare was formed in 1938 with ULAWS as its first branch.

Since then UFAW's most important achievement has been on behalf of laboratory animals. This success was due largely to the co-operation of many scientists, including biologists, and to the publication in 1947 of the first edition of the *UFAW Handbook on the Care and Management of Laboratory Animals*. Subsequent editions in 1957, 1967 and 1972 further advanced UFAW's influence in promoting the humane care of laboratory animals and gave it an international status. The demand for this book has been such that the fifth edition can now be published. It is very fitting that its appearance should coincide with the celebration of UFAW's fiftieth anniversary.

The forewords to the first, second and third editions were written by distinguished scientists: Sir Thomas Dalling, Sir Harold Himsworth and Sir Peter Medawar, all of whom have paid generous tribute to UFAW's ideals. The Council of UFAW have honoured me by asking me to contribute a foreword to the fourth and now to the fifth edition. I make no claim to any particular contribution to biology but as UFAW's founder and President I feel proud indeed that UFAW continues to have provided for nearly 30 years a standard publication on the welfare of the laboratory animals that have contributed so much to the welfare of human beings.

Today there is no longer any need to argue the case for the technical value of good animal husbandry, but a word may be in place as to the reason why this work is published by an organization that exists for promoting the welfare of animals for their own sakes. The motive will be readily understood by anybody who has seen both good and bad husbandry. It fortunately happens that the animals most suitable for scientific work are those that are healthy, tame, comfortable and contented, and that wastefulness or irresponsibility in dealing with them is as discreditable on scientific as on humanitarian grounds. We can feel sure that along with increasing knowledge and understanding of the physical requirements of the animals, of their mentality, and of the effects on them of their environment, will go increasing consideration and sympathy for them and a growing determination to use them economically and compassionately and only in situations where other scientific procedures cannot provide valid results.

This work has been written by experts in the various fields of laboratory-animal science, and edited by the staff of UFAW. It seeks to put the case for the humane treatment of laboratory animals in a rational way. UFAW does not necessarily sanction all the procedures that are described and hopes that as knowledge accumulates alternative and more humane methods may be developed where necessary.

UFAW is most grateful to the many contributors to this *Handbook* for the work which they have undertaken, to members of the Editorial Board for their planning, and to the members of the staff concerned with the production, particularly Mrs. Catherine Brockhurst for assistance with the general editing, Mrs. Joan Lee for coping with the arduous task of typing and re-typing manuscripts, and Mr. Michael Prichard, A.R.P.S. and Mr. Michael Cope, for the line drawings and many of the photographs.

1976   C.W.H.

List of Contributors

ADAMS, C. E.
  Agricultural Research Council, Animal Research Station, 307, Huntingdon Road, Cambridge.

ASHBY, G. J.
  Deceased.

BALL, D. J.
  Zoological Society of London, Regent's Park, London, N.W.1.

BARNETT, S. F.
  University of Cambridge, Department of Animal Pathology, School of Veterinary Medicine, Madingley Road, Cambridge.

BELLAIRS, A. d'A.
  Department of Anatomy, St. Mary's Hospital Medical School (University of London), Paddington, London, W.2.

BLACKMORE, D. K.
  Faculty of Veterinary Science, Massey University, Palmerston North, New Zealand.

BLEBY, JOHN
  Laboratory Animals Centre, M.R.C. Laboratories, Woodmansterne Road, Carshalton, Surrey.

BOTERENBROOD, ELZE
  Hubrecht Laboratory of the Royal Netherlands Academy of Sciences and Letters, Uppsalalaan 1, Universiteitscentrum De Uithof, Utrecht, The Netherlands.

BOWEN, W. H.
  National Institute of Dental Research, Westwood Building, Room 532, Bethesda, Maryland 20014, U.S.A.

CHESTERMAN, F. C.
  Deceased.

COATES, MARIE E.
  The National Institute for Research in Dairying, Shinfield, Reading, Berks.

COID, C. R.
  Clinical Research Centre, Animal Division, Watford Road, Harrow, Middlesex.

COLE, J. H.
  Department of Agronomy & Applied Entomology, Huntingdon Research Centre, Huntingdon.

COOPER, D. M.
  Houghton Poultry Research Station, Houghton, Huntingdon.

DINSLEY, MARJORIE
  Barnet College, Wood Street, Barnet, Herts.

EISENBERG, J. F.
  National Zoological Park, Washington D.C. 20009, U.S.A.

FALCONER, D. S.
  Institute of Animal Genetics, West Mains Road, Edinburgh 9, Scotland.

FESTING, MICHAEL F. W.
  Laboratory Animals Centre, M.R.C. Laboratories, Woodmansterne Road, Carshalton, Surrey.

FIENNES, R. N. T-W-
  Zoological Society of London, Nuffield Institute of Comparative Medicine, Regent's Park, London, N.W.1.

FRAZER, J. F. D.
  The Nature Conservancy, 19, Belgrave Square, London, S.W.1.
HAMMOND, J. JNR.
  Cambridge University, School of Agriculture.
HARRISON, F. A.
  A.R.C. Institute of Animal Physiology, Babraham, Cambridge.
HARRY, E. G.
  Houghton Poultry Research Station, Houghton, Huntingdon.
HIME, MALCOLM J.
  The Hospital, Zoological Society of London, Regent's Park, London, N.W.1.
JORDAN, A. M.
  Tsetse Research Laboratory, Department of Veterinary Medicine, Langford House, Langford, Bristol.
KEYMER, I. F.
  Zoological Society of London, Regent's Park, London, N.W.1.
LANE-PETTER, W.
  University of Cambridge Animal Holding & Breeding Unit, Laundry Farm, Barton Road, Cambridge.
MACARTHUR, JUDY A.
  Universities Federation for Animal Welfare, 8 Hamilton Close, South Mimms, Potters Bar, Herts. EN6 3QD.
MAHONEY, C. J.
  Wisconsin Regional Primate Research Center, University of Wisconsin, 1223 Capitol Court, Wisconsin 53706, U.S.A.
MARSTON, J. H.
  Department of Anatomy, University of Birmingham Medical School, Birmingham 15.
NASH, T. A. M.
  Tsetse Research Laboratory, Department of Veterinary Medicine, Langford House, Langford, Bristol.
PAYNE, L. N.
  Houghton Poultry Research Station, Houghton, Huntingdon.
RACEY, P. A.
  Department of Zoology, University of Aberdeen, Tillydrone Avenue, Aberdeen.
RAY, PHYLLIS M.
  Universities Federation for Animal Welfare, 8 Hamilton Close, South Mimms, Potters Bar, Herts. EN6 3QD.
REDFERN, R.
  Pest Infestation Control Laboratory, Ministry of Agriculture, Fisheries & Food, Hook Rise South, Tolworth, Surbiton, Surrey.
ROBINSON, M. H.
  M.R.C. Laboratories, Laboratory Animals Centre, Woodmansterne Road, Carshalton, Surrey.
ROCHFORD, REV. P. J., O.S.B.
  Ampleforth College, York.
ROWE, F. P.
  Pest Infestation Control Laboratory, Ministry of Agriculture, Fisheries & Food, Hook Rise South, Tolworth, Surbiton, Surrey.
SCOTT, PATRICIA P.
  Royal Free Hospital School of Medicine, University of London, 8 Hunter Street, London W.C.1.

SCOTT, W. N.
> Universities Federation for Animal Welfare, 8 Hamilton Close, South Mimms, Potters Bar, Herts. EN6 3QD.

SPARROW, STEPHEN
> MRC Laboratories, Laboratory Animals Centre, Woodmansterne Road, Carshalton, Surrey.

SPENCER, K. E. V.
> Pharmacology Dept., Smith, Kline & French Laboratories, Welwyn Garden City, Herts.

STONES, L. C.
> Cooper Technical Bureau, Cooper Research Station, Berkhamsted Hill, Berkhamsted, Herts.

TAVERNOR, W. D.
> Surgery Department, Royal Veterinary College, Hawkshead House, Hawkshead Lane, North Mimms, Hatfield, Herts.

TOWNSEND, G. H.
> M.R.C. Laboratories, Laboratory Animals Centre, Woodmansterne Road, Carshalton, Surrey.

TREXLER, P. C.
> Department of Pathology, Royal Veterinary College, Royal College Street, London, N.W.1.

VERHOEFF-DE FREMERY, ROMEE
> Hubrecht Laboratory of the Royal Netherlands, Academy of Sciences and Letters, Uppsalalaan 1, Universiteitscentrum De Uithof, Utrecht, the Netherlands.

VEVERS, H. G.
> Zoological Society of London, Regent's Park, London, N.W.1.

WEIR, BARBARA J.
> Zoological Society of London, Wellcome Institute of Comparative Physiology, Regent's Park, London, N.W.1.

WHEELER, MARSHALL R.
> University of Texas at Austin, Austin, Texas 78712, U.S.A.

WRIGHT, C. A.
> Department of Zoology, British Museum (Natural History), Cromwell Road, London, S.W.7.

Contents

1 U.K. legislation relevant to the keeping of laboratory animals

P. M. RAY AND W. N. SCOTT

The principal Act of Parliament which regulates the use of animals for experiment is the Cruelty to Animals Act of 1876. There are, however, several other Acts which touch on the subject to a greater or less degree, and which can apply to those animals not actually under experiment, e.g. breeding stock. These will also be described in this chapter.

Cruelty to Animals Act, 1876

This Act regulates the carrying out, on living vertebrate animals (excluding man), of experiments calculated to cause pain. Such experiments may only be carried out at places registered for this purpose by the Home Secretary and by persons holding the necessary licence and certificates.

When a licence is used by itself, the following restrictions must be observed.

1. The animal must throughout the experiment be under the influence of an anaesthetic of sufficient power to prevent the feeling of pain. Urari, curare and similarly paralysing drugs are specifically excluded from being used as anaesthetics; they may not be used in experiments without the special permission of the Home Secretary.
2. The animal must be killed at the end of the experiment while it is still anaesthetised.
3. The experiment must be performed with a view to the advancement by new discovery of physiological knowledge, or of knowledge which will be useful for saving or prolonging life or alleviating suffering.
4. The experiment must not be performed for the purpose of attaining manual skill, nor to illustrate lectures in medical schools, etc.

Some of these restrictions may be lifted if the appropriate certificate is held.

For example, an anaesthetic is unnecessary for some experiments; the possession of certificate A allows for them to be carried out without anaesthetisation. Such experiments are subject to a limitation condition which states that no operative procedure more severe than simple inoculation or superficial venesection may be adopted. Under certificate B animals can be allowed to recover from an anaesthetic, provided they are killed as soon as the object of the experiment has been obtained, while certificate C permits the use of anaesthetised animals in demonstrations to students or to learned societies, provided the animal is killed immediately afterwards either by, or in the presence of, the licensee.

Horses, asses and mules cannot be used for any experiment unless certificate F has been granted; and cats and dogs cannot be used for experiments under certificates A or B unless the respective certificates E or EE are also held.

Probably the most important condition attached to licences is that known as the pain condition. This states that:

(a) If an animal at any time during any of the said experiments is found to be suffering pain which is either severe or is likely to endure, and if the main result of the experiment has been attained, the animal shall forthwith be painlessly killed;

(b) If an animal at any time during any of the said experiments is found to be suffering severe pain which is likely to endure, such animal shall forthwith be painlessly killed;

(c) If an animal appears to an Inspector to be suffering considerable pain, and if such Inspector directs such animal to be destroyed, it shall forthwith be painlessly killed.

Report of the Departmental Committee on Experiments on Animals

Ever since its introduction the 1876 Act has been criticised as being insufficiently far-reaching in its requirements. In 1963 the Government set up a Departmental Committee under the Chairmanship of Sir Sydney Little-wood 'to consider the present control over experiments on living animals, and to consider whether, and if so what, changes are desirable in the law or its administration'. The Committee submitted its Report in 1965 and made 83 recommendations, some calling for new legislation, some for improved administration, and others for the giving of statutory force to requirements already imposed by the Home Office. Regrettably, these recommendations have never been implemented.

Dogs Act, 1906

This Act, which is mainly concerned with the disposal of strays, prohibits the giving or selling 'for the purposes of vivisection' of any dog seized by the police.

Protection of Animals Act, 1911

This contains provisions for the protection of both domestic and captive animals. Acts amounting to cruelty include:
(a) cruelly to beat, kick, ill-treat, over-ride, over-drive, over-load, torture, infuriate or terrify any animal;
(b) to cause unnecessary suffering by doing, or omitting to do, any act;
(c) to convey or carry any animal in such a way as to cause it unnecessary suffering;
(d) to perform any operation without due care and humanity;
(e) the fighting or baiting of any animal or the use of any premises for such a purpose;
(f) the administering of any poisonous or injurious drug or substance to any animal.

Apart from the very wide specific instances defined in the Act as positive acts amounting to cruelty, it should be noted that it is also an offence to cause suffering by omission, such as by not killing or otherwise treating an animal in pain. There may be cruelty even when someone with a perfect right to kill an animal does it in a cruel and painful way. So if a man starts to kill an animal he must do so with as much dispatch as possible, for to allow it to linger in pain is cruelty.

The provisions of the Act do not render illegal any act lawfully done under the Cruelty to Animals Act, 1876.

Improvement of Livestock (Licensing of Bulls) Act, 1931

It is unlawful to keep a bull or a boar which has reached a certain age unless a licence to keep it for breeding or a permit to keep it for any other purpose is in force in respect of the animal.

Destructive Imported Animals Act, 1932

The Minister of Agriculture, Fisheries and Food is given power to prohibit and control the importation or keeping of destructive animals that are not native to this country. Orders are at present in force in respect of the importation into, or the keeping within, Great Britain of coypus, mink, musk rats, grey squirrels and rabbits other than the European variety. These prohibitions are absolute and it is an offence to import them, keep them or turn them loose. The Minister has, however, power to license their import and keeping in certain cases, e.g. for experimental purposes.

Docking and Nicking of Horses Act, 1950

The docking and nicking of horses is prohibited except where certified to be a necessity because of disease or injury in the tail. The import of docked horses is prohibited except for re-export or for breeding purposes.

Diseases of Animals Act, 1950

Under this Act the Minister of Agriculture, Fisheries and Food is empowered to make Orders having the force of law. The animals included in its scope are cattle (which include bulls, cows, oxen, heifers and calves), sheep, goats, and all other ruminating (or cud-chewing) animals, swine and poultry (which include domestic fowls, turkeys, geese, ducks, guinea-fowls, pigeons, pheasants and partridges). The Minister has power by order to extend these definitions to include any kind of four-footed beast or bird and to exclude pheasants or partridges from the definition 'poultry'.

It is the duty of everyone having in his possession or under his charge an animal affected with one of the diseases in respect of which an Order is in force to keep that animal so far as is practicable separate from all animals not so affected, and with all practicable speed to give notice of the fact to a police constable. The latter must in turn notify the necessary authorities as directed by any Order in relation to that particular disease. A person is presumed to know that such an animal is diseased unless he satisfies the court that he had not the knowledge and could not with reasonable diligence have obtained it.

Orders are in force in respect of the following diseases: foot and mouth, cattle plague, pleuropneumonia, epizootic lymphangitis in horses, asses and mules, sheep pox, swine fever, tuberculosis in bovine animals, anthrax, sheep scab, virus hepatitis, glanders in horses etc., parasitic mange in horses etc., fowl pest or plague, and rabies in four-footed animals.

Under the Diseases of Animals Act, 1950, the Minister of Agriculture, Fisheries and Food has power to regulate, authorize or prohibit the import of animals into this country. Such control is exercised by various Orders having the force of law, and since many of these are prolific in their provisions and regulations and technically specialized their repetition here would be out of place. They refer to dogs, cats, horses, cattle, sheep, goats, all other ruminating animals, swine, poultry and eggs for hatching, and various exotic animals.

Transport orders. There have been several Orders which regulate inland transit and sea transit of farm animals. A more recent Order—the *Transit of Animals (General) Order, 1973*—introduced a number of general measures intended to safeguard the welfare of all species of animals during their carriage by sea, air, road and rail. The definition of animal given in this Order includes all mammals (other than man) and any kind of four-footed beast which is not a mammal, together with all fish, reptiles, crustaceans and other cold-blooded creatures. The Order also covers birds of any species. Insofar as farm animals are concerned, the provisions of the Order apply only to carriage by air. Previous legislation covering their carriage by sea remains unaffected (although new regula-

tions are expected shortly), while the new *Transit of Animals (Road and Rail) Order, 1975* makes extensive provision for the carriage, by road and rail, of cattle, sheep, swine, goats and horses.

Control of rabies. It is worth noting that the *Rabies (Importation of Dogs, Cats and Other Mammals) Order, 1974* imposes a general prohibition on the importation of a large number of species of mammals, belonging to ten different orders. Such animals may be imported under licence, when they will be subject to strict quarantine conditions. Except in exceptional circumstances, they may be landed only at certain ports and airports, from which they must be removed without delay to authorised quarantine premises. The Order provides for the majority of animals to which it applies to be detained for a period of at least six months after landing, but in the case of vampire bats their entry into the country is only permitted on the basis of their being quarantined for life. Animals which are landed illegally may be destroyed.

Protection of Animals (Anaesthetics) Acts, 1954 and 1964

Subject to an important list of exceptions, these provide that the carrying out of any operation, with or without instruments, involving interference with sensitive tissues or bone structures upon any animal (excluding a fowl or other bird, fish or reptile) without the use of an anaesthetic so administered as to prevent any pain during the operation is deemed in law to be performed without due care and humanity. The exceptions are:

(a) any experiment authorized under the Cruelty to Animals Act, 1876;
(b) the rendering of emergency first aid to save life or relieve pain;
(c) the docking of dogs' tails before the eyes are open;
(d) the amputation of dogs' dew claws before the eyes are open;
(e) castration of a horse, mule or ass under twelve months; of a bull or sheep under three months; or of a goat or pig under two months. The practice of castration by means of a rubber ring or other device is permissible only if this is applied within the first week of life;

(f) various minor operations, exceptions to these being: the de-horning of cattle; the dis-budding of calves, except by chemical cauterization within the first week of life, or the docking of lambs' tails by rubber ring or other device unless applied in the first week of life;

(g) injections and extractions by means of a hollow needle.

Protection of Birds Acts, 1954

There are various offences in relation to killing or taking wild birds that arise through the protection given by this Act.

Horse Breeding Act, 1958

The keeping of stallions over the age of two years is regulated by licence issued to the owner by the Minister of Agriculture, Fisheries and Food.

Slaughter of Animals Act, 1958

The Act applies to horses, cattle, sheep, swine and goats and by it and the regulations made under it the slaughter of animals in both slaughterhouses and knackers' yards is controlled. In these places no animal may be slaughtered unless it is instantaneously killed or stunned by means of a mechanically operated instrument so as to be insensible to pain until it dies. Exceptions are made in respect of animals killed by the approved Jewish or Mohammedan methods. No animal may be slaughtered save by a person duly licensed by the local authority and over the age of eighteen.

There are certain exceptions to the Act's provisions when slaughter is for research or medical purposes or for preventing suffering or injury in emergency to any person or animal. The regulations made under the Act deal in detail with the conditions of lairages, etc. so as to prevent unnecessary pain and suffering to animals awaiting slaughter.

Animals (Cruel Poisons) Act, 1962

This Act contains provisions prohibiting the killing of animals by means of cruel poisons, 'animal' in this connection being defined as any mammal. It is no defence to claim that the poison was intended to kill pests and that reasonable precautions were taken to exclude access by domestic animals. The Animals (Cruel Poisons) Regulations, 1963, prohibit the use of yellow phosphorus and red squill and limit the use of strychnine to the destruction of moles.

Slaughter of Poultry Act, 1967

Domestic poultry should either be slaughtered instantaneously by decapitation, or dislocation of the neck or some other approved method, or be stunned by an approved instrument immediately prior to slaughter. Exceptions are made in respect of animals killed by the approved Jewish or Mohammedan methods.

Slaughter for commercial purposes may be carried out only on premises registered with the local authority.

Agriculture (Miscellaneous Provisions) Act, 1968

This Act makes it an offence for a person to cause or to permit unnecessary pain or distress to livestock situated on agricultural land and under his control.

Provision is made for the introduction of regulations, as follows:

(a) with respect to the dimensions and layout of accommodation for livestock, the materials to be used in constructing any such accommodation and the facilities by way of lighting, heating, cooling, ventilation, drainage, water supply and otherwise to be provided in connection with any accommodation;

(b) for ensuring the provision of balanced diets for livestock and for prohibiting or regulating the use of any substance as food for livestock and the importation and supply of any substance intended for use as food for livestock;

(c) for prohibiting the bleeding of livestock and the mutilation of livestock in any manner specified in the regulations, and for prohibiting or regulating the use of any method of marking or restraining livestock or interfering with the capacity of livestock to smell, see, hear, emit sound or exercise any other faculty.

In accordance with this Act, codes of recommendation for the welfare of cattle, pigs, domestic fowls and turkeys have been prepared and have received Parliamentary approval.

Provision is also made for the amendment of the law regarding the use of anaesthetics, as laid down in the Protection of Animals (Anaesthetics) Acts, 1954 and 1964, which at present allows certain minor operations to be performed without anaesthetics.

Some regulations have already been introduced, as follows:

The Welfare of Livestock (Docking of Pigs) Regulations, 1974 prohibit the docking of the tails of pigs which are agricultural livestock and over seven days old except where the operation is performed by a veterinary surgeon on health grounds or to prevent injury from tail-biting. It also prohibits any method of docking pigs otherwise than by the quick and complete severance of the part of the tail to be removed.

The Welfare of Livestock (Cattle and Poultry) Regulations, 1974 prohibit the performance on agricultural livestock of the following operations:

(a)   The docking of the tails of cattle;

(b)   The surgical castration of male birds;

(c)   Operations on birds (other than feather-clipping) to impede flight;

(d)   The fitting of blinkers to birds by a method involving mutilation of the nasal septum.

Veterinary Surgeons Act, 1966

This Act states that, subject to certain provisions, no individual shall practice, or hold himself out as practising or as being prepared to practise, veterinary surgery unless he is registered in the register of veterinary surgeons or the supplementary veterinary register. However, the Act authorises treatment given to an animal by the owner thereof, by another member of the household of which the owner is a member, or by a person in the employment of the owner or any other member of such a household. Although this would appear to give permission to a licensee or other member of staff to give remedial or prophylactic treatment to an experimental or stock animal, the Research Defence Society (1974) is of the opinion that, as a general rule, it would be advisable for disease or injury, other than any of a minor nature, to be treated by a veterinary surgeon.

Medicines Act, 1968

This Act controls the use of tests on animals for the establishment of the purity and potency of therapeutic substances in cases where the tests cannot be adequately carried out by chemical methods.

Animals Act, 1971

The Act makes provision with respect to civil liability for damage done by animals and with respect to the protection of livestock from dogs.

Breeding of Dogs Act, 1973

The act makes it unlawful for anyone to keep a breeding establishment for dogs unless they have obtained from the appropriate local authority a licence to do so. For the purposes of the Act a breeding establishment means 'any premises (including a private dwelling) where more than two bitches are kept for the purpose of breeding for sale.'

Before a licence is granted the local authority will have to be satisfied:

(a)   that the dogs will at all times be kept in accommodation suitable as respects construction, sizes of quarters, number of occupants, exercising facilities, temperature, lighting, ventilation and cleanliness;

(b)   that the dogs will be adequately supplied with suitable food, drink and bedding material, adequately exercised, and (so far as necessary) visited at suitable intervals;

(c)   that all reasonable precautions will be taken to prevent and control the spread among dogs of infectious or contagious diseases;

(d)   that appropriate steps will be taken for the protection of the dogs in case of fire or other emergency;

(e)   that all appropriate steps will be taken to secure that the dogs will be provided with suitable food, drink and bedding material and adequately exercised when being transported to or from the breeding establishment.

Local authorities may authorise in writing any of their officers or any veterinary surgeon or veterinary practitioner to inspect licensed premises in their area at reasonable times and to look not only at the premises themselves but also at the animals kept there.

Further information on the law relating to animals is contained in the Proceedings of the recent UFAW Symposium *Animals and the Law* (UFAW, 1975).

REFERENCES

Research Defence Society (1974). *Guidance Notes on the Law Relating to Experiments on Animals*. London: R.D.S.

UFAW (1975). Proceedings of Symposium *Animals and the Law*. Potters Bar: UFAW.

Copies of Acts of Parliament and Statutory Instruments may be obtained from H. M. Stationery Office, Holborn Viaduct, London, E.C.1.

2 Genetic aspects of breeding methods

D. S. FALCONER

The maintenance of a stock of laboratory animals requires only the replacement of old breeding animals by younger ones. It is the way in which these young animals are chosen and the way in which they are paired together that constitute the breeding method. Use of the right method enables the breeder to maintain the inherited characteristics of the stock, or to alter them to make the stock more suited for any particular purpose. The genetic theory on which different breeding methods are based is not at all simple, but fortunately it is not necessary to understand the genetics fully in order to apply any particular method with success. Only the simplest ideas of genetics need therefore be introduced here in explanation of the objects and the practice of breeding methods.

The basic ideas and terminology of genetics are briefly described in the following section, which leads to an explanation of the breeding methods applicable to the maintenance of stocks carrying inherited pathological defects. Though the maintenance of such stocks may not be an important aspect of breeding laboratory animals, the appropriate methods are described at the beginning because they depend only on very simple genetics, and their explanation serves to illustrate the basic laws of inheritance. The subsequent sections deal with the maintenance of inbred and random-bred strains, and with selective breeding. There are many elementary textbooks of genetics where the simpler aspects of the subject can be pursued; one that can be recommended is *The Science of Genetics* (Burns, 1972). The more complicated theory underlying inbreeding and selection is explained in *Introduction to Quantitative Genetics* (Falconer, 1960a), where references to the original sources are given. *An Introduction to Animal Breeding* (Bowman, 1974) gives a short account of the basic theory.

The mechanism of inheritance

Every individual receives from its parents a double set of genes, one set from each parent.

The genes guide the developing embryo so that it becomes a mouse if its parents were mice, or a rat if its parents were rats; and they also provide the adult animal with the ability to carry out all the many chemical processes that constitute living. There are probably some tens of thousands of genes in each set and they are all exceedingly stable, passing from generation to generation unchanged by time or circumstances. So long as all the genes are functioning normally, we have no means of recognizing them individually. Occasionally, however, one or another suffers a change known as a *mutation* and fails to perform its normal function. The mutated gene is also very stable and is transmitted from generation to generation in its defective form so that we see an inherited defect which gives us the opportunity of studying the mechanism of inheritance. An animal exhibiting a defect caused by a mutated gene is known as a *mutant*. When a hitherto unknown mutant is found we can infer from it the existence of a hitherto unknown gene. The mouse is the best known genetically of the laboratory mammals and there are at present well over 300 genes known in mice from their mutations. But this is only a very small fraction of the genes necessary for a mouse to develop and function normally.

The nature of the defect caused by a mutated gene may provide important evidence about the function that the normal gene performs, and in this way the study of mutants can contribute to the study of physiology and pathology. Mutated genes are given names that are more or less descriptive of the defect shown by the mutant animal. The names are contracted to form a symbol consisting of one or more letters, usually but not always the first letter or letters of the name. These symbols make it much easier to express the mechanism of inheritance in writing. A symbol is needed also for the normal form of the gene and the symbol + is widely, though not universally, used for this. The + symbol can stand for the normal form of any gene, and it is always to be taken to refer to the normal form of the particular mutant gene under discussion.

The genes are transmitted from parents to offspring in the germ cells or gametes. Every egg (female gamete) and every sperm (male gamete) contains one set of genes, so that the fertilized egg has two sets of genes. The egg divides, and by growth and repeated division it develops into the adult individual. At each division all the genes of both sets duplicate themselves so that every part of the adult individual still has a double set of genes, but when this individual itself forms germ cells it divides up, shuffles, or *segregates* the genes so that each gamete carries only one set. This is the basic mechanism from which the laws of inheritance derive.

Let us now consider just one of all the many genes that may have revealed its existence by a mutation. This gene then exists in two forms: the normal form $(+)$ and the mutated form (m). Every individual has two representatives of this gene, one received from its father and one from its mother. There are therefore three possible *genotypes* (i.e. arrangements of the pair of representatives of this gene). The individual may have received a normal gene from each parent, in which case its genotype will be $+ +$; or it may have received a mutated gene from each parent and so have the genotype mm; or finally it may have received a normal gene from one parent and a mutated gene from the other and have the genotype $+m$. It makes no difference which of the genes came from the father and which from the mother, and this genotype can equally well be written as $+m$ or as $m+$. Individuals with the genotypes $+ +$ or mm are both called *homozygotes*, and individuals with the genotype $+m$ are called *heterozygotes*. (These terms derive from the term *zygote*, which is the name given to the fertilized egg formed by the union of a sperm with an egg: 'homozygote' means that the uniting egg and sperm were alike and 'heterozygote' that they were unlike with respect to this gene.)

When an individual breeds and produces offspring it transmits one or other of its genes, but never both, to any one of its offspring. It cannot transmit both because the gametes (eggs or sperm) carry only one of the pair of genes. The homozygous genotypes transmit the same form of the gene to all their offspring. The $+ +$ genotype produces gametes that all carry $+$ and so transmits the $+$ gene to all offspring. The mm genotype produces gametes that all carry m and so transmits m to all offspring. The outcome of matings between homozygotes is thus easy to predict. Matings between two mutant homozygotes $(mm \times mm)$ breed true and produce nothing but mm offspring. Matings between two normal homozygotes $(+ + \times + +)$ also breed true and produce nothing but $+ +$ offspring. Matings between mutant and normal homozygotes $(mm \times + +)$ produce nothing but heterozygotes, $+m$. Heterozygous individuals $(+m)$, when they breed, do not produce gametes that are all alike but produce two sorts of gamete, one sort carrying the m gene and the other sort carrying the $+$ gene. They therefore transmit the m gene to some of their offspring and the $+$ gene to others. The two sorts of gamete are produced in equal numbers and each offspring has an equal chance of receiving one or the other. This is exactly analogous to tossing a coin. The coin has an equal chance of coming down heads or tails, and if a large number of tosses are made, about 50 per cent are heads and 50 per cent are tails. If a heterozygote is mated to a homozygote, for example $+m \times mm$, all the offspring get m from the homozygous parent, and chance only enters into which gene they get from the heterozygote. They have an equal chance of getting m or $+$ so, in the long run, 50 per cent of the offspring of this mating will be mm and 50 per cent $+m$. The words 'in the long run' are necessary because of the chance element. A coin seldom comes down heads in exactly half the tosses, and if a small number of tosses is made the total score may be very far from 50 per cent. In the same way, among a small number of offspring it very seldom happens that there are exactly 50 per cent of one genotype and 50 per cent of the other. But the larger the number of offspring, the closer the ratio comes to 50 per cent.

Matings between two heterozygotes $(+m \times +m)$ are more complicated because both parents produce two sorts of gamete and chance enters in twice. To pursue the coin-tossing analogy, we now have to toss two coins simultaneously, one representing the gene received from the father and the other the gene received from the mother. If a large number of double tosses are made, about 25 per cent will be double heads and 25 per cent double tails. These represent mm and $++$ genotypes respectively among the offspring. The remaining 50 per cent of tosses will give heads on one coin and tails on the other. These represent the heterozygotes, $+m$, among the offspring. Thus matings between two heterozygotes give in the long run 25 per cent of mm, 50 per cent of $+m$ and 25 per cent of $++$ among the offspring.

because very often the heterozygote cannot be distinguished by its appearance from one or other of the homozygotes. Usually a single mutant gene is not enough to cause any visible abnormality, and the single normal gene in a hererozygote suffices to make the heterozygote normal in appearance. The mutant gene is then said to be *recessive* to the normal gene, and by convention the symbol for the mutant gene is written with a lower-case initial letter. Thus if a genotype is written as $+m$ this signifies a hetero-

TABLE I

Types of mating, and genotypes expected among the offspring

Genotypes of parents	Genotypes of offspring	Genetic ratio mm : $m+$: $++$
$mm \times mm$	all mm	1 : 0 : 0
$mm \times ++$	all $m+$	0 : 1 : 0
$++ \times ++$	all $++$	0 : 0 : 1
$m+ \times mm$	50% mm, 50% $m+$	1 : 1 : 0
$m+ \times ++$	50% $m+$, 50% $++$	0 : 1 : 1
$m+ \times m+$	25% mm, 50% $m+$, 25% $++$	1 : 2 : 1

(m stands for any mutant gene, and $+$ for the normal form of this gene)

The results of all the different types of mating are summarized in Table I. Each type of mating has its characteristic ratio of genotypes among the progeny, as shown in the last column of the table. The genetic proof that an inherited abnormality is caused by a single defective gene consists in demonstrating the expected ratio among the progeny of known types of mating. With small numbers of offspring the ratios actually observed are seldom exact because of the chance element already explained. Usually the more offspring there are the closer does the ratio approach to the theoretical expectation; but often, even with a large number of offspring, there appear to be too few of the mutant animals simply because, if the defect is a severe one, some mutants die before they can be classified, or, if the defect is very trivial, some mutants may escape detection and be classified as normal.

What we see in the animals examined is not, of course, the genotype itself but the appearance of the animal, whether it has or has not the defect characteristic of the gene in question. The appearance of an animal in this sense is called its *phenotype*. The distinction between phenotype and genotype is a necessary one

zygote and also conveys the meaning that the mutant gene m is recessive and that the appearance of the heterozygote is therefore normal. A familiar example of a recessive is the *albino* gene of mice, rats and other rodents. The three genotypes and their phenotypes are as follows:

genotype: cc $+c$ $++$
phenotype: *albino*   normal

(Contrary to the general rule the symbol for the *albino* gene is c, standing for 'colour' which is the characteristic of the normal gene rather than of the mutant gene.)

When a mutant gene produces a detectable abnormality in the heterozygote the gene is said to be *dominant* over the normal gene, and the gene symbol is written with a capital initial letter. Sometimes the two mutant genes present in the homozygote MM produce no more severe abnormality than the single gene present in the heterozygote $M+$. The *Rex* gene in mice is an example and gives these phenotypes:

genotype: $ReRe$ $+Re$ $++$
phenotype: *Rex*   normal

Genes referred to as dominant are, however, more often only partially dominant and exhibit a more severe or even a different abnormality in homozygotes. A simple example of such a gene is *Naked* in mice. Heterozygotes ($N+$) lose their body hair soon after it has grown and thereafter show patches of bare skin. Homozygotes (NN) grow almost no visible hair and are bald all over. The defect in homozygotes of dominant mutants is, however, often so severe that the homozygous genotypes die before birth and are not seen except in embryological investigations. The *Yellow* gene of mice is an example: *Yellow* homozygotes die at a very early embryonic stage and only the heterozygotes survive.

Reference to the genetic ratios of Table I will show what phenotypic ratios are expected with recessive mutants or with any condition of dominance. Examples of the expected phenotypic ratios will be given later by reference to known genes of mice and explained in relation to the problems of stock maintenance. There is, however, one further feature of mutated genes that must be explained first.

Sometimes there is more than one different mutant form of the same gene. For example, *albino* and *chinchilla* in mice, rabbits, and other animals, are different forms of the same normal gene, *chinchilla* being less severely abnormal than *albino*. In discussing cases of this sort the term 'gene' becomes confusing because it can have two meanings. We can talk about the '*albino* gene' the '*chinchilla* gene' and the 'normal gene'; or we can talk about *albino*, *chinchilla* and normal as being three alternative forms of the same gene. So the term 'gene' can refer to the *albino*, *chinchilla* and normal forms separately, or to all three together. Furthermore, if we talk about a 'normal gene' we have no simple way of distinguishing between the normal form of the *albino* gene and the normal form of any other gene. In order to distinguish between the two meanings of the 'gene', the term *allele* is used for the separate forms, and the term *locus* for all the forms together. 'Allele' means 'alternative form', and 'locus' means position or site, referring to the physical location of the gene. Thus, we say that *chinchilla* is an allele of *albino*, or that *chinchilla* and *albino* are alleles. And we refer to the gene, of which they are alternative forms, as the '*albino* locus'.

Chinchilla and *albino* are not the only alleles known at the *albino* locus. There are *light chinchilla* and *dark chinchilla* in rabbits, *extreme dilution* in mice, and *himalayan* in both rabbits and mice. The same letters are always used as symbols for all the alleles at one locus, and the different alleles are distinguished by superscripts. Thus *albino* is symbolized by c, *chinchilla* by c^{ch}, *himalayan* by c^h.

The genetic behaviour of multiple alleles is very simple. Because each individual has two representatives of each gene (i.e. locus), it can carry two and no more than two alleles at one locus. Thus if, for example, a stock carried the three alleles *albino*, *chinchilla* and normal, the individuals could have the following genotypes and no others:

$++$ — homozygous normal
$c^{ch}c^{ch}$ — homozygous *chinchilla*
cc — homozygous *albino*
$+c^{ch}$ — heterozygous for normal and *chinchilla*
$+c$ — heterozygous for normal and *albino*
$c^{ch}c$ — heterozygous for *chinchilla* and *albino*

The ratios expected in the offspring of any sort of mating between these genotypes can easily be worked out. For example, the mating:

$$+c^{ch} \times c^{ch}c$$

would give:

$$+c^{ch} \ : \ +c \ : \ cc^{ch} \ : \ c^{ch}c^{ch}$$
$$\tfrac{1}{4} \ : \ \tfrac{1}{4} \ : \ \tfrac{1}{4} \ : \ \tfrac{1}{4}$$

Maintenance of mutant stocks
Simple recessive

By a simple recessive is meant a recessive gene whose homozygote exhibiting the abnormality lives and breeds satisfactorily. Most of the common colour genes of rodents are of this sort: *albino* (c) will be taken as an example. A stock is most simply maintained by matings between *albinos*:

$$cc \times cc \quad \rightarrow \quad \text{all } cc \ (albino)$$

These give nothing but *albino* offspring and there is no difficulty. Occasionally, however, it may be necessary to outcross to another strain in order to maintain vigour. If the other strain is normal, i.e. not *albino*, the outcross mating gives all heterozygous offspring which are normal in phenotype, thus:

$$cc \times ++ \quad \rightarrow \quad \text{all } +c \ (normal)$$

If the heterozygous offspring are then mated together they give one-quarter *albino* and three-quarters normal among their offspring, thus:

$$+c \times +c \;\rightarrow\; \underset{(\frac{1}{4}\,albino)}{\tfrac{1}{4}cc:}\; \underset{(\frac{3}{4}\,normal)}{\underbrace{\tfrac{1}{2}+c:\tfrac{1}{4}++}}$$

The *albinos* are then taken to repeat matings of the first type. Note that by the laws of chance it may be necessary to raise 10 or more offspring from the matings between heterozygotes before even one *albino* appears. (Roughly one in 20 families of 10 offspring are expected to contain no *albino*.)

Instead of heterozygotes being mated together, a heterozygote can be backcrossed to an *albino* from the original strain. This gives half *albino* and half normal among the offspring, thus:

$$+c \times cc \;\rightarrow\; \underset{(albino)}{\tfrac{1}{2}cc:}\; \underset{(normal)}{\tfrac{1}{2}+c}$$

In this case the normal offspring are all of one genotype, namely heterozygotes. This type of mating, known as a *backcross,* has several advantages. It yields the genotypes of the parents in equal numbers among the offspring and so can be repeated generation after generation; and since it yields normals and abnormals in equal numbers it is often the most suitable mating when normal litter-mates are required as controls.

Recessives with one sex a poor breeder

Mutants of physiological interest often produce defects severe enough to impair seriously the ability of homozygotes to breed. The recessive *hairless* (*hr*) of mice, for example, cannot be maintained by matings between homozygotes because homozygous (*hr hr*) females cannot suckle their young. If, however, one sex breeds satisfactorily the stock can be maintained without difficulty by matings of the backcross type. *Hairless* males breed satisfactorily and backcross matings can be made thus:

$$+ hr\,♀ \times hr\,hr\,♂ \rightarrow \tfrac{1}{2}\,hr\,hr : \tfrac{1}{2}+hr$$

Recessives with both sexes unable to breed

If homozygotes of both sexes are unable to breed the maintenance of a stock of the gene gives some trouble. The procedure will be explained by reference to the *pituitary dwarf* gene (*dw*) of mice as an example. Homozygous *dwarfs* (*dw dw*) lack pituitary growth hormone and both sexes are sterile. *Dwarf* homozygotes can therefore be produced by only one type of mating, namely matings between heterozygotes. These produce one-quarter *dwarfs* and three-quarters normals among the offspring, thus:

$$+ dw \times + dw \;\rightarrow\; \underset{(\frac{1}{4}\,dwarf)}{\tfrac{1}{4}\,dw\,dw}\; \underset{(\frac{3}{4}\,normal)}{\underbrace{:\tfrac{1}{2}+dw::\tfrac{1}{4}++}}$$

Among the normal offspring, two-thirds are heterozygotes (+*dw*) and suitable for continuing the strain but the remaining one-third are normal homozygotes (+ +) and so unsuitable for further breeding. The two genotypes cannot be distinguished by appearance and, until proved by breeding test, the normal animals are conveniently designated as '+*dw*?'. There are two possible breeding plans. One is to mate untested normals to an older animal already proved by a breeding test to be a heterozygote, thus: +*dw*? × +*dw*. One-third of such matings will, on the average, prove to be + + × +*dw* and will produce no *dwarf* offspring. Two-thirds will prove to be +*dw* × +*dw* by producing at least 1 *dwarf* offspring. The newly proven heterozygote can then be used to test another untested normal. This method gives the lowest proportion of failures, but it has the disadvantage of needing care in the exchange of partners in the tests. The second method, which gives a higher proportion of failures but causes less trouble, is to mate the untested offspring together at random, thus: +*dw*? × +*dw*? Since each member of the pair has a two-thirds chance of being a heterozygote the pair together has a $\frac{2}{3} \times \frac{2}{3}$ or $\frac{4}{9}$, chance of being of the desired genotypes, +*dw* × +*dw*. Thus on the average nearly one half of these random matings will be successful in producing *dwarf* offspring. The pairs that fail to produce *dwarfs* are discarded and the stock continued from the pairs that succeed. If a pair fails to produce *dwarf* offspring this means that one or other, or both members of the pair, is a normal homozygote, + +. It is therefore uneconomical to try these animals in matings with different partners unless there is a shortage of untested animals, because an untested animal is of more value than one with even partial evidence against its being suitable.

The two procedures both rest on making test matings to distinguish between the two genotypes within the normal phenotype. A single *dwarf* among the offspring is enough to prove with complete certainty that both parents are heterozygotes, but the absence of *dwarf* offspring is not absolutely certain proof that the parents are not both heterozygotes, because they could be heterozygotes which have failed by chance to produce a *dwarf* offspring. This raises the question how many normal offspring are to be taken as sufficient evidence for discarding the parents as unsuitable. There is no simple answer to this question and the decision rests on how often one is willing to discard suitable parents on false evidence. Supposing the mating is one between heterozygotes: each offspring then has a three-quarter chance of being normal. Two offspring have a $\frac{3}{4} \times \frac{3}{4}$ or $\frac{9}{16}$ chance of both being normal. Table II shows the chance, expressed as a percentage, that a family of *n* offspring will contain no *dwarf* even though both parents are heterozygotes. Suppose for example that a mating produces six young in its

fails to produce a mutant homozygote among the offspring is calculated on the assumption that the ability of the mutants to survive up to the time of classification is not impaired. If a large proportion of mutants die before they can be recognized it is necessary to have a large number of offspring before a mating can be regarded as adequately tested.

Dominants

The only difficulty about maintaining a stock of a dominant mutant is that it is necessary to use the mutant animal as one, at least, of the parents in every mating. Since no normal animal $(++)$ carries or can transmit the gene, a mating between two normal animals cannot produce mutant offspring. The possibility of maintaining a stock therefore rests on the ability of the mutant animals to breed. If homozygotes of both sexes are able to breed, as for example with *Rex* (*Re*) in the mouse, then it is possible to maintain a true breeding strain by mating homozygotes together, thus:

$$Re\ Re \times Re\ Re \quad \rightarrow \quad \text{all } Re\ Re$$

TABLE II
Probability (per cent) that a pair of heterozygous parents will produce no mutant homozygote among their offspring

No. of offspring (n)	1	2	3	4	5	6	7	8	9	10	11	12	14	16
Chance per cent of no homozygote	75	56	42	32	24	18	13	10	8	6	4	3	2	1

(The mutants are assumed to be fully viable)

first litter with no *dwarf* among them. If it is decided to discard the parents, the probability that this will be a wrong decision is 18 per cent. The probability of 18 per cent means that if 100 matings are discarded, each on the evidence of 6 offspring, 18 of them will have been suitable matings wrongly discarded. In practice the decision is usually how many litters, rather than how many offspring, are needed for a reasonably certain test. As a general rule applicable to mice it is probably best to let every mating have two litters. If a mating does not produce enough young in two litters to test it with reasonable certainty, then it is not worth keeping for reasons of productivity. The chance given in Table II that a mating between heterozygotes

It is, however, seldom possible to breed from homozygotes of dominant genes.

The most generally useful type of mating for the maintenance of a dominant gene is the mating between the heterozygous mutant and a normal animal. This produces mutant and normal offspring in equal numbers, thus:

$$Re+ \times ++ \quad \rightarrow \quad \begin{array}{cc} \frac{1}{2}Re+ & : \quad \frac{1}{2}++ \\ (Rex) & (\text{normal}) \end{array}$$

If two heterozygotes are mated together, and the gene is fully dominant, a ratio of 3 mutant to 1 normal is expected, thus:

$$Re+ \times Re+ \quad \rightarrow \quad \begin{array}{c} \frac{1}{4}Re\ Re : \frac{1}{2}Re+ \ : \ \frac{1}{4}++ \\ (\frac{3}{4}\ Rex) \qquad (\frac{1}{4}\ \text{normal}) \end{array}$$

There are then two different genotypes among the mutant progeny, which is often undesirable if they are indistinguishable by appearance.

If the mutant homozygotes die before birth all the living mutants must be heterozygotes, and matings between two mutants will produce the ratio of 2 mutants to 1 normal among the live young. *Yellow* (A^y) is a gene of this sort in mice, and matings between two *Yellows* produce *Yellow* and normal offspring as follows:

$$A^y + \times A^y + \rightarrow \underset{\text{(Dies)}}{\tfrac{1}{4}A^y A^y} : \underset{(\tfrac{2}{3}\ Yellow)}{\tfrac{1}{2}A^y +} : \underset{(\tfrac{1}{3}\ normal)}{\tfrac{1}{4} + +}$$

This type of mating is necessary if homozygous embryos are needed for study, but for stock maintenance it has the disadvantage of the possible disturbance to reproduction caused by resorbing embryos. It is better to mate mutants to normal animals as exemplified by *Rex* above.

Normal variation

To see the inheritance of a gene displayed in the phenotypic ratios described in the previous section, it is necessary to have a mutant gene with an effect serious enough to cause a visibly abnormal individual. Gene differences (i.e. differences between the normal and mutant form of a gene) with so large an effect are very

rarely found except when deliberately maintained by an appropriate breeding method. If these gene differences were the only causes of inherited differences between individuals, most stocks would exhibit no genetic variability at all. This, however, is not the case. No two individuals of any one stock are exactly alike; they differ in countless ways—in size, shape, fertility, behaviour and so on. These differences between entirely normal individuals may be referred to as 'normal variation' in distinction from the abnormalities discussed in the previous section.

Some of this normal variation is caused by environmental factors such as nutrition, but every genetic analysis has shown that, except in highly inbred strains, the differences between individuals are to some extent inherited. The inherited differences in normal characteristics, such as differences of size or shape, are moreover caused by gene differences subject to the same mechanism of inheritance as are the inherited abnormalities described in the previous section. The only distinction is that the visible effects of the gene differences are so small that they cannot be individually recognized. The variation that can be seen or measured is caused by a great many gene differences acting together, each with a small effect, instead of by a single gene.

TABLE III

Heritabilities of various characters in mice and rats

Character	Heritability per cent	Reference
Mice		
Body weight at 6 weeks	40	Falconer (1973)
Weight gain, 3–6 weeks	25	LaSalle *et al.*, (1974)
Tail length at 6 weeks	60	Falconer (1954)
Litter size, first litters	20	Falconer (1960b)
Ovulation rate	30	Land and Falconer (1969)
Blood pH	15	Wolfe (1961)
Number of lung tumours induced by urethane	20–50	Falconer and Bloom (1962, 1964)
Antibody responses to various antigens	0–80	Claringbold *et al.* (1957) Sobey and Adams, 1961)
Rats		
Body weight at 9 weeks	35	Zucker (1960)
Amount of white in hooded pattern	40	data of Castle and Wright (1916)
Ovary response to gonadotrophic hormone	35	Chapman (1946)
Age of puberty in females	15	Warren and Bogart (1952)
Cortical cholinesterase activity	80	Roderick (1960)

(The figures quoted are rounded to the nearest 5 per cent because heritabilities vary from strain to strain and are seldom known with greater accuracy) difference with a large effect. This variation among normal individuals can be controlled or utilized by application of the appropriate breeding methods, and these methods form the subject of the rest of this chapter.

To study the variation among the individuals of any stock it is necessary to measure the individuals. The characteristic measured is called the *character* or *trait* under study. This might be, for example, the weight at a fixed age, the growth over a fixed period or the number of young born per litter. The variation that is directly measurable is the *phenotypic* variation caused by both genetic and environmental factors in conjunction. Breeding methods influence only the genetic portion of the variation, so the efficacy of any breeding method in changing the phenotypic characteristics of the strain depends on the extent to which differences between individuals are inherited. This differs according to the character, some characters being more strongly inherited than others. It is difficult to measure experimentally the relative amount of genetic variation, but part of the genetic variation can be easily measured from the degree of resemblance between relatives. The relative amount of genetic variation measured in this way is called the *heritability* of the character. Some examples of heritabilities in laboratory animals are given in Table III. The heritability sets a minimum to the relative importance of genetic factors as causes of phenotypic variability.

The breeding methods to be discussed in this section are inbreeding and random breeding. The first is designed to reduce or remove the genetic variation and the second to preserve it.

Inbreeding

Inbreeding is the mating together of individuals which are related to each other through having one or more ancestors in common. The offspring of such a mating are inbred to a degree dependent on the closeness of the relationship between their parents. It is the relationship between the parents that makes the offspring inbred: either or both of the parents may be inbred themselves, but if they are not related to each other the offspring are not inbred. The primary consequence of inbreeding is to reduce the number of individuals that are heterozygous for any one gene pair, and to increase the number that are homozygous for one or other member of the gene pair. The reduction in the number of heterozygotes and increase in the number of homozygotes can be worked out mathematically and provides a measure of the degree of inbreeding known as the *coefficient of inbreeding*. The coefficient of inbreeding ranges from 0 per cent at the start of inbreeding to 100 per cent when inbreeding is complete. The most rapid increase in the coefficient of inbreeding that can be achieved with animals is by mating full brothers and sisters, known as *full-sib* mating. Figure 2.1 shows the inbreeding co-efficient in successive generations of full-sib mating. It will be seen that the inbreeding coefficient increases rapidly at first, but more and more slowly as time goes on and, in principle, it never quite reaches 100 per cent. It is, therefore, never strictly true to say that a strain is completely inbred and the decision as to whether a strain is sufficiently inbred for any particular purpose is a purely arbitrary one. The Committee on *Standardized Nomenclature for Inbred Strains of Mice* (1952), in defining what is meant by an 'inbred strain', laid down 20 generations of consecutive full-sib mating as being the minimum required. This gives an inbreeding coefficient of 98·6 per cent which means that on the average any individual is homozygous for 98·6 per cent of the genes that were heterozygous originally, or that 1·4 per cent of the originally heterozygous genes are still heterozygous. Whether a gene that affects the outcome of an experiment remains heterozygous or not is a matter of chance, the chance being 1·4 out of 100 that it will be heterozygous. Strictly speaking the term 'inbred' has little meaning without a statement of the degree of inbreeding, but an 'inbred strain' is generally understood to be a strain with an inbreeding coefficient of at least 98·6 per cent.

There are two practical consequences of inbreeding, both of which follow from the reduction of heterozygotes and increase of homozygotes. The more obvious of these is the *inbreeding depression* well known to everyone who has practised inbreeding. The animals become generally less healthy and more susceptible to

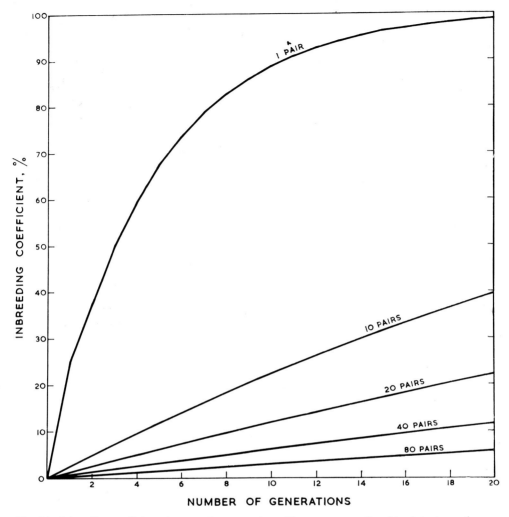

Fig. 2.1. Inbreeding coefficients in successive generations. The upper curve (1 pair) relates to continuous full brother × sister mating. The other curves relate to random breeding in stocks maintained by different numbers of pairs of parents, as indicated.

disease and their reproductive capacity is reduced. This effect of inbreeding follows from the fact that most deleterious genes are recessive. In a non-inbred strain these genes are present mainly in heterozygotes where, being recessive, they do not show in the phenotype. As inbreeding proceeds, however, they appear more and more often in homozygotes where they exert their full deleterious effect on the phenotype. The decline of reproductive capacity that follows inbreeding may be a serious disadvantage in the use of inbred animals; it usually makes inbred strains troublesome to maintain and inbred animals more costly to produce. There are some steps that can be taken to diminish the extent of the inbreeding depression; these will be explained later.

The second practical consequence of inbreeding is to change the amount of genetic variability among the animals. This is the purpose for which inbreeding is practised, but whether the change is an increase or a decrease depends on how the inbreeding is carried out. The practical aspects of inbreeding are, therefore, of the greatest importance to the consequences that follow.

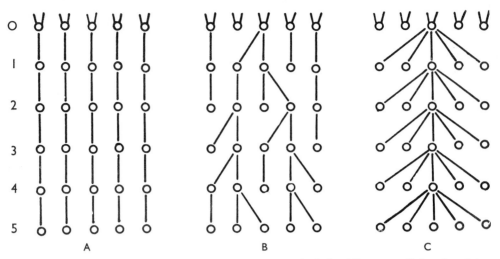

Fig. 2.2. Diagrammatic illustration of the parallel-line (A) and single-line (C) systems of inbreeding. B is a simplified example of the combination of the two systems that occurs in practice. Each circle represents one pair-mating, and the connecting lines indicate the ancestry of each pair. Successive generations are numbered on the left.

There are two contrasting systems of inbreeding which are illustrated schematically in Figure 2.2. The first, Figure 2.2A, which may be called the 'parallel-line' system, is as follows. The stock to be inbred may be assumed to consist of a number of families, each family being the progeny of one single-pair mating. Five foundation families are represented at generation 0 in the figure. One pair of brother and sister are taken from each family and mated. These mated pairs, in turn, produce families and one pair is again taken from each family. This is repeated in subsequent generations. The consequence of this system is that the inbred stock consists of different lines of descent, each line being separated from all the others in ancestry right back to the foundation generation.

The second system, which may be called the 'single-line' system, is illustrated in Figure 2.2C. Here several pairs of full-sibs are taken from only one of the foundation families and are mated to produce the first inbred generation. Five matings are shown in the diagram. In the next generation all the matings are again made from only one family and the remaining families are discarded. This procedure is repeated in subsequent generations. The consequence of this system is that the inbred stock consists of a single line of descent, all the animals of the stock tracing back in ancestry to a single pair in the immediately preceding generation, and to a single pair in the foundation generation.

In practice, neither of these two contrasting systems of inbreeding can be carried out exactly as described, and any practical programme consists of a mixture of the two systems. Under the parallel-line system several sib pairs out of each family will be mated in order to allow for pairs that failed to reproduce. Even then some lines will be lost and will have to be replaced by offshoots from another line. Under the single-line system one mating will not always provide enough offspring to ensure the continuation of the line, and pairs for mating will often have to be taken from several families. Moreover the single-line system, if strictly followed, would not produce enough animals for use and at some stage the line would have to be expanded. In practice then, a combination of the two systems such as that shown in Figure 2.2B is followed. Here the foundation matings of five lines are shown. One (the second counting from the left) failed to reproduce and was replaced by a sub-line taken from the third line. The fourth line failed after the first generation and was replaced by another sub-line from the third line. Two other lines failed later and were replaced by sub-lines also from the third line.

Thus in the fourth and fifth generations the entire stock consists of one line tracing back to a single foundation pair, but this line is divided into two major sub-lines with separate ancestries back to the first generation. It is the structure of the strain when fully inbred—the number of lines and major sub-lines that remain—that is of importance to the genetic consequences.

The effect of inbreeding on the genetic variation can now be explained. As inbreeding proceeds, more and more of the animals become homozygous for one or other of the genes of each gene-pair that was originally heterozygous; but the animals of some lines become homozygous for one member of any gene-pair, while those of other lines become homozygous for the other member. For example, if the original stock contained both albino and non-albino animals, some lines would come to have nothing but albino homozygotes (cc) while other lines would have nothing but non-albino homozygotes $(++)$. Thus the lines become more and more different genetically from one another, while animals belonging to the same line become more and more alike genetically. This is the reason why the genetic variability can be either increased or decreased by inbreeding. Inbreeding by the parallel-line system, with the retention of many lines, leads to an increased genetic diversity, while inbreeding by the single-line system leads to a decrease of genetic diversity.

The genetic diversity between different lines is not usually one of the objects of inbreeding but it gives the breeder the opportunity to combat, to some extent at least, the worst effects of inbreeding depression. If a large number of lines are present at the beginning, the ones that suffer most from inbreeding depression can be eliminated and replaced by expansion of the better lines. This process of replacement may delay the progress towards genetic uniformity within the surviving lines, so that the proportion of heterozygotes may be greater than is indicated by the calculated inbreeding coefficient. The extent of this delay cannot be precisely predicted, but in any case it cannot be avoided if a high degree of inbreeding is to be achieved without too serious a loss of reproductive capacity.

The chief object of inbreeding is generally to produce a strain of genetically identical animals.

When inbreeding is virtually complete any two animals are genetically equivalent to identical twins. For this purpose, therefore, an inbred strain must consist of a single line, all the animals tracing back to a single pair at some stage after the inbreeding is virtually complete. Sub-lines that are separated in the early stages of inbreeding become genetically differentiated from one another. Therefore during the inbreeding process, up to at least 20 generations, there must be a continued elimination both of the original lines and also of the sub-lines, until in the end only one sub-line is left. Sub-lines that are separated after inbreeding is virtually complete cannot become genetically different from each other as a result of continued inbreeding, but they do become slowly differentiated as a result of mutation. Therefore, though a highly inbred strain may safely be expanded into sub-lines for immediate use, if the sub-lines remain separated for more than a few generations (five would probably be safe) there may be some loss of uniformity.

The uniformity of animals from the same inbred strain is a uniformity of genetic constitution. For some purposes, such as for example work involving tissue transplantation, genetic uniformity is all that matters. For many other purposes, however, it is the phenotypic uniformity that is important. The genetic variation is only a part of the total variation and, though the genetic part of the variation is effectively removed by inbreeding, the environmentally caused part of the variation remains. Therefore, the improvement of uniformity achieved by inbreeding depends on what character is required to be uniform, and on how much of the original variation of this character was genetic. If 50 per cent of the variation is genetic, inbreeding cannot do better than halve the variation; and if 10 per cent is genetic, inbreeding cannot reduce the phenotypic variation by more than one-tenth.

The advantage in respect of uniformity gained by inbreeding is sometimes even less than the foregoing considerations suggest because the environmental part of the variation does not always remain unchanged. In some respects inbred animals are more sensitive to environmental factors and show more environmental variation than non-inbred animals. Consequently the environmental variation of some characters increases with inbreeding, and this may more than counterbalance the reduction of genetic variation. There is no generally accepted theory to account for the increased sensitivity to environmental factors and the experimental evidence is insufficient to allow any generalization about what characters are affected in this way.

The use of F_1 crosses

For some purposes, animals produced by crossing two highly inbred strains may offer important advantages over the inbred animals themselves. The first generation, called the F_1, of a cross between two highly inbred strains consists of animals that are genetically uniform but are not inbred. The inbred parents are all homozygous for all genes, so the F_1 animals are all heterozygous for the genes that differ between the two parent strains. There are, therefore, no genetic differences between one F_1 animal and another of the same cross; and since the F_1 animals are homozygous for no more genes than non-inbred animals are, they do not suffer from inbreeding depression. The reproductive capacity of inbred parents when mated to a different strain is usually improved, and the F_1 animals are improved in viability. So the F_1 animals are less costly to produce than inbred animals. For the continued production of F_1 animals, however, the inbred strains must be maintained, since each batch of F_1 animals must be the immediate product of a cross between the inbred strains. If the F_1 animals are bred and offspring obtained from them, the genetic uniformity is immediately destroyed.

The F_1 animals of a cross have the same degree of genetic uniformity as the parent inbred strains, but their phenotypic uniformity may be greater or less, because they may be more or less resistant to environmental influences. Experimental evidence shows that F_1 animals are more uniform than inbreds for many characters, but not for all. It is therefore not possible to predict in advance whether F_1s are likely to be superior to inbreds in this respect. The evidence on this matter is discussed by Chai (1962) and Becker (1962).

Maintenance of inbred strains

The maintenance of inbred strains is a responsible matter which should not be left to untrained workers. The only guarantee that an inbred strain is what it purports to be is the reputation of those who have maintained it in the past. The years of labour that have gone into the making of the strain can be undone at a stroke by the introduction of a single extraneous animal. One mouse looks very much like another even to a trained eye, and an error may all too easily pass unnoticed. Many of the existing inbred strains of mice are albino, and with them the risk of error is at its greatest because albinism masks all other coat colours and effectively covers up any evidence of error that might be provided by other coat-colour genes. Some of the inbred strains of guinea-pigs, in contrast, are marked with characteristic spotting patterns, and with them an error would be more easily detected. The rule for maintaining an inbred strain should be, therefore, carefully kept records and careful scrutiny of the animal-house routine for possible sources of mistaken identity or illegitimacy.

New genetic variation is continually produced by mutation. In consequence, it is necessary to maintain an inbred strain by continued inbreeding in order to eliminate the new variation as it arises. At the same time, selection for productivity should be applied to the strain in order to eliminate any deleterious new genes that may arise. The selection consists of elimination of the sub-lines with the worst reproductive performance, and their replacement by new sub-lines derived from the best of the existing ones. The sub-lines should not be allowed to run too long in parallel or they may become differentiated genetically. In other words, all the animals in the strain at any one time should trace back to one pair of parents mated not more than about, say, five generations previously.

New genetic variation may arise rapidly enough to cause detectable differences between sub-lines after as little as ten generations of separation (Grewal, 1962). The widely distributed inbred strains maintained in different laboratories become split into many sub-strains separated by many generations of independent ancestry; these sub-strains may become quite seriously differentiated and they cannot be regarded as being genetically equivalent (Bailey, 1959).

Random breeding

Random breeding means that the young animals for breeding are chosen without regard to their parentage, and mated together without regard to their relationships: the breeding is thus 'random' with respect to pedigree. It is, however, understood that matings are only made between animals of the same stock, or in other words that the stock is closed to introductions from outside. The purpose of random breeding is to avoid the consequences of inbreeding; that is to say to maintain the genetic variability and to prevent inbreeding depression. Random breeding, however, is not by itself a means of avoiding inbreeding. What random breeding does is to prevent the stock from becoming subdivided into separate lines of descent. The rate of inbreeding is determined by the number of parents that leave offspring to continue the stock. If, for example, each line of descent is propagated from only two parents in each generation the matings are necessarily sib-matings and inbreeding is rapid. If there are four parents in each generation, and mating is made at random, some of the matings will be sib-matings and others will be between less close relatives. In general the larger the number of parents in each generation the more distant is the average relationship between pairs mated at random; and since the rate of inbreeding depends upon the closeness of the relationship between mated pairs, the larger the number of parents the slower is the rate of inbreeding. Therefore, to reduce the inbreeding to a minimum one must maintain the stock by the largest possible number of parents in each generation. The number of parents does not have to be excessively large to reduce the inbreeding to a very small amount. The theoretical rate of inbreeding, when there are equal numbers of male and female parents, is approximately equal to 1/2N per generation where N is the number of parents. The progressive increase in the inbreeding coefficient with different numbers of parents is shown for the first twenty generations in Figure 2.1, and as a further aid to assessing the number of parents required, Table IV shows the number of generations that would elapse before the inbreeding coefficient reached the level produced by one generation of brother by sister mating, i.e. 25 per cent.

It is not easy to decide what is a tolerable rate of inbreeding and therefore how many parents must be used. The decision depends partly on the length of time over which the stock is to be maintained. Whereas 10 pairs might be adequate for a stock that was only required for a short time, it would not be adequate for a stock to be maintained over a number of years. If a stock is to be maintained for an indefinite period of time one should probably aim at a rate of inbreeding of not more than about 1 per cent per generation. With random mating this would require the use of offspring from 25 pairs of parents. Figure 2.1 and Table IV are based on theoretical rates of inbreeding. What actually happens to the genetic properties of the stock depends also on the selection, both natural and artificial, applied. If the offspring of the less productive matings are discarded, the selective pressure in favour of higher productivity will tend to counteract the effect of inbreeding on productivity.

The rates of inbreeding given in Figure 2.1 and Table IV are based on the assumption of equal numbers of male and female parents, and for this reason the number of parents is reckoned in pairs rather than individuals. Having equal numbers of males and females gives the slowest inbreeding for a given number of individual parents. If each male is mated to several females, the resultant inbreeding depends more on the number of males than on the number of females. For example, a stock maintained by 10

TABLE IV

Inbreeding in a closed stock with random mating

No. of pairs of parents	Rate of inbreeding per generation (per cent)	No. of generations equivalent to 1 full-sib mating
10	2·5	11
20	1·25	23
40	0·625	46
60	0·42	69
80	0·31	93
100	0·25	116

male and 40 female parents would be equivalent in respect of inbreeding to one maintained by 16 single-pair matings. Therefore, if the avoidance of inbreeding is the main objective, as many males as can be accommodated in the space available should be used as parents.

Minimal inbreeding

Random breeding amounts to a complete absence of any system in the propagation of the stock. There are two ways in which planned breeding can be utilized so as to reduce the rate of inbreeding below the rate resulting from strict random breeding with the same number of parents. The first and most important concerns the choice of animals to be used as parents, and the second concerns the choice of mates. Planned breeding designed to give the lowest possible rate of inbreeding with a given number of parents may be referred to as the system of 'minimal inbreeding'.

Not all the young animals produced are used as parents of the next generation. Random breeding implies that the animals to become parents are chosen at random from all the available young animals. This results in some families contributing more future parents than other families. In other words the parents of one generation do not, under random mating, contribute equally to the propagation of the stock. The rate of inbreeding can be reduced by making the contributions more nearly equal, and this can be very easily achieved by taking two offspring from each of the pairs of parents and using them as parents of the next generation. It does not matter if two females are taken from some families and two males from others. All that matters is that each pair of parents in one generation should contribute as nearly as possible the same number of offspring to be parents of the next generation. If the numbers contributed are exactly the same—two for each pair—then the rate of inbreeding is half the rate under random mating with the same number of parents. Therefore if 50 pairs were thought to be necessary with random mating only 25 are necessary with minimal inbreeding. In practice, some pairs will prove to be sterile and some families will be discarded because their parents, though not sterile, have been too unproductive.

In order to keep up the numbers of parents, substitutes must be found and these should be chosen from the most productive pairs but taking only one substitute from each pair. In this way, while most of the pairs contribute two offspring to be future parents, the least productive pairs contribute none and the most productive contribute three.

The second modification of random breeding is less important, but may be incorporated as an additional feature of the system of minimal inbreeding. Under strictly random mating there are different degrees of relationship between the mated pairs. Some pairs have the least possible relationship to each other, some are cousins, and occasionally matings between full sibs are made. Random mating therefore results in contemporaneous families having different degrees of inbreeding. Furthermore, the average degree of relationship between the mated pairs varies from generation to generation, so that the average inbreeding coefficient fluctuates erratically from generation to generation. For example, if by chance several sib-matings were made in one generation the inbreeding coefficient would rise steeply, and would fall again in the next generation if this happened to include no sib-mating. The second modification of random breeding is therefore the avoidance of matings between close relatives, such as sibs and cousins, and the object is to reduce the differences of the inbreeding coefficients, both within the stock at any one time and between successive generations. Avoidance of closely related matings in this way does not reduce the average rate of inbreeding. Its advantage lies in the greater uniformity of the inbreeding coefficients and consequently the greater uniformity also of the genetic characteristics of the stock.

In practice it is often rather troublesome to arrange matings at random with the avoidance of sibs or cousins, and it is easier to follow a regular system based on a prearranged schedule of matings. It seems natural therefore to look for a regular system of mating that would be easy to operate and would give uniformity of the inbreeding coefficients both within and between generations. Table V gives the mating schedules for two such systems. The notation in the table is as follows. Each mated pair is given a number and the young animals carry with them the number of the mating that produced them, until they themselves are given a new number when they are mated. A new series of numbers, starting from 1, is given in each successive generation. Two offspring are taken from each mating and used as parents of the next generation. The way the new matings are made up is written as, for example, '$1 = 1 \times 2$', which means that the new mating No. 1 is made between an animal produced by mating No. 1 in the previous generation and an animal produced by mating No. 2 in the previous generation. The '2' used for $2 = 2 \times 3$ is a full sib of the '2' used in $1 = 1 \times 2$.

TABLE V
Regular systems of mating with minimal inbreeding

A			B
	Generations		All
1 & 8	2 & 9	3 & 10	generations
$1 = 1 \times 2$	$1 = 1 \times 3$	$1 = 1 \times 4$	$1 = 1 \times 2$
$2 = 2 \times 3$	$2 = 2 \times 4$	$2 = 2 \times 5$	$2 = 3 \times 4$
$3 = 3 \times 4$	$3 = 3 \times 5$	$3 = 3 \times 6$	$3 = 5 \times 6$
$4 = 4 \times 5$	$4 = 4 \times 6$	$4 = 4 \times 7$	$4 = 7 \times 8$
$5 = 5 \times 6$	$5 = 5 \times 7$	$5 = 5 \times 8$	$5 = 1 \times 2$
$6 = 6 \times 7$	$6 = 6 \times 8$	$6 = 6 \times 1$	$6 = 3 \times 4$
$7 = 7 \times 8$	$7 = 7 \times 1$	$7 = 7 \times 2$	$7 = 5 \times 6$
$8 = 8 \times 1$	$8 = 8 \times 2$	$8 = 8 \times 3$	$8 = 7 \times 8$

One regular system is the cyclical system given in Table VA. The matings are shown for a stock maintained by eight pairs in every generation, but the system can be applied to any number of pairs. The matings for three generations are shown, and the succeeding generations follow the same pattern, the numbers moving up one notch each time. When the turn of matings $1 = 1 \times 1$, $2 = 2 \times 2$, etc. comes round at the eighth generation these are omitted, because they would be sib-matings; the matings listed for generation 1 are made instead and the cycle repeats itself. The consequences of this system are that all the mated pairs in any one generation have the same relationship, so the object of having uniform inbreeding coefficients within each generation is achieved. But the relationships are not the same in successive generations and the object of having uniformity between generations is therefore not achieved.

The other system, shown in Table VB, has the great practical advantage that there is only one mating schedule, which is repeated in every generation. It will be seen that each mating is duplicated: for example $1 = 1 \times 2$ and also $5 = 1 \times 2$. Different individuals are, of course, used for the two matings, and the sexes can be arranged according to convenience. This system works very well if the number of mated pairs is a power of 2; i.e. if there are 4, or 8, or 16, or 32, or 64, etc. pairs. Then the relationship between the mated pairs are the same, both within each generation and in successive generations. The objectives of uniformity of inbreeding coefficients among contemporaries and regularity in successive generations are therefore both achieved. If, however, the number of pairs is not a power of 2 the uniformity within generations is lost.

The system given in Table VB has been investigated, along with other systems, by Kimura and Crow (1963). It and the cyclical system of Table VA both give rates of inbreeding very close to the theoretical rates for random breeding with equal representation of the families, i.e. a rate equal to $1/4N$, where N is the number of individual parents, or twice the number of pairs. It is questionable, however, whether the irregularities of the inbreeding coefficients which the regular systems are designed to avoid are of much practical significance because they are rather small. Ease of application is therefore probably the chief advantage of a regular system, and the only really important aspect of minimal inbreeding is to ensure that each mated pair contributes equally to the parents of the next generation; i.e. to take as far as possible two offspring from each pair to be parents of the next generation.

One further point of importance in connection with minimal inbreeding should be mentioned. Since the inbreeding coefficient at any time depends on the number of generations that have been passed through since the strain was established, it is obviously an advantage to propagate the strain as slowly as possible. Parents should therefore be kept breeding for as long as their reproductive life permits, and the new parents should be taken from the later litters of the old parents.

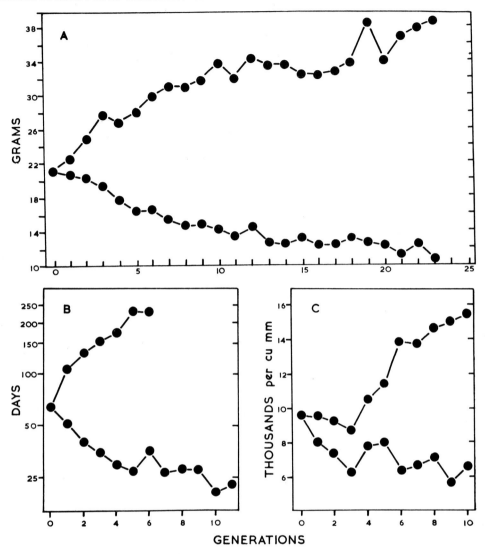

Fig. 2.3. Examples of the response to selection. The mean level of the character selected is shown on the vertical scales. In each case the stock was divided at the beginning into two strains, one of which was selected upwards and the other downwards. A: Body-weight of mice at 60 days of age (MacArthur, 1949); B: Resistance of rats to induced dental caries—number of days till appearance of first cavity (Hunt *et al.*, 1944); C: Total leucocyte count in the blood of mice (Weir *et al.*, 1953. See also Weir and Schlager, 1962).

Selection

Selective breeding provides the means of changing the inherited characteristics of a strain. It consists of choosing the individuals to be parents according to the characteristics that it is desired to change. If, for example, it is desired to increase the body weight, then the largest individuals are chosen to be the parents of the next generation; or if it is desired to increase response to a certain drug, then the individuals showing the greatest response are chosen as parents. Even if there is no desire to change any characteristic of the strain, it is usually advisable to apply selection for productivity in order to reduce the adverse effects of inbreeding as explained in the previous section. Selection is therefore an important aspect of any inbreeding method.

Selection can be applied to any character in which the individuals of the strain differ and in which the differences can be measured or recognized. The practicability of a selective breeding programme depends, however, on how easily the character to be selected can be measured. There have been many experimental studies of selection in laboratory mammals and almost all have proved selection to be effective in changing the mean level of the character selected. Most of the characters listed in Table III have been proved to respond to selective breeding. The changes produced in some typical experiments are illustrated graphically in Figure 2.3. The questions about selective breeding to be discussed here concern the procedures by which the greatest rate of response can be obtained.

Rate of progress

The rate at which the selected character changes in response to selective breeding depends on two factors: the heritability of the character and the average superiority of the selected parents. If, for example, selection is applied to body weight with a heritability of, say, 25 per cent and the individuals chosen as parents average 4 g heavier than all the animals out of which they were chosen, then the response will be 25 per cent of 4 g, i.e. 1 g. Under these circumstances the strain would increase in body weight by 1 g per generation. Experiments have shown, how-ever, that the response is usually rather erratic, as can be seen from Figure 2.3. A regular improvement of the character selected is there-fore not to be expected and the rate of improvement cannot be accurately assessed until several generations of selection have been made.

Both of the factors that determine the rate of progress are to some extent under the control of the breeder. The heritability, which is the proportion of the variation that is inherited, can be increased by careful control of the environmental conditions. If the animals are subjected to widely different environmental conditions the non-genetic variation will be increased and the heritability reduced. Control of the environmental conditions, so that every individual experiences as nearly as possible the same conditions, therefore increases the rate of response. The average superiority of the selected parents can be increased by choosing the parents out of a larger number of individuals, or in other words by reducing the proportion of individuals kept for breeding. The actual number of parents used is decided by considerations of the rate of inbreeding discussed in the previous section, so the proportion kept as parents can be reduced only by increasing the number of offspring obtained from the previous generation out of which the future parents are to be chosen. If, on the average, only two offspring are obtained from each pair, obviously no selection can be applied because all the offspring must be used to replace their parents. If four offspring are obtained per pair, 50 per cent of them must be selected as future parents; if eight offspring are obtained only 25 per cent must be selected. The superiority of the selected individuals increases as the percentage needed for replacement decreases. So selecting 25 per cent and dis-carding 75 per cent is better than selecting 50 per cent and discarding 50 per cent. Increasing the number of offspring obtained, however, introduces the further complication of time. To raise more offspring before selection is made takes a longer time, and this increases the interval between generations. Though the progress made per generation will be increased, the progress made per unit of time may be reduced, and it is the progress in terms of weeks or months that matters in practice. The best compromise between raising a large number of offspring and keeping the generation interval short depends on the reproductive charac-teristics of the animal and the stock. With mice it is usually not worth while waiting for second litters unless the number raised and measured is fewer than six per litter. Later litters can, of course, be raised to supply animals for use. The general principle is that the future breeding stock should be selected from among the off-spring of young parents.

Relation of inbreeding to selection

The selection of parents under selective breeding governs the choice of which individuals are to be used as parents, but it does not govern the way the selected individuals are mated together. It is, therefore, possible to practise inbreeding or random breeding at the same time as selection, and the method of breeding has an important bearing on the outcome of the selection. If inbreeding is practised the response to selection is likely to be more rapid at the beginning but to come to an end sooner and to yield less improvement in the end. Experiments have shown that selection with random breeding or with minimal inbreeding continues to yield progress, though at a diminishing rate, for about twenty generations. With rapid inbreeding by brother-sister matings, progress is likely to cease after about five generations. For the sake of rapid initial progress, however, inbreeding might sometimes be worth while. It is then of the greatest importance that the inbreeding should be by the parallel-line system because members of the same line quickly become genetically alike while members of different lines become genetically more unlike. The selection should be applied chiefly by elimination of the worse lines and retention of the best, and when the strain is reduced to a single line little further progress can be expected from the selection.

Unless there is any special reason to inbreed with the selection, inbreeding should be avoided. It is therefore important to know how the selection is likely to affect the rate of inbreeding. Selection applied with random mating tends to reduce the number of families that contribute parents of the next generation because the selection concentrates the choice of parents among the better families. Therefore if the rate of inbreeding is to be kept to a certain level, a larger number of pairs must be selected and mated than would be required under random mating without selection. This effect of selection is greater when the heritability of the character selected is high, because then there is a higher degree of resemblance between members of the same family and selection is more likely to lead to a choice of parents from among a very few families.

If the selection is to be continued for a very long time and the avoidance of inbreeding is an important consideration, then the selection can be applied with the system of minimal inbreeding. This system requires that two individuals shall be chosen as parents from each family and the choice open to selection is thereby restricted. Instead of choosing the best individuals out of all the offspring, the best two have to be chosen out of each family. In this way the genetic variability between the families is not utilized in the selection. The average superiority of the selected individuals is consequently reduced and the rate of progress is approximately halved.

Difficult characters

A character such as body weight, which can be measured on both sexes and early in life, presents few problems of selection. Other characters, however, present a variety of problems of which the most important are presented by (a) characters measurable on one sex only, (b) characters that require the killing of the animal for measurement, (c) characters measurable only late in life, and (d) characters of an all-or-none sort. The procedures that may be adopted for selection of such characters are briefly as follows:

(a) If a character refers to only one sex, such as for example the age of vaginal opening, the simplest procedure is to disregard the other sex, taking the individuals to be used as parents at random. The rate of progress is then half what it would be if both sexes could be selected. It is possible, however, to apply some selection to the sex that cannot itself be measured. If it is the male sex that cannot be measured then males can be selected on the basis of their sisters' values. The mean measurement of each family of sisters is calculated and males are taken from the families with the best averages. Some care must be exercised to ensure that the males do not all come from a very few families, so that the rate of inbreeding is not too high.

(b) If the measurement of the character to be selected involves the death of the animal, as in the case of the ovarian response to gonadotrophic hormone, there are two procedures that can be adopted. The first is to make measurements on some members of each family and to select the unmeasured animals on the basis of the average values of their brothers and sisters. Measurements might be made on all the offspring born in first litters, and the second-litter offspring be selected as parents. If the measurement can be delayed until after sexual maturity there is an alternative procedure. A large number of animals can be mated and the measurements made only after the litters are weaned. The litters are then selected on the basis of their parents' values. This method requires a great deal of space because all the offspring must be mated before they can be selected.

(c)   Characters that can be measured only late in life are troublesome to select because of the long interval between generations, if matings are not made until after measurements have been made. The progress can be speeded up by mating all the animals before measurement in the manner described in the previous paragraph.

(d)   There are some characters that cannot be measured but can only be grouped into qualitative classes such as 'alive' or 'dead', 'resistant' or 'susceptible'. The nature of the class to be selected may present very serious difficulties, but assuming the desired class are survivors and can be bred from, there is one device that can speed the progress of selection. If the desired class represents a large proportion of the animals, say 50 per cent or more, then selection will not be very effective for the following reason. Only about 25 per cent of the offspring are needed to replace their parents; but there is no means of knowing which are the best 25 per cent among the desired class. Conversely if the desired class represents a very small proportion, then some individuals of the other class will have to be used as parents. The most effective use of selection can therefore be made when the desired class amounts to exactly the percentage that must be selected for replacements. It is sometimes possible to adjust the treatment so as to give the required percentage in the desired class. As selection proceeds and progress is made, the proportion of the desired class increases and the treatment should be further adjusted so as to keep the proportion always equal to the requirement for replacement.

Selection for productivity

Selection for productivity has been mentioned previously as a means of minimizing the undesirable effects of inbreeding. Productivity is, however, an important character in its own right because the economic value of a strain depends very largely on its productivity. Selection for productivity, therefore, merits special consideration.

The first decision that has to be made is how productivity is to be measured. The simplest measurement is the number of offspring weaned, or the number raised to maturity. This may be an adequate measure when only one or two litters are raised and productivity is of secondary importance. When the improvement of productivity is the primary object, however, it is desirable to take account also of the time factor, or the rate of production, and also of the quality of the offspring produced. (The inclusion of quality as an aspect of productivity can be justified on the ground that the quality of the young influences their productivity in the next generation.)

The simplest way to take account of the time factor is to count the number of offspring produced in a fixed period of time after mating. This method is, however, a little inflexible because it does not allow an assessment to be made before the fixed time has elapsed, and it does not allow credit to be given to pairs that maintain a high rate of production over a longer period. The best method is to divide the number of young produced by the number of days since mating, multiplying by 100 to give a percentage index (Lane-Petter *et al.*, 1959). The index of productivity measures the rate of production as the number of offspring produced per 100 days. It can be calculated at any time for any pair and selection can be made whenever it is convenient.

The quality of the offspring can be assessed from their weight at weaning, and the total weight of the offspring weaned or raised can be used as a measure of productivity that combines both quantity and quality. The total weight of offspring attaches more importance to numbers than to weights because the pairs with the largest number of young nearly always come out best. It does, however, provide a simple means of ensuring that the quality of the young does not suffer seriously by any increase in the number produced.

Selection for productivity cannot be applied without some sacrifice of other objectives of the breeding system. If selection is also being applied to some other character, then selection for productivity restricts the choice of individuals open to selection for the other character. And if the avoidance of inbreeding is an objective, then selection for productivity reduces the number of families that contribute parents to the next generation and so increases the rate of inbreeding. The breeding programme, therefore, has to be based on the decision of which objectives are the more important.

REFERENCES

Bailey, D. W. (1959). Rates of sub-line divergence in highly inbred strains of mice. *J. Hered.,* **50,** 26–30.

Becker, W. A. (1962). Choice of animals and sensitivity of experiments. *Nature, Lond.,* **193,** 1264–6.

Bowman, J. C. (1974). *An Introduction to Animal Breeding.* Studies in Biology No. 46. London: Arnold.

Burns, G. W. (1972). *The Science of Genetics.* London: Collier-Macmillan.

Castle, W. E. & Wright, S. (1916). Studies of inheritance in guinea-pigs and rats. *Publs Carnegie Instn.,* No. **241.**

Chai, C. K. (1962). Choice of animals and sensitivity of experiments. *Nature, Lond.,* **193,** 1266.

Chapman, A. B. (1946). Genetic and nongenetic sources of variation in the weight response of the immature rat ovary to a gonadotrophic hormone. *Genetics,* **31,** 494–507.

Claringbold, P. J., Sobey, W. R. & Adams, K. M. (1957). Inheritance of antibody response. III. Heritability of response to sheep red cells. *Aust. J. biol. Sci.,* **10,** 367–73.

Falconer, D. S. (1954). Validity of the theory of genetic correlation. An experimental test with mice. *J. Hered.,* **45,** 42–4.

Falconer, D. S. (1960a). *Introduction to Quantitative Genetics.* Edinburgh: Oliver and Boyd.

Falconer, D. S. (1960b). The genetics of litter size in mice. *J. cell. comp. Physiol.,* **56** Suppl. 1: 153–67.

Falconer, D. S. (1973). Replicated selection for body weight in mice. *Genet. Res.,* **22,** 291–321.

Falconer, D. S. & Bloom, J. L. (1962). A genetic study of induced lung-tumours in mice. *Br. J. Cancer,* **16,** 665–85.

Falconer, D. S. & Bloom, J. L. (1964). Changes in susceptibility to urethane-induced lung tumours produced by selective breeding in mice. *Br. J. Cancer,* **18,** 322–52.

Grewal, M. S. (1962). The rate of genetic divergence of sub-lines in the C57BL strain of mice. *Genet. Res.,* **3,** 226–37.

Hunt, H. R., Hoppert, C. A. & Erwin, W. G. (1944). Inheritance of susceptibility to caries in albino rats (*Mus norvegicus*), *J. dent. Res.,* **23,** 385–401.

Kimura, M. & Crow, J. F. (1963). On the maximum avoidance of inbreeding. *Genet. Res.,* **4,** 399–415.

Land, R. B. & Falconer, D. S. (1969). Genetic studies of ovulation rate in the mouse. *Genet. Res.,* **13,** 25–46.

Lane-Petter, W., Brown, A. M., Cook, M. J., Porter, G. & Tuffery, A. A. (1959). Measuring productivity in breeding in small animals. *Nature, Lond.,* **183,** 339.

LaSalle, T. J., White, J. M. & Vinson, W. E. (1974). Direct and correlated responses to selection for increased post-weaning gain in mice. *Theoretical and Applied Genetics,* **44,** 272–277.

Roderick, T. H. (1960). Selection for cholinesterase activity in the cerebral cortex of the rat. *Genetics,* **45,** 1122–40.

Sobey, W. R. & Adams, K. M. (1961). Inheritance of antibody response. IV. Heritability of response to the antigens of Rhizobium meliloti and two strains of influenza virus. *Aust. J. biol. Sci.,* **14,** 588–93.

Standardized Nomenclature for Inbred Strains of Mice. (1952). *Cancer Res.,* **12,** 602–13.

Warren, E. P. & Bogart, R. (1952). Effect of selection for age at time of puberty on reproductive performance in the rat. *Ore. agric. exp. Sta. Tech. Bull.,* No. **25.**

Weir, J. A., Cooper, R. H. & Clark, R. D. (1953). The nature of genetic resistance to infection in mice. *Science,* **117,** 328–30.

Weir, J. A. & Schlager, G. (1962). Selection for leucocyte count in the house mouse and some physiological effects. *Genetics,* **47,** 1199–217.

Wolfe, H. G. (1961). Selection for blood-pH in the house mouse. *Genetics,* **46,** 55–75.

Zucker, L. M. (1960). Two-way selection for body size in rats, with observations on simultaneous changes in coat colour pattern and hood size. *Genetics,* **45,** 467–83.

3 The nutrition of laboratory animals

MARIE E. COATES

Animals living in the wild can be loosely classified according to their feeding habits into herbivores, which subsist mainly on vegetable foods, carnivores, which eat flesh, and omnivores, which choose a mixed diet. Under the artificial conditions of the laboratory most of the species commonly in use have become adapted to the omnivorous state. The need for uniformity among experimental animals demands that they be fed on a nutritionally adequate diet of known and repeatable composition. In consequence, standardized diets based on cereals, with supplements of good quality protein, vitamins and minerals, have come into general use. Much remains to be learned about the nutrient requirements of laboratory animals. This chapter outlines some general principles of nutrition and offers a few comments that may serve as a guide to the efficient formulation and practical use of diets in the animal laboratory. For those who seek more detailed information a list of relevant textbooks has been appended.

Before considering the subject of animal nutrition, and its specialised aspects relating to laboratory animals, it might be wise to go back to first principles and ask the basic questions— why does an animal need food and what are the conditions likely to influence the amount and kind of food it needs? Food is the raw material from which all animal tissues are built, whether they be new tissues formed during growth, or replacements for losses due to wear and tear. The animal needs minerals to form the skeleton and proteins from which to build the vast number of cells that make up the blood, muscles and other soft tissues of the body. Food is also the source of energy needed to generate heat and to support the chemical reactions on which all physiological processes depend. Many of these reactions are catalysed by vitamins or some inorganic elements which must, therefore, be provided in the diet. Finally, since they proceed in an aqueous medium, water is vitally necessary.

From these precepts it follows that the animal's need for food will be greatest when new tissues are being formed. Young animals in a state of rapid growth and breeding females during gestation will have the highest requirement for all types of nutrients; during lactation, also, the mother must increase her own food intake to compensate for the large amount she passes on to her young in her milk. The need for the energy components of food will be highest during periods of great physical activity or in a cold environment, where the heat lost must be replaced through an increase in metabolism. The resting, non-breeding adult, particularly if it is maintained in a warm environment, will have the least need for all classes of food since it has only to replace worn tissue and maintain its basic bodily functions.

Quantitative requirements, however, are not the only consideration. Although the food in the hopper may be a carefully calculated mixture of the required ingredients in the correct proportions it will be useless if the animal refuses to eat it. Furthermore, having entered the gastrointestinal tract, the ease with which the components of the diet can be digested and absorbed determines its true nutritive value. Thus palatability, digestibility and the physical nature of the diet are important characteristics that must be taken into account. Microbiological quality is another factor that should be considered. The food can be a vehicle for the spread of disease among laboratory animals unless care is taken to ensure that the raw materials used in its preparation are free from pathogens and parasites, and that the diet after mixing does not become contaminated.

Components of animal diets
Proteins

It is not strictly true to say that an animal needs protein as such. Proteins are complex substances made up of amino acids, the number, type and arrangement of which determine the nature of the protein. What the animal really needs is a source of amino acids from which to build the proteins appropriate to its own tissues. Some amino acids can be synthesized in the animal body from ammonia and simple carbon compounds. Others (the so-called 'essential' or 'indispensable' amino acids) cannot be synthesized, at least at a fast enough rate to satisfy the demand, and these must be supplied in the diet. The animal's digestive processes break down food proteins first to smaller chains of amino acids (peptides) and finally to free amino acids. The free acids and, possibly, the smaller peptides are absorbed and then transported by the blood to the tissues where they are rebuilt into the proteins characteristic of the animal concerned. The mechanism of synthesis demands that all the component amino acids of the particular protein being formed must be present simultaneously and in the right proportions. They cannot be stored. The amino acid in lowest concentration in relation to need will thus determine the rate at which protein can be formed in the tissues; other amino acids present in greater amounts than are needed will be wasted for this purpose, although they may be utilized as a source of energy.

It is clear, therefore, that a good dietary protein must not only contain the essential amino acids in sufficient quantity and the right proportion, but it must also be readily digestible in the alimentary tract so that the amino acids it contains are made available to the animal eating it. In other words, the *quality* as well as the *quantity* of protein in an animal diet is important. Amino acids can be destroyed by oxidation and, during prolonged heat treatment of a protein-containing food, may react among themselves or with carbohydrates to form compounds from which they cannot readily be released by the digestive enzymes. Thus protein quality may deteriorate during bad storage conditions, or as a result of overheating in the course of pelleting or sterilization of a diet. Such deterioration is not easily detectable and it would clearly be desirable to have some simple test of protein quality that could be applied to the diet or its protein-containing ingredients.

Evaluation of protein quality. Proteins consist of carbon, hydrogen, oxygen, nitrogen and, usually, sulphur. The nitrogen contributes about 16 per cent of the total weight, and the traditional method of assessing the quantity of protein in a food is to determine its nitrogen content and multiply by 100/16, or 6·25. This gives a value for the crude protein content of the food, but tells us nothing about the properties of that protein. With modern methods of physico-chemical analysis the individual amino acids can be identified and measured, but even this information does not adequately describe the nutritive value since it tells us nothing of their availability to the animal. The most realistic way to assess the nutritional quality of a protein is by feeding trials on animals—usually rats or chicks. Several biological measurements have been proposed as indicators of protein quality. The simplest is the *protein efficiency ratio* (PER), an expression of the weight gained by a growing animal per unit of protein eaten. Others involve nitrogen balance studies, which are measurements of the amounts of nitrogen eaten and excreted, the difference between the two values being a measure of the nitrogen retained, i.e. protein utilized, by the test animal. From this type of data several indices of protein quality can be calculated: *true digestibility* (TD) expresses the proportion of the food nitrogen that has been absorbed; *biological value* (BV) measures the proportion of the absorbed nitrogen that the animal has retained; and *net protein utilization* (NPU), a product of TD × BV, measures the proportion of the food nitrogen that has been retained. However, these tests are elaborate and expensive to perform, and they have limited value since the results should strictly be applied to the test substance only when it is fed as the sole source of protein under the conditions of the test. When it becomes part of a mixed diet in which other ingredients may make good its deficiencies, these rather artificial indices are not necessarily helpful.

A simpler procedure, though subject to similar criticism, is the microbiological test with *Streptococcus zymogenes*, devised by Ford (1960). The amino acid requirements of this organism can be completely satisfied, under the conditions of the assay, by supplements of casein. By measuring the extent to which the organism can grow in response to supplements of other proteins, their *relative nutritive value* (RNV), compared with casein as 100 per cent, can be determined. This technique has been found to give results that correlate well with those of the more cumbersome animal tests.

One of the amino acids most readily damaged during food-processing is lysine, which may become chemically bound and hence nutritionally unavailable. It is also the one most often present in only marginal amounts in animal diets. The amount of nutritionally available lysine is therefore likely to be a useful index of protein quality. Two relatively simple chemical tests, based on the reaction between fluorodinitrobenzene and undamaged lysine, measure the so-called 'available lysine' (Carpenter, 1960; Roach *et al.*, 1967). They have been applied to a range of protein components of human and animal foods and have proved a reliable indicator of nutritional value.

TABLE I

Amino acid composition and nutritive quality of two fish meals

Test		Meal A	Meal B
Amino acid composition (g/16 g N)			
Lysine		7·3	6·8
Histidine		1·6	1·7
Arginine		5·5	5·4
Aspartic acid		9·7	9·7
Threonine		4·4	4·3
Serine		3·9	3·8
Glutamic acid		13·8	14·4
Proline		3·9	3·6
Glycine		5·3	5·1
Alanine		6·4	6·3
Valine		4·9	4·9
Cystine		0·9	0·8
Methionine		2·6	2·4
Leucine		7·4	7·4
Isoleucine		4·2	4·3
Tyrosine		3·3	3·1
Valine		3·9	4·1
Tryptophan		1·0	0·8
Protein efficiency ratio (chick)		2·45	2·01
Net protein utilization (rat)		59	29
Relative nutritive value	Microbiological test with	78	44
Available methionine (g/16 g N)	*Strep. zymogenes*	2·5	1·6
Available lysine (g/16 g N)	Fluorodinitro- benzene test	6·5	4·8

(From data of Bunyan and Woodham, 1964.)

The data in Table I illustrate the need for some means of assessing the nutritive quality of protein foods and the usefulness of simple microbiological and 'available lysine' tests as indicators of nutritional value. The amino acid composition of two samples of fish meal, measured chemically, was very similar. However, biological assays with chicks and rats showed that meal B was nutritionally inferior to meal A. The chemical estimation of available lysine and microbiological assays of RNV and available methionine both confirmed the poorer protein quality of meal B.

Choice of proteins for animal diets. In view of the problems of digestibility, availability and amino acid balance it is obviously a practical impossibility to state precisely the amount of protein that should be supplied in an animal's diet to support optimal performance. Protein is usually the most expensive ingredient of a diet and economy demands that it should be used to best advantage. To ensure its most efficient utilization it must be easily digestible and should contain the optimal proportions of essential amino acids.

Table II gives the amino acid compositions of some of the common ingredients of laboratory animal diets. For comparison, values have also been included for whole hen's egg, which has been shown to be virtually ideal as a protein source for the rat. Not only does it supply the correct balance of amino acids but it is also completely digestible and has frequently been used as a standard of reference against which to compare the nutritive value of other protein foods. There are very few proteins that supply the ideal selection of amino acids and, in designing a diet for laboratory animals, it is customary to prescribe a mixture in which the deficiencies of one will be compensated for by excesses in others. As a very rough guide, a level of 20% crude protein should be adequate in most circumstances provided that it includes a proportion of good quality protein concentrates such as dried milk, fish meal, or soya. Diets based entirely on cereals are inadequate for growth and reproduction of most species. The crude protein content of cereals is usually only about 10–15 per cent and, furthermore, they are seriously deficient in lysine. Addition of some fish meal or dried milk will not only increase the quantity but will also improve the quality by supplying extra lysine. Apart from lysine, the sulphur-containing amino acids, methionine and cystine, are the ones most

TABLE II

Amino acid composition (g/16 gN) of some common ingredients of laboratory animal diets

	Barley*	Wheat*	Fish meal*	Ground-nut*	Soya-bean*	Skim milk powder†	Whole egg†
Total N (%)	1·50	2·15	10·33	8·08	8·64	6·21	5·60
Aspartic acid	6·47	4·76	8·79	10·96	10·74	7·68	10·36
Threonine	3·69	2·72	4·16	2·76	3·73	5·17	5·14
Serine	4·67	4·33	4·76	4·92	4·70	5·98	7·72
Glutamic acid	28·10	27·82	14·25	23·46	17·38	25·19	14·68
Proline	11·23	10·77	4·78	4·35	4·58	—	—
Glycine	4·62	3·54	9·98	5·78	4·11	2·19	3·59
Alanine	5·52	3·06	6·76	4·20	4·28	3·54	6·11
Valine	5·07	4·41	4·60	4·13	4·77	6·40	7·54
Methionine	1·41	1·75	2·79	1·04	1·30	2·36	3·01
Isoleucine	3·87	3·56	3·91	3·30	4·64	5·88	5·76
Leucine	7·25	6·62	7·07	6·39	7·31	10·66	8·90
Tyrosine	3·29	3·00	2·95	4·36	3·58	5·59	3·63
Phenylalanine	4·79	4·38	3·85	5·01	4·87	5·30	6·69
Lysine	3·79	2·96	6·69	3·38	6·15	7·76	6·65
Histidine	2·41	2·53	2·22	2·40	2·90	3·50	2·54
Arginine	5·15	4·57	6·43	11·18	6·49	3·15	6·15
Tryptophan	1·18	—	1·19	1·00	1·53	1·47	1·49
'Available lysine'	—	2·96	6·22	2·86	5·90	—	—

* Values kindly supplied by A. P. Williams, National Institute for Research in Dairying, Shinfield, Reading.
† Eggum (1968).

likely to be below the optimal level. Some of the amino acids, including methionine and lysine, are now produced fairly cheaply on a commercial scale, so it is economically feasible to make good their deficiencies in an otherwise adequate diet by supplementation with the pure substances.

Energy components

In a nutritional context, the term 'energy' is used to describe the components of the diet that provide the fuel an animal needs for the basal metabolic processes such as heart beat and respiration, for the production of heat to maintain body temperature and for the work done during physical activities such as running, gnawing or burrowing. Dietary energy is supplied mainly by carbohydrates and fats and, to a lesser extent, by the non-nitrogenous portions of amino acids that have escaped being built into proteins.

Carbohydrates, as the name implies, consist of carbon, hydrogen and oxygen. They are mostly of vegetable origin and include the starches, sugars and cellulose, but they must be broken down to simple sugars before they can be absorbed and utilized by animals. In diets for laboratory animals cereal starches form the chief source of carbohydrate. When they come

into contact with the digestive enzymes they are converted to dextrin, then maltose and finally to glucose. The glucose is absorbed into the blood stream and carried to the tissues, where it is oxidized to carbon dioxide and water. It is during the course of this oxidation that energy is released and used for other chemical reactions on which the bodily functions depend. Two other simple sugars from which energy is derived are fructose and galactose. Fructose and glucose arise on hydrolysis of sucrose, galactose and glucose by the hydrolysis of lactose, the sugar in milk. Galactose is thus an important source of energy for the suckling animal.

Immediately after a meal, more glucose is available than the animal can utilize and the excess must be stored until it is needed. The short-term storage product is glycogen, a compound built up from a large number of glucose molecules. It can be conveniently stored in the liver and muscle and readily utilized when required. If the intake of carbohydrate is greater than can be used or stored as glycogen the excess glucose is converted to fat and transported to the adipose tissue, usually the subcutaneous and abdominal depots. Thus an animal that is consistently taking in more carbohydrate than its current energy requirements demand will ultimately become obese.

The fats are less easily mobilized than is glycogen, but they provide a long-term form of storage and, in periods of starvation, are metabolized to provide energy.

Cellulose, hemicellulose and lignin make up much of the hard, structural parts of plants— the husks of grains, for instance, and the stems of grasses. Cellulose, like starch, consists of chains of glucose molecules but they are arranged in such a way that they cannot be released by enzymes of the mammalian or avian digestive tract. Hemicelluloses are made up of combinations of other simple sugars with uronic acids. Although more easily hydrolysed than cellulose they are also resistant to digestion in the alimentary tract. Cellulose and hemi-cellulose can, however, be degraded by some micro-organisms. Ruminants derive an important contribution to their energy requirements from cellulose after it has been subjected to bacterial action in the rumen. In animals like the guinea-pig and rabbit, which have a large caecum with a dense microbial population, some digestion of cellulose does go on. In other species it is of little nutritional value except as bulk, or roughage. Lignin, the structure of which is unknown, is completely indigestible.

The fatty components (lipids) of animal diets also consist of carbon, hydrogen and oxygen, but, as the proportion of oxygen is lower than in carbohydrates, they are a correspondingly richer source of energy. They occur in animal diets as constituents of seeds, e.g. soya bean, groundnut, linseed, or of animal products, e.g. fish and meat meals, whole milk. They may be deliberately added as, for instance, fish liver oil, tallow or vegetable oils. The most abundant of the lipid components of animal diets are the neutral fats, that is, esters of glycerol with fatty acids. The characteristics of a fatty acid are determined by the number of carbon atoms it contains, and whether or not they are fully combined with hydrogen atoms. If they are, the fatty acid is described as 'saturated'. Un-saturated fatty acids lack two, four, or more hydrogen atoms and are, in consequence, less stable and become readily oxidized on exposure to air. The unpleasant odour and flavour of rancidity that develops in fish liver oils under adverse conditions of storage is the result of oxidative changes in the highly unsaturated fatty acids they contain. It is common practice for manufacturers to add an antioxidant to such products to prevent deterioration of the oils during storage or after they have been mixed into a diet. Unsaturated fatty acids have lower melting points than the corresponding saturated compounds of the same chain length; maize oil, for instance, contains mainly glycerides of unsaturated fatty acids and is liquid at room temperature whereas coconut oil, which has a high proportion of glycerides with saturated fatty acids, is solid.

In the small intestine fats are very finely emulsified, with the aid of bile salts. Fatty acids are released from some of them by the action of the pancreatic enzyme, lipase. These are absorbed, and recombined with glycerol in the intestinal wall to form neutral fats. The rest of the fats are absorbed directly, as a fine emulsion, by way of the lymphatics. They may be oxidized in the muscles or liver to provide energy or they may be transported to the storage depots. There, along with the fat formed from surplus carbohydrate, they are retained until the demand for energy increases above the dietary intake. Meanwhile, the fat stores afford a useful form of insulation that helps to conserve body temperature.

Measurement of dietary energy. Until recently it has been customary to measure dietary energy in terms of heat. The unit is the small calorie, defined as the amount of heat required to raise one gram of water through one centigrade degree. Since this is a very minute amount, it is more convenient to use the kilocalorie (kcal), a thousand small calories. However, the calorie is not an absolute unit, and can vary slightly according to circumstances, e.g. with altitude. For this reason, under the recently agreed International System of Units, the joule (J) has superseded the calorie as a unit of energy. The joule is a measure of work done* and, for nutritional purposes, one small calorie is equivalent to 4·184 joules.

The *gross energy* (GE) content of a diet, or component of a diet, is measured in terms of the heat generated when a given quantity is completely combusted in a bomb calorimeter.

* A joule is the work done when a force of one newton acts through one metre. A newton is the force required to accelerate one kilogram by one metre per second per second.

However, not all the gross energy of a diet is necessarily available to the animal that eats it; some may not be digested (e.g. cellulose) and some that has been digested may be lost in the urine, flatus, sweat etc. or dissipated in heat that cannot be used by the body. The *digestible energy* (DE) takes into account the faecal losses; it is the gross energy of the diet minus the gross energy (i.e. heat of combustion) of the corresponding faeces. The *metabolizable energy* (ME) is a value frequently used with reference to laboratory animal diets. It is probably the most meaningful index of nutritive value since it represents that part of the dietary energy that is transformed into energy useful to the animal. It is determined by subtracting the energy lost in the urine and combustible gaseous products of digestion from the digestible energy. A further value, the *net energy*, takes into account all energy lost, including that wasted in unnecessary heat. It is principally of importance in farm animal production, because it is a measure of the proportion of dietary energy that is converted into body tissue, milk, eggs, etc. However, it varies with the species, the physiological state of the animal and the environmental temperature and has little relevance in laboratory animal practice.

Determinations of the digestible, metaboliz-able and net energy values of feedingstuffs are made by direct experiments on animals. The results, strictly speaking, apply only to the particular species of animal on which the test was done, but variation between species is not great. Metabolizable energy values of many of the common ingredients of animal diets have been measured on farm livestock, but little or no work has been done with laboratory animals. The most comprehensive range of feedingstuffs has been tested on poultry, and a selection of average values based on published results is given in Table III. Although they may not be strictly valid for other species, in the absence of direct information on laboratory animals they offer a workable approximation.

Relationship of energy to other dietary constituents. It is impracticable to state a quantitative energy requirement for any species of animal because the need is influenced by variable factors such as physical activity, ambient temperature, rate of growth and physiological state. In practice, most laboratory species are fed *ad libitum*, and usually in this circumstance the animal will voluntarily regulate its food intake to satisfy its requirement for energy. It is of little importance whether the energy is obtained from carbohydrate or fat. There is no evidence that fat *per se* is required,

TABLE III

Metabolizable energy values (kcal/g dry matter) of some animal feedingstuffs

Barley	2·97	Casein	4·50	Glucose	3·70
Maize	3·90	Fish meal	2·87	Starch	4·10
Oats	2·97	Groundnut meal	3·05	Sucrose	3·74
Wheat	3·42	Meat meal	2·46	Maize oil	8·76
Wheat bran	1·41	Soya bean meal	2·62	Animal fat	7·10

Values kindly provided through the courtesy of Dr. W. Bolton (ARC Poultry Research Centre, Edinburgh.)

TABLE IV

Comparison of normal and high-fat rat diets

Ingredient	ME value (kcal/g)	Normal diet Quantity (g)	Normal diet ME contributed (kcal/kg diet)	High fat diet Quantity (g)	High fat diet ME contributed (kcal/kg diet)
Casein	4·5	230	1035	230	1035
Maize starch	4·1	490	2009	390	1599
Sucrose	3·7	170	629	170	629
Maize oil	8·8	50	440	150	1320
Salt mixture	negligible	50	—	50	—
Vitamin supplement	negligible	10	—	10	—
Total		1000	4113	1000	4583

except as a source of essential fatty acids (see below) and as a vehicle for the fat-soluble vitamins. However, it can be seen from Table III that fats have a higher calorific value than carbohydrates and, in fact, provide about twice as much energy per unit weight. On the assumption that an animal eats enough food to satisfy its energy needs, it follows that the weight of food eaten will be determined by its ME content. If a diet contains a lot of fat its ME value will be higher, weight for weight, than a similar diet with less fat. In Table IV two variations of a simple experimental rat diet are considered.

In the high-fat diet, 10 per cent of the starch has been replaced by an equal weight of maize oil. The ME values for the protein, carbohydrates and fat have been taken from Table III and the amount of energy each contributes has been calculated. (Strictly speaking, an allowance should have been made for moisture content, since the values in Table III are expressed per gram of dry matter.) The sum of the individual contributions gives the total ME in a kilogram of diet. It can be seen that substitution of some of the starch by maize oil has led to an increase in ME of about 11 per cent, from 4·1 kcal (17·2 kJ)/g for the normal diet to 4·6 kcal (19·4 kJ)/g for the high-fat diet. Thus the food intake of rats given the high-fat diet is likely to be about 11 per cent less than that of those given the normal diet and, inevitably, their intake of protein, vitamins and minerals will be concomitantly reduced. It is wise, therefore, to think of quantitative requirements for essential nutrients as a proportion of the energy content, rather than of the weight, of a diet. For instance, the dietary standards for rats and mice recommended by the Laboratory Animal Science Association (1969) refer to diets with ME contents of between 3·6 and 3·8 kcal (15 to 16 kJ)/g, and the suggestion is made that in diets of significantly higher energy content the levels of other nutrients should be proportionately increased.

Inorganic elements

The importance of many of the inorganic elements to the proper growth and functions of the animal body has long been recognised. The two needed in greatest amounts are calcium and phosphorus, which together constitute the major parts of the skeleton and teeth. Sodium and potassium are quantitatively next in importance since they are largely responsible for maintaining the correct osmotic pressure and acid/base balance of the tissue fluids. Others, often referred to as trace elements, are required in much smaller quantities but are no less essential. Iron, for example, is a vital constituent of the blood pigment, haemoglobin, and molybdenum plays an important role in several enzyme systems. Some of the trace elements are required in such minute amounts that there is no need to add them to the diet. They occur in sufficient quantity in the major dietary components, in the drinking water, or as impurities in the mineral supplements. A brief discussion follows of the elements of most importance in the nutrition of laboratory animals.

Calcium and phosphorus. These two elements are generally considered together because they are very closely linked in their physiological functions. Their major role is in bone and tooth formation; almost all the calcium and 80 per cent of the phosphorus in the animal body are found in the skeleton and teeth. The remainder occurs in the soft tissue, where both elements are concerned in the buffering reactions that maintain equilibrium in the tissue fluids. Phosphorus plays an important part in many enzyme systems, particularly those concerned in the release of energy. The presence of calcium ions is essential in the mechanism of blood coagulation.

When considering dietary allowances of calcium and phosphorus it is important to ensure that there is not only enough of each, but also that they are in the correct ratio to one another. If there is a gross excess of either element, the other cannot be efficiently utilized. For most species the optimal dietary ratio of calcium to phosphorus lies somewhere between 2:1 and 1:1. Vitamin D is necessary for the proper absorption of calcium from the intestine and deposition of calcium phosphate into bone. A deficiency of vitamin D can be partially, but not entirely, overcome by increasing the levels of calcium and phosphorus in the diet.

Young growing animals have a great demand for calcium and phosphorus while the skeleton is being laid down. If either element is lacking the animal suffers from rickets, a condition in which the bones fail to ossify and become stunted and malformed. In non-breeding adults the need for calcium and phosphorus is relatively low, but gestating and lactating females must be given a plentiful supply of these elements in the diet to compensate for the amounts they transfer to the developing embryos or secrete in the milk. Laying birds need both elements, but particularly calcium, for eggshell formation; hence their dietary requirement is high and is related to their rate of egg-production. If the dietary intake is inadequate the animal is said to be in negative calcium balance and, under the influence of the parathyroid hormones, calcium and phosphorus are mobilized from the bones in order to satisfy the demands of other physiological functions. If a negative balance persists in adults, osteoporosis develops, a condition in which the bones are gradually demineralised until their structure deteriorates and they become brittle and easily fractured.

Cereals are a poor source of calcium and, unless a diet contains a generous quantity of animal products such as dried milk or meat-and-bone meal, extra calcium must be given as, for instance, limestone flour, sterilized bone meal or calcium phosphate. Phosphorus is more abundant in plant foods, particularly in bran and other roughage, but much of it is in the form of phytic acid, with which calcium and some other elements form insoluble salts. In this form the phosphorus is not readily absorbed and, furthermore, the calcium or other elements combined with it are also unavailable to the animal. The ability to utilize phytin phosphorus seems to differ between species. One reason for variation is that phytin can be broken down by microbial enzymes. Ruminants can therefore utilize it satisfactorily, but for non-ruminants it is far less effective than inorganic phosphates. The availability of phytin phosphorus is increased if the diet contains generous amounts of vitamin D. Excessive amounts of calcium and phosphorus in the diet can be harmful because, through the formation of double salts, the absorption of some other minerals can be impaired.

Magnesium. Magnesium is another element that occurs mainly in the skeletal tissue and which, like calcium and phosphorus, can be mobilized from bone if the dietary intake is inadequate. Its presence is essential to a number of very important enzyme systems. A young animal deprived of magnesium grows poorly, with derangements in bone formation. Severe deficiency at any age results in neurological disturbances, leading finally to death in convulsions.

The absorption and excretion of magnesium closely follow those of calcium and phosphorus. It has been shown in several species that a high dietary level of calcium and phosphorus increases the need for magnesium, as a result either of impaired absorption, accelerated excretion, or a combination of both. Conversely, an excess of magnesium can disturb the animal's ability to retain calcium. At normal levels of calcium and phosphorus intake the magnesium content of animal foods is usually adequate to satisfy requirements without further supplementation.

Sodium, potassium and chlorine. These three elements are together mainly responsible for maintaining the correct water balance in the tissues. Potassium is located chiefly inside the cells, sodium in the extra-cellular fluids. The concentration of sodium, potassium, chloride and bicarbonate ions controls the osmotic activity into and out of the cells, and regulates the pH of the body fluids. Calcium and magnesium are also concerned in these functions. There is enough naturally-occurring potassium in the conventional animal foods to satisfy requirements but it is customary to add common salt to provide sodium and chlorine. This practice has the added advantage that salt stimulates salivary secretion and thus acts as an appetiser. If there is an unlimited water supply a high dietary salt content is unlikely to be harmful because an animal will simply drink more and rapidly excrete the excess salt in the urine. With a restricted water intake, however, the dietary level of sodium chloride should not grossly exceed the requirement.

Iron and copper. Iron is a constituent of haemoglobin, the pigment of the blood cells that carries oxygen to all tissues of the body. At any stage in an animal's life inadequate intake of iron leads to anaemia. It is only required in relatively small amounts which, in most circumstances, are present in the natural ingredients of laboratory animal diets. Surprisingly, milk has a very low content of iron and the suckling animal has to rely on the stores it accumulates during gestation to carry it over the first few weeks of life. Thus iron is of particular importance in the nutrition of breeding females, who must receive in the diet enough iron to ensure that the offspring are born with ample stores and that the mother's own reserves are not seriously depleted.

Copper is required in even smaller amounts and is also essential for haemoglobin synthesis. Without it iron cannot be incorporated into haemoglobin, hence an apparent deficiency of iron may, in fact, be a lack of copper. Milk contains very little copper which, like iron, should therefore be supplied to breeding females. Copper is also concerned in the synthesis of keratin and of melanin, hence the hair of animals deprived of copper is sparse and poorly pigmented.

The copper content of natural foodstuffs varies. In some areas of the world the soil is extremely low in copper and the crops and pasture do not contain enough to satisfy the requirement of animals feeding on them. The condition of sway-back in lambs was shown to be due to this cause. Care should be exercised in supplementing diets, however, since an excess of copper can be toxic. Although in young pigs a level of 250 ppm copper in the diet has been shown to promote growth, and even twice that level can be tolerated with no harmful effects, chicks and rats develop a haemolytic anaemia at comparable levels of supplementation.

Cobalt. Cobalt deficiency has long been recognized as a wasting disease in cattle and sheep grazing on pasture in areas where cobalt is lacking in the soil. Recent evidence shows, however, that the direct cause of the condition is a lack of vitamin B_{12}, which contains cobalt. The vitamin is synthesized by micro-organisms in the alimentary tract, and in ruminants, which depend for many of their vitamin requirements on products synthesized by rumen micro-organisms, a lack of cobalt leads to a deficiency of vitamin B_{12}. In single-stomached animals vitamin B_{12} is synthesized by bacteria in the lower gut but, except in coprophagous species, little or none of the resulting product can be utilized. Such animals require a direct source of vitamin B_{12} and, as far as can be shown, they have no need for cobalt except as a constituent of the vitamin.

Manganese. Manganese functions in several enzyme systems and a serious deficiency is apparent in adverse effects on the reproductive system of both males and females. Delayed sexual maturity, irregular oestrous cycles, resorbed foetuses and weak offspring have been reported. Manganese is also concerned in normal bone growth, although the precise mechanism is not understood. In chicks, manganese deficiency leads to perosis, in which swollen hock joints permit the Achilles tendon to slip, and the bird becomes lame. In mammals malformations of bone resulting from manganese deficiency take the form of crooked legs and enlarged hocks. A high dietary intake of calcium and phosphorus can interfere with the availability of manganese and so increase the requirement.

Zinc. Zinc is an essential component of the respiratory enzyme, carbonic anhydrase. It also appears to play a part in the process of keratinization and in bone formation. A deficiency therefore results in generally retarded growth, poor hair development and malformed bones. Its availability from the diet can be adversely affected by high levels of calcium and by phytic acid. Soya bean meal has a particularly adverse effect on the availability of zinc, partly because of its phytin content but also because there appears to be another zinc-binding factor in soya bean meal. For these reasons the level of zinc naturally present in animal foods may be inadequate in some circumstances. Laboratory animals can take in a significant quantity of zinc from galvanized cages or food and water vessels. High levels of zinc can be toxic; anaemia has been reported in rats, as a result of antagonism between zinc and copper.

Iodine. The thyroid gland produces a hormone, thyroxine, which contains iodine. If the dietary intake of iodine is not adequate, thyroxine production cannot be maintained, with the result that the animal becomes sluggish, overweight, and sexually retarded. The thyroid gland becomes enlarged, a condition known as goitre. Iodine deficiency is not common, except in areas where the drinking-water, the soil and consequently the crops, have an exceptionally low iodine content. In these circumstances it is usual to provide iodine in the diet, frequently in the form of iodized salt.

Selenium. Although selenium can take over some of the functions of vitamin E there is still some doubt as to whether or not it is an essential element. It cannot prevent the reproductive disturbances in animals deprived of vitamin E, but it effectively counteracts the liver necrosis in rats and exudative diathesis in chicks that are observed under certain specialized dietary conditions (see reviews by Schwartz, 1961 and Scott, 1962). Crops grown on soil rich in selenium may contain enough to cause toxic effects in animals fed on them. Although there are well-documented reports of selenium toxicity in farm animals in some areas of the world, the condition is unlikely to arise under the conditions of laboratory animal feeding currently prevailing in the UK.

Vitamins

There is no hard and fast definition of a vitamin but it can be described as an organic substance which, although required in only very small amounts in the diet, is nevertheless essential to animal health and well-being. From the relatively minute amounts that are needed it seems highly likely that vitamins function as biological catalysts and, in fact, many of them have already been shown to take part in one or more of the enzyme systems that catalyse the chemical reactions continuously proceeding in living matter. In general, the effects of deficiency of a particular vitamin are broadly similar between species, although some manifestations may be peculiar to a particular species or class of animal (Coates, 1968).

When the existence of vitamins as accessory food factors was first recognized their chemical identity was unknown and they were given names based on letters of the alphabet, allotted in order of their discovery. Their activity was measured in units which were defined in terms of national or international standard reference preparations. Now that the chemical constitution of all the established vitamins is known there is no need for trivial names or arbitrary units, except where a group of closely related substances all have vitamin activity. In Table V the vitamins are listed with their accepted chemical names and a brief description of the major effects of deprivation.

Vitamins A, D, E and K are fat-soluble. They occur in the fatty components of foods and are absorbed with them. A diet should always contain at least a small quantity (say, 1 per cent) of fat to act as a vehicle for these vitamins. The fat-soluble vitamins can all be stored in the animal body, mainly in the liver. It is possible for an animal to live for relatively long periods without a dietary source so long as it has ample tissue reserves on which to draw. The vitamins B and C are water-soluble. Some of the B vitamins occur in foods in combination with large molecules, frequently proteins, from which they are usually readily released by the animal's digestive processes when they reach the alimentary tract. These bound forms are not so readily available to micro-organisms, a fact of some significance in the estimation of the vitamin B content of foods. A very convenient method of measuring vitamins of the B complex is by microbiological assay. This technique is based on the fact that the growth of a vitamin-requiring micro-organism is quantitatively related to the concentration of vitamin in the culture medium. In applying microbiological assays to dietary materials it is essential to ensure that the chosen method will detect bound forms of the vitamin, otherwise the result may be a considerable underestimate of the value of the food for higher animals. The water-soluble vitamins, with the exception of vitamin B_{12}, are not stored to any great extent and signs of deficiency rapidly develop if the dietary supply is inadequate.

Vitamin A. Preformed vitamin A occurs in animal tissues in the form of its alcohol (retinol), retinyl esters or its aldehyde (retinal). In plants it is present as its pro-vitamins, i.e. those carotenoid pigments that can be converted to vitamin A in the animal body.

Fish-liver oils are a rich, but not very stable, form of vitamin A. The polyunsaturated fatty acids in the oils readily become oxidized on exposure to air and the resulting products accelerate destruction of vitamin A. It is unwise to use fish-oils as a source of vitamin A in diets unless they have been stabilized by the addition of an anti-oxidant. Vitamin A can be prepared synthetically on a commercial scale and dry products, usually retinyl acetate or palmitate, are often used to supplement animal diets. Even in this form the vitamin is not stable when it has been dispersed in the diet and thus exposed in a finely divided state to atmospheric oxygen. To obviate losses by oxidation, preparations have been developed in which the particles of vitamin A are dispensed in a protective wax or gelatin coating. In this form the vitamin remains stable in the diet throughout long storage periods and is fully available to the animal if the diet contains some fat. A disadvantage is that the coated beadlets are rather large, and great care must be taken during preparation of the diet to ensure that they are evenly dispersed throughout.

Several carotenoids have vitamin A activity. The most significant is β-carotene, an orange pigment occurring in maize, dried grass and carrots. It is made up of two molecules of vitamin A and, theoretically, it can be broken down in the small intestine to yield half its weight of vitamin A. In practice, however, the conversion of carotene to vitamin A is influenced by a variety of factors and the ideal conversion ratio is seldom achieved. It is of interest to note that the cat is unable to effect the conversion of carotenoids, and must receive preformed vitamin A in its diet. β-carotene is more stable than vitamin A, but is subject to some oxidative destruction during storage.

Vitamin A can have toxic effects if given in great excess. Adverse signs of overdose have been demonstrated in many species of laboratory animals, and have included skeletal deformities, anaemia, retarded growth and poor hair development (Deuel, 1957a). However, the toxic effects have only been observed with doses several thousands of times in excess of normal dietary intakes and are unlikely to occur under ordinary conditions of feeding.

Vitamin D. Several forms of vitamin D occur in nature. They arise as the result of ultraviolet irradiation of their precursors, which are all sterols of related chemical structure. The two of most importance in animal nutrition are ergocalciferol (vitamin D_2), which is formed on irradiation of the vegetable sterol, ergosterol, and cholecalciferol (vitamin D_3) which results when the animal sterol, 7-dehydrocholesterol, is irradiated. Fish-liver oils are an excellent source of vitamin D_3. Both cholecalciferol and ergocalciferol are produced commercially and are used to supplement diets, either in oily solution or as dry powders. The vitamins D are relatively stable substances. Birds cannot utilize ergocalciferol very effectively, hence synthetic cholecalciferol or fish-liver oils should always be given in their diets. In general, mammals use both forms of vitamin D indiscriminately, although recent evidence suggests that New World monkeys may have a specific requirement for vitamin D_3. When an animal is exposed to strong sunlight it can synthesize vitamin D from precursors in its superficial tissues and its need for a dietary supplement is consequently reduced.

High intakes of vitamin D can be toxic, resulting in demineralization of the bones and deposition of calcium in the soft tissues. Evidence with rats and chicks suggests that at least one hundred times the normal requirements (and probably even more) can be tolerated without harmful effect (Deuel, 1957b; Taylor *et al.*, 1968).

Vitamin E. The general name, vitamin E, is applied to the tocopherol compounds, the most biologically active of which is α-tocopherol. The vitamins E are widely distributed in the vegetable components of diets, and are particularly abundant in the germ of cereal grains. Wheat germ oil has long been used as a dietary supplement of vitamin E, although it is now more usual to provide the synthetic dl-α-tocopheryl acetate in oily solution or as a dry powder.

Vitamin E is concerned in a variety of physiological functions, but the effects of deficiency differ markedly between species. In certain circumstances vitamin E behaves as an antioxidant in the tissues, and it can be quite successfully replaced by other antioxidants such as DPPD (diphenyl-p-phenylenediamine). As previously mentioned, in some of its effects it is interchangeable with selenium (see reviews by Moore, 1962 and Green, 1962). In laboratory animals the most important function of vitamin E is in the maintenance of normal reproduction. Deficiency leads to infertility in the male and, in the female, failure to carry her litter successfully to term. There is no evidence that this role of vitamin E can be satisfactorily taken over by other substances.

Table V

Effects of vitamin deficiencies in laboratory animals

Vitamin	Chemical name	Effects of deprivation	Comments
A	Retinol (1 iu ≡ 0·30 μg retinol or 0·34 μg retinyl acetate)	Poor growth. Reproductive failure. Visual disturbances—impaired dark adaptation. Keratinization of epithelia—particularly the cornea. Xerophthalmia (rat). Increased susceptibility to infection.	Some carotenoids have vitamin A activity. Theoretically, 0·6 μg β-carotene = 0·3 μg retinol.
B_1	Thiamine	Retarded growth. Polyneuritis, resulting in incoordinated movements (ataxia), neck retraction (opisthotonus) in birds, convulsions. Cardiac failure.	Thiamine is concerned in carbohydrate metabolism. Requirement is reduced if dietary energy is supplied as fat.
B_2	Riboflavin	Retarded growth. Loss of hair. Opacity of the cornea. Neurological abnormalities. 'Curled-toe' paralysis (chick). Dermatitis (dog, Rhesus monkey).	
—	Nicotinic acid	Loss of appetite. Depressed growth rate. Black tongue (dog). Rough coat, anaemia (pig). Perosis (chick).	Nicotinic acid can be synthesized in the animal body from tryptophan. Severe deficiency is only encountered on diets low in tryptophan.
—	Pantothenic acid	Loss of hair pigment (achromotrichia). Adrenal necrosis. Loss of appetite. Blood-stained whiskers, 'spectacle eye' (rat). Dermatitis (chick).	
B_6	Pyridoxine	Dermatitis. Microcytic anaemia. Hyperexcitability, convulsions. Marked decrease in growth. Decreased antibody production.	Pyridoxol, pyridoxal and pyridoxamine and their corresponding phosphates are all equally active.
—	Biotin	Dermatitis. Paralysis. 'Spectacle eye' (rat). Reproductive failure. Alopecia (mice). Perosis (chick).	Biotin forms a complex with avidin, a factor in raw egg white. Severe deficiency is only observed if raw egg white is included in the diet.
—	Folic acid complex. Pteroylmonoglutamic acid (and allied compounds with more than one glutamic acid residue)	Poor growth. Macrocytic anaemia. Leucopenia.	Many species appear to derive some folic acid from microbial synthesis in the gut. In such animals, severe deficiency only occurs if sulphonamides are included in the diet.
B_{12}	Cobalamin	Retarded growth. Reproductive failure.	Requirement for vitamin B_{12} is reduced on diets containing ample methionine or choline.
C	Ascorbic acid	Scurvy (primates, guinea-pig).	Most other species can synthesize enough in their tissues and do not need a dietary source.
D_2	Ergocalciferol (1 iu vitamin D ≡ 0·025 μg cholecalciferol)	Rickets in young animals. Osteoporosis in adults. Black feather pigmentation (chick).	Vitamins D_2 and D_3 equally effective for most mammals. Birds utilize vitamin D_2 very inefficiently.
D_3	Cholecalciferol (1 iu = 1 D ≡ 0·025 μg cholecalciferol)		Animals and birds synthesize vitamin D if exposed to UV radiation. Dietary requirement is therefore reduced if they are maintained in bright sunlight.
E	The tocopherols (1 iu = 1 mg synthetic dl-a-tocopheryl acetate)	Muscular dystrophy. Reproductive failure (except in cattle and sheep). Encephalomalacia. Exudative diathesis. (chick) Liver necrosis. Yellow fat (rat, pig, mink).	Some signs of deficiency (e.g. muscular dystrophy, encephalomalacia) can be prevented by antioxidants and others (e.g. exudative diathesis, liver necrosis) by selenium.
K_1 K_2 K_3	Phylloquinone Menaquinone-7 Menaphthone	Haemorrhagic disease prolonged blood clotting time.	Vitamin K_2 is synthesized by micro-organisms in the alimentary tract.

Vitamin E is relatively stable except in the presence of rancid fats, so if a diet contains a high level of unsaturated fatty acids it should be liberally supplemented with vitamin E.

Vitamin K. The two main naturally-occurring forms are vitamin K_1 (phylloquinone, or 2-methyl-3-phytyl-1, 4-naphthoquinone), which is found in green plants, and vitamin K_2 (menaquinone-7, or 2-methyl-3-difarnesyl-1, 4-naphthoquinone) which is synthesized by bacteria. A simpler compound, vitamin K_3 (menaphthone or 2-methyl-1, 4-naphtho-quinone) can be synthesized chemically on a large scale and, being much cheaper than the natural forms, is commonly used to supplement diets. It has the added advantage that it forms a water-soluble complex with sodium bisulphite or sodium diphosphate.

Vitamin K is concerned in the mechanism of blood clotting and a lack of it results in a haemorrhagic disease in which bleeding occurs very readily as a result of slight abrasion. Its significance in the nutrition of rats and mice has been reviewed by Hacking and Lane-Petter (1968). Vitamin K is synthesized by intestinal bacteria, and coprophagous animals derive at least part of their requirement by ingestion of their excreta. A recent report of haemorrhagic disease in a specified-pathogen-free rat colony (Gaunt and Lane-Petter, 1967), which was cured by administration of menaphthone, suggests that the contribution of vitamin K from micro-organisms in conventionally reared animals may be nutritionally very significant. In SPF environments the microflora of the alimentary tract is usually limited and the bacteria responsible for production of vitamin K may be absent. In this event it seems likely that the dietary requirement of SPF animals will be many times higher than that of their convention-ally kept counterparts. Germ-free animals certainly depend entirely on their diet for a supply of vitamin K and there is evidence that they utilize the synthetic vitamin K_3 less well than vitamin K_1 (Gustafsson *et al.*, 1962; Wostmann *et al.*, 1963).

At many times the physiological level vitamin K_3 has been shown to cause growth inhibition and anaemias in several species. No toxic effects have been observed with similar doses of vitamin K_1.

The vitamin B complex. The vitamins B have been linked together by accident of history rather than by any close association of structure or function. As far as is known, they all play the part of co-factors in one or more enzyme systems. They occur together in nature in liver and, with the exception of vitamin B_{12}, in cereals and yeast. It was in the course of attempts to isolate the anti-beriberi and anti-pellagra factors from such starting materials that the existence of further vitamins was recognized.

In diets of natural ingredients the cereal components will contribute largely to the animal's requirements of the B vitamins, although it should be noted that nicotinic acid occurs in cereals in a bound form that is only poorly utilized by most species (Chaudhuri and Kodicek, 1960). Dried yeast and liver extracts are excellent natural sources of the B vitamins, but as most of them can now be produced commercially at reasonable cost it is becoming common practice to supplement diets with concentrates of the pure vitamins.

There is little doubt that all the vitamins of the B complex are synthesized microbially in the gastro-intestinal tract. Ruminants do not need a source in their diet because they are able to subsist on the vitamins produced by rumen organisms. Rabbits, guinea-pigs, rats and mice are likely to obtain at least part of their require-ments when they eat their faeces; it is well-known, for instance, that folic acid deficiency cannot be demonstrated in the rat unless sulphonamides are added to the diet to suppress the activity of the gut microflora. As has been discussed earlier with reference to vitamin K, there is a danger that animals maintained under SPF conditions may not be associated with the organisms responsible for vitamin synthesis, and such animals should be given ample vitamin supplements in the diet. Vitamin B_{12} is some-what remarkable in that a lack of it is seldom observed under practical feeding conditions, even though it is present only in substances of animal origin (liver is a rich source). The reasons are twofold; first, it is synthesized microbially in the gut and second, it is efficiently stored in the liver, kidneys and other tissues. Significant reserves are passed over from dam to foetus and, as it is needed in very minute amounts, are sufficient to last the offspring until well into adult life.

Requirements of some of the B vitamins can be influenced by the constituents of the diet. Nicotinic acid is synthesized in the tissues from its precursor, tryptophan. However, tryptophan is one of the essential amino acids and will be available for synthesis of nicotinic acid only if it is in ample supply. Experiments with rats and chicks suggests that the minimal requirement for tryptophan is about 0·15 per cent of the diet. Levels in excess of this can be used to form nicotinic acid and so reduce the need for it in the diet. One of the functions of vitamin B_{12} is as a catalyst in the formation of methyl groups which are subsequently incorporated into methionine. Thus if a diet contains plenty of methionine, or of methylating compounds such as choline, the need for vitamin B_{12} is spared. Vitamin B_{12} cannot be completely replaced by methionine, however, as it is also necessary for other unrelated reactions. There is a special relationship between thiamine and the source of energy in a diet. Thiamine plays an important role in the metabolism of carbohydrates, but is not specifically concerned in the production of energy from lipids. In consequence the need for thiamine is considerably greater when the dietary source of energy is carbohydrate than when fat is the major energy component.

Vitamin C. Vitamin C occurs in nature as ascorbic acid and its corresponding dehydro-compound. It is not produced by the gut micro-organisms but very many species of animal can synthesize vitamin C in their own tissues and so have no need of a dietary source. The two important exceptions are primates and guinea-pigs, which lack the necessary enzyme to perform the last step in the synthesis and must therefore obtain their vitamin C from an external source. Vitamin C occurs abundantly in fresh fruits and vegetables but is not present to any great extent in the ingredients of laboratory animal diets as they are currently formulated. It can be given in the drinking water but, as it is liable to destruction by oxidation, it is better added as a supplement to the diet.

Other essential nutrients

Essential fatty acids. Although the animal body is capable of synthesizing fats from carbohydrate (cf p. 31), the products are mainly saturated fatty acids, or oleic acid, with one double bond; the ability to form fatty acids with two or more double bonds in the molecule appears to be lacking. It has been shown in several animal species that diets completely devoid of fat do not support optimal growth. Scaly skin, poor hair growth and impaired reproductive performance have been observed after long periods on a fat-free diet. The effects can be prevented by including linolenic, linoleic or arachidonic acids in the diet, which have thus become known as the 'essential' fatty acids. It now seems generally agreed that arachidonic acid is the biologically active compound and that linoleic acid is converted to it in the tissues. Linolenic acid is not convertible to arachidonic, and its metabolic role is still unclear. Arachidonic acid is found only in animal fats. The other two are constituents of animal and vegetable fats and are particularly abundant in maize, soya bean and groundnut oils.

The biological role of the essential fatty acids has been discussed by Deuel (1957c), but there is little information on the quantitative requirements of any species. Holman (1960) recommends for rats that 1 per cent of the dietary calories should be supplied in the form of linoleic acid, and several groups of workers have suggested a level of between 1 and 2 per cent of the total dietary calories for the chick. This level of requirement is likely to be achieved on the cereal-based diets in common use at present but there are indications that the need for essential fatty acids is increased on diets containing a high level of saturated fats.

Choline. Choline differs from the vitamins in that it is not a catalyst but instead provides the basic material from which several biologically important substances are formed. Its ester, acetylcholine, acts as a chemical transmitter of nerve stimuli. Choline contains methyl groups which can readily be removed and used in the synthesis of other important methyl-containing biological compounds such as methionine. Choline forms part of the phospholipid, lecithin, which is concerned in the transport of lipids. In animals given diets low in choline, fat accumulates in the liver. The ability to prevent fatty infiltration of the liver, described as a 'lipotropic' action, is shared by several other compounds including methionine, betaine and inositol, but the lipotropic mechanism is not properly understood. In the chick, choline is one of the factors concerned in prevention of perosis but, again, the mechanism of action is unknown.

Choline is synthesized in the tissues, but not at a rate fast enough to fulfil all its functions, and some must be provided in the diet. If there is a good supply of other methylating compounds such as methionine or betaine, the need for choline is correspondingly less. Choline deficiency is unlikely on diets of natural ingredients since it occurs in very many foods, particularly in the phospholipid fraction of animal and vegetable fats.

Inositol. Although inositol is sometimes listed among the essential nutrients, the evidence to justify its inclusion is not convincing. Structurally it is similar to the simple sugars. It is found in animal tissues and, in the form of a phospholipid, is a constituent of brain. In certain circumstances it has a lipotropic action and has been claimed to prevent alopecia in mice. As far as is known, it can be synthesized in the tissues at a rate fast enough to satisfy ordinary requirements. Furthermore, it is formed by microbial action in the alimentary tract. It is widely distributed in plant and animal products. It occurs in plants as its hexaphosphate, phytin (see p. 34) and it is also found in wheat germ, soya, groundnut and yeast.

Water

The physical and chemical properties of water combine to make it the most important constituent of living matter. It is also the most abundant, comprising at least 70 per cent of most soft tissue. An animal can withstand starvation for long periods but deprivation of water is lethal in a very few days.

Water is the vehicle in which most of the other constituents of the body are dissolved or suspended. About half the total body water is inside the cells; the rest is extracellular. The products of digestion are transported in an aqueous medium from their site of absorption in the gastro-intestinal tract to the tissues, and many of the consequent waste products are excreted in solution in the urine. Water is an essential constituent of any metabolic reaction involving hydrolysis. It plays an important role in heat regulation of the body. Because of its high heat capacity it can absorb much of the heat generated during metabolism, so that body temperature does not rise excessively. Heat is dissipated through the skin and in the water vapour exhaled through the lungs. The high latent heat of evaporation of water means that relatively large heat losses are achieved by the loss of a small volume of water.

Water is being continually lost from the body in the urine and faeces, by exhaled breath and through the skin. In order to maintain the correct fluid balance on which cell function depends, the lost water must be replenished. Most of an animal's water supply is taken in directly as fluid or as a component of the food, but a small part of the requirement arises as the product of certain metabolic processes. Metabolic water is formed, for instance, during the utilization of glucose when, in order to provide energy, it is converted in the tissues to carbon dioxide and water.

Bruce (1950) estimated the average daily water intake of several species of laboratory animal and calculated a requirement in terms of body weight. However, an animal's need for water is influenced by environmental, physiological and dietary factors. In a warm environment the animal will need to lose heat and will consequently pass out a great deal of water in its breath and through its skin. (In animals with thick fur and few sweat glands, the lungs are a much more important means of losing heat.) Growing or gestating animals will use water for the formation of new tissue, lactating females will pass out a large volume in the milk. The volume of urine excreted varies with the composition of the diet. Proteins give rise to nitrogenous endproducts, particularly urea, that must be eliminated in the urine. If the concentration of electrolytes in the intracellular fluid rises the animal becomes thirsty and drinks water to restore the balance. Thus diets high in protein or in salts increase the need for water. In view of the variable nature of an animal's water requirement, and of the serious consequences of inadequacy, it is highly advisable that animals should have constant access to a supply of fresh water. No detrimental effects result from excessive intake, because over-consumption of water simply results in a more copious excretion until the fluid balance is readjusted to its optimal.

Fibre

The fibrous portion of a diet consists of substances like cellulose and lignin that have little or no nutritive value. It has, however, some desirable properties which depend on its physical nature. When a diet is eaten, fibre (roughage) helps to distribute it in the alimentary canal, preventing the formation of a solid mass that would be impenetrable by digestive enzymes. The addition of roughage to a diet usually increases its volume per unit weight. Within reasonable limits this is considered an advantage because bulky diets tend to distend the digestive tract and thereby increase its efficiency. Furthermore, the presence of undigested fibre in the lower gut has a laxative effect that aids the removal of food residues. For some species fibre contributes to the subtle qualities that make a diet acceptable.

Although a certain amount of fibre in the diet is advantageous, it must be remembered that it occupies space at the expense of more nutritive components. If excessive quantities are present there is a danger that, although an animal eats to capacity, it will be unable to take in enough diet to satisfy its nutrient needs.

Non-nutrient additives

It is well known that addition of minute amounts of antibiotics or other chemotherapeutic agents to the diets of young farm animals, particularly chicks and pigs, increases their growth rate. The generally accepted explanation is that antibiotics suppress some so far unidentified organism(s) that prevent the animal from realising its full growth potential. Although the judicious use of antibiotics as growth promoters has undoubtedly been of economic advantage, their indiscriminate addition to animal feeds incurs serious risk to human and animal health by encouraging the emergence of resistant strains of organisms that cause disease in man or animals. The pros and cons of antibiotic usage are extensively discussed in a report by a Joint Committee on the Use of Antibiotics in Animal Husbandry and Veterinary Medicine (1969). There seems no justification for the inclusion of antibiotics in the diet of laboratory animals. Rats, mice and rabbits given nutritionally adequate diets seem to derive little benefit from the addition of antibiotics to their diet and guinea-pigs may be positively harmed (see review by Braude et al., 1953). Medicated feeds should never be relied upon to compensate for poor husbandry. In the event of an outbreak of disease, it may be convenient to administer therapeutic agents in the diet or drinking water (coccidiostats or anthelmintics, for example), but this should only be a short-term expedient. Laboratory animals given an adequate diet and kept under proper conditions of hygiene have no need of drugs to maintain them at their optimal level of performance.

Formulation of diets for laboratory animals

In designing a diet for any type of animal the first requisite is a knowledge of the nutrient requirements of the species concerned. Such knowledge is regrettably sparse for all but the commonest species of laboratory animal. The available information has been summarized by the National Research Council (1962) and the Laboratory Animal Science Association (1969). The nutrient requirements of animals of agricultural importance have been more thoroughly studied and are discussed in publications by the Agricultural Research Council (poultry, 1963; ruminants, 1965; pigs, 1967). In the absence of experimentally determined estimates of requirement a system of trial and error, until optimal performance is attained, is the only approach. An intelligent guess can be based on the known nutrient requirements of a closely allied species, but it can be nothing more than a reasonable starting point. There is sufficient evidence of variation in nutrient requirements between strains (Porter, Lane-Petter and Horne, 1963; Porter and Lane-Petter, 1966) and even between individuals within a strain (Williams and Pelton, 1966; Williams and Deason, 1967) to point the danger of drawing too close an analogy between species.

Choice of ingredients

The choice of materials will largely be determined by availability and cost, the protein-containing foods being the most expensive. In stock diets, the amino acid balance can be made up from various mixtures of cereals and protein concentrates and it is common practice for manufacturers to ring the changes among these components in order to get the optimal nutritive value for the least cost. However, it is false economy to use poor quality raw materials, even at a lower price. Protein foods that have been damaged by over-heating or bad storage conditions will not contribute the calculated amounts of available amino acids; in cereal grains from which the husks have been inadequately separated the proportion of nutrients to fibre will be too low. The quality of raw materials varies widely between samples, as Cassidy (1957) has pointed out. The protein content of wheat, for example, can vary from as little as 8 per cent in poor quality soft wheat to as much as 19 per cent in a good quality hard wheat. Such differences in composition are not easily detectable without elaborate chemical or biological analyses. Nevertheless, the supplier should give some indication of the nutrient quality of his products wherever possible. Figures such as the crude protein level in meat or fish meals, calcium and phosphorus in bone meals or β-carotene in dried grass meal are all useful guides to the quality of the product to which they refer.

When the cereals and protein concentrates have been chosen, the energy value of the diet will inevitably have been fixed within fairly narrow limits. In farm livestock, if a fast rate of growth is desirable it is customary to increase the energy concentration in the diet by the addition of animal or vegetable fats. In laboratory animal practice there is no great advantage in high energy diets, since neither rapid growth of the young nor the laying down of fat is usually called for.

In calculating the amounts of vitamins and minerals that must be added, account should be taken of the quantities that may already be present in the major ingredients. On one hand, there is no need to incur additional expense by, for instance, adding to a cereal diet a complete supplement of vitamins of the B complex. On the other, positive harm can be done by over-supplementation with minerals—by, for example, addition of the total requirement of calcium to a diet containing a high level of fish meal.

The object of formulating a balanced diet is to provide the animal with its full requirement of nutrients in the right proportions and without unreasonable excess. It follows that, to attain this object, the animal must be fed solely on the diet, and no other food must be offered. If, in the interests of economy or variety, it is allowed to eat other foods of lower nutritional value—a handful of oats, a tidbit of carrot—its intake of the balanced diet will inevitably be reduced and it will not eat enough to satisfy its requirements of all nutrients. Should there be some good reason to offer supplementary foods their quantity and nutritive value should be estimated; the proportions of nutrients in the basal diet should then be adjusted (e.g. by increasing the protein content) so that the intake of diet plus supplementary food together cover the animal's needs.

Acceptability

Having achieved a nutritionally satisfactory formula, the diet has to be presented to the animal in a form that it will eat. Sensory characteristics like odour, flavour, texture and even colour all influence the acceptability and palatability of food but it is not always easy to decide which is likely to attract an animal (see review by Lane-Petter, 1957). Texture is an important consideration, from the point of view both of acceptability and ease of handling. Powdered diets are not easy to eat and are readily spilt and wasted. Wet mashes are more attractive but quickly go sour and mouldy. Pelleted or cubed diets do not suffer from either of these disadvantages and have become the preferred form of feeding. Commercially they are prepared by injecting steam into the mixed diet and then forcing it under pressure through metal dyes of appropriate size. Small batches of pellets can be successfully prepared in the laboratory by mixing the diet with an aqueous solution of methyl cellulose, forcing the resulting paste through a mincing machine and drying the cylinders of diet that emerge. Pellets are a hygienic and economical method of feeding because they can be supplied in suspended wire baskets from which they are not easily scattered and where they are out of reach of fouling. Their consistency is important. If they are too soft, powder rubs off in the storage sack or feeding basket and diet is wasted, but if they are too hard, the animals may not be able to eat them. Porter and Lane-Petter (1960) quote an example of a cubed diet that supported satisfactory performance of a hybrid strain of mouse, whereas a weaker inbred strain failed to thrive because the animals had difficulty in eating the pellets. The authors point out, however, that hardness and 'gnawability' are not necessarily synonymous.

The keeping quality of diets and dietary ingredients

When a balanced diet is formulated it is assumed that the recommended amounts of nutrients will be present when the diet is eaten by the animal. It must, therefore, retain its original level of nutrients throughout preparation and storage, or an allowance must be made for likely losses. Destruction of the nutritive quality of a diet usually results from interactions among its individual components, from atmospheric oxidation, or a combination of both; some dietary constituents (e.g. riboflavin, vitamin A) are sensitive to light. Deterioration is likely to be accelerated by the presence of moisture and by a rise in temperature. It follows that, to attain optimal keeping quality of a diet, the major considerations are an intelligent choice of the original ingredients and, after mixing, storage in a cool dry place, preferably not exposed to direct sunlight.

Some examples of relatively unstable dietary constituents have been quoted earlier in this chapter. The vitamins are, in general, the most labile. It is almost inevitable that some losses will occur of vitamins A, C and E, even under good conditions of storage, and it is wise to allow a reasonable excess in the original formula. In natural foods vitamins of the B complex frequently occur bound to proteins, in which form they are more stable than are supplements of the pure vitamins. The composition of the salt mixture can significantly alter the pH of a diet and thus have a marked effect on the stability of other components; thiamine, for instance, is less stable at alkaline than at acid pH.

With the increasing use of sterilized diets it is important to know the extent to which the nutrient content of diets is affected by the different processes of sterilization (see review by Coates, 1970). The damaging effects of heat on foods has long been recognized. The degree of damage is, in general, proportional to the temperature and the time of exposure. Many of the vitamins are labile to heat but, as they are relatively cheap, a diet can be generously supplemented to allow for possible losses; alternatively, a vitamin supplement sterilized without heat (e.g. by filtration) can be fed separately. Of more consequence is the damage done by heat to the nutritive value of protein foods. This is much less readily compensated for, since the degree of destruction is difficult to assess (see p. 43) and very expensive to make good with supplements of amino acids. In this respect gamma-irradiation appears to offer some advantage. From the information so far available (see review by Ley et al., 1969) doses of up to 7 Mrad, which is considerably in excess of the dose needed for sterilization, have little or no effect on the protein quality of laboratory animal diets. Some destruction of vitamins, particularly vitamins A, E, B_6 and thiamine, has been observed at doses of 5 Mrad (Coates et al., 1969). There is less information about the effects of ethylene oxide fumigation on the nutritive quality of animal diets. Although some losses of vitamins and amino acids have been recorded, Charles et al. (1965) reported successful maintenance of five generations of mice without evidence of dietary insufficiency.

Microbiological quality of diets

One of the easiest ways by which disease can be spread is through contaminated food or water. In the course of harvesting, handling and processing, the raw ingredients of animal diets may acquire a heavy load of microbial contaminants, among which may be organisms that cause disease in animals. In order to achieve a diet with a low count of organisms, the first essential is to pay careful attention to the microbiological quality of the original ingredients. Substances that have been subjected to heat treatment in the course of preparation, like fish meal, steamed bone flour and dried milk, usually have a low bacterial count. Cereals carry a heavier load, and animal products that have not been heated are very likely to be contaminated with undesirable organisms. After a diet is mixed it must be protected from extraneous contamination, particularly from the excreta of wild birds and rodents, which are highly likely to carry pathogens. It should be stored in an area into which they cannot penetrate, preferably in bins or strong paper sacks. When the diet is eventually fed to the animals it should be offered in quantities small enough to be eaten before it goes stale. The hoppers should be designed to prevent the food being fouled by excreta or flooded with the drinking water. Dry diet is preferable to wet mash, which presents an excellent medium for the growth of moulds. From the point of view of food hygiene, the feeding of green food is bad practice. It is rarely eaten completely, and wilted cabbage leaves lying in a cage encourage moulds and other undesirable microbial contaminants.

To obviate the risk of introducing disease via the food, it is not unusual to treat diets to some form of sterilization process to reduce their microbial load. Diets may be infected with bacteria, fungi, protozoa and helminths, many of which can be easily destroyed by, for instance, a mild heat treatment. However, spore-forming organisms are difficult to eradicate because the spores are very resistant to heat and other sterilizing agents. Sterility implies complete absence of any detectable living matter, and if a diet is described as sterile it should be demonstrably free from all microbial contaminants. The methods in common use for the sterilization of laboratory animal diets are autoclaving (i.e. the application of steam at pressures higher than atmospheric), fumigation with ethylene oxide gas and exposure to gamma-radiation (see review by Coates, 1970). Pasteurization, the application of mild heat (usually about 80°C), is a treatment that destroys vegetative organisms but not spores. A pasteurized diet therefore cannot be sterile, but the number of contaminants it carries is likely to be low.

Whatever the process of sterilization, it inevitably increases the cost of the diet, and may harm its nutrient content. Furthermore, it is not good enough for a diet to be sterile when it leaves the production plant—it must remain sterile through the vicissitudes of transport and storage until it is fed to the animals. This implies special packaging, which further adds to the cost, so it is as well to consider for what classes of animal the extra precaution is really justified. It is self-evident that gnotobiotic animals must be provided with sterile diet. Many workers take the precaution of using sterile diets also for SPF animals. Although the entry of non-pathogenic organisms into an SPF area is permissible, it is impossible to be completely sure that unwanted pathogens and parasites are absent from a diet unless it has been sterilized. For ordinary conventional animals such elaborate precautions are unnecessary. They need food that is free from the common pests, parasites and pathogens, an object that has a good chance of being achieved with less costly procedures. The Laboratory Animal Science Association (1969) recommends microbiological standards for diets for rats and mice which can equally well be applied to most other laboratory species. The recommendations are that (1) salmonellae should be absent, (2) presumptive coliforms should be not more than 10 per gram of diet, and (3) the total viable organisms should not exceed 5,000 per gram. The appropriate tests for checking that these standards have been attained are also set out. If the raw materials are of first quality and the milling and mixing conditions very clean, the recommended standards may be attained without further processing. For extra safety, pasteurization or very mild autoclaving can be applied. In the case of the pelleted diets, the heat generated during the pelleting process is sufficient to reduce the bacterial count significantly.

Water is a ready vehicle for the spread of disease, particularly if it becomes fouled with excreta. For this reason bottles or automatic drip services are preferable to open pans for the provision of water. All water containers must be frequently cleaned to prevent a build-up of contaminants. The water supply can be a source of risk to an SPF colony. Chlorination or acidification will control most of the likely water-borne contaminants, but it may be necessary to filter it if there is any risk of infestation with parasites or protozoa.

REFERENCES

Agricultural Research Council. (1963). *The Nutrient Requirements of Farm Livestock. No. 1. Poultry.* London: A.R.C.

Agricultural Research Council. (1965). *The Nutrient Requirements of Farm Livestock. No. 2 Ruminants.* London: A.R.C.

Agricultural Research Council. (1967). *The Nutrient Requirements of Farm Livestock. No. 3. Pigs.* London: A.R.C.

Braude, R., Kon, S. K. & Porter, J. W. G. (1953). Antibiotics in nutrition. *Nutr. Abstr. Rev., 23,* 474–95.

Bruce, H. M. (1950). The water requirement of laboratory animals. *J. Anim. Techns Ass., 1* (3), 2.

Bunyan, J. & Woodham, A. A. (1964). Protein quality of feedingstuffs. 2. The comparative assessment of protein quality in three fish meals by microbiological and other laboratory tests, and by biological evaluation with chicks and rats. *Br. J. Nutr., 18,* 537–44.

Carpenter, K. J. (1960). The estimation of available lysine in animal-protein foods. *Biochem. J., 77,* 604–10.

Cassidy, J. (1957). The commercial manufacture of compressed diets for laboratory animals. *Proc. Nutr. Soc., 16,* 63–6.

Charles, R. T., Stevenson, D. E. & Walker, A. I. T. (1965). The sterilization of laboratory animal diet by ethylene oxide. *Lab. Anim. Care, 15,* 321.

Chaudhuri, D. K. & Kodicek, E. (1960). The availability of bound nicotinic acid to the rat. 4. The effect of treating wheat, rice and barley brans and a purified preparation of bound nicotinic acid with sodium hydroxide. *Br. J. Nutr., 14,* 35–42.

Coates, M. E. (1968). Requirements of different species for vitamins. *Proc. Nutr. Soc., 27,* 143–8.

Coates, M. E. (1970). The sterilization of laboratory animal diets. In *Symposium on Nutrition and Disease in Experimental Animals,* ed. Tavernor, D. London: Baillière, Tindall & Cassell.

Coates, M. E., Ford, J. E., Gregory, M. E. & Thompson, S. Y. (1969). The effect of gamma irradiation on the vitamin content of diets for laboratory animals. *Lab. Anim., 3,* 39–49.

Coates, M. E., O'Donoghue, P. N., Payne, P. R. & Ward, R. J. (1969). *Dietary Standards for Laboratory Rats and Mice. Laboratory Animal Handbook No. 2.* London: Laboratory Animals Ltd.

Deuel, H. J. (1957a). In *The Lipids,* vol. 3, p. 605. New York: Wiley.

Deuel, H. J. (1957b). In *The Lipids,* vol. 3, p. 681. New York: Wiley.

Deuel, H. J. (1957c). In *The Lipids,* vol. 3, p. 816. New York: Wiley.

Eggum, B. O. (1968). *Amino Acid Concentration and Protein Quality.* Copenhagen: Stongaards. (In Danish.)

Ford, J. E. (1960). A microbiological method for assessing the nutritional value of proteins. *Br. J. Nutr., 14,* 485–97.

Gaunt, I. F. & Lane-Petter, W. (1967). Vitamin K deficiency in SPF rats. *Lab. Anim., 1,* 147–9.

Green, J. (1962). Metabolic effects of vitamin E and selenium. *Proc. Nutr. Soc., 21,* 196–202.

Gustafsson, B. E., Daft, F. S., McDaniel, E. G., Smith, J. C. & Fitzgerald, R. J. (1962). Effects of vitamin K-active compounds and intestinal micro-organisms in vitamin K-deficient germ-free rats. *J. Nutr., 78,* 461–8.

Hacking, M. R. & Lane-Petter, W. (1968). Factors influencing dietary K requirements of rats and mice. *Lab. Anim., 2,* 131–41.

Holman, R. T. (1960). The ratio of trienoic to tetraenoic acid in tissue lipids as a measure of fatty acid requirement. *J. Nutr., 70,* 405–10.

Lane-Petter, W. (1957). Modern ways of feeding laboratory animals. *Proc. Nutr. Soc., 16,* 59–62.

Lane-Petter, W. & Porter, G. (1963). Compound mouse diets. *Nature, Lond., 198,* 1013–4.

Ley, F. J., Bleby, J., Coates, M. E. & Paterson, J. S. (1969). Sterilization of laboratory animal diets using gamma radiation. *Lab. Anim., 3,* 221–54.

Moore, T. (1962). The history of selenium-vitamin E inter-relationships. *Proc. Nutr. Soc., 21,* 179–85.

National Research Council (1962). *Nutritional requirements of domestic animals. No. 10. Nutritional requirements of laboratory animals.* Publ. No. 990. Washington: N.R.C.

Porter, G., Lane-Petter, W. & Horne, N. (1963). Assessment of diets for mice. 1. Comparative feeding trials. *Z. Versuchstierk., 2,* 75–91.

Porter, G. & Lane-Petter, W. (1965). Observations on autoclaved, fumigated and irradiated diets for breeding mice. *Br. J. Nutr., 19,* 295–305.

Roach, A. G., Sanderson, P. & Williams, D. R. (1967). Comparison of methods for the determination of available lysine value in animal and vegetable protein sources. *J. Sci. Fd. Agric., 18,* 274–8.

Schwarz, K. (1961). Development and status of experimental work. In *Symposium on nutritional significance of selenium (Factor 3). Fedn. Proc. Fedn. Am. Socs. exp. Biol., 20,* 666–702.

Scott, M. L. (1962). Antioxidants, selenium and sulphur amino acids in the vitamin E nutrition of chicks. *Nutr. Abstr. Rev., 32,* 1–8.

Swann, M. M. (1969). *Report of Joint Committee on the Use of Antibiotics in Animal Husbandry and Veterinary Medicine*. Cmnd. 4190. London: H.M.S.O.

Taylor, T. G., Morris, K. M. L. & Kirkley, J. (1968). Effects of dietary excesses of vitamins A and D on some constituents of the blood of chicks. *Br. J. Nutr.*, **22**, 713–21.

Williams, R. J. & Pelton, R. B. (1966). Individuality in nutrition: effects of vitamin A deficient and other deficient diets on experimental animals. *Proc. nutn. Acad. Sci. U.S.A.*, **55**, 126–34.

Williams, R. J. & Deason, G. (1967). Individuality in vitamin C needs. *Proc. natn. Acad. Sci. U.S.A.*, **57**, 1638.

Wostmann, B. S., Knight, P. L., Keeley, L. L. & Kan, D. F. (1963). Metabolism and function of thiamin-naphthoquinones in germ-free and conventional rats. *Fedn Proc. Fedn Am. Socs. exp. Biol.*, **22**, 120–4.

Gyorgy, P. & Pearson, W. N. (1967). *The Vitamins*, vols 6 & 7. London & New York: Academic Press.

Maynard, L. A. & Loosli, J. K. (1962). *Animal Nutrition*, 5th ed. New York: McGraw-Hill.

Munro, H. N. & Allison, J. B. (1964). *Mammalian Protein Metabolism*, vols 1 & 2. London & New York: Academic Press.

Munro, H. N. (1969). *Mammalian Protein Metabolism*, vol 3. London & New York: Academic Press.

Robinson, F. A. (1966). *The Vitamin Co-factors of Enzyme Systems*. Oxford: Pergamon Press.

Sebrell, W. H. & Harris, R. S. (1968). *The Vitamins*, 2nd ed., vols 1 & 2 (3, 4 & 5 in press). London & New York: Academic Press.

Shütte, K. H. (1964). *The Biology of Trace Elements*. London: Crosby Lockwood.

Tyler, C. (1964). *Animal Nutrition*, 2nd edn. London: Chapman & Hall.

FURTHER READING

Comar, C. L. & Bronner, F. (1960–64). *Mineral Metabolism*, vols 1A, 1B, 2A, 2B. London & New York: Academic Press.

4 The selection and supply of laboratory animals

John Bleby and Michael F. W. Festing

The planning of any experiment or trial involving the use of animal material requires the most detailed and careful consideration of the wide variety of species and strains that are now available. Unfortunately, although many research workers carefully define their requirements with respect to chemicals or complicated laboratory equipment, often they appear to be unaware of the specifications that should be similarly applied to laboratory animals, and end up by simply asking for a white mouse! The aim of any user of laboratory animals should be to achieve maximum accuracy with the minimum number of animals, and it is, therefore, essential that careful consideration should be given to the choice of animal.

The first step when selecting an experimental animal is to specify the type needed. Such specification must take into account the following factors:

1. Selection of species
2. Selection of breed or strain
3. Selection of animals of the requisite quality, especially with respect to health status.

Selection of species

A number of factors need to be taken into account in the choice of a species for any given research project.

Probably the most important factor is the nature of the target population, i.e. the population about which inferences are to be drawn. Few problems arise when the investigator is content to restrict his conclusions to one of the commoner species of laboratory animals. However, when the results of an experiment are to be extrapolated to man much more serious consideration must be given to the choice of the experimental animal. In many cases more than one species will be required, though it is usual to carry out the experiments in a sequential manner, starting with the cheapest species (or tissue or organ culture), and continuing with the series only if the results are satisfactory at each stage. With this type of experiment, rats, mice and other rodents can often be used in the preliminary screening experiments, but data on rabbits, dogs and possibly primates may be necessary before the safety of the treatment can be assessed for clinical trials.

In some cases the target population consists of humans suffering from specific diseases such as diabetes. If so, an animal model of the disease may be wanted and this may dictate the choice of species. Cornelius (1969) discusses animal models of various human diseases that are now available. It must be stressed that in most cases the aetiology of the disease is not the same in animals as in man, but, nevertheless, such animal models may suggest possible approaches in tackling the disease in humans.

In cases where the target population is undefined, and the research worker is content to restrict his conclusions to the species actually investigated, a wide choice may be available. Table I lists thirteen important characteristics of seventeen species of laboratory animal and may be used to obtain a preliminary evaluation of the merits of different species for any given project.

One of the most important characteristics is size, which largely determines both the cost of the animal and the amount of space needed during the experimental period. Size also determines the ease with which operative procedures can be performed. A wide range of small rodents are available at the lower end of the scale, and farm animals are readily available at the higher end of the scale, but there is limited choice in the 10 to 30 kg weight range. As soon as the miniature pig is fully developed and readily available it may become a useful addition to the dog in this weight range.

TABLE I

RODENTS AND LAGOMORPHS

Character	Mouse	Hamster Chinese	Hamster Syrian	Rat	Guinea-Pig	Rabbit
Typical adult weight (g)	25–30	30–40	100–140	250–400	500–800	1000–7000
Average longevity (years)	1–2	3	1–2	2–3	6–8[8]	5–6
Chromosome number (2n)	40	22	44	42	64	44
Age at puberty (days)	35	45–60	45–60	45–75	45–75	150–210
Usual breeding age (days)	50	80	50	80	80	150–210
Usual breeding method [1]	PM	HM	HM	PM/H	PM	HM
Length of oestrus cycle (days)	4–5	4	4	4–5	14–16	—
Gestation period (days)	20	21	16	21–23	65–72	31–32
Average litter size	6–9	4–6	5–7	6–10	3–4	6–8
Weight at birth (g)	1–2	1–2	1–2	5–6	85–90	100
Weight at weaning (g)	10–12	6–8	40	35–40	250	1000
Age at weaning (days)	19–21	21	21	21	14–21	50
Productivity (young per breeding unit per year [3]	50–100	20–30	20–50	50–100	12–18	15–20

[1] PM = Permanently mated; HM = Hand mated; H = Harem mated
[2] At 22 weeks of age
[3] A breeding unit is the space occupied by 1 breeding female
[4] Incubation period
* These figures should only be used as a guide in the choice of a species. Specific strains or breeds may differ substantially from the figures given.

The productivity of the species is another characteristic which has an important influence on total costs. Productivity can vary enormously within a species, and the figures given in Table I should be used only as a guide. For example, it will usually be lower in inbred animals than in non-inbred ones.

The other characteristics given in Table I may be important under some conditions, though otherwise of minor importance. Further details of each of these species may be obtained by reference to the relevant chapters of this handbook.

Principles for pre-clinical testing of drug safety

This problem has been examined by the World Health Organization (1966) who arrived at the following general conclusions:

The effects of all drugs should be studied on the various systems, such as the cardiovascular, respiratory, central and autonomic nervous systems, and on neuromuscular functions.

In appropriate cases, evidence should be obtained from experiments that have been devised to resemble conditions in which the drug will be used for clinical purposes.

Toxicity should be studied on several species —three or more in acute experiments, and two or more in long-term investigations. One of the species should be non-rodent; both sexes should be studied. For long-term toxicity studies, the selection of species should be guided by evidence obtained from acute toxicity tests and from metabolic studies. Wherever possible, species should be chosen that are sensitive to the particular substance under test as well as those that have a metabolic pattern resembling that of man.

CARNIVORES			AVES	UNGU-LATES	PRIMATES					
Ferret	Cat	Dog	Quail	Mini-pig	Macaca Mltta.	Macaca Irus	Vervet	Baboon	Marmoset	Squirrel
750–800	3500–4500	10000–30000	110–160	20000[2]	7000 ♀ 6000 ♂	6000 ♀ 4000 ♂	4700 ♀ 2600 ♂	28000 ♀ 13000 ♂	250 ♀ 240 ♂	800 ♀ 550 ♂
5[8]	13–17	13–17	?	>6	15–20	10–15	~20	25–30	8–12	10–20
40	38	78	?	38	42	42	60	42	44–48	44
180–270	180–240	180–240	42–50	60–70	800	900	?	1200	?	?
180–270	270	270	42–50	80–100	1200	1200	?	1500	400	1000
H	HM/H	HM	PM	H	H/HM	H/HM	H/HM	H/HM	PM	H
—	15–28	22	—	21	28	28	30	36–44	?[6]	28
42	63	63	16[4]	114	146–186	146–186	180–213	154–183	134–140	168–182
6–7	3–4	3–6	200[5]	5–7	1	1	1	1	2	1
10	110	200–500	7	400–600	380–450	280–350	250–350	750–1050	30–35	45–65
450	700–800	?	48[7]	4500–6000	?	?	?	?	?	?
50	50	42–50	—	56	120–180	120–180	90–150	~180	80–140	100–170
10–12	10–12	6–12	50–150	10–15	1	1	1	1	4	1

[5] Eggs per breeding female (total in a year)
[6] Post-partum mating occurs
[7] Weight at 3 weeks
[8] This figure is of doubtful accuracy.

Animals used for carcinogenicity tests

This problem has also been considered by the World Health Organization (1969), who reached the following general conclusions:

Rats, mice and hamsters are regarded as being suitable; hamsters may be more suitable than the other two species for the testing of certain aromatic amines.

Guinea-pigs are known to be refractory to the action of certain carcinogens and are considered unsatisfactory.

Rabbits are of limited use in carcinogenicity tests.

Desert rats and steppe lemmings may prove useful for test purposes in the future but information on their suitability is limited at present.

Dogs are still recommended for testing suspected bladder carcinogens of the aromatic amine group, though studies now in progress may show that they are not indispensible for the purpose. Periods of up to 7 years have been found necessary for tumour-production by bladder carcinogens. In general the use of dogs is not practical.

Monkeys. Recent studies indicate that monkeys are sensitive to a variety of carcinogens and are likely to prove valuable in the testing of certain hormonal preparations.

Other non-rodent mammals. None can at present be recommended.

Trout are susceptible to the induction of hepatoma but are otherwise of limited use.

Birds. Budgerigars, ducks, domestic fowl and quail are susceptible to carcinogens, but more background information is needed before general recommendations for testing can be made.

Principles for the testing of drugs for teratogenicity

The World Health Organization (1967) has reached the following general conclusions regarding test animals:

Mice, rats and rabbits are the test animals most frequently used. Otherwise the choice is arbitrary in the absence of any indication, as yet, of a species with a susceptibility close to that of man.

Chick embryos are too sensitive for this purpose. Other species less extensively used are pigs, dogs, cats and monkeys. The choice of certain monkeys appears logical because of their phylogenic proximity to man and studies now being made may indicate that these primates are more useful than rodents.

Susceptibility of the monkey embryo to teratological agents resembles that of the human more closely than other species, but studies in primates require special skills, financial investment and close co-operation between disciplines. Such work could best be carried out in suitably co-ordinated research centres.

Selection of breed or strain

In some cases, the selection of the breed or strain of animal within the species selected will be determined by the characteristics which dictated the selection of species. If, for example, the species was selected because of a specific hereditary disease, then usually only certain strains, breeds or mutant types of the species will exhibit the disease. Similarly, if size is an important characteristic, a breed or strain of the necessary size should be selected.

In other cases there is considerable freedom in the selection of the strain of animals, and the main problem is to decide whether or not to use inbred animals (if they are available). Festing (1969) considered the conditions under which inbred strains, F_1 hybrids and other strains should be used, and came to the following conclusions:

1. Inbred strains and F_1 hybrids should be more widely used in general research, to replace non-inbred strains.

2. The argument that non-inbred strains should be used because they are widely representative of the species is not generally valid. Most of the so-called non-inbred stocks of laboratory animals probably do not contain a great deal of genetic variation. In any case, the variation

within a stock is usually considerably less than the variation between different inbred strains and, in order to ensure that a range of genetic material is used, studies should incorporate a range of different strains.

3. The best experimental material would be a mixture of different genotypes (such as inbred strains or F_1 hybrids) arranged in a factorial design of experiment, since these would be representative of several genotypes and would also give results of high precision together with information on the importance of genetic factors in response to the treatment.

In certain types of research, the use of inbred strains or F_1 hybrids is essential and immunology and cancer research utilizes a wide variety of inbred strains of mice and other species. On the other hand, inbred strains are not widely available for species other than the smaller laboratory rodents. Table II shows the number of inbred strains available for the main species listed by Jay (1963), although additional inbred strains have undoubtedly been bred in the intervening period.

TABLE II

Species	No. of inbred strains** listed	No. of strains in development
Mouse	200	?
Syrian hamster	5	17
Chinese hamster	—	9
Rat	49	15
Guinea-pig	3	4
Rabbit	2	17

*Compiled from Jay (1963). **Includes different sub-lines.

Health standards

Once the species and strain or type of animal has been decided, the next step is to determine the standard of health that is required. The advantages and disadvantages of specific pathogen free (SPF) and gnotobiotic animals are discussed in Chapters 10 and 11. However, it is now well established that the use of disease-free animals can often lead to a substantial reduction in the number needed for any given experiment, and this factor should be borne in mind when deciding the standard of health. Godwin *et al.* (1964) carried out haematological observations on SPF rats and presented evidence that accepted standards for blood counts in the rat are based on infected animals, and it seems evident that these standards should be revised. Fletcher *et al.* (1969) studied the survival rate of two groups of dogs which underwent experimental heart valve replacement, namely pure bred labrador retrievers and healthy, conditioned pound dogs. They found that survival of the labradors was 93 per cent compared with only 73 per cent for the pound dogs, and after listing the costs of each operation they concluded that the use of the healthy dogs was economically justified, as well being ethically more acceptable. Costings on animals undergoing other types of experiments, particularly long-term experiments, would often show similar results.

Tracing the source of animals

Having decided on a species and particular type of animal within the species, the next step is to locate an appropriate source.

Several publications are available which can be used to locate specific types of animals, and the most useful of these are:
International Index of Laboratory Animals (Edited by Festing and Butler, 1975). This Index, which is published by the MRC Laboratory Animals Centre with the support of the International Committee on Laboratory Animals (ICLA) is designed specifically to enable the research worker to trace a source of animals, wherever they may be held throughout the world. The Index mainly covers the more common laboratory species, but some information is also given on the less common species and animals obtained from the wild.

The information is given under three headings: strain, breeding system (i.e. whether the animals are inbred or non-inbred) and whether they are SPF (specific pathogen free), CD (caesarian derived) or germfree, and the code of the location of the stocks. This code consists of two parts, the first of which indicates the country, and the second the holder of the stock.

A few mutant stocks are included in the *International Index* but coverage is by no means complete. Reference to stocks of *Drosophila* has deliberately been avoided as these can be traced more efficiently through the Drosophila Information Service (see Ch. 42). Copies of the *International Index* may be obtained from the *MRC Laboratory Animals Centre, Medical Research Council Laboratories, Woodmansterne Road, Carshalton, Surrey, England.*

Animals for Research (Eds. Sundborg and Yager, 1968). This publication is described as a 'directory of sources of laboratory animals, fluids, tissues, organs, equipment and materials', but it is mainly confined to sources on the North American continent. Some details of the origin of the stocks is given, and the booklet contains a section on equipment and materials which could be useful. Copies may be obtained from *The Institute of Laboratory Animal Resources*, 2101 *Constitution Avenue, N.W., Washington 25, DC, USA.*

Genetic Strains and Stocks (Jay, 1963), is a listing of the name, origin and main characteristics of inbred strains or strains in development, as shown in Table II. Holders of each of these strains are also listed. Although this paper was published in 1963 it is still of great importance, being the only list giving origin and characteristics of such a wide range of species. It is a most useful adjunct to the two publications mentioned previously.

Standardized Nomenclature for Inbred Strains of Mice. Fourth Listing (Staats, 1968), provides a world-wide list of all officially recognized inbred strains of mice, brief details of the history and characteristics, and a list of holders of the strains.

Inbred Strains of Mice No. 6 (Staats, 1969) is a companion issue to *Mouse News Letter* (see below) giving brief details of the inbred strains of mice maintained at 101 laboratories throughout the world. The amount of information given varies, but a useful feature is the inclusion of a list of names of individual workers at each laboratory. Consultation of this publication provides a convenient way of following up the history of any strain held by another laboratory.
Mouse News Letter, edited by Dr. A. G. Searle and distributed by the MRC Laboratory Animals Centre, Carshalton, Surrey, England,

is published twice a year. It gives an up-to-date listing of the known mutants of the mouse and, in some issues, a cross classification system showing the holders. In other issues a reference describing the mutant is given. Occasional information is included on inbred strains of mice. The term mouse also extends to occasional references to *Peromyscus* stocks and mutants.

Standardized Nomenclature for Inbred Strains of Rats (Festing & Staats, 1973). This listing gives a brief description of the correct methods of describing inbred rat strains and lists a total of 100 strains, together with brief details of their origin and characteristics and the location of known colonies.

Genetics of the Norway Rat (Robinson, 1965). This is the standard text book on the genetics of the rat and gives a full listing of both mutants and inbred strains known at that time. The book should be used as a means of *tracing* a supply of a strain or mutant only as a last resort, although as a source of information it is excellent.

Biology of the Laboratory Mouse (Green, 1966). This book covers all aspects of the biology of mice and is an excellent source of *information*, but as stocks can be traced more efficiently by other methods it should be used in conjunction with the publications listed above.

The supply of experimental animals

There are two ways in which a supply of animals can be obtained: either they can be bred by the user, or they can be purchased from a commercial breeder. These two sources may not be available on all occasions, but where either course is open the following factors need to be considered:

Breeding by user

(a) Physical facilities available, such as animal-rooms and equipment. Both quality and quantity should be carefully considered.
(b) Staff available, including numbers and quality.
(c) Knowledge of breeding methods for the species concerned.

These facilities should be considered in relation to the numbers and quality of the animals needed. Chapter 6 will help in deciding the area needed to breed a given number of animals, and Chapter 10 should be consulted about the type of facilities required to keep a colony disease-free.

Even if the accommodation, staff and expertise are available, some consideration should be given to the alternative uses that could be made of the space. In a research laboratory, it would be bad policy to use valuable laboratory space for breeding animals. Further, animals bred by the user are usually more expensive than those bred by a commercial breeder who knows how to keep costs down.

Commercial breeding

If the experimental animal selected is a standard type produced by commercial breeders on a large scale, it is usually more economical to buy in supplies. Even if the animals are not currently available, it may be possible to arrange with a breeder to provide them specially at an economic price. When negotiating such an arrangement, the following factors should be taken into account:

(a) The reliability and reputation of the breeder concerned.
(b) The continuity of supplies. The breeder should be asked to give some indication of his plans to produce the desired level of output. The adequacy of these plans can then be judged according to the information given in Chapter 5.
(c) Physical facilities available to the breeder. Wherever important negotiations are to be carried out, the breeder's facilities should be inspected, with a view to judging the quality of animals which are likely to be produced. If such an inspection cannot be carried out, other customers should be contacted and asked for an assessment of the quality of animals and service supplied by the breeder.
(d) Location. Ideally, this should be as near as possible, and all animals should be delivered to the user's premises by road.

If this is not possible, the nearest breeders should be given preference, other things being equal.

(e) Price. Good quality animals will cost more than animals of a poor quality. However, it will seldom be in the interests of the research worker to sacrifice quality for the sake of price. For this reason, it is helpful if the quality can be judged in some objective way, such as by the microbiological status of the animals.

(f) Quality. The most important aspect of quality is usually the health status of the animals. However, classification according to microbiological status will usually require facilities that are not available to the average user.

In order that the above factors may be met, almost all the commercial breeders in the United Kingdom belong to the MRC Laboratory Animals Centre's Accreditation and Recognition Schemes for suppliers of laboratory animals. Under these Schemes samples of animals from each commercial breeder are screened at least twice a year, and their products are classified according to the results of this screening.

In other countries it may be possible in some cases to obtain a listing of the microflora and fauna from the breeder, but such information should be treated with a certain amount of reserve and the competence of the screening laboratory should be carefully checked.

The accreditation and recognition schemes for suppliers of laboratory animals*

The original accreditation scheme was started in 1950, and has developed into a comprehensive system whereby serious and reliable suppliers of laboratory animals can be distinguished from the remainder.

As the breeding and procurement of all species cannot be encompassed within one basic set of rules, two schemes are available:

* *The Accreditation and Recognition Schemes for Suppliers of Laboratory Animals*, 2nd edn. (1974) published by the MRC Laboratory Animals Centre.

The Accreditation scheme for the breeders of mice, hamsters, rats, guinea-pigs, rabbits, cats and dogs.

The Recognition scheme for breeders and suppliers of vertebrate species who, for various reasons, are unable to fulfil the requirements of accreditation.

In both cases, a breeder or supplier has to sign a statement that he accepts the rules of accreditation or recognition, and apply to have his premises inspected by the Laboratory Animals Centre.

The Rules

The rules of accreditation and recognition are strict but practical, so that the smooth flow of high quality animals is ensured. No breeders can be considered for accreditation unless they are in a position to produce annually not less than 20,000 mice, or 5,000 hamsters, or 10,000 rats, or 2,000 guinea-pigs, or 5,000 rabbits, or 400 cats, or 300 dogs. One of the most important rules is that they must supply animals of their own breeding, and the resale of animals brought in from other premises is absolutely prohibited. Such strict measures are absolutely essential if dealing and the consequent spread of disease is to be prevented.

As the terminology regarding the health status of laboratory species is somewhat ambiguous, the species bred by accredited breeders are screened at regular intervals by the Centre and divided into five categories, depending upon their microbiological flora.

Categories

Briefly, the five categories are as follows:
Category 1 are comparable to non-barrier maintained animals traditionally called conventional and should be free from all evidence of infectious diseases, especially those communicable to man, on both clinical and post-mortem examination, with special reference to the following organisms: all salmonellae and shigellae, *Mycobacterium tuberculosis, Pasteurella pseudotuberculosis*, pathogenic dermatrophic fungi, *Sarcoptes scabiei.*
Category 2 are comparable to conventional animals maintained under high standards of management and should be of the same status

as Category 1 and, in addition, should be free from the following additional organisms: intermediate stages of cestodes and all obligatory parasitic arthropods. Further species-specific demands are the absence of: the viruses of ectromelia in mice and myxomatosis in rabbits.
Category 3 are comparable to those animals previously designated caesarian derived. They should be of the same status as Category 2 and, in addition, should be free from the following organisms: *Bordetella bronchiseptica*, all pasteurellae, all species of mycoplasma (excluding hamsters and guinea-pigs), all coccidia and all pathogenic helminths. Further species-specific demands are the absence of *Streptobacillus moniliformis* in mice and rats, *Corynebacterium muris* in mice, all types of pneumococcus in guinea-pigs and rabbits, and rabbit syphilis in rabbits.
Category 4 are comparable to those animals previously described as specific pathogen free and should be of the same status as Category 3 and, in addition, should be free from the following organisms: all pneumococci, *Klebsiella pneumoniae*, *Listeria monocytogenes*, all helminths, all pathogenic protozoa. Further species-specific demands are: the absence of all species of mycoplasma in hamsters and guinea-pigs and *Fusiformis necrophorus* in rabbits.
Category 5 are animals which result from the use of closed-system sterile techniques and are free of all demonstrable organisms.

Recently, genetic monitoring has also been introduced into the Accreditation Scheme.

The Recognition Scheme caters for those species that for various reasons are unable to fulfil the requirements necessary for accreditation. It incorporates species such as lemmings, gerbils, domestic fowl and primates, as well as other more unusual species such as, for example, amphibia and reptiles.

Maintaining standards

Laboratories can look to accredited breeders for the supply of animals of a certain minimum standard of quality—although it need hardly be added that no one can guarantee that this standard will invariably be maintained. But where an accredited animal fails to come up to standard the user has some redress, and the sanction of withdrawal of the certificates of accreditation is a powerful incentive to the breeder to maintain high standards.

An important part of the operation of this scheme is the technical advice and help made available to breeders. They have access to laboratory examination of their animals and, in the course of visits made from time to time by the staff of the Centre, or at any time upon inquiry, to detailed advice and help in practical matters.

The Accreditation and Recognition Schemes have survived during a period when there has been a steadily rising demand for quality in laboratory animals, and they have demonstrated that private breeding is susceptible to a useful measure of control. Furthermore, today a very high proportion of purchasers of laboratory animals buy *only* from accredited breeders.

It is significant that in 1968, the breeders and suppliers formed their own trade association, The Laboratory Animal Breeders Association, and have published a code of ethics.

Laboratory animal centres

It will always be necessary for laboratory workers to be able to look to some central office for information about the availability of animals they require and, if they breed some themselves and have a surplus, to enquire where this surplus may be needed by other users. Thus the situation where one laboratory is short of, say, mice, while another has to destroy a surplus for which it has no immediate use may be avoided by the central publication of a circular such as *Parade State* which lists surplus animals produced by Accredited and Recognised suppliers.

From time to time, unusual species or strains are requested and these may have to be imported from overseas. A Centre's experience in handling such demands is often useful, not only in finding a source of the animals required, but in facilitating their procurement. In this way contact is established with many sources in other countries and it is possible for data and information to be compiled covering the sources of almost all animals used for research purposes.

Since 1946 several countries have established

central offices to ensure that adequate supplies of high-quality laboratory animals are available for their national biological research programmes. As these programmes advance, so does the standard of the laboratory animal required, and these standards can be reached only if there is some form of central organization able to state what the minimum standards must be.

Above all, national centres make possible international liaison and collaboration either directly between one another or indirectly through the International Committee for Laboratory Animals (ICLA), whilst at the same time maintaining direct contact with the laboratory worker and the animal breeder. Without some form of central organization, therefore, the supply and procurement of laboratory animals is severely handicapped, leading to an unnecessary waste of time and money.

In addition to the U.K. Laboratory Animals Centre, the following countries have established either a centre of their own or an office/bureau: Argentine, Australia, Austria, Belgium, Bulgaria, Canada, Czechoslovakia, Denmark, France, German Democratic Republic, Germany—Federal Republic, Greece, Hungary, Iceland, India, Ireland, Israel, Italy, Japan, Lebanon, Netherlands, Nigeria, Norway, Poland, Rumania, South Africa, Spain, Sweden, Switzerland, Turkey, USA, USSR. Further details can be obtained from the Secretary-General of ICLA, Dr. S. Erichsen, National Institute of Public Health, Geitmyrsveien 75, Oslo 1, Norway.

REFERENCES

Cornelius, C. E. (1969). Animal models—a neglected medical resource. *New Engl. J. Med.*, **281**, 934–44.

Festing, M. F. W. & Butler, W. (1975). *International Index of Laboratory Animals*. MRC Laboratory Animals Centre, Carshalton, Surrey.

Festing, M. F. W. (1969). The use of inbred strains, F_1 hybrids and non-inbred strains in research. *Proc. IV ICLA Symposium, Washington, D.C.*

Festing, M. F. W. & Staats, J. (1973). Standardized nomenclature for inbred strains of rats. *Transplantation*, **16**, 221–245.

Fletcher, W. S., Herr, R. H. & Roberts, A. L. (1969). Survival of purebred labrador retrievers versus pound dogs undergoing experimental heart valve replacements. *Lab. Anim. Care*, **19**, 506–8.

Green, E. L. (1966). *Biology of the Laboratory Mouse.* New York: McGraw-Hill.

Godwin, K. O., Fraser, F. J. & Ibbotson, R. M. (1964). Haematological observations on healthy (SPF) rats. *Br. J. exp. Path.*, **45**, 514–24.

Jay, G. (1963). Genetic strains and stocks. In *Methodology in Mammalian Genetics*, ed. Burdette, W. J. San Francisco: Holden-Day.

Medical Research Council Laboratory Animals Centre. (1974). *The Accreditation and Recognition Scheme for Suppliers of Laboratory Animals*, 2nd edn. MRC Laboratory Animals Centre, Carshalton, Surrey.

Queinnec, G. & Chemin, M.C. (1968). Les souches de rats disponibles pour la recherche. *Expérimentation animale*, **1**, 193–210.

Robinson, R. (1965). *Genetics of the Norway Rat.* Oxford: Pergamon.

Searle, A. G. *Mouse Newsletter.* Published twice yearly by MRC Laboratory Animal Centre, Carshalton, Surrey.

Staats, J. (1968). Standardized nomenclature for inbred strains of mice: fourth listing. *Cancer Res.*, **28**, 391–420.

Staats, J. (1969). *Inbred Strains of Mice No. 6. Companion Issue to Mouse Newsletter No. 41.* Bar Harbour, Maine: Jackson Laboratory.

Sundborg, M. B. & Yager, R. H. (1968). *Animals for Research.* Publ. No. 1678. Washington: Nat. Acad. Sci.

World Health Organization. (1966). *Principles for Pre-Clinical Testing of Drug Safety.* Technical Report Series No. 341. Rome: W.H.O.

World Health Organization. (1967). *Principles for the Testing of Drugs for Teratogenicity.* Technical Report Series No. 364. Rome: W.H.O.

World Health Organization. (1969). *Principles for the Testing and Evaluation of Drugs for Carcinogenicity.* Technical Report Series No. 426. Rome: W.H.O.

5 Production methods

MICHAEL F. W. FESTING

Good quality experimental animals can be expensive, but viewed in the context of the total cost of a research project the cost of the animals is usually low. However, because of both the expense and the moral obligations involved, animals should be used in a research project likely to cause pain only if no alternative is available.

The purpose of this chapter is to examine the ways in which the total cost of research and the number of animals used can be reduced. The four main ways are as follows:

1. Production of laboratory animals as economically as possible, in sufficient numbers to satisfy experimental requirements.
2. Avoidance of the production of surplus animals.
3. Production of animals of the requisite quality, with regard to the disease status and genetic background.
4. Assistance from a qualified statistician to ensure that the proposed experiment is correctly planned and fully analysed. This will frequently result in fewer animals being needed to achieve the objectives of the experiment.

The *operational* techniques which can be used by the breeders of laboratory animals to achieve objectives 1 and 2 above are considered.

The supply/demand system

The relationships between the demand for and the production of laboratory animals are shown diagrammatically in Figure 5.1.

As a result of an initial demand, a *breeding colony* is set up. In due course weanling animals are produced, and these go into the *growing stock* colony until they reach the minimum usable age (or weight), when they go into the *stock inventory,* and are available for use. The

animals are then either *issued* or exceed the maximum usable age and have to be *culled as surplus*. The time taken to grow from the minimum usable age (or weight) to the maximum usable age is known as the *shelf-life* of the animal.

If large numbers have to be culled as surplus, a decision may be taken to reduce the size of the breeding colony. Conversely, if there is a chronic shortage so that a queue constantly develops, a decision may be taken to increase the size of the colony.

In any system such as the one shown in Figure 5.1, it is useful to have a single criterion of success to judge the effectiveness of different management methods. It can be assumed that the animals' contribution to the overall cost of a research project is made up of three parts:

(*a*) The cost of the animals used in the experiment.

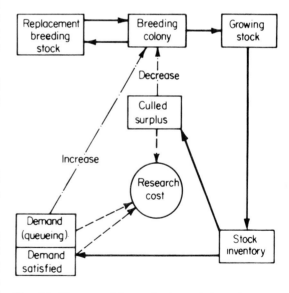

Fig. 5.1. Diagram of the production and demand system for laboratory animals.

(b) The cost of animals produced for the experiment but not actually used.

(c) The cost of failing to produce sufficient animals for the experiment.

In many cases it would be logical to assume that the best management policy would be the one that minimized this total cost. Thus, if a large surplus is produced, the total cost of animals actually used must rise; conversely, if insufficient animals are produced, considerable costs may be incurred by the under-employment of scientists and equipment.

A closed system such as that shown in Figure 5.1 in which there is a *feedback* of information which controls the level of production is known as a *servo system*. This system is commonly encountered in manufacturing industries, biology and electronics, and there is considerable information about its characteristics. In the production of laboratory animals, three points are important:

1. If there is a sudden and unforeseen increase in demand, there is inevitably a time-lag of several weeks before production can be increased, since gestation periods and growth rates cannot be altered. This time-lag often causes the size of the breeding colony to be over-corrected in order to relieve the pressure of demand that builds up when the supplies are short.

2. If the demand is continually rising a chronic shortage may arise, due again to the time-lag in increasing production.

3. Weekly output and demands for animals will vary, due to natural biological variation and chance factors. If feedback is too sensitive these normal fluctuations may suggest over- or under-production, and steps may be taken to alter the size of the colony. This will lead to continual changes in the number of breeding females, and output will tend to fluctuate as a result of these changes.

Conversely, if the feedback mechanism is too insensitive, real changes in demand or output may not be recognized. This in turn will lead either to excessive wastage of animals, or to shortages. It is important, therefore, that a recording system should be used which can identify any trends in demand that may occur.

Records

Good records are essential in controlling the production process. They are usually kept for two distinct purposes:

1. To gather information of scientific importance, including pedigree and performance records of individuals or groups of animals.

2. To monitor and control the production process.

Dickie (1966) gives details of recording systems suitable for gathering scientific information, but such systems are seldom suitable for colony control.

A simple system which can be used to control the production process is shown in Figure 5.2, but more sophisticated systems could be devised with weekly rather than monthly recording.

This type of recording gives most of the information needed to control the colony efficiently and estimate performance levels, although in small colonies the individual month's figures will be subject to large chance fluctuations. However, over a period of time a pattern will emerge, and the records may then be used to monitor the relationship between supply and demand, and to show whether the size of the breeding colony needs to be altered.

The most important information given by such records is:

1. Output in terms of young weaned/breeding female/week.

2. Output per unit area of animal room per week (if the area stays constant this will be directly related to item 1 above).

3. Percentage of usuable production not issued (percentage wastage).

4. Trends in production, issues, percentage wastage and other factors listed.

This information can be used as feedback to control breeding activities. For example, the monthly issues can be graphed, and if a trend is shown, this can be taken into account in the future.

MONTHLY STOCK INVENTORY AND PRODUCTION RECORD

	Jan.	Feb.	Mar.	Apr.	May	June	July	Aug.	Sept.	Oct.	Nov.	Dec.	Year's Summary
1. No. of breeding ♀♀ in colony													
2. No. of breeding ♂♂													
3. No. reserved for future breeding													
4. Growing stock on hand													
5. Total size of colony													
6. No. of litters weaned in month													
7. Total young weaned in month													
8. Number culled as surplus to requirements ♂													
♀													
9. Number issued for use ♂													
♀													
10. Total usable production (8 + 9)													
11. Usable young/breeding ♀/WK $\frac{1}{4}(10 \div 1)$													
12. Average litter size, weaned (7 ÷ 6)													
13. % pre-weaning mortality*													

* Obtained separately from a random sample of, say, 50 litters born during the month.

Fig. 5.2. Example of monthly recording system.

THE COMPONENTS OF THE PRODUCTIVE SYSTEM

Demand

Demand is usually expressed as the mean number of animals of a specified type required in a given period of time. However, for efficient production planning, additional information is required if costs are to be minimized. This falls into the following three categories.

Changes in the mean level of demand. These changes may occur both as long-term increases (or decreases) in demand, and also as cyclical effects such as seasonal patterns. In addition, sudden changes in the demand for any one type of animal may occur as a result of changes in research emphasis or personnel. The production manager who is aware of these changes in mean demand is clearly in the best possible position to plan future production.

The distribution of demand. Superimposed on the mean level there are usually weekly or monthly fluctuations in demand of an essentially random nature. The magnitude and nature of these are of great importance in deciding the level of output that is required.

Flexibility of demand. If demand is so inflexible that the specified number of animals has to be produced each week without fail, then a higher level of production will be required than if the research worker is prepared to accept shortages in some weeks, provided they are made up in other weeks. The flexibility of demand is closely related to the shelf-life of the animals. In some cases it may be essential to have the specified number of animals, but the exact age or weight may not be important. In practice, this amounts to a flexible demand, since surpluses produced one week can be stored and used the following week. A knowledge of the flexibility of demand is essential if the animal colony is to be run efficiently.

Estimating average future demand

In cases where the production colony supplies only a few research workers, it is sometimes possible to get a firm estimate of future demands by close liaison with the users of the animals.

In many cases, however, the users are either not known or are not in a position to give firm

orders so far in advance, or such liaison may be completely impracticable. Moreover, in some cases it may not be desirable to tie the research worker down to long-term predictions.

Demand has sometimes to be estimated for a considerable time in the future. Where, for example, production facilities are already fully utilized it may be necessary to estimate demand for a period of five or six years, in order to decide upon the size of any new production facilities that may be needed.

In all these cases, the only basis for the estimation of future demand may be the pattern of past demand. Estimation of future trends from past records can be carried out by methods of analysis known as the *Analysis of Time Series*. A simple explanation of these methods is given by Thirkettle (1968). Briefly, it separates the three causes of fluctuations in demand: long-term trends; chance fluctuations; and cyclical (such as seasonal) trends.

In many cases the chance fluctuations are so great that they completely obscure any long-term trend. A method commonly used to eliminate these and illustrate the trend is that of moving averages. With this method a moving average of, say, five months' sales (or demand) is calculated, and this is then graphed. A five-month moving average of the sales of one of the LAC mouse strains is shown in Figure 5.3. This shows the trend clearly, and a trend-line can be drawn in by eye. In this graph the trend is for an approximately constant or slightly declining

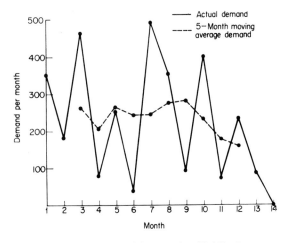

Fig. 5.3. Monthly demand for a strain of LAC mice.

demand, so that there would be no need for dramatic changes in the numbers of breeding stock provided that past sales truly represented demand. Care must be taken not to extrapolate too far, as the greater the extrapolation the less reliable is the estimate obtained.

Seasonal effects can be estimated by considering average demand over a period of several years in January, February, etc., or the demand in each quarter of the year. Carter (1957) obtained an estimate of the seasonality of *production* in the Harwell mouse colonies by this method, but estimates of seasonality of *demand* can be obtained in the same way provided that sufficient data are available.

Production methods

The success of a breeding operation may be examined under four main headings:
1. Output per breeding cage or per unit area of animal house (either in terms of numbers or value).
2. Output per unit of labour.
3. Quality of the animals produced.
4. How supply equates with demand.

Since in some cases success under one of these headings may be antagonistic to success under another, a correct balance must be found.

The factors relevant to success under each of these headings are shown in Table I, and are discussed below.

Maximizing output per cage

The breeding method. This will be determined largely by the biological characteristics of the species and by the type of animal wanted. In general, breeding systems may be divided into two main groups, each of which can be sub-divided into two classes:

(a) *Permanently mated groups,* in which the female is allowed to litter down in the presence of the male and other members of the group.

There are two main sub-divisions of this system:
  (i) *Monogamy,* i.e. a single male and female housed together.
  (ii) *Polygamy,* i.e. one or more males are housed with several females.

(b) *Temporarily mated groups,* in which the females are separated from the group before parturition.
  (i) *Harem systems,* in which the males and females are run together but separated prior to parturition.
  (ii) *Hand-mating systems,* in which the male and female are left together only for sufficient time for mating to occur.

T ABLE I
Sub-objectives of a breeding system

Sub-objective	Criterion of success	Relevant factors
1. Maximize output (a) Per breeding cage.	No. weaned/cage/week.	*Basic Husbandry:* Breeding method. Culling and stock turnover. Genetic potential. Husbandry and environment. Disease status.
(b) Per unit area.	No. weaned/unit area/week.	*Basic husbandry* plus stocking density.
2. Maximize output per unit of labour.	Mins. of labour/animal produced/week.	*Basic husbandry* plus labour saving devices, work routine, method of recording, staff quality plus training. Even work load.
3. Maximize quality.	Absence of disease. Uniformity. Genetic reliability. Weight for age.	*Basic husbandry* plus methods of genetic control. Manipulation of litters. Barrier maintenance.
4. Equate supply with demand.	Low wastage. Short queue length.	*Basic husbandry* plus variations in demand. Variations in output. Consequences of failure to meet demand.

TABLE II

Common breeding systems for laboratory animals

Species	Permanently mated		Temporarily mated	
	1♂:1♀ (Monogamy)*	1♂:>1♀ (Polygamy)	Harem (♀ removed when pregnant)	Hand mated
Rodents, Lagomorphs				
Mouse	+	+		
Deer mouse	+			
Vole	+	+		
Chinese hamster	+([2])	+([1])		+
Steppe lemming	+			
Gerbil	+			
Norway rat	+		+	
Guinea-pig		+		
Rabbit				+
Carnivores				
Ferret			+	
Cat			+	+
Dog				+
Primates				
Marmoset	+			
Rhesus				+
Baboon				+
Aves				
Quail		+		
Chicken		+		

* This system commonly used for inbred matings of many species.
([1]) Breeding by collar method, Belcic & Weihe (1967).
([2]) Some genetically adapted colonies only, Porter & Lacey (1969).

The breeding systems most usually employed in the production of the common laboratory animals are given in Table II.

With some species several different breeding systems may be used, in which case the choice may have a considerable influence on the output per unit area, labour costs and quality.

Mice are usually permanently mated, but the ratios of males to females can vary from 1 : 1 up to 1 : 6 or 1 : 7. Output per unit area is usually higher with the wider ratios, though output per female is lower. Levels of output with different ratios of females to males found by Heine (1965) are given in Table III.

The advantage of the wider ratio can be judged in relation to the overall economic consequences of the various systems. Walker and Stevenson (1967) listed the recurrent costs incurred in running a specific pathogen free building (Table IV).

If these figures are typical, then it is clear that the overheads should be spread over as large an output as possible if the cost of production is to be kept low. This implies that the wider ratios would be more economical, provided that they did not result in a higher labour cost per unit of output. This would be unlikely, since labour costs are probably more closely related to the number of boxes than to the number of animals per box.

TABLE III

Output from different polygamous ratios in mice

	Output	
No. ♀♀ per box	Young/♀/week	Young/box/week
1	1·98	1·98
2	1·35	2·70
3	1·12	3·36
4	1·00	4·00

TABLE IV

Recurrent costs of running a specific pathogen-free unit

Item	% Cost
Overheads	37
Labour	37
Food	12
Heat & Lighting	9
Bedding	4
Sterilizing	1

(Reproduced with permission from Walker & Stevenson, 1967)

Thus, unless there are other considerations (such as keeping a strict account of pedigrees) the wider ratios would generally be more economical. Unfortunately, there is scant information on the productivity of systems using wider ratios than 1 : 4, though such systems are used commercially. However, it is probable that larger cages are needed for such mating systems.

Rats are commonly bred either in monogamous pairs or in harems where about six females and one or two males are housed together, the pregnant females being removed prior to parturition. A female is returned to the harem only after weaning her young, and therefore does not conceive at the postpartum oestrus.

A sophisticated form of the harem system has been described by Lane-Petter *et al.* (1968), in which the young are cross-fostered two days after birth. Uniform litters of 12–14 pups of a single sex are made up, and surplus animals (usually unwanted females) are culled at this stage. The deprived mothers are returned to the harem, where they conceive about two weeks later. This results in an overall higher level of productivity of animals than is obtained from the conventional harem system, and the young may be more uniform than animals from normal litters.

The relative merits of monogamous pairs and harems for breeding rats are compared by Tucker & Peacock (1974). Production per breeding female was higher in the monogamous pairs than in the harems (1·4 and 0·8 young/female/week respectively), but a higher stocking density could be achieved with the harems as considerably fewer males were needed. Total output per m² of animal room was not very different between the two colonies (532 young in the monogamous pairs and 468 in the harems). It was concluded that, as monogamous pairs were easier to manage, this method of production was preferable for small colonies. For large colonies, on the other hand, the harem system may be more efficient provided that careful attention is given to details of colony management.

Further details of the breeding systems used for other species may be found by reference to the relevant chapters of this book.

Stock turnover policy.   Under certain conditions both the culling method and the stock turnover policy can have a strong influence on total productivity per unit area. For example, Carter (1957) concluded that productivity was maximized if CBA/Ca strain mice were allowed to produce five litters before being replaced. Unfortunately, expressing the breeding life of a mouse in terms of the number of litters produced can be dangerous, since if this is adhered to too strictly, sterile matings may stay in the colony indefinitely.

Festing (1968) examined the productivity of nine inbred strains of mice, bred as monogamous pairs, after they had been mated for 10, 17 and 23·3 weeks. The results are shown in Table V.

TABLE V

*Productivity (No. of young/breeding ♀ /week)
of nine inbred strains of mice
bred for various periods*

Strain	Weeks in Production		
	10 wks.	17 wks.	23·3 wks.
AKR	0·60	0·63	0·55
A2G	1·05	1·11	1·14
BALB/c	0·63	0·74	0·79
CBA	0·98	1·17	1·11
C57BL	0·41	0·62	0·64
C57BR	0·30	0·36	0·36
DBA/1	0·85	0·81	0·69
DBA/2	0·58	0·58	0·54
NZW	0·58	0·59	0·67
Average	0·66	0·74	0·72

Although there were strain differences in the optimum length of the breeding period (cf. BALB/c and DBA/1), the differences in productivity between the 10, 17 and 23·3 week periods were relatively slight. On average, therefore, it makes little difference to total productivity whether the colony is allowed to breed for, say, 17 weeks or 23 weeks, although the longer period would usually be preferred for practical reasons. Similar results were obtained by Heine (1965).

TABLE VI

*Productivity of a single strain of rats after
various lengths of breeding life (based on
data from 40 breeding pairs)*

Weeks in production	Productivity (Y/ ♀ /wk)
5	1·56
10	1·85
15	1·70
20	1·57
25	1·53

For rats there appears to be little published information on the optimum breeding period, though some original data on the Porton/Lac strain are given in Table VI.

Here, productivity would be at a maximum after only 10 weeks of breeding, but a policy of culling at this stage would tend to be impractical and breeding cycles of about six months are more usual.

Another important aspect of stock turnover policy is the establishment of a routine for saving future breeding stock, making up new matings, and culling old breeding stock. Such routines, when carried out on a weekly basis, help to spread the work-load by evening out fluctuations in output; they ensure that stock shall be mated at the correct age and culled after an optimum period of breeding, and that the age-structure of the population is kept constant.

Mating schedules for rats and mice that can be applied to large colonies are given in Tables VII and VIII. Such schedules ensure a routine stock turnover policy, and should not be confused with the randomization schemes described by Falconer in Chapter 2 which describe *how* the future breeding stock should be chosen.

The use of the schedules, which are suitable for permanently mated or harem systems, can be illustrated by reference to the routine carried out in one week. For example, in the mouse schedule (Table VIII) the following operations would be carried out in week 10:

1. The breeding females in cages labelled J1, J2, etc., would be culled.
2. A new set of matings labelled J1, J2, etc., would be made up from future breeding stock labelled J. This stock would have been reserved five weeks previously (week 5) and would therefore be eight weeks old. The

TABLE VII

A mating schedule for rats suitable for colonies of more than 130 breeding groups
(Harems or Monogamous pairs)

Week[1]	Young stock mated in cages labelled:[2]	Weanlings saved for breeding and labelled:[3]	Males removed from:
1	A1, A2, . . . etc	J	D1, D2, . . . etc
2	B1, B2, . . . ,,	K	E1, E2, . . . ,,
3	C1, C2, . . . ,,	L	F1, F2, . . . ,,
4	D1, D2, . . . ,,	M	G1, G2, . . . ,,
5	E1, E2, . . . ,,	N	H1, H2, . . . ,,
6	F1, F2, . . . ,,	O	I1, I2, . . . ,,
7	G1, G2, . . . ,,	P	J1, J2, . . . ,,
8	H1, H2, . . . ,,	Q	K1, K2, . . . ,,
9	I1, I2, . . . ,,	R	L1, L2, . . . ,,
10	J1, J2, . . . ,,	S	M1, M2, . . . ,,
11	K1, K2, . . . ,,	T	N1, N2, . . . ,,
12	L1, L2, . . . ,,	U	O1, O2, . . . ,,
13	M1, M2, . . . ,,	V	P1, P2, . . . ,,
14	N1, N2, . . . ,,	W	Q1, Q2, . . . ,,
15	O1, O2, . . . ,,	X	R1, R2, . . . ,,
16	P1, P2, . . . ,,	Y	S1, S2, . . . ,,
17	Q1, Q2, . . . ,,	Z	T1, T2, . . . ,,
18	R1, R2, . . . ,,	A	U1, U2, . . . ,,
19	S1, S2, . . . ,,	B	V1, V2, . . . ,,
20	T1, T2, . . . ,,	C	W1, W2, . . . ,,
21	U1, U2, . . . ,,	D	X1, X2, . . . ,,
22	V1, V2, . . . ,,	E	Y1, Y2, . . . ,,
23	W1, W2, . . . ,,	F	Z1, Z2, . . . ,,
24	X1, X2, . . . ,,	G	A1, A2, . . . ,,
25	Y1, Y2, . . . ,,	H	B1, B2, . . . ,,
26[4]	Z1, Z2, . . . ,,	I	C1, C2, . . . ,,

[1] The rats breed for 26 weeks with this schedule.
[2] 1/26th of the breeding groups are mated each week.
[3] Rats 12 weeks old at mating. Number saved should correspond with the numbers mated each week, plus some to allow for selection at mating time.
[4] Begin again at week 1.

TABLE VIII

A mating schedule for mice, suitable for large colonies
(over 100 breeding cages)

Week[1]	Young stock mated in cages labelled: [2]	Weanlings saved for breeding and labelled:[3]	Males removed from:
1	A1, A2, . . . etc	F	D1, D2, . . . etc
2	B1, B2, . . . ,,	G	E1, E2, . . . ,,
3	C1, C2, . . . ,,	H	F1, F2, . . . ,,
4	D1, D2, . . . ,,	I	G1, G2, . . . ,,
5	E1, E2, . . . ,,	J	H1, H2, . . . ,,
6	F1, F2, . . . ,,	K	I1, I2, . . . ,,
7	G1, G2, . . . ,,	L	J1, J2, . . . ,,
8	H1, H2, . . . ,,	M	K1, K2, . . . ,,
9	I1, I2, . . . ,,	N	L1, L2, . . . ,,
10	J1, J2, . . . ,,	O	M1, M2, . . . ,,
11	K1, K2, . . . ,,	P	N1, N2, . . . ,,
12	L1, L2, . . . ,,	Q	O1, O2, . . . ,,
13	M1, M2, . . . ,,	R	P1, P2, . . . ,,
14	N1, N2, . . . ,,	S	Q1, Q2, . . . ,,
15	O1, O2, . . . ,,	T	R1, R2, . . . ,,
16	P1, P2, . . . ,,	A	S1, S2, . . . ,,
17	Q1, Q2, . . . ,,	B	T1, T2, . . . ,,
18	R1, R2, . . . ,,	C	A1, A2, . . . ,,
19	S1, S2, . . . ,,	D	B1, B2, . . . ,,
20[4]	T1, T2, . . . ,,	E	C1, C2, . . . ,,

[1] Mice breed for 20 weeks only on this schedule.
[2] 1/20th of the total number of breeding cages needed are mated each week.
[3] This results in matings at 8 weeks of age.
[4] Begin again at week 1.

number of breeding groups would be $\frac{1}{20}$ of the total number of groups in the colony.

3. A group of weanling mice would be saved as future breeding stock and labelled O.
4. Males would be removed from cages labelled M, since these could no longer sire progeny before the group was culled.

Once such a schedule has been set up it is easy to administer.

Flexibility in the size of the breeding colony can be obtained by varying the number of breeding groups (e.g. pairs or harems) mated each week. However, such alterations are necessary only when there has been a change in demand or when a seasonal pattern has been identified, although it is advisable to keep a slight surplus of future breeding stock to allow for sudden increases in demand.

Culling. Culling as a routine health-control procedure should be carried out in all colonies to reduce the exposure of healthy members to sick or diseased animals.

Culling may also be carried out to increase productivity per unit area. An effective culling programme will be based on an individual or box record of performance. Usually an arbitrary standard of achievement is set up for each breeding unit, taking account of the strain, the time the unit has been in production, and the operational difficulty of applying the standard. Any breeding group that fails to reach this standard is culled. The general experience at the Laboratory Animals Centre has been that culling can have a dramatic effect on output. This is pronounced in colonies of poor inbred strains, but has less effect in the better breeding colonies of non-inbred strains, especially in polygamous breeding systems where it is usually difficult to identify poor breeding females.

A routine which has been applied successfully at the LAC is to examine the breeding records of all breeding pairs (the animals are inbred mice and rats bred as monogamous pairs) every three weeks, and the standards given in Table IX are applied.

The system is arbitrary, and screens only for fertility. On the other hand, it is so simple that it can be applied routinely to large numbers of breeding pairs, and in practice it seems to be highly effective. It would be possible to devise a

TABLE IX

Culling standards for rats and mice applied in the LAC colonies

Weeks since mating	Replace unless the following No. of litters have been produced
6	1 or more
9	2 ,,
12	3 ,,
18	4 ,,
21	replace anyway

similar system based on the cumulative number of young weaned from each breeding female, that would, theoretically, be better than the above system.

Genetic potential. The genetic potential of the stock is one of the most important factors in determining output. For example, Festing (1968) found that strain differences accounted for 87 per cent of the variation in productivity among a group of inbred lines of mice within a single building. Similarly, Roberts (1961) found striking strain differences in the life-time productivity of five different strains of mice ranging from 18·6 to 93·4 young per breeding female. Almost certainly similar differences exist between strains and breeds of other species.

Unfortunately, the relative importance of genetic factors in determining output when a wider range of environments is encountered has not been studied. It would be interesting to know the range of variation in reproductive performance of a single strain of animals housed in a number of different normal laboratory environments.

Husbandry and environment. Although there have been a number of studies of the effects of different environmental factors on the reproductive performance of laboratory animals, much further work remains to be done.

The most important factors likely to influence productivity are temperature, light, diet, cage-design, bedding and nesting material, noise, humidity and a group of intangible factors summed up under the term stockmanship. A full evaluation of these is beyond the scope of this chapter and the reader is referred to the relevant chapters for details. However, if any one of these factors differed far from the optimum the effects on breeding-performance could be disastrous.

Disease status. Although animals showing clinical signs of disease will usually have a reduced reproductive performance, there is little information available on the benefits, in terms of increased reproductive performance, to be obtained from specific pathogen free conditions over clean conventional conditions. Many conventional colonies, however, show overt disease of one kind or another, and even where overt disease is absent, environmental stress may cause a sub-clinical disease to become manifest. Experimental stress can have similar effects. Thus, the breeding of animals under SPF conditions is usually advantageous. An additional advantage of SPF colonies is that the stocking densities within such colonies can be much higher than in conventional colonies because there is less risk of an outbreak of disease. For some species, the reduced overhead costs in high stocking-densities, coupled with low pre-weaning mortality, probably makes the total cost of production competitive with conventional methods. With cats, for example, where pre-weaning mortality can be well over 30 per cent in conventional colonies, it is usually only about 5–8 per cent in SPF colonies (Festing and Bleby, 1969). High stocking-densities, however, should not be used unless the ventilation rate is adequate, and the temperature is controlled within certain limits.

Maximizing output per unit area

The factors influencing productivity that have been discussed so far are those relating to individual cages. However, the overhead cost of any *building* must be spread over the total output. Thus, at the same time as output per cage is maximized, the number of cages that can be accommodated should also be maximized, provided that the desired quality standards can be maintained and disease outbreaks avoided. Even if the building is well designed and the animals are disease-free, the standard of husbandry must be higher with higher stocking-densities. It is not, therefore, possible to recommend any specific stocking levels, but Table X lists some successful levels that have been reported in the literature, and Table XI lists some rule-of-thumb stocking-densities that can be used for some common species.

TABLE X

Some stocking densities actually reported

Species	No. Breeding ♀	Area of Animal room (ft.²)	Area of Cage (ft.²)	Breeding Method (¹)	S.Ft.² per Breeding ♀	Reference
Mouse	196	86	0·54	P 1 : 1	0·44	2
	300	170	—	P 1 : 1	0·57	3
	1,600	160	—	P 1 : 5	0·10	5
	1,200	800	1·00	P 1 : 2	0·67	7
	600	200	0·66	P 1 : 3	0·33	7
	1,000	1,600	0·58	P 1 : 2	1·60	7
	2,000	700	1·04	P 1 : 5	0·35	7
Hamster	350	140	0·70*	H 1 : 5	0·40	5
	100	200	1·00*	HM	2·00	7
Rat	400	396	1·34	P 1 : 1	0·99	1
	80	86	1·34	P 1 : 1	1·08	2
	300	150	0·62*	H 1 : 6	0·50	5
	800	800	1·00*	H 1 : 6	1·00	7
	400	600	1·00*	H 2 : 15	1·50	7
	3,000	4,000	1·17*	H 1 : 5	1·33	7
Guinea-pig	336	960	16·00	P 1 : 12	2·90	6
	2,200	2,100	12·00	P 1 : 8	0·95	7
	100	130	8·00	P 1 : 6	1·30	7
	300	750	16·00	P 1 : 17	2·50	7
Rabbit	200	2,400	8·00*	HM	12·00	7
	100	800	—	HM	8·00	7
	60	480	8·00*	HM	8·00	7
	4,000	24,000	—	HM	6·00	7
Cat (SPF)	30	360	—	H 1 : 5	12·00	1

* Littering cages.   (¹) P = Permanently mated, H = Harems (♀ removed prior to parturition), HM = Hand mated.
1. Unpublished data, LAC colonies.
2. Walker & Poppleton (1967).
3. Findlay (1967).
4. Coid, Box & Flynn (1962).
5. Bacharach *et al.* (1958).
6. Weitz (1954).
7. Robinson (1969) unpublished data on LAC accredited commercial breeders in the UK.

TABLE XI

'Rule of thumb' estimates of the area of animal room needed for a breeding female

Species	Area of animal room/breeding ♀ Ft.²	cm.²
Mouse	0·5	450
Hamster	0·8	700
Rat	1·0	900
Guinea-pig	2·0	1,800
Rabbit	8·0	7,500
Cat	12·0	11,100

The annual output that can be expected from one breeding unit (i.e. the space occupied by one breeding female) of most of the common species of laboratory animals is given in Table I of Chapter 4.

Finally, Table XII gives some levels of output actually achieved under various conditions.

All this information can be summarized to give a guide to the output that can be expected per unit area for some of the common small species of laboratory animals. This is given in Table XIII.

The problem of deciding in more detail the amount of space required to produce a given output of laboratory animals has been considered by Festing and Bleby (1968).

Maximizing output per unit of labour

Labour costs account for approximately 30–35 per cent of the total cost of a commercially bred animal. Hence, any savings in labour will have a significant effect on total costs.

Assuming that standards of basic husbandry allow high output per unit area, labour efficiency can be increased by attention to the following:

The establishment of standard work routines. Inevitably, much of the work involved in running a breeding colony is routine. Small savings in the amount of time spent (e.g. in servicing a cage) can amount to considerable savings over a week.

Frequently work routines develop naturally, rather than as an act of conscious thought, so

TABLE XII
Annual output reported per ft² of animal room

Species	Strain	Breeding System	Output (No. weaned/Ft²/ year)	Reference
Mouse	A2G (Inbred)	Harem 1 : 5	225	2
	GFF ,,	Polygamous 1 : 5	300	
Mouse*	?	Pairs	136	3
Mouse	CDI (Non-inbred)	Pairs	463	5
Mouse	BDFI (i.e. Inbred C57BL ♀ ♀)	Harem 1 : 5	60	6
Hamster	Syrian	Harem 1 : 5	47	2
Rat	Porton	Pairs	75	1
Rat	WAG, PVG (Inbred)	Harem 1 : 5	56	2
Rat*	?	Pairs	46	3
Rat	Charles River	Harem 1 : 4	58	6
Cat	'Mill Hill'	Hand mating	0·56	4

1. Festing (1969) (Unpublished data from LAC colonies).
2. Bacharach, Cuthbertson & Flynn (1958).
3. Walker & Poppleton (1969).
4. Lamette & Short (1966). Calculated on last three years' data only.
5. Foster *et al.* (1963).
6. Zibas & Oleson (1960).
* Calculated on stated productive capacity, assuming $\frac{1}{2}$ building devoted to each species.

TABLE XIII
'Rule of thumb' guide to the output that can be expected per unit area

Species	Approx. output expected	
	Young/ft²/year	Young/1000 cm²/yr.
Mouse	150–400	160–430
Hamster	∼50	∼54
Rat	45–60	48–65
Guinea-pig	6–9	7–10
Rabbit	1·5–2·0	1·6–2·2
Cat	∼·06	∼·07

there is often scope for streamlining. An elementary knowledge of work-study methods is helpful in analysing current procedures, so that modifications can be suggested, but even the application of common sense will often give satisfactory returns. Procedures such as changing water bottles, cleaning cages, weaning young stock and feeding should all be examined to see whether a change of routine could save labour.

The use of labour-saving devices. In strict economic terms it is probably worth spending £8000—£10,000 to save one man's labour (assuming that it costs at least £1000 to employ one man for a year). However, considerable savings can be made from modest expenditure. Although these will rarely make it possible to use fewer animal technicians, they may allow the staff more time for looking after the stock.

Large savings can also be made from the installation of cage- and bottle-washing machines, automatic watering and wire mesh floors in the cages. Other labour-saving devices such as automatic feeding and cleaning have also been used successfully, and the use of disposable items may be more important in the future, though the desirability of any single device should be examined critically.

The elimination of unnecessary records. The collection of unnecessary and frequently unusable records seems to be particularly common in laboratory animal production. What frequently happens is that someone starts a recording system, often without any clear thought about what the system is designed to achieve, and when he eventually leaves, the recording system stays on. In this way, records proliferate, and the time spent recording, processing and filing can be out of all proportion to the value of information gathered. A careful examination of all recording systems should be made about every two years, and unnecessary record-keeping eliminated.

Levelling out peak work loads. Non-seasonal breeding systems result in an even work-load throughout the year, and consequently are more efficient in using labour than are highly seasonal systems. Since demand for animals is usually only slightly seasonal, animal rooms should be kept at uniform temperatures and humidities. Uniform light cycles throughout the year will also reduce the seasonality of production.

Maximum use of labour. Bleby and Porter (1969) give some figures on the work-load that one animal technician can be expected to carry under various conditions. These figures should be helpful as a standard, against which labour efficiency in specific cases may be judged.

Improving quality

Freedom from disease is essential if high-quality animals are to be produced. Production and maintenance of disease-free colonies are discussed in Chapters 10 and 11.

Genetic uniformity can be achieved by the use of inbred strains, while long-term stability of non-inbred strains may be obtained by maintaining closed colonies with the maximum avoidance of inbreeding and no selection. For further details see Chapter 2.

Attempts should always be made to standardize the environment for all animals. Even within a single animal room there are considerable variations in the physical environment, particularly between cages at different heights.

Equating supply and demand

Meeting a single order

Breeders are frequently required to meet a single order which requires the setting-up of special matings. The order might be, for example, for 500 weanling mice born within a specified time.

The decision on the number of females that should be mated for this requires a knowledge of the effective fertility and the average litter-size of the animals concerned, the effective fertility being the number of females likely to conceive in the allowed period. Beilharz (1968) listed the percentage of females littering down on the 20th to 23rd day after pairing (Table XIV, mice).

These figures are largely a reflection of the number of female mice that conceive on the first, second, etc., day after pairing. The effective fertility then would depend on the requirement specified in the order. Supposing the order was for mice born over a three-day period, then the effective fertility would be $13\cdot4 + 13\cdot4 + 35 = 61\cdot8$ per cent (approximately).

Similar figures for rats (original data from 127 mated females) are also given in Table XIV.

TABLE XIV
Time of littering after introduction of the male

Days after male introduced	Per cent of females littering Mice	Rats
20	13·4	—
21	13·4	—
22	35·0	9·4
23	17·7	26·0
24		22·8
25	20·5	17·4
26		4·7
Later or infertile		19·7

Estimates of litter-size can be obtained from recent records of the average of a random sample of approximately 60 litters. With estimates of both effective fertility and litter-size it is possible to calculate the number of females that would need to be mated to produce the numbers required. This estimate does not, however, leave any margin for safety. This margin is necessary because purely by chance an unusually low number of females may be fertile, or many small litters may be produced. The statistical calculations of the number of females that have to be mated to be confident of producing enough animals are complicated, so Tables XV and XVI have been compiled for this purpose.

The tables make certain assumptions, which may not be fully valid under all circumstances. These are:

1. All environmental conditions remain the same as when the estimates of litter-size and fertility were obtained.
2. Litter-size is normally distributed, and fertility is binomially distributed (these assumptions are not too critical if large numbers are involved).
3. The standard deviation of litter-size is 2·5 young per litter, for all values of mean litter-size.

If these assumptions are not met, a higher safety margin should be allowed.

The use of Tables XV and XVI will be illustrated with an example: Suppose an order comes in for 250 weanling mice, all born within a four-day period. How many females would need to be mated, assuming litter-size at

TABLE XV

*Number of litters required to give a 97·5% probability of achieving the number of animals needed**

No. of Animals Needed	Average effective litter size													
	2	3	4	5	6	7	8	9	10	11	12	13	14	15
25	25	15	10	8	7	5	5	4	4	3	3	3	3	2
50	40	25	18	14	11	9	9	7	7	6	5	5	5	4
75	56	35	25	19	16	13	12	10	9	8	8	7	7	6
100	71	44	32	25	21	17	15	13	12	11	10	9	8	8
125	86	54	39	31	25	21	18	16	15	13	12	11	10	9
150	100	63	46	36	29	25	22	19	17	16	14	13	12	11
175	114	73	53	41	34	29	25	22	20	18	16	15	14	13
200	128	82	60	47	38	33	28	25	22	20	19	17	16	15
225	142	91	66	52	44	37	32	28	25	23	21	19	18	16
250	156	100	73	58	48	40	35	31	28	25	23	21	20	18
275	170	109	80	63	52	44	38	34	30	27	25	23	21	20
300	184	118	87	68	56	48	42	37	33	30	27	25	23	22
350	211	136	100	79	66	55	48	43	38	35	32	29	27	25
400	239	154	113	90	74	63	55	48	43	39	36	33	31	29
450	266	172	127	100	83	70	61	54	48	44	40	37	34	32
500	293	190	140	110	92	78	68	60	54	48	45	41	38	35
550	320	207	153	121	100	85	74	66	59	53	49	45	42	39
600	346	225	166	132	108	93	81	72	64	58	53	49	45	42
650	373	242	179	142	117	100	87	77	69	63	57	53	49	46
700	400	260	192	152	126	107	94	83	74	67	62	57	53	49
750	427	278	205	163	134	115	100	89	79	72	66	61	56	52
800	453	295	218	173	144	122	106	94	85	77	70	65	60	56
900	506	330	245	194	161	137	119	106	95	86	79	72	67	63
1000	559	365	271	215	177	152	132	117	105	95	87	80	74	69
1250	691	452	335	266	222	188	164	145	131	119	109	100	93	86
1500	821	539	400	318	262	225	196	174	156	142	130	119	111	103
2000	1082	711	529	421	350	298	260	230	207	188	172	159	147	137
2500	1342	883	657	523	434	371	324	287	258	234	214	198	183	171
3000	1600	1054	785	625	519	443	387	343	309	280	257	237	219	204
3500	1857	1225	913	727	605	516	451	400	359	326	299	276	256	238
4000	2116	1396	1040	829	686	589	514	456	410	372	341	314	292	272

* Assuming standard deviation of litter size = 2·50 young used in conjunction with Table XVI there will be a 95% confidence of getting the specified number of animals.

weaning is 8·9 young and 80 per cent of the females litter down on the 20th to 23rd day?

From Table XV an estimated 31 litters will be needed (litter-size 9, number needed 250). From Table XVI with an 80 per cent effective fertility, 44 females should be mated to produce 30 litters, so approximately 46 females would be needed to produce 31 litters. Working backwards as a check, 46 females, 80 per cent of which would be expected to produce litters of 8·9 young would be expected to produce about 328 young mice, so in this case a safety margin of 78 extra mice (31 per cent extra) would be needed to give reasonable confidence that the correct number of animals would be produced. On average, too few animals would be produced one time out of every 20.

Continuous production

Given an estimate of the average productivity (young/female/week) of a group of animals under a particular set of environmental conditions, it is easy to calculate the number of breeding females needed to achieve a given *average* level of output. Problems arise, however, when the average level of output has to be related to the actual demand. This is because:

1. Normal biological variations cause weekly fluctuations in productivity.
2. Weekly variations in demand can occur.
3. The shelf-life of the animals produced may range from a day to several weeks.
4. The flexibility of demand can vary greatly.
   Because of item 1 above, it is clear that

TABLE XVI

Number of breeding females to be mated to give a 97·5% (1) probability of achieving the specified number of litters(2)

No. Litters Wanted	Percent 'Effective Fertility'															
	15	20	25	30	35	40	45	50	55	60	65	70	75	80	85	90
10	119	87	69	57	48	41	36	31	28	25	22	20	18	17	15	14
15	161	118	94	77	65	56	49	43	38	35	31	28	26	24	22	20
20	201	149	118	97	82	71	62	55	49	44	40	36	34	31	28	26
25	241	178	141	117	99	85	75	66	60	54	49	45	41	37	35	32
30	280	207	165	136	115	100	88	78	70	63	57	52	48	44	41	37
35	319	237	188	155	131	114	100	89	80	72	66	60	55	51	48	43
40	357	266	210	174	148	128	112	100	90	81	74	68	63	58	53	49
45	396	292	233	193	164	142	125	111	100	91	83	76	69	64	59	55
50	433	320	256	211	180	156	137	122	110	100	91	83	77	71	66	61
55	471	350	278	230	195	169	149	133	120	109	100	91	85	78	72	67
60	508	376	300	248	211	182	162	144	130	118	108	99	90	85	78	72
65	545	404	323	267	227	196	174	155	140	127	116	106	98	90	84	78
70	583	433	345	285	242	210	186	166	150	136	124	114	106	98	90	84
75	620	462	367	303	258	225	199	177	159	145	132	122	112	104	96	90
80	656	488	389	322	274	237	210	188	169	154	141	130	119	110	102	95
90	730	543	432	358	305	266	234	209	189	171	157	144	135	124	115	107
100	803	595	476	395	336	293	259	230	208	190	173	159	146	137	127	118
120	948	708	563	466	398	346	306	273	246	225	206	191	174	163	151	141
140	1092	812	649	538	458	400	353	315	285	259	235	219	205	188	177	164
160	1236	924	734	609	520	454	400	358	324	296	270	250	231	216	200	188
180	1378	1030	820	680	580	506	445	400	362	331	303	279	259	240	225	210
200	1520	1136	906	751	641	557	493	442	400	365	335	310	286	266	249	231
250	1874	1399	1116	927	791	692	610	547	493	450	416	384	354	331	309	289
300	2226	1665	1326	1101	941	818	729	651	590	538	493	458	424	396	369	346
350	2577	1927	1537	1276	1090	949	841	755	686	625	571	529	493	458	429	400
400	2925	2190	1746	1450	1239	1080	961	858	778	708	655	605	562	525	489	458
450	3274	2450	1953	1623	1388	1210	1076	962	870	795	735	676·	630	587	550	515
500	3622	2704	2162	1797	1536	1340	1184	1064	967	876	812	751	697	650	609	571
600	4315	3226	2577	2142	1832	1598	1414	1271	1156	1050	967	894	835	778	729	681
700	5007	3745	2990	2486	2126	1858	1648	1477	1339	1225	1129	1043	967	900	847	795
800	5696	4264	3399	2830	2421	2112	1875	1682	1521	1391	1282	1183	1109	1031	967	906
900	6382	4775	3816	3173	2714	2372	2098	1886	1714	1568	1436	1332	1241	1159	1087	1024
1000	7069	5285	4225	3516	3007	2626	2333	2091	1901	1731	1600	1482	1377	1286	1206	1136

(1) Used with Table XV there will be a 95% probability of getting the specified number of animals.
(2) Calculated from the usual formula for the lower limit of a confidence interval.

average *production* should be higher than average *demand*, unless demand is completely flexible or the shelf-life of the animals is so long that a large pool of animals can be built up to even-out the fluctuations in output.

Where these conditions do not exist, the consequences of failure to meet the demand in any one week must be assessed. This is usually difficult, but some idea could be obtained if it were assumed, for example, that failure to produce enough animals might make it necessary to buy in animals from an outside source, at a cost which would usually be higher than that of home-produced animals. Alternatively, a subjective approach would be to say that sufficient animals must be produced, on average, say, nine weeks out of ten. Table XVII is based on this latter approach with the additional assump-

tion that the demand is constant though output fluctuates about a mean value with a standard deviation of 3·0 young/female/week. This figure is based on the data of Porter and Festing (1969) for mice, and Festing (unpublished) for rats.

In Table XVII it is assumed that a shelf-life of three weeks, for example, means that the demand will be satisfied as long as the average production over any three-week period is above the demand level. Such a situation would occur if demand was either flexible or inflexible, but age at use did not matter within three weeks.

In using Table XVII, the first priority is to obtain an estimate of the productivity of the breeding females in terms of usable young/female/week. This can be obtained from past records, as before. The next step is to decide on the shelf-life of the animals. This should be

Table XVII
*Number of breeding females needed (demand constant)**

Demand per Week	Shelf Life (wks)	0·50	0·75	1·00	1·25	1·50	1·75	2·00	2·25	2·50
					Productivity (Y/♀/Wk)					
10	1	97	50	32	23	18	14	12	10	9
	3	53	30	20	15	12	10	9	7	7
	6	40	24	16	13	10	9	7	6	6
25	1	144	80	54	40	31	26	22	19	16
	3	94	56	39	30	24	20	17	15	13
	6	78	48	34	26	21	18	16	14	12
50	1	214	125	86	65	52	43	37	32	28
	3	156	96	69	53	43	36	31	28	25
	6	137	87	62	49	40	34	29	26	23
75	1	282	167	117	90	72	60	52	45	40
	3	216	135	97	76	62	52	45	40	36
	6	193	123	90	70	58	49	42	38	34
100	1	352	208	147	112	92	77	66	58	51
	3	275	173	126	98	80	68	59	52	46
	6	250	159	117	92	76	64	56	50	45
125	1	404	249	177	137	110	93	80	70	62
	3	332	210	153	121	98	83	72	64	57
	6	303	196	144	113	93	79	69	61	55
150	1	471	288	206	159	130	109	94	82	74
	3	388	246	180	141	116	99	85	75	68
	6	358	231	172	135	110	94	83	72	66
175	1	529	327	235	182	149	125	108	95	84
	3	445	282	207	163	134	114	99	87	78
	6	412	269	199	156	128	110	96	85	76
200	1	590	366	263	205	166	141	122	107	95
	3	502	320	234	185	152	129	112	99	88
	6	467	303	222	177	147	124	108	96	87
250	1	708	441	320	249	204	172	149	131	117
	3	610	392	289	228	187	159	138	122	109
	6	571	373	276	218	182	153	135	119	106
300	1	824	520	375	292	240	203	176	155	138
	3	718	465	342	269	222	189	164	147	130
	6	681	445	328	259	216	183	161	142	128
350	1	936	595	431	339	276	234	203	178	160
	3	830	538	396	313	257	219	190	169	151
	6	790	515	380	303	250	213	185	164	149
400	1	1056	666	486	380	313	265	230	202	181
	3	936	607	450	353	292	250	216	192	172
	6	894	586	433	342	282	243	210	188	169
450	1	1170	740	541	424	350	296	256	225	202
	3	1043	678	502	396	327	276	243	215	192
	6	999	655	484	384	320	272	237	210	188
500	1	1281	818	595	467	384	328	282	250	223
	3	1153	749	552	437	362	310	269	238	213
	6	1102	724	538	424	350	303	262	234	210
750	1	1832	1177	864	681	562	475	412	365	328
	3	1681	1100	812	645	535	458	396	352	316
	6	1624	1069	795	630	524	445	392	346	310
1000	1	2381	1537	1129	894	735	625	548	482	433
	3	2209	1444	1076	853	708	605	525	466	419
	6	2143	1413	1050	835	692	595	520	458	412

* Table gives number of breeding females needed to fulfil the stated demand on average nine weeks out of ten. Numbers obtained from the formula:

$$PN - \sigma / \sqrt{\left(\frac{N}{L}\right)} - DP = 0,$$ where N is the number of breeding females, P = productivity, L = shelf life, D = demand per week, and σ = standard deviation of productivity = 3·0 (assumed).

estimated according to both the true shelf-life and the flexibility of demand.

The tables should, of course, be used only as a guide to planning a colony. As soon as it is up to full strength the normal feedback mechanisms will come into action and the size can be altered according to size of surplus and the success with which demand is met.

Fluctuating demand

Demand can both fluctuate widely, and vary so much in nature, that it is impossible to construct tables to take account of this. The best that can be achieved is to state some general principles. On the one hand, if demand fluctuates widely and is inflexible, then considerably more breeding females will be needed; but if the demand is very flexible, then not many more females would be needed than if demand were constant. In practice, for a given flexibility of demand fluctuations will require more breeding females to be used than if the demand were constant.

In all cases of fluctuating demand the best approach would probably be to use the tables as a guide and add a rough estimate of the additional numbers needed. No amount of sophisticated mathematics can replace the use of common sense and experience in adjusting a breeding colony to meet the demand.

REFERENCES

Bacharach, A. L., Cuthbertson, W. F. J. & Flynn, G. W. (1958). The economics of laboratory animal breeding—rats and mice. *Lab. Anim. Cent. coll. Pap.*, **7**, 31–44.

Beilharz, R. G. (1968). Effect of stimuli associated with the male on litter size in mice. *Aust. J. biol. Sci.*, **21**, 583–5.

Belcic, I. & Weihe, W. H. (1967). Applications of the collar method for breeding Chinese hamsters. *Lab. Anim.*, **1**, 157.

Bleby, J. & Porter, G. (1969). Timed assessment of the duties of animal technicians. *J. Inst. Anim. Techns.*, **20**, 12–24.

Carter, T. C. (1957). Breeding methods II. Economic considerations. In *UFAW Handbook on the Care and Management of Laboratory Animals*, 2nd edn. London: UFAW.

Coid, C. R., Box, P. G. & Flynn, G. (1962). The density of experimental animal populations. *Lab. Anim. Cent. coll. Pap.*, **11**, 29–32.

Dickie, M. M. (1966). Keeping records. In *The Biology of the Laboratory Mouse*, 2nd edn. Ed. Green, E. L. New York: McGraw-Hill.

Festing, M. F. W. (1968). Some aspects of reproductive performance in inbred mice. *Lab. Anim.*, **2**, 89–100.

Festing, M. F. W. & Bleby, J. (1968). A method for calculating the area of breeding and growing accommodation required for a given output of small laboratory animals. *Lab. Anim.*, **2**, 121–9.

Festing, M. F. W. & Bleby, J. (1969). Breeding performance and growth of SPF cats (*Felis catus*). *J. small Anim. Pract.*, **11**, 533–42.

Findlay, G. H. (1967). Factors considered in designing an animal house. *J. Inst. Anim. Techns.*, **18**, 29–36.

Foster, H. L., Foster, S. J. & Pfau, E. (1963). Large-scale production of caesarian originated, barrier sustained mice. *Lab. Anim. Care*, **13**, 711–8.

Heine, W. (1965). Problems of large-scale production. *Fd. Cosmet. Toxicol.*, **3**, 223–8.

Lamotte, J. H. & Short, D. J. (1966). The breeding and management of cats under laboratory conditions. *J. Inst. Anim. Techns.*, **17**, 85–95.

Lane-Petter, W., Lane-Petter, M. & Bowtell, C. W. (1968). Intensive breeding of rats. 1. Crossfostering. *Lab. Anim.*, **2**, 35–39.

Porter, G. & Festing, M. F. W. (1969). A note on breeding performance of mice in four different types of plastic cage. *J. Inst. Anim. Techns.*, **21**, 78–83.

Porter, G. & Lacey, A. (1969). Breeding the Chinese hamster (*Cricetulus griseus*) in monogamous pairs. *Lab. Anim.*, **3**, 65–68.

Roberts, R. C. (1961). The lifetime growth and reproduction of selected strains of mice. *Heredity, Lond.*, **16**, 369–81.

Thirkettle, G. L. (1968). *Whelden's Business Statistics*, 6th edn. London: Macdonald & Evans.

Tucker, D. K. & Peacock, W. (1974). A comparison of two rat-breeding systems. *J. Inst. Anim. Techns*, **25**, 15–20.

Walker, A. I. T. & Poppleton, W. R. A. (1967). The establishment of a specific pathogen free (SPF) rat and mouse breeding unit. *Lab. Anim.*, **1**, 1–5.

Walker, A. I. T. & Stevenson, D. E. (1967). The cost of building and running laboratory animal units. *Lab. Anim.*, **1**, 105–9.

Weitz, B. (1954). A new small animal unit. *Lab. Anim. Bur. coll. Pap.*, **2**, 11–20.

Zibas, V. & Olson, J. J. (1960). The development and operation of a rat and mouse breeding laboratory. *Cancer Chemother. Rep.*, **7**, 13–8.

6 The animal house and its equipment

W. LANE-PETTER

Introduction

The three clients

The architect designing an animal house has three clients to satisfy: the investigators who are to use the animal house, the animal technicians whose place of work it will be, and the animals who will live all their lives there. The claims of these three clients are sometimes in conflict, and a compromise has to be worked out that will not completely override the claims of any one of them. When the architect has arrived at the best possible compromise with his clients' demands, he can then turn his attention to the more aesthetic side of his design. But it cannot be too strongly stressed that, in a laboratory animal house, functional efficiency is paramount.

Flexibility

In most laboratories a variety of species of animals will be in use, and the relative numbers of each will not remain constant from year to year. As far as possible, therefore, this fact should be taken into account, so that a room that is used for rabbits one year can be converted easily to house mice or rats the next. Certain species, such as dogs and monkeys, have such special requirements that their rooms cannot be easily turned over to other species. Yet even in these cases the special requirements should not be so permanently built into the structure of the house that it can never be dismantled or adapted for any other purpose.

Planning

The design team must include members representing all three clients, together with administrative and financial members. The architect will bring his own expert consultants on heating and ventilation, on electrical and plumbing services and so on. The clients will be well advised to include on their side an engineer who will have the responsibility for maintaining the animal house in good order when it is built, for it is not unknown for an animal house to meet all the requirements that have been specified but to present almost insuperable problems to the engineer who has to keep it operating.

All members of the design team have much to contribute. The demands of the investigator must be taken very seriously, because without his work there is no *raison d'être* for the house; but his demands must be realistic, and if they are not, this must be explained to him. For example, it is unrealistic to demand, and difficult or impossible to justify on scientific grounds, an animal room with a large window and a temperature requirement of $\pm 1°$ all the year round. There is no reason why the investigator's demands should not be challenged, but it is no part of the architect's or the administrator's job merely to overrule him.

The senior animal technician who is to work in the animal house must have a place in the design team. He or she will contribute much that is of practical value, and may also represent the needs of the animals. Once again, the animal technician's demands may be challenged, but not lightly overruled. The same is true of the maintenance engineer.

At this stage the architect should have a fairly clear idea of what he is being asked to produce. He is entitled to be told what animals he has to accommodate, and in what environmental conditions: what their special needs are, and what, outside normal architectural experience, is peculiar to laboratory animal accommodation. From then on it is his task to produce sketch plans, and discuss them freely with his colleagues in the design team.

A useful publication is *The Design and Function of Laboratory Animal Houses* (Hare and O'Donoghue, 1968). Other general sources of information are Animal Welfare Institute (1966); Institute of Laboratory Animal Resources (1968); Jonas (1965); Hill (1963); Lane-Petter (1963); Lane-Petter and Pearson (1971); Nuffield Foundation (1961); Ottewill (1968);

74

Short and Woodnott (1969). Reference should also be made to the Guidance Notes (1974) published by the Research Defence Society, which include Home Office recommendations and requirements.

General principles of design

Subdivisions

Any animal house, from the smallest to the largest, has a number of functional subdivisions, which may be thought of as separate departments linked together by operational considerations.

Since all animal colonies are always at risk of unwanted infection coming in from outside, there must always be a more or less efficient peripheral barrier against such invasion; and there may also be subsidiary barriers within the periphery, as an added protection.

The basic subdivisions of an animal house are:

1. Accommodation for animals destined for experiment: that is, a breeding colony, and a room for stock.
2. Accommodation for animals under experiment.
3. Store rooms.
4. An area for cleaning cages and other equipment, disposing of soiled bedding, carcases etc.
5. An office for administration.
6. A room for mechanical plant.
7. Corridors giving access to all these areas.

The relative proportions of the various areas will vary to some extent. Large animal houses will be able to allot a higher proportion of the total area to animal accommodation, but only if the planning of the whole house is carefully worked out. An idea of relative areas is given in Figure 6.1.

From this it will be seen that not more than about half the total floor area of the house can be used for accommodating animals, together with operating theatres, X-ray rooms, laboratories and other places where animals may be held or used.

A surprisingly large proportion (about 15 per cent) is needed for corridors, and about 10 per cent for each of the following: stores, cage cleaning and the group of miscellaneous purposes included in the term administration.

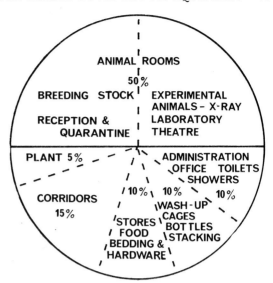

Fig. 6.1. Approximate relative areas of the subdivisions of an animal house.

These subdivisions will be discussed in more detail shortly, but it is first useful to consider two other aspects of design, namely operational needs and provision of barriers.

Operational needs

Traffic. A substantial traffic circulates in any animal house, which partly accounts for the high proportion of the total floor area occupied by corridors. Staff have to move from place to place; food, bedding, clean cages and other materials and equipment have to be brought to the animal rooms, while dirty cages and equipment, soiled bedding and other materials have to be taken away; and animals may be moved both to and from rooms. Much of this movement is on wheeled vehicles.

The main traffic movements are shown diagrammatically in Figure 6.2.

The design team should make a reasonably detailed calculation of the volume, in terms of cubic metres and kilograms, of each item of traffic: not only its total volume, but how the total is distributed over the days of the week and the hours of the day. It will be found that the flow is not even; there are peak flows, now in one direction, now in another. Dirty traffic (soiled cages and bedding) should not mix with

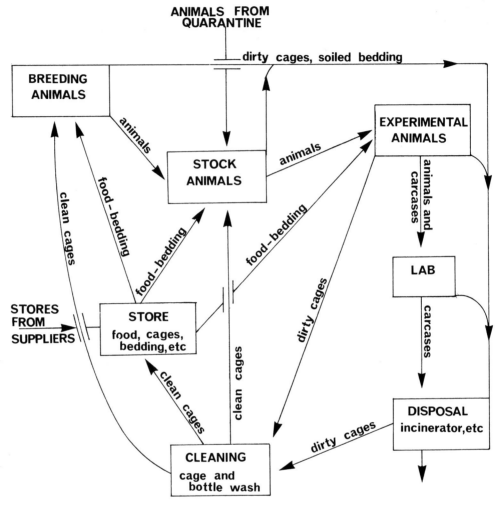

Fig. 6.2. Diagram of traffic flow within the animal house.

clean traffic (materials and clean equipment being delivered to rooms). They should be separated by space or by time; that is, they should use different corridors, or they should flow at different times of the day.

These calculations are necessary in order to decide how large the corridors should be. If they are too small, there is a periodical traffic jam. If they are too large, valuable space is wasted and more maintenance is incurred.

Not the least important part of the traffic circulation is that of the staff. They have to move themselves, and also a considerable volume of material, from place to place.

Unnecessary steps mean extra work and effort. Much can be done in the organization of the work of the staff to reduce unnecessary steps, especially within rooms, but there is still scope in the design of the animal house to see that the most frequent journeys are not the longest ones. A frequent journey is from the cage rack to the weighing, treatment or examination table, and back: these should be close to one another. Less frequent is the journey from the semi-bulk store to the animal rooms; the distance between these can be greater. But everywhere steps up or down, or steep slopes, are to be avoided, because wheeled traffic will not negotiate them.

Lay-out. A study of the traffic will give many pointers to an ideal layout. There are, of course, other considerations.

Animal rooms are usually grouped together because it is more convenient from the point of view of plumbing, ventilation ducts and heating, and for other mechanical reasons. Grouping may also be preferable for operational reasons; but on the other hand, some animal rooms, such as quarantine rooms, or rooms for infected animals, may need to be isolated from the rest.

An early decision has to be made about access corridors. Animal houses are built on either the one-corridor or the two-corridor system. In the one-corridor system all access is from the same corridor, staff and materials as well as animals passing either way through the same doorway. In the two-corridor system all traffic entering the room uses one doorway, and all traffic leaving it another doorway, thus creating a sort of one way traffic through the room. The argument in favour of the two-corridor system is that it reduces the risk of cross-infection within the animal house. Perhaps it does, although there is no convincing evidence that it makes any substantial contribution to this; there are so many other pathways of cross-infection in an animal house that the question of one or two corridors is probably not very important. Two corridors are certainly more expensive than one, and thus need some justification in terms of operational efficiency.

The relationship of stock or breeding animal rooms to the experimental suite needs careful study. Generally speaking, to move a cage of animals from one room to the next is as disturbing as to move them a greater distance, and they will take just as long to settle down in their new room. Therefore, the advantage of proximity of stock to the experimental room does not rest on the argument of minimal disturbance, and it may be countered by the added risks or inconveniences of having the two animal departments too close together. In most cases it will be found preferable to arrange complete separation of the two departments, perhaps with separate buildings, or separate staff, entrances, stores and services.

Where there is a quarantine room, for the reception and treatment of animals coming from outside sources, it should be well separated from all other animal rooms, preferably with a separate entrance. For mammals imported into Great Britain the requirements of the quarantine rooms must conform with the Regulations made by the Ministry of Agriculture, Fisheries and Food under the Rabies (Importation of Dogs, Cats and Other Mammals) Order, 1974, details of which are obtainable from the Animal Health Division of the Ministry at Hook Rise South, Tolworth, Surbiton, Surrey.

Storerooms should be near the reception area for materials, as well as being conveniently placed for distribution to the animal rooms. They should be protected from undue heat from sterilizers and other equipment; from steam or dampness from washing areas; and from effluent ventilation with its burden of moisture, heat, dust and possibly infection.

Relation to laboratory. There are two schools of thought about the relation of animal rooms to laboratories. One demands that animal facilities be always conveniently available to the laboratory workers; it wants an animal room in every laboratory suite. The other points out that animal care of a high standard is possible only when the animal rooms are more or less grouped together, and their operation integrated.

Both points of view have something to be said in their favour. It can be very difficult for the investigator to be separated from his animals by considerable distances, and he may waste a great deal of his time going to and from them. Moreover, he may have in his laboratory materials and apparatus that cannot conveniently be taken to a distant animal room, making it necessary to bring the animal to his bench, and observe it throughout the experimental period.

On the other hand, animal care is still necessary during the experimental period (except in the case of short-term non-recovery experiments), and the immediate vicinity of the laboratory may be quite unsuitable for providing a high standard of care.

The way this conflict is to be resolved will depend much on the nature of the work, and the proposed organization of the laboratory and animal house complex. The solution will be, as usual, a compromise, with some concessions to experimental demands or convenience, and some to the demands of animal care.

A common compromise will be to site the breeding or stock animal house on its own, strictly isolated and operated as a unit in its own right. Experimental animal facilities will be provided, more or less in conjunction with laboratory facilities, depending on the nature of the proposed work. The breeding or stock animal unit has the responsibility of providing animals for experimental use; it should also provide animal care, since this is a skilled task, and the laboratory unit cannot be expected to provide the necessary expertise. The laboratory side will supply special equipment and apparatus, as well as technical help in experimental procedures that are outside the normal range of skills of an animal technician.

Barriers

In the context of laboratory animal houses, a barrier is a physical structure, or complex of structures, that separates animals from the general environment, and thus provides them with a special environment in which they enjoy more or less protection from the hazards of the outside world. According to this definition, a barrier may be anything from the shelter of a conventional building to the sophistication of a germfree isolator.

It is important to remember that there is a continuous spectrum of sophistication, from a lock on the door to the absolute exclusion of all germs, and that any particular levels of sophistication are purely arbitrary. An example illustrating categories of barriers will be found in Parrott and Festing (1971). Provided that the arbitrary nature of these is not forgotten, such an example is useful and a comparable classification will be used in what follows.

'Conventional' barrier.   Ever since scientists have kept laboratory animals, they have provided 'conventional' animal houses for them. Such accommodation provides more or less protection from the weather; from a roof to keep the rain off rabbit hutches to a fully air-conditioned building. It will also provide more or less protection from insects, wild rats and mice, and other pests; be compatible with a more or less high standard of ordinary cleanliness; and offer more or less encouragement to the staff to observe the sort of hygienic standards that are to be found in a good institutional kitchen. Food, bedding and other materials coming into such an animal house will be examined, controlled and treated in such a way as to exclude the likelihood of their being vectors of dangerous infection: in fact, the analogy with the institutional kitchen is a close one.

The frequent entry of those who do not work in the house, and who have no good reason to visit it, will be discouraged, for visitors may be vectors and thus constitute a danger to the animals. A close and continued watch will be kept on the animals' health and incipient epidemics will be nipped in the bud as far as possible.

A 'conventional' animal house, run in accordance with sound principles of ordinary hygiene, has served laboratories well for many decades, and is likely to continue to do so. Its hygienic standards are compatible with the perpetuation of a high standard of general health in the animals, provided their more detailed care is adequate.

Detailed care includes nutrition; badly fed animals will not maintain good health in any circumstances. It also includes such factors as density; high population densities increase the risk of disease quite markedly. Poor standards of handling, coupled with a relaxation of general cleanliness, will also be likely to lead to a breakdown in the health of the colony.

One great advantage of a 'conventional' animal house is its convenience. It is possible to enter it with only minimal ritual: a change of coat and footwear, perhaps, and the donning of a mask and disinfection of the hands. Movement and communications are simple. Indeed, a good 'conventional' animal house may be the only one compatible with many kinds of experimental work.

'SPF' barrier.   'SPF' stands for 'specific (or specified) pathogen free', and has been defined in the following terms:

'Animals that are free of specified microorganisms and parasites, but not necessarily free of the others not specified' (*ICLA Bulletin*, 1964).

However inadequate this definition, it has the merit of simplicity, and it applies only to animals. But there has grown up around the term, whether spelt out in full or contracted to its three initial letters, 'SPF', a mystique which has led to the letters being applied to methods, buildings and barriers. In the present discussion the term 'SPF' is used, in inverted commas, to indicate a somewhat more rigorous barrier than has been described as 'conventional'.

'Conventional' animal houses are regarded as more or less open systems. 'SPF' units should likewise be regarded as open, but not so wide open.

The main feature that distinguishes an 'SPF' animal house is that entrance and exit of staff and of materials is made much more troublesome than in a 'conventional' animal house. Staff will be expected to change, shower and put on sterilized working clothes each time they enter, which will normally be twice a day. The casual entry of visitors, or investigators, will be strongly discouraged, but if they do enter they will have to observe the same changing and showering routine.

Materials will enter through some kind of sterilizing lock: an autoclave, a sterilizing gas chamber, a disinfectant entry (dunk) tank. This applies to food, bedding, cages and other equipment, instruments, tools, and everything else that has to come within the barrier where the animals are kept. It is thus hoped that none of these materials will be a vector of pathogenic infection.

Air coming into such an 'SPF' building will be filtered. Opinions differ greatly about the degree of filtration that is necessary and commensurate in rigour with the other features of the barrier, from the total exclusion of particles larger than 5 μm in diameter, to that of particles a tenth of this size or less. The incoming air must at least be free of insects and dust: it will also need to be treated in some way, up to full air-conditioning.

Water from a main supply should normally have a very low bacterial count, and often contains free chlorine. But it cannot be relied upon to be sterile, and may require further treatment before passing through an 'SPF' barrier.

Finally, there are the animals. No amount of cleaning up the physical environment within the barrier is of use if the foundation colony introduced into it already carries an inventory of organisms that may cause disease. The first animals to be introduced will have been separated from their normal burden of microorganisms and parasites by being delivered by hysterectomy and being either hand-raised or fostered (see Chapter 10).

Isolators. An isolator is a container for laboratory animals, the periphery of which is an absolute barrier to the passage of microorganisms (including viruses). It is thus, unlike the foregoing 'conventional' and 'SPF' animal houses, a closed system, and it provides for the animals within it an environment whose microflora is, or can be, known in its entirety. It is a germfree or gnotobiotic system, and it is dealt with in detail in Chapter 11.

Because an isolator is an absolute barrier system in itself, it does not need to be in a rigidly barriered room. An ordinary working room is the right place for it, but it is advisable to keep down the level of dust in the rooms, so that the air filters in the isolators will last longer; and to exclude gross pests, which might damage the fabric of the isolators.

Accessibility and communication. In no animal house should there be complete freedom for all to come and go as they please. In general, only those who work in the animal house, or who have a specific reason for going there, should be allowed in. In the case of experimental animal houses in Great Britain, this will be a condition laid down by the Home Office. But those who have legitimate business in the house must be able to enter, and leave, without unreasonable difficulty. If, for example, showering and changing is made too uncomfortable or inconvenient, staff will in time stop observing this precaution; and if investigators are asked to do more than is seen to be necessary in the way of changing footwear and outer garments, they will refuse to co-operate. Every precaution is, in fact, troublesome. Every precaution that is demanded, therefore, must be supported by good evidence of its necessity.

Many animal rooms are windowless, for reasons which will be discussed later (p. 85). It is then not always possible to look into animal rooms from outside, and if access is restricted, visual and direct auditory communication may be prevented. Yet there is constant need for those inside the animal house to be in communication with those outside.

A telephone is an obvious necessity. Telephone bells, if loud and ringing in an animal room, can seriously disturb breeding rats and mice; they should therefore be located in corridors or offices, rather than in animal rooms. In a large house there should be enough telephone extensions to ensure that too much time is not wasted by staff walking long distances to answer the telephone.

The lack of visual communication, especially in an animal house with a strict barrier, may be largely overcome by the provision of windows at various points, giving visual access across the peripheral barrier. If these windows have inserted in them an auditory diaphragm (Fig. 6.3) normal direct conversation can be carried on between the two sides of the barrier without breaking its integrity.

The barrier concept. The peripheral barrier (apart from an isolator) is, emphatically, not an absolute bar to invasion, and internal barriers (between rooms for example) are an added hindrance to the spread of infection.

Bricks and mortar, stainless steel, plastics and glass are not the only materials for a barrier. In recent years the usefulness of flowing air has been exploited as a most efficient barrier against invasion by germs. The surgeon in the operating

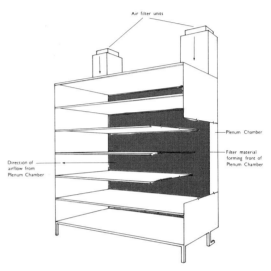

Fig. 6.4. Filter rack. Air is drawn through the filter units into the plenum chamber and passes out to the rack through the filter material at the front of the chamber.

Fig. 6.3. Window with auditory diaphragm. This diaphragm can be fitted to allow conversation to take place in normal speaking tones. (Photograph: UFAW).

theatre, under a vertical laminar flow hood, relies on flowing air to carry away from the operation site particles that might otherwise settle and bring infection to the site. The laminar air flow clean bench enables aseptic manipulations to be carried out under apparently open conditions. The filter rack (Fig. 6.4) described by Lane-Petter (1970) makes use of the same principle. Flowing air, when properly used, is in fact, a very efficient barrier. It has the advantage that as often as it is violated, it reforms.

The peripheral barrier, short of the absolute conditions of an isolator, is always a relative concept. For this reason, there is very little meaning in such terms as 'barrier-maintained', 'SPF' and 'conventional', unless their use is accompanied by a definition *ad hoc*. But experience has shown that such terms are often used without defining them, and that they acquire a semblance of precision which they cannot possess. It is perhaps time to think of dropping them.

Departments

There are seven departments or subdivisions of an animal house, as was outlined at the beginning of this chapter (p. 75). Their relative proportions are shown in Figure 6.1. They are respectively for normal animals, for experimental animals, for stores, for cleaning cages and disposing of soil, for administration, for plant, and for corridors. The animal departments, part of the stores, part of the cleaning facilities, part of the administration facilities, and some corridors, will all be within whatever peripheral barrier is provided. Parts of the stores, cleaning and administration facilities, some corridors, and the plant room will be outside the barrier.

The seven departments will be considered individually.

Normal animal department

To talk about normal animals begs a definition of the word 'normal'. In this context it is used for animals that are not under experiment, but are destined to be so used, including breeding colonies.

Breeding colonies are usually kept as isolated as possible. In any animal house complex the strictest peripheral barrier should be round these colonies. The animals within a breeding colony are self-recruiting: that is, they do not normally receive new members from outside. The only exception to this rule is when a new colony has to be founded, or the gene pool in an existing one enlarged by an introduction from outside. When this happens, full precautions have to be taken to prevent the accidental introduction of infection.

However, the normal animals may be obtained from another source, such as a breeding centre or a commercial breeder, and in this case the normal animal department has to provide only holding facilities for them, after a suitable period of quarantine.

There are certain inherent dangers in obtaining animals from any outside source, however reliable. The outside animals may appear perfectly healthy, but a change of environment may light up an infection that has been latent and produce disease, even of epidemic proportions. Also, if animals (especially of the same species) are obtained from different sources, each group may carry, with no damage to itself, an infection that may be pathogenic to the others. For this reason, to mix animals from different sources is to ask for trouble.

There is, in these circumstances, a need for some sort of quarantine area, where incoming animals can be held for a sufficient length of time to ensure that their introduction into the main holding area will not be followed by an outbreak of disease. Quarantine rooms need to be small and multiple, isolated as far as practicable from each other and above all from the main holding area. The length of stay in them depends entirely on what infections are likely to be present in the imported animals that could cause trouble in the holding area or the experimental animal house. Animals coming from good breeding centres or commercial breeders may need little or no quarantine, although groups of animals from two such sources, however good, should not be mixed. But sometimes the incoming animals need a long period of observation and cleaning up before they can be released. It is not possible to state any definite length of time for the quarantine period.

There must also be a reception area in the normal animal department. If animals are being brought in regularly from outside, this area should be adjacent to the quarantine facilities, and it should be possible to hold incoming animals there for a few hours, so that they may be examined before being accommodated. If importations of animals are infrequent the same reception area that is used for food, bedding and other materials will probably serve to receive animals.

To sum up, the normal animal department will have provision for receiving animals and materials; for quarantine of incoming animals; for breeding animals; and for holding both animals that have been bred there and those obtained from outside until they are needed for experiment.

Experimental animal department

After transference from the normal animal department to experimental quarters animals may need a period of acclimatization or conditioning in their new environment before the experiment proper begins, and for this purpose the same sort of accommodation as serves for holding animals in the normal animal department will usually be suitable.

For much experimental work, however, special accommodation is necessary, such as climatic chambers, isolation cages, metabolism cages, or provision for pre- or post-operative care.

In the experimental animal department there will also be special experimental or observation areas, such as an operating theatre suite, an X-ray room, sound-proof or other special rooms, a treatment room for carrying out simple manipulations such as injections, and often a laboratory with special experimental apparatus.

Stores

Every animal house needs an adequate area for storage of materials. Food and bedding are usually delivered at intervals of seldom less than a week (apart from fresh foods such as vegetables, fruit, meat and fish and even these may not come more often than once a week) and therefore there must be storage for not less than 2 or 3 weeks' supply. Most of the materials come in bulk, and have to be broken down into semi-bulk before they arrive at the animal rooms. By bulk is meant a two-weeks' supply of feed, in 25 kg bags for easy handling and stacking in the store. These are transferred, as required, to bins, usually on wheels in semi-bulk, and the bins go to the animal rooms.

Food in pellet form should be stored in a cool, dry, airy room, and in such conditions ought to keep well for 3 or 4 weeks. Fresh or frozen foods will often have to be stored in a refrigerator. Bedding, in bags, can be stored with the pelleted food, and its keeping life is almost indefinite.

Cages, cage tops, bottles, hoppers and other hardware always present a storage problem because they are so bulky. A hardware store should use every available space, and, if at all possible, equipment that is not in regular use should be removed from the animal house and stored elsewhere.

There is always a need to store small-bulk, valuable or perishable materials, such as drugs, chemicals, and food supplements. Some of these may require special conditions of storage, such as refrigeration. There will also be instruments to store, and these, like drugs and some chemicals, need to be under lock and key. And there should always be a first aid box, kept where all staff are aware of its existence, and regularly inspected.

Some of the stores will need to be kept within the peripheral barrier, thus necessitating storage facilities on both sides of the barrier.

Cleaning area

Cages, cage tops, bottles and other equipment in regular use need regular cleaning; and soiled bedding, dead animals and other waste material have to be disposed of.

By far the biggest cleaning load is cages and cage tops. It is today unusual, and always inefficient, for cages to be cleaned inside the animal rooms. Normally, the dirty cages and their soil are removed from the animal rooms and cleaned out and washed in an area set aside for this task; often mechanical cage-washers are used for this purpose.

The cleaning area is usually outside the peripheral barrier, and after cleaning the cages are returned through some sterilizing portal of entry. Sometimes, however, a mechanical cage-washer is installed within the barrier, thus saving a certain amount of transportation and sterilization. A third possibility is to install a tunnel type of cage-washer in the peripheral barrier, so that dirty cages come outside, their soil is removed, and they re-enter via the cage-washing machine.

Whatever method is chosen, provision must be made for stacking dirty cages awaiting cleaning, disposing of soiled bedding and other waste, and stacking clean cages ready for re-use. The necessary stacking space can be quite considerable.

Bottle-washing is more frequently carried out within animal rooms, especially if a type of bottle that can be chemically sterilized is used. But in many animal houses bottles are treated in the same way as cages.

Soiled bedding must be removed from the animal house as soon as possible. It is often put into paper or plastic bags for easy disposal. It may be taken away and incinerated elsewhere, or it may be burnt in an incinerator in the animal house.

The cleaning area is the dirtiest part of the animal house, and the work done there the least pleasant. It is likely to be steamy and often smelly. Everything should be done to make it as little unpleasant as possible, and the cleaning as easy as possible. In this way those who are employed there may be persuaded to take a pride in their work, thus ensuring that the area is not objectionable to the senses or a danger to the health of animals or staff.

Administration

Every animal house, however small, needs an office: at the very least, a quiet corner of a room, with a desk or table, where the chief technician can keep his records and do his thinking. For records always have to be kept—records of animals in and out, bred, bought, delivered and died; housekeeping records; staff records; and, not least important, records of observations and research which all good animal technicians will want to make from time to time.

Staff also need rest periods, where they can sit down for ten minutes morning and afternoon, and have a cup of tea or coffee. In 'conventional' animal houses the laboratory canteen facilities may serve the animal house staff, but in 'SPF' units, and in many more open animal houses, a rest and refreshment room may have to be provided. It need not be large, and it should not be too comfortable, but it does help if it has a window on the outside world.

Showers and lavatories are provided today in most animal houses, and are a necessary part of those with more rigorous barriers. The location of these facilities inside and outside the barrier is a matter of dispute; probably they are needed on both sides. Showers should be of the best type, from the point of view of plumbing; if they are not, sooner or later they will not be used. Many shower mixer-valves and sprays are not capable of giving long trouble-free service, and particular attention should be paid to choosing the most robust and reliable fittings, regardless of cost.

The plumbing also demands attention: it is disconcerting, when taking a shower, to be scalded or frozen because someone elsewhere has turned on a tap that diverts water from the shower supply.

Plant

Heating, ventilating, air-conditioning, steam-raising, water-treatment and other mechanical plant, and electrical installations and switchgear, need to be accommodated somewhere. The contractors will always demand a generous area for all the plant, to facilitate installation and make it simpler to maintain. As far as possible, this demand for space should be met, all the more so because it will usually be cheap space that is not useful for other purposes. In Figure 6.1 some 5 per cent of the total area is allotted to plant. This must be regarded as a minimum, and should perhaps be read in terms of cost rather than area.

Corridors

Corridors are essential for communication, but they are unproductive areas and therefore should not be larger than necessary. They are the main pathways of traffic within the animal house, and it is the volume of this traffic, at peak periods, that must determine the width of corridors.

Corridors may also be used as parking areas for some of the wheeled traffic—food and bedding bins, cage trolleys, and so on. If this is permitted—and it is usually a great convenience —then the width of the corridor must take it into account; alternatively, there must be parking alcoves at intervals for the purpose.

Wheeled traffic tends to knock the surface of walls, and hit salient corners. The walls and corners of corridors should, therefore, be appropriately protected.

Summary of design

The subjects so far discussed will enable a first sketch plan to be drawn, and it is useful at this stage to attempt such a drawing. Few people, even those with appropriately trained minds, can visualize an abstract animal house, or even one they see in plan, or a drawing, until they have studied it and thought about it for a long time.

The rest of this chapter will deal with the construction, equipment and furnishing of the animal house, and certain aspects of its operation.

Construction

Animal houses are very liable to shift the emphasis of their occupancy from one species to another, and from one purpose to another—perhaps to species and purposes that were not foreseen when the house was built. Moreover, there have been and will be revolutionary developments in the breeding, management and use of laboratory animals, and these developments will lead to rapid obsolescence of methods.

The designers must, therefore, take into account the fact that the needs are almost certain to change, probably within the first five years of the building of the house and perhaps radically within 15 years. It should therefore have a planned lifetime of, say, ten years, after which time it may expect to be pulled down and rebuilt, or at least completely refashioned internally.

If this sounds impractical or extravagant, past follies may be brought to mind. There are today too many animal houses which were built to last too long: they cost too much in the first instance and therefore cannot, without unacceptable expense, be brought up to date. They are usable, in that they accommodate animals; but those who work in them are too often aware of their shortcomings and the inefficiency that is unavoidable in operating them.

Because of all these considerations, the construction of an animal house must be simple and not unnecessarily costly; and it should allow for subsequent alterations to be made as required.

Materials

Walls. The external shell of the animal house is governed by its site and surroundings. Any appropriate building material can be used for external walls and roof, bearing in mind that it must be proof against rodents and other pests (which will certainly try to get in), and that it should have a high level of thermal insulation.

External walls will usually be load-bearing, and there may also have to be some load-bearing points inside. These are fixed points, and should be carefully located, because a load-bearing pillar in the middle of an animal room is a permanent nuisance.

Internal walls will, for the most part, not have to be load-bearing, and can therefore be of lighter construction. They may, however, be required to carry cage-racking. The economics of such construction should be carefully weighed. A light wall could be so much cheaper to construct that floor-supported cage-racks, with their increased mobility and adaptability, would be both more convenient and more economical.

Ceilings and roofs. Ceilings offer two choices: solid and suspended. On a solid ceiling all light fittings are mounted in the ceiling and have to be serviced from the animal room. Suspended ceilings offer a chance of recessing fittings into the ceiling, and servicing them from above, but this may be a small advantage to set against loss of vertical space and the danger of wild rodents, insects and general dirt and dust getting into the space above the suspended ceiling.

For a single-storey building a flat roof can be very effective. One type that works well consists of layers, from below upwards, of ceiling board, glass wool insulation, air space, strawboard or similar insulating board, roofing felt and felspar chips. Provided all is rendered pest-proof, there is no danger of infection coming in through such a roof. Variants of the same system utilize preformed concrete and foam plastic sections.

For a warm climate, solar heat gain through the roof can be serious. Polished aluminium sheeting will reflect most of the solar heat.

Pitched roofs can also introduce a serious problem of solar heat gain, unless the roof space is well ventilated.

Floors. There is no ideal floor material for an animal house.

Asphalt provides a continuous surface, but it deforms on point loading, and it can crack as a result of thermal shock. Quarry tiles have joints that can collect dirt, and they are very difficult to lay absolutely level. Terrazzo and other slabs, being larger, have fewer joints, and are easy to clean, but they are, like quarry tiles, rather expensive. Well trowelled granolithic cement is an acceptable surface, but is not cheap, and it sometimes cracks. Various surface treatments such as rubber or plastic sheeting, and various combinations of plastics and other materials have been described and tried; all are rather expensive, and none is without other disadvantages.

Perhaps one of the most serviceable types of floor, and certainly one of the cheapest, is plain cement, laid level and power-floated treated with a suitable floor varnish, such as epoxy-resin or polyurethane. If the right varnish is chosen, the floor will be smooth, waterproof and easy to clean, wet or dry; it will not be unreasonably slippery; it will be very durable, with an expected life of many years in spite of heavy wheeled traffic and frequent washing down; and it can be very easily repaired or re-coated. If the cement develops crazing or large cracks, these can be filled in with varnish.

Opinions vary about floor drains. It seems useful to be able to swill down floors and wash all into a floor gully, but to get a good fall into the gully a slope of about 7 per cent is necessary, and this is awkward for racks and wheeled trolleys. Drains and gulleys are also dirt traps and possible portals of entry for pests. On the whole, level floors, without drains, are to be preferred in animal rooms and corridors. Mopping down is an easy and efficient way of cleaning them. Floor drains will, however, be useful around autoclaves and cage-washers.

Doors, windows, etc. The type of door to be used is a matter of choice, but metal frames, with either metal or metal-clad doors, are to be preferred. In some doors inspection windows will be necessary. Hinged doors can be made mouse-proof (mice are accomplished escapists); sliding doors, although often more convenient from the point of view of space and traffic, are very difficult to make mouse-proof. In some circumstances doors may also need to be airtight.

Windows are not always necessary in animal rooms. Animal technicians are divided in their preference for windows; about half of them

want them, the other half do not, and there seems to be no correlation with age or sex in these preferences. For most animal rooms, and for all with controlled ventilation, windows if present must be small, and should not open. They are to look through only. The extra expense of double-glazing small windows is negligible, and this should be provided.

Finishes. Wall surfaces have to be washable, often with detergents or disinfectants, occasionally with hoses. They must also be resistant to the air of an animal house, which may contain ammonia. Glazed bricks and tiles have been much used; they are resistant but are expensive, and they have numerous joints, and are therefore not so hygienic as they appear. Gloss paint has also been much used, but it does not seem to have any advantage over a good-quality emulsion paint; the latter is surprisingly durable, and can be renewed with the least trouble. Emulsion paint is also suitable for ceilings.

Other finishes, such as plastic panels or continuous sheeting, are much more expensive and seldom justifiable.

Floor finishes have already been mentioned above.

Miscellaneous details. An animal house should be as free as possible of pipes, conduits, buttresses, re-entrants, skirting, cracks and any projections which provide opportunities for the accumulation of dirt or interference with cleaning. If pipes must be carried along the wall they should not be tucked away tidily under benches, where they will never be cleaned; they should be placed in full view, so that their cleaning can hardly be avoided.

Electrical conduits can harbour insects, and be a portal of entry from outside. They should be plugged with glass wool at the inner end.

Benching should be simple in design, and as far as possible free of harbourage for dirt. Wood has been regarded as an unhygienic material for animal rooms. It is true that badly constructed benches of untreated softwood can provide harbourage for insect pests in a very dirty animal house, and that such wood can also absorb water and rot, but the use of resin-bonded plywood, or wood treated to make it waterproof, is perfectly permissible.

Services

Electricity. Electric supply is needed for light and for power outlets. The lights may have to be on a timeswitch, perhaps one for each room. Lighting intensity should, for most purposes, be suitable for normal working. Fluorescent tubes produce much less heat than filament bulbs and are usually preferred.

Power outlets will be needed for many purposes—weighing machines, vacuum cleaners, instruments etc.—and should be plentiful.

All electric fittings may have to be waterproof.

Water. Each square metre of animal room floor may contribute up to 2 kg of water vapour per 24 hours, via the animals, and this must be supplied in the form of drinking water. In addition water is needed for washing of all kinds.

The water used must be either sterile, or with a very low and completely harmless bacterial count. Provision will have to be made for water-treatment in accordance with the client's specification: by chlorination, acidification, filtration, heating and cooling, ultra-violet irradiation, or de-ionization.

Hot water, as well as cold, will be needed wherever hands have to be washed, but the volume of hot water needed is not great, except in the cage cleaning area; it can often be supplied most conveniently by small electric heaters over sinks, which save a lot of pipework.

Air and vacuum. There may be a need for a compressed air supply. A built-in vacuum system is often used for cleaning cages and floors. Many systems are on the market; it is important to choose one that can handle the difficult material of a soiled cage, and that permits blockages to be cleared without dismantling part of the building.

Gas. A gas supply may be needed.

Heating and ventilating

The subject of heating and ventilating is far too big to be treated more than cursorily in this chapter, and it is a subject of study in its own right. It is, however, important to realize that ventilating an animal room is very different from ventilating a room for human occupancy. In an animal room there is a much greater problem in the need for a uniform degree of air-change and ventilation in all parts of the room where animals are located; and this is made more difficult because the animals are in cages, and the cages on shelves, both of which interfere with the free circulation of air. The heating and ventilating consultant must take these matters into account and he must realize that the provision of a specified number of air changes to the room does not automatically ensure that every cage in the room has the benefit of these changes.

With this serious caution in mind, it may be said that for most animal rooms some 15 air changes per hour are necessary; but figures both above and below this are sometimes specified, and often for good reasons. Indeed, the amount of ventilation may well be better expressed in some other way than the number of room air changes per hour.

Recirculation of air is not an economy, because the dust produced in any animal house (mainly by the animals themselves) blocks filters quickly and necessitates very fine re-filtration of the recirculated air; 100 per cent fresh air is the rule.

All temperature control, both heating and cooling, can be provided by the air at 15 changes so that there should seldom be a need for radiators, heated panels, or similar equipment. But in certain types of animal house, such as dog-kennels and stalls for farm animals, the number of air changes can be much less than 15 per hour, and local heat may have to be supplied. Underfloor heat (for dog-kennels), infra-red lamps, black body heaters, and hot water coils may all find a place in such circumstances.

The temperature of the animal house needs to be kept within a specified range. It should not be forgotten that all animals (apart from the very young in nest) have a considerable capacity to adjust to a wide range of ambient temperature, and that to deprive them of the opportunity of making such adjustments may not be in the best interests of their health (this has been discussed by Lane-Petter, 1963). Unless, therefore, there are good reasons for a narrower range, a specified temperature plus or minus 3°C is acceptable. If a narrower range in the immediate surroundings of a cage is required permanently a thermostatic cabinet will usually be required. A steady reading on the thermometer or thermograph in the animal room is no guarantee that every cage enjoys such uniformity of temperature.

Humidity is another variable that requires some control. In cold weather relative humidities of heated air may be very low, and humidification will be necessary. Most laboratory animals demand a range of relative humidity of 40–60 per cent.

Incoming air needs to be free of insects and gross dust. Parrott and Festing (1971) suggest filtration down to 5 μm particle size, except for gnotobiotic systems, where 0.5 μm filters are needed. Others suggest that, even for open systems ('SPF' or 'conventional'), filtration should be down to 0.5, 0.3 or even 0.1 μm particle size.

The reasoning behind these excessive demands is obscure, and there is little or no practical evidence that infections have been transmitted to animals from air that has not been so finely filtered. The cost of excessively fine filtration rises rapidly, especially below 5 μm, and the benefits may well be illusory. A ventilating engineer has a right to challenge a very expensive specification, especially when other features of the animal room are not of a comparable standard.

However, there are two exceptions to this stricture on excessively fine filtration of air. If air is to be recirculated, the burden of fine particles to be removed will be far greater than is to be found in almost any fresh air, even in the centre of a large city. All filters may allow a percentage—perhaps a very small percentage, but a certain proportion of the total—of particles through, and a heavily burdened air coming into the filter will be matched by a correspondingly higher number of particles emerging.

The other exception concerns climatic cabinets. Here the volume of air being passed is much less than in a room, and very high efficiency filtration is a negligible extra burden.

For germfree isolators, absolute filtration is obligatory.

All these requirements, for heat and humidification as well as for cooling, de-humidification and filtration, can be met by full air-conditioning. Air-conditioning plants are expensive to install; they may be as much as 40 per cent of the cost of the building. They also require a certain amount of skilled maintenance. On the other hand, a good plant may solve so many other problems that its expense is justified. It should be neither demanded nor rejected without a careful balancing of the pros and cons.

Reference has already been made to the dust produced in an animal room, mainly by the animals. It can be considerable, and it comes from food, bedding, skin detritus and dried excreta. It is scattered from the cages by the movements of the animals, and conveyed round the room by air currents. Dust is an important vector of infection, for bacteria and viruses attach themselves to dust particles.

Odours are always present in an animal house, but with good standards of hygiene and management they should be neither obtrusive nor excessive. Ammonia is an irritant odour formed by bacterial activity on urine. An acid bedding (such as peat moss or sugarbeet pulp) will suppress ammonia formation; a neutral or alkaline, relatively less absorbent bedding (such as sawdust) will permit it.

Apart from hygiene, good ventilation is the answer to odour; but if a very high number of air-changes is found to be needed in order to remove unpleasant odours, a fault in care and management should be sought.

Fixed equipment

Apart from the heating, ventilating and water-treatment plant, electrical switchgear and other machinery (which are all likely to be located in the plant room) an animal house has a number of other items of fixed equipment which must be functionally placed. They should therefore be considered in the early design stage, since both the space they occupy and the services they call for will have a bearing on the general lay-out.

Items of fixed equipment include sterilizers, dunk tanks, entry and exit locks, mechanical delivery systems, incinerators and macerators.

Sterilizers. The most useful type of sterilizer is an autoclave, preferably of the high-vacuum rather than the downward-displacement type. This can be used for sterilizing cages and other hardware, food, bedding and most of the other things that have to come into the animal house. A complete sterilizing cycle, even in a large autoclave, will take from 30 to 60 minutes (seldom outside these limits) so that, allowing for a warming-up period at the beginning of the day, 7 to 14 cycles can be run in an ordinary working day.

Steam cabinets at atmospheric pressure are not often used today, but their cycle time is about the same as that of an autoclave. Their sterilizing efficiency, and in particular their penetration, is much less than that of an autoclave.

Gas chambers use a sterilizing gas, such as ethylene oxide, or sometimes formalin, in conjunction with steam. Ethylene oxide is very penetrating, and will pass through thin sheets of plastics which are impervious to water and air. It will sterilize at temperatures as low as 20°C, using a mixture containing about 90 per cent ethylene oxide, which is flammable, even explosive. Alternatively, a mixture of about 10 per cent ethylene oxide with an inert gas, which is not flammable, will sterilize at about 60°C. Both methods entail a long cycle of several hours duration, so that only one or perhaps two loads can be put through in 24 hours. Moreover, the load must be warmed up beforehand and aired afterwards, which entails the use of considerable space. An ethylene oxide sterilizer, therefore, has to be several times larger than an autoclave if it is to handle the same volume of material, and there has to be more space around it. This factor must be considered in the planning.

Dunk Tanks. Most barriered animal houses possess a dunk tank; that is, an entry disinfectant tank partially divided by a vertical partition, under which are passed materials for surface sterilization. The size of the tank will depend on the volume of material to be passed through it. It will require a cold water supply and a drain.

Locks. Various entry and exit locks, for staff and perhaps for materials, may be required in addition to the sterilizer and dunk tanks. Sometimes they may need to be equipped with ultra-violet lights, insecticidal sprays, disinfectant sprays, or other special devices, according to the client's specification.

Delivery systems. In some large animal houses the volume of food and bedding that is used is considerable, and its delivery from store to point-of-use (the animal room) may entail a substantial load of work. In such cases delivery may be gravity-fed from high-level hoppers, through chutes, or by a pneumatic device that delivers the material through a tube. If such equipment is to be installed, it must be taken into account early in the planning.

Incinerators. The disposal of large quantities of soiled bedding and other refuse is another problem, and chutes or pneumatic devices may have to be provided to remove the refuse from the cleaning area. If it is to be burnt on site an incinerator will be needed, and this occupies space, creates heat which has to be removed, and makes a mess.

Macerators. For the disposal of small carcases and small quantities of soiled bedding (when they have a low level of radioactivity) a macerator may be used. This needs a water supply, a drain, and a source of electrical power.

Movable equipment

This includes cages, racks, trolleys, instruments and other movable items.

Cages

Standards and recommendations for cage sizes have been published by the Institute of Laboratory Animal Resources (1960–65), Porter, Scott and Walker (1970) and the Research Defence Society (1974). These are useful guides, but they rest almost entirely on empirical findings and views.

In designing or choosing a cage for a particular purpose, attention must be directed to the four functions which it has to fulfil. Any detail which cannot fairly be included under one of these four headings is a fad, and should be rigorously discarded (but see Wallace, 1963).

A cage must confine the animal. This is an absolute requirement. The cage must be made of a material which the animal cannot break, distort or destroy, e.g. by gnawing or pulling apart. The mesh of the wire or bars must be small enough to prevent the animal, or its young, from escaping; in this respect particular attention must be paid to well-fitting doors and food baskets or other openings. Door-fastenings must be secure, for many animals are persistent and ingenious fiddlers, and will worry apart quite safe-seeming closing-devices. Monkeys are particularly liable to release themselves, and a padlock is the only really trustworthy answer; but rabbits also show a surprising aptitude for undoing catches. Rats and even mice can lift off the lids of boxes in which they are confined, unless these are held firmly in place either by their own sufficient weight or by fasteners.

A cage must allow the animal to live in health and comfort. The first consideration here is size. There is little or no definite evidence about ideal or minimum cage-size for most species, and the allowance of space (whether for single animals or groups) is nearly always a matter for intelligent conjecture. The length of time during which the animal is to be confined must also be considered. For example, it is permissible to go below the normally accepted limits of size when packing animals for transport, but for this very reason they should never be left in their travelling-boxes after arrival at their destination.

Probably the best advice that can be given to someone who wants to provide his animals with good accommodation is that he should see other people's animal houses in which the species in which he is interested are being kept in obviously good health and comfort. (For a fuller discussion of the animals' needs, see Lane-Petter, 1963.)

It is worth adding here that smaller animals (such as rats, mice and chickens) have a marked tendency to crowd. If large numbers are put into one cage, however capacious, they will huddle together and those underneath will be suffocated. Therefore, except for short periods, not more than 50 mice, rats or chicks should be caged together. For larger animals the figure will be smaller.

The animals' need for light, or darkness, is important. For example, rats and mice are better housed in boxes, particularly when breeding, and rabbits and ferrets should be provided with a darkened nest-box for kindling; but guinea-pig sows have no obstetrical shyness, and will farrow happily in an open pen among a group of their fellows.

A cage must be economical in first cost, accommodation and maintenance. Today, more than ever before, the first cost of cages is a considerable item in the animal-house budget. It is extravagant to order cages in small numbers, and equally to avoid standardization. Simplicity of design should be the aim.

The durability of cages, however, can become an obsession. The stainless steel cage with an almost indefinite life can become a heavy burden on the children and the grandchildren of the person who introduced it. Although the steel may still be bright after half a century of use, the design will certainly be outdated. A good principle is to design cages, and many other articles of animal house equipment also, to have an expected life not exceeding five years, and to make provision for amortization of their cost during that period. Repairing of cages is often money wasted. This policy will, of course, increase the numerical demand for cages, thus enabling manufacturers to arrange long runs of popular patterns with consequent reduction in cost. With the present tendency to get away from the craftsman-built metal cages to moulded plastic boxes and trays, or other mass-produced articles, this reduction in cost is likely to be considerable.

The advice of the manufacturer is often useful in suggesting ways in which simplification can facilitate manufacture without prejudice to the animal. A slight alteration in dimensions may, in the case of sheet metal, effect considerable economy in material, for sheet metal is made in sheets of standard size.

Cost aside, the accommodation needed for the cage, whether in use or in store, is important. The smaller the cages the more can be accommodated, but this consideration must be matched against the comfort and health of the animals kept in them. Large sizes also aggravate the problems of sterilizing and storing, and it is here that a collapsible cage has the greatest advantage. Many large collapsible cages have been designed, with considerable success.

Maintenance includes the attention required while the cage is occupied by an animal, and the prevention of undue deterioration while in use, undergoing cleaning, or in store. Inconvenient catches, doors badly placed or shaped, awkward corners, ill-conceived trays or other parts which are not readily interchangeable, are some of the faults that add to the burden of day-to-day work. Crevices and inaccessible corners encourage the accumulation of soil and consequent rapid corrosion. But no cage is perfect in this respect, and the need for periodical overhaul must be accepted as a current charge in any animal house. It should be the rule that damaged and corroded cages must be immediately withdrawn from use and either put in order or thrown away.

A cage must satisfy experimental requirements. For many experiments an unstressed animal is required, and a cage which satisfies the foregoing requirement is sufficient. But for certain work there are special considerations that demand special cages, and some of these considerations may be at cross-purposes with the other requirements. A cage which under the present heading is satisfactory may not meet the animal's requirements for health and comfort, but here the duration of the experiment is relevant. An animal confined in an anti-coprophagy cage, for example, is not likely to remain unstressed for long.

Experimental requirements are so varied that in many cases it is necessary to design a special cage for the purpose, and standardization, even of metabolism cages, is seldom possible. But in working out such a design, it is well to bear in mind that the needs of the experiment may be in conflict with the foregoing requirements, and that where the conflict is resolved in favour of the experiment, it may be at the expense of the other requirements, in particular the health and comfort of the animal. This in turn may have an important bearing on the experiment itself. To describe all the special types of experimental cage which have been designed for different purposes would be an endless task and not very profitable. Moreover, this is not the place to deal with points which more properly belong to the description of the experiments in which animals are taking part.

Many useful patterns of cage are illustrated in Short and Woodnott (1969). Special types of experimental cages may be found in Gay (1965a & b, 1968) and in numerous other publications describing experiments that require such equipment. All the main manufacturers of animal house equipment advertise in the appropriate journals, and most of them exhibit at the annual meetings of the Institute of Animal Technicians, the Gesellschaft für Ver-

suchstierkunde, the American Association for Laboratory Animal Science and some other organizations. A study of the catalogue, and a visit to a trade exhibition, will be more useful than pages of description.

Shelving

Cages have to stand on shelves, or be suspended from them. There is an endless variety of shelving that is used for this purpose, most of it useful, some very good. It is only necessary to say that, apart from being convenient in use and able to carry the required weight, shelves should be durable and resistant to spilled water or urine; adjustable, if necessary; and easy to clean and dismantle.

Shelves may be attached to the walls of the animal room; this is often an economical way of utilizing space, but it has the disadvantage that the cleaning of the wall behind the shelves can be troublesome. It also demands a strong wall, which may add to the cost of construction of the animal house.

They may also be disposed as free-standing or suspended racks. Suspended racks need anchor points above, usually in the ceiling, which may raise constructional problems. They leave the floor clear, but the anchor structures accumulate dirt. Free-standing racks, often on wheels for mobility, are on the whole the more practical, for they make no special demands on ceilings or walls, and they introduce maximum flexibility in the animal house. The whole rack can, if necessary, be cleared out of the room for cleaning or repair. There are many patterns of mobile racks on the market, as a study of manufacturers' catalogues or a visit to a trade exhibition will reveal. Particular attention may be drawn to a recent innovation, in which individual cage ventilation is combined with a mobile cage rack (Fig. 6.4, see also Lane-Petter, 1970).

Watering equipment

Water should be available at all times to all laboratory animals.

Except for some farm animals, water is now always offered either by bottle or by some sort of automatic watering device. Troughs are for the most part obsolete, on grounds of inconvenience and bad hygiene.

Water bottles. The inverted bottle and drinking spout depends for its operation on certain simple physical principles (Lane-Petter, 1952). In practice it is found that, with bottles of up to 500 ml capacity, leakage caused by agitation or change in temperature or pressure is not generally a serious problem, although spontaneous emptying may take place even in smaller bottles if the dimensions of the tube differ markedly from those given below, if there is any leakage round the bung, or if the spout touches bedding or other material in the cage. Where the water consumption of a pen of animals is greater than can be met by one 500 ml bottle, two or more such bottles should be provided in preference to one bottle of larger size.

The bottle is best held outside the cage, the tube being bent so as to be accessible to the animal within the cage. It may be laid on top of the cage, slung in a wire frame at the side or in front, fixed with a spring clip, or accommodated in a suitable compartment built into the cage. The shape is immaterial: blood-transfusion bottles, medical flats, and ginger-beer bottles are all cheap and serviceable. Pathological specimen bottles of 125 or 250 ml are especially suitable, for their wide mouths greatly facilitate cleaning. Medical flats are useful for laying on the tops of mouse boxes, slightly tipped up at the end, but they are rather troublesome to clean.

The internal diameter of the drinking-tube should be 6 to 9 mm and that of the terminal aperture about 3 mm. The length of the tube and whether it be straight or bent, is a matter of convenience, and it may be made of glass, metal or hard plastic. It may be integral with bung or cap, whether made of hard plastic, or metal, or it may be separate. Glass tubes are very widely used, for anybody can cut and flame a piece of soda-glass tube. However, such tubes may have serious disadvantages: soda-glass is fragile (more so if worked by people unskilled in glassblowing) and not only may it be chewed by the animals but it can give rise to accidents among staff during removal from bottles. Glass, however, being transparent, does not conceal dirt, nor does it corrode; it holds a big drop and is cheap.

Metal tubes have been used in many laboratories to avoid breakages and accidents. Aluminium tubes usually prove too soft, and are rapidly chewed by many species; brass and monel metal have a longer life, but even non-ferrous metals corrode or collect accretions to some extent. Stainless steel is the best choice for metal tubes.

A variant of the drinking tube on a water bottle is the conical cap. This was first introduced in Germany, by the Altromin Company, and consists of a stainless steel cone fitted on to a plastic bottle by means of a rubber washer. The cone is perforated at its apex, and the animal licks or sucks water from the perforation.

One modification of this has been described by Barber (1970); it has no washer, and the cap presses on to a machined taper on the neck of the bottle.

Another development is used by Carworth Europe, and consists of a similar stainless steel conical cap which fits on to the neck of the bottle with a sort of bayonet fixing. A quarter of a turn of the cap is enough to release it. Like Barber's cap it has no washer. (Fig. 6.5.)

Automatic watering systems.   There are several automatic watering devices on the market. For larger animals (such as dogs and monkeys as well as for rabbits and guinea-pigs) they work well, because they can be large and robustly constructed to easy tolerances and there is little problem of deposition of mineral matter from the water, leading to stoppage or leakage. But for rats and mice the problem is more difficult, because these animals lick, rather than suck, their water, and the drinking nipple has to be actuated by a very small effort, and therefore has to be made to fine tolerances. Minute deposits of grit can cause the valves either to stop delivering water or to flood.

No automatic device functions properly all the time, although some have a very high level of efficiency. If the cages being used have grid bottoms and the trays underneath can lead excess water away, a flooding drinking nipple is of little consequence. But if the cage is a shoe box, or the tray underneath the grid does not drain away, a flooded cage is inevitable, with drowned animals.

In either case, a blocked system means thirsty animals, and this may go unnoticed for some time. Demineralizing of the water supply may reduce the formation of calcareous deposits in

Fig. 6.5. Cage and water bottle. The plastic water bottle has a stainless steel cone cap. This cap has no washer and therefore both bottle and cap can be sterilized chemically. The bottle locates automatically in the cage top. The bedding in the cage is dried sugarbeet pulp. (Photograph: UFAW.)

the watering system, but cannot totally eliminate blockage or floods.

There is no doubt that an automatic watering system can be a great labour-saver, for bottle watering can occupy 10–30 per cent of the total work load in the animal room. But no automatic system is perfect, and probably never will be. Against the convenience of an automatic system must be set the occasional flood or stoppage, and the labour of keeping clean the tubes and plumbing associated with it. There is a third point that is often overlooked. A glance at a water bottle will often tell the technician something about the state of the animals in the cage; if it is fuller than expected, why are the animals not drinking? or if it empties too soon, is it a leaking bottle, or are the animals drinking too much? Automatic systems do not provide this information, and they may also cause either drowning or thirst. Their use is not necessarily advantageous and, when they go wrong, they carry the risk of avoidable suffering for the animals.

Trolleys and tables

There is much material that has to be moved round the animal house, and as far as possible it should be put on wheels. Bins for food and bedding are easily mounted on bin boys, and working tables may be wheeled. In some animal houses water-carriers are used to take water to the cages, so that bottles need not be brought to the tap (Fig. 6.6).

Vacuum cleaners and perhaps also spraying and mopping equipment should run on wheels.

Instruments

All animal houses have a need for some instruments, such as balances (for weighing animals), a microscope (for examining smears, see Fig. 6.7), syringes and surgical instruments. Pro-

Fig. 6.6. A simple way of dispensing acidified water. This water carrier, which holds 60 litres, can be moved around the animal room and the bottles filled from the automatic valve. (Photograph: UFAW.)

vision has to be made for the safe storage of instruments.

Gnotobiotic equipment (see Chapter 11).

Furniture and housekeeping equipment

There will be some necessary furniture such as tables, chairs, cupboards, lockers (for changing rooms), brooms, buckets, waste bins, and so on.

In animal houses any place that can accumulate dirt or discarded objects is a danger to hygiene and cleanliness. This applies particularly to cupboards. It should be a rule that cupboards are not permitted in an animal room, and should rarely be found in an animal house: if they are asked for, a very good case has to be made for them.

First-aid

Every animal house must have a suitably equipped first-aid point.

Lethal chamber

There is from time to time a need to kill animals, and suitable methods are described in Chapter 14 and in the relevant specific chapters. There will be a need in the animal house for the necessary apparatus in one or more lethal chambers.

Hygiene

The principles and practice of hygiene are dealt with in Chapter 7. It is only necessary here to draw attention to those aspects that may make special calls on the design or the equipment of the animal house.

Cleaning makes no special claim, apart from the possible provision of a built-in vacuum system and floor drains near the cage-washer and autoclave. The sterilizing requirements have been referred to on p. 87 and in Chapter 7.

When a new animal house is taken into use it will need an initial fumigation, to sterilize all the exposed surfaces within the area. From time to time this will have to be repeated, and methods of fumigation are described in Chapter 7. It is necessary to point out here only that rooms must be capable of being sealed and made

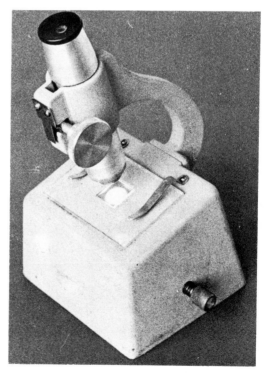

Fig. 6.7. A single microscope which can be used for examining vaginal smears, etc. (Photograph: UFAW.)

airtight on such occasions as fumigation is necessary.

Disinfestation—the destruction of pests, whether insect or rodent—should seldom if ever be necessary, but it may occasionally happen that there is an unwelcome invader. The possible need for disinfestation should be borne in mind by the architect.

Lastly, the use and disposal of radioactive substances make special demands, which are indicated in various official publications and regulations. The subject has been well covered in Short and Woodnott (1969).

REFERENCES

Animal Welfare Institute (1966). *Comfortable Quarters for Laboratory Animals*. New York: Animal Welfare Institute.

Barber, B. R. (1970). Development of a one-piece cap for animal drinking bottles. *Bull. Inst. Anim. Techns*, **6**, (7), 8.

Gay, W. I. (1965a and b, 1968). *Methods of Animal Experimentation*, vols. I–III. New York & London: Academic Press.

Hare, R. & O'Donoghue, P. N. (1968). *The Design and Function of Laboratory Animal Houses*. London: Laboratory Animals.

Hill, B. F. (1963). Symposium on research animal housing. *Lab. Anim. Care*, **13**, 219–467.

International Committee for Laboratory Animals (1964). Terms and definitions. *ICLA Bulletin*, **14**, Annex 1.

Institute of Laboratory Animal Resources (1960–1965). *Standards for the Breeding, Care and Management of Laboratory Animals*. Cats (1964), Chickens (1966), Dogs (1964), Guinea-pigs (1964), Hamsters (1960), Mice (1962), Rabbits (1965), Rats (1962). Washington: ILAR.

Institute of Laboratory Animal Resources (1968). Guide for laboratory animal facilities (revised). *Publ. Hlth Serv. Publs, Wash.*, No. 1024.

Jonas, A. M. (1965). Laboratory animal facilities. *J. Am. vet. med. Ass.*, **146**, 600.

Lane-Petter, W. (1952). Mechanics of the animal water bottle. *Nature, Lond.*, **169**, 465.

Lane-Petter, W. (1963). The physical environment of rats and mice. In *Animals for Research—Principles of Breeding and Management*, ed. Lane-Petter, W. London & New York: Academic Press.

Lane-Petter, W. (1970). A ventilation barrier to the spread of infection in laboratory animal colonies. *Lab. Anim.*, **4**, 125–34.

Lane-Petter, W. and Perason, A. E. G. (1971). *The Laboratory Animal—Principles and Practice*. London & New York: Academic Press.

Nuffield Foundation (1961). *The Design of Research Laboratories*. London: Oxford University Press.

Ottewill, D. (1968). Laboratory animal houses. *Architects Journal*, **147**, 1247 and *Information Sheets* 1597–1602.

Parrott, R. F. & Festing, M. F. W. (1971). Standardised laboratory animals. *MRC Laboratory Animals Centre Manual Series*, **2**, 4–7. Carshalton: MRC Laboratory Animal Centre.

Porter, G., Scott, P. P. & Walker, A. I. T. (1970). Caging standards for rats and mice: recommendations by the Laboratory Animal Science Association Working Party on caging and penning. *Lab. Anim.*, **4**, 61–6.

Research Defence Society (1974). *Guidance Notes on the Law Relating to Experiments on Animals*, 2nd edition. London: Research Defence Society.

Short, D. J. & Woodnott, D. P. (1969). *The I.A.T. Manual of Laboratory Animal Practice and Techniques*. London: Crosby Lockwood.

Wallace, M. E. (1963). Cage design principles, practice and cost. *J. Anim. Techns Ass.*, **14**, 65.

7 Hygiene

D. K. BLACKMORE

Introduction

By definition, a chapter in this book entitled 'Hygiene' should be concerned with any factor which relates to health and prevention of disease in laboratory animals. Obviously space does not permit such a wide approach to the problem, and other chapters will cover certain aspects of disease-control in the more common species of laboratory animal maintained under different systems of management. The emphasis of this contribution will be on the theoretical aspects of the prevention of the introduction or build-up of infectious agents in any animal house.

To formulate a logical plan to prevent infection within an animal unit, four factors must be considered and appreciated: the nature of the infectious agents to be precluded; the probable routes of entry to the animal house of such agents; the means by which such agents may be destroyed or prevented from entry; and finally, methods by which the efficiency of a given hygienic measure can be assessed.

This chapter will not contain descriptions of apparatus such as autoclaves, which are often employed for sterilization purposes, neither will it contain long lists of various chemicals, some of which may inhibit the growth of or even kill certain organisms. Technical information on apparatus and its use can be found elsewhere (Sykes, 1958; Short, 1969), while lists of so-called disinfectants are of very doubtful value. It is intended to provide some of the background theory, coupled with more practical advice, which will allow readers to evaluate critically the problems of hygiene specifically associated with their own units, and thus to formulate a logical plan of action.

Infectious Agents and Their Sensitivity

For the purpose of this chapter, the infectious agents will be divided into four broad groups: the viruses, rickettsia and allied organisms; the bacteria and mycoplasmata; the fungi, including yeasts; and the higher parasites, including the protozoa and metazoa. Although each of these groups of infectious agents have certain similarities in relation to their resistance to physical or chemical factors, wide generalizations must be treated with caution, and obviously exceptions occur.

Viruses

The viruses are the smallest of the infective agents, and the size of the actual elementary particle or viron varies from around 20 mμ for some of the picornaviruses such as the virus of murine encephalomyelitis, to 300mμ for some of the pox viruses, a group which includes the agents of rabbit myxomatosis, and mousepox or ectromelia. Although the majority of viruses are extremely small, individual virus particles do not occur naturally in the environment as they are always associated with some other organic matter. Infective particles of less than 1 μ in diameter are unlikely to occur, and it is improbable that the majority of airborne diseases are associated with particles of less than 5 μ. These facts are most important when considering the practical design of filtration systems for either air or water. Viruses are strict intracellular parasites and tend therefore to be somewhat unstable outside the host, but their survival is dependent on many factors and subject to considerable variation. Some may be inactivated within a few hours, while others may survive for several months. As a generalization, viruses are particularly susceptible to heat (50°–60°C for 30 mins), ultra-violet irradiation, oxidizing agents and formalin vapour. The effects of dessication and other chemical agents tend to be rather variable according to the group of viruses concerned.

Bacteria

The smallest bacteria are in the region of 0.5–1.0 μ, but some spirochaetes may be over 10 μ in length and most methods of destroying these organisms depend on destruction of one or more essential enzyme systems within the cell. These enzymes, being protein in nature, can therefore be destroyed by any of the physical or chemical means which denature protein. Many species of bacteria can produce much more resistant spores, the walls of which protect the essential enzyme systems. Some of the spores can withstand temperatures of 120°C for more than 30 minutes. Even among the vegetative forms of bacteria there are considerable differences in resistance. This is perhaps best exemplified by the mycobacteria, with their protective wax envelopes, but even amongst the organisms without obvious protective cell walls certain species are particularly resistant to chemical denaturation, typical examples being *Pseudomonas aeruginosa* and *Staphylococcus aureus*. Any organic matter surrounding an organism also provides protection.

Fungi

Fungi, like the bacteria, can produce spores and their destruction depends on the inactivation of one or more of their enzyme systems, but these spores are often not as resistant to heat as are bacterial spores and few survive a moist heat of 80°C. This group of organisms is less sensitive to high osmotic pressures than are the bacteria, whose growth is suppressed by such solutions.

Higher parasites

The wide spectrum of the many different types of higher parasite can create particular problems. Many of these parasites produce at some stage in their development thick-walled cysts or ova which are highly resistant to many chemical compounds and can be effectively destroyed only by physical methods, such as steam sterilization or gamma irradiation. Methyl bromide is one of the few chemical agents effective against the majority of arthropod parasites and many helminth ova and coccidial occysts. Oocysts are also sensitive to heat, especially if the humidity is low, and can be destroyed easily with 10 per cent ammonia, a chemical which is considerably cheaper and more practical to use than methyl bromide. A knowledge of the life cycle of certain parasites will help to develop hygienic measures of control.

Obviously, the cycle of tapeworm development can be prevented by eliminating all suitable intermediate hosts, while thorough animal-house cleaning would make it more difficult for flea larvae to survive as they feed on organic waste material and are not strict parasites during this stage of their development.

The factors governing the infectivity of a given agent are extremely complex, and include such factors as the natural immunity of the potential host, the virulence of the organism and the number of organisms with which the animal is challenged. For all practical purposes, a single bacterial organism will be incapable of causing infection. Therefore, although a certain technique may not destroy every potentially infective particle, if the numbers of such agents are sufficiently reduced, infection will often not occur. Many of the parasitic conditions must be excluded from this wide generalization, for a single oocyst is capable of producing a lethal infection in a fully susceptible host.

Whether or not an animal unit is managed under strict barrier principles, hygienic measures must be taken to minimize the chance of animals becoming infected. The techniques employed should be designed either completely to destroy any agent, whether pathogenic or non-pathogenic, or to reduce the numbers of any potential pathogen to such a level that the challenge to the animals at risk is minimal. It must be ensured that the few organisms present are prevented from increasing in numbers until they might constitute a hazard.

Some Theoretical Concepts of Sterilization and Disinfection

The absolute process of destroying all organisms is known as sterilization, and the less absolute process of destroying only specified organisms is known as disinfection. Many of the disinfectants, although reducing the number of potential pathogens, are by no means absolute in their action.

Before considering the various practical technicalities of sterilization and disinfection, it is important that certain theoretical concepts concerning the destruction of microbial populations are appreciated. These concepts apply to both physical and chemical methods of sterilization, and therefore to all methods likely to be employed in the maintenance of an animal unit.

Chick (1908) was one of the first workers to study the problems of sterilization in detail and suggested that the sterilization of a specific microbial population by a particular method could be compared to a unimolecular chemical reaction. This concept suggests that the rate of killing is proportional to the concentration of the sterilizing agent, and if the logarithm of the number of viable organisms exposed to the agent is plotted against the time of exposure, a straight line will be obtained. The slope of this line can be expressed as a mathematical constant, as shown in the equation below, and is known as the velocity constant of sterilization reaction.

$$K = \frac{1}{t} \log_{10} \frac{n_1}{n_2}$$

K = velocity constant; t = time of reaction; n_1 = initial number of viable organisms; n_2 = final number of viable organisms.

This linear expression is obtained when excessive amounts of the sterilizing agent are available and the rate of reaction is relatively rapid. In other conditions, and especially in relation to certain chemical substances, a sigmoid curve can occur, with an initial lag phase and a tailing-off effect. Obviously, reactions that have a tailing-off effect are to be avoided. If the rate of kill is sufficiently rapid, these initial and final lag phases will become insignificant and the reaction can be considered as represented by a straight line (Rubbo and Gardner, 1965). Whether a particular process with a known velocity-constant is likely to result in absolute sterility will depend on the number of organisms present, and the time for which they were subjected to the process. The degree of reduction of a microbial population by a particular treatment is known as the inactivation factor. This factor is calculated for a given process by using a test organism which is known to be highly resistant to the process, dividing the initial count of viable organisms by the final extrapolated count after the treatment.

Once the inactivation factor has been calculated for a given process, the degree of sterility likely to be achieved can be calculated by dividing the inactivation factor by the average number of organisms per article. As particularly resistant organisms are used for calculating the inactivation factor, the type contaminating the articles to be sterilized need not be identified when assessing the degree of sterility.

Many animal units incorporate a dunk tank containing a chemical disinfectant. Assuming that the inactivation factor of a particular tank was 10^5 when tested against *Pseudomonas aeruginosa*, and that the tank was being used for the transference of bottles of sterile milk and bags of sterile mouse pellets, and that the outside of the bottles was contaminated with an average of 100 organisms and the bags with 10,000 organisms, the degree of sterility for the milk bottles would be $\frac{10^5}{10^2}$ or 1,000, and the degree of sterility for the bags of food would be $\frac{10^5}{10^4}$ or 100. Thus there is a chance of one in every 1,000 bottles and one in every 100 bags of food being contaminated after passing through the dunk tank.

The resistance of organisms to a particular treatment can be affected considerably by the chemical or physical environment immediately surrounding the organism. Organic matter can have a profound effect on certain chemical disinfectants and can insulate organisms from heat sterilization processes. For certain organisms acid environment often greatly reduces the inactivation factor of a steam sterilization process—a matter of considerable importance in the manufacture of canned foods.

The various methods of sterilization and disinfection applicable to methods of laboratory animal hygiene can be classified conveniently under the headings of 'physical' and 'chemical'. The more important physical methods of sterilization are by heat, ultra-violet or gamma radiation, and filtration, whilst chemical methods involve the use of liquid vapours or gases. The effectiveness of any of these techniques depends on their power to coagulate protein or to destroy a vital enzyme system.

Sterilization by heat

Dry heat is essentially an oxidative process; wet heat results in the more effective process of protein coagulation, and so less time at a given temperature (thermal death time) is required to kill an organism. Although the spores of *Clostridium botulinum* can be destroyed by moist heat at 120°C in less than twenty minutes, they would need over two hours at the same temperature if subjected to a dry heat; alternatively, the temperature would have to be increased to 180°C. The usual methods of dry heat sterilization include the use of hot-air ovens, naked flames for surface sterilization, and incinerators. Items sterilized in the conventional cheaper ovens are usually subjected to temperatures of 180°C for 20–30 minutes. More modern high-vacuum infra-red ovens reach temperatures of 280°C (Darmody *et al.*, 1961), and thus the time of exposure can be reduced to 15 minutes. Brief exposure to temperatures of over 300°C will result in sterilization, and such temperatures can be achieved by applying an open flame to the surface of the objects to be sterilized.

Sterilization by moist or wet heat can be achieved by heating an object in an aqueous solution either below or at its boiling point, by the use of live steam at 100°C, or by the use of steam under pressure at temperatures above 100°C. Very few vegetative bacteria can withstand temperatures above 80°C for more than a few minutes, but the spores of some thermophilic bacteria require up to 35 minutes at 120°C.

Sterilization by radiation

In terms of animal house hygiene, radiation sterilization is either by gamma rays or by the longer electromagnetic wave radiation of ultraviolet light; these processes interfere severely with the metabolism of the organisms by inducing ionization of vital cell components, particularly the deoxyribonucleic acid (DNA). Gamma radiation, usually produced from a Cobalt 60 source, has good powers of penetration and never induces radioactivity in items subjected to the process. Apart from destroying microbes, gamma radiation can affect certain molecules of the item being sterilized, and in the case of foodstuffs palatability may be affected. The lethal activity of gamma radiation is enhanced by the presence of oxygen and reduced by the removal of water. Vegetative bacteria and larger viruses require a dose of only 0.05–0.5 M rads for their inactivation, while organisms such as *Micrococcus radiodurans*, spores of *Bacillus pumilus* and some of the smaller viruses, including certain phages, may require between 2·0 and 4·0 M rads (Rubbo and Gardner, 1965).

Non-ionizing ultra-violet (U.V.) radiation has virtually no penetrative powers and the most bacteriocidal wavelengths are those of between 2800–2600 Å. The majority of commercially produced quartz tube low-pressure mercury vapour lamps emit 95 per cent of their radiation at 2·537 Å. The dosage of U.V. radiation is measured as the product of intensity and time of exposure in microwattseconds per cm² or ergs per cm². Organisms vary considerably in their resistance to U.V. light, and the dose required to produce an inactivation factor of 90 per cent varies from 1,000 microwattseconds per cm² for influenza virus to 150,000 for *Aspergillus niger* spores. The intensity of radiation produced by an emitter varies inversely with distance from the lamp. It must also be remembered that lamps have a limited life, and no U.V. system should be used which does not have some time recorder device incorporated. Ultra-violet radiation can cause severe retinal damage and precautions must be taken to ensure that the eyes of animals or staff are shielded from direct ultra-violet radiation.

Filtration

Porous filters for water sterilization are usually manufactured from porcelain, kieselguhr, or asbestos, and their efficiency depends on the actual pore size and the length of the channels in the material and electrostatic properties, while membrane filters act purely as mechanical sieves.

The majority of air-filters are constructed from glass-wool or fibre-glass paper. Glass-wool filters do not act as sieves but depend on four different factors: interception, or direct impact; inertial impaction; electrostatic precipitation; and diffusion.

Interception (or direct impaction of a particle with a fibre of the filter) is independent of the airflow, but only larger particles will be held back by this mechanism. Inertial impaction is a process dependent on the air velocity and will trap larger particles travelling at relatively high speed. Electrostatic precipitation is also independent of air velocity and tends to trap charged particles between 10 and 40 μ in diameter. Diffusion is the process by which particles of less than 0·3 μ at comparatively low air velocities diffuse out of the air stream by Brownian movement and impinge on the filter material.

The efficiency of a filter is usually expressed either as a percentage retention or penetration of particles of a given size, and estimation of this is usually carried out by either a sodium flame, a methylene blue, or a dioctyl-phthalate (D.O.P.) test. In all these tests, particles of an average size of less than 0·5 μ are generated and used to challenge the filter, using different air velocities. HEPA or ultra-high efficiency filters have a retention efficiency of 0·003 per cent for particles of 0·3 μ in diameter and can produce an absolutely sterile air flow. When considering fibre-wool filters, it should be remembered that as the efficiency rises there is usually an associated increase in resistance to the air flow, so that the problem of a forced air supply becomes greater.

Chemical disinfectants

There are a multitude of different chemical agents which can be used to prevent the growth of, or destroy, organisms if conditions are ideal. However, few should be relied upon as being entirely satisfactory, and in practice these agents are subject to considerable abuse and misuse, and to misleading claims by the manufacturers. Very few chemical agents are capable of killing bacterial spores in the concentration and under the conditions in which they are likely to be used, and their value is mainly in those areas where some physical method of sterilization cannot be employed.

New disinfectants are marketed each year, and it is impossible and probably undesirable to attempt to list each one. However, these chemical agents can be classified under several general headings, such as phenols, soaps, alcohols, dyes, quaternary ammonium compounds, halogens, heavy metals, ampholytic agents and miscellaneous substances, including formaldehyde; their various properties have been well reviewed (Sykes, 1958).

It is not possible to select an agent that will be effective in all circumstances. Peracetic acid is a most effective bacteriocidal, fungicidal and anti-viral agent, but it has obnoxious fumes and is ineffective against the majority of higher parasites. The halogens, especially the chlorine-based compounds, are effective against the majority of viruses but are rapidly inactivated by the presence of organic matter. The choice of a satisfactory product may become a matter of chance unless its activity is assessed both in the laboratory and later, while actually in use, against the specific organisms whose growth is to be controlled.

Sterilization by gas or vapour in a laboratory animal facility is usually by means of formaldehyde (Darlow, 1958), ethylene oxide (Kelsey, 1961a) or beta-propiolactone (Spiner and Hoffman, 1960) and these agents, if used correctly, have a wide spectrum of activity against all the infective agents. Formaldehyde can be generated either by heating a solution of Formaldehyde B.P., which is a 37 per cent solution of the agent in water, stabilized with methyl alcohol (formalin), or by mixing two parts of formalin with one part of potassium permanganate and allowing the oxidative reaction to volatilize the remainder of the formaldehyde. An increased temperature increases the efficacy of the treatment, and it is important to have, if possible, a relative humidity above 75 per cent. Similar conditions are required when beta-propiolactone is used. For fumigation of a room, 500 ml of formalin are required for every 27 m^3 of space. Polymerization of the formaldehyde, and the coating of exposed surfaces with a white deposit, will be reduced if the wall temperatures are above 18°C.

Ethylene oxide, although expensive, can be an effective sterilizing agent, but its use must be carefully controlled to achieve its efficient use (Ernst and Shull, 1962a and 1962b), and to avoid accidents associated with its explosive properties. The gas must, therefore, be generated only in a specially designed chamber in which the relative humidity can be kept to about 33 per cent, which is the optimal level to achieve the maximum effects of the agent.

The exposure-time for both formalin and ethylene oxide is rather prolonged—up to 24 hours. The exposure-time for ethylene oxide can be reduced according to the concentration of the gas, but for practical purposes a cycle will always take several hours.

Potential Sources of Infection and Practical Preventive Measures

Infectious agents can gain access to an animal colony by a wide variety of methods. Irrespective of whether animals are maintained under a strict barrier system or by so-called conventional methods, attention should always be paid to good hygiene in order to minimize the hazard of infection. The main potential portals of entry for infectious agents in a colony of laboratory animals are:

1. The ventilation system and the incoming air.
2. The water supply.
3. The food and bedding.
4. Contaminated equipment, especially cages, racking, water bottles, etc.
5. The animals themselves.
6. Infected uteri if hysterectomy techniques are being carried out.
7. Vermin, especially wild rodents and various arthropods.
8. Staff.

The relative hazard of each of these potential sources of infection will vary considerably between different units, and is especially related to the species of animal at risk and the proximity of the unit to sources of potential infection. Before formulating logical hygienic measures, an attempt must be made to assess the relative hazard of each of the factors mentioned, as the final success can be only as good as the weakest link in the chain of disease preventive measures.

Air supplies

In most circumstances, the hazard of airborne infections from the air supply to a building does not constitute a major problem. As the majority of infectious agents are carried on particles greater than 5μ in diameter, relatively simple filters will reduce the hazard considerably. If the building is situated away from other animal houses the number of significant pathogens in the incoming air will be extremely small. In modern barrier-maintained units, high-efficiency filters are often installed and these expensive filters are protected from undue clogging by cheaper and disposable prefilters, which extract the majority of particles above 5μ from the incoming air.

Electrically-heated air incinerators are part of the air filtration system of the Gustafsson lightweight stainless steel isolator (Gustafsson, 1959), and air incinerators might also be a useful adjunct for the air outlets of rooms containing experimentally infected animals.

If animals housed within a building develop an infection that is disseminated via the respiratory route, a major problem is created. Apart from the destruction of infected animals, the only alternative is to increase the ventilation rate to such an extent that the infectious agent is diluted beyond an infective level. Such action is usually impossible in the normally constructed animal house.

Water supplies

Although mains water supplies are not completely sterile, they can certainly be assumed to be free of obligate pathogens. It is theoretically possible for a water supply to become transitorily contaminated if a water main is fractured, but the probability of such an occurrence is remote. Small numbers of saprophytes, including species of pseudomonas, can be recovered from most water supplies, and such organisms may proliferate in storage tanks and around the outlets of taps. If it is important to exclude organisms such as *Pseudomonas aeruginosa*, some form of water sterilization or disinfection must be carried out and precautions taken to avoid contamination from within the unit, of the initially sterilized water.

It is difficult to recommend any practical method for producing large quantities of sterile water. A heat interchange pasteurization plant would probably be the most reliable, but obviously the capital costs of such a plant are very high. Filtration, using kieselguhr or Berkefield filters, is an easy system to install and maintain, and is of relatively low capital cost. Although on theoretical grounds the system would appear to be efficient, there is some divergence of opinion on the use of such filters for animal house facilities. Large multiple-membrane or paper filters have also been used with variable results. Finally, systems for the sterilization of water by ultra-violet radiation are available. In view of the lack of experience of really well-tried methods for the sterilization of large quantities of water for animal houses, and the ever constant hazard of contamination of water supplies subsequent to initial sterilization, acidification and chlorination of water may be a more realistic procedure.

Food and Bedding

In most circumstances, non-sterilized or non-pasteurized food and bedding constitutes the greatest disease hazard for an animal unit. The raw materials from which many of the rations are compounded, especially proteins of animal origin, are often contaminated with pathogenic organisms that are able to survive the manufacturing process. Manufacturing and storage areas are frequently heavily infested with wild rodents which may contaminate the raw materials, rations, or bedding material with a variety of pathogens, especially salmonellae and various higher parasites. Any animal unit must have a storage area which is vermin-proof and of a suitable environment to suit the material being stored.

Diets and bedding that are not produced by a process which results in a packaged sterilized or pasteurized product must be subjected to some form of suitable treatment before use. Green food cannot be sterilized before being fed, and as it can be contaminated easily by wild birds (which often excrete salmonella, *Yersinia pseudotuberculosis* and other possible pathogens) every effort should be made to avoid feeding unprocessed raw materials.

Although gamma radiation is theoretically the best method for sterilizing animal foodstuffs it is a rather expensive process, and when transport and handling charges are included the cost is virtually doubled. If a moist-heat system is being used for sterilization a modern high-vacuum autoclave is recommended, so that the sterilization cycle is as short as possible and the degradation of heat-labile substances is minimal. Ethylene oxide has been used for sterilizing diets, but (apart from the problems already mentioned in using this substance) there is some divergence of opinion on its persistence in treated food.

Commercially prepared canned cat and dog foods and long-life sterile milk are examples of pre-treated products that can be used in a strictly barrier-maintained unit after disinfection of the outside of the containers, and for this purpose a chlorine-based dunk tank, which will be discussed later, is probably ideal.

Equipment and other inanimate objects

An animal house has to be supplied constantly with cages, racks, water bottles, instruments, paper, etc. The chance of a particular object being contaminated with pathogens varies enormously, and a poorly cleaned cage from a potentially infected area is a much greater hazard than a recently purchased item of scientific equipment. The proper assessment of the potential hazard of an object, and the type of material from which it is manufactured, will dictate the hygienic measures necessary to be taken before admitting it to the unit. Although wet heat, as produced by an autoclave, is an ideal treatment for most cages, it would be a disastrous one for most pieces of scientific apparatus. A fumigation technique (especially one using ethylene oxide) would be much more suitable for the latter. Dry heat, or the use of hot-air ovens, probably has little place in the modern laboratory animal house, except for the sterilization of glassware needed for work on pyrogen-testing, or for sterilizing heat-stable powders, oils and waxes.

The sterilization of the actual fabric of a building is probably best carried out by formalin fumigation, after the sealing of any outlets. Prior treatment with an open flame of grossly contaminated surfaces might be indicated when a particularly virulent or resistant infection has occurred in a colony of animals, an obvious example being the treatment of an area known to have been contaminated by a primate infected with tuberculosis. All objects must be thoroughly cleaned before sterilization is attempted, as the presence of organic matter reduces the efficiency of most techniques of sterilization or disinfection.

A wide variety of gas-fired or electrically-heated incinerators is available, and such equipment is essential for the rapid disposal of infected carcasses and waste material from an animal room. Such apparatus must be situated well away from animal houses, as smoke or hot air from inefficiently operated incinerators may contain infective particles.

Stock animals

In a non-barrier-maintained unit the animals themselves may provide a continual source of infection. Every attempt should be made to eliminate animals with obvious clinical disease, as they are likely to disseminate large numbers of organisms. Ideally, all infected animals should be eliminated, whether or not infections are apparent, but such absolute measures are impractical in the absence of complete barrier facilities. There can be no excuse for maintaining colonies with diseases which can be transmitted to man, such as lymphocytic choriomeningitis (LCM) or ringworm. Neither should colonies affected by diseases likely to cause lethal epidemics in the animals or seriously to interfere with experimental procedures be maintained. Any animal unit must, therefore, have facilities for *post-mortem* examination of animals which die or are culled, and for the examination of random samples of animals in order to check for evidence of pathogenic or potentially pathogenic organisms.

If the unit is non-barrier-maintained and extra conventional animals are to be introduced to the colony, a quarantine period is essential, during which a sample of the animals can be screened for diseases, such as latent ectromelia, which if introduced, could decimate the original colony.

The issue of animals from a strictly barrier-maintained unit can create certain problems. If the autoclave is not in continuous use it may be used as an exit lock between each sterilization run, given sufficient time for it to cool. A double-sided ultra-violet light chamber can be most useful, but after use it must be left for some time so that the air can be resterilized. More recently, exit locks (designed almost as a miniature wind tunnel) supplied with clean air, have been used with success. The flow of air is outwards, and such devices can be used continuously for the despatch of animals in filtered containers.

Hysterectomy derivation

Hysterectomy or caesarean techniques are often employed in the establishment and maintenance of strictly barrier-maintained colonies such as germfree or specific-pathogen-free (SPF) colonies. Unless strict hygienic measures are employed, these techniques can constitute a real threat to the health of the colony. It is normally assumed that the full-term embryos in the uterus are sterile, but some infectious agents such as LCM, reo virus III and certain parasites can infect the young animals before birth. Other organisms such as pasteurella and mycoplasma are able to infect the apparently healthy uterus (Flynn *et al.*, 1968) and foetuses can become contaminated with infective maternal fluids during their removal from the uterus.

The proper hygienic control of a hysterectomy programme should, therefore, include the initial screening of the donor female for evidence of potential uterine or trans-placental infections, strict observance of aseptic principles during all surgical procedures, and the isolation and screening of offspring and maternal tissues before the young are admitted to the colony. The use of plastic film positive-pressure isolators (Trexler and Reynolds, 1967) can provide most useful temporary quarantine facilities. The introduction of gravid uteri to an isolator or animal house requires the use of a chemical dunk tank. Such tanks are also useful for the introduction of other packaged presterilized objects with minimal superficial contamination. A variety of chemical solutions can be used in such tanks, and probably those which are based on either a halogen, especially chlorine, or on formalin are most suited for these purposes. If the tank is being used for the introduction of pre-sterilized inanimate objects they must be immersed for periods long enough for the agent to exert its lethal effects. Uteri in separate containers passed through such a fluid lock will not have sufficient time in the tank for the fluid to have any germicidal effect, and it will only wash the container mechanically and provide a fluid lock to the unit. If bacteria and viruses are considered the most likely source of contamination, a chlorine-based agent is probably indicated. If higher parasites are also a likely hazard, then a formalin-based preparation might be more suitable.

Vermin

Wild rodents are frequently infected by a wide variety of bacterial, viral and parasitic conditions which can be transmitted to their laboratory-housed relations, and great care must be taken to prevent wild rodents from gaining access to animals, food stores or any other ancillary buildings. Apart from acting as the intermediate hosts for various cestodes, arthropods can also mechanically transmit certain other infections. As well as using rodent barriers at all doorways and constructing vermin-proof buildings, the wild population of rats and mice in the vicinity of the unit should be reduced by ensuring that all waste is disposed of quickly and efficiently.

Animal attendants and other staff

The personnel entering an animal unit can constitute a disease hazard to the animals, either by the mechanical carriage of strict animal pathogens on their skin or hands, or by dissemination of their own organisms, especially from the upper respiratory tract. The mechanical carriage of organisms by staff is usually the greater hazard and washing and showering facilities, together with the use of good protective clothing, will be of value, although showering tends to increase the shedding of organisms indigenous to the skin of humans (Bethune et al., 1965; Speers et al., 1965). A technician who keeps as a hobby a colony of conventional rabbits, which would undoubtedly have endemic pasteurellosis and coccidiosis, is likely to be contaminated and should not be allowed to work in an SPF unit.

The majority of viruses seem to be rather host specific, so human respiratory viruses are usually a serious hazard only to higher non-human primates. On the other hand, staff can often be symptomless carriers of pneumococci, which are potential pathogens for rabbits and guinea-pigs, and the author has recovered *Pasteurella multocida* from the throat of an apparently healthy animal technician. There is also evidence that staphylococci of human origin can cause infection in barrier-maintained colonies (Blackmore and Francis, 1970). Staff can contract ringworm from domestic pets and become a hazard for animals under their care. Although the incidence of human salmonellosis in this country is relatively uncommon, carriers can occur. It is therefore sensible to instigate a regular microbiological screening programme for staff working in an SPF unit and to prevent them working in the unit if they have clinical evidence of respiratory or enteric infection.

Assessment of Sterilization or Disinfection

The most carefully planned system of hygiene will be of little value unless its efficiency is continually checked, and for this a wide variety of methods is available.

These methods can be divided into three main groups: physical, chemical, and microbiological. Generally, physical and chemical methods are the most practical for the assessment of sterilization techniques; they are usually more easily carried out, and are subject to a minimum of experimental variation.

Physical methods

Apparatus designed for sterilizing objects by heat should be fitted with automatic time and temperature recorders, but although such a record might suggest a satisfactory sterilizing cycle, poor packing or operation of an autoclave can result in pockets of unsatisfactory heat penetration. It is therefore important to check on such occurrences by the use of movable thermocouple probes and a suitable separate time and temperature recorder. The in-use efficiency of filters can be checked physically by Royco particle counters, but they are expensive either to buy or to hire. The intensity of ultra-violet and gamma radiation can be checked by suitable intensity meters, but it is most unlikely that an animal facility would ever have its own Cobalt 60 gamma radiation plant.

Chemical methods

A wide variety of useful chemical indicators can be used as a constant check on a variety of sterilizing processes. Although many of these methods may not be particularly accurate, they are based on simple changes in colour and require very little skill to interpret. They therefore provide a useful guide as to whether an object has been subjected to some sterilization process and are suitable for all levels of staff.

A wide variety of chemicals which change colour when exposed to certain temperatures for a given period of time are used to assess heat sterilization processes. These chemical indicators include: four different types of Brown's tubes which change colour from red to green; the 'Steam-Clox' indicator, which changes from purple to green; 'Diack' controls which contain a wax tablet which melts at temperatures of 121°C after 5–8 minutes; autoclave tape which develops brown streaks; polyvinyl chloride indicator containing an azo dye, which when subjected to gamma radiation changes from yellow to red; Royce's sachets, which change from yellow to purple if subjected to ethylene oxide; and paper indicator strips which turn black when dipped in solutions containing more than 200 ppm of available chlorine. These methods are well described by Rubbo and Gardner (1965).

Microbiological methods

Microbiological testing falls into two distinct areas: the examination of an object after sterilization or disinfection, and tests which are used merely to indicate that an agent or process is apparently efficient. The former is a purely microbiological problem and will not be discussed here. For the latter problem, the spores of particularly resistant organisms can be used to assess certain sterilization processes. Paper strips are impregnated with a known number of such spores and then included with items that are being sterilized; they are subsequently incubated to ensure than no viable spores remain. Examination of a strip which originally contained 10^5 viable spores would provide as much information as the examination of 1,000 items contaminated with 100 spores which had been subjected to the same sterilization process. The spores of *Bacillus stearothermophilus* are particularly resistant to moist heat, certain *Clostridia* spores to dry heat, *Bacillus globigii* to ethylene oxide and *Bacillus pumilus* to gamma radiation. The number of spores per strip should never be less than 10^5. Kelsey (1961b) has described the use and production of thermophilic spore papers.

It is most important that disinfectants should be tested in the laboratory against the organisms they are needed to control, and later by 'in-use' tests in the areas in which they are actually used. Although most text books refer to Rideal-Walker, Chick-Martin and other phenol-coefficient tests, it is suspected that they are now of almost only historical interest as more efficient

and reproduceable 'Capacity use-dilution tests' (Kelsey *et al.*, 1965) have been introduced. In general principle, three or four selected test organisms are grown in nutrient broth, with or without added yeast, to simulate 'dirty' or 'clean' conditions. Various dilutions of the disinfectants are then prepared and 1 ml of the bacterial suspension added every ten minutes. Two minutes before each addition of further bacteria, the disinfectant solution is tested for sterility. The test is fully described by Kelsey and Sykes (1969).

In conclusion, if the people responsible for the hygiene of an animal colony formulate their policies on a proper understanding of the infectious agents likely to challenge their animals, and the principles of how the agents should be destroyed, a satisfactory system of hygiene should be obtained. However, it must be remembered that circumstances can change, and policies and methods must be re-evaluated and checked continually.

REFERENCES

Bethune, D. W., Blowers, R., Parker, M. & Pask, E. A. (1965). Dispersal of *Staphylococcus aureus* by patients and surgical staff. *Lancet*, i, 480–3.

Blackmore, D. K. & Francis, R. A. (1970). The apparent transmission of staphylococci of human origin to laboratory animals. *J. comp. Path.*, **80**, 645–51.

Chick, H. (1908). Quoted by Sykes, C. (1958) in *Disinfection and Sterilization*. London: Spon.

Darlow, H. M. (1958). The practical aspects of formaldehyde fumigation. *Mon. Bull. Minist. Hlth Lab. Serv.*, **17**, 170–3.

Darmody, E. M., Hughes, K. E. A., Jones, J. D., Prince, D. & Duke, W. (1961). Sterilization by dry heat. *J. clin. Path.*, **14**, 38–44.

Ernst, R. R. & Shull, J. J. (1962a). Ethylene oxide gaseous sterilization. I. Concentration and temperature effects. *Appl. Microbiol.*, **10**, 337–41.

Ernst, R. R. & Shull, J. J. (1962b). Ethylene oxide gaseous sterilization. II. Influence of method of humidification. *Appl. Microbiol.*, **10**, 342–4.

Flynn, R. J., Simkins, R. C., Brennan, P. C. & Fritz, T. E., (1968). Uterine infections in mice. *Z. Versuchstierk.*, **10**, 131–6.

Gustafsson, B. E. (1959). Lightweight stainless steel system for rearing germfree animals. *Ann. N.Y. Acad. Sci.*, **78**, 17–28.

Kelsey, J. C. (1961a). Sterilization by ethylene oxide. *J. clin. Path.*, **14,** 59–61.

Kelsey, J. C. (1961b). The testing of sterilizers. 2. Thermophilic spore papers. *J. clin. Path.*, **14,** 313–9.

Kelsey, J. C., Beely, M. M. & Whitehouse, C. W. (1965). A capacity use-dilution test for disinfectants. *Mon. Bull. Minist. Hlth Lab. Serv.*, **24,** 152–60.

Kelsey, J. C. & Sykes, G. (1969). A new test for the assessment of disinfectants with particular reference to their use in hospitals. *Pharm. J.*, **202,** 607–9.

Rubbo, S. D. & Gardner, J. F. (1965). *A Review of Sterilization and Disinfection.* London: Lloyd-Luke.

Short, D. J. (1969). Sterilization and disinfection. In *The I.A.T. Manual of Laboratory Animal Practice*, 2nd ed., Short, D. J. & Woodnott, D. P. London: Crosby Lockwood.

Speers, R., Berhard, H., O'Grady, F. & Shooter, R. A. (1965). Increased dispersal of skin bacteria into the air after shower-baths. *Lancet*, i, 478–80.

Spiner, D. & Hoffman, R. K. (1960). Method of disinfecting large enclosures with ß-proprolactone vapour. *Appl. Microbiol.*, **8,** 152–5.

Sykes, C. (1958). *Disinfection and Sterilization.* London: Spon.

Trexler, P. C. & Reynolds, C. J. (1957). Flexible film apparatus for the rearing and use of germfree animals. *Appl. Microbiol.*, **5,** 406.

8 Transport of Laboratory animals

G. H. TOWNSEND AND M. H. ROBINSON

The production of laboratory animals in specialised breeding units, commercial or otherwise, is highly preferable to breeding them in buildings designed primarily for housing animals under experiment (Lane-Petter, 1969). Some species cannot be bred in captivity in sufficient numbers to satisfy demand and so must be captured in their native habitats. Therefore, at some stage in their lives the majority of laboratory animals need to be transported from their original breeding area to a holding unit or experimental room. Transport may involve movement from room to room, building to building or from site to site.

The ultimate ideal in the transport of animals is their arrival at their destination in the same physiological, psychological and health status as at the point of departure. Procedures must be simplified and minimized and each transfer should involve a check of all the steps. The factors which may influence the outcome of the journey include:

1. The quality of the animals.
2. The species, age, sex and number of animals.
3. The animal container itself.
4. The environmental conditions experienced at the outset, during and after transportation both within and around an animal container (temperature, humidity, noise, light and ventilation).
5. The quality and quantity of litter, bedding, food and water.
6. The speed, mode and efficiency of transport.
7. The care and attention experienced by the animals before, during and after transport.
These factors will be discussed in some detail.

The quality of the animals

Only healthy animals in first-class condition from a colony with a known disease-free history should be selected for transport. Prior to packing, each animal should be thoroughly examined by a person *experienced in the signs of ill-health exhibited by that species.* Any deviation from normal behaviour or condition should warrant rejection of the animal.

The increasing importance of animal experimentation, particularly long term investigations (Research Defence Society, 1963) focusses attention on the need for uniform, high-quality and definable animals (Carworth Europe, 1968).

Descriptive terms such as conventional, caesarian-derived and specific pathogen free are no longer adequate and recourse must be made to other forms of definition based on the animals' microbiological and parasite burdens (MRC Laboratory Animals Centre, 1969) (see Chapter 4) as well as its environment and husbandry. The animal that is disease-free is more likely to arrive at a destination in a healthy state and, what is equally important, more likely to remain so.

Transport invariably involves *stress and often reveals sub-clinical infection.*

Even when animals are transferred in a single journey from a breeding to a holding area in the same building, or from holding area to experimental premises as part of a daily routine, they should be examined before the movement takes place and given the same care and attention as if the journey was complex and of long duration.

Species, age, sex and number of animals

Confinement within a container, variation in environmental conditions, and noise and movement affect all species to a varying degree. Inbred animals tend to be most disturbed. Carnivores are affected more than rodents, and primates more than both. During transport the greatest stresses experienced are by newly captured wild animals, especially primates, from their native areas to laboratory quarters. All wild animals require a period of adjustment before removal from their native area; this not only conditions the animals to confinement but serves as a period of quarantine, accustoms them gradually to an unnatural diet and provides opportunity for the feeding of supplementary proteins, vitamins and minerals to protect them on the journey.

The transport of primates is highly specialized and consultation with an expert is of value. Good liaison must exist between exporter and recipient for some primates do not travel well during certain seasons (Medical Research Council, 1959).

Species should never be mixed and all animals in a consignment should be approximately the same age and from the same room or staffing unit. This is important to avoid fighting, bullying and cross-contamination. Predators should not be transported in the same vehicle as their natural prey.

The sexes should not be mixed unless they are litter mates or established breeding groups. Pregnant females should not be transported unless absolutely necessary and certainly not during the last third of the gestation period.

The maximum size of the container is usually governed by factors such as shippers' regulations or ease of handling. Therefore, the number of animals per container or compartment may be limited but in any case should not exceed the maxima laid down in Appendix 1.

It is of great advantage to select the animals several days before transport and cage them together. Especially is this important in the case of hamsters, for it will enable them to develop a social order and become accustomed to each other. In the case of hamsters young animals from mixed litter groups are compatible but at over five weeks great care should be exercised and excess bedding used. Adults, other than breeding pairs, should be caged individually.

The animal container

Design and materials

The standard design for an animal transport container has for many years taken the form of a rectangular box but a cylindrical shape is stronger, safer and marginally cheaper. For extremely timid species, such as the guinea-pig, which tend to panic and crowd into corners, a circular shape is particularly advisable. In rectangular boxes the sides should slope slightly towards the top.

Materials for the container may be wood, metal, cardboard or plastic.

Wood is warm, provides a good insulation and is less expensive than metal but it is almost impossible to sterilize or clean wood properly and such containers can only be used once for the carriage over long distances of mice, hamsters, rats, guinea-pigs, rabbits, reptiles, amphibia and small primates. Wood in the form of crates may be used for cats, dogs, lambs, birds, calves or poultry. Depending on the species, economies can be made by the use of hardboard or plywood but wickerwork hampers and boxes are not recommended.

Metal conducts heat and is expensive. Non-returnable boxes are desirable but if metal containers have to be returned then they should be sterilized by the recipient before return to the consignor and again sterilized before re-use.

Crates made wholly of metal or partly of wood may be used for the larger animals mentioned. In box form metal may be used for the smaller species but crates or wire cages should never be used.

Cardboard containers are usually the cheapest but are difficult to sterilize and involve stapling and folding in construction. They are not as strong as those constructed from other materials.

Fig. 8.1. Typical disposable cardboard boxes for category 1 and 2 rodent stock. (Approximate sizes 46 cm × 28 cm × 13 cm, and 31 cm × 13 cm × 10 cm.) (Photograph: MRC Laboratory Animals Centre.)

The grade and thickness (or grammage) of board is critical and should be checked with the box manufacturers. Corners and the base normally suffer the worst effects of urination and these can be protected by spraying with a plastic or wax coating (F. H. Evans, personal communication).

Mice, hamsters, rats, guinea-pigs, rabbits, day-old chicks, and other varieties of small birds can be transported in cardboard containers on journeys up to twelve hours. Beyond this period there is danger of damage to the base caused by urine, with subsequent escape of the animals (Figs, 8.1, 8.2 and 8.3).

Fig. 8.2. A range of cardboard rabbit despatch boxes for category 1 and 2 stock. (Approximate sizes 41 cm × 36 cm × 23 cm; 37 cm × 25 cm × 25 cm; 46 cm × 25 cm × 25 cm.) (Photograph: MRC Laboratory Animals Centre.)

Fig. 8.3. Two forms of internal fittings used in rodent boxes. (Approximate sizes 37 cm × 36 cm × 15 cm; 31 cm × 13 cm × 10 cm.) (Photograph: MRC Laboratory Animals Centre.)

Plastic materials are becoming more widely used in the construction of animal transport containers but at the time of writing they are expensive, for the relatively few purchased must bear the total cost of the moulds. Some commercial breeders have designed plastic transport boxes for rodents which can continue to be used in the experimental rooms (Fig. 8.4).

Fig. 8.4. Plastic transport boxes suitable for categories 1, 2, 3 and 4 animals. (Approximate sizes 53 cm diameter × 36 cm; 31 cm × 31 cm; 23 cm × 23 cm.) (Photograph: MRC Laboratory Animals Centre.)

Since the economically priced containers (wood and card) are difficult to sterilize they should never be returned to the animal suppliers for re-issue but destroyed after unpacking.

In many cases the choice of container is dictated by factors other than those of economy and sterility. For instance large primates (over 25 kg) may be transported in metal cages. Smaller primates, dogs and cats are usually contained in specially designed wooden crates of light weight construction when shipped by air.

The British Standards Institution and the International Air Transport Association have laid down standards concerning the transport of animals by air; these include the size, construction and packing of suitable containers (British Standards Institution, BS.3149 and International Air Transport Association, 1975).

As a general rule rodents and other small fur-bearing mammals are rather shy and inclined to panic; therefore see-through windows are not to be recommended. However, larger species should always be given an opportunity to view their surroundings.

The containers for Category 3 and 4 animals (Fig. 8.4) (MRC Laboratory Animals Centre grading scheme, see Chapter 4) must be air-tight and sealable. The choice of materials is therefore limited to metal or plastic.

Category 5 (germfree animals) pose special problems during transport such as the packaging of the animals inside the isolator or through an attached sleeve, and the provision of a sterile environment and air supply. The transport container must again be air-tight and so the materials commonly used in construction are metal, plastic or glass (Fig. 8.5). Clear plastic or glass are preferable but if metal must be used then a window should be provided. The condition of the animals must be easily discernible at all times so that adjustment to the filtered air flow can be made to prevent discomfort or asphyxiation. Prior to use the container should be irradiated or autoclaved for double the normal time and enclosed in a double layer of plastic film during sterilization. Bedding, food and source of moisture may be supplied from inside the isolator.

Kilner-type jars fitted with a modified lid are suitable containers for the transport of a few germfree mice (Fig. 8.6). A satisfactory litter or bedding material for such a container is absorbent lint (D. K. Blackmore and T. van de Klee, personal communication).

Fig. 8.6. Small filter container suitable for short journeys. (Photograph: MRC Laboratory Animals Centre, by courtesy of Blackmore and Van der Klee.)

Density

Laboratory animals, mainly rodents, lagomorphs, and carnivores, need space to move. This requires careful attention to packing density. These densities should be fixed according to carefully compiled tables (MRC Laboratory Animals Centre, 1957 and Appendix 1) and not left to the discretion of anyone who happens to be available for the task. Several authorities have written on this subject (Animal Welfare Institute, 1965; National Academy of Sciences, 1961). In hot weather the number of animals in each compartment should be reduced by one quarter.

Environmental conditions

Ventilation is of vital importance and the system adopted should be dictated by the grade of the animals concerned.

The M.R.C. Laboratory Animals Centre's grading scheme (L.A.C., 1974) (see Chapter 4 for details) provides two categories of animal that can be maintained under conventional

Fig. 8.5. Germfree transport isolator with outer container. (By courtesy of Charles River, Inc., U.S.A.)

conditions; namely Categories 1 and 2. Such animals require a clean container but a filtered air supply is not necessary. Categories 3, 4 and 5, on the other hand, require containers fitted with air filters.

Air vents are best sited on the sides of containers to avoid asphyxiation caused by piling in transit and occlusion of lid vents. If lid vents are used they must be protected from occlusion by the use of adequate spacers and 50 per cent more vent area should be allowed. With filter-covered vents twice the area should be allowed as for open vents. If there is any possibility of blockage of ventilation, cleats should be fastened on the outside of the box.

Assuming maximum packing densities as laid down in Appendix 1 the ratio of vent to container surface area (neglecting the base) given in Table I should not be reduced.

TABLE I
Ratio of vent to surface area of transport containers

	Conventional on sides	Conventional on top	Filter containers
Mice	1:30	1:20	1:10
Hamsters	1:30	1:20	1:10
Rats	1:30	1:20	1:10
Guinea-pigs	1:50	1:30	1:20
Rabbits	1:50	1:30	1:20
Cats	1:50	1:30	1:20
Dogs	1:50	1:30	1:20

These ratios are calculated for travel in temperate zones. Where ambient temperatures exceed 27°C double the ventilation area should be allowed.

Large numbers of small holes are better than small numbers of large holes over a given area. Vents are best placed high on the sides of the container to avoid draughts. Circular or oval vents are preferable to rectangular or square designs. Where a vent is composed of a series of holes punched in a metal sheet the ventilation must be calculated on the area of the holes and not that of the total window.

The holes or vents in conventional containers should be covered on the inside by wire mesh of such a size that the animals' muzzles or feet cannot protrude.

Filter material varies in its porosity. For transport of Categories 3, 4 and 5 animals, filters nearly 100 per cent efficient at 3 μ, 95 per cent efficient at 1 μ and 82 per cent efficient at 0·5 μ will suffice. Fibre-glass filter media meeting such requirements are easily obtainable. The actual fixing of filter material is of prime importance and it is best sandwiched between sheets of expanded metal. Alternatively, metal gauze may be used on the inner face and nylon or plastic gauze on the outer (Fig. 8.7). The sealing of the edges is essential. Filters are sterilized with the transport container and should not be used again under any circumstances.

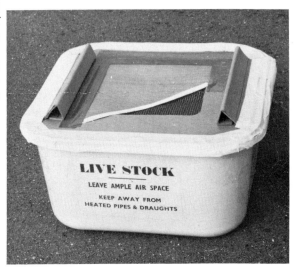

Fig. 8.7. Container to show filter protection and fixture; and basic gauze vent for conventional use. (Photograph: UFAW.)

Air can be supplied to containers in transit if the ventilation port is circular, i.e. top of glass jar or part of transit isolator. Either a rubber puffer operated manually or an electric pump may be attached to a cap over the filter and a valve supplied to adjust the pressure to some few inches of water (Fig. 8.8). Prevention of high humidity and heat stroke are particularly important in filter containers and one half the packing density recommended in Appendix 1 is desirable.

The quality and quantity of bedding, food and water

Bedding and litter should always be provided. Such material should be clean, wholesome and acceptable to the animals. Since moisture may

Fig. 8.8. Small germfree transport jar with system of forced ventilation for use during decontamination of isolator entry block. (Photograph: MRC Laboratory Animals Centre, by courtesy of Blackmore and Van der Klee.)

affect the strength of animal containers the litter must be able to absorb urine efficiently. Several substances used for this purpose also have deodorant powers, although deodorants, per se, must never be used.

The commonly used materials include coarse softwood sawdust, granulated peatmoss or proprietary mineral or wood products (sterilized for Category 3, 4 and 5 animals). Such substances are usually provided to a depth of one to two inches.

The quantity and type of bedding offered will vary with the species, type and size of container, and also should be influenced by the ambient temperatures to be expected during the journey. A liberal supply of hay should be offered to rabbits, guinea-pigs and hamsters; other rodents may be given woodwool or shredded paper (MRC Laboratory Animals

Centre, 1957). Amphibia require moist moss or leaf mould to a depth of several inches.

Food and water

It is most important that animals should have had access to food and water up to half an hour before packing for despatch. Hungry and thirsty animals do not travel well and the disturbance may deprive them of their appetite, even if food is available. Animals should not be offered food to which they are unaccustomed and, of course, those originating from other than conventional colonies may receive only pasteurised or sterilized foodstuffs. Food should always be available to the animals although it is possible to restrict cats, dogs, and larger animals to one feed per day. Amphibia, fish and reptiles do not require food during the journey.

Dry pelleted diets together with sliced fruit, root vegetables and hay are satisfactory for rodents and lagomorphs.

The use of water containers fixed on to, or placed within, travelling boxes or cages is seldom satisfactory as they are liable to leak through jolting. Cats and dogs must be watered every 12 hours and primates every six hours. Where ambient temperatures exceed 32°C these periods must be significantly reduced.

Smaller species may be provided with water in the form of fruit, or root vegetables. One per cent reconstituted agar in small containers may also be used and is suitable for Category 3, 4 and 5 animals.

Where several animals are confined in the same container every effort must be made to provide several sources of water to prevent fighting. Alternatively the animals should be watched and any indication of bullying suppressed. All animals, including the timid, should be given every opportunity of slaking their thirst.

Before the journey begins transport agencies must be made aware of the feeding and watering requirements of the animals and the consignor should satisfy himself that the agencies have proper staff and facilities to carry out the instructions.

Speed, mode and efficiency

It is to be remembered that animal transport commences at the door of the parent animal quarters and ends within the walls of the receiving animal quarters. All stages in the transport of a batch of animals must be taken into consideration and merit as much care and attention as the major portion of the journey.

Although much can be accomplished in minimizing the effects of adverse environmental factors and the stresses of travelling, probably the greatest single aid is speed. Even when movement is from room to room in the same building animals should not be left in corridors or on tables for collection.

Most authorities agree that transport by air is best—certainly over long distances. Greater judgement is required in relation to shorter journeys—the availability of suitable flights; distances and the facility of travel between animal quarters and airfield; the efficiency of documentation and movement facilities at air terminals.

In the case of transport between states and countries consignors of animals should realize that legal requirements will differ. Obviously, statutory regulations must be obeyed and consignors should always contact the appropriate authorities of the receiving states or countries for guidance. Information may be sought from the appropriate consulate, embassy, trade attaché or government veterinary authorities.

Regulations may refer not only to animals but also to their packing, transport and items of diet, bedding and litter. Furthermore some authorities prohibit importation of certain species from other countries or insist on periods of quarantine (see Appendix 4).

Short distances are best covered by road transport and it is advisable, wherever possible, to use one's own vehicle and staff. Thus there is control and supervision from door to door.

Ideally the driver should receive instruction in laboratory animal care and management and the vehicle be modified for the purpose. It should be insulated, and fitted with an air-intake coupled to a thermostatically controlled heating or air-conditioning device. Removable racking should be provided and constructed in such a way that movement of the animal containers is reduced to a minimum. The interior of the vehicle must be so designed as to allow ease of cleaning and disinfection. Lighting should be provided and a window situated between the animal compartment and the driver to allow frequent visual inspection.

The problem of middle-distance transport should never be solved solely on the basis of economics. Due consideration must be given to all available methods and those involving in-transit delays rejected. Each journey will have weak points such as overnight delays, changes of carrier and cross-terminal movements. Climatic and weather changes should also be borne in mind.

The recipient is in the best position to ensure an ordered journey because he will know best his own lines of communication. The producer can cross-check the journey, suggest improvements to the recipient and ensure attention to weather, personnel, quality of container, bedding, food, water, marking, labelling and documentation. Confirmation of the date and time of arrival by letter or telephone is advisable. The invoice should correspond exactly with the consigned animals.

In many cases the arrangements and documentation associated with animal transport are complex and require considerable experience. When exportation is involved due consideration should always be given to the employment of a specialized shipping agency.

Care and attention

Animals should be selected and grouped at least 24 hours before transport. They should be examined thoroughly and crated not more than 30 minutes before the journey commences.

Inspection of the animals should take place as near to packing time as possible and any necessary veterinary certificates signed. Care should be taken that the signatory and his qualifications are acceptable to the authorities of the country of import. The usual form of certification is shown in Appendix 2.

The container should be examined before use to ensure satisfactory construction with special regard to the air vents and filters (if any).

A package relies for its arrival largely upon its labels. These must be firmly attached before the animals are crated. More than one is necessary, preferably of different types, i.e. one adhesive and one tie-on. Writing must be distinct, using large and legible lettering bearing in mind dim lighting or strange eyes.

If a veterinary certificate of health is required a copy should be attached to the container and protected by a plastic envelope.

Labels should show: Name, complete address and telephone number of consignee.

Name, complete address and telephone number of consignor.

Purchase order number.

Date and time packed.

Description of contents, number, sex, species, strains, age.

An outline illustration of an animal, preferably of the species involved.

Indication if box is one of a number, i.e. 1/6.

Instructions on handling, feeding, watering, etc. easily accessible.

The value of the box (for Customs information).

Abbreviations should be avoided. The design of the label should include in bold print such words as: URGENT—LIVE ANIMALS FOR RESEARCH. KEEP AWAY FROM HEAT, COLD, RAIN OR DIRECT SUN. ON NO ACCOUNT OPEN IN TRANSIT. Figure 8.9 shows a selection of labels in use.

Care should be taken to ensure that the animals themselves, the information on the labels, veterinary certificates and invoices match-up accurately. Confusion, loss or delay may result if there are discrepancies.

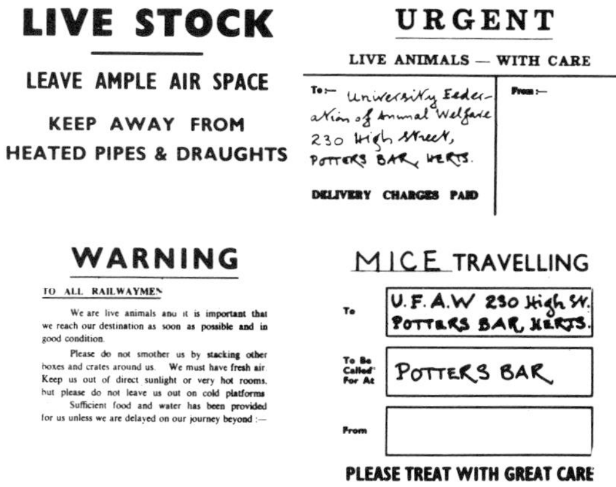

Fig. 8.9. Suitable adhesive labels, used in the United Kingdom. (Photograph: MRC Laboratory Animals Centre.)

Departure procedure

The consignee should receive full details of the shipment (including estimated time of arrival) *before* transportation begins.

Once the animals leave the breeding or rearing area they should never, on any pretext, be allowed to return.

Transit

The care and attention of hundreds of hours can be destroyed by a few minutes' carelessness or ignorance. Personnel handling animals in transit need either to be specially trained or specially briefed. When using commercial passenger or cargo carriers choice should be made, wherever possible, among those specializing in animal transportation. If this is not possible then the services of a specialized shipping agency should be employed.

During transit a number of points must be borne in mind. These are summarized in Appendix 3, a copy of which might well be displayed prominently in despatch, transit and reception areas. Attendants should carry out speedily and diligently all in-transit instructions. If containers are to be opened then every effort must be made to ensure that animals do not escape. Since such accidents do sometimes occur, then the attendants must be skilled in, and equipped for, re-capture.

Replenishment of food and the offering of water presents excellent opportunities for a quick examination of the animals themselves. Injured or dead animals should be removed, the former speedily and humanely destroyed and the latter packed in plastic bags and refrigerated prior to pathological examination. Here again it is evident that the attendant must be fully trained and briefed in all aspects of the work. A method of euthanasia suitable for the species should also always be at hand especially in the case of air transportation.

Where the consignee is responsible for collection of the animals from an air terminus, freight depot or railway station, an attendant should be waiting for their arrival. Animals should not be left in such surroundings to be collected when convenient.

On arrival at the recipient's animal quarters, the containers should be speedily opened in a closed room away from other animals, a thorough check of the consignment made and the animals placed in previously prepared cages, fed, watered and rested.

Loss of body weight in transit is not uncommon. It varies from species to species and depends upon conditions of transit. Ten per cent loss of weight is not impossible on long journeys and efforts to rectify this depletion should be attempted by making water (at room temperature) freely available to all newly arrived animals.

Sick or injured animals should be destroyed and (plus any dead on arrival) sent to a pathology laboratory for examination.

Any complaints should be made immediately to the suppliers who will normally replace sick or dead animals if this is desired by the recipient. If there is any question of the liability of the transporter a claim should be made by the supplier. This liability usually involves death, injury or escape of animals, or an act or default on the part of the carrier or company. Claims should be made in writing as early as possible supported by:

A copy of the original receipt.
A copy of the original invoice.
A claim form with full details.
An inspection report, duly signed.
A post-mortem report, if appropriate.

Even when carriers or agents are not involved, complaints should always be supported by an inspection report and/or a complete post-mortem report. Such information will be of value to the consignor who must, of course, make every effort to prevent future mishaps or eradicate a disease problem.

The consignor should be willing to accept some liability for the animals for seven days after arrival on the consignee's premises. However, this liability will vary depending on the recipient's standards of husbandry and quarantine.

After the animals have passed the first examination and rested in the reception area they should be housed away from the main colonies in a state of quarantine for a period of twenty-one days. This is not always practical but even so every effort must be made to avoid direct and indirect contact between the new animals and any other animals on the premises.

Consignments of animals, even of the same species from the same source, should never be mixed until the full period of quarantine or isolation has been served by the latest arrivals.

To obtain the best results from newly acquired animals recipients would do well to liaise with the suppliers and maintain their colonies under identical systems of housing, caging, husbandry and nutrition. Alternatively, purchasers should explore the sources of animals and choose those adopting a routine closely akin to their own.

Costing

It is common practice for the recipient to pay both for the transport container and for the carriage. This places responsibility upon the supplier to use the most economical yet satisfactory container while the user should decide, in consultation with the supplier, the best method of carriage.

It is essential to realize that the initial expense of an animal, container, carriage, etc., is but a fraction of the cost of the experiment itself. Even in acute investigations a high-quality animal efficiently transported will amply repay the higher costs in more accurate, reproducible results, which in turn can lead to fewer animals being required.

Investigators and animal technicians should be aware of the relative value of purchase, passage and experiment because this will conclusively show that every care spent on the first two will be repaid in the third.

APPENDIX 1

As a general guide all animals must be able to stand, sit, lie down, turn around and stretch out to their full length within a container. Ruminants, pigs, equines, etc., however, should not be allowed room to turn around.

A useful formula for constructing crates or single compartments for individual animals is to allow:

Length = from point of nose extended, to set on tail + 1/3.

Width = width of animal at shoulder × 2.

Height = head raised to full extent.

Species	Weight of Animal	Maximum Number per Compartment	Space per Animal (cm²)	Height of Box (cm)
Mice	15–20 g	25	20	10
	20–35 g	25	26	13
Hamsters	30–50 g	12	32	13
	50–80 g		88	13
	80–100 g	refer to page 106	136	13
	over 100 g		160	13
Rats	35–50 g	25	40	13
	50–150 g	25	52	13
	Adult	12	100	13
Guinea-pigs	170–280 g	12	90	15
	280–420 g	12	160	15
	over 420 g	12	230	15
Rabbits	under 2·5 kg	4	770	20
	2·5–5 kg	2	970–1160	25
	over 5 kg	1	1160–1400	30
Cats	Adult	1 or occasionally 2	1400	38
Dogs	Adult	1 or occasionally 2	see notes above	
Monkeys	4·5 kg or under	total weight of animals not to exceed 23 kg or 12 animals.	4050 90 × 45	48
Day-old Chicks	—	50	1225	12
Quail	—	20	762 × 457	18

APPENDIX 2

Export of (.....................) to (.....................)

THIS IS TO CERTIFY that on ....
on behalf of the .... (despatching body)
.... at the above address,
I have examined the undermentioned:

....

Breed ....
Sex ....
Age..
Distinctive markings ....

The animals show no signs or symptoms of rabies or other infectious or contagious disease, and are in good bodily condition and fit to travel.

Signed......................... ....
A. N. OTHER (qualifications.........).
Veterinary Inspector.

APPENDIX 3

List of points which might well be displayed in places where packing, transit changes and reception of animals take place.

Packaging:
1. Selection and examination of animals (preparation of veterinary certificate).
2. Preparation of invoices, etc. Notification to consignee.
3. Feeding and watering prior to packing.
4. Examination and preparation of container.
5. Addition of litter, bedding, food and moisture.
6. Fixing of labels (and veterinary certificates, feeding and watering instruction).
7. Entry of animals into container.
8. Sealing of container.

Transportation:
1. Immediate delivery to shipping agent, transporter or consignee.

2. Instructions to agent or transporter concerning in transit feeding, watering and inspection (if required).

Reception:

1. Consignee or agent to terminus to await arrival of animals.
2. Collection of animals from terminus.
3. Inspection of container for evidence of damage and/or escape.
4. Transportation to reception room.
5. Examination, re-caging, feeding, watering and resting animals.
6. Removal to quarantine facilities.
7. Lodging of complaints (if any) with consignor.
8. Daily examination of animals.
9. Removal from quarantine after 21 days and satisfactory reports.

APPENDIX 4

There is a considerable volume of legislation concerning the transport of live animals to, within, in transit via and from the UK. Advice should be sought from the Ministry of Agriculture, Fisheries and Food, Hook Rise, Tolworth, Surrey. (See also Chapter 1).

The following Acts and Orders embody many of the legal requirements:

Protection of Animals Act, 1911

Transit of Animals (General) Order, 1973

Rabies (Importation of Dogs, Cats and Other Mammals) Order, 1974.

REFERENCES

Animal Welfare Institute (1965). *Basic Care of Experimental Animals*, revised edition, Appendix 4. New York: Animal Welfare Institute.

British Standards Institution (1961–1966). *BS.3149: Recommendations for the Carriage of Live Animals by Air*.

Part 1. Monkeys and other primates for laboratory use.

Part 2. Small and medium-sized seed-eating birds.

Part 3. Rodents, rabbits and small fur-bearing animals.

Part 4. Dogs and cats.

Part 5. Day-old chicks and turkey poults.

Part 6. Reptiles.

Part 8. Birds other than chicks, turkey poults and small and medium-sized seed-eaters.

Part 9. Fish, amphibia and invertebrates.

Part 10. Larger carnivores.

Carworth Europe (1969). Uniformity. *Collected Papers*, vol. 3. Alconbury, Huntingdon: Carworth Europe.

International Air Transport Association (1975). *IATA Live Animal Regulations*. Fourth Edn. Geneva: IATA.

Lane-Petter, W. (1969). The rising quality of laboratory animals. *J. Sci. Technol.*, **15** (5), 8.

MRC Laboratory Animals Centre (1957). *Laboratory Animals Bulletin*, consolidated edition. Carshalton: MRC Laboratory Animals Centre.

MRC Laboratory Animals Centre (1974). *The Accreditation and Recognition Scheme for Suppliers of Laboratory Animals*, 2nd edn. Carshalton: MRC Laboratory Animals Centre.

National Academy of Sciences (1961). *Guide for Shipment of Small Laboratory Animals*. Publ. No. 846. Washington: Nat. Acad. Sci.

Research Defence Society (1963). *New Drugs*. Conquest Pamphlet No. 17.

Taylor, G. B. (1974). Carriage of animals at high altitudes. *Vet. Rec.*, **92**, 204–205.

FURTHER READING

Coates, M. E. (1968). *The Germfree Animal in Research*. London & New York: Academic Press.

Council of Europe (1969). European Committee for the Protection of Animals during International Transport. *European Treaty Series*. No. 6.

Short, D. J. and Woodnott, D. P. (1969) *The I.A.T. Manual of Laboratory Animal Practice and Techniques*, 2nd ed. London: Crosby Lockwood.

9   Training of Laboratory Animal Personnel

Judy A. MacArthur

Introduction

There are wide differences in the requirements of laboratory animal science, both between countries and between institutions within a country. These differences can be explained, not only by the function of the institution and the direction of its research, but also by environmental conditions and other local factors.

Because of this wide variation, the term laboratory animal science can have no precise definition. In addition, entrants to this discipline will come from a wide range of educational and professional backgrounds, a fact which must be taken into account when devising a training programme. However, despite these difficulties there is every reason for collaboration between individuals, institutions and countries for the purpose of raising the standards of training of the personnel involved in the production and use of laboratory animals.

Organisation of training

To justify the provision of education in the subject is a simple task; laboratory animal science is undergoing remarkable growth and development and, as the needs of the research scientist become more demanding, the operators responsible for the production and care of his laboratory animals must be of a higher standard. To produce a training programme which embraces and satisfies the requirements of all grades of operators is much more difficult.

When the pyramidal structure on which most institutions of any size will operate is examined it can be seen that the staff of an animal house can be graded by responsibilities into three groups (see Fig. 9.1).

The administrative head of the animal house should be a scientist of sufficient standing to hold the respect of the research scientists whom the animal house serves. He must know the capacity and limitations of his facility and be able to allocate available space fairly to satisfy, as far as possible, the demands of all his clients and research colleagues. He must also be able to refuse animals when necessary, and to advise scientists both on specific techniques and on licensing procedures, where these apply.

The senior technicians should each have their own area of responsibility, reporting either directly to the operational head or through a chief technician. This latter is usually a very experienced and highly trained senior technician, but he may occasionally be a graduate gaining experience in the field of laboratory animal science. Both chief and senior technicians should be good managers in addition to having wide technical experience, since they are responsible for the day-to-day running of the animal facili-

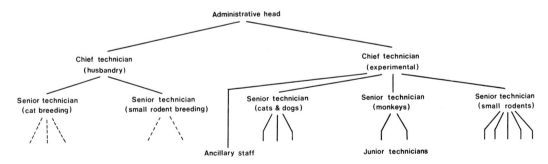

Fig. 9.1. Hypothetical organizational structure for an animal facility. Adapted, with permission, from Lane-Petter and Pearson, 1971).

ties, either as a whole, or, in the case of unit heads, as a specific aspect of the work of the facility.

Unit heads work through technicians, junior technicians, and ancillary staff. The latter carry out simple operations such as general cleaning and do not require a specific training programme. The technicians and junior technicians comprise the third group which requires training, the former being basically more experienced and perhaps with additional examination successes to their credit. Each animal technician is responsible for the daily running of one or more animal rooms and is also required and expected to be able to identify the unusual on his own initiative.

Thus the three groups to be considered in any training programme are:

1. Administrative head or scientist
2. Chief or senior technician
3. Junior technician

Training of the scientist

In order to run an animal facility both efficiently and economically, the administrative head must have an adequate background in modern laboratory animal science. He should also have the experience required to run such a facility, making full use of his scientific and technical staff. Such training and experience is no longer to be gained by apprenticeship but requires some sort of formal training. In addition to the general science of laboratory animals, specialised sciences such as comparative biology, genetics, animal nutrition etc., have developed. However, for the moment only the general science will be considered, it being assumed that trainees completing this course may go on to specialise.

The entrance requirements for such a course have been the subject of some debate. Previous education to degree level in veterinary medicine would ideally be a minimum requirement. Although candidates with a degree in medicine or a biological science might be suitable, they would benefit from a year's experience working with laboratory animals prior to commencing the formal training. The course content would obviously have to be varied to take account of the educational attainments of the trainees.

During the training programme, emphasis should be placed on the most commonly used vertebrates: mice, rats, hamsters, guinea-pigs, rabbits, cats, dogs, non-human primates and poultry. Veterinary graduates would obviously be at an advantage with regard to other species such as farm animals.

The ideal course would probably last about two years with the award, on graduation, of a diploma or M.Sc. degree. The training would be divided more or less equally into:

1. Theory: lectures and formal classes;
2. Practical: experience working in various sections of an animal facility;
3. Research.

The first two sections should be sandwiched to run concurrently; the research project could either be carried out simultaneously with these or separately in the latter stages of the course. Such research, resulting in a short thesis, is most important from the point of view of enabling the laboratory animal scientist to appreciate the needs of the research scientist and to understand methods of compilation of data and bio-statistics. In addition, it would make him more fitted to carry out later his own research projects or possibly prepare him to specialise in a particular aspect of laboratory animal work.

The practical work should provide a back-up for the theoretical training and should cover as wide a field as possible. For this reason, only a small number of students should be accommodated on a course, thus allowing the practical training of each student to be considered individually. In some cases, it may be possible for the practical training of all students to be carried out at the same institution as the theoretical training. This can be done by allowing trainees to participate directly, under supervision but at increasing levels of responsibility, in the daily operation of the animal facilities in an institution.

The formal training course should cover the following subjects:

1. Biology of laboratory animals, including anatomy, physiology, ethology, genetics and nutrition;
2. Selection of research animals, including an understanding of animal models for both spontaneous and experimentally induced diseases;
3. Husbandry of laboratory animals, including maintenance of conventional and germ-

free colonies, environmental requirements and quarantine;

4. Organisation of the animal house, including design and administration, routine equipment, and special facilities for research;

5. Administration of staff, including education of animal attendants;

6. Legal and moral considerations, including appropriate legislative controls and conservation of endangered species;

7. Techniques in animal experimentation, including anaesthesia, handling, injections, euthanasia and hysterectomy derivation;

8. Experimental surgery including aseptic technique, pre- and post-operative care and common surgical procedures;

9. Clinical medicine, including diseases of laboratory animals, their diagnosis and control;

10. Clinical pathology, including laboratory aids to diagnosis and pathology of diseases of laboratory animals.

Once an institution has been established to provide the above course it could easily expand to fill two other needs: firstly, as a training centre for graduates wishing to specialise in a particular aspect of laboratory animal science, and thus as a reference centre for all people working in this field and, secondly, to provide a short course for research scientists who intend to use laboratory animals. Eventually, completion of such a course might become an essential requirement for obtaining a licence to perform animal experiments. Training need last only a few days, covering the basic elements of the scientific course, including selection and acquisition of animal models, legislation and conservation, simple techniques of handling, anaesthesia, removal of body fluids, euthanasia, and sources of further information. To ensure that animal wastage is reduced to a minimum, principles of experimental design should also be included. The Report of the Departmental Committee on Experiments on Animals (Littlewood Report, 1965) recommended that applicants for a licence to perform animal experimentation be 'suitably qualified in training, knowledge and skill, to perform the procedures proposed'. The course described above would certainly go a long way towards satisfying these criteria and generally improving the standard of the applicant.

Training of technicians

During the last ten to twenty years there has been a world-wide increase in, and appreciation of, the need for adequately trained animal technicians. The International Committee on Laboratory Animals has always been conscious of this need and one of its stated aims is: 'To encourage the training of animal technicians by the provision of training manuals in several languages and by the establishment of a specimen programme of courses for laboratory animal technicians. . . .' A complete set of lecture notes has now been made available, and the subject is under continual review.

Such training leads to improvements in the productivity and quality of laboratory animals and hence to increased reliability of experimental results. The investigator can entrust the technician with far more responsibility regarding the acquisition of animals and the execution of experiments. All this is to the real benefit of the scientist and renders such a technicians' course of undoubted value. In return, the animal technician must be appreciated by the investigator and used in the context of his training and experience, and should, therefore, be incorporated into the research team on the same basis as any other technician with specialist knowledge. His advice during the planning stage of the experiment should be sought and, if he is kept properly informed, his ability to observe clinical and behavioural changes during the experiment is invaluable.

The education of animal technicians should be suitably graded. The best-known courses are at two levels: one for the junior and the other for the senior animal technician. However, further grades can be introduced into this basic format and the presence of a good career structure will have the added advantage of attracting a more conscientious trainee.

The junior technician

The duties of the junior technician include the routine care of one or more animal species. In order to gain as much experience as possible, it is important to practise rotation of staff between units with different functions. This will result in a flexible staff capable of meeting the changing needs of the research workers. Rotation should not be considered until the junior technician is adequately trained in his initial unit and has completed a basic theoretical training.

The simplest approach to the preliminary course is to start with a part-time training given either in the evenings or during working hours. The trainee should already be employed in an animal house in order that he may gain concurrent practical experience. The entrance qualifications for the course should relate to both academic and practical ability. Obviously the ability to learn is important, but equally so is a genuine interest in animal welfare and a rational approach to animal work. The duration of the training will vary according to the proportion of time available for the teaching of theory but, in general, a course extended over one to two years will provide a better training than a more concentrated course of three to six months.

The main species studied should be those vertebrates which are commonly used: mice, rats, guinea-pigs, hamsters and rabbits—although information on some which are less widely used—cats, dogs and non-human primates—should also be included. The course should be constructed along the following lines:

1. Zoology, including classification of the animal kingdom and elementary structure and function of the various organs;
2. Environment of the animal, including the animal house, animal cages and the apparatus used to record and control the environment;
3. Breeding and breeding administration, including systems and methods and record-keeping with simple analysis;
4. Nutrition, including dietary constituents and the preparation of diets;
5. Diseases and their control, including signs of normal health and behaviour, nature and recognition of disease and prophylaxis;
6. Hygiene, including sources of infection and precautions, disposal of wastes and practical aspects of personal hygiene;
7. Handling and sexing animals, packing for transportation, despatch, inspection on receipt, and quarantine;
8. Layout of the animal house and equipment, including routine operation and care of washing machines, autoclaves and animal and food balances;
9. Experimental procedures, including preparation of animals for experiment and post-operative care, maintenance and sterilisation of instruments, anaesthesia and minor techniques;
10. Routine care and humane killing of animals;
11. Welfare legislation as applicable and the ethics and purpose of laboratory animal investigation;
12. Simple arithmetic and composition in the national language.

It is very difficult to draw a line between the subjects suitable for the junior course and those suitable at a more senior level. In those countries where a three-tier course is already in existence, the preliminary course may exclude some of the above subjects. However, where only two tiers exist, the junior technician may often be involved in direct support to experiments and be expected to perform, under supervision, simple biotechnical operations.

The senior technician

The nature of the activities of the senior animal technician may also vary. He may be responsible for giving support to animal experiments which will differ according to the field of experimentation. Hence he may be required to give injections, take body fluids, anaesthetise, assist in operations, or perform necropsies and remove organs for further examination. He may have to supervise the animals under test and produce written reports on his observations. Alternatively, he may be responsible for animal accommodation and breeding programmes. In this case he will be required to analyse records and select future breeding stock, to purchase animals, feedstuffs and other necessary materials and to maintain the health status of the colony by environmental control and suitable prophylactic measures.

In any case, a high standard of technical expertise and experience will be needed, together with the ability to manage personnel and accept responsibility. For these reasons, the qualifications for entry to the senior technicians' course should be much more demanding. Academic ability must be rated highly and an ordinary degree in a biological science might be a suitable qualification. On the other hand, the trainee may already have gained the junior qualification and wish to further his career. In that case, it is essential that at least two years should be spent working as a junior technician. The final decision must, of course, be left to the interviewing panel and, for this reason, these recommendations should be considered only as guidelines.

The training of a senior technician must cover, in much greater depth, all the subjects in the junior course. In addition, the number of animal species studied should be increased to include detailed study of cats, dogs and non-human primates, together with the elements of breeding and maintenance in several other species, including more rodents (gerbil, Chinese hamster and cotton rat), birds (duck, chicken and turkey), farm animals (horse, cow, sheep, goat and pig) and cold-blooded species (frogs, toads, fish and reptiles). New aspects which must be added to the junior course are as follows:

1. Elementary genetics, including principles of inbreeding and random breeding and selection of breeding stock;
2. Nutrition, including comparative aspects, design of special diets and quality control of commercial diets;
3. Management of experimental animals, including experimentally infected animals, radiobiology experiments and surgically prepared animals;
4. Management of germfree animals, including hysterectomy derivation, barrier operation, and routine monitoring;
5. Anaesthesia of laboratory animals, including principles, methods and signs, and euthanasia;
6. Signs of ill-health and disease, including a knowledge of the transmissible diseases and infestations, and methods of control;
7. Experimental techniques (including sampling of blood, urine, faeces), post-mortem procedures and collection of tissues, calculation of doses and dilutions and random division of test groups;
8. Applied pathology, including treatment of wounds and understanding of inflamma-

tion and immune responses;
9. Elementary microbiology, including normal flora, pathogenic organisms and microscopy;
10. Safety measures, including prevention of accidents and first-aid for laboratory hazards, zoonoses, and precautions with toxic substances;
11. Discipline in the animal house and supervision of staff, including work planning and personnel management;
12. Routine ordering and purchase, supervision of record-keeping, elementary statistics and report writing;
13. Sources of further information, references, and reference centres.

The current situation

In April 1970, the governing board of the International Committee on Laboratory Animals (ICLA) decided to assemble information from its member countries relating to the various training courses available in the field of laboratory animal technology. Twenty-nine national members provided documents relating to their training schemes during the twelve-month period ending June 1972 and a survey of the situation was published in 1973. The results were quite encouraging. Thirteen countries already had an established course at one or more levels and leading to formal examinations. These courses varied in duration from three months to three years and most students attended on a part-time basis. In general, the courses were organised either by a government body, or by an educational institute, or by a national laboratory animal organisation such as A.A.L.A.S. (American Association for Laboratory Animal Science) in the U.S.A. or I.A.T. (Institute of Animal Technicians) in the U.K. One of these courses was held by correspondence. All courses required concurrent practical experience.

Three other countries were proposing to start new courses in the near future, and, indeed, some of these may be established by now. Eleven countries had no regular formal training programme at all. However, four of these were making some effort, usually along the lines of occasional short lecture courses.

In the remaining two countries, some laboratory animal training was given, either as a small part of a general technical course, or else privately by concerned individuals within their own establishments.

Conclusions

The scheme of training in laboratory animal science which has been described here probably envisages an ideal situation. To organise the entire scheme from conception to completion is obviously a mammoth task. The ICLA survey, however, does show that there is an increasing awareness of the need, in this field, for training on a national scale. Of the 29 countries in the survey, only seven had no training scheme whatsoever and almost all of these expressed dissatisfaction with this situation. It seems to be important to know what to aim for, but, even if it is impossible to achieve this aim in the forseeable future, it is imperative for every country to start a scheme of some kind on at least one level. It is to be hoped that this will then lead to the emergence of more training courses at other levels, with a consequent raising of standards.

REFERENCES

Institute of Animal Technicians (1970). Examination Syllabuses and Rules and Conditions of Examination (revised 1973). Oxford: I.A.T.

International Committee on Laboratory Animals (1959–1973). ICLA Bulletins Nos. 5(1959), 7(1960), 11(1962), 33(1973).

International Committee on Laboratory Animals (1976). Proceedings of the 6th Symposium held at Thessaloniki.

Lane-Petter, W. and Pearson, A. E. G. (1971). *The Laboratory Animal—Principles and Practice*. London & New York: Academic Press.

National Research Council (1972). *A Guide to Post-doctoral Education in Laboratory Animal Medicine*. Washington: Nat. Acad. Sci.

Report of the Departmental Committee on Experiments on Animals (1965). Chairman: Sir Sydney Littlewood. Cmnd. 2641. London: HMSO.

SOURCES OF FURTHER INFORMATION

American Association for Laboratory Animal Science. 4, East Clinton Street, Box 10, Toliet, Illinois 60434, U.S.A.

Gesellschaft für Versuchstierkunde. Secretary, W. Rossbach, Institut für Biologisch-Medizinische Forschung AG, CH-4414 Füllinsdorf, Switzerland.

Institute of Animal Technicians, 16, Beaumont Street, Oxford OX1 2LZ, England.

Institute of Laboratory Animal Resources. National Research Council, 2101 Constitution Avenue, Washington, D.C. 20418, U.S.A.

International Committee on Laboratory Animals. Secretary, S. Erichsen, National Institute of Public Health, Postuttak Oslo 1, Norway.

Japan Experimental Animal Research Association. Laboratory Animal Science Office, Kyonan 1-7-1, Musashino-shi, Tokyo 180, Japan.

Laboratory Animal Science Association. Secretary, P. Eaton, Charing Cross Hospital Medical School, Fulham Palace Road, London, W6 9HH, England.

10 Disease-free (SPF) animals

John Bleby

Specific-pathogen-free (SPF) animals were defined by the International Committee on Laboratory Animals (ICLA, 1964) as 'animals that are free of specified micro-organisms and parasites, but not necessarily free of the others not specified'. This term is somewhat controversial but it is likely to continue in general use until a more appropriate definition is found. The term SPF was the first one used for animals originally obtained by hysterectomy and maintained in a controlled but non-sterile environment (Foster, 1963). Other terms such as 'clean animals', 'disease-free', 'pathogen-free', 'caesarian-derived' are also in use, but in every case the meaning is identical. In fact, SPF animals are conventional animals that are free of their common and troublesome diseases.

SPF animals are, therefore, the disease-free animals that have been sought for so many years. Their derivation and maintenance may be summarized as follows:

(1) drawn from stock that had been originally established by hysterectomy;
(2) maintained within a barrier designed to discourage the entry of undesirable micro-organisms;
(3) maintained on diets, bedding and nesting materials known to be free of pathogenic micro-organisms;
(4) cared for by skilled technical staff.

The vast majority of conventional animals (those maintained in non-barrier buildings) harbour a collection of pathogens including ecto- and endo-parasites, fungi, protozoa, bacteria and viruses which result in overt or potential disease. For many years all those who work with laboratory animals have been frustrated by the presence of disease which has often led either to premature death of the animals or to gross interference with experimental results.

Improvements in the environment within laboratory animal houses and the rising standard of laboratory animal husbandry have done much to reduce the incidence of disease. On the other hand the incidence of most virus, protozoan (coccidiosis), ecto- and endo-parasitic infections has remained unaltered. Chemotherapy has not been very successful in the eradication of commonly recurring infections, and only in one or two cases has success been recorded. Wicker *et al.* (1958) for example showed that animal colonies can be freed of a single specific parasitic or bacterial infection, but only after tedious testing and extensive eradication programmes.

Experience has shown, therefore, that animal colonies cannot be freed from most of their common pathogens and parasites by chemotherapy and improvement in environment or husbandry, but with the arrival of germfree studies and the production of animals entirely free from all micro-organisms it was soon realized that an alternative way had been found. The key to this development was the aseptic removal of full-term foetuses from the mother by hysterectomy and their introduction and maintenance inside a sterile environment. Thus by utilizing germfree techniques involving caesarian section or hysterectomy, an infected colony could be freed from its common pathogens and parasites.

There is evidence to show that the placenta acts as a very efficient filter and protects foetuses from almost all bacterial and viral agents carried by the mother. The common pathogens such as salmonella, pasteurella, the causal organism of Tyzzer's disease, ectromelia, infantile diarrhoea, chronic respiratory disease, tapeworms, roundworms, and all other flora and fauna of a conventional colony do not appear to cross the placenta. However, this generalization should be treated with caution because a number of parasites do penetrate the placenta during the migration of their larval stages and the virus of lymphocytic choriomeningitis (LCM) also crosses the placental barrier (Traub, 1939). Furthermore, recent evidence (Casillo & Blackmore, 1972) has shown that some other undesirable micro-organisms which have been implicated in the contamination of SPF buildings, can invade the uterus.

It would seem prudent, therefore, for the hysterectomized or caesarian-derived young to be quarantined in a germfree isolator (see chapter 11) until tests have shown them to have a satisfactory flora and fauna. The young can then be introduced aseptically into the SPF building without the risk of contamination.

Hysterectomy

The first stage of the operation, i.e. the removal of the gravid uterus from the mother, is performed in a suitable room outside the main barrier. The operator and assistant are gowned and masked in accordance with the full sterile procedures practised in an operating theatre. The ventral surface of the pregnant female animal is shaved before she is humanely killed by a method which has the minimum effect on the foetuses and which, therefore, automatically precludes such agents as barbiturates. In the case of the smaller rodents, such as mice, the use of cervical fracture (Fig. 10.1) is quick and humane if done by a skilled technician. This method is not applicable to larger species, such as rabbits and cats, but a suitable alternative such as a halothane/oxygen mixture can be used for inducing anaesthesia, following this with direct intra-cardiac injection of barbiturate once hysterectomy has been completed. Ether, cyclopropane, and other inflammable or explosive

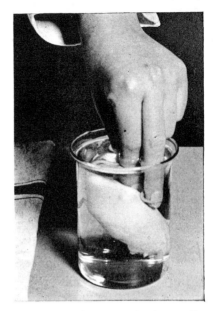

Fig. 10.2. Complete immersion in a sterilizing fluid. (Photograph: MRC Laboratory Animals Centre.)

agents are, of course, contraindicated when diathermy is being used.

The smaller species are completely immersed (Fig. 10.2) in a suitable sterilizing solution, such as a newly prepared solution of 2 per cent hypochlorite and 2 per cent cetrimide maintained at body temperature, as soon as they have been killed or anaesthetized. The animal is then pinned out on a sterile cloth as shown in Figure 10.3, the abdomen is covered with an adhesive sterile drape such as is used in human surgery,

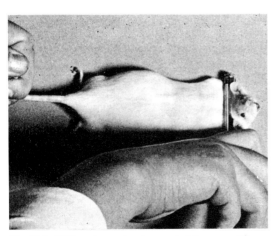

Fig. 10.1. Killing the mother by cervical fracture. (Photograph: MRC Laboratory Animals Centre.)

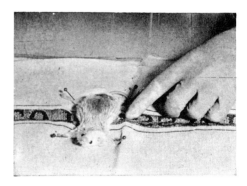

Fig. 10.3. Pinning out prior to covering with sterile plastic drape. (Photograph: MRC Laboratory Animals Centre.)

and an incision is made with a diathermy cutting needle simultaneously through the plastic drape and the skin (Fig. 10.4). One of the main advantages of using a diathermy cutting needle is that its method of cutting makes a sterile incision, leaving sterile edges to the wound, and little or no haemorrhage occurs to obscure the field. The muscle layers are then incised (Fig. 10.5) and after sectioning has been effected above the ovaries through their attachments and immediately below the cervix, the gravid uterus is removed intact from the abdomen (Figs. 10.6 to 10.8). During this process the uterus is protected from contact with the unsterile outside surface of the mother by the sterile plastic drape. When the body wall is being incised care must be taken not to puncture the fragile gravid uterus or the gut. Should this happen the whole effective barrier is invaded and the carcase, uterus, surgical instruments and drapes in use must be disposed of immediately. It is important to stress that a completely separate set of sterile cloths and surgical instruments must be used for each operation in order to prevent any risk of cross contamination. Furthermore, the operator and assistant must wash their gloved hands between each operation in a suitable sterilizing fluid.

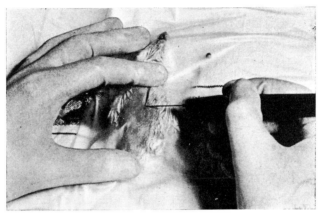

Fig. 10.4. Incision by diathermy through drape and skin layers. (Photograph: MRC Laboratory Animals Centre.)

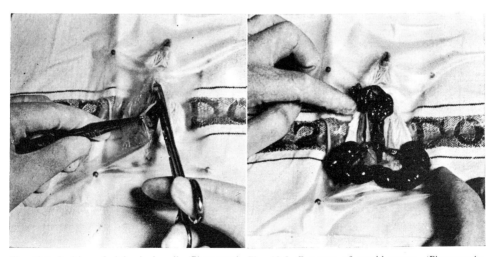

Fig. 10.5. Incision of abdominal wall. (Photograph: MRC Laboratory Animals Centre.)

Fig. 10.6. Exposure of gravid uterus. (Photograph: MRC Laboratory Animals Centre.)

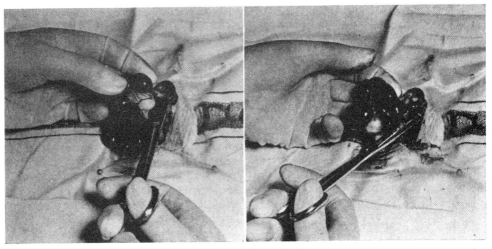

Fig. 10.7. Sectioning through the ovarian attachments. Photograph: MRC Laboratory Animals Centre.)

Fig. 10.8. Sectioning immediately below the cervix. (Photograph: MRC Laboratory Animals Centre.)

The intact gravid uterus is placed into a small sterile plastic box or similar receptacle fitted with a lid and containing the same sterilizing fluid at a temperature of approx. 38°C in which the uterus floats free (Fig. 10.9). In order to achieve a high survival rate of foetuses the fluid in which the uterus is floating must be maintained at blood heat, otherwise foetal blood temperature will be considerably lowered. The covered box is quickly passed into the germfree isolator or SPF building through a dunk tank containing long-acting disinfectant (Figs. 10.10 and 10.11). The box is fitted with a lid for two important reasons: firstly to prevent the cold disinfectant in the dunk tank from gaining access to the warm fluid jacket around the uterus, and secondly, to prevent any pressure from being exerted on the foetuses owing to the depth at which they are passed through the dunk tank. Inside the isolator or SPF building the uterus is opened with fine scissors, and the foetuses dried and resuscitated

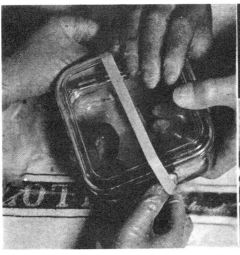

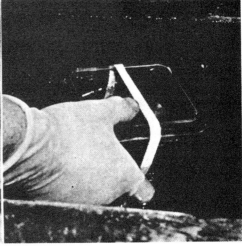

Fig. 10.9. Securing lid to sterile box containing gravid uterus floating in sterilizing fluid at 38°C.

Fig. 10.10. Passing of box into the dunk tank. (Photograph: MRC Laboratory Animals Centre.)

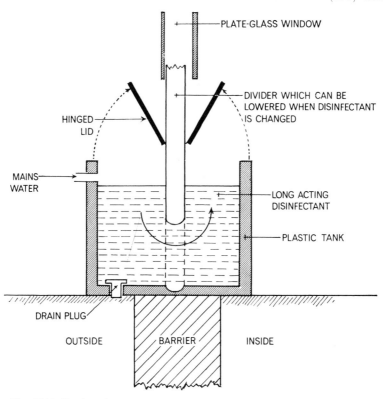

Fig. 10.11. Dunk tank.

(Fig. 10.12). When the foetal membranes have been removed from around the body of each pup, it is important to leave the placenta in contact at the umbilicus for a short time after respiration has been initiated, in order that the maximum amount of placental blood may be allowed to flow into the pup (Fig. 10.13). Eventually complete severance of the

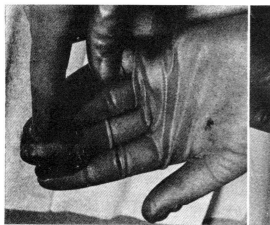

Fig. 10.12. Opening up of uterus. (Photograph: MRC Laboratory Animals Centre.)

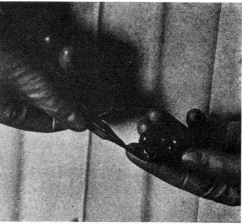

Fig. 10.13. Placenta still in contact with umbilicus. (Photograph: MRC Laboratory Animals Centre.)

umbilical cord is made with a stubby thermo-cautery needle which, by coagulating the blood, prevents any umbilical haemorrhage. The operation is best carried out on a small heated table because again it is essential to maintain body temperature. The young pups sometimes require stimulation by gentle rubbing with sterile gauze and drying of the nostrils until they are breathing well. Food is given to them either by hand, day and night, until they are old enough to feed themselves, or by fostering them on to lactating mothers already resident inside the isolator or SPF building. The latter method is much simpler but care has to be taken before-hand to bring about accurate synchronization of two breeding programmes, one on the outside of the isolator or SPF building, the other on the inside. In practice female animals that are to become foster mothers are mated so as to pro-duce their own litters some three days or so before they are due to receive foster pups.

with parity. A matter of hours in a short gesta-tion period of generally 21 days is critical, and if hysterectomy is carried out too early (possibly 24 hours before normal parturition) the pups will be immature and the survival rate will go down (Fig. 10.14). The gestation period for each strain must, therefore, be accurately deter-mined, and the females hysterectomized as near to normal parturition as possible. Another variable factor that has to be taken into account is the length of the period between the killing of the mother and the point when the foetuses begin to be affected. It has been found that in mice and rats some 10 to 15 minutes are avail-able, but in the case of guinea-pigs this period is reduced to 4 minutes, probably because of their advanced stage of development and high oxygen requirement at birth. However, in all cases (provided the hysterectomy team is well re-hearsed) there is sufficient time available for the gravid uterus to be removed, passed through the

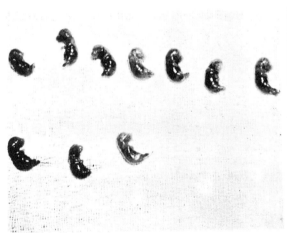

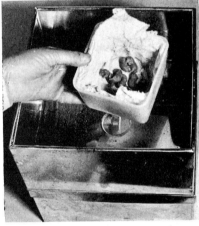

Fig. 10.14. Immature pups immediately after severance at the umbilicus. (Photograph: MRC Laboratory Animals Centre.)

Fig. 10.15. Mature pups, in warm receptacle containing foster-mother's bedding, being put into an incubator. (Photograph: MRC Laboratory Animals Centre.)

Meanwhile, on the outside, the females to be hysterectomized are mated so as to reach full term three days after the females on the inside have produced their litters. Complications can arise over the length of gestation periods; in mice these may vary from strain to strain and

dunk tank and opened up inside the isolator or SPF building.

For the foetuses to be removed under maxi-mum asepsis, the operation should be performed before first-stage labour has commenced and when the cervical seal is still intact. Some workers, in an attempt to obtain as mature pups as possible, wait until the first pup has been born, but once the birth canal has opened infection could gain access to the interior of the uterus.

Fostering

When the young pups appear pink in colour, and are kicking and breathing evenly, they are placed in a warmed receptacle containing some of their foster mother's bedding which contaminates them with the smell of their new mother. After about half an hour the foster mother is removed from her own box and her existing litter is replaced by the hysterectomized pups (Fig. 10.15). She is then returned and usually accepts her new litter without too much difficulty or suspicion.

During fostering, the mouse box or cage should be left undisturbed for about 72 hours and the foster mother carefully observed to see if she is accepting her new litter. It is useful to know that some mouse strains make better foster mothers than others and these should be used whenever possible.

entry of wild rodents and flying and crawling insects and special attention must be paid to drains, entry and exit locks for personnel and ventilation. The barrier must be adequately sealed throughout and plastic materials which never completely harden are invaluable for sealing off roof areas, window sections and other similar places through which insects and dust may penetrate.

Drains and gulley-traps that are fitted with mud-traps should have a sufficient depth of water containing a disinfectant to prevent the entrance of crawling insects. To obviate any likelihood of their drying up in hot weather, the seals should be topped up at regular intervals. As a general rule, solid concrete floors and ceilings are to be recommended. Ideally, there should be no windows in the animal rooms, but if there are, then they should be permanently sealed.

It is essential that the design of the building allows for the maximum amount of maintenance to be undertaken from outside the main shell of the building, i.e. outside the barrier. A useful way of facilitating this is to have a false roof 1·5–1·8 metres above the ceilings of the animal rooms

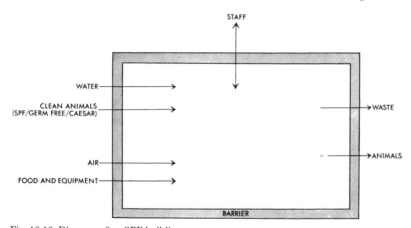

Fig. 10.16. Diagram of an SPF building.

The SPF building

An SPF building (Fig. 10.16) is basically a simple affair and can be considered as a tank with the minimum number of holes cut in it to facilitate the entry and exit of personnel, food, water, air, waste materials and animals. The building must be designed so as to exclude the

to act as a crawlway and allow the maintenance of such installations as heating, lighting, ventilation and other services.

As staff are the main vectors of reinfection it is necessary to keep the number of persons who enter the clean area down to an absolute minimum. If maintenance has to be undertaken from within the barrier, difficulties can arise from the possible lack of appreciation by maintenance staff of the standard of cleanliness required. Sterilizing chambers such as the autoclave and ethylene-oxide fumigator should, therefore, have as many operating controls as possible outside the barrier.

Operation

The main discipline of the whole operation of an SPF building is that of preventing reinfection and invasion of the clean area by pathogens. There are many carriers, of which the staff working inside are the most important.

Staff

The staff of an SPF building are in a specialized position; they must fully understand and appreciate the principles involved, and at all times be conscious of their duty to safeguard and maintain the barrier. They need to possess standards of intelligence and education above those needed for work in a conventional animal house and it is essential for the building to be under the general supervision of a person of suitable qualifications, and of adequate seniority, to maintain strict discipline at all times.

Upon entering the building the personnel are required to undress completely and shower. Entry/exit locks should be designed so that it is impossible to cross the barrier without actually passing through a shower area. Once through the shower, the staff put on a sterile working uniform which has already been brought into the building through either the autoclave or the gas chamber. The virtues of very simple clothing have been considered against spaceman-like outfits designed to prevent any possible infection of the animals by the staff. The latter measures are probably extreme and not justified as long as the hair is covered and an efficient face mask is worn, plus rubber gloves during work in the animal rooms. In this connection, it is helpful if persons working in an SPF building do not keep pets at home; furthermore, they should never be allowed to enter the building if they have been in contact with conventional animals until some considerable period of time (at least one week) has elapsed. Similarly, no one should enter the building unless in good health, and any illnesses such as sore throats, diarrhoea or suppurating skin areas should be reported immediately to the

person in charge, who will arrange alternative work away from the SPF building.

Ventilation

All incoming air should be filtered free from dust particles in order to remove the main sources of airborne contamination. It is not necessary to use sterilizing filters, which are very expensive and need frequent changing as they become blocked. A further refinement can be the introduction of an ultra-violet pathway within the ducting; this ensures some sterilization of the air. Both electrostatic precipitators and electronic filtration are quite effective but they have the disadvantage that their initial installation is expensive. Mechanical filters are usually the most practical, and as they are standard in most air systems they can be easily maintained. Generally, 0·8 to 0·9 μ appears to be the best size for particle-removal but it is advisable to consult ventilation specialists for detailed information concerning the different types and methods of filtration which are available and best suited to local climate and requirements. Because of the extremely closed environment within an SPF building, full air conditioning including humidity and temperature control is recommended in all climates. The aim should be to maintain a temperature within the barrier which can be controlled between 10° and 21°C including individual temperature control by thermostat in each animal room giving a temperature of about 19° to 21°C in the case of rodents. The number of air-changes per hour per room should be within a range of 15 to 20 so as to provide adequate ventilation, irrespective of the type, species, and density of animals in the room. One of the most important aspects of the ventilation of an SPF building, apart from the filtration of dust, is to ensure that a positive pressure with respect to the corridor outside is maintained within each room, and there should be a positive pressure inside the corridors with respect to the pressure outside the barrier. Thus, there is a falling pressure gradient from the animal rooms to the corridors, to the washing-rooms, to the outside, which ensures that any airborne contamination tends to be carried away from the animal rooms towards the entry/exit locks. Furthermore, should any leaks or apertures appear in the barrier, unfiltered air cannot pass in since filtered air is being forced out by the positive pressure.

Entry ports and sterilization

No food or equipment must be allowed to pass through the barrier and enter the building unless it is sterilized by a suitable method, of which there are several.

Dunk Tank. A dunk or dip tank consists of a rectangular plastic or fibre glass tank placed in the wall of the barrier and kept filled with disinfectant to such a height that it floods around a simple division at the top of the tank, thus making a fluid barrier (Fig. 10.11). There are many suitable and highly efficient disinfectants, such as phenolic derivatives and quaternary ammonium compounds, that do not require frequent changing but retain adequate potency for some time. The provision of a divider which can be lowered to the floor of the tank before emptying will allow the disinfectant to be changed without breaking the barrier. When the tank has been emptied by removing the drain plug at its base and refilled, the divider is raised again to its original position below the surface of the disinfectant. The dunk tank is a very useful and simple entry-and-exit lock and is ideal for introducing such items as metal cage-tops and other impermeable equipment that can be left overnight or over a week-end in the tank. The dunk tank also provides a rapid method of entry for gravid uteri when new animals enter the buildings, although entry via an isolator is now favoured.

Steam. This is the most commonly used method of sterilization and is carried out by means of a double-ended ultra-high vacuum autoclave situated in the barrier with one door, i.e. the dirty end, facing outside, and the other door, the clean end, on the inside of the barrier. The autoclave should be able to pull high vacuum and maintain temperatures of 121 to 134°C for considerable periods, and ideally should incorporate facilities whereby formalin vapour can be used instead of steam. It is operated in the usual manner, except when being used for sterilization of diets. This presents technical problems which are dealt with below.

Ethylene oxide Sterilization by fumigation with ethylene oxide has distinct advantages in, for example, the sterilization of books and papers that cannot be passed through an autoclave or dunk tank. Ethylene oxide is highly toxic and in the pure form is violently explosive in all mixtures with air ranging from 3 to 80 per cent or even more (Phillips and Kaye, 1949). The gas is, therefore, supplied commercially mixed with 10 per cent of carbon dioxide, to reduce its explosive nature. But full precautionary procedures such as the provision of flash-proof switches and lamps, must be observed to avoid any possible risk of explosion, and expert advice must be taken when a fumigator is installed. In an SPF building a double-ended fumigator is placed across the barrier in the same way as an autoclave, and consists of a reinforced chamber which can be completely evacuated. After the fumigator has been loaded a full vacuum is drawn and vaporized ethylene oxide of a concentration directly proportional to the amount of load being sterilized is then run into the evacuated chamber and left for some hours. The optimum period of exposure to the gas is in the region of seven to eight hours for adequate sterilization of diet or four to five hours for such items as paper, depending upon the concentration of gas being used. The gas is then drawn from the chamber and filtered air is run from the inside of the SPF building followed by two further evacuations and air rinsings.

The most likely hazard of using ethylene oxide is the possibility of residues being left behind unless adequate air washing takes place after fumigation. These residues can cause irritation and death to animals because of their highly toxic nature and this substance is not, therefore, recommended for the sterilization of diets (Porter & Bleby, 1966). Nevertheless, ethylene oxide affords a useful adjunct to other methods, especially in the sterilization of delicate equipment such as microscopes and cameras.

A fumigator, like an autoclave, has to be operated correctly and requires responsible staff. It is not essential for an SPF building to have both an autoclave and an ethylene-oxide fumigator, but it is useful to have both methods of sterilization available. Thus an alternative port of entry is provided if one method breaks down, and both double-ended chambers provide suitable means of exit for soiled bedding, faeces, and other waste matter.

Sterile diets

The provision of a sterile diet for SPF animals is difficult. When food is sterilized, some of its nutritive value is destroyed, but two methods have been developed which have largely overcome this problem. These are ultra high vacuum autoclaving and the use of gamma radiation. High vacuum autoclaves have the advantages of being relatively simple to operate and requiring only steam and electricity supplies. For the sterilization of food supplies they must be capable of operating on high-temperature short time-cycles so that the destruction of the thermolabile constituents is kept to a minimum. Examples of acceptable cycles are 134°C for 4 minutes or 121°C for 10 minutes. Pre- and post-vacuum cycles are always necessary.

To counterbalance the destruction of thermolabile substances, such as vitamins, some commercial companies have produced pelleted diets fortified with an excess of these substances, so that after sterilization by heat the food retains a correct balance of essential nutrients. However, antivitamins can be produced from vitamins by heat treatment (Porter & Lane-Petter, 1965).

The sterilization of diets by gamma radiation is more complex and requires the convenient availability of an appropriate radioactive source, usually cobalt 60. Recent work (Ley *et al.*, 1969) has shown that gamma radiation is a suitable method for the treatment of a variety of laboratory animal diets intended for SPF colonies. Owing to the high penetrating power of this radiation the diets can be packed before treatment in a manner which prevents recontamination during transport and storage. A radioactive dose of 2·5 Mrad has proved effective for the control of contaminating organisms and irradiated diets appear to be nutritionally satisfactory. Because of the advantages of having sterile diet available when required, it is likely that irradiated diets will become increasingly used.

It is worth noting that although SPF units are the primary users of sterile diets, in the course of time the breeders of all colonies are likely to follow suit. Ultimately, there is little doubt that all laboratory animal colonies will be fed on sterile diets and one further hazard to the animal and research worker will have been removed.

Water supplies

Ideally, all water supplies should be sterilized before passing through the barrier. However, water sterilization can present certain technical difficulties and may, therefore, not be warranted if the human drinking water supply is shown to be pure enough for the purposes for which the disease-free animals are required. Nevertheless, the use of unsterilized water places a barrier building continually at risk and is therefore a factor which should be most carefully considered.

Exit of animals and materials

Animals which are being issued from the SPF building and are to be maintained in an SPF condition, are prepacked within the barrier in suitable containers fitted with filters. The containers are, like everything else, introduced through the appropriate sterilizing entry port. They can be passed out of the barrier through an ultra-violet lock consisting of a simple chute with doors at each end which are never opened together. As there is positive pressure within the barrier the gradient flow of air is from the inside out, but because of the possibility that outside air might intermingle and get within the barrier in the air-lock before the inner doors are once again opened, the ultra-violet light is left on for 20 minutes to sterilize any unfiltered air. As was mentioned previously, the best exit for soiled bedding, faeces, and other waste material is through the autoclave or fumigator, and a routine can be established whereby waste material is bagged up in disposable plastic bags and tied off within the animal rooms and then stacked somewhere within the vicinity of the inner door of the chamber. After a sterilizing run has been made for the purpose of introducing new material into the building, and before the inner door is closed, the bags of waste material are placed in the autoclave or fumigator, to be removed from the outside at the appropriate safety signal from inside and disposed of in the usual manner. The inner door of the chamber is not reopened until after another sterilizing run has been made.

Screening and control of pathogens

It is impossible to operate an SPF building satisfactorily without regular microbiological screening of the animals within the building. Screening includes the regular checking of diets for sterility, whatever the method of sterilization, and, in fact, anything non-human which enters the building. It also includes sampling of water and of dust from the air filters of the ventilating system and from all other possible sources of entry of pathogens.

Nose and throat swabs should be taken at regular intervals from all staff in order to detect those which harbour undesirable micro-organisms.

Contamination and breakdown

Contamination and breakdown of the system by the introduction of pathogens into the building is inevitable. However, if all possible pre-cautions are taken and vigilance is maintained, SPF buildings can run for some years without a breakdown. In order to be able to take action in the event of breakdown the building should be designed so that each animal room is a complete entity in itself with no possible risk of cross-contamination. The extent to which this can be carried out depends largely upon the size of the building and on local conditions but, if each animal room is complete in itself, then if infection occurs in one room it is possible to kill out the whole of that room and fumigate and resteri-lize it without upsetting the rest of the working of the building. This is made easier by the incorporation in the outside wall of each animal room of a small sealed metal door which can be opened in such emergencies, facilitating the entry and exit of staff to carry out sterilization and fumigation from outside the barrier (Spiegel, 1965). If the building is very small then one might have to consider a complete kill-out and re-stocking. In this connection it is useful to have nearby the services of a small germfree unit (Chapter 11) where from time to time the animals from certain rooms or of certain strains may be re-cycled to ensure, as far as is humanly possible, that they are clean.

Uses and functions of SPF animals

Much has already been written on the use and function of SPF stock. Godwin *et al.* (1964) have shown in haematological observations on healthy

(SPF) rats that accepted standards for blood counts are based on infected animals and it seems evident that these standards should be revised. It has been shown that disease-free colonies can be established and that they cost no more in terms of time, labour, food and money than conventional colonies (Innes *et al.*, 1957).

The ultimate test for any experimental animal is its usefulness in the scientific world, and many users have until recently appeared somewhat dubious of the advantages of SPF animals. Nelson & Collins (1960) have shown that the most favourable change in an SPF colony is increased vigour, and that the removal of the parasite burden is followed by a higher produc-tion rate. Further, there is no significant differ-ence between SPF and conventional stock in their susceptibility to infection and other agents. SPF mice are now known to be superior in experiments lasting over an extended period, and fewer unexplained deaths occur.

The disease-free or SPF animal is rapidly becoming the minimum accepted standard, and furthermore is likely to be demanded in the future for the statutory testing of compounds destined for public use.

REFERENCES

Casillo, S. & Blackmore, D. K. (1972). Uterine infections caused by bacteria and mycoplasma in mice and rats. *J. comp. Path.*, **82**, 477–482.

Foster, H. L. (1963). Principles of breeding and management. In *Animals for Research*, ed. Lane-Petter, W. London & New York: Academic Press.

Godwin, K. O., Fraser, F. J. & Ibbotsen, R. M. (1964). Haematological observations on healthy (SPF) rats. *Br. J. exp. Path.*, **45**, 514–24.

International Committee on Laboratory Animals (1964). *Terms and Definitions*. Bulletin No. 14. London: I.C.L.A.

Innes, J. R. M., Donati, E. J., Ross, M. A., Steufer, R. M., Yevich, P. R., Wilson, C. E., Farber, J. F., Pankevicius, J. A. & Downing, T. O. (1957). Establishment of a rat colony free from chronic murine pneumonia. *Cornell Vet.*, **47**, 260–80.

Ley, F. J., Bleby, J., Coates, M. E. & Paterson, J. S. (1969). Sterilisation of laboratory animal diets using gamma radiation. *Lab. Anim.*, **3**, 221–54.

Nelson, J. B. & Collins, G. R. (1960). Establishment and maintenance of a specific pathogen-free colony of Swiss mice. *Proc. Anim. Care Panel*, **11**, 65–71.

Phillips, C. R. & Kaye, S. (1949). Sterilizing action of gaseous ethylene oxide. *Am. J. Hyg.*, **50,** 270–9.

Porter, G. & Bleby, J. (1966). Ethylene oxide sterilization. (Observations on the use of food, cages and nesting material in the breeding of mice). *J. Inst. Anim. Tech.*, **17,** 160–6.

Porter, G. & Lane-Petter, W. (1965). Observations on autoclaved, fumigated, irradiated and pasteurized diets for breeding mice. *Br. J. Nutr.*, **19,** 272.

Shafik, A. & Khiskin, L. (1950). Determination of ethylene oxide and methods of its recovery from fumigated substances. *J. Sci. Fd Agric.*, **1,** 71.

Spiegel, A. (1965). In *Food and Cosmetics Toxicology*. London: I.C.L.A.

Traub, E. (1939). Epidemiology of lymphocytic choriomeningitis in a mouse stock observed for four years. *J. exp. Med.*, **69,** 801–17.

Wickert, W. A., Rosens, S., Dawson, H. A. & Hunt, H. R. (1958). *Brucella bronchiseptica* vaccine for rats. *J. Am. vet. med. Ass.*, **133,** 363–4.

11 Gnotobiotic animals: I

History and terminology

Animals are usually associated with a great variety of micro-organisms and parasites, a few of which may be acquired directly from the parent during the embryonic or foetal state but the majority of which come from the environment. Some of these are responsible for infectious disease, some, such as the rumen flora, are necessary for the normal development of the host, while the effects of the vast majority are relatively unknown. Though the gut flora of the laboratory mouse has been studied intensively, in 1968 Savage, Dubos and Schaedler reported a previously undescribed bacterium in the intestine of the mouse. Not only had this organism attracted little attention prior to this time, but it was also present in larger numbers than any other species.

The sterile techniques of the microbiological laboratory are not applicable to the rearing of laboratory animals because of the more involved manipulations required and the larger area needed for maintaining the animals. In 1895, Nuttall & Thierfelder described an enclosed cage sterilized by autoclaving in which guinea-pigs were reared for about 2 weeks without bacterial contamination. In 1915 Küster described apparatus in which a goat was maintained free of bacterial contamination for 34 days. This apparatus was the first capable of maintaining a sterile environment indefinitely while permitting manipulations through arm-length rubber gloves, the introduction of sterilized supplies and the removal of waste materials and various samples (see, Luckey, 1963, for the early history). This type of apparatus, later termed an isolator, is similar to the equipment used today for the rearing and study of germfree animals. The demonstration that a laboratory animal could reproduce in a sterile environment made it possible to consider the more extensive use of these methods.

At the present time, germfree rats and mice are reared in many laboratories and are even available commercially; colonies have been maintained in sterile environments since 1954. Reproduction has been obtained from guinea-pigs, rabbits, dogs, domestic fowl and Japanese quail. Primary stock (that is, animals derived from parents carrying the usual microbial contaminants) have been reared for varying periods of time and include the following: monkey, cat, pig, lamb, goat, calf and both live- and egg-bearing fish.

As with most new technologies, the development of terminology has been a problem. The early workers used terms such as *microbe-free* and *germfree*, and the latter term is still widely used. However, the term *germ* can mean either a pathogen or any micro-organism. For this reason, the terms *germfree* and *pathogen-free* are occasionally confused. Furthermore, *germfree* is not an appropriate description for animals which have been inoculated with one or more pure cultures. Since many animals are used in this condition, the need for another term is obvious. The designation *gnotobiotics*, meaning 'the field of investigation involved in the production and use of organisms isolated from all others or in the presence of known kinds of organisms' (Reyniers *et al.*, 1949), and its derivative 'gnotobiote' are widely used. This refers to an animal that is free of all microbes or one that is associated with any combination of organisms derived from pure cultures. Terms such as *axenic, monoxenic, dixenic*, etc., may be used to designate the number of species in an association (Dougherty, 1953). Animals having an unmodified microbiota are usually described as *conventional*, though the scientific term *holoxenic* has been used (Raibaud *et al.*, 1966), and *heteroxenic* if the microbiota has been deliberately altered.

Uses

The apparatus and methods available at present make it possible to undertake in a sterile environment almost any laboratory-animal investigation that can be accomplished in the ordinary open room. However, the use of the sterile environment involves considerably more effort and expense which must be equated with the probable advantage of such a procedure. A true comparison of costs involves the entire operation from the initial design of the experiment to the final analysis of results. Almost always the most expensive item is the cost of the investigator's time rather than the more obvious cost of the animal and isolation equipment.

Some of the uses of the gnotobiotic animal may be grouped as follows:

(a) The axenic animal may be studied by itself in order to determine its development and reactions upon a diet that can be defined chemically more completely than would be possible if micro-organisms were present. The animal also serves as a source of sterile organs and tissues for cultivation. In the absence of micro-organisms, the reduction in the varieties of antigens present is of significance for the study of defence mechanisms.

(b) The gnotobiote, when associated with one or more pure cultures, gives an opportunity for a more critical study of the aetiology of infectious disease than would otherwise be possible. In addition, animals with no previous contact with the agent studied have a more uniform susceptibility than those which have had a previous infection. Infections with a single agent also produce antisera with greater specificity than those produced in conventional animals.

(c) The role of various components of the normal flora can be studied by infecting the gnotobiotic host with either pure or mixed cultures from the normal flora.

(d) Gnotobiotic animals are frequently used for the production of colonies or herds of animals free of one or more specific agents.

Gordon & Pesti (1971) have reviewed the use of the gnotobiotic animal in the study of host microbial relationships, while Heneghan (1973) has reviewed the whole field of research using gnotobiotes.

Standard biological data

Two types of data should be examined carefully when working with gnotobiotes: microbio-logical evidence as to the gnotobiotic status (freedom from contamination) and a description of the morphology and physiology of the animal since this may differ from the conventional.

Gnotobiotic status

The limitations of laboratory procedures used to demonstrate the gnotobiotic status of an animal are the same as those of sterility testing in that the absence of all other living forms cannot be rigorously demonstrated. The microbiological condition of the animal must be inferred from the results of a series of tests for the presence of various living agents. While it is necessary to establish some minimal standards in order that results from different laboratories and at different times can be compared, it is desirable to make as thorough an examination as possible. Procedures should be modified from time to time on the chance that some heretofore undetected micro-organisms will be noticed. New methods should be tried as they become available. It is quite possible that in the future all animals will be shown to carry some form of symbiote. The presence of such organisms, intimately associated with the host and obligately transmitted from generation to generation, would be of less importance for most studies than the usual micro-organisms transmitted from the environment or diseased parents.

Monitoring procedures are of two types: routine tests to detect a defect in contamination control procedures and more elaborate tests designed to determine whether or not contaminants have been vertically transmitted from parent stock or have escaped the routine test procedures. The more elaborate tests require the services of a competent microbiologist and parasitologist. Routine tests can be performed by a technician with adequate training in microbiology (Wagner, 1959). Experience has shown that both microscopic examination and cultivation on a variety of media are required routinely. Stained faecal smears should be examined, preferably always by the same person, in order to detect a variation in the numbers and types of organisms which may reflect either a change in the food or the presence of a living contaminant. Dead organisms, of course, are found in the food and will appear in the faeces. Aerobic and anaerobic culture tests must be made at room temperature and 37°C. Routine examinations should also occasionally include samples of bedding and smears from the nose and mouth of the animal.

A committee established by the Institute for Laboratory Animal Resources, Washington D.C. has studied procedures for monitoring gnotobiotes and recommended standards (Anon, 1970). These include recommendations for elaborate as well as routine examinations.

Characteristics of the gnotobiote

Comparative studies have been made of rats, mice, and chickens in gnotobiotic and conventional environments (Pleasants, 1968; Gordon, 1968). Rats, mice, and Japanese quail appeared to reproduce equally well with and without microbiota. The growth rate of rats, mice and chickens can be within the normal range in a sterile environment while the growth rate of guinea-pigs and rabbits under axenic conditions does not equal those under comparable conditions in the open room.

As expected, the humoral and cellular defence systems do not develop as well as in animals receiving a high level of antigenic stimulus from the microbiota. However, both systems respond in a characteristic fashion following stimulation with antigens.

The most spectacular difference in rodents between conventional and most gnotobiotic animals is the greatly distended caecum. In some animals, the caecum may be 30 per cent of total body weight. A distended caecum impairs reproduction in at least the rabbit and guinea-pig and may rupture or become twisted in older rats and mice, causing death. The contents of the enlarged caecum and the large bowel have a higher moisture content in the absence of microbiota. Somewhat similar caecal distention can be produced by feeding animals large quantities of antibiotics.

Comparisons of animals with and without micro-organisms in the gut have shown that bacteria are responsible for the degradation of endogenous protein, mucopolysaccharides and other substances which are produced by glands and cells lining the alimentary canal and by the normal sloughing of cellular elements. In the conventional animal, bacteria apparently break down this material in the small intestine and

caecum so that it may be utilized by the host. In the absence of the appropriate microbes the animal cannot utilize this material (Loesche, 1968). Bacteria are also responsible for many reactions within the alimentary canal involving bile pigments, lipids and sterols. Several workers have found the metabolic rate in the gnotobiote was substantially lower than in the holoxenic counterpart.

Husbandry
Isolation
Isolators now in use for rearing gnotobiotes consist of chambers with either flexible or rigid walls. The air used for ventilation passes through filters made of fine glass-wool that removes all suspended micro-organisms. Air leaving the isolator chamber may pass either through protective filters or through traps that permit movement in one direction only. Manipulations are made through arm-length gloves attached to the wall of the isolator. In large isolators the gloves may be supplemented with or replaced by half-suits which permit the attendant to insert the upper portion of his body into the work chamber while still remaining microbiologically isolated. A few sterile rooms have been built in which the operator dons an impervious garment, the surface of which is sterilized with germicide so that he may then enter the room without introducing microbiological contamination. However, at present this does not seem to be of much practical importance though it may become so with the increased size of gnotobiotic installations.

By far the majority of isolators used at present are made of flexible plastic film and are modelled after the isolator described by Trexler & Reynolds (1957) (Figs. 11.1 and 11.2). A great variety of films are available, differing in flexibility, strength and optical properties. Polyvinyl chloride either 0·3 or 0·5 mm thick is the most commonly used material for construction of the chambers. Either peracetic acid or ethylene oxide is used for sterilization (Sacquet, 1968). Arm-length rubber gloves or shorter gloves attached to flexible film plastic sleeves can be used for manipulation. Materials are introduced or removed from the isolator chamber without microbial contamination by means of either a sterile lock or a germicidal bath or trap. The sterile lock may be a small double-

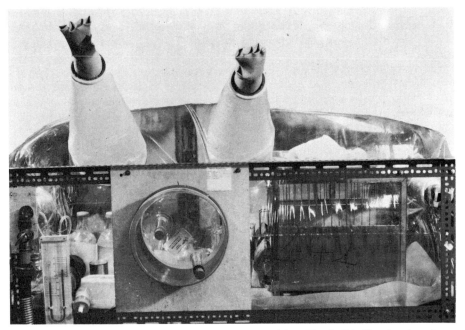

Fig. 11.1. Front view of an isolator used for the rearing of piglets. The flexible film chamber, sleeves and gloves are disposable. Note the entry port in the centre foreground. The port is 30 cm in diameter and 15 cm deep, and is closed with flexible film caps on the inside and the outside. (Photograph: UFAW.)

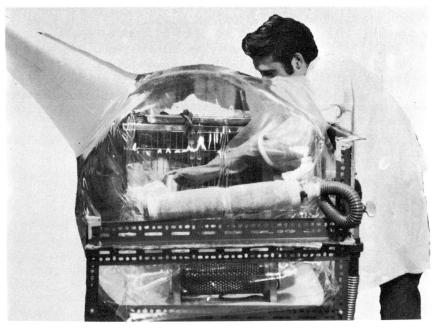

Fig. 11.2. Side view of piglet isolator. Note the air filter in the centre foreground and the heater in the lower foreground inside of the isolator chamber. The chamber has two pairs of gloves so that the attendant can have assistance in working with the animals. (Photograph: UFAW.)

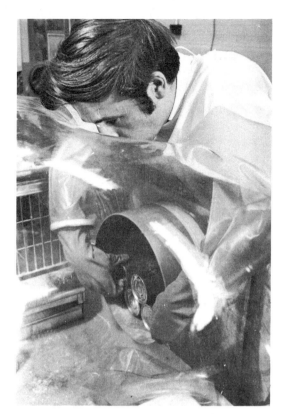

doored autoclave, penetrating the chamber wall, or more often a chemically sterilized lock. A 2-per-cent aqueous solution of peracetic acid with a small amount of nonionic detergent is the germicide of choice since it is active both as a liquid and a vapour, acts rapidly and leaves no toxic residue. Because peracetic acid is so corrosive, glass, stainless steel, pure aluminium or plastics are the only common compatible materials. The chemically sterilized lock consists of a rigid cylindrical entry port attached to the wall of the chamber, the inside of which is sealed with a removable flexible film cap (Fig. 11.3). A cap, sleeve, or bag may be attached to the outer end of the entry port making it possible to introduce loads of different sizes or to attach one isolator to another.

Flexible film isolators are adaptable to a variety of tasks in the laboratory because the chambers can be made in almost any size or shape (Fig. 11.4). Small chambers may contain

Fig. 11.3. Entry port in use. The attendant has removed the inside cap which lies on the floor of the chamber in the foreground. Empty cans are being placed in the entry port. The inside cap will be replaced before the outside cap is taken off to remove the cans. (Photograph: UFAW.)

Fig. 11.4. Isolators are placed on racks in order to conserve space in the laboratory. A single attendant can service the animals while the cages are on racks. However, if two individuals are required for manipulations, the isolator is removed from the racks by means of a fork lift truck and placed on a trolley. (Photograph: UFAW.)

Fig. 11.5. Animal cage within the isolator. The supplies, tinned milk and sterile water are stored on the floor of the isolator. The tray on top of the isolator contains syringes, sponges, thermometers, etc. (Photograph: UFAW.)

but a single mouse cage; larger isolators have been made to take an entire rack of cages or single pens that house animals as large as a calf (Figs. 11.5 and 11.6). The danger of contamination with film isolators is no greater than with those made of rigid material since the glove,

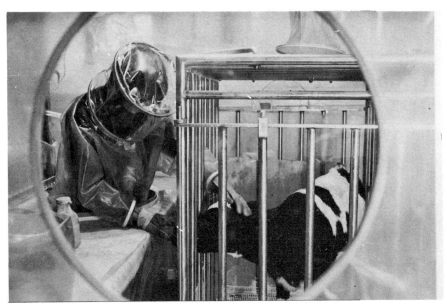

Fig. 11.6. Calf isolator. The wall of the chamber is made of polyethylene which can be readily sealed but is translucent. The clear window is made from 'Melinex', supported by a rigid ring. The attendant is in a half suit, which is attached to the bottom of the rigid entry ring. (Photograph: UFAW.)

common to both, is more vulnerable to puncture than the chamber itself.

Flexible film chambers can be readily altered and repaired in the laboratory using cement or simple heat-sealing equipment. Experimental apparatus can easily be used with the flexible film chambers because tubes, wires and mechanical probes can be passed through the walls without the risk of contamination. Delicate procedures such as thymectomy of mice less than 24 hours old can be performed routinely (Wilson *et al.*, 1966). More elaborate procedures such as are required by human surgery can also be accomplished in these isolators (Levenson *et al.*, 1964). (See also McLaughlan *et al.*, 1974).

Sterilization of apparatus and supplies

The sterilization of apparatus and supplies should be managed so as to provide a negligible probability of contamination. If an accidental contaminant is detected it is important that the probable sources of contamination are reduced to a minimum so as to simplify locating the cause. For this reason excess sterilization treatment should be used with all items that are not so damaged. Minimum sterilization times should be determined for each type of load so that all sterilization can be performed with an appropriate safety factor. The load to be tested should be inoculated with a known quantity of resistant organisms to provide a resistant population at least equal to the worst condition encountered in practice.

A high-vacuum autoclave, though not essential, simplifies sterilization procedures by the rapid removal of air from filters, bedding, dry feed and other supplies. The removal of air to ensure penetration of saturated steam into the load is more critical for supplies to be inserted into an isolator than for ordinary purposes because the sturdy package needed to prevent contamination upon removal from the autoclave retards the flow of air and steam. A metal drum with filters similar to those on the isolator is commonly used for this purpose though sealed packages of heat resistant plastic film vented with the same type of filter are also satisfactory provided they are handled carefully.

All the sterilization required for the maintenance of gnotobiotic animals can be accomplished by the use of a high-vacuum autoclave and peracetic acid. However, ethylene oxide can be used for the sterilization of completely assembled isolators. Radiation is also satisfactory for sterilizing food, bedding and a variety of apparatus. The choice is a matter of convenience and economics.

Environmental conditions

The environmental conditions should be the same for the gnotobiotic animal as for the conventional animal. An animal relieved of its microbial burden may tolerate greater extremes of temperature and humidity. However, it seems advisable to keep the environmental conditions within the isolator suitable for maintaining either type of animal.

Requirements for the care of gnotobiotic animals are the same as those for ordinary animals as far as their comfort and well-being are concerned. There are advantages in using the same caging and husbandry practice for gnotobiotic as for ordinary animals because this will simplify the making of comparisons between the two groups.

Difficulty is frequently experienced in maintaining ventilation within the isolator comparable to that in the open animal room. Ventilation in the animal room is expressed in terms of air changes per hour but the amount of fresh air an individual animal receives depends upon the population density and the pattern of air movement. Within the isolator, turbulence produced by bringing in high velocity air through a small opening will avoid dead spots but may result in draughts on some cages. These can be avoided by the use of a diffuser within the isolator or by placing the filter within the isolator to provide low-velocity air movement.

Air-flow through an isolator can be conveniently determined by the use of an orifice gauge on the inlet side of the air system. Temperatures can be determined by means of mercury thermometers or other temperature sensing devices within the isolator. Humidities can be determined by measurements taken on the effluent air after it leaves the outlet filter or the outlet trap, provided that non-aqueous fluids are used in the trap. The air velocity is usually sufficient for wet-bulb determinations.

Derivation of gnotobiotic animals

Mammals are obtained by removing the young from the uterus by sterile surgical procedures. If the young are free of microbial contamination while in the uterus they can be introduced into the isolator without contamination, hence the desirability of selecting breeding-stock that is free of infections transmissible to the foetus. This may not be possible with mice, since Pollard (1966) reported one or more viruses in all gnotobiotic mice adequately tested. Dogs are often infected with *Toxocara canis* so that the puppies may be reared in the isolator free of micro-organisms but harbouring these parasites. They may be removed by dosing the puppies with a suitable vermifuge (Heneghan *et al.*, 1966).

Either hysterotomy or hysterectomy can be used for obtaining young from the dam before introducing them into the isolator. Hysterectomy involves the excision of the entire uterus and its passage into the isolator through a germicidal bath. The young are then removed and passed into another isolator for rearing away from the maternal tissue. For removal of the young by hysterotomy or Caesarean section the isolator is attached directly to the abdomen of the dam. A thin plastic membrane forming part of the floor of a surgical isolator is attached to the surgically prepared skin of the abdomen or flank of the gravid animal. An incision is made through the plastic film and skin by means of a hot wire cautery to prevent contamination by organisms in the hair follicles and sebaceous glands. Standard surgical procedures are used for the rest of the operation.

Gnotobiotic birds are obtained by passing the eggs through suitable germicidal solutions and permitting them to hatch inside a sterile isolator. Mercuric chloride or peracetic acid is used to treat the surface of the egg. The eggs must be selected carefully, avoiding shells with faecal stains (Harrison, 1969). Since there are many egg-transmitted infections—bacterial, mycoplasmal and viral—it is necessary to select flocks that are free of these agents. For hatching germfree turkey poults, Mohamed & Bohl (1969) have indicated the importance of obtaining eggs from a flock known to be free of salmonellae, mycoplasmas and any other known egg-transmitted micro-organisms. This is especially true for *Mycoplasma meleagridis* in the United States, where infection of the reproductive tracts of both turkey hens and cocks is widespread and often results in the infection of embryos. These workers maintain a specified-pathogen-free turkey flock which serves as a source of eggs for the hatching of germfree poults.*

Handling

The handling of animals within the isolator is somewhat more difficult than in the open room because of the necessity of gloves (Fig. 11.7). Furthermore, a bite is not only annoying to the attendant but results in contamination of the

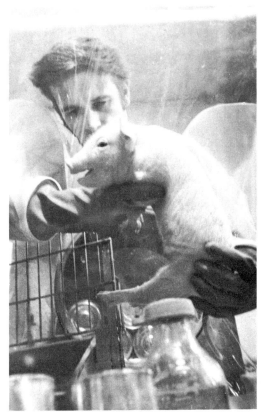

Fig. 11.7. The attendant handling animals through rubber gloves attached to the chamber wall. The flexible walls permit considerable movement of the attendant. (Photograph: UFAW.)

* *Editors' note.* For an account of the production of gnotobiotic chickens and quail see Coates, M. E. (1968) in: *The Germ-free Animal in Research*, edited by M. E. Coates. London & New York: Academic Press.

isolator. Mice are frequently handled by their tails, using long forceps. The forceps should have rubber covered tips and, to avoid injury, the tail of the animal should be held near the base. When handling a rat protective gloves are usually worn over the isolator gloves. These are frequently of a heavy plastic variety which can be steam sterilized. Isolators can be readily equipped with two pairs of gloves so that the operator may have an assistant to hold the animal for marking, injection and other procedures.

The handling of animals within the isolator is facilitated by good visibility. It is essential that the plastic in front of the face has good optical properties and is kept clean.

Transport

Gnotobiotic rats and mice have been shipped all over the world without contamination. They may be sent in ordinary flexible film isolators placed in a protective box with air supplied by a battery-powered blower. This package is quite heavy and expensive to transport. Animals may also be placed in light-weight flexible film isolators with sufficient filter area to provide adequate ventilation without the use of mechanical assistance (Trexler, 1968). These small isolators can be placed in protective boxes made of cardboard and shipped in much the same manner as ordinary animals.

Rats and mice, as well as the larger animals such as pigs and calves, can be shipped in an isolator carried within a motor vehicle. Air can be supplied by means of a battery-operated blower using either a special battery or the electrical system of the vehicle. Care must be taken to maintain adequate temperature control. It may be necessary to install additional heaters if the outside temperature is low. On the other hand, care must be taken that the inside of the isolator does not become overheated in direct sunlight. The sun shining through the windows of the vehicle can heat up the interior of the isolator very rapidly.

Laboratory procedures

Experimental procedures

In general, any procedure that can be used with ordinary laboratory animals can be carried on within an isolator provided that all the materials required can be sterilized. Experience has shown that isolators that do not interfere with visibility and manipulations can be constructed. Press-polished polyvinyl chloride sheeting has good optical properties and either the entire isolator may be made of this material or a panel can be inserted for viewing. Some distortion is associated with the flexibility of the PVC sheeting, but this can be eliminated by the use of rigid acrylic panels or a thin, transparent film, such as Melinex* which can be stretched over a circular rigid supporting ring to serve as a viewing port in the flexible wall of the chamber (Trexler, 1971). Dissecting microscopes which remain on the outside of the isolator have been used to examine materials within the isolator by extending the isolator wall to form a small pouch which is placed on the microscope stage. Gloves can be placed in the wall of the pouch to assist in manipulations.

Many studies require the use of materials that are sensitive to various sterilizing procedures, so that a combination of treatments may be needed during the course of a study. Many materials, such as syringes and a variety of injectables, may be purchased pre-sterilized by an appropriate process. Usually this treatment is adequate, though the requirements for sterility are more rigorous for gnotobiotics than for medical purposes. It is advisable to remove labels from ampoules in order to ensure surface sterilization. Ampoules sealed by the fusion of glass are preferred to those with screwed caps or rubber bungs since it is difficult to ensure sterilization in crevices. Liquids can be sterilized by filtration, with the filter serving as a sterilizing entry orifice. The filter may be sterilized in the conventional manner and the sterile end passed through an opening into the interior of the isolator. However, less risk of contamination is incurred if the sterilized filter assembly is passed into the isolator chamber through an entry port and the tube for receiving the non-sterile material forced out through an opening in the wall of the isolator and secured to prevent leakage.

Food, water, bedding and similar items are normally introduced into an isolator through some form of sterilizing entry port. For short-

* Imperial Chemical Industries Ltd.

term experiments, procedures with a higher risk of contamination may be used provided adequate microbiological monitoring is maintained. The occasional loss due to contamination can be compensated by the increased number of experiments that may be performed. Thus sterile materials can be passed into the isolator chamber against a flow of sterile air either from the chamber or another sterile filter protecting the entry port. Small specimens and other materials can be readily removed from the isolator through a small tube penetrating the isolator wall. Contamination is prevented by the outward flow of air. After passage of the material, the outer opening is closed by a rubber bung dipped in peracetic acid. Wires, tubing, etc., can be readily passed through the wall of the isolator chamber by first inserting them in a rubber bung in order to provide a larger surface for attachment either to the film directly or to a rigid tube sealed to the wall of the chamber. In this way, sterile constant temperature baths have been maintained within the isolator without sterilizing the control instruments. Tubing, carrying water from an external temperature-controlled bath, passes through the isolator wall and serves as a heat exchanger for the sterile bath. Wires and tubing can be passed through the wall after the chamber has been sterilized if they are first introduced into the isolator and then forced out through the wall of the chamber. This can be readily accomplished provided the wire or tubing is coiled within a plastic test tube and the mouth of the test tube is closed by a rubber bung through which the wire or tubing passes. The bottom of the tube is then forced against the wall of the chamber and a small circular hole made by a cutter operated from outside the isolator. The test tube is then forced through the small opening so that the flexible film of the chamber wall stretches and forms a flare around the body of the test tube. The flare is then taped to the test tube and the bottom cut off so that the end of the tubing or wire can be removed.

Anaesthesia and euthanasia

Anaesthesia and euthanasia can be induced within an isolator in much the same way as under conventional conditions except that sterile agents must be used and the hazard of explosion restricts the use of flammable materials. such as ether. Many of the barbiturates can be obtained for sterile injections. Pentobarbitone sodium* can be sterilized in the autoclave provided it is suspended in a solution of 10 per cent ethyl alcohol and 20 per cent propylene glycol (Pilgrim & D'Ome, 1955). Halothane and other gaseous agents can be introduced into the isolator as a gas passing through a sterilizing filter (Cook & Dorman, 1969). For anaesthesia it is useful to provide two filtering systems, one carrying the agent, the other air, so that the level of anaesthesia can be controlled.

The management of gnotobiotes infected with pathogens

The use of pathogens with gnotobiotic animals may require confinement as well as exclusion isolation. The infection of a gnotobiotic animal with some pathogens may present a greater hazard than in a conventional animal because the lack of competition from the normal microbiota may enable the agent to grow in larger numbers and in places such as the waste material where it would ordinarily not thrive. For this reason, it is advisable to use great care in working with pathogens in gnotobiotes.

Such work may impose a variety of constraints on the design of appropriate isolators and accessory apparatus. Rigid isolators are sometimes used in preference to those with flexible walls because there is less danger of punctures. However, gloves are the most vulnerable portion of the isolator and these will be the same whether flexible or rigid chambers are used. Flexible chambers can be used with negative pressure which prevents the leakage of air from the interior of the chamber through a puncture or other orifice. This is a valid consideration provided the agent can be airborne. However, fluids can leak from a hole in the isolator whether the pressure is positive or negative. Negative pressure requires the use of a rigid support either outside or inside the chamber. Glove movements with negative pressure isolators are almost as restricted as with rigid walled chambers.

Isolators harbouring pathogens are usually equipped with an outlet filter rather than an outlet trap, in order to prevent infective

* Nembutal: Abbott Laboratories.

aerosols from emerging from the apparatus. For additional protection, should the filter become damaged, the effluent air can be incinerated or vented to the atmosphere for dilution.

Materials that are removed from an isolator harbouring dangerous pathogens must be treated with a decontaminating agent. This can usually be accomplished by the same means as that used to sterilize the entry port. Waste materials, dead animals, etc., may be placed in plastic bags, sealed and passed into the sterile lock where the surfaces of the bags are decontaminated by a suitable germicide. The bags are then removed to the autoclave or incinerator. Plastic bags can be attached directly to the entry port for removal of materials (Trexler & Reynolds, 1957). After the materials are placed in the bag a secure closure can be made by means of a heat sealer. The seal can be cut and the bag removed for disposal. As an alternative, the bag may be tied-off and the entry port and stub-end of the bag sprayed with a germicide before removal.

Necropsies can be performed within the isolator to prevent the dispersal of pathogens during the procedure. Animal boards, instruments, fixatives, slides and other apparatus may be introduced into the isolator as needed. A special isolator may be used for necropsy if the isolator containing the animal is crowded or if many animals must be processed. A tubing passing through the wall of the isolator is convenient for either obtaining quantities of blood or removing waste fluids. If suction is used the effluent air should be filtered. It is advisable to heat the air so as to reduce the humidity before filtration.

At the termination of the study, the entire isolator must be decontaminated with an appropriate agent. A decontaminant such as hypochlorite solution is frequently introduced into the isolator through the entry port in some spraying device. Perhaps the best method of decontamination is to use disposable isolators which may be autoclaved in toto.

The use of isolators for the confinement of pathogens is an important consideration whether or not gnotobiotes are involved. Isolators can provide more security than the traditional isolation techniques and are less subject to human error because they provide secure mechanical barriers. Usually an isolator system will be more economical to install than the traditional quarters. Flexible film isolators have a considerable advantage over more permanent installations because they occupy little storage space when not in use and can be altered to meet changing requirements.

REFERENCES

Anon. (1970). *Gnotobiotes.* Washington: National Academy of Sciences.

Cook, R. & Dorman, R. G. (1969). Anaesthesia of of germ-free rabbits and rats with halothane. *Lab. Anim.,* **3,** 101–6.

Dougherty, E. C. (1953). Problems of nomenclature for the growth of organisms of one species with and without associated organisms of other species. *Parasitology,* **42,** 259–61.

Gordon, H. A. (1968). Is the germ-free animal normal? A review of its anomalies in young and old age. In *The Germ-Free Animal in Research,* ed. Coates, M. E. London & New York: Academic Press.

Gordon, H. A. & Pesti, L. (1971). The gnotobiotic animal as a tool in the study of host microbial relationships. *Bact. Res.,* **35,** 390–429.

Harrison, G. F. (1969). Production of germ-free chicks: a comparison of the hatchability of eggs sterilized externally by different methods. *Lab. Anim.,* **3,** 51–9.

Heneghan, J. B. (ed.) (1973). *Germfree Research.* London & New York: Academic Press.

Heneghan, J. G., Floyd, C. E. & Cohn, I., Jr. (1966). Gnotobiotic dogs for surgical research. *J. surg. Res.,* **6,** 24–31.

Küster, E. (1915). Die Gewinnung, Haltung und Aufzucht keimfreier Tiere und ihre Bedeutung für die Erforschung naturlicher Lebensvorgange. *Arb. K. GesundhAmt.,* **48,** 1–70.

Levenson, S. M., Trexler, P. C., Laconte, M. & Pulaski, E. J. (1964). Application of the technology of the germ-free laboratory to special problems of patient care. *Am. J. Surg.,* **107,** 710–22.

Loesche, W. J. (1968). Accumulation of endogenous protein in the cecum of the germ-free rat. *Proc. Soc. exp. Biol. Med.,* **129,** 380.

Luckey, T. D. (1963). *Germ-free Life and Gnotobiology.* London & New York: Academic Press.

McLaughlan, J., Pilcher, M. F., Trexler, P. C. & Whalley, R. C. (1974). The surgical isolator. *Brit. med. J.,* **1,** 322–324.

Mohamed, Y. S. & Bohl, E. H. (1969). Personal communication.

Nuttal, G. H. F. & Thierfelder, H. (1895). Thierisches Leben ohne Bakterien im Verdauungskanal. *Z. physiol. Chem.*, **21**, 109–21.

Pilgrim, H. I. & d'Ome, K. B. (1955). Intraperitoneal pentobarbital anaesthesia in mice. *Expl Med. Surg.*, **13**, 401–3.

Pleasants, J. R. (1968). Characteristics of the germ-free animal. In *The Germ-free Animal in Research*, ed. Coates, M. E. London & New York: Academic Press.

Pollard, M. (1966). Viral status of germ-free mice. *Nat. Cancer Inst. Monogr.*, No. 20. p. 166.

Raibaud, P., Dickinson, A. B., Sacquet, E., Charlier, H. & Mocquot, G. (1966). La microflore du tube digestif du rat. IV. Implantation controlée chez le rat gnotobiotique de différents genres microbiens isolés du rat conventionnel. *Annls Inst. Pasteur*, **111**, 193–210.

Reyniers, J. A., Trexler, P. C., Ervin, R. F., Wagner, M., Luckey, T. D. & Gordon, H. A. (1949). The need for a unified terminology in germ-free life studies. *Lobund Rep.*, **2**, 151–62.

Sacquet, E. (1968). Equipment design and management. Part I. General technique of maintaining germ-free animals. In *The Germ-free Animal in Research*, ed. Coates, M. E. London & New York: Academic Press.

Savage, D. C., Dubos, R. & Schaedler, R. W. (1968). The gastrointestinal epithelium and its autochthonous bacterial flora. *J. exp. Med.*, **127**, 67–76.

Trexler, P. C. (1968). Equipment design and management. Part II. Transport of germ-free animals and current developments in equipment design. In *The Germ-free Animal in Research*, ed. Coates, M. E. London & New York: Academic Press.

Trexler, P. C. (1971). Microbiological isolation of large animals. *Vet. Rec.*, **88**, 15–20.

Trexler, P. C. & Reynolds, L. I. (1957). Flexible film apparatus for the rearing and use of germ-free animals. *Appl. Microbiol.*, **5**, 406–12.

Wagner, M. (1959). Determination of germ-free status. *Ann. N.Y. Acad. Sci.*, **78**, 89–101.

Wilson, R., Bealmear, M. & Sjodin, K. (1966). A technique for thymectomizing germ-free mice. *J. appl. Physiol.*, **21**, 279.

12 Gnotobiotic animals: 2

Marjorie Dinsley

Feeding

The nutritional requirements of germfree animals are similar to those of conventional animals and the same types of diet may be used for both. However, a supplement must be added to the diet of germfree animals to compensate for the absence of those nutrients (notably vitamin K) normally synthesized by intestinal micro-organisms and also for the changes which occur in the diet during sterilization and storage.

Sterilization of diet inevitably causes some destruction of nutrients, the amount of damage depending on the sterilization method. The method chosen will depend on the facilities available but should not be one which causes major damage to the diet.

Diet is commonly sterilized by steam under pressure. This process results in loss of certain vitamins, especially thiamine, and has an adverse effect on the nutritive value of proteins. The degree of destruction increases with the temperature and duration of treatment. Clearly, the treatment must be sufficient to guarantee sterility but excessive temperatures and sterilization times must be avoided because of their effects on heat-labile nutrients. A compromise must be made, and a temperature of $121°C$ for 25 or 30 minutes is usually recommended. It is advisable to sterilize diet in an autoclave able to draw a high vacuum (approximately 28″ Hg). A precycle vacuum prevents the formation of air pockets and facilitates steam penetration thus causing a rapid rise in temperature, while a vacuum at the end of the cycle quickly reduces the temperature. This means that the diet is subjected to high temperatures for no longer than is necessary to ensure sterilization.

In this country, diet is often sterilized by gamma-radiation from a ^{60}Co source. During this process there is a temperature rise of only a few degrees and, although some vitamins are lost, it is less destructive than steam sterilization. An added advantage is that the cubes do not stick together as happens during autoclaving.

Diet for irradiation is double-wrapped in sealed plastic bags; it is advisable to evacuate the inner bag to reduce the loss of vitamin E during sterilization. Vacuum-packing produces a compact, easily-handled package and if the vacuum is subsequently lost it is an immediate indication that the bag has been punctured and that the package must be rejected. Diet for gnotobiotic animals is usually irradiated at 4–5 Mrads. Many plastics become brittle and crack at this high dose but laminates, such as cellulose/polythene and nylon/polythene, are less severely affected and are recommended for wrapping.

During storage of diet, there is a progressive loss of certain nutrients and this raises a problem in small units where prolonged storage of diets at room temperature sometimes occurs. In such units, diet must be bought in small quantities and regular checks must be made on the age of the diet in each isolator.

Reddy *et al.* (1968) discuss in more detail the factors affecting the nutritional adequacy of diets for germfree animals and describe diets which have been used for many species. Diets for germfree mice and rats are easily obtained in this country where vitamin-supplemented diets suitable for sterilization are commercially available. Germfree rabbits and guinea-pigs are not widely used at the present time and there is less information about their dietary requirements. In general, under germfree conditions, these species have shown poor growth, enlarged caeca and poor reproduction. However, diets reported to be adequate for growth and reproduction have been described for rabbits (Reddy *et al.*, 1965) and guinea-pigs (Pleasants *et al.*, 1967). Diets which support good growth in germfree chickens are mentioned by Coates (1968). Techniques for hand-rearing small laboratory mammals from birth through weaning are described by Pleasants (1968).

Large mammals are sometimes reared germfree for short periods of time and diets adequate for growth of pigs are described by Whitehair

147

(1968). Trexler (personal communication, 1970) has found that young pigs and calves may be reared on reconstituted evaporated cows' milk, which, in the case of pigs, is fortified with a mineral supplement.

Germfree mice and rats have been raised on chemically-defined, water-soluble diets sterilized by filtration (Reddy *et al.*, 1968). Such diets are potentially valuable in those fields of nutrition where dietary intake must be precisely controlled. They may also prove useful in immunology because their components are of low molecular weight and therefore of low antigenicity.

Genetic management

Generally, the aim of genetic control is to produce a colony which is uniform and stable, so that animals are comparable in different parts of the population and at different times. This control is necessary in both inbred and random-bred strains (Dinsley, 1963). Proper genetic management is sometimes difficult to achieve under gnotobiotic conditions and it is tempting to ignore the need for planned breeding programmes. But the full value of gnotobiotic animals will not be realized unless their genetic standards are equal to those of conventional animals.

Random-bred colonies

These strains are probably heterozygous at a large number of loci and their characteristics can be changed by genetic drift, selection and inbreeding. The strains may be stabilized, to some extent, by using a planned breeding programme based on the following concepts:

Colony size. The effective colony size is not the total number of animals in the colony at any one time but the number of breeding animals which will contribute offspring for breeding in the next generation. If the colony is too small, inbreeding will occur and the stock will begin to show signs of inbreeding depression.

Selection of breeding stock. This must be planned to ensure that all parts of the colony are represented. A great deal of genetic variation will be lost if some sections of the colony do not contribute stock for future breeding. Ideally, each breeding pair should contribute two animals for breeding in the next generation.

Mating system. Genetic variation must not only be retained in the colony, but must also be spread evenly throughout. Breeding stock must be mated in a regular pattern so that isolated units are not formed within the colony. Such units would quickly diverge and the colony would cease to be a single, uniform population.

Maintaining a colony of random-bred animals in isolators raises some problems. The amount of isolator space available for breeding is usually limited and thus the colony tends to be too small. If small isolators are in use, a number of them will be needed to house a breeding stock of adequate size. However, the colony must not be allowed to separate into isolated sublines which do not interchange genetic material. To prevent this, breeding stock must be regularly exchanged between isolators, and the frequency of the exchange depends on the number of breeding animals in each. Usually, it is necessary to do this every second or third generation. The disadvantage of this practice is that a contamination, present but undetected, in one isolator may be spread to others. To reduce the risk, reliable methods must be used to maintain and monitor gnotobiotes.

Another problem which may arise is loss of part of the colony by contamination of one of the isolators. Each time this happens, the colony is reduced in size, part of the genetic variation is lost and the colony becomes more inbred. Some inbreeding will also occur if insufficient animals are used to start the gnotobiotic colony. Frequently, this foundation stock is too small and the animals are chosen haphazardly so that they are not representative of the colony from which they were derived.

These difficulties do not arise if new gnotobiotic stock can be derived routinely from an existing conventional colony. In these circumstances, the gnotobiotic stock is genetically part of the conventional population and is not itself a permanent breeding colony which will be continued indefinitely.

It is clear that the sort of breeding programme necessary for the proper management of a random-bred colony is not well-suited to the small isolators of the gnotobiotics unit and there is a trend towards the use of inbred strains which are more easily controlled under gnotobiotic conditions.

Inbred colonies

Genetic management of gnotobiotic inbred strains presents no special problems. The colony may be small, and complicated mating systems are unnecessary because matings are always between littermates. Only one litter is needed to start the colony.

However, there are occasions when it is necessary to transfer animals from one isolator to another. A breeding stock of a gnotobiotic inbred strain is usually spread over a number of isolators and consists of sub-lines which are quite separate from one another. These sub-lines will eventually show some degree of divergence, because of the accumulation of mutations, and they should not be allowed to continue indefinitely. Sub-lines are normally pruned after three or four generations so that all the animals in the current generation have a common pair of ancestors three or four generations back. With the exception of the last few generations, a single line can be traced back to the foundation generation of the colony.

Many people routinely take isolators out of use for overhaul and this provides a good opportunity for pruning sub-lines. Instead of automatically transferring animals from the old isolators to the new, all the sub-lines should be inspected and animals from the best sub-line used to stock the new isolator. If isolators are allowed to run indefinitely, the stock will have to be replaced at intervals by animals from whichever isolator houses the line chosen to continue the colony. Clearly, it is desirable to increase the generation interval of the strain so that transfer of stock does not occur too frequently. This can be done by delaying pairing of animals until they are adult and by choosing future breeding stock from later rather than from earlier litters.

Comparative studies using gnotobiotes and conventional animals

Valid comparisons between gnotobiotic and conventional animals are not easy to make. Such comparisons are usually complicated by environ-mental variables which may contribute to the difference between the two groups. Genetic differences may also cause complications when the two colonies have been separated for many generations. In the course of time, genetic changes will occur and the gnotobiotic colony will become increasingly different from the parent colony maintained in conventional conditions. Random-bred strains are particularly liable to this kind of divergence. Clearly, the two groups of animals should have a common genetic background and this may be achieved in one of two ways: either by routinely creating new germfree stock from the conventional colony, or by giving a conventional flora to some of the gnotobiotes and using these animals as the conventional control group.

REFERENCES

Coates, M. E. (1968). Chickens and quail. In *The Germ-free Animal in Research*, ed. Coates, M. E. London & New York: Academic Press.

Dinsley, M. (1963). Inbreeding and selection. In *Animals for Research. Principles of Breeding and Management*, ed. Lane-Petter, W. London & New York: Academic Press.

Pleasants, J. R. (1968). Small laboratory mammals. In *The The Germ-free Animal in Research*, ed. Coates, M. E. London & New York: Academic Press.

Pleasants, J. R., Reddy, B. S., Zimmerman, D. R., Bruckner-Kardoss, E. & Wostmann, B. S. (1967). Growth, reproduction and morphology of naturally born, normally suckled germ-free guinea-pigs. *Z. Versuchstierk.*, **9**, 195.

Reddy, B. S., Pleasants, J. R., Zimmerman, D. R. & Wostmann, B. S. (1965). Iron and copper utlization in rabbits as affected by diet and germ-free status. *J. Nutr.*, **87**, 189.

Reddy, B. S., Wostmann, B. S. & Pleasants, J. R. (1968). Nutritionally adequate diets for germ-free animals. In *The Germ-free Animal in Research*, ed. Coates, M. E. London & New York: Academic Press.

Whitehair, C. K. (1968). Large mammals. In *The Germ-free Animal in Research*, ed. Coates, M. E. London & New York: Academic Press.

13 Anaesthesia

W. D. TAVERNOR

The term laboratory animal is impossible to define, for at the present time a very wide range of animal species are used for experimental purposes, ranging from the conventional laboratory rodents to large farm animals, as well as birds and fish. The scope of this chapter must be limited therefore to general considerations, and the basic principles, of anaesthesia in animals. For more detailed aspects the reader is referred to the other chapters of this volume and to specialised text books.

From both the humanitarian and scientific standpoints, the production of satisfactory anaesthesia is a pre-requisite for the performance of any surgical or other painful interference upon animals. It is impossible to undertake precise surgery without adequate anaesthesia, but the method of anaesthesia to be adopted may vary considerably according to the procedure to be carried out, the species of animal to be used, and the conditions under which the operator is working. The preoperative preparation and examination of the patient should be the first consideration of the anaesthetist, and the presence of disease will certainly modify his approach. In the case of surgery on experimental animals, however, it would be foolish to utilize animals that are not in good health. In order to minimize the possibility of vomition, and also to avoid respiratory embarrassment as a result of pressure on the diaphragm, it is essential that the patient should not have received a meal within three hours of the administration of a general anaesthetic. In adult ruminants it is impossible to empty the rumen, and although food may be withheld for a longer period of time, and the water intake reduced, it is also essential to insert a cuffed endotracheal tube during general anaesthesia to prevent inhalation of regurgitated ruminal contents.

Recent years have seen the introduction of a large number of drugs that may be used as part of a balanced anaesthetic technique. Modern anaesthetic methods involve the production of adequate levels of analgesia and narcosis as well as relaxation of voluntary muscles. Whilst it may be possible to meet all three requirements with a single drug, it can be equally practicable, and in some cases preferable, to use one or more drugs to fulfil the requirements of each aspect. For convenience the subject of anaesthesia will be considered under the general headings of *Local Anaesthesia; Premedication; Intravenous Anaesthesia; Inhalational Anaesthesia;* and *Muscle Relaxants.* It should be stressed at this stage that, whatever anaesthetic technique is adopted, it is absolutely essential to meet the oxygen requirements of the patient.

Local anaesthesia

This term is usually employed to describe the effect of drugs that block the conduction of an impulse along a nerve fibre, thereby eliminating sensory nerve impulses. In addition, the blocking of motor nerve fibres will produce relaxation of the innervated muscles. Although a large number of agents have been considered for their local anaesthetic activity, only a few compounds are used in clinical practice. The most widely known drugs are cocaine, procaine and lignocaine. Today, lignocaine is probably the most frequently used, being extremely stable in solution and producing minimal tissue damage or irritation. Compared with procaine, it has a shorter period of onset, a more intense and longer duration of action, and is more rapidly absorbed from tissues and mucous surfaces. Lignocaine is generally used as a 2 per cent solution of the hydrochloride salt, to which is added adrenaline hydrochloride in a strength of 1 in 100,000. The local vasoconstrictor action of adrenaline approximately doubles the time for complete absorption.

Local anaesthetics may be employed by a number of methods, including surface application, intra-articular injection, infiltration, and regional nerve block. Such techniques may be suitable for brief manipulations when restraint is not an important factor, or as part of a technique involving the production of narcosis or light general anaesthesia with other drugs. Generally speaking, such techniques are not suitable for the majority of surgical interferences, although in adult ruminants the use of local anaesthetics alone may be the method of choice.

Premedication

The term premedication is used to describe the administration of a drug before the induction of general anaesthesia. Such drugs may be administered to produce sedation or basal narcosis (a state of central nervous depression short of anaesthesia), or to reduce the excessive flow of mucus and saliva that may occur when irritant anaesthetics such as ether are to be used. Sedative drugs are used to reduce the irritability of the central nervous system, thereby enhancing the effects of a subsequent anaesthetic agent. Many sedative drugs produce respiratory depression and this must always be borne in mind, especially if further respiratory depressant drugs such as thiopentone are to be used to induce anaesthesia. Sedation can be utilized in order to allay apprehension and fear and to minimize resistance to the induction of anaesthesia. Although this form of premedication may have a limited application in the smaller laboratory animals, it is particularly important when dealing with species or individuals that are difficult to restrain. If a painful condition is already present, or likely to develop, an analgesic drug should be administered, and if excessive salivation and mucus secretion is likely to occur, atropine sulphate should be given. Suitable premedication invariably smooths the induction and course of anaesthesia, reduces the quantity of anaesthetic required, and produces a quiet recovery period. This latter point is particularly important in animals that have undergone extensive surgical interference.

A large variety and number of drugs may be used in the pre-anaesthetic preparation of an animal. As mentioned above atropine sulphate may be administered for its blocking effect on cholinergic nerves, in order to control the secretion of saliva and mucus. It should be remembered, however, that the drug may also reduce vagal tone to the heart, and this may be manifest as an increase in heart-rate and a small rise in blood-pressure. The drug also produces mydriasis, which may interfere with the ocular signs associated with subsequent anaesthesia.

Morphine and the synthetic morphine-like drugs are widely used for the pre-anaesthetic medication of some species. The drugs in this group have marked analgesic and, to a lesser extent, narcotic effects. Morphine itself has a number of undesirable side effects: for example, its stimulation of the vagal centre in the medulla oblongata. These effects include vomition, defaecation, excessive bronchial secretion, slowing of the pulse, and respiratory depression. The use of this drug is contra-indicated in cats, as in this species it may produce marked excitement. The synthetic derivatives of morphine have been developed in an attempt to retain the analgesic and narcotic properties of the drug whilst eliminating the untoward side-effects. Pethidine is probably the analgesic drug of choice for most species, having a central analgesic action similar to morphine but no marked hypnotic action, and little effect on the blood-pressure. Pethidine does, however, produce some degree of respiratory depression. Other morphine-like drugs include thiambutene (useful in dogs), etorphine (Reckitts M99—mixed with acepromazine and known as Immobilon—is particularly useful in large animals of both domestic and wild species), phenoperidine and fentanyl (these latter two drugs being used in combination with the so-called neuroleptic drugs to produce the condition described as neuroleptanalgesia—see below).

The term ataractic has been applied to drugs that produce sedation without producing drowsiness. The terms tranquillizer and sedative are often interchanged indiscriminately with ataractic, and in animal species, where it is difficult to define the subjective response to a drug, there may be very little difference between the terms. The main drugs that may be considered under this heading are those that have been derived from phenothiazine and include agents such as chlorpromazine, promethazine, promazine, trimeprazine, acepromazine, and very many more drugs of a similar chemical structure. The agents in this group in general produce central sedative effects along with a degree of antihistamine or antiadrenaline (α-adrenergic receptor blocking agent) activity that varies from one compound to another. The pharmacological effects of the phenothiazine derivatives are well documented, and they are suitable for the pre-anaesthetic sedation of most species of animal. It is, of course, advisable to check on their detailed pharmacological effects before utilizing them in a particular experimental procedure. Another group of drugs that are used clinically as major tranquillizers are the butyrophenone derivatives such as haloperidol and droperidol. Their properties resemble those of chlorpromazine and include anti-emetic activity.

The term 'neuroleptic' has been used to describe ataractic agents that antagonize the emetic action of morphine-like drugs, and includes compounds such as chlorpromazine and the butyrophenone derivatives. When these compounds are combined with morphine-like analgesics the resultant state has been described as one of neuroleptanalgesia. Various combinations of drugs have been developed (usually an analgesic such as pethidine, phenoperidine or fentanyl with either haloperidol, droperidol or chlorpromazine) in an attempt to meet the requirements of a particular species of animal. So far the usefulness of these drugs has been demonstrated particularly in dogs, cats and lower primates, although much more information is still needed in this field.

The barbiturates may be given in low dosage to produce sedation or narcosis prior to the induction of anaesthesia, but they are mainly used as general anaesthetic agents. Drugs such as chloral hydrate, chloralose, ethyl carbamate (urethane) and bromethol may also be used to produce narcosis and in some cases general anaesthesia. The three latter agents are not generally suitable for periods of anaesthesia followed by recovery but are useful in non-recovery experiments involving the pharmacological investigation of drugs.

Another group of compounds that has aroused interest is the cyclohexamine group, of which phencyclidine has been found to be extremely useful in the lower primates and also in the pig. More recently investigations with another compound in this series, ketamine, have shown promising results in man and in some laboratory animals. These drugs can be used to produce a state of catalepsy (in which the animal remains motionless and relaxed in any position in which it is placed, although it is not anaesthetized)

upon which it is possible to superimpose general anaesthesia, for example, by the administration of a mixture of 50 per cent nitrous oxide and 50 per cent oxygen. More recently a thiazine compound, xylazine hydrochloride, has been shown to be an effective sedative in cattle, horses and certain laboratory animals. Other drugs have been developed for the hypnosis and sedation of birds and fish. Of these, metomidate has been shown to be an effective hypnotic when given by intramuscular injection to birds, and propoxate effectively quietens and reduces the oxygen requirements of fish, when added to the water.

The usefulness of premedication techniques in laboratory animals will vary according to the ease with which individual animals or species can be handled and injections given, the procedure to be carried out and the facilities available.

General anaesthesia

General anaesthetics cause a depression of the cells of the central nervous system resulting in a reversible loss of consciousness, together with variable degrees of analgesia and muscular relaxation that will allow surgical interventions to be carried out. General anaesthetic drugs may be gases or volatile liquids that are administered by inhalation, or non-volatile drugs that are given by intravenous or intraperitoneal injection. The difference between the concentration of an anaesthetic drug that will depress the activity of the cerebrum and that which produces paralysis of the medullary centres gives an indication of safety of that agent. The speed of onset, depth, and duration of anaesthesia will depend upon the particular drug used, the dosage, and the route by which it is administered.

It is important to be able to assess the depth of anaesthesia and, although there are no clear-cut lines of demarcation between them, it is customary to describe four stages of anaesthesia: induction; narcosis; anaesthesia; and bulbar paralysis. The induction stage covers the transition from consciousness to the stage of narcosis. During this period, voluntary excitement may be seen. In narcosis the animal has lost consciousness but may show exaggerated reflexes in response to stimuli. The third stage, that of surgical anaesthesia, may be divided into planes (in man three planes are usually described) of light, medium and deep anaesthesia. As anaesthesia progressively deepens, so the response to external stimuli becomes sluggish and disappears; the respiratory rate may increase slightly, whilst the tidal volume decreases; and muscular relaxation becomes more profound. Observation of these responses and of movements of the eyeball may be used to assess the plane of anaesthesia, although it must be remembered that the responses will vary according to the species of animal and the anaesthetic agent employed.

The choice of anaesthetic will depend upon a number of factors including the operative procedure or investigation to be performed, the species of animal, the condition of the patient, and the laboratory conditions and facilities available. For certain operative procedures, where the animal is not allowed to recover, suitable anaesthesia may be obtained by using a drug such as chloralose, or by repeated injections of barbiturates. On the other hand, where recovery is required such techniques are contraindicated and inhalational drugs are preferable. For operations around the head or neck an endotracheal tube is essential, and for open chest surgery some form of positive pressure ventilation and a closed anaesthetic circuit are necessary. In pharmacological investigations of new drugs it is, of course, necessary to utilize an anaesthetic agent with known pharmacological effects, and a technique that is readily reproducible. The use of certain electrical equipment will preclude the use of inflammable or explosive agents.

The size of the patient and its temperament will vary according to species. Intravenous injection is feasible only in animals with reasonably large and readily accessible veins, and the ability to apply a face mask, with or without prior sedation, will be an important consideration in the use of an inhalational agent for induction. The idiosyncrasies of a species to a particular drug must also be considered.

Whilst it is illogical to utilize sick or diseased animals for experimental purposes, on certain occasions it will be necessary to anaesthetize animals that are not in good health. Toxaemic conditions will increase the susceptibility of animals to barbiturates, and disease of the liver or kidneys will interfere with the ability of the animal to detoxify or excrete certain drugs and will therefore affect the dosage employed. Disease of the respiratory system will be an important consideration when inhalational agents are used.

The equipment and assistance available in a laboratory will influence the choice of anaesthetic technique. For inhalational agents a minimal amount of equipment is essential, and for some species of animal a certain amount of assistance will be necessary. Whilst it is usual to consider the cost of an anaesthetic agent, in most experimental procedures the cost in skilled man-hours involved and the equipment utilized will in the majority of instances far outweigh the cost of any anaesthetic drug.

Non-volatile agents

These agents may be administered, usually in aqueous solution, by either intravenous injection, intra-peritoneal injection, rectal injection, or by mouth. The intravenous route is to be preferred in most instances, since it is possible to keep direct control of the amount of drug given and of the depth of anaesthesia induced. The induction time is also far more rapid when the drug is given by the intravenous rather than the oral route, but in some species of laboratory animals intravenous administration may be difficult or impossible. It is usually possible to use the intraperitoneal route, although occasionally the rectal route may be adopted.

For intravenous injections it is essential to use a sharp needle of a gauge related to the size of vein in the recipient animal. Where repeated injections will be necessary it is advisable to use an indwelling intravenous needle (such as the Mitchell needle) or cannula. When very small animals are anaesthetized it is preferable to use a very dilute solution so that strict control can be kept of the quantity administered. The drug should be given slowly so that the depth of anaesthesia can be assessed by the elimination of reflex responses and by close observation of respiration.

Drugs such as chloral hydrate, chloralose, ethyl carbamate (urethane) and bromethol may be used to produce general anaesthesia, but owing to the slow and protracted recovery from anaesthesia these agents are used normally only for non-recovery procedures. They are widely used in the investigation of new drugs in pharmacology laboratories, and chloral hydrate is occasionally used to produce anaesthesia in horses and cattle. Chloral hydrate is available in crystalline form and is usually prepared as a 10 per cent solution in water which is irritant if injected perivascularly. Chloralose is prepared by heating a mixture of anhydrous glucose and anhydrous chloral. For injection it is made up as a 1 per cent solution by boiling in 0·9 per cent sodium chloride solution. The solution is administered intraperitoneally at a temperature of 30–40°C before the chloralose comes out of solution. It has been suggested that chloralose produces narcosis with hypnosis rather than true anaesthesia. Urethane is usually made up as a 25 per cent solution in water and given as either an intravenous or an intraperitoneal injection. Bromethol is soluble in water to 3·5 per cent at 40°C, but may also be used as a 10 per cent solution in amylene hydrate (Avertin), these preparations usually being given by rectal injection. These compounds have been largely replaced by the barbiturates for most species.

The most widely used intravenous anaesthetics are the sodium salts of the short-acting barbiturates—principally thiopentone, but also pentobarbitone, thialbarbitone, thiamylal and methohexitone. The agents are fairly stable in solution but, apart from pentobarbitone, solutions are normally freshly prepared. Thiopentone is irritant in solution and should not be used in strengths above 2·5 per cent. It is probably the most frequently used of the barbiturates, and is normally the drug of choice to induce anaesthesia prior to maintenance with an inhalational agent. Pentobarbitone is a useful non-irritant agent for the anaesthesia of small laboratory animals particularly for non-recovery experiments. Methohexitone has a limited usefulness in that it produces anaesthesia of only a few minutes duration. It is non-irritant, however, and it is the drug of choice for the induction of anaesthesia in ruminants during the first few weeks of life.

Other, non-barbiturate, intravenous drugs have been investigated, but these have a very limited application in laboratory animal anaesthesia. They include hydroxydione, a steroidal compound which is extremely irritant and

produces venous thrombosis, and propanidid, an ultra short-acting agent that is used to produce very short periods of anaesthesia in out-patient clinics and dentistry. More recently a mixture of the steroids alphaxolone and alphadolone acetate (Saffan) has been shown to be a very effective anaesthetic in cats and primates. It may also prove to be of use in laboratory rodents. This agent is non-irritant and may be given by either the intravenous or intramuscular route to produce anaesthesia of between 10 and 12 minutes' duration.

Inhalational agents

The advantages of inhalational agents lie in the fact that anaesthesia is rapidly induced and rapidly regulated or reversed. However, their use requires constant supervision and the availability of a certain amount of equipment. If a gas is to be used it is normally available under pressure in a cylinder, in either the gaseous or liquid state. It is therefore necessary to have a reducing-valve to bring the pressure down to a manageable level, and to have some form of flow-meter to control the flow of gas. If a volatile liquid is to be used some form of vaporizer is necessary and, whilst there are a number of types avail-

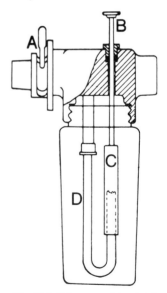

Fig. 13.1. Ether vaporizing bottle
A. Inlet bypass control valve
B. Cowl control
C. Cowl
D. 'U' tube.

able, the most commonly used is the Boyle's bottle. This type of vaporizer (Fig. 13.1) may be used for ether, chloroform, trichlorethylene and halothane. In the case of ether, the vaporizing gas or gases enter the bottle through a by-pass valve at A. It is possible by means of this valve to regulate the amount of gas passing through the bottle. The gases pass down the 'U' tube (D) and emerge from the lower end of the cowl (C), the position of which may be adjusted by moving the rod (B). The operator is thereby enabled either to bubble the gases through the ether or to pass them through the bottle above the fluid level. Using the controls (A) and (B), any required concentration of ether vapour may be obtained. Vaporizers for chloroform and trichlorethylene are essentially the same, whereas for halothane the diameter of the bottle is smaller and the gases are never bubbled through the liquid anaesthetic. The sophisticated forms of anaesthetic machine incorporate methods for the supply and control of anaesthetic gases and the vaporization of volatile liquids.

Administration of an inhalational agent to the animal requires the use of a mask or endotracheal tube, together with suitable tubing to connect the component parts of the equipment. It is usual to incorporate a rebreathing bag into the circuit, and if a closed circuit is used a soda lime canister is also necessary to absorb the exhaled carbon dioxide. The basic equipment for the supply and administration of inhalation anaesthesia are illustrated in Figure 13.2. In experiments of long duration, or where open-chest surgery is performed, equipment for positive pressure ventilation of the lungs will be required. This may take the form of a Starling pump or a specially designed anaesthetic ventilator, but in any event the output of the ventilator should be related to the minute volume requirements of the patient.

The most important aspect of any anaesthetic technique is to ensure adequate supplies of oxygen. Whilst an animal may be left to breathe air either by spontaneous respiration or via a ventilator, it is preferable to supply pure oxygen. Oxygen is available under pressure in cylinders of varying size, to which it is usual to attach a pressure-gauge (to indicate the content) and a reducing-valve, the supply of gas to the patient being controlled by a flow-meter. The minimum requirement of oxygen in any mixture is 20 per cent but it is preferable to use a concentration of 25 per cent or more.

Nitrous oxide is a gas with analgesic properties but only a weak narcotic effect, as a result of which it is normally used to supplement

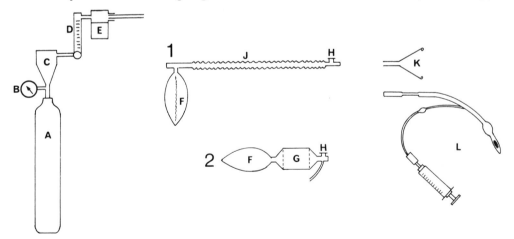

Fig. 13.2. Basic components of equipment for inhalational anaesthesia.
1. Magill attachment to face mask (K) or endotracheal tube (L).
2. Closed to and fro circuit with carbon dioxide absorption for attachment to endotracheal tube.

A. Oxygen cylinder	E. Vaporizer	J. Corrugated tubing
B. Contents gauge	F. Reservoir bag	K. Face mask
C. Reducing valve	G. Soda lime canister	L. Cuffed endotracheal tube
D. Flow meter	H. Expiratory valve	

some other anaesthetic drug. It is available under pressure in cylinders in a liquid phase, and a reducing-valve and flow-meter are used in a similar manner to that adopted for oxygen cylinders. It is usual to mix nitrous oxide with oxygen (never less than 20 per cent oxygen), which may be administered directly to the patient or used to vaporize a volatile anaesthetic liquid.

Cyclopropane is a gas with potent anaesthetic properties, but since it is highly inflammable, and because mixtures of the gas with air or oxygen are explosive, it is not widely used nowadays. This gas is available in cylinders in a liquid form under low pressure, and no pressure-reducing valve is required. It is normally administered as a 20–25 per cent mixture with oxygen but, as the level of anaesthesia is very easily altered, the use of this gas requires constant supervision.

Of the volatile liquids, chloroform has been widely used for more than 100 years. It is a potent anaesthetic agent, but because of its toxic effect on the myocardium, liver and kidneys when administered for any length of time it is not widely used in animal anaesthesia. If it is to be used, it is advisable to use oxygen as the vaporizing gas and to restrict the administration to a short period of time.

Diethyl ether is probably the volatile agent that is most commonly used in laboratories. It has a low toxicity but, owing to its high inflammability and the fact that it forms explosive mixtures with air or oxygen, its use is precluded when electrical equipment or naked flames are present. This drug also has an irritant action on mucous surfaces, with a consequent increase in mucus secretion which may be reduced by premedication with atropine. Ether is widely used for the anaesthesia of laboratory rodents; a flow of oxygen is used to vaporize the drug and a simple mask or funnel for administration. It may also be used in a closed anaesthetic circuit.

Ethyl chloride is another liquid with anaesthetic properties. It is non-irritant to the respiratory mucosa and induction of anaesthesia is rapid. This drug is essentially a single-dose anaesthetic; over-dosage occurs easily and circulatory failure is not uncommon during induction. Ethyl chloride is used to induce short

periods of anaesthesia in laboratory animals by placing them in a closed glass jar containing a vapour concentration of 20 per cent of the drug. The animal is removed from the container before respiration fails, and anaesthesia lasts for a further two to five minutes.

Trichlorethylene is an anaesthetic liquid which resembles, but is much less potent and less toxic than, chloroform. It is non-irritant, non-inflammable and non-explosive, and induction is smooth provided that the drug is vaporized in a flow of oxygen. It has good analgesic properties, but if it is used to produce full muscular relaxation cardiac irregularities may be seen. It is usually administered in a semi-closed anaesthetic circuit with a mixture of nitrous oxide and oxygen but, owing to the formation of toxic products in the presence of soda lime, it cannot be used in closed circuits.

The halogenated hydrocarbon, halothane, has become the most widely used inhalational anaesthetic agent in recent years. This drug is non-inflammable and non-explosive, and it may be used safely with soda lime. It is a potent anaesthetic with low toxicity and a wide margin of safety. It is a liquid that is readily vaporized and is normally used in a 2–4 per cent mixture with oxygen for induction, and in a 0·5–2 per cent mixture for maintenance of anaesthesia. Halothane may be administered by a face mask or via an endotracheal tube in semi-closed, or closed, anaesthetic circuits. Probably its main disadvantage is that it produces a moderate degree of hypotension, and cardiac irregularities have been recorded in some animals during its use. It can however be used effectively and with a high degree of safety in all animal species.

Methoxyflurane is another halogenated hydrocarbon that has been less widely used in animal anaesthesia. This drug also is neither inflammable nor explosive and it can be used in conjunction with soda lime. It is less readily vaporized than halothane and it is, moreover, difficult to obtain concentrations of more than 3·5 per cent at normal temperatures. Consequently methoxyflurane is not satisfactory for induction, particularly in large animals. However, following induction with a short-acting barbiturate, methoxyflurane is a very safe agent with which to maintain anaesthesia.

Muscle relaxants

All anaesthetics are central nervous system depressants and, if administered to a level to produce adequate muscular relaxation, there are dangers of respiratory or cardiac depression and of toxicity. To overcome this difficulty, modern anaesthesia has been developed so that the anaesthetic agent is used to produce loss of consciousness and analgesia, and muscular relaxation is obtained by the administration of a drug that prevents the normal transmission of a nerve impulse at the neuromuscular junction. Such drugs are termed muscle relaxants.

Normal neuromuscular transmission is effected by the release, at the neuromuscular junction, of acetylcholine which reacts with a receptor site on the muscle fibre and produces a depolarization of the surface of the fibre followed by a contraction. Any drug that interferes with this action will produce neuromuscular blockade, which will be manifest by general muscular relaxation. There are two main types of clinically effective muscular relaxant drugs.

Firstly, there are the drugs that have a curariform action (e.g. d-tubocurarine, gallamine) and which compete with acetylcholine for the receptor sites, subsequently blocking the access of acetylcholine to the sites. There is no depolarization or muscle contraction and the duration of the block depends upon the dose. Such a block may be termed a competitive or non-depolarizing block and can be reversed by the administration of an anticholinesterase drug, such as neostigmine.

Secondly, drugs such as suxamethonium and decamethonium also compete with acetylcholine for the receptor sites; they produce depolarization and muscular contraction. Such a block is termed a depolarizing block and is not reversed by the administration of neostigmine. Generally speaking the block produced by non-depolarizing agents is of longer duration (20–40 minutes) than that produced by depolarizing agents (3–10 minutes), but the effect of these drugs will vary according to the species of animal in which they are used.

There is a third group of drugs that produces muscular relaxation by a central action, whereby reflex activity is diminished by depressing conduction across the synapses of the internuncial neurones in the spinal cord. Drugs in this group are mephenesin and the guaiacol glycerol ethers, but they have a very limited use.

It cannot be emphasized too strongly that these agents *do not produce either anaesthesia or analgesia* and, if it is not adequately anaesthetized and can still feel pain, the animal will not be able to respond to the painful stimulus in a normal manner because of the general muscular relaxation. It is essential that unconsciousness and analgesia should be maintained throughout the operation. Muscle relaxants are useful as part of the anaesthetic technique for poor-risk patients, but they should be used only by skilled anaesthetists. There would seem, however, to be little use for these drugs in the surgery of laboratory animals.

Electrical anaesthesia

The passage of an alternating current across the brain in order to produce anaesthesia has been reported in a variety of species. The effectiveness of this technique has not yet been proven to the satisfaction of many anaesthetists and, from the present state of knowledge, this form of anaesthesia cannot be recommended for general use.

Dangers associated with anaesthesia

Apart from the dangers associated with the use of muscle relaxants, which have been emphasized already, there are a number of other points that need to be stressed. It is most important that respiration should be closely observed throughout anaesthesia and, in many cases, monitoring of cardiac function, with an electrocardiograph or a pulse monitor, is warranted.

If a long-acting barbiturate is being used administration should extend over at least 4–5 minutes, and close attention should be paid to the loss of reflexes. When irritant solutions are being administered it is essential to have the needle correctly positioned within the vein, and for intraperitoneal injections care should be taken to avoid injection into the abdominal viscera.

When positive pressure ventilation is employed, the respiratory rate and tidal volume must be adjusted to meet the requirements of the particular patient; over-ventilation can be almost as undesirable as under-ventilation. Every endeavour should be made to avoid respiratory or circulatory embarrassment due to the positioning of the patient, or by pressure on the animal by the operator, instruments or recording equipment.

Other dangers are associated with overdosage of a particular anaesthetic agent, or with circulatory collapse during anaesthesia. The first sign of overdosage with most anaesthetic agents is usually respiratory depression. Overdosage by inhalational agents can usually be reversed rapidly by terminating the supply of anaesthetic and ventilating with pure oxygen. In the case of intravenous anaesthetics it may be possible to administer an antidote. For barbiturate overdosage, bemegride sodium can be given intravenously, and other analeptic drugs such as leptazol, picrotoxin and nikethamide may also be used in cases of respiratory depression. The specific morphine antagonists (e.g. nalorphine,

cyprenorphine) may be given by intravenous injection to counter the respiratory depression produced by overdosage with morphine or morphine-like drugs, and the neuromuscular block produced by a non-depolarizing relaxant drug may be reversed by the administration of neostigmine.

Any blockage of the airway must be cleared immediately, and when this is caused by blood, mucus or vomitus, suction is essential. It must also be remembered that cuffed endotracheal tubes can be occluded by kinking during repositioning of the patient. Cardiac arrest usually results from hypoxia and the treatment of this emergency should be by both ventilation and cardiac massage.

Care should be taken to avoid heat-loss both during and whilst recovering from anaesthesia. Similarly any blood-loss should be replaced by whole blood or dextran solutions. In long surgical procedures it is desirable to monitor the acid-base and electrolyte balance of the animal and to use replacement therapy where necessary.

Finally, it is most important to stress the

Glossary of drugs referred to in the text

Approved or Pharmacopoeia name and salt	Trade name	Supplier in U.K.
Acepromazine maleate	Acetylpromazine	Boots
	Notensil	Fisons
Alphaxolone and alphadolone acetate	Saffan	Glaxo
Atropine sulphate B.P.		
Bemegride sodium	Megimide	Nicholas
Bromethol (tribromethanol)	Avertin	Bayer
Chloral hydrate	Chloralhydrate	B.D.H.
Chloralose	Chloralose	B.D.H.
Chloroform	Anaesthetic Chloroform	May & Baker
Chlorpromazine hydrochloride	Largactil	May & Baker
Cocaine hydrochloride B.P.		
Cyprenorphine hydrochloride	Revivon	Reckitt & Colman
Dexamethonium iodide		Koch-Light Labs.
Diethyl-thiambutene hydrochloride	Themalon	Burroughs Wellcome
Droperidol	Dehydrobenzperidol	Janssen Pharmaceutica
Ether (anaesthetic diethyl ether)	Anaesthetic ether	May & Baker
Ethyl chloride	Ethyl chloride	Bengué & Co.
Etorphine hydrochloride	M99	Reckitt & Colman
Etorphine hydrochloride and acepromazine maleate	Immobilon	Reckitt & Colman
Fentanyl dihydrogen citrate	Fentanyl	Janssen Pharmaceutica
Gallamine triethiodide	Flaxedil	May & Baker
Haloperidol	Haloperidol	Janssen Pharmaceutica
Halothane	Fluothane	I.C.I.

Approved or Pharmacopoeia name and salt	Trade name	Supplier in U.K.
Hydroxydione	Viadril	Pfizer
Ketamine hydrochloride	Ketalar	Parke Davis
Leptazol	Leptazol B.P.	B.D.H.
Lignocaine hydrochloride	Xylocaine	Astra Chemicals
	Xylotox; Lidothesin	Willows Francis
	Lignostab	Boots
Mephenesin	Myanesin	B.D.H.
Methohexitone sodium	Brietal Sodium	Lilly
Methoxyflurane	Penthrane	Abbott
Metomidate hydrochloride	R7315	Janssen Pharmaceutica
Morphine hydrochloride B.P.		
Nalorphine hydrobromide	Lethidrone	Burroughs Wellcome
Neostigmine methyl sulphate	Prostigmine	Roche
Nikethamide	Coramine	Ciba
Pentobarbitone sodium	Nembutal	Abbott
	Sagatal	May & Baker
Pethidine hydrochloride	Pethidine	Roche
Phencyclidine hydrochloride	Sernylan	Parke Davis
Phenoperidine hydrochloride	Phenoperidine	Janssen Pharmaceutica
Picrotoxin	Picrotoxin	Koch-Light Labs.
Procaine hydrochloride	Planocaine	May & Baker
Promazine hydrochloride	Sparine	Wyeth
Promethazine hydrochloride	Phenergan	May & Baker
Propanidid	Epontol	Bayer
Propoxate	R7464	Janssen Pharmaceutica
Suxamethonium chloride	Scoline	Allen & Hanbury
	Anectine	Burroughs Wellcome
Thialbarbitone sodium	Kemithal	I.C.I.
Thiamylal sodium	Surital	Parke Davis
Thiopentone sodium	Pentothal	Abbott
	Intraval	May & Baker
Trichlorethylene	Trilene	I.C.I.
Trimeprazine tartrate	Vallergan	May & Baker
d-tubocurarine chloride	Tubarine	Burroughs Wellcome
Urethane (ethyl carbamate)	Ethyl carbamate	B.D.H.
Xylazine hydrochloride	Rompun	Bayer

importance of post-anaesthetic care. This is particularly important following major surgical interferences, when the post-operative care is of major significance in the recovery of the animal.

REFERENCES

Croft, P. G. (1964). *An Introduction to the Anaesthesia of Laboratory Animals*, 2nd ed. London: UFAW.

Hall, L. W. (1971). *Veterinary Anaesthesia and Analgesia*, 7th ed. London: Baillière, Tindall & Cassell.

Lumb, W. V. & Jones, E. W. (1973). *Veterinary Anaesthesia*. Philadelphia: Lea & Febiger.

Sawyer, D. C. (1965). *Experimental animal anaesthesiology*. Proceedings of Symposium held at USAF School of Aerospace Medicine Texas: USAF School of Aerospace Medicine.

Soma, L. R. (1971). *Veterinary Anaesthesia*. Baltimore: Williams & Wilkins.

Westhues, M. & Fritsch, R. (1961). *Die Narkose der Tiere*, vols I & II. Berlin & Hamburg: Paul Parey.

14   Euthanasia

W. N. Scott and P. M. Ray

Introduction

The word euthanasia means a painless death. This is obviously not achieved if animals struggle or writhe in terror during the operation, or regain consciousness after mutilation. True euthanasia depends on the rapidity with which unconsciousness is achieved and the maintenance of this state until death occurs. Even humane methods may cause suffering if used by nervous or unskilled individuals. Those willing, and of suitable temperament, to undertake the task should first practise on dead animals, preferably under the supervision of an experienced person. Young people should not be allowed to kill animals.

Techniques of killing vary according to the species and the purpose for which the animal is being killed. The more sophisticated methods used for domestic animals are seldom practicable in the case of wild animals or animals unaccustomed to confinement or handling by man. When an animal is killed for food its flesh must be wholesome and fit for consumption, so lethal chemicals which might leave undesirable or harmful residues in the carcase cannot be used; the gas carbon dioxide is the only chemical suitable for such use. The animal must also be properly bled to ensure the keeping quality of the carcase; this is achieved by cutting the large blood vessels in the neck after the animal is unconscious.

Euthanasia of laboratory animals may be necessary because of prolonged or severe pain resulting from an experiment, or it may be an integral part of an acute experiment, or it may be essential because the animal has had an accident, is diseased or is surplus to requirements. Techniques of killing vary according to the species and the purpose for which the animal is being killed. For example, substances such as pentobarbitone sodium, which have an effect on the foetus, are contra-indicated when carrying out hysterectomies for the purpose of producing SPF or gnotobiotic animals. Physical methods of killing involve considerable bruising and haemorrhage at the site of the injury and for this reason may be unsuitable in certain cases.

Physical methods include shooting, stunning and breaking the neck. Chemical methods include the use of such substances as carbon dioxide, and the barbiturate pentobarbitone sodium.

Physical methods of killing when properly carried out are quick, humane and practicable and in certain circumstances provide the only means possible. They are, however, often repellent to sensitive people and should not be undertaken except with deliberation and purpose.

Physical methods

Dislocation of the neck, decapitation and stunning

These methods should be used only to kill animals that can be handled easily and have relatively thin skulls. Some manual dexterity is essential but once the operator is familiar with the technique apprehension on the part of the animal is minimal, death is quick and suffering slight. At the same time it is recognised that many people cannot steel themselves to these procedures and prefer to use other methods. Human feelings, however, should as far as possible not be allowed to influence the use of the most humane techniques. Although these may be distasteful to the operator they are often less distressing to the animal than more complicated procedures involving chemicals and other means.

Poultry.

(a)   The usual method is by dislocation of the neck. The bird is held by the shanks with the left hand, and kept close to the hips of the operator. The head is held immediately behind the skull between the first two fingers of the right hand, with the thumb under the lower beak. The neck is fully extended by moving the right hand downwards; then by means of a strong downward thrust at the same time pulling the bird's head backwards over the neck, the latter is dislocated at the junction of the cervical region and the occipital part of the skull.

(b) The bird may be stunned by damaging the brain. An effective stunner (Fig. 14.1) for this purpose can be made thus: a piece of wood 40 cm long × 5 cm wide × 1 cm thick is required; of this 13 cm is shaped into a handle; a metal bar 28 cm long × 0·5 cm wide × 1 cm thick is nailed to each edge of the remaining 28 cm of wood. The bird is held up with the hand by the legs, tail and the long feathers of the wings and given a very sharp rap at the back of the head (fowls just behind the comb) with the edge of the stunner.

Fig. 14.1. Poultry stunner.

(c) The head may be cut off with a heavy and sharp chopper. An assistant should hold the bird by its legs, wings and breast, placing the head and neck sideways on a block with the bird facing away from the operator.

(d) Burdizzo forceps or bloodless castrators may be used for killing large fowls such as old cocks, turkeys, geese, etc. They are applied immediately behind the head. When the forceps are closed the jaws dislocate the vertebrae and cut the spinal cord and jugular vein. The skin remains intact.

After stunning or killing poultry by any of the methods described the throat should be cut diagonally as near to the head as possible.

Convulsive movements in poultry follow all methods of killing; decapitation usually results in a spray of blood.

Small Birds.

(a) Holding the bird with the left hand, take the neck between the thumb and forefinger knuckle of the left hand. Apply the thumb and forefinger knuckle of the right hand to the neck, close to those of the left hand. Grip hard with both hands simultaneously and give a sharp twist; the head should come off immediately.

(b) Hold the body in the left hand and with a swinging motion strike the head smartly on a hard object. Add a second blow for safety's sake. This method is suitable for birds up to the size of a pheasant.

Rabbits.

(a) Hold the hind legs in the left hand as in Figure 14.2. With the fingers of the right hand extended and rigid, use the heel of the hand—the back edge as in Karate—and strike the back of the rabbit's neck

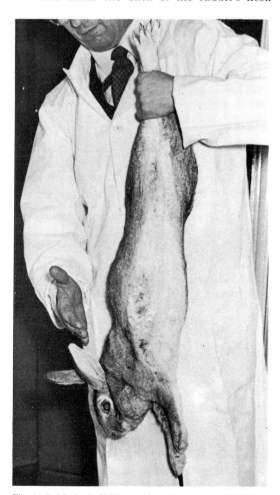

Fig. 14.2. Method of killing rabbit. (Photograph: UFAW.)

violently downwards. Instead of the hand a heavy stick or a poultry stunner may be used.

(b) Place the rabbit on a table; with the left hand lift it by the ears until its forepaws are just clear of the table, then hit it a hard blow behind the ears with a poultry stunner.

Guinea-Pigs. These may be killed either by preliminary stunning followed by cutting the throat or by dislocation of the neck. These methods are described in detail in Chapter 19.

Laboratory Rats. If the animal can be handled, place it on a duster and wrap the duster firmly round the body, including both front legs; kill by concussion thus:

(a) Hold head downwards and strike very hard behind the ears with a stout wooden stick or poultry stunner.

(b) Hold firmly, belly upwards, and strike the back of the head *very hard* against a hard horizontal surface such as a sink or a table.

Laboratory Mice. Dislocation of the neck is a simple and humane method of killing mice (Fig. 14.3). The animal is held by its tail and placed on a surface that it can grip, when it will stretch itself out so that a pencil or similar object can be placed firmly across the neck. A sharp pull on the tail will then dislocate the neck and kill at once.

Alternatively, the animal may be held as already described and hit hard behind the ears with a stunner (see rat, above).

Reptiles and Amphibia. These should be killed by a sharp blow on the head or by decapitation with bone forceps, secateurs or a sharp pair of scissors.

Fish. Small fish under 250 g in weight are best killed by holding the fish in the left hand, placing the right thumb in the mouth and the right forefinger at the junction of the head and neck, and bending the head sharply until a crack is felt.

Larger fish are killed by striking the back of the head with a stone or heavy metal object. Eels may be decapitated with a sharp knife.

Electrocution

Electrocution can be either painless or very painful and both skill and technical knowledge are essential for its proper use. If electrocution is to be humane, it is essential that a sufficiently large alternating current shall pass through the *brain* of the animal; methods which kill by passing current through the heart but not the brain are inhumane.

Shooting

General Considerations. A humane-killer (for which a police permit is required) should be used if possible. There are two types: those which fire a free bullet and those in which a blank cartridge propels a captive bolt; the latter type is the safer to operate but is much heavier than the former and is not suitable for horses. Particulars of humane-killers available in England and overseas and directions for their use, are obtainable from UFAW.

In all cases the instrument must be fired with the muzzle placed firmly against the animal's head, pointing in the required direction. It is essential that the shot should penetrate the brain, which is of comparatively small size and situated in the upper part of the skull; adequate restraint of the animal is therefore necessary. Stunning with the humane-killer destroys the brain and results in immediate unconsciousness. The animal will still be alive, however, in the

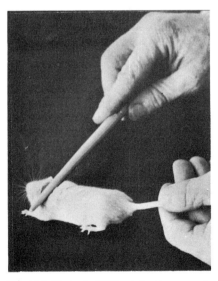

Fig. 14.3. Method of killing laboratory mouse. (Photograph: UFAW.)

sense that the heart will continue to beat until weakened by haemorrhage.

In the case of cats and average-sized dogs the initial head wound involves some of the larger blood-vessels and a considerable amount of bleeding will occur. After stunning larger animals the rate of bleeding should be accelerated by cutting through the large blood vessels in the neck with a sharp knife. This increased loss of blood will then stop the action of the heart more quickly.

Dogs: captive bolt humane-killer. If the dog is nervous it may be necessary to have it muzzled; this is done by making a loop of a strong piece of bandage or soft cord. The loop is slid over the middle of the nose and the ends drawn up gently but firmly on top. The ends are then firmly knotted twice under the chin and brought together and tied behind the neck (Fig. 14.4A). The lead should be fastened to a solid object.

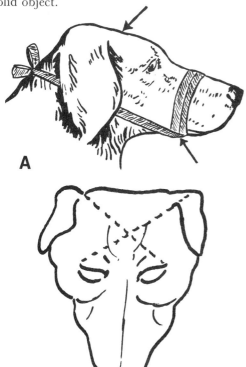

Fig. 14.4. Method of killing dog with humane-killer
  A. Method of applying muzzle.
  B. Target area—1 cm above the point where the dotted lines cross and a little to one side.

The dog should be either sitting or standing but not lying down. Stand on the dog's right side and a little in front of it. Steady the head by holding the scruff of the neck with the left hand, but see that the hand is at least 10 cm away from the pistol muzzle piece. With the right hand place the muzzle of the loaded pistol firmly on the dog's head midway between the level of the eyes and the base of the ears, but a little to one side so as to miss the bony ridge which runs down the middle of the forehead (Fig. 14.4B). Aim slightly across the dog and towards the spine. If the animal is restless wait patiently for a quiet interval before squeezing the trigger and firing.

Dogs: free bullet humane-killer. Take the dog out of doors to soft ground; the background of the line of fire should be earth or some substance in which the bullet will bury itself. Make sure that nobody is standing in the line of fire.

Steady the dog's head gently with the left hand clear of the line of fire. Place the muzzle of the loaded pistol flat on the target area as already described. Aim slightly across the dog towards the spine so that the bullet will travel partly on the side of the body opposite to the side on which it entered the head. Wait until the animal is quiet, then squeeze the trigger.

Ungulates. Figure 14.5 illustrates the method by which each species should be shot in order to produce complete and immediate unconsciousness. The black spots show where the instrument should be placed, and the arrows show the direction in which it should be fired. It is important that the muzzle should be held firmly but gently against the animal's head.

If the carcase is intended for human consumption it must be bled immediately after shooting. This is done by suspending it by the hind legs and cutting through the large blood vessels in the neck.

In all animals there may be convulsive movements after shooting, but if the eyes are fixed, dull, wide open and insensitive to touch the animal will be unconscious.

Until weakened by haemorrhage the heart beat may continue for some minutes after shooting. If it continues too long or rebreathing starts shoot again and cut the main blood vessels in the neck to encourage bleeding. For the easy removal of blood after slaughter it is best to allow the animal to bleed near a drain or on a hard floor sloping towards a gutter. The blood is easily washed away with *cold* water and a scrubbing brush. When this is not possible the animal should be bled over an armful of straw or a heap of loose earth which, after absorbing the blood, can be burnt or buried.

Cattle. Using either a captive bolt or free bullet humane-killer, place the muzzle of the gun firmly and at right angles to the head (Fig.

14.5A); extra care is required if the animal's head is tied close to the ground.

When shooting bulls or old and hard-headed beasts, place the muzzle very firmly 1 cm to the side of the ridge that runs down the centre of the face. A humane-killer firing a free bullet is the most efficient way of killing old and hard-headed animals. Never fire while the animal is moving its head; wait patiently for a quiet interval before squeezing the trigger.

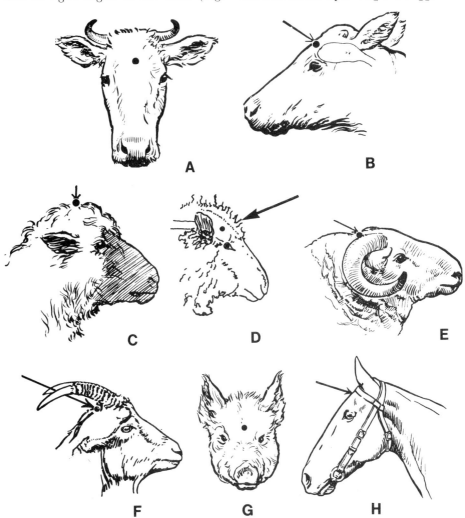

Fig. 14.5. Methods of killing ungulates with a humane-killer. The black spots show where the instrument should be placed; the arrows show the direction in which it should be fired.

A. Cattle
B. Calves
C. Hornless sheep (captive bolt)
D. Hornless sheep (free bullet)
E. Horned sheep
F. Goats
G. Pigs
H. Horses

Immediately after firing, a cane, or piece of strong wire, should be inserted into the hole made by the bolt or bullet and moved vigorously backwards and forwards to ensure complete destruction of the nerve centres before the animal is bled.

Hornless Sheep and Rams. If using a captive bolt humane-killer raise the animal's muzzle gently by hand and then place the instrument firmly against the tip of the head as shown in Figure 14.5c and aim in the direction indicated by the arrow towards the gullet.

If using a free bullet humane-killer place the muzzle of the pistol in the hollow (the spot where the horn grows; the hollow may be felt for with a finger) on the upper and right side of the face about equal distance from the eye and the ear as shown by the circle at point of arrow (Fig. 14.5D). Steady the head of the sheep by holding the nose, *not the neck*; point the instrument towards the animal's left shoulder and fire.

Horned Sheep and Rams. If using a captive bolt humane-killer place the muzzle just behind the middle of the ridge which runs between the horns and aim towards the gullet (Fig. 14.5E). When shooting rams or very heavy sheep, always use the strongest cartridges for the captive bolt.

If using a free bullet humane-killer shoot just above the eyes, pointing towards the spine and steady the head by holding the nose.

All sheep are best held by an assistant, who should be out of the line of fire when a free bullet is used. The practice of steadying sheep by a stick between their hind legs is cruel and should not be followed.

Goats. Using either a captive bolt or a free bullet humane-killer, place the muzzle of the instrument behind the horns and aim towards the animal's mouth (Fig. 14.5F). When using a free bullet take care that no one is in the line of fire.

In the case of kids, shoot from the front as for calves.

Pigs. Using either a captive bolt or a free bullet humane-killer place the muzzle of the instrument about a finger's width above the level of the eyes, halfway across the forehead, aiming well up into the head (Fig. 14.5G).

Horses, Mules and Donkeys. A free bullet humane-killer should always be used as captive bolt pistols are not suitable.

A head collar or bridle should be put on the animal and held by an assistant. If the animal is restless it should be blindfolded.

It is essential to shoot above the eyes. The brain is in the upper part of the head and particular care must be taken to place the muzzle high up as shown in Figure 14.5H, and to fire the bullet where the arrow points (halfway across the forehead). If the muzzle of the instrument is sloped place it flat on the forehead; the bullet will then be directed a little upwards. If the muzzle of the instrument is not sloped, tilt it very slightly by lowering the cartridge chamber end so that the bullet will travel in the direction required. Immediately after shooting pith with a rod or wire and cut the main blood vessels.

Chemical methods

UFAW has carried out exhaustive tests on the most suitable chemical methods of euthanasia and the recommendations that follow have been approved by its Scientific Advisory Committee.

Carbon dioxide

Although in the past various volatile agents have been used for the euthanasia of small animals, UFAW firmly recommends carbon dioxide as the most suitable chemical from the point of view of both the animal and the operator.

The gas carbon dioxide is $1\frac{1}{2}$ times heavier than air; it does not burn and will in fact extinguish flames; it has no smell and no colour. It is stored in cylinders as a liquid under pressure, which when slowly released turns into gas. If the outside temperature is very cold and the liquid is quickly released then carbon dioxide ice may be formed around the outlet.

Most cylinders are supplied with a simple locking device which is replaced by a reducing valve before use. A short length of plastic or rubber hose is used to connect the valve outlet with the container or cabinet to which the gas is to be introduced.

For the use of carbon dioxide a special euthanasia cabinet has been developed by UFAW (Fig. 14.8). This is described with details of construction in the Appendix.

Carbon dioxide euthanasia using a plastic bag. The animal is placed in a cage and allowed time to get accustomed to its new environment. In the case of mice, rats, hamsters, rabbits, budgerigars, pigeons and chickens several animals (of the same species) may be placed in the same cage provided there is sufficient floor space and the animals do not fight. Cats (with the exception of young kittens), mink and other small carnivores should be caged singly.

When the animal has settled down the cage is placed inside a clear polythene bag approximately five times the volume of the cage. The bag is pressed close to the sides of the cage and through the mouth of the bag the end of the hose from the carbon dioxide cylinder is introduced (Fig. 14.6). The operator holds the mouth of the

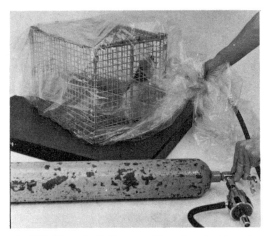

Fig. 14.6. Method of killing rat with carbon dioxide, using a plastic bag.

bag tightly around the hose with one hand and with the other turns the cylinder key and releases a slow flow of gas. The plastic bag will begin to inflate and by the time it is about half full the animal will have become narcotised or unconscious. The time to unconsciousness varies according to the height of the animal's head from the floor of the cage. Mice usually keep their nostrils very close to the ground and become unconscious in 10–20 seconds. Cats may take as long as 2 minutes. When the polythene bag is full of gas, but not under pressure, the hose is withdrawn, the gas turned off and the mouth of the bag sealed by tying with a rubber band or tape. By this time the animal will be unconscious, but it should still be kept under observation until all signs of breathing have

ceased. When this occurs leave the animal in contact with the gas for another 10 minutes, after which open the bag and remove and empty the cage.

If it is necessary to kill small wild animals caught in box-traps the trap can be placed in a polythene bag 5 times the volume of the trap and carbon dioxide introduced as described.

Carbon dioxide euthanasia using UFAW euthanasia cabinet. Place the animal or animals in a cage as already described. Allow them sufficient time to settle down and get accustomed to the surroundings.

Make sure that the dirt tray of the cabinet is firmly inside. Open the sliding lid of the cabinet and introduce the gas delivery hose so that the end rests on the dirt tray. Turn on the gas so that it flows gently; an excessive rush of gas will cause turbulence. As the cabinet fills with carbon dioxide the air is displaced upwards. The level of the gas in the cabinet can be judged by inserting a lighted match or taper; carbon dioxide does not support combustion and the level at which the light is extinguished indicates the height which the gas in the cabinet has reached.

When the level of the gas is within 5 cm of the top of the cabinet carefully place the cage containing the animal inside the cabinet and resting on the dirt tray. Move the sliding lid of the cabinet up against the hose; the gap allows for the escape of air. Watch the animal closely when it becomes unconscious, let the gas flow for another 15 seconds and then turn it off; withdraw the hose and shut tight the sliding lid of the cabinet. Keep the animal under observation until breathing stops and leave the cage in the cabinet for another 10 minutes, after which it may be removed and emptied.

A considerable amount of gas will remain in the cabinet and if more animals are to be killed it is only necessary to top this up to the required level and then continue as described above.

When operations are finished the gas may be emptied from the cabinet by pulling out the dirt tray. If the cabinet is on a table above floor level the gas will sink to the lowest level available and become harmless by dispersal and dilution in the atmosphere.

If the dirt tray becomes fouled with urine and faeces it should be removed and cleaned before any further operations take place. Generally, however, this does not occur when carbon dioxide is used and it is possible to continue a series of operations without having to release the gas already in the cabinet and the amount of gas required from the cylinder is therefore less. When operations have finished for the day the dirt tray should be withdrawn and cleaned and the inside of the cabinet wiped with a cloth soaked in water containing a mild detergent. *Strong-smelling disinfectants should not be used for cleaning.*

Destruction of Birds with Carbon Dioxide.∗

The method usually employed for killing fowls is the dislocation of the cervical vertebrae either by hand or with Burdizzo forceps (page 159). The biggest drawback to this method is the considerable amount of feather loss and production of dust resulting from flapping of the wings. Since dust may well carry disease organisms, particularly when birds are destroyed at the end of an experiment involving a pathogen, any method of destruction which will reduce the amount of dust produced deserves consideration. Carbon dioxide has been suggested as a suitable killing agent since its inhalation causes little or no distress to the bird, it has a reasonably fast action, and it depresses nervous activity. At the instigation of the Universities Federation for Animal Welfare the value of carbon dioxide for the destruction of birds was investigated.

A lidless destruction chamber (Fig. 14.7) was constructed of 12 mm exterior grade plywood reinforced with 2·5 cm × 5·0 cm battens, and mounted on 7·5 cm castors. In order to ensure an even distribution of the gas, a 1·25 cm copper pipe drilled with 3·6 mm holes at 10 cm centres was fitted to the inside of the chamber at levels of 5 cm and 65 cm from the bottom. A cylinder of carbon dioxide with regulating valve (B.I.G. Model 90∗) was attached to the pipe by means of a length of plastic tubing. The birds were placed in a polypropylene crate† with a grid floor, since solid-floored crates offer an obstacle to the diffusion of carbon dioxide through the chamber.

The procedure that has been developed is as follows:—The chamber is filled with carbon

∗ Reproduced with permission from the article by Miss D. M. Cooper in *Vet. Rec.* (1967).

∗ British Industrial Gases Ltd.

† Plasson Agricultural Plastics, Baytree Farm, Manley, Cheshire.

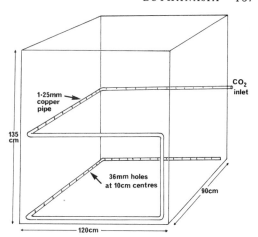

Fig. 14.7. Portable carbon dioxide destruction chamber designed for poultry.

dioxide to within 30 cm of the top. This is achieved by allowing the gas to flow until a lighted match held at this level is extinguished. The crate containing birds up to a total weight of 20 kg is hoisted by block and tackle, and after the chamber has been moved underneath the crate is lowered into the chamber and moved up and down through a distance of 30 to 60 cm to help displace any air pockets between the birds and the feathers. The birds become anaesthetised within 30 to 45 seconds and a very small amount of flapping which may occur is inhibited within a minute. To ensure that all birds are killed before the crate is removed it is necessary to allow five minutes to elapse from the time the crate enters the chamber. When the birds have been in the chamber for two minutes, more carbon dioxide is introduced for 30 to 45 seconds to replace the gas which has been used for destruction. It has been observed that even though the chamber is filled with gas initially, the time taken for destruction of the first crate of birds is at least 50 per cent longer than that for the second and subsequent crates. This may be due to retention of small air pockets in the chamber. On removal of the crate the birds are laid on the floor to allow a check to be made that all are dead.

This technique may be modified for the destruction of single adult birds or small numbers of young chicks by placing the birds in a clear polythene bag (preferably 500 gauge) and then filling the bag with carbon dioxide. A rubber band placed about the neck of the bag prevents the gas from escaping.

It has been found that with chickens of all ages, and adult Japanese quail, the period of exposure to the gas should not be less than five minutes. The time taken to carry out this method of destruction is similar to that involved in the dislocation technique, but the former is less objectionable to poultry attendants who have to carry out the task, especially when large numbers of birds are destroyed, and the virtual elimination of the dust hazard is an added advantage. The cost of this technique is quite small.

Pentobarbitone sodium

Pentobarbitone sodium is a medium-acting barbiturate of general value in both anaesthesia and euthanasia (see Chapter 13). The lethal dose may be taken as three times the anaesthetic dose.

Administration. The solution may be administered by the intravenous, intracardiac or intraperitoneal routes. Owing to skin absorption and irregular results the subcutaneous administration of pentobarbitone sodium is not recommended.

For ease of administration, however, and particularly in the case of intractable and nervous animals, the intraperitoneal route is best. Pentobarbitone sodium powder contained in capsules may be administered by mouth, but the effect is often uncertain. Narcosis may take two or more hours to develop and even then be incomplete.

Dosage for intraperitoneal administration. The rate of dosage should not be less than 100 mg pentobarbitone sodium per kg bodyweight of the animal to be destroyed.

The time taken for the drug to act varies, but the full effect should be seen in 10–15 minutes. The drug produces very deep narcosis which to those with little experience might appear to be death. A very careful examination should be made for any signs of breathing, and if doubt exists an additional dose should be given. IN NO CASE SHOULD A BODY BE SENT FOR DISPOSAL UNTIL RIGOR MORTIS HAS SET IN.

APPENDIX I

THE UFAW EUTHANASIA CABINET—for use with carbon dioxide (Fig. 14.8). Description:

A box $60 \times 60 \times 60$ cm with a solid block-wood base and solid back and sides.

A sliding dirt tray lined with formica is fitted over the base to pull out from the front.

The top consists of a perspex panel set in a wooden frame sliding forward to open.

There should be no projections inside the cabinet. Jointing should be neat and sliding surfaces as close fitting and airtight as possible.

Fig. 14.8. UFAW Euthanasia Cabinet, for use with carbon dioxide.

Information on more recent work is contained in the Addendum.

15 Post-mortem Techniques for Laboratory Animals

STEPHEN SPARROW

Introduction

The post-mortem examination of any animal necessitates the adoption of procedures which are designed: a) to provide maximum information about the condition of the animal immediately prior to death and, b) to protect animals and personnel from any infectious disease that may be transmitted from the dead animal. The importance of this second requirement cannot be overstressed, particularly in the case of primates where the hazards involved should preclude examination unless full facilities and experienced staff are available (see Chapter 31: Disease Control).

There are three possible reasons for post-mortem examination of laboratory animals:

(a) to determine the pathological changes that might result from an experimental procedure;

(b) to determine the cause of disease not associated with experimental procedures;

(c) to examine apparently healthy animals on a routine basis in order to determine the microbiological status of the colony from which they originated. (Accreditation Microbiological Advisory Committee, 1973).

Although the fundamental principles of examination and the equipment required are the same in all cases, the different objectives require different procedures to be adopted. The procedures described below are those designed to determine the cause of disease not associated with experimental procedures.

Facilities and Equipment

A room which has been set aside for post-mortem examination and which is used for no other purpose is required. It should have surfaces that are easily cleaned and disinfected and be fitted with a table (preferably stainless steel) with a lip and central drain hole. There should also be a basin with hot and cold running water. An adjustable light and lens mounted over the table are particularly useful for the examination of smaller animals. Species up to the size of rabbits are more conveniently examined if pinned on to boards and a number of these of various sizes are required.

The following instruments are essential and should be maintained in good working order:

Scalpels (with disposable blades)
Straight-pointed scissors
Bowel scissors
Rat-tooth forceps
Dressing forceps
Bone forceps
Bone saw
Syringes and needles (disposable)
Pins
Enamel trays

There is also a need for pots and jars in which to put specimens for histological and parasitological examination, swabs and media for microbiological examination, some method of recording the findings, and an adequate disposal system for contaminated materials and carcases.

Protection of personnel is essential; all staff in the post-mortem room should wear gowns and any persons handling tissue should wear rubber gloves.

The Examination

It is always an advantage to receive animals alive, in which case an ante-mortem inspection may be carried out. In addition, if correct euthanasia is carried out, the pathological picture is less likely to be confused by the presence of post-mortem changes. In any event, post-mortem examination should be carried out as soon as possible after death. If delay is inevitable the autolytic and putrefactive changes may be slowed considerably by placing the carcase in a refrigerator.

Euthanasia

Any of the methods described in Chapter 14 may be used, but minimal post-mortem changes will occur if carbon dioxide is used for the destruction of the smaller rodents and sodium pentobarbitone for animals of the size of guinea-pigs and above.

Blood samples for haematological and parasitological examination should be taken by cardiac puncture immediately after death.

External Examination

The condition of the skin and coat and the presence of any external parasites are noted; it is useful to have a dissecting microscope for this purpose (Owen, 1972). All orifices are examined for the presence of discharges and any lesions which are either palpable or visible on the surface of the body are recorded.

Internal Examination

The animal is pinned through the extended legs on to the board, ventral surface uppermost. The whole ventral surface, from nose to anus, is swabbed with disinfectant to lay dust and hair.

The skin is incised from the mandibular symphysis to the pubic symphysis and reflected from the thorax and abdomen. The texture of the skin, the amount of subcutaneous fat and the colour of the underlying tissues are noted.

The abdominal muscles are incised in the midline from the xyphisternum to the pubic symphysis and along the line of the last ribs, taking care not to damage or disturb the abdominal organs. The abdominal musculature is pinned back.

The colour and position of the abdominal organs, the gut motility and the amount of peritoneal fluid are noted.

The thoracic cavity is exposed by lifting the xyphisternum and cutting the costal cartilages and intercostal muscles along each side. In the smaller rodents it is possible to use scissors for this, but bone forceps will be necessary for rabbits and the larger species.

If it is necessary to culture for microbiological examination from any site, this should be done before the viscera are disturbed.

Removal of the viscera is achieved by cutting through the mandibular muscles, lifting the tongue with forceps, dissecting the pharynx and larynx and continuing down the neck removing the trachea, oesophagus, lungs and heart. The diaphragm is cut around its periphery and the dissection completed against the dorsal surface of the abdomen to the brim of the pelvis. The visceral mass is then completely removed by severing the rectum. The urinogenital organs are then dissected free from the body and placed with the viscera on an enamel tray.

Special examination of the organs may now be carried out.

Special Examination

During this part of the examination it may be necessary to take samples for histological, parasitological or biochemical examination.

A routine for parasitological examination is described by Owen (1972). Histology sections should be 5 mm thick and fixed in 10 per cent formal saline for 24 hours. Specimens for toxicology, for example, liver, kidney and stomach contents, are put into chemically clean jars (Randall, 1954). Samples for investigation of nutritional status should be deep-frozen as soon as possible if immediate analysis is not to take place.

The larynx, trachea and bronchi are slit with scissors, the incision continuing to the apex of both lobes of the lung, and the presence of froth, haemorrhages, congestion or other lesions noted.

The pericardium is slit open and the heart exposed. The heart is cut from the base of each ventricle through to the auricle to facilitate examination of the endocardium and cardiac valves.

It is essential to look for lesions and changes in the cut surfaces of the liver, spleen, adrenals and kidneys, as well as on the surface. The kidney is slit longitudinally through to the pelvis and the capsule should strip off cleanly. The bladder is slit open and the mucosa inspected. Abnormalities of the urine or the presence of calculi are noted.

The intestine is displayed by cutting it from the mesenteric attachments; it is convenient to examine the lymph nodes at this stage and to look for the presence of gas at various points in its length. The alimentary tract is slit from oesophagus to rectum, noting the colour, consistency and amount of the contents, the thickness of the wall and the character of the mucous membranes. It is important to remember that the mucosa of the gut, particularly the stomach, undergoes very rapid post-mortem changes. Samples of the gut contents may be taken for parasitological examination.

Central Nervous System

This is not usually examined unless either the history, or the ante-mortem or post-mortem examination suggest neurological abnormality.

In small rodents, the bones of the skull may be cut round with scissors after reflecting the skin from the top of the head. The brain is then easily removed by cutting through the hind brain and the attachments at the base of the cranial cavity with scissors (Olson, 1954).

To remove the spinal cord, the animal is pinned on its ventral surface, the skin and dorsal muscles are dissected away from the occiput to the sacrum, the neural arch is cut on each side, with scissors in the smaller species

and bone forceps in the larger, the spinal roots are cut and the cord lifted clear.

Muscular-Skeletal System

This again is only examined in detail if there is evidence of disturbance of this system.

The post-mortem routine is completed by examination of the organs of special sense if earlier examination shows this to be necessary.

All those parts of the body which are not required for further examination are placed in strong plastic bags and incinerated.

REFERENCES

Accreditation Microbiological Advisory Committee (1973). *Microbiological Examination of Laboratory Animals*. Carshalton: MRC Laboratory Animals Centre.

Olson, C. (1954). Necropsy procedure for laboratory animals. In *Veterinary Necropsy Procedures*, Ed. Jones, T. C. & Gleiser, C. A. Philadelphia: Lippincott.

Owen, D. (1972). *Common Parasites of Laboratory Rodents and Lagomorphs*. MRC Laboratory Animals Centre Handbook No. 1. London: HMSO.

Randall, R. (1954). Selection and preparation of specimens for laboratory examination. In *Veterinary Necropsy Procedures*, Ed. Jones, T. C. & Gleiser, C. A. Philadelphia: Lippincott.

16   The Rabbit

C. E. ADAMS

General biology

The rabbit (*Oryctolagus cuniculus*) is classified in the small order, Lagomorpha, which also includes the pikas (*Ochotona*), the hares (*Lepus*) and the cotton tails (*Sylvilagus*).

According to Nachtsheim, the rabbit was domesticated by the first century B.C., but another authority (Zeuner, 1963) suggests that the first experiments in domestication took place in French monasteries between the sixth and tenth centuries A.D. A detailed account of its history and spread in Europe has been given by Thompson & Worden (1956), who noted that the first undoubted records of rabbits in Britain occur in the 13th century.

The rabbit is adapted to a wide variety of conditions from semi-arid desert with less than 17·8 cm rainfall to sub-tropical with 187 cm annual rainfall but thrives best under temperate pastoral conditions. Habits vary according to the environment so that on sandy heathland rabbits live in burrows, whereas on moorland or under wet conditions they resort to runs or galleries or utilize hedgerows or hollow trees for cover.

Wild female rabbits reach sexual maturity at 4–6 months of age and the males somewhat later. In the British Isles the breeding season is at a peak in the spring and early summer, though pregnancy may occur at any time of the year.

Brambell (1944) estimated the mean litter size at birth to be 4·87 compared with 5·38 corpora lutea. Wild rabbits will not breed in captivity except when given plenty of space and cover (Assheton, 1910; Wilson, 1936). It is notable that female wild rabbits born in the laboratory as a result of egg transfer to tame hosts retain their shy behaviour and fail to mature sexually. In the bucks the testes descend and spermatogenesis occurs; however, only a few have mated under observation. According to Stodart & Myers (1964), liberated domestic rabbits are much more susceptible to disease and predators than are wild rabbits. They lack an urge to go underground, have a tendency to a diurnal activity pattern and are relatively inefficient in protecting their young. It is believed that all the successful mainland invasions (Europe, Australia, New Zealand and South America) have sprung from wild stock.

Contributions of general interest include those of Sandford, 1957; Templeton, 1962; Napier, 1963; Casady, Sawin & van Dam, 1966; and Robinson, 1958. A very comprehensive bibliography is available (Makepiece, 1956) which lists nearly 3,500 references on all aspects of the rabbit. For an account of the behaviour of rabbits reference should be made to Denenberg *et al.* (1969). One point of special interest is the habit of coprophagy, or pseudorumination, involving the re-ingestion of the special soft faecal pellets which are excreted mainly in the early hours of the morning and taken directly from the anus by the lips. The significance of this process lies in the bacterial synthesis of certain B vitamins in the caecum; rabbits prevented from practising coprophagy die within three weeks.

Breeds and strains

In Great Britain there are some 35 breeds and more than double that number of varieties, whilst the American Rabbit Breeders Association lists standards for 28 breeds and nearly 80 varieties. Most of these are attributable to the rabbit's popularity as a show animal. The British Rabbit Council 1969 Yearbook lists 36 National Specialist Clubs. Liveweights vary from about 1 kg (Netherland Dwarf) to 6 kg or more (Flemish Giant) with the intermediate weights especially well represented. The female is heavier than the male. There is extreme variation in body conformation and ear size, and a wide range of colours exists. Very few specialized strains have been developed so far for laboratory purposes, but the International Index of

Laboratory Animals (1971) provides a useful reference on the location of inbred and non-inbred strains. In the past most laboratories obtained their supplies through dealers whereas now, with a growing emphasis on stock of known origin and background, accredited breeders supply increasing numbers. Another source is the commercial rabbitries which specialize in two meat breeds, Californians and New Zealand Whites.

Uses

There appears to be little general agreement with regard to the most suitable types of rabbit for various purposes except that for venous injection and bleeding, breeds with large ears (lops, half lops and large breeds) are often preferred.

Except for studies on size inheritance or other specialized purposes, the Netherland Dwarf and Polish breeds are too small; moreover, the ears are extremely short, making injection difficult, especially as these breeds tend to be very lively. On the other hand, the biggest breeds though placid can be inconveniently heavy to handle, especially for female assistants, and they require greater cage space and consume more food. The Dutch breed, being of medium size, is ideal for many purposes. Unless does are specifically required, males are equally suitable for most purposes. However, from puberty onwards males cannot be kept in groups (see p. 186).

Research uses include cardiac surgery and studies of hypertension, infectious diseases, virology, embryology, etc. The species is used routinely in serological work and, latterly, for screening embryotoxic agents and teratogens. Other interesting applications are in the breeding of tsetse flies and the development of oral contraceptives. The rabbit is also important for

teaching purposes in anatomy and experimental physiology. It is especially suitable for research on reproduction since ovulation is non-spontaneous, there is no seasonal anoestrum, gestation is short and semen is easily collected. By-products, including blood, plasma, cells, complement, liver, and brain powder are available commercially for laboratory purposes. One of the few areas where the use of the rabbit has declined is for pregnancy-testing, since the Friedmann test has been superseded.

Standard biological data

There is a wealth of data on the body fluids, including blood, urine and milk; cerebrospinal, synovial, serous and perivisceral fluids; digestive, dermal and reproductive secretions, and aqueous and vitreous humour. Values for the above are to be conveniently found under one cover (Dittmer, 1961), whilst other useful sources include those of Albritton (1952) and Kaplan (1962). Some of the commoner blood values are summarized in Table I (see also Little, 1970) and the composition of milk is shown in Table II. The volume of urine excreted is 50–75 ml/kg bodyweight/day. The volumetric composition of the rabbit skeleton has been analysed by Gong & Ries (1970).

The embryology of the rabbit is particularly well documented. The egg, which measures 160 μ in diameter, is the largest so far recorded in mammals and also the fastest developing, reaching the early blastocyst stage by the time it enters the uterus 72–80 hr p.c. It can be fertilized in vitro (Chang, 1959) and is the easiest of mammalian eggs to culture through the cleavage stages (for references see Adams, 1970a). The blastocyst is especially large, expanding to 5 mm or more before implantation, which occurs from 160 hr onwards. The blasto-

TABLE I
Selected blood values

Plasma volume ml/kg bodyweight	38·8 (27·8–51·4); 43·5 (35·1–49·8)
Erythrocyte volume (ml/kg)	16·8 (13·7–25·5); 17·5 (13·4–22·8)
Whole blood volume (ml/kg)	55·6 (44–70); 69·4 (57·6–78·3)
Erythrocyte count million/mm³ blood	4·8 (3·3–5·5)[a]; 6·3[b]
Erythrocyte packed volume, haematocrit (ml/100 ml blood)	44·1 (32–50)[a]; 39·8[b]
Haemoglobin concentration g/100 ml blood	14·2 (11·0–15·7)[a]; 12·8[b]

[a] = newborn (2–18 hr);   [b] = adult female
Reproduced with permission from Dittmer (1961).

TABLE II
Composition of rabbit's milk
(values expressed as g/100 g whole milk)

Ash	Water	Protein	Fat	Carbohydrate	Source
2·6	69·5	15·5	10·5	2·0	Grimmer (1925)
2·0	73·1	10·9	12·1	1·8	Bergman & Turner (1937)
—	—	10·4	16·7	2·0	Davies (1939)

cyst stage has proved useful for the screening of embryotoxic substances (Adams *et al.*, 1961). Detailed measurements of the egg from 24–144 hr and of the blastocyst from 140–160 hr have been given by Adams (1958) and Kodituwakku & Hafez (1969) respectively. The normal stages of embryonic development from 9–21 days gestation are described and illustrated by Minot & Taylor (1905).

The length of pregnancy is usually 31–32 days with a tendency for the smaller litters to be carried longer. In a group of 35 albino does, in which both the times of mating and parturition were known, the mean length of gestation was $31·0\pm0·8$ days with a mean litter size of 6·3. Birthweights vary from less than 30 g in the smallest breeds to more than 70 g in the largest breeds. Litter size has a marked effect on birth weight; in an albino strain weighing 3·5 kg, the mean birth weights varied from 35 g in litters of 10 young to 70 g in those of 2 young. In the smallest litters birth weights are exaggerated by the extended gestation. In rabbits autopsied on day 28 litter size had no demonstrable effect on foetal weights, indicating the importance of the last few days of pregnancy during which a 50 per cent gain in foetal weight may occur (Adams, 1960b).

The diploid number of chromosomes is 44.

The rectal temperature in a healthy animal is usually within the range of 38·5°–40°C, the average being 39·5°C (Dukes, 1955). Rectal temperature may be modified by excitement, handling or environmental disturbances.

The respiration rate in the normal adult animal is within a range of 38 to 65 per minute with an average of about 50. In the young rabbit the rate is much higher and in the infant it may exceed 100 respirations per minute. The volume of air taken in per breath is of the order of 20 ml (Spector, 1956).

A description of spontaneous malformations in the New Zealand White rabbit has been given by Palmer (1968); in more than 18,000 foetuses examined the incidence of major malformations was 0·66 per cent, intermediate between that reported for the rat and the mouse.

Lifespan
Comfort (1956) states that the lifespan can almost certainly exceed 15 years, and quotes several authenticated cases of 10–14 years, all of which were males. Under laboratory conditions, both sexes show signs of ageing by 6 to 7 years. In our colonies most stud males have been culled by 6 years of age, though they may retain their fertility beyond this age. Three 6 to 7-year-old bucks, at present under observation, are still producing semen but the quality is low with less than 10 million sperm/ml. The reproductive lifespan of the doe rarely exceeds 5 years, owing to the occurrence of uterine tumours, the incidence of which is strongly age-dependent, rising from 4·2 per cent of does aged 2 to 3 years to 79 per cent or more in those aged 5 to 6 years (Greene, 1941; Adams, 1970b).

Husbandry
Housing
General recommendations on the design and function of rabbit accommodation have been made by Napier (1963).

Large rooms accommodating 200–300 animals offer certain advantages but equally have disadvantages, particularly with regard to disease-control. Smaller rooms enable one not only to separate different classes of stock more easily but also permit individual workers to have their own accommodation, thus often eliminating conflicting demands. Consequently, rooms designed to accommodate units of 50–60 animals may be the best for laboratory purposes. They offer the further advantage that entire rooms can be closed down, sterilized and left empty in rotation. Where surgery is carried out, a special room for pre- and post-operative cases is also useful.

At one time wood was used almost exclusively in the construction of cages, but nowadays it has been largely superseded by metal, Examples of a typical wooden hutch and a bank of metal cages are shown in Figures 16.1 and 16.2. The dimensions are as follows: *Metal* cages: 48 cm wide, 61 cm deep and 46 cm high; *wooden* cages: 58 cm wide, 61 cm deep and 51 cm high. The wooden cage is particularly suitable for nursing does.

In the third edition of this Handbook the recommended standard quoted for the floor area of a breeding hutch was 0·23 m² per kg of bodyweight. Translated into actual sizes, this gives a floor space of 61 × 76 cm for small breeds, 61 × 91 cm for medium sized breeds, and 61 × 122 cm for larger breeds. The minimal size recommended for any hutch is 61 cm square and 45 cm high. In practice, it is inconvenient to have cages of many different sizes and, therefore, where various breeds are kept some compromise has to be made. In this case one should err in favour of the largest rabbits because cramped quarters, besides being inhumane, quickly lead to an

Fig. 16.1. A wooden rabbit hutch. (Photograph: R. Patman.)

Fig. 16.2. A bank of metal cages for rabbits. (Photograph: R. Patman.)

increase in the incidence of disease and to un-thrifty animals. All cages should be of standard-ized design wherever possible.

For the floors of metal cages the size of the wire mesh is important; we have found that 16 mm mesh, 14 gauge wire is very satisfactory; no problems with sore hocks have arisen on such floors, except in Rex rabbits. For metal trays used in wooden hutches aluminium is more economical than galvanized metal, as the latter is not so resistant to urine and, therefore, has a shorter life.

At weaning we transfer a proportion of young stock to colony-type runs measuring 1·2 m × 1·5 m and 0·7 m high (Fig. 16.3). Each pen is capable of taking 20–25 weanlings, which are reduced in number at intervals to avoid over-crowding. Straw is used as bedding on a concrete floor which slopes gently for drainage.

Breeding does kept in wooden hutches that are provided with straw bedding do not need a nesting-box, but does kept in metal cages must be given a nesting-box some days before parturition. For medium-sized breeds boxes measuring 25 × 30 cm high and wide and 38–40 cm long are suitable, but they are relatively expensive and difficult to disinfect. Disposable nest-boxes should be considered. For the accommodation of rabbits for short periods a

Fig. 16.3. Colony pens for rabbits. (Photograph: R. Patman.)

small wire cage (45 × 45 cm) is useful. Wicker baskets are unsuitable because they are easily damaged by gnawing.

Food and watering utensils

A great variety of food and watering utensils are available. Within a laboratory every effort should be made to standardize equipment as far as possible. Open food and water bowls are less hygienic than hoppers and water bottles. Hoppers should be so designed that the ration cannot be easily scratched out. As the filling and cleaning of water bottles is very time-consuming the use of automatic watering systems is becom-ing more common.

Bedding

No bedding is required on mesh floors, but saw-dust is used for the trays. For solid-floored hutches straw is most satisfactory, though saw-dust or wood chips can be used if straw is not available.

Environmental conditions

Light. Until quite recently it has been common practice to rely upon natural lighting, although this is subject to wide seasonal variation, depend-ing upon latitude. However, a trend is develop-ing towards giving supplementary lighting when the day-length falls below 12–14 hours. (Lights on at 6 a.m. and off at 8 p.m. is a recommended schedule.) The sudden switching on of lights during the hours of darkness may sometimes lead to rabbits in double cages fracturing the spine in the lumbar region, as a result of leaping. *Temperature.* Although rabbits are tolerant of fairly wide fluctuations in temperature, the desired range is within 10–18°C. In England the summer temperatures are rarely high enough to cause heat stress, but if ventilation is poor dis-comfort can be caused. The low and high critical air temperatures are -7°C and $28\cdot3$°C respec-tively, with the thermoneutrality range of -5°C to 30°C (Spector, 1956). Good ventilation is most important, and draughts should be avoided. *Humidity.* No specific information is available covering the optimum humidity for the rabbit house. The general recommendation is 50 per cent relative humidity (Mach & Lane-Petter, 1967), while Napier (1963) suggests 75 per cent.

Noise. Noise should be avoided as it can have a disturbing effect on rabbits, interfering with copulatory or maternal behaviour, etc.

Identification

Permanent methods of marking include the use of rings, tattoing in the ear, and ear-notching. Very young rabbits, from day-old, may be ear-notched or marked with dyes such as acriflavine or gentian blue, and then tattooed after weaning. Rings, which are placed over the hock joints before 8–12 weeks of age, are available in a wide variety of sizes to suit different breeds.

Handling

All handling should be firm but gentle; rabbits sensing insecurity are more likely to struggle. Young rabbits can be lifted by grasping them firmly over the loins, the fingers on one side, the thumb on the other. For lifting or carrying larger animals, the ears together with the fold of skin over the shoulder are grasped with the right hand whilst the left hand is placed under the rump to support the animal's weight; if restraint is necessary the animal can be held

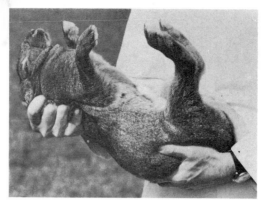

Fig. 16.5. Position for examining or carrying a rabbit. (Photograph: Miss M. Abbott.)

under the left arm (Fig. 16.4). Other positions, including one suitable for examining the animal, are shown in Figures 16.5 and 16.6.

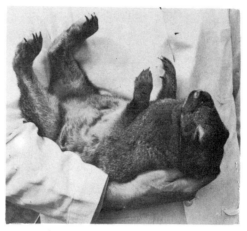

Fig. 16.6. Position for carrying a rabbit. (Photograph: Miss M. Abbott.)

For injection or other treatment rabbits should always be placed on a non-slippery surface such as towelling. When one is working single-handed, an animal that will not remain still can be restrained by means of a special box (Thorpe, 1944; Casady *et al.*, 1966) (Fig. 16.7). Care should be taken, because some rabbits struggle so violently when restrained that their backs may break.

Transport

Sexually immature rabbits can be transported in crates or, like adults, in specially constructed

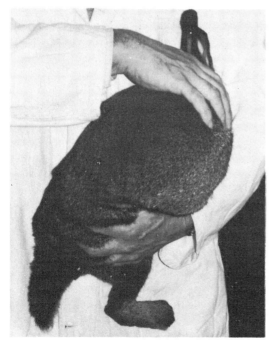

Fig. 16.4. Position for carrying a rabbit. (Photograph: Miss M. Abbott.)

travelling boxes, which should be light, strong, and waterproof. Resin-bounded plywood is suitable. Measurements should be: height 30–35 cm; width 20–30 cm; and length 35–45 cm, according to the size of the animal. Ventilation should be through the lid and sides, with baffle boards to prevent any direct draught.

Feeding*

Under natural conditions the rabbit's diet consists mainly of green herbage and other plant material. Before compounded rations were developed variable diets were fed, including cereals, freshly-cut grass, sainfoin, lucerne, vetches, cabbage, mangolds, kale, herbs and hay, according to season and availability. The digestibility of such feeds has been reviewed by Jarl (1944), whilst a table of coefficients of digestibility of commonly used rabbit foods is given by Aitken & Wilson (1962), whose publication contains a detailed review of the literature on nutrition. Bräunlich's (1965) booklet entitled 'Feeding Rabbits' contains much useful information and data.

Fig. 16.7. Rabbit positioned in a restraining box. (Photograph: C. E. Adams.)

* Much of the material in this section is adapted from the relevant section in the third edition of this Handbook.

Energy requirements

Maintenance. The basal or standard energy requirements of rabbits of different weights have been measured (Benedict, 1938; Lee, 1939a, b, c). Requirements for 24 hours, in round figures (calculated from the equation of Lee (1939b)), increase from 60 kcal for rabbits weighing 1 kg by increments of 20 kcal per 0·5 kg to 220 kcal for rabbits weighing 5 kg. According to Brody (1935) for mammals in general the energy-value of the total digestible nutrients for maintenance should be about twice the basal expenditure. On this basis the energy-requirements of rabbits for maintenance range from 120 kcal at a body-weight of 1 kg by increments of 40 kcal for each additional 0·5 kg body-weight to 440 kcal at a body-weight of 5 kg.

Reproduction. The energy-requirements of pregnant does and breeding bucks have not been established. Recommended increases are usually expressed as a percentage of maintenance allowances. Scandinavian authorities recommend increases of from 5 to 40 per cent for both does and bucks.

Lactation. Estimates have been made of the additional energy of digested food required for peak milk-production by does, based on available information on yield and energy value of milk and assuming that the doe converts food-energy into milk-energy as efficiently as a good cow (Sandford, 1957; Aitken & Wilson, 1962). They are in agreement with recommended allowances for lactating does, viz. two or three times the maintenance allowance by the third or fourth week of lactation. Inclusion of the litter's needs of solid food in the recommended allowances for a doe in late lactation raises these to four or more times the maintenance allowance.

Growth. Axelsson (1949) gave standards for energy requirements in terms of metabolizable energy for weight-gain of growing rabbits at different ages. Axelsson's standards, converted to energy of digested food (Aitken & Wilson, 1962), together with average weights and gains in weight of rabbits of medium-sized breeds as given by Sandford (1957), are shown in Table III.

Protein requirements

Detailed recommended allowances of protein for maintenance, reproduction, lactation and

TABLE III

Estimated energy requirements for growth of rabbits of medium-sized breeds

Age (weeks)	8	10	12	14	16	18	20	22	24
Approx. weight (kg)	0·96	1·28	1·62	1·87	2·10	2·27	2·41	2·55	2·67
Weekly gain (g)	156	170	128	114	85	71	71	57	—
Energy of digested food (kcal per g gain)	4·2	5·1	5·9	6·8	7·6	8·5	9·4	10·2	11·1

Compiled from data given by Axelsson (1949a), Aitken & Wilson (1962) and Sandford (1957).

growth are available from Scandinavia and the United States (Axelsson, 1949a; National Research Council, 1954). The recommendations are summarized in Table IV.

It has been calculated from the National Research Council's examples of rations supplying recommended allowances of protein that the N.R.C. allowances in terms of grams of protein per 1,000 kcal of energy of digested food are 46 for maintenance, 58 for growth, 57 for pregnancy, 63 for lactation and 64 for growth and fattening (Aitken & Wilson, 1962).

According to the N.R.C. report, for normal growth and fattening of rabbits between 1·8 and 3·2 kg live-weight the requirement of digestible protein is 16 per cent of the air-dry ration. The expected average daily gain is said to be 34 g per head. According to results of feeding trials made at Cornell University (Smith *et al.*, 1960) such an allowance of protein would seem to be excessive. New Zealand White rabbits, weighing about 1·4 kg at the beginning and about 2·7 kg at the end of the trials, were given pelleted diets containing 12 to 19 per cent of crude protein in the air-dry diet. Judged by weekly weight-gain and efficiency of food-utilization the requirement of crude protein was between 13 and 14 per cent of the air-dry ration. A digestibility study showed that one of

the diets which had 14 per cent of crude protein contained 9·4 per cent of digestible crude protein. With that amount of protein, average weekly gains equalled or exceeded the average expected gain postulated in the N.C.R. report. Rabbits given a diet containing 12 per cent of crude protein ate less, gained less, and utilized food less efficiently than rabbits given diets with 13 per cent or more of protein. In Denmark good growth and fattening were reported in rabbits at a progeny-testing station where the two breeds principally represented were White Landrace and French Voedder weighing on average about 3 and 4 kg respectively at 22 weeks (Nielsen *et al.*, 1958). It has been calculated that the Danish fodder plan provided 39 g of digestible crude protein per 1,000 kcal of energy of digested food and a nutritive ratio of 1:3·4 or 1:3·5 compared with 64 g per 1,000 kcal and a nutritive ratio of 1:1·8 by the N.R.C. diet for growth and fattening (Aitken & Wilson, 1962).

Fat requirements

It may be seen from the table of composition of rabbit feeds given by Aitken & Wilson (1962) that the feeds commonly included in dry rations contain from about 1 to 8 per cent crude fat. The effects on growth of diets containing

TABLE IV

Recommended allowances of digestible protein

	g per 1,000 kcal of metabolizing energy	Per cent of air-dry ration
Maintenance	30 to 36	12
Pregnancy	40 to 44	14
Lactation	48 to 52	20
Growth ⎫ Growth and ⎬ fattening ⎭	46 or 48 at 2 to 3 months decreasing to 36 to 38 in 8th or 9th month	16

Summarized from recommendations of Axelsson (1949a) and NRC (1954).

different levels of fat have been studied only with purified diets (Wooley and Mickelsen, 1954; Thacker, 1956). Wooley and Mickelsen were unable to evaluate the effect of the fat-content of diet on growth. Thacker's results suggested that in purified diets 10 to 25 per cent of fat was better than 5 per cent. Rabbits given 10 per cent or more of fat ate more and made better live-weight gains than those given 5 per cent of fat. Fat is a relatively expensive component of the diet.

Fibre requirements

A review of the literature yields little evidence likely to be helpful in the choice of fibre content of diets for rabbits. The best guide would seem to be the fibre content of diets upon which rabbits thrive. It has been calculated that the fibre content of commonly recommended diets ranges from about 10 to 16 per cent of dry matter for does with litters and about 15 to 26 per cent for other rabbits. A pelleted feed which is said to be satisfactory for all classes of rabbits contains about 13 per cent (Aitken & Wilson, 1962).

Water requirements

Records of the water intake of laboratory rabbits, mainly of the Dutch breed, maintaining normal growth on a pelleted diet at an environmental temperature of 21°C were kept by Cizek (1961). Mean values and standard deviations, based on 5,000 daily observations of each sex, were for males $235 \cdot 2 \pm 95 \cdot 61$ ml per day, or $104 \cdot 1 \pm 37 \cdot 79$ ml per kg per day, and for females $240 \cdot 0 \pm 103 \cdot 82$ ml per day, or $99 \cdot 5 \pm 38 \cdot 57$ ml per kg per day. In the study made by Johnson et al. (1957) New Zealand White rabbits drank about 120 ml per kg at 70 days of age, the amount decreasing to about 64 ml at 340 days in an environmental temperature of 28°C. When the temperature dropped to 9°C the corresponding figures were about 76 ml decreasing to 46 ml per kg.

Water should always be supplied ad lib.

Salt

Lack of an adequate supply of salt causes loss of water from the body, a low rate of growth, and low milk-production. Utilization of food is impaired and the health and condition of the animals is profoundly affected. Recorded salt-consumption of New Zealand White rabbits fed on concentrates and hay with a small allowance of green stuff and a salt lick was on average 0·1 g daily for bucks and 2·4 g daily for does with litters (Templeton, 1938). Rabbits given salty feed and offered unsalted water increase water consumption when salt intake is greater than about 0·3 g daily (Gompel et al., 1936). In North America salt is commonly added at a level of from 0·25 to 0·5 per cent of the total ration or as 1·0 per cent of mash. Complete pellets made from plant foods usually include 0·5 or 1 per cent of salt.

Vitamins and minerals

Bräunlich (1965) states that there is little information available concerning the mineral requirements of rabbits. However, certain con-

TABLE V
Estimated requirements for certain elements

	Requirement
Calcium	0·561–0·672 % of the feed
	1·0 % of the feed
	0·73 % of the feed
Phosphorus	0·513 % of the feed
	0·5 % of the feed
Potassium	0·8 % of the feed
	0·6 % of the feed
Manganese	0·3 mg/animal daily
Magnesium	40·0 mg/100 g of the feed
Cobalt	less than 0·1 mcg/animal daily

* Compiled by Bräunlich (1965) from estimates made by other research workers. For a complete list of references see Bräunlich's original paper. Reproduced by courtesy of Roche Products Ltd.

ditions, e.g. anaemia, resulting from mineral deficiencies are recognized. Estimated requirements for a few elements appear in Table V.

According to Bräunlich rabbits depend upon a dietary supply of a number of the vitamins but precise requirements are known for only a few of them. Information appears to be especially lacking on the requirements of the water soluble vitamins. Various conditions are known to occur as a result of a deficiency, e.g. of vitamin A (xerophthalmia, nervous disturbances etc.); of vitamin D (a rachitic condition has been produced experimentally though it is not known whether this vitamin is actually required); of vitamin E (muscular dystrophy, and myocardial damage) and of vitamin K (disturbed reproductive function).

Bräunlich makes the following recommendations for the fortification of compounded feeding stuffs:

Vitamin A	9000 i.u./kg
,, D	900 i.u./kg
,, E	40 i.u./kg
,, K_3 (menadione)	1 mg/kg
Nicotinic Acid	50 mg/kg
Choline	1300 mg/kg

Though certain vitamins of the B complex are synthesized in the caecum and re-ingested, it is also suggested that some supplementation with B_1, B_2, pantothenic acid and B_6 is justified. No supplementation with vitamin C is required.

Pelleted diets

Pelleted, compounded diets have obvious advantages and are now universally used. No supplements are necessary. Apart from facilitating feeding, such diets minimize the hitherto characteristic seasonal variations and enable results to be compared more reliably both between laboratories and from year to year.

In the United Kingdom, those most commonly used are diet 18 (Bruce & Parkes, 1946), modified diet 18, and SG1 (Short & Gammage, 1959). The composition and analysis of diet 18 and SG1 are given in Table VI.

Modified diet 18 differs from the published formula in that soya meal is substituted for the ground-nut cake and a trace-elemented mineral supplement for the salt and chalk. A vitamin supplement containing 16,000,000 i.u. of vitamin A, 4,000,000 i.u. of vitamin D_3, both stabilized, 32,000 mg of riboflavin and 100,000 mg of vitamin E per 50 kg may be added to the SG1 diet at the rate of 2·5 kg/100 kg.

Examples of compounded diets used in Denmark, France, Germany, Italy and the U.S.A. are given by Bräunlich (1965). Recommendations concerning the amount of pelleted diets to be fed are not readily available; they are usually fed *ad libitum*, although adult rabbits fed in this way often become excessively fat. Some pelleted diets tend to 'powder' and therefore result in considerable waste. If this happens attention should be given to the composition and size of the pellet.

Breeding

The rabbit has no oestrous cycle, but some authors have suggested that there is a certain rhythm in sexual activity. Under favourable conditions, sexually mature does remain on heat for long periods during which Graafian follicles are continuously developing and regressing in such a way that fairly constant numbers are available for ovulation. At certain times, for example during lactation, or if nutrition is poor,

TABLE VI
Percentage composition of Diet 18 and Diet SG1
Calculated analyses

Diet 18		Diet SG1			Diet 18	Diet SG1
Bran	15	Bran	10	Crude protein	24·8	20·33
Barley	20	Middlings	18	Crude fibre	10·3	9·05
Linseed Cake, Exp.	10	Sussex Ground Oats	12	Oil (Ether extract)	3·8	4·18
Groundnut Cake, Exp.	15	White Fish Meal	10	Calcium	1·4	0·89
Meat Meal (60% protein)	8	Dried Grass Meal	20	Phosphorus	0·8	1·05
Grass Meal (18% protein)	30			Chloride }	1·4	0·23
Chalk	1			Sodium }		0·15
Salt	1					

or when moulting, does may show no desire to mate, owing to a suppression of follicular activity.

The vaginal smear technique is not used for the detection of oestrus; instead one utilizes the appearance of the vulva which under the influence of oestrogens enlarges and becomes reddish-purple in colour. However, some does will mate even when the vulva is relatively small and pale. Owing to the sometimes marked divergence between the physiological and behavioural aspects of oestrus it is recommended that the doe's reaction to a buck should be observed; copulation can easily be prevented if it is not desired. The typical behaviour-pattern of a doe in oestrus is unmistakable, as she exhibits a pronounced lordosis. The intensity of behaviour, however, seems to vary between breeds.

Mating should be arranged to take place in the male's quarters; if moved elsewhere bucks tend to show more interest in their new surroundings than in the doe. The characteristic sequence of mounting, with exploratory pelvic thrusts culminating in copulation and orgasm, often occupies less than 30 seconds. If mating does not take place readily, there is no point in allowing a buck to exhaust himself. However, before assuming that a doe is anoestrous, she should be tried with a second male since her response may vary between males. When several bucks are available, it is possible for one person to obtain 40 matings per hour.

Sometimes does, although oestrous, may not permit mounting, especially if the male is young and inexperienced. In such cases forced mating may be attempted. The doe is restrained with one hand whilst the rear quarters and tail are raised with the other hand. When the buck mounts, the doe if oestrous will show lordosis.

Ovulation is generally non-spontaneous, requiring the stimulus of mating for its induction. However, under certain conditions conducive to intense sexual excitement, ovulation may occur spontaneously, even in does occupying separate cages. Mechanical stimulation of the vagina may also lead to ovulation. Variations exist in individual does of both the time of onset and duration of the ovulation process; the limits are from about 9 to 13 hours (Harper, 1961). The number of ovulations, which is not influenced by increasing the number of copulations above one, is strongly related to bodyweight (Gregory, 1932). Recently-obtained data on this aspect are presented in Table VII.

Does mate readily *post partum*, and contrary to earlier findings the pregnancy thus initiated can be maintained when large litters are suckled, depending upon the level of feed intake (Adams, 1967). Does suckling 8 young required 250 g of diet 18 daily to day 10 (thereafter they received 400 g) to maintain pregnancy. When the feed intake was reduced to 75 g daily, pregnancy succeeded only in does suckling no more than one young. When pregnancy fails it does so about 7 days *post coitum*, owing to regression of the corpora lutea; this condition can be overcome by treatment with follicle-stimulating hormone (FSH). Does fed *ad libitum* and mated *post partum* (p.p.) can become pregnant repeatedly and lactate continuously for months. This practice is being adopted commercially to intensify production. When mating takes place 5 days p.p. the litter is weaned just before the expected parturition; some breeders, however, prefer to remate their does 21 days p.p.

A proportion of does, sometimes 25 per cent or more, fail to ovulate after mating; this is probably due to a deficiency of luteinizing hormone (LH) in their pituitary gland. Ovulation failure is influenced by season, an important factor being the number of daylight hours (Farrel *et al.*, 1968); in the Northern hemisphere (lat. 50°–60°N) the late summer and autumn

TABLE VII

The relationship between ovulation rate and bodyweight in three breeds of rabbits

Breed	No. does observed	Mean no. ovulations	Mean liveweight	g weight/ovulation
Polish	55	4·58±0·14 (3–7)	1431±29	312
Dutch	40	6·95±0·26 (4–11)	2304±34	331
'A'	57	11·25±0·29 (8–18)	3713±55	330

period is generally considered to be the most difficult and the spring the least so.

In naturally-mated does examined in August, September and October 1968, 0, 3 and 36 per cent respectively failed to ovulate. However, in does treated with human chorionic gonadotrophin (HCG) ovulation failure is extremely uncommon; thus, of 157 Polish, Dutch and Albino does treated intravenously with 25 i.u. HCG between late August 1968 and January 1969, only one failed to ovulate, and that one had infantile ovaries. Even does that refuse to mate will ovulate and may become pregnant after treatment with HCG so that if they are artificially inseminated their oestrous behaviour need not be tested. The major disadvantage of HCG is that after repeated treatments no response is obtained, owing to antibody formation; from the 3rd to the 6th treatments the proportion of does failing to produce litters rose from 21 to 100 per cent (Adams, 1961a). Fortunately, in such does ovulation may still occur after mating. Various copper salts, e.g. gluconate, sulphate and acetate, can also be used to induce ovulation but they are not as effective as HCG; fewer does ovulate, and those that do shed fewer eggs (Suzuki & Bialy, 1964).

Induction of ovulation by mating with a vasectomized male or by treatment with HCG will lead to pseudopregnancy, in which functional corpora lutea develop and the uterus and mammary glands undergo changes characteristic of pregnancy. Nest-building activity, such as carrying straw and plucking fur, is associated with the termination of pseudopregnancy which happens in 75 per cent of does on days 17–19 (range 10–23).

Does may accept the male at any time during pregnancy or pseudopregnancy, except for a brief period from 33 to 39 hours after the first mating, but ovulation does not follow when active corpora lutea are present (Hammond, 1925).

Using the technique of abdominal palpation (Suitor, 1946), pregnancy can be diagnosed as early as 9 days after mating, when the uterine swellings measure only 12 mm (Adams, 1960b). As the swellings increase in size, reaching 20 mm at 13 days, pregnancy diagnosis becomes easier. At the Animal Research Station, Cambridge, does are palpated routinely on the tenth day

after mating and if there is doubt the examination is repeated a few days later. Early diagnosis of pregnancy is important because it facilitates better management of the pregnant does and at the same time permits early remating (day 20) of non-pregnant does. Palpation also enables one to estimate the stage of pregnancy in does whose breeding history is unknown. The technique is particularly useful in following the course of pregnancy in does that have been treated experimentally with drugs whose action may interfere with pregnancy. Test-mating is not a reliable procedure for diagnosing pregnancy.

Parturition generally occurs during the early hours of the morning and for this reason it usually passes unobserved. Cross's (1958) observations on induced labour in full-term does showed that the whole process took only 7 to 20 minutes. In untreated does parturition is normally completed within 30 minutes. Occasionally parturition is split: that is, part of the litter is born several hours, even a day or more, before the remainder. Routine palpation of does *post partum*, which is strongly recommended, will reveal if any foetuses are retained. Such does, and those which have not littered by the thirty-fourth day, may be successfully treated with oxytocin* to induce labour. Young retained *in utero* beyond the thirty-fifth day die, and if they are not expelled the uterus becomes fouled with decaying tissue which prevents further pregnancy. Normally each foetus is expelled together with its placenta, which is almost always immediately eaten by the doe who continues to lick the young. The young move to the teats and suckle whilst the birth of their siblings proceeds. During the height of lactation which occurs in the third week p.p. milk yield may reach 35 g/kg liveweight.

Sometimes does, more especially primiparae, mutilate part or all of their litters. When cannibalism occurs on more than two occasions offending females should be culled. Similarly, does that do not prepare adequate nests, but scatter their newborn young about the cage are hardly worth retaining as breeding-stock.

In 62 natural parturitions of primiparae, which produced 490 young, only 5 young were

* Pitocin brand: Parke Davis. 200 mμ in 0·5 to 1 ml of normal saline given intravenously.

stillborn while 11 others died shortly after birth (Adams, 1960a, b). In superovulated does the incidence of stillbirths is significantly greater, reaching 25 per cent or more. Abortion is extremely uncommon, and occurs only after the twentieth day of pregnancy. If the entire litter dies before day 20 elimination takes place by resorption, which is very rapid.

The sexing of newborn young is not difficult; differences can be seen when slight pressure is applied to the external genitalia with the thumbs, causing an eversion of the rudimentary penis or vulva (Fig. 16.8). In experienced hands the technique is very accurate and rapid. Before examining the litter it is advisable to remove the mother; she should not be returned until the young have settled down again in the nest.

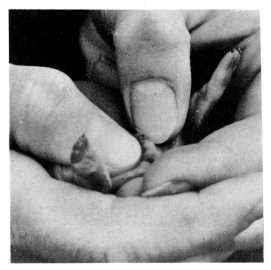

Fig. 16.8. Technique for sexing a day-old rabbit. (Photograph: C. E. Adams.)

If only male or only female offspring are required the unwanted sex can be destroyed at birth and litters can be made up to the required size by fostering young from does that produced the smallest litters. These does can then be re-mated. Fostered young should not differ in age by more than a few days from the foster mother's litter.

The two sexes are produced with equal frequency. In 1,024 litters containing 9,854 young there were 4,928 males and 4,926 females, giving a sex ratio of 100·04. It must be emphasized however that when only a few litters are considered the sex ratio may show very wide variations.

Sexual development in the doe is influenced by many factors, including breed, nutrition and season of birth. The smaller breeds reach sexual maturity earlier than the larger breeds, and a high plane of nutrition also hastens the onset of sexual function. It is not possible to state precisely when does should be mated for the first time; the limits vary from 3–4 months in small breeds to 8–9 months in large breeds.

Ideas are changing regarding the number of litters that breeding does should be required to produce annually, particularly under commercial conditions where profitability is much influenced by this factor. With early weaning and/or concurrent pregnancy and lactation some does can produce 6–10 litters annually. In large breeds a lifetime production of 80–120 young over 3 years represents a satisfactory performance. The smaller breeds, of course, will produce relatively fewer young. Observations on reproductive performance are given in Table VIII. The most outstanding doe produced 163 young in 24 litters by 58½ months of age.

The male

It is hardly possible to make firm recommendations concerning the frequency of use of males, which show wide individual variations. It has proved possible to ejaculate bucks as frequently as once a day for as long as 10 months without impairing their libido, semen production, or

TABLE VIII

Reproductive performance of Strain A does, weighing 3·5–4 kg

No. does	No. litters	Total no. young	Mean Age at last parturition (months)
24	12·2±0·9 (3–21)	77·3±6·0 (18–133)	37·1±2·0 (16–52·5)
15	11·2±1·6 (4–24)	83·5±11·9 (17–163)	34·0±3·9 (13·8–58·2)

Reproduced with permission from Adams (1970b).

fertility (Gregoire *et al.*, 1958). In exhaustion tests up to 20 collections have been made over a period of 10 hours. We have records of a buck that copulated 37 times in three 45 minute periods between 11.35 hr and 17.43 hr. Even the 35th mating produced a litter and litter-size was normal throughout. When used the following morning on three does, every one became pregnant. Thus, over limited periods, it may be possible to exploit bucks more than was formerly thought practicable.

The ejaculate varies in volume from 0·2–1·0 ml or more, depending upon breed, and in density from 100–1,000 million sperm/ml. Above a certain density, 350–400 million sperm/ml, ejaculates show wave motion. In three different regimes of semen collection, viz. once weekly, once daily except Sundays, or twice in succession on Monday, Wednesday and Friday the mean sperm output was 273, 519 and 619 millions (Desjardins *et al.*, 1965). Increasing the frequency of collection from once weekly to once daily resulted in a 50 per cent decrease in sperm concentration and a 40 per cent decrease in ejaculate volume, but a 40 per cent greater weekly sperm output (Gregoire *et al.*, 1958) (see also Amann, 1966).

A simple test for male fertility consists of obtaining a smear from the vulva immediately after copulation, when motile sperm should be present. However, the finding of motile sperm does not alone furnish proof of fertility. Ideally, a more critical analysis of male fertility should include a thorough examination of one or more semen samples collected with an artificial vagina, combined with a series of test matings.

Selection and discarding of breeding stock

There are certain primary requirements for all breeding stock. Sawin & Curran (1952) have shown that the development of strains with good reproductive efficiency involves rigorous selection for six major factors: fertility, fecundity, milk-production, maternal behaviour, growth rate, and viability. Although environmental factors affect these characteristics, they are basically inherited and details should be included in the recording system. The only characteristic which cannot be determined directly is milk-production. An index of this characteristic can be obtained by using the weight of the litter at 21 days, or when the youngsters first leave the nest if they do so earlier. Youngsters, particularly of the larger types, which leave the nest before the age of about 21 days do so mainly through hunger, thus indicating poor milk-yield in the doe.

Few does can be profitably kept much over 3 years of age and in commercial units culling after 12 matings or 108 weeks' production is recommended. Apart from ill-health, the main reasons for culling are low or declining fertility (small litters or failure to conceive regularly) and failure to rear young. Fertility declines long before old age owing to defects in the uterus which tend to precede ovarian changes (Adams, 1970).

Inbreeding and outbreeding

For a general consideration of breeding systems see Chapter 2. Choice of stock, establishment of a colony, breeding programme and stock improvement relative to the rabbit have been described by Napier (1963). There is a need for inbred strains, but although several attempts have been made to establish them no completely isogenic strain (20 generations) yet exists. This must reflect the common experience that within 3 or 4 generations of close inbreeding serious difficulties arise in maintaining the line. For example, in one of our strains, up to 80 per cent of the young died at about 8–12 weeks of age, showing symptoms of enteritis (diarrhoea). Very considerable facilities are required to maintain the number of stock necessary for selection through the early generations. For example, Napier (1963) suggests that approximately 100 cages will be required if a strain is closed to all outside blood.

We have found the hybrid is extremely useful for many laboratory purposes but little work appears to have been done on the hybridization of more than two breeds. It is expected that the growing need for rabbits for specialized experimental purposes will stimulate the production of inbred lines and special strains.

Breeding does are normally caged individually and mating, which can be precisely controlled, is arranged as described on p. 182. One male is required per 12–15 does.

Artificial insemination

Artificial insemination gives results at least as good as those obtained from natural mating, over which it has several advantages. The equipment and techniques used in semen collection, preparation of the doe and deposition of semen into the vagina have been described and the literature reviewed by Adams (1961b, 1962). For experimental purposes insemination can also be carried out surgically or intra-peritoneally by injection (Adams, 1969).

Fertility is well maintained down to 1 million sperm, which means that semen can be much diluted. Insemination can be carried out with good results several hours before or up to 4 hours after the ovulating stimulus is given, though in practice the two procedures should be arranged near together, if possible. Recently, the successful long term storage of rabbit spermatozoa by deep freezing has been reported (Stranzinger et al., 1971); at 10°C fertility is retained for up to 120 hours.

Weaning and rearing

There is a growing tendency to wean stock at progressively younger ages, down to 4 weeks under commercial conditions. However, in the laboratory weaning at 6–8 weeks is still favoured because the young are better grown and more resistant to any nutritional disturbances that may follow weaning.

After weaning, youngsters are best kept in pairs or trios or in colony groups of 15–20, depending upon the area available. All young stock should be separated into sexed groups before 3 months of age, when males may need to be separated into individual cages. About this time the bucks begin to show aggressive behaviour and attack each other, particularly in the scrotal region causing severe injury to the testes, so that ultimately only one intact male remains. This provides a severe example of natural selection.

Does may be left together until they are 18–20 weeks of age.

Laboratory procedures

Selection of animals

Emphasis should be placed on selecting only healthy well-grown animals, preferably of known breed and age. For studies of reproduction it is usually necessary to separate the females for at least three weeks before an experiment, which may also require proof that an animal was fertile immediately beforehand.

Anaesthesia

Nembutal* (pentobarbitone sodium 60 mg/ml B. Vet.C.) is widely used as an anaesthetic. For complete sedation the recommended rate is 1 ml/2·27 kg bodyweight. Alternatively, to shorten the period of hypnosis this dose may be halved and the anaesthesia completed with ether. It is most important to give the injection slowly. A peculiarity worthy of note is the hyper-sensitive state that may arise if the needle is accidentally withdrawn before injection is complete, making reinsertion most difficult. When ether is used, control of the depth of anaesthesia needs constant attention because an overdose quickly leads to respiratory failure, while if too little is given surgery is impossible. As an alternative to ether, we prefer to use Fluothane† (halothane B.P.) carried in oxygen by means of a simple anaesthetic machine fitted with a Trilene‡ (trichloroethylene) bottle. Nembutal is still used for induction at the rate of 0·8 ml (Polish), 1·0–1·2 ml (Dutch) and 1·5 ml for breeds of 3–4 kg body-weight. A technique for the anaesthesia of germfree rabbits with halo-thane has been described recently (Cook & Dorman, 1969). Very young rabbits may be anaesthetized with ether or halothane alone

Artificial respiration can be applied successfully if the heart-beat is still strong. The animal is placed on one side and the tongue is pulled forward and deflected to one side. The thorax is compressed 20–30 times per minute. For further information on anaesthetic practices the reader is referred to Croft (1964) and Hall (1966).

Euthanasia

Rabbits can be killed by dislocation of the neck or by carbon dioxide* (see Chap. 14). In the case of larger rabbits, not required for food, the method of choice is by the intravenous injection

---

* Abbott Laboratories Ltd.
† I.C.I. Ltd.
‡ British Oxygen Company.

* Author's note. Although I have not personally used carbon dioxide for euthanasia I am advised by UFAW that it is a very satisfactory method for rabbits.

of Expiral (pentobarbitone Sodium B. Vet. C. 200 mg/ml, Abbott). Given by *rapid* intravenous injection 1·5 ml is sufficient to kill a large rabbit. Older buck rabbits are difficult to kill cleanly by physical methods.

Experimental feeding procedures

The use of a pipette was found unsatisfactory by Hills and McDonald (1956) for hand-rearing; they were successful, however, with feeding-bottles fitted with small rubber teats. Capsules can be conveniently given by stomach tube.

Collection of specimens

Blood. Blood can easily be collected from the marginal ear vein, which is prepared as follows. The area around the vein is shaved and then bathed with 70 per cent alcohol before the vein is nicked with a razor blade. Blood may also be drawn into a syringe. After collection the flow often continues but it can be stemmed by placing a little cotton wool over the vein.

Large quantities of blood may be obtained by heart puncture. The donor is anaesthetized and placed in a dorsal position. It is not necessary to use a syringe to draw the blood, which will run very freely through a hypodermic needle if the puncture is well done. For a large rabbit the recommended needle size is 1·85–2·10 mm × 5 cm; the needle point should be really sharp.

Cook & Dorman (1969), using 6–8 week old rabbits, have taken 2 ml of blood on 6–8 occasions by heart puncture, with an interval of 7 days between each bleeding. They used 2 ml sterile disposable syringes and 0·51 × 16 mm sterile disposable needles.

Urine. Urine can often be obtained by applying slight pressure over the bladder or directly by insertion of an inseminating pipette. In insemination (p.186) the technique of introducing the pipette into the vagina is desired to avoid stimulation of the release of urine.

Faeces. The hard faecal pellets can be collected in a special cage, while the soft pellets, which are normally reingested early in the morning, can be collected directly from the rectum (before excretion) or from the stomach (after reingestion) at autopsy.

Milk. Small quantities of milk are obtainable by applying pressure to the teats of lactating does.

Oviduct secretions. A method for the continuous volumetric collection of oviduct secretions has been described by Clewe and Mastroianni (1960), who reported that secretion rates ranged from 0·02–0·13 ml/hour.

Semen. Semen is collected with the aid of an artificial vagina. The model described by Walton (1958) is very satisfactory. Bucks can be easily trained to ejaculate into the artificial vagina (Fig. 16.9). For success, it is important to employ the correct temperature (about 42°C) and pressure. Individual variations in male sex drive determine the degree of stimulus required, which varies from the artificial vagina alone to a doe. In practice, a rabbit skin is normally used.

Fig. 16.9. Semen collection using an artificial vagina. (Photograph: R. Patman.)

Dosing and injection procedures

Oral. Anything placed in the rabbit's mouth is usually rejected, so that a stomach tube must be used for oral dosing. Powdered materials are encapsulated.

Certain substances may also be incorporated into a pelleted feed, e.g. antibiotics, coccidiostats, etc.

Subcutaneous: convenient sites are the flanks or nape of the neck. A fold of skin is pinched up and the needle (No. 15) inserted in an anterior direction.

Intravenous: the marginal ear vein is used, and the needle (No. 19) is introduced nearer the tip than the base of the ear (Fig. 16.10). Injection is facilitated by bathing the vein with warm

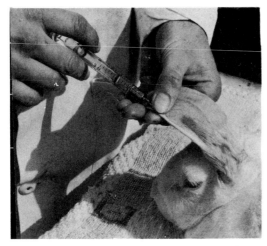

Fig. 16.10. Intravenous injection in the rabbit. (Photograph: C. E. Adams.)

water or alcohol; shaving is rarely necessary except possibly in breeds with dark coloured ears. It is important to ensure that the injection is not being made around the vein, because some vehicles lead to necrosis of the tissue.

Intramuscular: the hind leg is a convenient site.

Intraperitoneal: the animal is suspended by the hind legs so that the gut falls forward. The point of insertion is mid-line forward of the bladder.

Other procedures. Surgical procedures, e.g. laparotomy in connection with ovariectomy, egg transfer, etc., are facilitated if food and water are withheld from the previous evening.

Disease control

Disease aspects have been very fully dealt with by McDougal (1929), Blount (1945) and Ostler (1961).

A recent survey of disease in laboratory animals (Seamer & Chesterman, 1967), which included more than 33,000 rabbits, showed that in breeding colonies about one fifth of all rabbits born (14,851) died before use, and that approximately half of the deaths occurred before weaning. Five per cent of bought-in rabbits died before use. The most commonly reported diseases were intestinal (including enteritis, mucoid enteritis, intestinal impaction, scours and diarrhoea), coccidiosis and 'respiratory' (snuffles and pneumonia). A disease not mentioned specifically in the survey, namely pregnancy

toxaemia, can account for deaths that occur suddenly, especially in late pregnancy (Greene, 1937).

According to Lev (1963), germfree rabbits have been reared with only limited success: he noted that the problems lay in the nutrition of the weaned animal, for which no satisfactory diet had been described. Pleasants (1959) reported on the weaning period of germfree rabbits. SPF barrier-maintained rabbits are now available commercially.

The production of rabbits free from lesions associated with *Encephalitozoon cuniculi* has been described recently (Howell & Edington, 1968).

Except for a few diseases, e.g. ear canker or mange, treatment is difficult and frequently of no avail, as sick rabbits show little power of recovery. Moreover, prolonging the life of infected animals increases the risk of spreading the disease. Consequently, the immediate humane destruction of ailing animals is recommended, whilst the destruction or isolation of immediate contacts has to be considered. The adoption of prophylactic measures is, therefore, very important. These include:

1. Provision of adequate accommodation which should be dry and well-ventilated but not draughty.

2. Cages should be cleaned out regularly and sterilized at intervals. They should not be occupied continuously, but allowed to stand empty periodically.

3. Feeding should be regular and even.

4. Avoid the use of drinking bowls. Remember that certain infections may be spread by automatic watering systems which, therefore, require regular attention.

5. Avoid overcrowding, especially of nursing mothers and weanlings.

6. Examine carefully all bought-in stock and reject any showing evidence of snuffles (sneezing, nasal discharge or soiled fur on inner aspect of fore leg), scouring, ear canker, or vent disease. Isolate others for 2–3 weeks.

7. Prevent other animals, particularly dogs and vermin, from coming into contact with feeding stuffs.

8. Inspect breeding stock regularly for signs of vent disease. Snuffles can also be transmitted at coitus as it is possible for the

organism to be transferred from the nares to the genitalia during coprophagy. Kyaw (1945) described female sterility resulting from the organisms (*Pasteurella septica* and a *Brucella bronchiseptica*-like organism) invading the genital tract.

9. Carrying boxes, crates and utensils should be regularly sterilized.
10. New stock or weanlings should be introduced only into clean cages.
11. Practise treatment for coccidiosis as a routine measure. Sulphamezathine* (sulphadimidine solution 16 per cent) is added to the drinking water at the rate of 60 ml/4½ litres. On premises where the infection is known to be present, pregnant does should be given a five day course of treatment two to three weeks before they litter. This should be repeated at weaning time for another five days.
12. As outbreaks of myxomatosis still occur amongst wild rabbits, it may be considered prudent to vaccinate stock against this disease. Immunity lasts 9–12 months after which a second injection is advisable for breeding stock. In the case of an outbreak, the vaccine† is injected intradermally at the base of the ear; protection is then afforded within two or three days. Vaccination has no adverse effect on pregnant does.

REFERENCES

Adams, C. E. (1958). Egg development in the rabbit: the influence of post-coital ligation of the fallopian tube and of ovariectomy. *J. Endocr.*, **16**, 283–93.
Adams, C. E. (1960a). Prenatal mortality in the rabbit (*Oryctolagus cuniculus*). *J. Reprod. Fert.*, **1**, 36–44.
Adams, C. E. (1960b). Studies on prenatal mortality in the rabbit (*Oryctolagus cuniculus*): the amount and distribution of loss before and after implantation. *J. Endocr.*, **19**, 325–44.
Adams, C. E. (1961a). Artificial insemination in the rabbit. *J. Reprod. Fert.*, **2**, 521–2.
Adams, C. E. (1961b). *Artificial insemination in the rabbit.* Tech. Bull. No. 1. Commercial Rabbit Assoc.
Adams, C. E. (1962). Artificial insemination in rodents. In *The Semen of Animals and Artificial Insemination*, ed. Maule, J. P. Commonwealth Agricultural Bureaux: Farnham Royal.
Adams, C. E. (1967). Concurrent lactation and pregnancy in the rabbit. *J. Reprod. Fert.*, **14**, 351–2.
Adams, C. E. (1969). Intraperitoneal insemination in the rabbit. *J. Reprod. Fert.*, **18**, 333–9.
Adams, C. E. (1970a). The development of rabbit eggs after culture in vitro for 1–4 days. *J. Embryol. exp. Morph.*, **23**, 21–34.
Adams, C. E. (1970b). Ageing and reproduction in the female mammal with particular reference to the rabbit. *J. Reprod. Fert. Suppl.*, **12**, 1–16.
Adams, C. E., Hay, M. F. & Lutwak-Mann, C. (1961). The action of various agents upon the rabbit embryo. *J. Embryol. exp. Morph.*, **9**, 468–91.
Aitken, F. C. & Wilson, W. K. (1962). *Rabbit feeding for meat and fur*, 2nd ed. Tech. Comm. No. 12. Commonw. Bur. Anim. Nutr., Aberdeen.
Albriton, E. C. (1952). *Standard Values in Blood.* Philadelphia: Saunders.
Amann, R. P. (1966). Effect of ejaculation frequency and breed on semen characteristics and sperm output of rabbits. *J. Reprod. Fert.*, **11**, 291–3.
Assheton, R. (1910). *Guy's Hosp. Rep.*, **64**, 313–42.
Axelsson, J. (1949). Standards for nutritional experiments of domestic animals in the Scandinavian countries. *13th Int. Congr. Zool.*
Baker, F. (1944). Stability of the microbial populations of the caecum of guinea-pigs and rabbits. *Ann. appl. Biol.*, **31**, 121–3.
Benedict, F. G. (1938). *Vital energetics. A study in comparative basal metabolism.* Carnegie Inst. Washington Publ. 503.
Bergman, A. J. & Turner, C. W. (1937). *J. biol. Chem.*, **120**, 21. (Quoted by Dittmer, 1961.)
Blount, W. P. (1945). *Rabbits' Ailments: A Short Treatise on the Domestic Rabbit in Health and Disease.* Idle & London: Watmoughs.
Brambell, F. W. R. (1944). The reproduction of the wild rabbit, Oryctolagus cuniculus (L). *Proc. zool. Soc. Lond.*, **114**, 1–45.
Bräunlich, K. (1965). *Feeding Rabbits.* Basle: Hoffmann La Roche.
Brody, S. (1935). The relation between feeding standards and basal metabolism. Report of Conference on Energy Metabolism held at State College, Pennsylvania, p. 12.
Bruce, H. M. & Parkes, A. S. (1946). Feeding and breeding of laboratory animals. II. Growth and maintenance of rabbits without fresh green food. *J. Hyg., Camb.*, **44**, 501–7.
Casady, R. B., Sawin, P. B. & van Dam J. (1966). *Commercial Rabbit Raising Handbook No. 309.* Washington: U.S. Dept. Agric.
Chang, M. C. (1959). Fertilization of rabbit ova in vitro. *Nature, Lond.*, **184**, 466.

* I.C.I. Ltd.
† Weyvak: Mansi Laboratories.

Cizek, L. J. (1961). Relationship between food and water ingestion in the rabbit. *Am. J. Physiol.*, **201**, 557–66.

Clewe, T. H. & Mastroianni, L. (1960). A method for continuous volumetric collection of oviduct secretions. *J. Reprod. Fert.*, **1**, 146–50.

Comfort, A. (1956). *The Biology of Senescence*. London: Routledge & Kegan Paul.

Cook, R. & Dorman, R. G. (1969). Anaesthesia of germ-free rabbits and rats with halothane. *Lab. Anim.*, **3**, 101–6.

Croft, P. G. (1964). *An Introduction to the Anaesthesia of Laboratory Animals*. London: UFAW.

Cross, B. A. (1958). On the mechanism of labour in the rabbit. *J. Endocr.*, **16**, 261–276.

Davies, W. L. (1939). *The Chemistry of Milk*. New York: Van Norstrand. (Quoted by Dittmer, 1961.)

Denenberg, V. H., Zarrow, M. X. & Ross, S. (1969). The behaviour of rabbits. In *The Behaviour of Domestic Animals*, 2nd ed., ed. Hafez, E. S. E. London: Baillière, Tindall & Cassell.

Desjardins, C., Kirton, K. T. & Hafs, H. D. (1968). Sperm output of rabbits at various ejaculation frequencies and their use in the design of experiments. *J. Reprod. Fert.*, **15**, 27–32.

Dittmer, D. S. (1961). *Blood and Other Body Fluids*. Biological handbooks. Washington: Fedn Am. Socs exp. Biol.

Dukes, H. H. (1955). *The Physiology of Domestic Animals*, 7th ed. London: Baillière, Tindall & Cox.

Farrel, G., Powers, D. & Otani, T. (1968). Inhibition of ovulation in the rabbit. Seasonal variation and effect of indoles. *Endocrinology*, **83**, 599–603.

Gompel, M., Hamon, F. & Mayer, A. (1936). Individual differences in feeding habits of the domesticated rabbit. *Annls Physiol. Physicochim. biol.*, **12**, 471–503. (Cited in 3rd ed. *UFAW Handbook on the Care and Management of Laboratory Animals*.)

Gong, J. K. & Ries, W. (1970). Volumetric composition of the rabbit skeleton. *Anat. Rec.*, **167**, 79–86.

Greene, H. S. N. (1937). Toxaemia of pregnancy in the rabbit. 1. Clinical manifestations and pathology. *J. exp. Med.*, **65**, 809–32.

Gregoire, A. T., Bratton, R. W. & Foote, R. H. (1958). Sperm output and fertility of rabbits ejaculated either once a week or once a day for forty-three weeks. *J. Anim. Sci.*, **17**, 243–8.

Gregory, P. W. (1932). The potential and actual fecundity of some breeds of rabbits. *J. exp. Zool.*, **62**, 271.

Grimmer, W. (1925). *Tabulae biol.*, Berlin, **2**, 536. (Quoted by Dittmers, 1961.)

Hall, L. W. (1966). *Wright's Veterinary Anaesthesia and Analgesia*, 6th ed. London: Baillière, Tindall & Cassell.

Hammond, J. (1925). *Reproduction in the Rabbit*. Edinburgh: Oliver & Boyd.

Harper, M. J. K. (1961). The time of ovulation in the rabbit following the injection of luteinizing hormone. *J. Endocr.*, **22**, 147–52.

Hills, D. M. & McDonald, I. (1956). Hand rearing of rabbits. *Nature, Lond.*, **178**, 704–5.

Howell, J. McC. & Edington, N. (1968). The production of rabbits free from lesions associated with *Encephalitozoon cuniculi*. *Lab. Anim.*, **2**, 143–6.

Jarl, F. (1944). *Digestibility experiments with rabbits*. Lantbrukshogskol. Husdjursforsoksanst. Rep. No. 16. (English summary.)

Johnson, H. D., Cheng, C. S. & Ragsdale, A. C. (1957). Environmental physiology and shelter engineering with special reference to domestic rabbits. *Res.Bull.Mo.agric.Exp.Stn.*, 646 (45), 648 (46).

Kaplan, H. M. (1962). *The Rabbit in Experimental Physiology*, 2nd ed. New York: Scholars Library.

Kodituwakku, G. E. & Hafez, E. S. E. (1969). Blastocyst size in Dutch belted and New Zealand rabbits. *J. Reprod. Fert.*, **19**, 187–90.

Kyaw, M. H. (1945). Note on a bacterial cause of sterility in the rabbit. *Vet. Rec.*, **57**, 502–3.

Lee, R. C. (1939a, b, c). (a) The basal metabolism of the rabbit. (b) Basal metabolism of the adult rabbit and pre-requisites for its measurement. (c) Size and basal metabolism of the adult rabbit. *J. Nutr.*, **17**, 19–20; **18**, 473–88, 489–500.

Lev, M. (1963). Germ-free animals. In *Animals for Research. Principles of breeding and management*, ed. Lane-Petter, W. London & New York: Academic Press.

Little, R. A. (1970). Changes in the blood volume of the rabbit with age. *J. Physiol.*, **208**, 485–97.

Mach, E. & Lane-Petter, W. (1967). Animal house design. In *UFAW Handbook on the Care and Management of Laboratory Animals*, 3rd ed. Edinburgh: Livingstone.

Makepiece, L. L. (1956). *Rabbits: A Subject Bibliography*. Colorado: Fort Collins.

McDougal, J. (1929). *The Rabbit in Health and Disease*. Idle & London: Watmoughs.

Minot, C. S. & Taylor, E. (1905). The rabbit. *NormTaf. EntwGsch. Wirbeltiere*, **5**.

Napier, R. A. N. (1963). Rabbits. In *Animals for Research*, ed. Lane-Petter, W. London & New York: Academic Press.

National Research Council (1954). *Nutrient requirements for domestic animals. 9. Nutrient requirements for rabbits*. Publ. No. 331. Washington: Nat. Acad. Sci.

Nielsen, J., Jensen, N. E. & Hansen, H. (1958). 7th report from the Rabbit Progeny Testing Station on Favrholm for the years 1955–56 and 1956–57. *Beretn. Forsøgslab.*, **307**.

Ostler, D. C. (1961). The diseases of broiler rabbits. *Vet. Rec.*, **73**, 1237–52.

Palmer, A. K. (1968). Spontaneous malformations of the New Zealand White rabbit. The background to safety evaluation tests. *Lab. Anim.*, **2**, 195–206.

Pleasants, J. R. (1959). Rearing germ-free caesarian born rats, mice and rabbits through weaning. *Ann. N.Y. Acad. Sci.*, **78**, 116–26.

Robinson, R. (1958). Genetic studies of the rabbit. *Biblphia genet.*, **17**, 229–558.

Sandford, J. C. (1957). *The Domestic Rabbit*. London: Crosby Lockwood.

Sawin, P. B. (1950). *The Care and Breeding of Laboratory Animals*. London: Chapman & Hall.

Sawin, P. B. & Curran, R. H. (1952). Genetic and physiological background of reproduction in the rabbit. *J. exp. Zool.*, **120**, 165–201.

Short, D. J. & Gammage, L. (1959). A new pelleted ration for rabbits and guinea-pigs. *J. Anim. Techns Ass.*, **9**, 62–9.

Seamer, J. & Chesterman, F. O. (1967). A survey of disease in laboratory animals. *Lab. Anim.*, **1**, 117–39.

Smith, S. E., Donefer, E. & Mathieu, L. G. (1960). Protein for growing fattening rabbits. *Feed Age*, **10**.

Spector, W. G. (1956). *Handbook of Biological Data*. Philadelphia: Saunders.

Stodart, E. & Meyers, K. (1964). A comparison of behaviour, reproduction and mortality of wild and domestic rabbits in confined populations. *C.S.I.R.O. Wildl. Res.*, **9**, 144–59.

Stranzinger, G. F., Maurer, R. R. & Paufler, S. K. (1971). Fertility of frozen rabbit semen. *J. Reprod. Fert.*, **24**, 111–3.

Suitor, A. E. (1946). Palpating domestic rabbits to determine pregnancy. *U.S. Dept. Agric., Bull. No. 245.*

Suzuki, M. & Bialy, G. (1964). Studies on copper induced ovulation in the rabbit. *Endocrinology*, **74**, 780–783.

Templeton, G. S. (1938). Salt requirements of rabbits. U.S. Dept. Agric. Bureau of Biological Survey. *Wildl. Res. Mgmt Leafl.*, pp. 35–102.

Templeton, G. S. (1962). Domestic Rabbit Production, 3rd ed. Illinois: Danville Interstate Publishers.

Thacker, E. J. (1956). The dietary fat level in the nutrition of the rabbit. *J. Nutr.*, **58**, 243–9.

Thompson, H. V. & Worden, A. N. (1956). *The Rabbit*. London: Collins.

Thorpe, R. H. (1944). A box for holding rabbits for ear bleeding and other minor operations. *J. Path. Bact.*, **56**, 270–1.

Walton, A. (1958). Improvement in the design of an artificial vagina for the rabbit. *J. Physiol., Lond.*, **143**, 26–8 P.

Wilson, W. K. (1936). Cited by Thompson & Worden (1956) in *The Rabbit*, pp. 83–4.

Wooley, J. G. & Mickelson, O. (1954). Effect of potassium, sodium or calcium on the growth of young rabbits fed purified diets containing different levels of fat and protein. *J. Nutr.*, **52**, 591–600.

Zeuner, F. E. (1963). *A History of Domesticated Animals*. London: Hutchinson.

ADDITIONAL BIBLIOGRAPHY

(This list contains recent references which are not mentioned in the text).

Adams, C. E. (1972). Induction of ovulation and A.I. techniques in the rabbit. *Vet. Rec.*, **91**, 194–197.

Alliston, C. W. & Rich, T. D. (1973). Influence of acclimation upon rectal temperature of rabbits subjected to controlled environmental conditions. *Lab. anim. Sci.*, **23**, 62–67.

Castor, G. B. & Zaldivar, R. A. (1973). Tattooing rabbits' ears for identification. *Lab. anim. Sci.*, **23**, 279–281.

Cheeke, P. R. (1974). Feed preferences of adult male Dutch rabbits. *Lab. anim. Sci.*, **24**, 601–604.

Chou, S. T. & Robinson, G. A. (1972). Formulation of a laboratory rabbit diet. *Lab. anim. Sci.*, **22**, 48–55.

Davis, N. L. & Malinin, T. I. (1974). Rabbit intubation and halothane anaesthesia. *Lab. anim. Sci.*, **24**, 617–621.

Faith, M. R. & Simmons, J. B. (1972). A safe and versatile rabbit restrainer. *Lab. anim. Sci.*, **22**, 907.

Fox, R. R. & Crary, D. D. (1972). A simple technic for the sexing of newborn rabbits. *Lab. anim. Sci.*, **22**, 556–558.

Garvey, J. S. & Aalseth, B. L. (1971). Urine collection from newborn rabbits. *Lab. anim. Sci.*, **21**, 739.

Griffiths, H. J. (1971). Some common parasites of small laboratory animals. *Lab. Anim.*, **5**, 123–135.

Hampton, G. R., Sharp, W. V. & Andresen, G. J. (1973). Long term rabbit restraint—A simple method. *Lab. anim. Sci.*, **23**, 590–591.

Jilge, B. (1974). Soft faeces excretion and passage time in the laboratory rabbit. *Lab. Anim.*, **8**, 337–346.

Jones, J. B. & Bailey, D. E. (1971). Diseases of domestic rabbits: a bibliography. *Lab. Anim.*, **5**, 207–212.

Kent, G. M. (1971). General anaesthesia in rabbits using methoxyflurane, nitrous oxide, and oxygen. *Lab. anim. Sci.*, **21**, 256–257.

McSheehy, T. W. (1972). Rabbit accommodation suitable for long term experiments in the field. *Lab. Anim.*, **16**, 343–349.

Morgan, D. R. (1974). Routine birth induction in rabbits using oxytocin. *Lab. Anim.*, **8**, 127–130.

Nims, R. M. & Reader, D. J. (1973). Production of hyperimmune serum with mature rabbits. *Lab. anim. Sci.*, **23**, 391–396.

Nixon, G. A. & Reer, P. J. (1973). A method for preventing oral ingestion of topically applied materials. *Lab. anim. Sci.*, **23,** 423–425.

Opeck, D. K. (1973). A collapsible restraint apparatus for cardiac bleeding of rabbits. *Lab. anim. Sci.,* **23,** 276–278.

Plant, J. W. (1974). Control of Pasteurella multocida infections in a small rabbit colony. *Lab. Anim.,* **8,** 39–40.

Skartvedt, S. M. & Lyon, N. C. (1972). A simple apparatus for maintaining halothane anaesthesia of the rabbit. *Lab. anim. Sci.,* **22,** 922–924.

Ward, G. M. (1973). Development of a Pasteurella-free rabbit colony. *Lab. anim. Sci.,* **23,** 671–674.

Weary, M. E. & Wallin, R. F. (1973). The rabbit pyrogen test. *Lab. anim. Sci.,* **23,** 677–681.

17 The Laboratory Mouse

W. Lane-Petter

General

The mouse is by far the most commonly used laboratory mammal. It owes this unique position to its small size and high fecundity and because like its larger relative the rat it resembles man in being largely unspecialized (apart from its highly specialized rodent teeth). It has adapted itself readily to domestication and cage life. It is in consequence the cheapest and the most readily available of laboratory mammals.

General Biology

Origin. The wild or house mouse (*Mus musculus*) belongs to the sub-family Murinae, family Muridae, sub-order Myomorpha of the order Rodentia. It has a world-wide distribution and, as its name implies, it is commonly found in association with buildings inhabited by man; for example, dwellings, warehouses, farms and food stores. It is also common in habitats apart from man, where food and shelter are plentiful.

All the existing laboratory strains of *Mus musculus* have been derived by selective breeding from the wild type. A number of sub-species have been described. The most important is found in the region of South East Asia, and has been designated *Mus bactrianus*, but it almost certainly does not merit specific status.

General description. The pelage of the wild mouse is agouti, with a lighter colour on the ventral surface. The eyes are black and the skin pigmented. Wild mice vary somewhat in size, but at 4 weeks of age they are likely to weight 18–20 g. They will continue to grow thereafter, but at a decreasing rate. Adult mice may reach 30–40 g at 6 months or more of age. Wild mice are omnivorous and like rats are experimental feeders; that is, they will eat anything edible and even try many things that are inedible, but unfamiliar substances will be eaten very tentatively, the mouse returning to take more of them only if the first bite produces no ill-effects. Wild mice will penetrate small holes, wall cavities and roof spaces; they can travel considerable distances in pipes as small as 2·5 cm in diameter, and even get past S-bends. They can easily climb brick walls. Although their thermotactic optimum is high they can nevertheless survive at low temperatures and Barnett (1960) has reported mice breeding in refrigerated meat stores.

Laboratory mice are similar in size to wild mice, but through selective breeding over more than three-quarters of a century a wide range of coat colours and a considerable range of sizes have been developed. Most laboratory mice are albinos.

For a detailed description of the laboratory mouse the reader is referred to Green (1966) for general biology; to Grüneberg (1952) for genetics; to Rugh (1968) for reproduction and development; to Cook (1965) for the anatomy of most of the systems; and, for a wealth of practical information, to Simmons & Brick (1970).

Size range. The mouse weighs about 1g at birth, and 8–12 g at weaning (18–21 days), some strains being consistently larger than others. Subsequent growth of a typical albino strain is shown in Table I.

Table I

Age (days)											
	20	24	28	32	36	40	44	48	56	63	70
Weight (g) ♂	13	17	21	25	28	31	34	35	38	39	40
♀	13	17	20	23	25	26	28	30	32	33	34

Weight gain of CFLP mice (from published records provided by Carworth Europe). Older mice may reach 50 g or more.

Generally speaking, inbred strains are smaller at all ages, animals from small litters are likely to be heavier than those from large litters, and increased age, in the absence of chronic disease, will be accompanied by increasing obesity.

Behaviour. Crowcroft (1966) has given an entertaining and informative account of the behaviour of wild mice. In the laboratory the mouse is timid, gentle and easily handled, photophobic, gregarious, more active at night than in the daytime, and a determined escaper. Male mice tend to be pugnacious when they mature, and some strains are vicious fighters, but females are usually docile and seldom fight except occasionally when they have a litter and are disturbed by a stranger.

The presence of man tends to inhibit activity in the mouse. This can be easily verified if a mouse-room is observed after normal working hours. Provided the mice are unaware of the presence of the observer their activity and the accompanying rustling and commotion will be considerable, but it will suddenly cease if the observer makes his presence known.

Mice do not tolerate well being singly caged. Such animals eat less food and gain less weight than those caged in groups. They tend to seek rest in dark corners, and will become very nervous and disturbed if compelled to remain in a bright light. When caged together male mice may attack each other savagely, especially if the cage has little or no bedding to afford the weaker ones a refuge; at the same time they will produce a strong odour of acetamide which can rapidly fill a room (Lane-Petter, 1967).

Mice will escape from their cage if they can, and (unlike rats) do not often return to captivity. Escapers will be found running over the tops of the cages, or hiding in dark corners on the floor. A strange mouse resting on the wire top of a cage may be attacked from below by the inmates and seriously wounded.

Strains.   There are more inbred strains of mice than of any other mammalian species, and these have been developed over many years by selective breeding in the fields of genetics and cancer research. The introduction of inbred strains of mice some 40 years ago marked a turning point in cancer research, because for the first time it became possible in transplantation experiments in mammals to distinguish between tissue and tumour immunity. A large number of mutants have appeared in mice and many of these are responsible for pathological conditions which may closely resemble congenital states in man. Grüneberg (1956) has catalogued the more interesting pathological mutants.

The present number of inbred strains of mice is somewhere between 100 and 200, depending on what degree of recognition is given to sub-strains. The most complete and up-to-date information about inbred strains will be found currently in *Mouse News Letter*. Further informa-

tion about non-inbred strains is listed in the *International Index of Laboratory Animals* (1971).

Uses in the laboratory

Laboratory mice account for 60–80 per cent of all mammals used as laboratory animals. Their main uses in the laboratory are in bio-assay and toxicity tests, screening of new compounds, microbiology, virology, radiobiology, cancer research (especially inbred strains) and behaviour research.

The greatest numbers are used in bio-assay, toxicity tests and the screening of new compounds; for all these purposes outbred albino mice are generally preferred. Albinos are also most commonly chosen in microbiological and virological studies. In cancer research, and to some extent in behaviour studies, a variety of strains are necessary, especially closely inbred strains and those with a natural incidence of spontaneous tumours. The recent development of the *nude* mouse, which is a hairless, homozygous recessive and lacks a thymus and is thus deficient in T-lymphocytes, has been of fundamental interest in the field of tissue immunity and transplantation research.

Life-cycle

Life-span. The life-span of the laboratory mouse is anything from $1\frac{1}{2}$ to $2\frac{1}{2}$ or more years. Some inbred strains, however, may develop tumours or conditions such as leukaemia, which will kill them before they are 12 months of age.

Development. Most mice weigh about 1g at birth and 8–12g at 20 days. Festing (1969) has reported an inverse relationship between litter size and weight at weaning.

Young are born hairless and with their eyes and ears closed. They are very active and vocal from the time they are born; indeed, if they are not so they are likely to be rejected by the dam. Okon (1970) has discussed the ultrasonic responses of mouse pups to tactile stimuli.

After about three days the hair begins to grow over the body, the vibrissae being particularly noticeable. By 10 days of age there is a good pelage and the external auditory meatus has opened. At 12 days of age the eyes are open, the pups are very active and able to run around rather stumblingly, and at 13 to 14 days of age they will eat any solid food they can reach. They will also learn to drink water from a bottle or drinking nipple, but it is said that they have to be taught to do this, either by another mouse or by the technician who looks after them.

Breeding cycle. The laboratory mouse is sexually mature at 3½ to 4 weeks of age. The gestation period is 17–22 days with a mean of 19–20 days. Young mice can be weaned at any time from 16 days onwards. The female comes into oestrus approximately every 4 days and will be receptive for perhaps 12 hours. If she is successfully mated she will go on to a normal pregnancy, but if she is mated by an infertile male she may develop a

pseudo-pregnancy which can last nearly 3 weeks. She also comes into oestrus immediately after parturition, but simultaneous gestation and lactation is likely to lead to a delay of up to 7 days in implantation.

A male laboratory mouse will successfully mate 2 or 3 females in one night, but he may become infertile through exhaustion while still able to copulate, and it is in these circumstances that he may induce a pseudo-pregnancy. Bruce (1963) has reported extensively on the part played by olfaction in mice and its effects on

Fig. 17.1 Plastic breeding box used at the MRC Laboratory Animals Centre. (Photograph: LAC.)

breeding and implantation. According to her observations, which have been amply confirmed by other workers, a female mouse successfully mated by one male may fail to become pregnant by that male if she is exposed to another male within 24 hours: and this failure is more likely if the second male is of a different strain from the first. Bruce has shown that the suppression of the first pregnancy depends on the sense of smell being intact; if it is destroyed (by cutting the olfactory nerve path) the Bruce effect is not observed.

Husbandry

Housing and caging (Figs. 17.1–17.3)

Mice may be housed in shoe-box type cages of almost any material or in cages in which walls and floor are of wire mesh. In shoe-box cages they require bedding and, for breeding, a nesting material. In grid cages, they will require a nest-box or compartment for breeding, or at the very least an area of solid floor with ample nesting material. Short and Woodnott (1969) recommend 310 to 390 cm^2 floor area for a breeding pair. Mice on long-term experiments can be satitfactorily accommodated in cages $22 \cdot 5 \times 10$ cm: these will hold 3 mice. Mice should not be held for more than a few hours in numbers greater than 30 per cage, however big the cage, because they tend to congregate in large heaps and the ones underneath may be suffocated.

Laboratory mice are now almost always fed on pelleted diets, and these are best offered in hoppers where the animals have to eat the pellets through the wire. Mice are particularly liable to pull half-eaten pellets through the wire and let them drop on the bottom of the cage while they go back for more. A friable diet therefore is likely to be pulled through in large quantities and the mice may half fill the cage with uneaten food.

Fig. 17.2. Mouse battery used at Allington Farm. (Crown copyright.)

Fig. 17.3. Drawer from the battery shown in Fig. 17.2. (Crown copyright.)

Water is given by bottle or automatic drinker. Mice can learn to drink from either, provided it is within easy reach.

Mice are naturally active animals, and a shoe-box type of cage with a wire grill on the top will give them ample opportunity for exercise. Even a heavily pregnant female will spend a lot of her time hanging by any or all of her feet from the wires of the cage top.

Plastic cages of almost any material will be readily gnawed if the mouse is able to find an edge where gnawing can start. The same is true of aluminium and even of pure zinc, but galvanized and stainless steel are gnawproof.

Bedding

Various types of bedding may be used including sawdust, wood shavings, cellulose wadding, peat moss, granulated clays, ground corncobs, dried wood chips, paddy husks, and dried sugar beet pulp (Lane-Petter & Lane-Petter, 1970). Whatever the material it is important that there is enough of it to absorb urine and other moisture in the cage, because mice do not tolerate wet bedding.

Nesting material may be wood wool, paper shavings or simply a paper towel which the animal will shred up and utilize efficiently.

Physical environment

The temperature of the mouse-room may be anywhere in the range 20–25°C. It will be found that in the lower part of this range the food consumption will go up in relation to weight gain, but because of their extreme adaptability the mice themselves will maintain excellent health. An optimal temperature for the mouse-room is 24–25°C.

The relative humidity should be held between 45 and 65 per cent. 12–15 changes of air per hour in the animal-room are recommended, but it is important that the room ventilation is such that every individual cage will get its share of air change. Since even in a small animal-room there are likely to be very many mouse cages it may sometimes be difficult to ensure their adequate ventilation.

A type of ventilated rack described by Lane-Petter (1970) will in many cases go far to ensure an adequate and uniform degree of ventilation (Fig. 6.4).

Mice are sensitive to noise, especially in the higher frequencies of the human auditory range and beyond it. Many people believe that background music has a calming effect upon mice and will to some extent blur sudden noises that are ordinarily disturbing, such as hammers banging on nails and metal utensils being crashed. Some observations about the possible effects of music and of noise on mice has been recorded in Woolstenholme & O'Connor (1967); see also Iturrian (1971) and Pfaff (1974).

The male mice of some species are extremely sensitive to certain chlorinated hydrocarbons (Christensen, Wolff, Matanic, Bond & Wright, 1963) and their sensitivity has been shown to be due to the presence of circulating androgens. The effect on the animals is to produce a massive renal necrosis, from which they will die within a few days. Only adult intact males of some strains are affected. Castrated males, females and immature mice of both sexes are not affected, but castrated males, or females injected with androgens, become susceptible.

The breeding performance of mice is influenced by the day-length. A winter light cycle—perhaps 8 hours of illumination to 16 hours of darkness—will lead to fewer pregnancies than a more prolonged illumination. For maximum breeding throughout the year the mouse-room

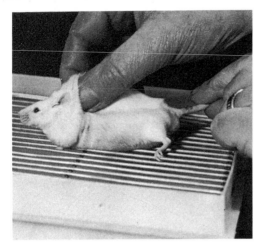

Fig. 17.4. Method of holding mouse by grasping the scruff of the neck and the tail. (Photograph: UFAW.)

should be illuminated for 12–16 hours in every 24, controlled by a time switch.

Identification

It is possible to run a colony of inbred mice in which every animal is fully recorded without marking a single one, but there are occasions when it is necessary to identify individuals, and this is most easily done by clipping or slitting the ears according to various patterns (see for example Short and Woodnott, 1969, for a code of markings that has been widely accepted). Unpigmented ears can also be tattooed.

For identification that does not have to last more than 2 or 3 weeks, painting the tail with a marker pen containing waterproof ink serves well.

Handling (Figs. 17.4–17.5)

Mice are provided by nature with a very convenient tail by which they may be picked up, and even a heavily pregnant female will come to no harm if held by the tail. It is important to remember, however, that the animals must never be held by the tip of the tail, but only by the proximal half.

To immobilize a mouse more securely, put it on a rough surface such as the top of the cage, and grasp a generous fold of the skin of the scruff of the neck between finger and thumb. The tail may be held and kept in place by the little finger of the hand holding the scruff. A mouse so held can breathe comfortably, and can be carefully examined and if necessary injected intramuscularly or intraperitoneally by the same person, using his free hand.

Mice may also be grasped by the whole hand, and young ones may be picked up by the handful without injury.

It is important to remember that some strains of mice are much more lively than others, and they will be at their liveliest at $2\frac{1}{2}$ to 3 weeks of age. At this age they may jump vertically out of an open cage, and when caught may attempt to bite. Some first-generation crosses between inbred strains are particularly mercurial at this age, and a bench with shutters round it may be necessary to prevent escapes.

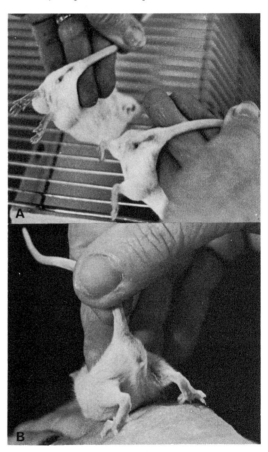

Fig. 17.5. Sexing of (A) female and (B) male mice. *Note:* (A) Short ano-genital gap and the relatively hairless line between the genital aperture and the anus. (B) Long ano-genital gap and presence of scrotum.

Mice are small enough to survive a free fall from almost any height, and they have a well-developed righting reflex; an accidental fall from the table will seldom if ever cause injury.

Transport

Mice, being small animals, have a high metabolic rate and they therefore produce a lot of heat. During transport their density within the cage, and the density of cages in a vehicle, are likely to be high, and therefore they are much more likely to suffer from heat associated with lack of ventilation than from cold. They can, however, suffer from cold on a journey, particularly where this is associated with draughts that not only remove heat but also cause dehydration. Therefore, whatever the container in which they are packed for a journey, they should be placed in a cool, ventilated, but not draughty place.

Mice will gnaw their way out of most cardboard containers within a few hours. No carrier likes to have escaped mice in his vehicles, and airline pilots will quite properly refuse to fly their planes if there is any suspicion that an escaped mouse is aboard. It is, therefore, necessary to see that transport containers are escape proof.

Containers made of cardboard should be lined with wire mesh. Wooden containers of material 1 cm or so in thickness will suffice for most journeys, unless the mice can find a loose knot. A travelling container that has proved very satisfactory consists of a plastic tray at top and bottom, in between which is a wire, or wire and sheet plastic, wall forming the periphery of the box. This wall can be lined with filter material, or it can be of open mesh with cardboard half way up the side to keep the bedding from coming through the mesh and to give the animals some protection from side draught. The corners of such a cage should be generously rounded so that it is impossible to stack the cages in such a way that no air can circulate through them. Experience has shown that these cages are very satisfactory in use and their cost is reasonable (Fig. 17.6).

Transport cages should contain a generous amount of bedding material and, if the journey is likely to last more than six hours, some feed. The mice will come to no harm if they do not get

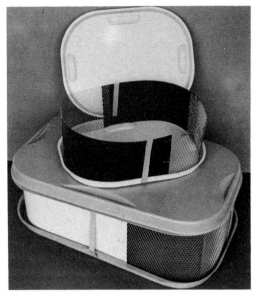

Fig. 17.6. A convenient type of transport cage for rats (larger cage), mice (smaller cage) or other small rodents. The top and bottom are of plastic, and the side a combination of plastic sheet and expanded metal. (Photograph: UFAW.)

drinking water for 24 hours during the journey. Suitably-packed mice will travel well by road, rail or air, and even audiosensitive strains seem to be unaffected by aircraft noise.

For recommendations for transport by air made by the British Standards Institution see British Standard 3149 Part 3 (1961).

Feeding

The wild house-mouse is omnivorous, but its preferred foods seem to be cereals, seeds, and many types of stored food-stuffs.

The mouse, like most rodents, has a simple digestive tract. A heavily lactating female may ingest her own body-weight of food and water in 24 hours, and this places an enormous burden on the alimentary system. Such a high metabolic rate makes exceptional demands on the diet, and it is therefore remarkable that the mouse has been used so little in nutritional research, and that in consequence so little is known in detail about its requirements.

Standard diets

In a set of indispensable recommendations (Coates, O'Donoghue, Payne & Ward, 1969) energy content, lipid content, total protein, and relative amino-acid content, as well as levels for vitamins and minerals, are proposed but no attempts made to formulate suitable diets. The recommended levels are in most cases related to observations on rats, on the assumption that mice are not very different in their dietary needs. Further detailed information will be found in Lane-Petter & Pearson (1971), Chapter 7.

Most laboratory mouse diets today are presented in either compressed or expanded pellet form 8–12 mm in diameter. There are many formulae in use, but they all have the same basic pattern. There is a large cereal element: either all one cereal, or a mixture of two or more. Most of the protein comes from fish meal, meat meal, dried milk or a mixture of any of these. Some may be supplied by soya bean flour. The fat will include tallow, corn oil or other vegetable fats. Yeast is often present. The vitamins and minerals are usually supplied in the form of a premix. Some of the formulae in use are open; that is, the total list of ingredients and their relative proportions are published. Others are the property of the manufacturer, who maintains strict secrecy about the formula, but will list the possible ingredients (without giving relative amounts) and guarantee minimum or maximum levels of certain components. He will also give a representative analysis; this may be based on analyses of batches of diet as produced, but in some cases it is merely theoretical, having been worked out from tables of average compositions of ingredients (see Tavernor, 1970).

Most scientists would prefer to know what precisely they are offering to their animals, but at the present time it may be preferable to feed a proprietary diet with a secret formula (which may have been the subject of some excellent feeding trials) that gives good results, rather than offer a diet made to some open formula that has not been adequately tested under intensive conditions, or in which the specification of the ingredients is too vague to ensure high quality and batch uniformity.

Pelleted diets give the impression of being standard diets, but too often this is an illusion. Many workers have had the disconcerting experience of their breeding or experimental programmes being disrupted because a new batch of diet—from the same supplier and made

to the same formula—has in use shown itself to be different from the one supplied previously. This applies particularly to mice, which are the most sensitive indicators of dietary inadequacy (Porter, Lane-Petter & Horne, 1963; see also Gärtner, Pfaff & Rode, 1974.

The food hopper should contain pellets at all times. Mice will gnaw them through the base, and when they are small enough to be pulled through may drop them in the cage and go back for more, thus wasting the half-eaten material; pellets should therefore be reasonably hard.

Water

Water must always be available, in a bottle with a drinking tube or in a satisfactory automatic drinker. Mice require about the same weight of water as of food. To prevent massive bacterial proliferation in the water bottle the water may be acidified with hydrochloric acid to give a pH of 2·0–2·5. This may be achieved by adding 2 ml hydrochloric acid (B.P.) to 3 litres of tap-water.

Prolonged administration of such acidified water does not appear to have any harmful effect on mice, nor would it be expected to, since hydrochloric acid is a normal ingredient of gastric secretion. On the other hand, the acid suppresses bacterial, fungal and algal growth in the bottle, and it also prevents the deposition of calcareous matter from the tap-water.

Supplementation

A complete diet should not need any supplementation, but it is evident that from time to time some trouble may arise with a particular batch of diet, leading to a crisis situation.

In such circumstances it may be necessary to supplement the diet temporarily, and this can often be done by dusting or spraying the pellets with a preparation containing the supplement. More exact dosing entails the making up of a new diet, or individual administration of the supplement to each animal—a considerable task in a large mouse colony.

Sometimes the supplement can be given in the drinking water. If this is to be done, it is necessary to ensure that the substance added will be stable in the bottle, and not destroyed by acid, light, or other influences.

Occasionally it may be desired to give an

antibiotic or other drug to the colony. The need for this is rare, but may be compelling. It must be emphasized that routine antibiotic administration is never permissible; the animals so treated are rendered quite unsuitable for experiment; they may also develop an abnormal and possibly pathogenic microflora. However, a serious infection may rapidly be brought under control by a short course of medication, and this is best given in the water at an appropriate dose level.

Breeding

Mice breed readily all the year round. Breeding is more prolific if they are kept in groups: either a number of females with a male, or a group of pregnant females together. A useful maternal group is three females, with their litters; the young will be raised in a single nest, and suckling is indiscriminate.

Litters may be of 20 or more, but in a good outbred strain an average of 12 is normal. The female will produce as many as 8 or 10 litters if she is given the opportunity, but after the first 5 litters her productivity falls; the litters are likely to be less frequent and smaller in size, and the weaning losses higher.

A breeding life-time of up to 9 months is usually enough. The male, too, should be retired after not more than 6 months of breeding; by this time he is likely to be getting obese, and his fertility will be falling off.

Selection

In any breeding colony of mice it is necessary to apply some sort of selection in order to ensure acceptable rates of productivity and low levels of mortality and morbidity. There are countless methods of selection, ranging from the most highly sophisticated methods employing random tables and numerous calculations to just picking out a few good-looking animals whenever some new breeding stock is required. The latter method is nearly always disastrous in the long run, because the largest and best looking animals are likely to come from small litters, and this will lead to an inevitable decline in productivity. On the other hand an intensive selection for the best 10 per cent from which may be chosen the next generation of breeders will in all but the

very largest colonies lead to too small a gene pool being carried over from one generation to the next and, in consequence, a rise in the coefficient of in-breeding.

In a non-inbred colony closely-related matings should be avoided, especially those as close as brother by sister. However, the occasional closely-related mating, if it is not repeated, is most unlikely to produce any observable or lasting damage to the colony, and therefore it is seldom if ever necessary to adopt a system so rigid that such a mating can never occur. It is sufficient to deal in high probabilities. Selection therefore should aim at eliminating the worst 10 per cent or so from the list of candidates for future breeding stock. Thus, females that produce small litters, fail to lactate well, or produce abnormal young, should not be given any further chance to breed. Ruthless culling of such poor performers will itself have a sufficiently strong selective effect on the whole colony to ensure that high standards are maintained and even improved. Quite a useful general system is to use this method of negative selection (that is the elimination of the worse end of the scale) for choosing female breeders, but to use a more positive selection for the males. If the very best litters are set aside to provide future male breeding stock, those same litters should not be used for providing females. This will avoid the possibility of brother by sister mating. The negative selection of the females will ensure the maintenance of an adequate gene pool. For a further discussion of this, see Lane-Petter & Pearson (1971), Chapters 4 & 10.

Culling of inferior animals is vitally important in all breeding colonies. Inferior animals include not only those that are obviously below weight, deformed in some way, or unthrifty, but also the males that take too long to get their mates pregnant and females that are inclined to delayed implantation, do not get themselves mated on time, produce small litters, or fail to raise all or nearly all of them to weaning.

Inbreeding

When it comes to inbreeding the situation is quite different. Here repeated brother by sister mating is the rule, at least for the primary colony. The definition of an inbred strain is to be found in Staats (1968): 'A strain should be regarded as inbred when it has been mated brother × sister (hereafter called b × s) for twenty or more consecutive generations. Parent by off-spring mating may be substituted for b × s matings provided that, in the case of consecutive parent × offspring matings, the mating in each case is to the younger of the two parents.'

Although there is no theoretical reason why b × s mating should not on occasion be bigamous, in practice it is very much easier to keep strictly to the principle of monogamy. Many systems of managing an inbred colony have been described, and the choice is usually a matter of convenience. However, there are certain unbreakable principles:

1. The definition of an inbred strain must never be departed from. A single breach of this rule can completely nullify many generations of inbreeding.
2. Care must be taken not to establish a number of parallel sub-lines in an inbred colony. This will happen in any pair that is separated from any other pair in the colony by more than a very few generations.
3. A sufficient degree of selection must be applied to prevent the strain from dying out.

For further information on inbreeding see Ch. 2.

Outbreeding

A number of breeding systems are used for the production of mice other than inbred strains. Permanent pair-mating may be employed, in which the male and the female are never separated, and the young that are born are weaned at 17–20 days. In many cases, the female will have been mated at post-partum oestrus and therefore she will be simultaneously lactating and gestating the next litter. Successful post-partum matings can be expected in at best 80 per cent of cases and in these there will be some delayed implantation. Therefore, in a permanent monogamous pair-mated colony, although the average inter-litter interval (assuming that post-partum mating always takes place) should be about 20 days, in practice it will seldom be found to be below 25 days.

If two females are placed permanently with one male, that male will beget nearly twice as many young. There is likely to be a slight rise in infant mortality, especially when the two females are out of step in their production and a newborn litter comes along ten days after the previous litter from the other female. In cases like this, the older pups are likely to strip the milk off the mother of the younger litter and the younger pups may not get enough milk. However, the decrease in productivity per female is only small, and the saving in space and cages very considerable.

A further step in the same direction may be made by putting three or more females, up to a maximum of about six, to each male. With these larger numbers, larger cages are required than are used for pairs, but the productivity per foot of shelf space will certainly go up. So will the infant mortality. Moreover, it will seldom, if ever, be possible to give an exact birth date to any young produced by this harem system. However, despite its disadvantages, undoubtedly it is the simplest system to operate.

All the foregoing systems take advantage of the fact that the mouse has a post-partum oestrus. Simultaneous gestation and lactation places a very big burden on the physiology of the dam. If the strain being bred normally produces large litters, say an average of 12 or more young, the dam that is pregnant by a post-partum mating may well be unable to carry a new litter and produce enough milk to give optimal growth to the litter she has just had. The result will be small weanlings or high infant mortality or both. With a very fecund strain, therefore, it may be expected that better results would be obtained by separating pregnant females before parturition and putting them either on their own or with two or three other females in a maternity cage and in fact this has sometimes proved to be the case. With discontinuous breeding (that is without the post-partum mating) it is found that the dams can raise the bigger litters that they are bearing to optimal weight at weaning, and that the system allows a great saving of shelf space and of labour. This underlines the fact that in a breeding unit productivity should be measured by at least three criteria, namely productivity per breeding female, productivity per pair of hands employed and productivity per foot of shelf space.

Whatever method of breeding is chosen, care must be taken to identify and cull all breeders that are not doing their fair share of work. Technicians who fail to cull their slow breeders, their infertile males and females, or their bad mothers, will soon find their colonies over-burdened with passengers and turning out far fewer animals than they should.

Cross-fostering, as described in Chapter 18, can be used as well with mice as with rats. Since mice tend to raise their young more readily in small communities, a mixed group of, say, 36 young mice should be placed in a single nest with three dams; it has been found that even dams previously unacquainted with each other will settle down without difficulty in these communal conditions.

The best time for cross-fostering mice is when the young are two days old, but it is possible at any time from 1 to 4 days.

Special requirements

From time to time users request timed pregnant females. The oestrous cycle of the mouse is about 4 days and oestrus itself lasts for a matter of hours only. For about 24 hours after mating the vagina of the female is filled with a very obvious plug of coagulated ejaculate. This coagulum is some 3 mm in diameter at the open end of the vagina, and thus can easily be seen by the naked eye. It may stay in for as long as 48 hours, but it is liable to shrink and come away. It also turns colour slightly during the second 24 hour period, becoming a sort of pearly grey. Recent mating can thus be positively confirmed by the presence and appearance of a vaginal plug.

Whitten (1958) found that if female mice of breeding age were placed together in a cage the majority of them tended to become an-oestrous. If after a few days the male was put among them, either in the same cage or placed in a separate wire cage where he could see and be seen and smelt, then the majority of the females would come into oestrus together some three days later. The same effect can be achieved by placing among the females a male with severed vasa deferentia.

The Whitten effect is useful in getting an appreciably higher proportion of timed preg-nant mice on a given day than is possible otherwise.

From time to time it is necessary to identify the stage in the oestrous cycle of the mice by vaginal smear. A sample of vaginal contents is removed with a suitable spatula and spread out on a clean glass slide. This can be examined under a simple microscope with a magnifica-tion of about x 100 either fresh or after drying in the air.

The appearance of vaginal smears is well illustrated in Short & Woodnott (1969). Only a very elementary microscope is necessary, and one with its own illumination is quite sufficient and much more portable than many larger instruments (see Fig. 6.7).

Weaning and growing on

From 14 days, mice begin to eat and drink independently, but they cannot eat very hard pellets until 18–20 days of age and although they may be weaned from 16 days onwards, weaning before 18 days will cause a severe check in growth from which they may not entirely recover; that is, they will always remain rather below normal weight. Mice should not normally be left with the dam beyond 21 days of age.

Young mice of weaning age are very active and quick moving, and an inexperienced tech-nician may have some difficulty in handling them. When in the cage and at rest they tend to huddle together, forming large heaps with some buried in the middle. There is a constant move-ment in the heap that tends to bring those inside to the surface. Not more than about 30 mice should be caged together for any length of time.

Growth after weaning will be influenced not only by the diet but also by the density of animals in the cage and by the ambient tempera-ture. The optimum density will vary with the strain, the type of cage and the room tempera-ture. Generally speaking, lower temperatures and reduced densities lead to slower weight gain, but too high densities also slow down growth and densities should be kept under close observation in order to achieve optimal growth.

Disease Control

A useful account of mouse diseases will be found in Tuffery & Innes (1963), and the pathology of mice is comprehensively dealt with by Cotchin & Roe (1967). This is not the place to discuss individual diseases, but something of disease control should be mentioned.

Factors favouring disease and its spread

Anything that places an extra burden on an animal's metabolism is likely to predispose it to disease. Poor nutrition, overcrowding, lack of hygiene, inadequate ventilation, bad handling, and exposure (directly or by indirect contact) to mice suffering from infectious disease will all favour an outbreak of disease in a colony.

A large colony of high density offers a more favourable environment for the spread of infection than does a small colony. Infrequent changes of bedding and cages will also favour the development of disease, and so will a poor general standard of cleanliness in the animal room. Poor ventilation, which is by no means uncommon, is another adverse factor; unskilled handling, by stressing the animals, also tends to predispose towards disease.

In principle, none of these environmental factors will lead to an outbreak of infectious disease if the specific infective organism is not present; but even in modern animal houses with efficient peripheral barriers it must not be assumed that such organisms can always be successfully and permanently excluded.

Hygiene

A wild mouse is probably the greatest risk to the health of a laboratory mouse colony. The mousehouse must, therefore, be absolutely proof against rodents, and as far as possible against insects and other pests.

The next important hazard is the technician, whose hands touch the mice and the food and bedding in the cages. These hands may have picked up dangerous infection elsewhere—from stroking a cat which hunts wild mice, for example. Careful and frequent handwashing is therefore essential, and it should be followed by rinsing the hands in 70 per cent alcohol (methylated spirits 3 parts, water 1 part).

Food and bedding may be sources of infection, especially if they have been exposed to the risk of contamination by wild rodents. Pelleted feeds in the bag are unlikely to contain dangerous organisms (although this possibility cannot be ruled out), and to offer them without further treatment is a calculated, but generally small, risk. Bedding materials may be more dangerous, especially wood products such as sawdust or shavings which are by-products of woodmills and are very likely to be contaminated not only by bacteria (such as salmonella) that may cause serious disease, but also by helminth eggs, and even arthropod parasites.

In short, good social cleanliness will go far to protecting a healthy colony of mice from disease.

Some common diseases

Tuffery (1959) has listed a number of mouse diseases, among which ectromelia and infantile diarrhoea (virus diseases), salmonellosis (a bacterial disease) and Tyzzer's disease (of uncertain aetiology but thought to be bacterial) were regarded as the four most serious. Today, ectromelia and salmonellosis are rare in healthy mouse colonies, infantile diarrhoea is only an occasional nuisance in breeding colonies, and Tyzzer's disease is sporadic and unpredictable, but seldom very serious. Less general virus infections, such as Sendai, mouse hepatitis, lymphocytic choriomeningitis, and a number of others, may interfere with certain types of investigation but do not produce devastating epidemics.

Ectoparasites are far less common than they used to be, although mites are still seen in some colonies. Mice may be dipped in a solution or suspension of a miticidal agent, such as monosulfiram (Tetmosol), benzyl benzoate, or gamma benzene hexachloride, and this is fairly effective, if tedious. Alternatively, and much less laboriously, a strip of material which slowly releases a volatile organophosphorus compound (for example, Vapona) may be placed on the top of the cage, and this will lead to a rapid and complete kill of mites. Fraser, Joiner, Jardine & Galvin (1974) have described the use of dichlorvos pellets in the bedding, which is both convenient and effective, while Green and Needham (1974) report the use of fine particulate sulphur in a drum in which the mice are rolled.

Endoparasites, particularly pin-worms, are often present but are well tolerated by the host and do not often interfere with the experimental work. Tapeworms are not often seen.

The only way to ensure that a mouse colony is free from most or all of these common infections and infestations is to obtain the animals by hysterectomy and foster them on to germfree or ex-germfree dams, and then ensure that they do not become recontaminated.

Culling

One of the earliest signs of infection in a colony is slight loss of condition in a few animals. A conscientious and experienced animal technician will be the first to notice this, and it is important that at the first evidence of ill-health the affected animals are vigorously culled. Tuffery (1962) has discussed the applications and usefulness of culling which in many cases will limit the spread of an infection.

Quality control

Apart from culling, which is a method of quality control, the amount of continuous monitoring of the health of a mouse colony depends on the resources available.

Routine microbiological screening of animals is probably the most expensive and least rewarding procedure. Careful examination of culled animals is always useful, and may give the first evidence of trouble. A search for virus antibodies, by microserological screening, will indicate the presence or absence of a range of murine viruses in the animals examined. Macroscopic examination of a sample of surplus animals, or of culls, followed by histological examination of tissues which show lesions, is always useful.

Quality control may be likened to the work of customs officers at a busy airport. It is not possible to examine all baggage, and even if it were, some well hidden contraband might get through. But spot checks, together with general precautions, by skilled customs officers keep smuggling at an acceptably low level. That is about the best that can be expected in most animal colonies.

LABORATORY PROCEDURES INVOLVING MICE AND RATS

K. E. V. SPENCER

Since similar general laboratory procedures are used for both rats and mice this account can be applied to either species: important differences are noted where they exist.

The importance of an adequate experimental design cannot be too strongly emphasised.

Every experiment, including those of the pilot 'look-see' variety, should, as far as is practicable, be designed so that it is capable of producing a statistically valid result. Powerful tools, such as the analysis of variance, quantal analysis and non-parametric methods, are now available for this. If the help of a competent statistician cannot easily be obtained then there are some excellent texts on the subject such as those by Smart (1970) and Rümke & De Jonge (1964).

The use of proper control groups is essential and the practice of comparing experimental results with values quoted in the literature can at the best give only re-assurance to the experimenter.

Careful standardization of the experimental conditions helps to reduce variability and to increase the reliability of the experiment. A factor often neglected is the effect of the laboratory environment on the results of experiments with mice and rats. Familiar examples are the increase in toxicity of amphetamine and the decrease in toxicity of alpha-chloralose as the environmental temperature is raised. For most work the experimental room should have the same conditions of lighting, temperature and humidity as the animal room in which the mice or rats are reared. These species are nocturnal and behavioural studies should be carried out on animals that are accustomed to reversed-daylight conditions of a dim red light. A night cycle of from 10 a.m. to 10 p.m. is often convenient as this allows feeding, watering and cleaning to be done before 'night-fall'.

The number of experimental animals housed together in a cage is important as this affects the temperature and humidity of the cage microclimate and also has a marked effect upon the endocrine balance of the animals.

A knowledge of the details of the technique of animal husbandry used is also important in assessing the causes of variability. For example, the occasional use of insecticide sprays can cause large variations in drug action; this is due to their ability to cause induction of liver microsomal enzymes which markedly alter rates of drug metabolism. The rat or mouse arriving in the laboratory for an experiment is not just a living test-tube; it has a history which must be understood by the experimenter through mutual trust and co-operation with the animal technician.

The saving in expensive experimental time and animal material, together with the increased confidence in the results obtained, makes the proper design of experiments the first priority in laboratory procedures.

Anaesthesia

For minor surgical procedures of short duration such as cardiac puncture, carbon dioxide or ether can be used to produce anaesthesia. Either solid carbon dioxide or ether is placed in the bottom of a glass desiccator, covered with a perforated platform and the mouse or rat put in the vessel which is then closed. The animal is removed when it has lost consciousness and the operative procedure quickly carried out. Additional ether can, if necessary, be given by a simple gauze face mask. For blood sampling by means of cardiac puncture, carbon dioxide anaesthesia is much to be preferred since it causes least change in the blood chemistry; for example, the blood glucose levels are similar to those present in the conscious animal.

Ether can also be used for experiments of longer duration, although it does cause troublesome and copious salivary and bronchial secretions, and is in addition highly inflammable. The volatile, non-inflammable anaesthetic halothane is much more suitable and with it the depth of anaesthesia can be readily controlled, recovery is very rapid with no ill-effects, and little or no salivary or bronchial secretions are produced. Halothane can be administered by passing oxygen through it in a simple vaporizer with an outlet tube wide enough to act as a face mask for the animal. Since so little of the substance is required to anaesthetise rats or mice it is doubtful if a closed-circuit apparatus is necessary for most work. However, Simmons & Smith (1968) describe a suitable closed-circuit apparatus which could easily be made in a workshop.

Non-volatile anaesthetics such as sodium pentobarbitone and sodium hexobarbitone are also suitable for recovery experiments. Suitable doses for sodium pentobarbitone are 45–60 mg/kg intraperitoneally or 35 mg/kg intravenously and for the shorter-acting sodium hexobarbitone 75 mg/kg intraperitoneally and 47 mg/kg intravenously. Supplementary doses can be given

as required. As has already been mentioned, the duration of anaesthesia may depend on the previous history of the animal such as its exposure to insecticide sprays.

For non-recovery experiments of a few hours duration, urethane (ethyl carbamate) is very suitable since it causes less cardiac and respiratory depression than the barbiturates. The normal dose is 1000–1250 mg/kg given intraperitoneally as a 25 per cent solution in water. For a 200 gram rat an initial dose of 0·6 ml is given followed 10 minutes later by a further 0·2 or 0·4 ml depending on the degree of narcosis already reached. This gives a depth of anaesthesia suitable for visceral surgery and blood vessel cannulation. The trachea should be cannulated to provide a free airway. The blood glucose level of a rat anaesthetised with urethane is often double that of the un-anaesthetised animal and this high level slowly declines towards normal over three to four hours. Urethane is not suitable for repeated anaesthesia in the same animal and care should be taken to avoid repeated skin contact.

Euthanasia

Carbon dioxide euthanasia, using a specially designed cabinet, is the recommended method, although sodium pentobarbitone, injected intraperitoneally at three times the anaesthetic dose can also be used. Nitrogen is not recommended.

Collection of faeces and urine

Several good metabolism cages in which the faeces and urine can be collected separately are available. An all-glass 'Metabowl' (Jencons, Hemel Hempstead) can be used for either rats or mice for the collection of urine and faeces and in addition it can be adapted for the collection of expired air in radio-active tracer studies. North Kent Plastic Cages (Dartford) make a very effective stainless steel metabolism cage with a polypropylene funnel urine collector and faeces separator. Some animals produce soft faeces until they have become acclimatised to the metabolism cage.

Collection of blood

The composition of a blood sample will depend on the method of sampling. In a long-term experiment, therefore, it is essential that the same method should be used throughout for both control and experimental animals. Analytical methods that will enable the volume of the blood sample withdrawn to be kept to a minimum should be chosen. If the volume of the sample withdrawn is an appreciable proportion of the total blood volume, then it is important to remember that the composition of the last part of the sample will differ from that of the initial portion because of the haemodynamic changes produced by the large haemorrhage.

Rat. Large blood samples should be removed by cardiac puncture (Burhoe, 1940). The anaesthetised rat is placed on its back and the heart located under the 5th to 6th ribs by palpation with the index finger of the left hand, the left thumb being held on the rat's right side. Using a 2–5 ml syringe with a 19–25 mm 24–26 gauge needle, the needle is inserted at 45° to the long axis of the body into the thoracic cavity at this point until the heart is felt throbbing against the needle, which is then pushed into the ventricle and the sample withdrawn. The syringe and needle can contain an anticoagulant if necessary.

To obtain small blood samples (0·2–0·3 ml) for haematological or microchemical examination, blood can be taken from a tail vein by snipping the tip of the tail. The animal, including the tail, should be kept warm by means of a lamp and gently restrained. The tip of the tail is cleansed with alcohol and when this has dried the tip is snipped with a clean scalpel and the blood collected in a pipette or directly on to a slide or tube, discarding the first few drops. The tail must not be massaged or the sample will be diluted with tissue fluids.

Mouse. Blood samples can be obtained from mice by cardiac puncture, tail vein puncture or tail clipping, but in the experience of many people collection of blood from the orbital sinus is the most efficient method and the one which causes least stress to the animal (Simmons & Brick, 1970). The method is fully described by Riley (1960) and his paper should be consulted. A capillary pipette or microhaematocrit tube with a polished tip is passed by the side of the eye into the orbit so as to puncture some of the fragile vessels of the ophthalmic venous plexus. The animal is held by the back of the neck and the loose skin of the head is tightened with the thumb and middle finger. With the aid of the index finger, the eye is made to bulge slightly by further traction of the skin adjacent to the eye. The tip of the pipette is then placed at the inner corner of the eye and slid alongside the eyeball to the ophthalmic venous plexus. The polished tip prevents damage to the eyeball. After puncture of the plexus vessels slight withdrawal of the pipette enables the orbital cavity to fill with blood and the pipette fills by capillary action. Bleeding usually stops on withdrawal of the pipette and re-establishment of the normal ocular pressure on the venous network. Until familiarity with the technique is obtained it is advisable to use animals anaesthetised with carbon dioxide. Mice have been bled from both orbits repeatedly for several months without evidence of blindness or other serious damage.

Collection of other fluids

For details of the specialised techniques for the removal of cerebrospinal fluid, lymph and ascites fluid the books by Farris & Griffith (1962), Simmons & Brick (1970) and Gay (1965) should be consulted.

Dosing and injection procedures

Substances can be administered to rats and mice by the oral, subcutaneous, intramuscular, intraperitoneal and intravenous routes. Convenient volumes for injection are 0·2–0·3 ml/100 g bodyweight for rats and 1 ml/100 g bodyweight for mice. Drug concentrations should be adjusted accordingly.

For all routes other than the intravenous, the rat should be held with the palm of the left hand flat along its back and the thumb and index finger around and below the lower jaw, care being taken not to squeeze the trachea. Mice should be held by the loose skin on the back of the neck, using the thumb and forefinger of the left hand and restraining the tail with the little finger. This gives good control of the animal.

Oral administration is best carried out with a syringe fitted with a long hypodermic needle with a bulb soldered on to the end. This cannula is inserted into the mouth and gently slid over the tongue and pushed downwards; it should slide easily into the oesophagus, after which it can be pushed into the stomach. If there is any resistance the cannula should be withdrawn and another attempt made to find the oesophageal opening. It is helpful to tilt the animal's head slightly to align the oesophageal and stomach orifices.

Sub-cutaneous injections are readily given into the loose skin in the neck region of rats and mice or under the loose abdominal skin of rats. For intraperitoneal injections the animal should be held on its back with the abdominal skin slightly stretched. The needle should be at an angle of 10° with the abdomen and slightly to the left or right of the midline so as to avoid the bladder. The injection site should bot be too high, for otherwise there is a danger of puncturing the liver.

Intramuscular injections should be given with a 6·4 mm 24 gauge needle into the posterior thigh muscles, but it should be borne in mind that in both species the volume of muscle is quite small.

Intravenous injections are best given in the tail vein, using a 12·7 mm 24 gauge needle. The animal is restrained in a plastic holder with the tail protruding. It is essential that both body and tail be kept warm with a lamp. Intravenous injections into rats other than young ones require considerable skill. It is much easier to inject into the tail vein of mice.

In addition to the books already mentioned, D'Amour & Blood (1954) describe several techniques using the anaesthetised rat.

REFERENCES

D'Amour, F. E. & Blood, F. R. (1954). *Manual for Laboratory Work in Mammalian Physiology.* University of Chicago Press.

Barnett, S. A. (1960). Effect of low environmental temperature on the breeding performance of mice. *Proc. R. Soc.,* Series B, **151,** 87–105.

Bruce, H. M. (1963). Olfactory block to pregnancy among grouped mice. *J. Reprod. Fert.,* **6,** 451–60.

British Standards Institution (1961). Recommendations for the carriage of live animals by air—rodents, rabbits and small fur-bearing animals. *British Standard* 3149, part 3.

Burhoe, S. O. (1940). *J. Hered.,* **31,** 445.

Christensen, L. R., Wolff, G. L., Mazanic, B., Bond, E. & Wright, E. (1963). Accidental chloroform poisoning of BALB/cAnNIcr mice. *Z. Versuchstierk.,* **2,** 135.

Coates, M. E., O'Donoghue, P. N., Payne, P. R. & Ward, R. J. (1969). *Dietary Standards for Laboratory Rats and Mice. Laboratory Animals Handbook No. 2.* London: Laboratory Animals.

Cook, M. J. (1965). *The Anatomy of the Laboratory Mouse.* London & New York: Academic Press.

Cotchin, E. & Roe, F. J. C. (1967). *Pathology of Laboratory Rats and Mice.* Oxford: Blackwell.

Crowcroft, P. (1966). *Mice all Over. Behaviour in the House Mouse.* London: Foulis.

Farris, E. J. & Griffith, J. Q. (1962). *The Rat in Laboratory Investigation.* New York: Hafner.

Festing, M. (1969). *Laboratory Animals Centre News-Letter,* **37,** 9.

Fraser, J., Joiner, G. N., Jardine, J. H. & Galvin, T. J. (1974). The use of pelleted dichlorvos in the control of murine acariasis. *Lab. Animals,* **8,** 271–274.

Gärtner, K., Pfaff, J. & Rode, B. (1974). Zur Konstanz wertbestimmender Bestandteile in pelletierten Alleindiäten für Laboratoriumstiere. *Z. Versuchstierk.,* **16,** 183–196.

Gay, W. I. (1965a & b, 1968). *Methods of Animal Experimentation,* vols. I–III. New York & London: Academic Press.

Graham-Jones, O. (1964). *Small Animal Anaesthesia.* Oxford: Pergamon Press.

Green, C. J. & Needham, J. R. (1974). A simple technique for sulphur dressing mice. *Lab. Animals,* **8,** 327–328.

Green, E. H. (1966). *Biology of the Laboratory Mouse,* 2nd ed. New York: McGraw-Hill.

Grüneberg, H. (1952). *The Genetics of the Mouse,* 2nd ed. The Hague: Nijhoff.

Grüneberg, H. (1956). An Annotated Catalogue of the Mutant Genes of the House Mouse. *M. R. C. Memorandum No. 33.* London: HMSO.

International Index of Laboratory Animals 2nd Ed. (1971). Carshalton: MRC Laboratory Animals Centre.

Iturrian, W. B. (1971). Effect of noise in the animal house on experimental seizures and growth of weanling mice. In *Defining the Laboratory Animal.* Washington, D.C.: National Academy of Sciences.

Laboratory Animals Centre Collected Papers (1963). The choice of the experimental animal. **12.**

Lane-Petter, W. (1967). Odour in mice. *Nature, Lond.,* **216,** 794.

Lane-Petter, W. (1970). A ventilation barrier to the spread of infection in laboratory animal colonies. *Lab. Anim.,* **4,** 125–34.

Lane-Petter, W. & Lane-Petter, M. E. (1970). Dried sugar beet pulp as bedding. *J. Inst. Anim. Techns,* **21,** 31.

Lane-Petter, W. & Pearson, A. E. G. (1971). *The Laboratory Animal—Principles and Practice*. London & New York: Academic Press.

Mouse News Letter. Published twice yearly by MRC Laboratory Animals Centre, Carshalton.

Okon, E. E. (1970). The ultrasonic responses of albino mouse pups to tactile stimuli. *J. Zool.*, **162**, 485–92.

Owen, D. G. (1972). *Common Parasites of Laboratory Rodents and Lagomorphs*. London: HMSO.

Pfaff, J. (1974). Noise as an environmental problem in the animal house. *Lab. Animals*, **8**, 347–354.

Porter, G., Lane-Petter, W. & Horne, N. (1963). Assessment of diets for mice. 1. Comparative feeding trials. 2. Diets in relation to reproduction. *Z. Versuchstierk.*, **2**, 75–79, 171–82.

Riley, V. (1960). *Proc. Soc. exp. Biol. Med.*, **104**, 751–4.

Rugh, R. (1968). *The Mouse*. Minneapolis: Burgess.

Rümke, C. L. & De Jonge, H. (1964). *Evaluation of Drug Activities: Pharmacometrics*, vol. I., ed. Laurence, D. R. & Bacharach, A. L. New York: Academic Press.

Short, D. J. & Woodnott, D. P. (1969). *The I.A.T. Manual of Laboratory Animal Practice and Techniques*, 2nd ed. London: Crosby Lockwood.

Simmons, M. L. & Brick, J. O. (1970). *The Laboratory Mouse: Selection and Management*. New Jersey: Prentice-Hall.

Simmons, M. L. & Smith, L. H. (1968). *J. appl. Physiol.*, **25**, 324–5.

Smart, J. V. (1970). *Elements of Medical Statistics*, 2nd edn. London: Staples Press.

Staats, J. (1968). Standardized nomenclature for inbred strains of mice: fourth listing. *Cancer Res.*, **28**, 391–420.

Tavernor, W. D. (1970). *Nutrition and Disease in Experimental Animals*. London: Baillière, Tindall & Cassell.

Tuffery, A. A. (1959). The health of laboratory mice. A comparison of general health in two breeding units where different systems are employed. *J. Hyg. Camb.*, **57**, 386–402.

Tuffery, A. A. (1962). Husbandry and health. In *Problems of Laboratory Animal Disease*, ed. Harris, R. J. C. London & New York: Academic Press.

Tuffery, A. A. & Innes, J. R. M. (1963). Diseases of laboratory rats and mice. In *Animals for Research*, ed. Lane-Petter, W. London & New York: Academic Press.

Whitten, W. K. (1958). Modification of the oestrous cycle of the mouse by external stimuli associated with the male. Changes in the oestrus cycle determined by vaginal smears. *J. Endocr.*, **17**, 307–13.

Wolstenholme, G. E. W. & O'Connor, M. (1967). *Effects of External Stimuli on Reproduction*. London: Churchill.

18 The Laboratory Rat

W. LANE-PETTER

General

General biology

The laboratory rat has been developed over the last 70 years or so from the wild brown or Norwegian rat. There are three main groupings.

1. The Wistar albino, developed at the Wistar Institute, Philadelphia, Pennyslvania. This is a quiet, moderately prolific strain which has spread to laboratories throughout the world; it is rather resistant to infection and has a low incidence of spontaneous tumours. The head is wide, especially in the male, and the ears are long. The tail-length is always less than the body-length.
2. The Sprague-Dawley albino, developed at the Sprague-Dawley Farms, Madison, Wisconsin. This is more rapidly-growing than the Wistar, and more prolific. It has a longer, narrower head and a longer tail, which may equal the body length. It is also less resistant to infection, especially respiratory disease.
3. The Long-Evans rat. This is somewhat smaller than the two foregoing, and it has a black hood over the head, is black at the back of the neck and has a black line down the back. It was the subject of a recent symposium (Weisbroth, 1969). A hooded rat of similar appearance, and possibly the same origin, is often referred to in Britain as the Lister hooded rat.

Origin. The wild brown rat (*Rattus norvegicus*) belongs to the sub-family Murinae, family Muridae, sub-order Myomorpha, order Rodentia. Like the house-mouse, it is found all over the world, particularly in association with human habitation. In many areas it has supplanted the black rat, *Rattus rattus*, because it is stronger and more aggressive. Unlike the black rat, it adapts readily to breeding and living in laboratory conditions.

General description. The wild brown rat may be any colour from light agouti to black, with lighter coloration on the ventral surface. The eyes are black, and the skin pigmented. Adult males may attain a weight of 500 g or more, but females will seldom grow as big as this.

The laboratory rat is of similar size. Albino strains have pink eyes, hooded strains more or less pigmented eyes.

Rats are omnivorous and, like mice, are experimental feeders; that is, they will take an unfamiliar food with great caution and return to it only if they have suffered no ill effects. They are very intelligent animals and can be taught to carry out quite complicated manipulations, and this has encouraged their use in behavioural research.

Standard works on the rat are by Griffiths & Farris (1962) (general), by Green (1935) (anatomy) and by Robinson (1965) (genetics).

Size range. The rat at birth weighs about 5g, and will attain 35–50g at weaning at 21 days. The following table gives typical figures of subsequent growth for strains of Wistar (CFHB) and Sprague-Dawley (CFY) origin.

TABLE I

Age (days)	21	24	28	31	35	42	49	56	63	70
Weight (g)										
CFHB ♂		45	63	87	105	129	171	213	255	295	330
♀		43	55	71	83	99	127	155	183	209	230
CFY ♂		52	73	101	122	150	206	262	318	365	399
♀		50	65	86	106	130	172	210	240	258	272

Weight gains of CFHB and CFY rats (from published records provided by Carworth Europe).

Older male rats may weigh 500g or more, but females seldom go beyond 350g.

Hooded rats are considerably smaller at all ages.

Behaviour. Broadhurst (1963) has reviewed the use of the rat in behaviour studies. Barnett (1963) in a long series of papers has studied the wild brown rat and compared its behaviour with the laboratory rat. Calhoun (1962) has outlined its ecology and sociology.

The laboratory rat is normally quiet and very easily handled. The males are much less prone to fighting than are male mice, but if they do fight the loser is usually killed.

Rats are less photophobic than mice, and less gregarious. If a rat escapes from a cage it will more often than not seek to return to the same cage. The activity of rats is less inhibited by the presence of man than that of mice. Rats tolerate single caging well (Gärtner, 1969), and female rats with litters will tolerate the presence of their mate, but not of another female, in the same cage.

Despite their normal docility, laboratory rats, if badly handled or suffering from certain nutritional deficiencies (e.g. hypovitaminosis A) can become savage and will attack the handler, inflicting a deep bite. This awakening of a normally dormant aggressiveness can be contagious; in a normally quiet room the careless handling of a few rats, causing them to squeal, may result in the other rats, which have not been mishandled, becoming savage. But gentle handling soon restores good behaviour. Indeed, the more rats are gently handled, the quieter they become.

Strains. Robinson (1965) has reviewed the genetics of the Norway rat at considerable length, and Jay (1963) has also listed inbred strains. Generally speaking, however, the development of inbred strains of rat has not been carried out to anything like the same extent as with the mouse.

Uses of rats in the laboratory

Next to mice, rats are the most commonly used laboratory mammals, accounting for 10–15 per cent of the total. They are used mainly for toxicity studies, including long-term tests; for nutritional research; for behaviour research; and for cancer research. A small number are used in physiological, and rather more in pharmacological, investigations and for teaching.

Life cycle
Life Span.

The laboratory rat will live for at least 3 years, provided it is in good health. But the great majority of laboratory rats develop chronic respiratory disease, which causes progressive damage to the lungs, so that by the time the animal is $1\frac{1}{2}$ to 2 years old a large proportion of the lung tissue will have been destroyed. Such animals will not achieve a life span of 3 years.

Development. The rat weighs about 5g at birth. It is hairless, with closed eyes and ears, and no erupted teeth.

The new-born rat is scarlet in colour, but at 2 days it is pink. It is active and will suck at the mother's nipples almost from the first hour of extra-uterine life.

The body hair begins to appear on the 4th day and the eyes and ears open on the 13th day. By the 10th day the animals are well covered with fur, and begin to wander away from the nest. On about the 16th day they will begin to eat any solid food they can find in the cage, and to drink from the water-bottle if they can reach it. They will also begin to nibble faecal pellets, for the rat practises refection or coprophagy as a normal part of its feeding behaviour.

Breeding cycle. The laboratory rat is sexually mature at $6\frac{1}{2}$ to $7\frac{1}{2}$ weeks of age, and the female will have her first oestrous cycle at 6 to 7 weeks. The testes of the male descend at about this time, but they are retractable throughout life.

The oestrous cycle in the female lasts for about 12 hours, and recurs every 4 or 5 days; there is also a post-partum oestrus.

The rat has usually been regarded as a spontaneous rather than an induced ovulator. It is true that a female rat will normally have a regular cycle, unless she is with a group of females, when oestrus may be suppressed. She will, too, normally be receptive to the male only at oestrus. But a vigorous and experienced male rat will forcibly mate a female that is not in oestrus and the female will frequently ovulate, and become pregnant as a result (see Ritchie, 1970).

The Bruce effect (see Ch. 17) is not seen in the rat, and the Whitton effect (see Ch. 17) is much less marked than in the mouse. Pseudopregnancy is also much rarer. But loss of fertility, in both the male and the female, is not uncommon; it may be due to vitamin E deficiency, which will cause sterility that is permanent in the male but reversible in the female; to too high a temperature in the cage, which depresses male fertility; or to other causes.

Husbandry
Housing and caging

Rats are most conveniently caged in a larger type of shoe-box cage, with a wire lid, in the same way as mice. Short & Woodnott (1969) recommend 900 cm² of floor area for a breeding pair, and 1080 cm² for a female rat with litter. These are generous allowances: we have found that a female with 14 young will do well in a cage of floor area of 825 cm² while a cage of twice this size will accommodate a breeding male and six adult females. But as the animals grow older and bigger, they will need more space.

In growing on after weaning the density of rats in a cage can have a marked effect on their weight gain. They do not grow, weight for weight, as rapidly as mice, and their point of thermal equilibrium is rather lower, but the young will lose heat very rapidly if exposed in a large airy cage with too few companions for huddling. On the other hand, rats (and especially those of 3–4 weeks of age) will not tolerate too many to a cage, chiefly because of the excessive production of heat. The rat has sweat glands only on its pads. As in the mouse, the tail is the main means of losing heat, and vaso-dilatation in the tail will cause it to dissipate surplus heat to the environment, in much the same way as a car radiator loses heat. If, through overcrowding or from excessive ambient heat, this fails to cool the rat it has one last defence; it will salivate copiously (the rat has relatively enormous salivary glands) and cover its fur with saliva. If this fails, the rat will die within minutes of hyperthermia. If, however, it is removed and can cool down it will recover rapidly, provided it can replace the large volume of water that it has lost.

Like mice, rats will gnaw pellets through a wire basket, but they tend to pull half-eaten pellets through the wire, and then pick them up in their forepaws to eat them (mice often reject them and go back to the hopper for more). Although rats can gnaw hard pellets—indeed they can make short work even of such an un-gnawable substance as glass-filled nylon—the young have difficulty with them; it is better therefore to offer them pellets that are not too compacted.

Rats will drink readily from water bottles or automatic drinkers, and they show much more intelligence than mice in learning to do so.

Bedding

The same type of bedding as is used for mice (see Ch. 17) will do equally well for rats. They do not make such elaborate nests as mice, and a small paper towel is quite enough to make a nest.

Physical environment

A temperature anywhere in the range 20 to 25°C will serve for rats, but rapid or frequent fluctuations should be avoided. If they are kept in open wire cages, the ambient temperature will have to be much more carefully controlled than if they are in solid-walled, plastic cages.

The relative humidity is important to rats. If it falls below about 45 per cent, and the temperature is also low, an outbreak of ring-tail may occur (Njaa et al., 1957; Totton, 1958; Flynn, 1967). The humidity should therefore always be in the range 50–65 per cent.

The susceptibility of the rat to respiratory disease suggests that ventilation may be a very important factor in its environment. This has been discussed in a paper by Lane-Petter (1970), who has introduced the idea of a filter rack to ensure adequate and uniform ventilation of cages in a rat-room. The rat's lungs are sensitive to the effect of irritant dusts (wood dust, debris from other animals, food dust) and gases (ammonia, sulphur dioxide) and one of the first responses to irritation is an increase in goblet cell activity in the epithelium lining the respiratory tract from trachea to bronchioles. More severe or chronic irritation leads to a steady accumulation of lymphoid tissue in the lungs, at the bronchial and bronchiolar bifurcations, and round the smaller blood vessels. At this stage a superimposed infection, with myco-plasma for example, can lead to massive and progressive lung disease, from which the animal will never completely recover.

The filter rack (see Ch. 6) offers the possibility of preventing this sequence of events, and even of mitigating it if it is once in train. In effect, it provides each cage with a steady flow of more or less dust-free air that is little if at all contaminated by air from other cages.

Rats are less sensitive to noise than mice, but loud banging will upset the nursing mothers, and cause them to kill their young.

Controlled illumination is as important to rats as it is to mice in the regulation of breeding.

12 to 16 hours of light in 24 are recommended for uniform year-round breeding.

Identification

Ear clipping or punching, such as is described by Short & Woodnott (1969), is a convenient way of marking individual rats, but the ear of the rat is rather more fleshy than that of the mouse and it may object to having its ears so marked. Tattooing will also serve to identify the animals.

For a less permanent method of marking, a waterproof marking-pen may be used on the tail. Various colours, and positions on the tail, can contribute a useful code, and colours will remain for about 3 weeks, when they can easily be renewed. The most useful colours, which are distinguishable one from another, are black, blue, red, green and brown.

Handling (Figs. 18.1 and 18.2)

Small rats may be picked up by the tail, like mice. Larger animals can also be similarly handled, but the proximal half, preferably the root, of the tail must be grasped. A heavy rat held up by the middle of the tail may slip the skin like a finger-stall.

Larger rats, especially pregnant rats, should not be picked up, but only caught, by the tail. The weight should be taken by a hand, either underneath the body, or grasping the rat over the shoulders and chest. A rat will not resent being grasped from above, but it will become

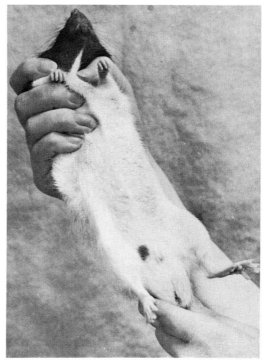

Fig. 18.2. Method of holding a rat for intramuscular (inner surface of thigh) or intraperitoneal injection. (Photograph: UFAW.)

nervous if driven into a corner and may bite. A sick rat, properly held, will not usually bite, but it may gently hold a finger or a fold of skin, as if to indicate that it wants to be very carefully handled.

Adult rats will not survive without injury a free fall on to a hard floor.

Transport

Rats, like mice, create a lot of heat, and therefore when packed in boxes for transport it is essential to ensure a good circulation of air round each cage. They will tolerate cold much better than heat.

For the type of box that should be used, see Ch. 8 and Figure 17.6.

Rats can safely go without water for 24 hours.

Feeding

The wild rat, like the wild mouse, is omnivorous. It is also an experimental feeder. It has a simple alimentary tract, and cannot vomit. The action of the salivary digestive enzymes, which occurs

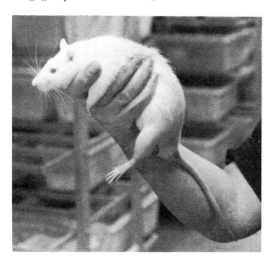

Fig. 18.1. Method of handling an adult rat. (Photograph: Lane-Petter.)

for the most part in the stomach, is important, and the stomach is normally never empty.

Standard diets

Reference should be made to *Dietary Standards for Laboratory Rats and Mice* (ed. Coates *et al.*, 1969). These standards are useful recommendations of dietary levels of all the major, and most of the minor, ingredients of a good diet.

Most of the remarks made in Ch. 17 in relation to the mouse are applicable to the feeding of the rat. Some strains of rat, however, seem to have especially high requirements for vitamin K (Hacking & Lane-Petter, 1968; Gaunt & Lane-Petter, 1967). The male rat is particularly susceptible to vitamin E deficiency and can undergo an irreversible atrophy of the germinal epithelium. For a further discussion of dietary requirements see Lane-Petter and Pearson (1971), Chapter 7.

Water

Water must be available at all times. Rats will drink more water than mice: up to 2 ml water for every gram of food eaten. Acidification of the drinking water with hydrochloric acid, as for mice (see Ch. 17), seems to cause no harm to rats, and has all the advantages already indicated for mice.

Supplementation

The remarks on supplementation made in Chapter 17 are equally applicable to rats.

Breeding

Rats breed easily all the year round, and do not have any particular breeding season. The number of hours of illumination in 24 almost certainly has much the same effect on rats as it does on mice.

Selection

The general principles of selection outlined in Chapter 17 apply equally to rats.

From time to time mutations are liable to occur in a rat colony. Two of the commonest ones are curly coat, and microphthalmia. The curly coat mutant is characterized by a curling and malformation of the vibrissae, a great reduction or absence of guard hairs, and a thickening of the undercoat, which shows a waviness. According to Robinson (1965) a number of genes cause this mutation, some of them dominant and all interacting.

Microphthalmia is not infrequently seen, and may also be genetically caused (see Robinson, 1965).

These and other genetic abnormalities should be bred out by negative selection. Since they will also, in most cases, be associated with general disadvantages such as depressed fertility, they are certain, in any event, to be discriminated against in the general breeding programme that selects for high productivity.

Inbreeding

Although inbreeding has been practised in laboratory rats (enough to demonstrate that its principles differ in no respect from those of mice, and that the results are often very similar) the great variety of inbred strains that have been developed with mice is not parallelled in rats. Jay (1963) has listed inbred strains of rats, and more up to date information will be found in the *International Index of Laboratory Animals*, 1975.

The general principles of inbreeding mice referred to in Chapter 17 apply equally to rats. It has been found that many colonies of laboratory rats that have not been consistently inbred are nevertheless so homozygous that when they are repeatedly sib-mated inbreeding decline either does not occur or is not serious.

Outbreeding

The rat, like the mouse, has a post-partum oestrus. It is possible to mate rats in permanent pairs, so that advantage may be taken of the post-partum oestrus, when a litter may be expected every 21–25 days; but this entails keeping as many breeding males as females, and male rats grow large, eat a lot of food and take up a lot of room. It is, therefore, not a very economical system.

Rats will mate readily in polygamous groups, with one male to a number of females. It is important to have a large enough cage, so that mating is not hindered by restriction of space. A cage of floor area 1550 cm² is large enough for a mating group of one male and six females.

A male rat will be able to serve satisfactorily a

harem of six to nine females, and if he is young and vigorous there will be few failures to impregnate. But a female rat, unlike a mouse, will not always tolerate another female in the cage once she has a litter; she prefers to be on her own, unless she is sharing a cage permanently with a male. If rats are bred in harems, therefore, the females have to be removed when heavily pregnant and placed on their own in maternity cages, where they remain until the young are weaned, which is usually at about 21 days of age. As a female is removed from the mating group or harem, she is replaced by another, either a maiden coming up for breeding, or one that has just weaned a litter. In this way, a stud male is kept constantly occupied with a full harem.

A female rat is capable of looking after as many as 14 young if she is well fed. Many rats are heavy milkers and if they have small litters they may suffer from engorgement mastitis, a condition that can resemble mammary tumour.

Lane-Petter, Lane-Petter & Bowtell (1968) and Lane-Petter & Lane-Petter (1971) have described a method of cross-fostering young rats, in order to standardize at an optimum level the sizes of litters raised by the dams. This method can be practised only in a large colony, where there are at least 500 breeding females.

At two days of age the young from several litters are removed from the dams and sorted out into groups, each group being an 'ideal' litter—that is, a litter of optimal size, say 14 young; each such group, which for further uniformity may be all of one sex, is then returned to one of the dams, who remains in her original cage. Such cross-fostering, when properly carried out, does not lead to rejection by the dam of the new litter, which she will raise exactly as if they were all her own.

The effect of cross-fostering is to produce standard litters of rats, whose growth rate is much more accurately predictable than that of natural, unadjusted litters. The optimal size of this cross-fostered litter may well be greater than the average litter size as born, which means that at every cross-fostering operation there will be a few surplus dams deprived of their young. These can safely be put back to the male, when they will come into oestrus again within about three days, and be mated.

Cross-fostering should normally be carried out when the young are 2 days old, but it may be done at 1, 2 or 3 days, provided all the young are of the same age.

Special requirements

Timed pregnancies are sometimes asked for, and there is no difficulty in obtaining rats with accurately timed matings. After mating, the rat will show a vaginal plug, as do mice, but it tends to fall out after a few hours. If the cage has a wire floor with a tray underneath, the plug may often be seen in the tray.

A more certain way to know if a mating has occurred is to examine a vaginal smear, which is much easier in the rat than in the mouse (see Ch. 17). If spermatozoa are seen in the smear, mating has occurred within the last 12–24 hours.

Weaning and growing on

Rats can be weaned at 17 days, but this will cause some check in growth. The normal age is 20–21 days, at which time they are well able to eat normal food and drink from a bottle or drinker. They are, however, more susceptible to heat loss than are older animals, and should therefore be caged at fairly high densities. A typical density, in a room of ambient temperature of 23–24°C, is 30 rats to a cage of floor area 1620 cm² during the first 7–10 days post-weaning. After that time the density needs to be reduced by up to 50 per cent over the next 14 days, and further still as the animals grow bigger. The actual density depends on the strain and the local conditions, and must be a matter for individual experiment: the above figures are given only as a general guide.

Gärtner (1970) has shown that the number of young adult male rats in a cage may have a profound effect on the corticosterone levels in the blood. He concludes that the lowest level of corticosterones, which is associated with maximal resistance to infection, is seen when a single male adult occupies a cage; the level rises to a peak at 5 rats per cage, and subsequently falls to the single-rat level when there are 10–12 rats together. Thereafter it rises steadily to 30 rats per cage, when a crowding effect begins to become evident. The common practice of placing 5 male rats in a cage for long term studies may therefore require re-examination.

Disease Control

Rat diseases are described in some detail in Lane-Petter (1963) and in Cotchin & Roe (1967).

Few virus diseases have been described in rats, and serious bacterial diseases are not common. Among them, however, should be mentioned leptospirosis, which is usually a mild infection in rats but very dangerous to man; and infection with *Streptobacillus moniliformis*, which can cause a number of lesions in rats, and is also dangerous to man.

But all rat infections are overshadowed by respiratory disease, which is by far the commonest and most serious condition seen in the species. It is nearly always associated with *Mycoplasma pulmonis*, but many years ago Nelson (1946) recognized that there was an associated causal factor, which he considered to be a virus. Even today the full aetiology of this disease has not yet been established. Lane-Petter (1970) has discussed it, and suggests that, whatever the role of mycoplasma in respiratory disease of rats, irritant dusts and perhaps gases in the air may play an important part.

Hygiene

The remarks made in Chapter 17 are equally applicable to rats.

Common diseases

Apart from respiratory disease and leptospirosis rats may carry an infection with *Mycoplasma pulmonis* in the middle or inner ears or in the naso-pharynx. This will lead to the characteristic signs of middle-ear disease and labyrinthitis, in which the animal carries its head awry and if picked up by the tail will twist and circle.

Sendai virus can cause pneumonitis (Tyrell & Coid, 1970), and a number of organisms may be associated with arthritis.

Rats are susceptible to pin-worm (*Syphacia obvelata* or *Aspicularis tetraptera*), but infestations are seldom sufficiently heavy to worry the animals. They are also susceptible to scabies of the ears and muzzle, in which the mite *Notoëdres muris* may be found. Treatment of scabies with monosulfiram (Tetmosol) or benzyl benzoate is usually completely effective. Fleas and lice will only be seen in the worst run colonies and are inexcusable.

For further information about rat diseases, see Cotchin & Roe (1967).

Culling

Culling of inferior animals is as important in a rat colony as in a mouse colony (see Ch. 17).

Quality control

See Chapter 17. The same methods are equally applicable to rats.

Laboratory Procedures

See Chapter 17.

REFERENCES

Barnett, S. A. (1963). *A Study in Behaviour. Principles of Ethology and Behavioural Physiology, Displayed Mainly in the Rat.* London: Methuen.

Broadhurst, P. L. (1963). The choice of animal for behaviour studies. *Lab. Anim. Cent. coll. Pap.*, **12,** 65.

Colhoun, J. B. (1962). A 'behavioural sink'. In *Roots of Behaviour*, ed. Bliss, E. L. New York: Harper.

Coates, M. E., O'Donoghue, P. N., Payne, P. R. & Ward, R. J. (1969). *Dietary Standards for Laboratory Rats and Mice. Laboratory Animal Handbook No. 2.* London: Laboratory Animals.

Cotchin, E. & Roe, F. J. C. (1967). *Pathology of Laboratory Rats and Mice.* Oxford: Blackwell.

Farris, E. J. & Griffiths, J. Q. (1962). *The Rat in Laboratory Investigation.* New York: Hafner.

Flynn, R. J. (1967). Note on ringtail in rats. In *Husbandry of Laboratory Animals*, ed. Conalty, M. L. London & New York: Academic Press.

Gärtner, K. (1970). Beziehung zwischen endokrinen System und soziologischen Situationen bei Massentierhaltung. *Zentbl. VetMed.*, **17,** 81.

Gaunt, I. F. & Lane-Petter, W. (1967). Vitamin K deficiency in 'SPF' rats. *Lab. Anim.*, **1,** 147–9.

Green, E. C. (1935). *Anatomy of the Rat.* New York: Hafner.

Hacking, M. R. & Lane-Petter, W. (1968). Factors influencing dietary vitamin K requirements of rats and mice. *Lab. Anim.*, **2,** 131–41.

Harris, J. M. (1965). Differences in responses between rat strains and colonies. *Fd. Cosmetics Toxicol.*, **3,** 199.

Harris, R. J. C. (1962). *Problems of Laboratory Animal Disease.* London & New York: Academic Press.

International Index of Laboratory Animals (1975). Carshalton: Laboratory Animals Centre.

Jay, G. E. (1963). Genetic strains and stocks. In *Methodology in Mammalian Genetics*, ed. Burdette, W. J. San Francisco: Holden-Day.

Lane-Petter, W. (1970). A ventilation barrier to the spread of infection in laboratory animal colonies. *Lab. Anim.*, **4,** 125–34.

Lane-Petter, W. & Lane-Petter, M. E. (1971). Towards standardized laboratory rodents: the manipulation of rat and mouse litters. In *Defining the Laboratory Animal*. Washington: National Academy of Sciences.

Lane-Petter, W., Lane-Petter, M. E. & Bowtell, C. W. (1968). Intensive breeding of rats. 1. Crossfostering. *Lab. Anim.*, **2,** 35–9.

Lane-Petter, W., Olds, R. J., Hacking, M. R. & Lane-Petter, M. E. (1970). Respiratory disease in a colony of rats. Part 1—The natural disease. *J. Hyg., Camb.*, **68,** 655–62.

Lane-Petter, W. & Pearson, A. E. G. (1971). *The Laboratory Animal—Principles and Practice*. London & New York: Academic Press.

Nelson, J. B. (1946). Studies on endemic pneumonia of the albino rat. *J. exp. Med.*, **84,** 7–23.

Njaa, L. R., Utne, F. & Braekkan, O. R. (1957). Effect of relative humidity on rat breeding and ringtail. *Nature, Lond.*, **180,** 290.

Ritchie, D. H. & Humphrey, J. K. (1970). Some observations on the mating of rats. *J. Inst. Anim. Techns*, **21,** 100.

Robinson, R. (1965). *Genetics of the Norway Rat*. Oxford: Pergamon Press.

Short, D. J. & Woodnott, D. P. (1969). *The I.A.T. Manual of Laboratory Animal Practice and Techniques*, 2nd ed. London: Crosby Lockwood.

Totton, M. (1958). Ringtail in new-born Norway rats, a study of the effects of environmental temperature and humidity on incidence. *J. Hyg., Camb.*, **56,** 190–6.

Tuffery, A. A. & Innes, J. R. M. (1963). Diseases of laboratory rats and mice. In *Animals for Research*, ed. Lane-Petter, W. London & New York: Academic Press.

Tyrrell, D. A. J. & Coid, C. R. (1970). Sendai virus infection of rats as a convenient model of acute respiratory infection. *Vet. Rec.*, **86,** 164.

Weisbroth, S. H. (1970). The Long-Evans rat in biomedical research. *Lab. Anim. Care*, **19,** 699.

19   Wild Rats and House Mice

R. Redfern and F. P. Rowe

The wild rodents most frequently maintained in the laboratory for research purposes are the common or Norway rat (*Rattus norvegicus* Berkenhout), the house mouse (*Mus musculus* L.), and, to a lesser extent, the ship or black rat (*Rattus rattus* L.). A great deal of the information given in Chapters 17 and 18 on the laboratory rat and the laboratory mouse is relevant to the wild forms, and it is the intention here to describe only those aspects concerning the care, maintenance and use of the wild animals that differ from those described for laboratory strains.

The techniques and equipment described below are mainly those used regularly by the authors in pest control research.

Hazards of working with wild rodents

When wild rodents are present in the animal house or laboratory it is essential to maintain a very high degree of general cleanliness and hygiene. This is because the common rat and house mouse (and probably all other wild rodents) are the vectors of several serious diseases which affect humans, and are also the hosts of parasites transmissible to man (see 'Introduction of wild rodents to animal house'). The most serious of these diseases is leptospirosis, or Weil's disease, for which the spirochaete *Leptospira icterohaemorrhagiae* is responsible. During the period 1940–1946, the human death rate was 22 per cent of reported cases (Broom & Alston, 1948). The common rat is the most important reservoir host of leptospirosis, although Twigg *et al.* (1969) have shown that there is an extensive reservoir of several strains of *Leptospira* in wild mammals in Britain. Broom (1958) records the results of a survey of 850 rats from different places in England and Wales, in which the infection rate was 43 per cent. The organism is harboured in the kidneys of the host and excreted in the urine. Entry into the human body can be through any abrasion in the skin, or via the mucous membranes of the eye, nose, throat or lungs during immersion in water. Opinions differ as to whether leptospires can pass through normal unabraded skin (Alston & Broom, 1958).

The main hazard, therefore, is for the technician to be bitten by a wild rat, but it is not difficult to imagine minor accidents from which infection might be contracted (e.g. cuts received from broken water bottles, accidental puncturing of the skin by a hypodermic needle, or an unnoticed abrasion coming into contact with a dirty rat cage).

Other diseases transmitted from wild rodents to man include rat-bite fever, salmonellosis, lymphocytic choriomeningitis, encephalomyocarditis and ringworm (for details see Somerville, 1961, and Hagan, 1955).

The possibility of the transmission of parasites from rodents to man is another reason for precautions to be taken by workers. Gibson (1967) states that cestode infection is one important hazard. The tapeworm *Hymenolepis nana* is common in wild rats and mice, and also affects man; another species *H. diminuta* is more rarely found in man. The parasitic nematode *Trichina spiralis* which spreads to man when infected pork is eaten, causing trichinosis, is also found in wild rats.

Precautions

The following precautions should be taken when animals or dirty equipment are handled:

(1) All cuts, scratches and other abrasions on the hands or arms must be covered.
(2) Protective clothing (e.g. overalls, laboratory coat) must be worn. Adequate laundering facilities should be available.
(3) Industrial gloves, or failing that, surgical gloves, should be used whenever possible (in some experimental procedures with mice this is not practicable).

(4) All dirty equipment should be washed thoroughly and sterilized after use.

(5) Samples of blood and urine taken from wild rodents should not be pipetted by mouth.

(6) Smoking and the consumption of food or drink in the animal house must be forbidden.

(7) Meticulous personal hygiene must be observed. Hands must be washed thoroughly on leaving the animal house, and before touching anything else outside.

Emergency medical treatment

To safeguard staff working with wild rodents, the following safety precautions are suggested:

(a) Every person should be given a booklet laying down basic rules about hygiene, instructions for dealing with wounds, etc. A section should be included giving information for medical practitioners about the symptoms of Weil's disease, and it is recommended that this be shown to a person's doctor if any unusual illness is contracted.

(b) First-aid equipment should be available in the laboratory.

(c) Staff should be urged to undergo a course of anti-tetanus injections.

(d) A person receiving a bite from a rat or mouse should be sent to hospital for immediate medical attention and the patient advised to inform his or her doctor thereafter.

Uses

The numerous and obvious advantages of working with domesticated strains of rats and mice preclude the use of the wild animals in most experimental studies. In certain fields of work however, particularly in rodent control, it is essential to examine wild rats and mice because of the physiological and behavioural differences between them and laboratory animals. Such research studies include the evaluation of rodenticides (Bentley, 1958), the investigation of food preferences and the response to traps, attractants and repellants.

Wild rats and mice have also been maintained under laboratory conditions for the study of population growth and intraspecific behaviour (Crowcroft & Rowe, 1957; Rowe & Redfern, 1969), the physiological effects of stress (Barnett, 1958) the epizootiology of diseases (Capel-Edwards, 1969), and genetic variation in different populations (Carter & Parr, 1969).

Caging

Breeding cages

Rats. Wild rats breed most readily in the animal house when they are given ample space, a dark nesting-box and are subjected to the minimum of human interference. Unrelated individuals, particularly males, often react aggressively towards one another when they are grouped, and it is advisable therefore to use only monogamous pairings (it is also much easier to deal with a small number of rats in a cage).

Lane-Petter (1963) states that, as wild rats are crepuscular or nocturnal by nature, their laboratory descendants show a preference for solid-walled cages rather than wire mesh ones. Notwithstanding this, large wire cages have proved satisfactory for breeding both species of rat and will withstand constant gnawing. Many types of plastic cage suffer from the disadvantage of having loosely fitting lids.

Figure 19.1 shows a family of *Rattus rattus* bred in a cage measuring $60 \times 36 \times 24$ cm, provided with a wooden nest-box $20 \times 20 \times 20$ cm. The box, which has a solid floor and one opening approximately 5 cm in diameter, is liberally supplied with nesting material.

A tray lined with sawdust and placed underneath the cage is used to absorb urine and to

Fig. 19.1. Breeding cage, with family of *Rattus rattus*. (Crown copyright.)

retain droppings. The sawdust can be changed at intervals, causing a minimum of disturbance to the rats.

Mice. Wild mice will breed in any of the cages at present available for laboratory strains, although the standard 'shoebox' design is too small for the best results. Successful breeding has been carried out with harem matings in polypropylene boxes ($43 \times 28 \times 13$ cm), as shown in Figure 19.10.

The Cambridge mouse cage is suitable for breeding wild mice.

Holding cages

Rodents caught in the field usually arrive at the animal house in colony cages or travelling boxes (see 'Transport', page 225). To prevent losses due to overcrowding, the animals should be caged individually without delay (see 'Handling'). It is advisable to allow a period of three weeks to elapse before they are used for experimental purposes; in this time any ailing individuals can be culled, while the remainder have time to recover from the stress of capture and transportation and any pregnant females will have dropped their litters (see also 'Introduction of wild rodents to animal house').

Rats. A cage that has been used to house rats during this quarantine period is shown in Figure 19.2. It is identical to the experimental cage (see below) except that food is held in a basket let into the top door. The small size of the mesh prevents rats taking and hoarding whole pieces of pelleted laboratory diet within the cage. The cage is rested on a tray, supported

Fig. 19.3. Experimental cage for wild rats. (Crown copyright.)

by angle pieces at the corners of the tray to raise it above the sawdust.

Mice. Mice which need to be tested immediately after their three weeks quarantine period is finished are usually kept in experimental cages, described below.

Animals may, of course, be held in any other type of mouse box.

Experimental cages

Much experimental work on wild rats and mice takes the form of free feeding tests in which rodenticides are presented to the animal to determine their efficacy in terms of toxicity and acceptance. It is often important to measure accurately the amount of food consumed and experimental cages and accessory equipment have been designed to meet this end.

Suitable cages for carrying out feeding tests on wild rats and mice are described below. Both types of cage are used, resting above metal trays. At the start of a feeding test the trays are lined with clean white absorbent paper, which facilitates the collection of any food spilled by the animal. In the case of rats, a wire grid resting on top of the tray prevents droppings from falling through.

Rats. The experimental cage (Fig. 19.3) measures $36 \times 28 \times 15$ cm and is strongly constructed from thick galvanized wire. There are two doors, one forming the rear end of the cage, and the other the front half of the roof. Both are secured with stout cage pins. The front door is used for introducing the rat into the cage and

Fig. 19.2. Holding cage for wild rats. (Crown copyright.)

the rear door for extracting it. When a food pot is inserted the rat should be imprisoned in the rear of the cage by a metal partition, which is inserted through a slot midway along the top of the cage.

Different types of water bottle can be used, including the globular type (as shown in Fig. 19.3) with the stalk fitting into a metal bracket at the rear of the cage. Other bottles may be supported on a stand on top of the cage with a bent tube protruding into the cage. Food pots, placed in the centre (or at the sides) of the front end of the cage are fixed in position by pot pins pushed through the two wire loops at the back of the pots. (It is essential to fasten pots in this manner to prevent them from being tipped over by the rat.) To prevent food spillage and contamination, pots are fitted with cross-wires which discourage the animal from climbing inside. *Mice.* An experimental cage designed for carrying out feeding tests using wild mice is shown in Figure 19.4, and measures $36 \times 28 \times 13$ cm. It is made of galvanized sheet metal with half of the top galvanized wire mesh (mice can gnaw perforated zinc). There is a transverse slot in the top to allow for the partition to be inserted. The bottom of the cage is made of wire mesh, and is removable to facilitate washing. The water-bottle fits into a small hole on top of the cage. Each mouse is provided with a small wooden nest-box into which it will readily retreat.

The same type of food pot is used as in the rat experimental cage, although there is no need to fasten it in any way.

Handling

Rats

Handling. Wild rats are among the most unpleasant and dangerous of experimental animals to handle. They will readily attack when cornered, bite viciously without provocation and are the vectors of several serious diseases (see 'Hazards of working with wild rodents'). It is possible to pick up young animals (up, to, say 100g) when wearing thick industrial gauntlets, but even this is inadvisable.

Different workers have devised and perfected their own methods for handling wild rats.

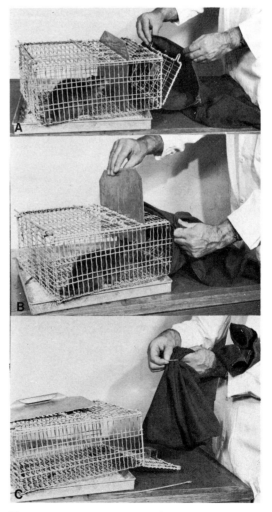

Figs. 19.5.A, B, C. Removal of wild rat from cage—(see text for description). (Crown copyright.)

Fig. 19.4. Experimental cage for wild mice. (Crown copyright.)

Evans *et al.* (1968), for example, describe a box in which rats can be manipulated, sexed and weighed. Information on other techniques is given by Davies & Grice (1962), Kilham & Belcher (1954), and Keighley (1966). Cisar (1973) advocates the use of butcher's boning-gloves for handling wild rats. Well-fitting gloves allow good manual dexterity and afford protection from bites; they are easily cleaned and sterilized.

A quick and very effective method that has been used to handle large numbers of rats is to persuade them to retreat into a strong, dark, closely woven cloth bag measuring about 75 × 45 cm. In the bag a rat will rest quietly and will very rarely attempt to bite through the cloth. This method may be used to extract a number of animals from a colony cage and to re-cage them individually, or to perform a specific operation (e.g. weighing or injection).

Removal from an experimental cage.   The technique for removing a single rat from a test cage is as follows: the animal is confined to the end of the cage that does not open by inserting a metal partition into the cage. The mouth of the

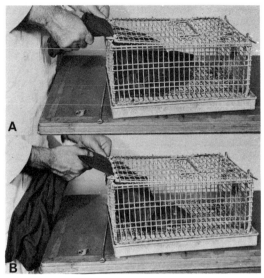

Figs. 19.7A, B. Release of wild rat into cage. (Crown copyright.)

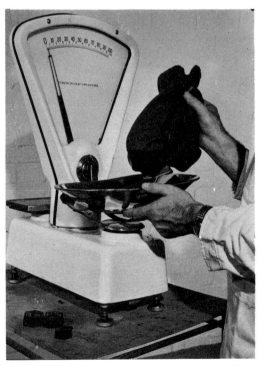

Fig. 19.6. Weighing wild rat in bag. (Crown copyright.)

bag is then stretched tightly over the side (or end) of the cage which opens, so that when the door is unfastened the opening is completely enveloped by the bag (see Fig. 19.5A). The partition is then removed (Fig. 19.5B). If this whole procedure is carried out calmly the rat will not get into a stressed condition and will enter the darkness of the bag with little delay. The mouth of the bag should then be closed and secured with a rubber band (Fig. 19.5C). The animal will then remain sufficiently quiet for it to be placed in a scale pan for weighing (Fig. 19.6).

It is a fairly simple matter to transfer the rat from the bag to its new cage. The bag is placed inside the cage through the top door (Fig. 19.7A). While this door is held firmly down the bag is drawn out carefully, forcing the animal to move towards the mouth of the bag and finally out into the cage (Fig. 19.7B). The door is then made secure by means of the pin.

Holding for experimental procedures.   It is possible to sex a rat when it is confined in a bag, and to carry out subcutaneous, intramuscular and intraperitoneal injections, and also oral intubation. To do so, it is necessary to have two technicians working together, one wearing thick gloves and holding the rat, and the other performing the practical operation. The rat should be manoeuvred to a corner of the bag with its tail towards the mouth of it. The first technician must use both hands to hold the animal down on the bench. Considerable pressure must be exerted to restrain the rat and prevent it from wriggling loose and turning towards the mouth of the bag. The second technician can then roll back the neck of the bag until the rat's tail and hindquarters are visible (Fig. 19.8).

Another method of holding individual rats is by means of an Emlen sleeve (Emlen, 1944). This is a collapsible wire cone into which a rat may be persuaded to run from a bag. When the

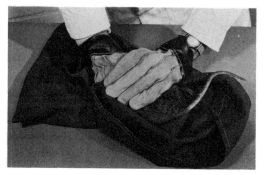

Fig. 19.8. Wild rat held in bag with hindquarters exposed. (Crown copyright.)

animal has entered, the back of the cone is gently squeezed shut. This immobilizes the rat (Fig. 19.9) and is an especially useful technique for dealing with the feet and tail, all of which protrude through the wire, or the urino-genital system (e.g. for taking vaginal smears). It is also possible to carry out injections. Lazarus & Rose (1975) have used an Emlen sleeve when freeze-marking both species of wild rat.

The rat is released by relaxing the grip on the

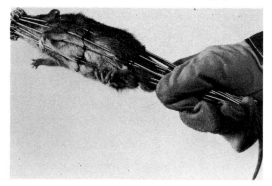

Fig. 19.9. Wild rat held in 'Emlen sleeve'. (Crown copyright.)

mouth of the cone, allowing it to turn round or to back out into a bag.

Mice

Handling. The wild mouse is a thoroughly intractable animal, quite different from its domesticated descendants. However, once the difficulties of handling have been overcome, some of the routine experimental techniques may be carried out with comparative ease by a single technician (e.g. oral intubation, withdrawal of blood from retro-orbital sinus (Riley, 1960)). For injection, the assistance of a second person is essential.

The necessary human attributes for dealing with wild mice are a gentle, determined approach, and a certain sleight of hand. The small size of this species precludes the use of protective gloves; any type of glove which allows the handler sufficient sensitivity of touch would not withstand the vicious attempts at biting which are usual. A few bites will inevitably be sustained at first; these should be treated in the manner described above.

Removal from experimental cages. The removal of one wild mouse from a colony is not easy and should be carried out with the cage in a

Fig. 19.10. Releasing wild mice in metal container. (Crown copyright.)

rectangular container with smooth vertical sides of not less than 45 cm in depth (Fig. 19.10). The lid of the cage is partially removed until a suitable animal emerges, whereupon the lid is quickly replaced (at this point care must be taken that mice do not run up the arm of the handler and leap to their freedom). The mouse may then be caught by placing the cupped hand over it while it is sheltering in a corner of the container (Figs. 19.11A and 19.11B). Provided the animal is not actually gripped at this stage it will not attempt to bite. The handler can then pick up the mouse by the tail with his other hand.

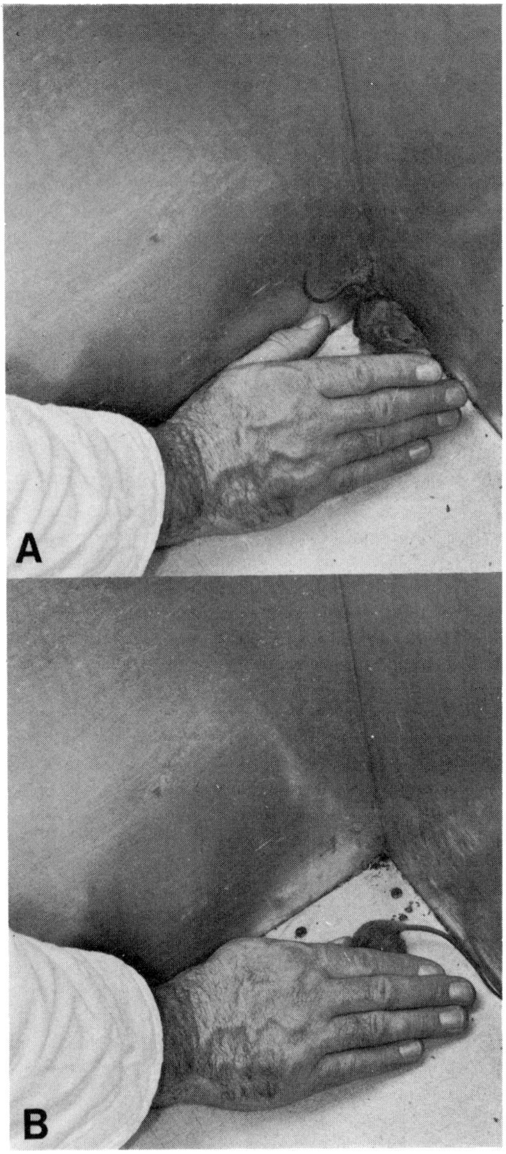

When mice are caged singly, handling is greatly facilitated by introducing a nest-box into the cage. A wooden box $10 \times 3.5 \times 3.5$ cm used for the postal transit of 7.5×2.5 cm glass specimen tubes serves excellently for this purpose. Mice will enter this readily and generally

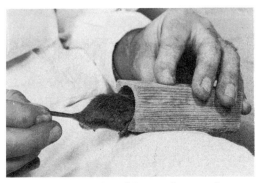

Fig. 19.12. Withdrawing wild mouse from wooden nest-box. (Crown copyright.)

sit with their tails protruding from the open end. They may then be pulled out by their tails (Fig. 19.12). If an animal is facing the entrance, a gentle tap on the front of the box will often make it turn round.

Once the mouse has been firmly gripped by the tail, it will not tolerate being suspended without support for more than a few seconds. If it is allowed to rest on the sleeve of the technician's laboratory coat (or on any rough surface on the bench) it will strain to escape rather than turn round on itself to bite its handler (Fig. 19.13).

Fig. 19.13. Wild mouse straining away from handler. (Crown copyright.)

Figs. 19.11A, B. Catching wild mouse in metal container. (Crown copyright.)

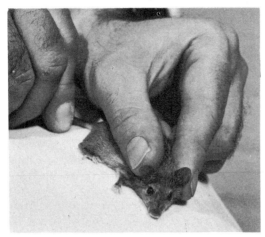

Fig. 19.14. Gripping wild mouse by scruff of neck. (Crown copyright.)

Holding for experimental procedures. Assuming that the technician is right-handed, the mouse must be transferred to the left hand before experimental techniques can be performed. With the mouse still straining against being held by the tail, the technician can quickly run his left hand up the mouse's back and grip it by the scruff of the neck (as far forward as between the

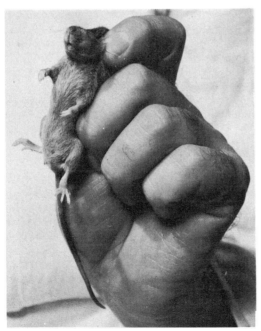

Fig. 19.15. Wild mouse held ready for experimentation. (Crown copyright.)

ears) with his thumb and first finger (Fig. 19.14). The remaining three fingers are then able to grip the loose skin right down the animal's back, rendering it immobile without causing injury (Fig. 19.15).

Wallace & Hudson (1969) have given details of the time taken to catch wild mice by hand for marking, etc. They concluded that this method is preferable to others where apparatus for catching and restraining is used.

Transport

Transit by road and rail

Rats. Because it is not difficult to obtain wild common rats by trapping in the field, it is rarely necessary to breed them in the animal house. When caught, batches of rats may be put together in strong wire colony cages for transit. A cage measuring $60 \times 38 \times 75$ cm will hold up to a dozen, medium-sized animals, provided they are released or thinned out immediately they reach their destination. Sufficient pelleted diet should be supplied in the cages to allow for unforeseen delays on the journey, together with a few cut potatoes, carrots or apples to provide moisture. (Soaked wheat is not suitable as it can fall through the bottom of the cage.) To reduce stress during travel, a large amount of hay or wood wool should be put into the cages. As a safety measure, it is advisable to secure all doors with stout wire.

Fig. 19.16. Wild rat travelling box with two colony cages. (Crown copyright.)

To limit fighting it is preferable to keep wild rats drawn from different sources separate during transit. It may then be convenient to use wooden travelling boxes 68 × 20 × 20 cm fitted with two colony cages (Fig. 19.16). Rats will become less agitated in dark conditions, although ventilation holes must of course be bored in the sides of the box. A padlock should be fitted.

When rail transport is used it is important to cut the journey time to a minimum, and therefore the person despatching the rats should notify the recipient of the expected time of arrival of the train. The travelling box must be marked with conspicuous 'Livestock' and 'To be collected' labels, the former being to prevent it being stacked in the middle of other luggage, thereby blocking the circulation of air.

Mice. A similar, but much smaller, wooden travelling box can be used to transport wild mice (Fig. 19.17). The box (45 × 30 × 30 cm) does not hold cages, but is lined with thin metal or asbestos sheet to prevent the mice gnawing the wood; asbestos has the advantage of insulating the box from extremes of external temperature. The box is divided into two compartments by a central partition, and has a sliding trap door at each end for introducing and removing the mice. The doors are secured by screws, and have ventilation holes (6 mm diameter) drilled in them.

To reduce the chances of fighting between mice during transit the box should be liberally supplied with hay or other nesting material.

Fig. 19.17. Wild mouse travelling box. (Crown copyright.)

Pelleted diet or wheat may be given (these should be put into the box before the mice are introduced!). The water requirements of mice are less than those of rats and therefore it is not so essential to include any vegetable matter in the travelling diet.

Details of rail transport described for rats apply for mice.

A dustbin can also be used for transporting mice. Provided ample nesting material is given thirty or so mice can be kept in a bin for a day or two. It is important that the inside surface of the bin, especially the seam where the metal is joined, is sufficiently smooth to prevent mice climbing out and for the same reason no long stalks of hay or straw should be left. The lid should, of course, fit tightly and have small air-holes drilled in it. It is wise to fasten the lid by tying string between the handles of the bin through that of the lid.

Transit by air

Experience has shown that wild rats may be transported satisfactorily by air and most of the details given in Chapter 8 apply also to the wild species. It is, however, imperative to confine rats in wire (e.g. weldmesh) cages which fit into an outer travelling box made of light timber or plywood (for details see British Standard 3149, part 3 (B.S.I., 1961). Wooden spacers at least 25 mm deep should be fixed to the top and sides of the box to prevent the air vents from being blocked in the course of stowage. A liberal supply of wood-wool will provide cover and reduce the space available for fighting.

Introduction of wild rats and mice to the animal house

It must be stressed that the introduction of wild rats and mice into the animal house is accompanied by the risk of bringing in parasites and disease transmissible to laboratory stocks. It is imperative, therefore, to isolate wild animals in their own rooms (if not their own buildings), and to take great care that cross-infection does not occur from dirty cages, sacks of food, or technicians' overalls etc. Ideally, complete sets of equipment should be kept for laboratory strains and the wild form, and different staff allocated to each.

On arrival at the animal house, wild rodents should be examined and treated for ectoparasites (for details see Tuffery & Innes, 1963). A routine procedure should be carried out in which new arrivals are, for example, dusted with pyrethrum powder to remove insect and mite parasites. This can be done by shaking dust through the wire mesh of the cage.

Laboratory procedures

Anaesthesia

The authors have used three methods of anaesthesia satisfactorily with wild rats:

Carbon dioxide and oxygen. A mixture of equal proportions of these two gases has been found most suitable for blood withdrawal from the retro-orbital sinus of the rat. This anaesthetic is also useful when bleeding by heart puncture is performed because it stimulates respiration and blood return to the heart (Payne & Chamings, 1964).

Ether. A mixture of ether and air is a better anaesthetic when blood samples are taken from the tail vein. Air is blown over, or bubbled through ether and into the jar containing the rat.

Methoxyfluorane. Taylor, Hammond & Quy (1969) state that in work where it is necessary to have rats unconscious for a considerable time (e.g. field routines of weighing, marking, etc.) methoxyfluorane is by far the most successful in comparison with chloroform, ether or halothane. It produces relaxed anaesthesia for long periods without any obviously harmful effect. No deaths occurred amongst several hundred rats anaesthetized.

Collection of blood

Different workers advocate a variety of techniques for blood sampling, including cardiac puncture, incision of the dorsal tail vein, and amputation of the tip of the tail. Karsel & Levitan (1953) describe a method especially suitable for mice in which the external jugular vein is used.

In the course of work on blood coagulation the following techniques have proved to be the most useful:

Rats. This method has been described by Ayres (1969). Blood is taken from the ventral tail vein with the animal under ether/air anaesthesia. The scaly tail of the wild species makes it impossible to see the vein (even with warming), but the technique can be perfected with little trouble. Disposable plastic hypodermic syringes are used with gauge 25 needles flushed out with sodium citrate solution to prevent clotting. The needle is inserted pointing towards the anterior end of the rat.

Mice. The orbital bleeding technique is most successful for wild mice, and has been used extensively. Holding the animal in the correct position is critical. For a full description of the method see Chapter 17 and Riley (1960).

Dosing and injection procedures

Oral administration. The hazards attached to handling wild rats make it difficult to administer substances orally. Redfern (1971) discusses the disadvantages of using anaesthesia and describes a safe technique for oral intubation, in which the rat is immobilised in a bag. Intubation in wild mice is less difficult; they should be held as shown in Figure 19.15. Intravenous cannulae, 5 cm in length, are suitable and should be used in conjunction with a hypodermic syringe.

Injection. Subcutaneous, intramuscular and intraperitoneal injections can be performed in both species. Two people are needed, one to hold the animal, and the other to manipulate the syringe. Rats should be held either in a bag (Fig. 19.8) or an Emlen sleeve (Fig. 19.9). Intravenous injection in the rat is performed in the same way as blood sampling from the ventral tail vein (see above). To make sure the needle is inserted in the vein, a very small amount of blood may be drawn into the injection fluid before the dose is given.

This method is not known to have been used with wild mice.

Acknowledgment. We are indebted to Mr R. Page for taking the photographs.

REFERENCES

Alston, J. M. & Broom, J. C. (1958). *Leptospirosis in Man and Animals*. Edinburgh: Livingstone.

Ayres, P. (1969). Private communication.

Barnett, S. A. (1958). Physiological effects of 'social stress' in wild rats. *J. psychosom. Res.*, **3**, 1–11.

Bentley, E. W. (1958). *Biological Methods for the Evaluation of Rodenticides*. M.A.F.F. Bull. No. 8. London: H.M.S.O.

British Standards Institution (1961). Recommendations for the carriage of live animals by air; rodents, rabbits and small fur-bearing animals. *British Standard 3149, part 3.*

Broom, J. C. (1958). Leptospiral infection rates of wild rats in Britain. *J. Hyg., Camb.*, **56**, 371–6.

Broom, J. C. & Alston, J. M. (1948). Weil's disease. Analysis of 195 cases in England. *Lancet*, **2**, 96.

Capel-Edwards, M. (1969). Spread of foot-and-mouth disease. *Lancet*, **2**, 901.

Carter, N. D. & Parr, C. W. (1969). Phosphogluconate dehydrogenase polymorphism in British wild rats. *Nature, Lond.*, **224**, 1214.

Cisar, C. F. (1973). The use of a butcher's boning glove for handling small laboratory animals. *Lab. Anim.*, **7**, 139–140.

Crowcroft, P. & Rowe, F. P. (1957). The growth of confined colonies of the wild house mouse (*Mus musculus* L.). *Proc. zool. Soc. Lond.*, **129**, 359–70.

Davies, L. & Grice, H. C. (1962). A device for restricting the movements of rats suitable for a variety of procedures. *Can. J. comp. Med.*, **26**, 62–3.

Emlen, J. T. (1944). Device for holding live wild rats. *J. Wildl. Mgmt.*, **8**, 264–5.

Evans, C. S., Smart, J. L. & Stoddart, R. C. (1968). Handling methods for wild house mice and wild rats. *Lab. Anim.*, **2**, 29–34.

Gibson, T. E. (1967). Parasites of laboratory animals transmissible to man. *Lab. Anim.*, **1**, 17–24.

Hagan, W. A. (1955). Diseases of laboratory animals transmissible to man. *Proc. Anim. Care Panel*, **6**, 26–9.

Kassel, R. & Levitan, S. (1953). A jugular technique for the repeated bleeding of small animals. *Science, N.Y.*, **118**, 563–4.

Keighley, G. (1966). A device for intravenous injection of mice and rats. *Lab. Anim. Care*, **16**, 185–7.

Kilham, L. & Belcher, J. H. (1954). A device for safe handling of wild animals in the laboratory. *J. Wildl. Mgmt*, **18**, 402.

Lane-Petter, W. (1963). The physical environment of rats and mice. In *Animals for Research*, ed. Lane-Petter, W. London & New York: Academic Press.

Lazarus, A. & Rowe, F. P. (1975). Freeze-marking rodents with a pressurized refrigerant. *Mammal Rev.*, **5**, 31–34.

Payne, J. M. & Chamings, J. (1964). The anaesthesia of laboratory rodents. In *Small Animal Anaesthesia*, ed. Graham-Jones, O. Oxford: Pergamon Press.

Redfern, R. (1971). Technique for oral intubation of wild rats. *Lab. Anim.*, **5**, 159–172.

Riley, V. (1960). Adaptation of orbital bleeding technique to rapid blood studies. *Proc. Soc. exp. Biol. Med.*, **104**, 751–4.

Rowe, F. P. & Redfern, R. (1969). Aggressive behaviour in related and unrelated wild house mice. (*Mus musculus* L.). *Ann. appl. Biol.*, **64**, 425–31.

Somerville, R. G. (1961). Human viral and bacterial infections acquired from laboratory rats and mice. *Lab. Anim. Cent. coll. Pap.*, **10**, 21–32.

Taylor, K. D., Hammond, L. E. & Quy, R. (1969). Private communication.

Tuffery, A. A. & Innes, J. R. M. (1963). Diseases of laboratory mice and rats. In *Animals for Research*. ed. Lane-Petter, W. London & New York: Academic Press.

Twigg, G. I., Cuerden, C. M., Hughes, D. M. & Medhurst, P. (1969). The leptospirosis reservoir in British wild mammals. *Vet. Rec.*, **84**, 424–6.

Wallace, M. E. & Hudson, C. A. (1969). Breeding and handling small wild rodents: a method study. *Lab. Anim.*, **3**, 107–17.

20 The Guinea-Pig

Michael F. W. Festing

Introduction

The guinea-pig is the fourth most widely used species of laboratory animal in the U.K. (after the mouse, rat and chicken), some 220,000 being used in 1972. This number appears to be relatively constant at the present time, with only a 15 per cent increase between 1956 and 1972, compared with a 94 per cent and a 308 per cent increase in the use of the mouse and rat respectively during the same period (Medical Research Council, 1974). Nevertheless, the guinea-pig remains an extremely important laboratory animal, with many features not found in other species.

The guinea-pig was introduced into Europe by the Spaniards in the middle of the 16th century, where it became a pet, being mentioned by Gessner in Zurich in 1554 (Mason, 1940). The first use of the species as a laboratory animal cannot be traced with certainty. As early as 1664, Robert Hook performed experiments on the effects of fixed air on mice, but the first use of the guinea-pig was probably at the beginning of the 19th century (Mason, 1940).

According to Rothschild (1961), guinea-pigs belong to the order Rodentia, sub-order Hystricomorpha and family Caviidae. The genera of cavies recognised by Huckinghaus (1962) are shown in Table I.

TABLE I

Cavy genera and species recognised by Huckinghaus (1962)

Common name	Genus	Species
Cavy, aperea, cobaye	Cavia	porcellus
		aperea
		fulgida
		stolida
Cuis	Galea	musteloides
		spixi
		wellsi
		palustris
Salt desert cavy	Microcavia	australis
		niata
		shiptoni
Moco	Kerodon	rupestris

The exact origin of the domestic guinea-pig, *Cavia porcellus*, is not known, though it may have been domesticated since pre-Inca times (i.e., before about AD 1450). *C. porcellus* no longer occurs in the wild, and it is not clear which of the existing wild species were its ancestors. The most likely candidates are *C. cutleri*, *C. rufescens* and *C. aperea*. Crosses between *C. porcellus* and *C. aperea* are fully fertile, though some individuals of later generations are infertile. Only the females of hybrids between *C. rufescens* and *C. porcellus* are fertile. On present evidence it seems probable that *C. porcellus* is derived from *C. aperea*, and that the other two species are really forms of *C. aperea* (Weir, 1972). Further information on *C. aperea* can be found in Chapter 24.

According to Walker (1964), the members of the genus Cavia are found in South America from Colombia and Venezuela southwards to Brazil and Northern Argentina. They inhabit a wide variety of habitats ranging from rocky regions to savanas, swamps and the forest fringes.

The combined length of the body and head is 225–355 mm, and the adult weight is usually between about 450 and 700 g. There is no external tail, and there are three digits on the hind feet and four on the front feet, all of which have claws.

In wild forms the pelage is usually coarse, long and greyish or brownish in colour. However, in the domesticated guinea-pig, the coat may be long and fine and may radiate in a series of rosettes. A wide variety of coat colours are known.

All members of the genus have stocky bodies with relatively short legs and ears.

In the wild, guinea-pigs live in small groups of five to ten individuals in burrows which they may excavate themselves or which may have been abandoned by other animals. They are very timid, and usually leave their burrows at nightfall to travel by paths to their feeding grounds. They eat a wide range of vegetation, and are reported to be particularly fond of alfalfa.

Behaviour in semi-captivity

Although the behaviour of wild guinea-pigs has not been studied in detail, King (1956) studied the social relations of the domestic guinea-pig living, under semi-natural conditions, in an enclosure with an area of 2,500 ft². A total of nine guinea-pigs from three different races, distinguishable by colour, were placed in this enclosure, and they and their progeny were studied for a period of one year.

It was concluded that the domestic guinea-pig is slow to adapt to semi-natural conditions, but once they have adapted, their social behaviour is chiefly characterised by:

(a) allelomimetic behaviour in which the young follow the adults, and the males follow the females;

(b) et-epimeletic behaviour in which the young call for attention from the adults, and the adults call for attention from the human caretakers, particularly for supplies of food;

(c) sexual behaviour which matures early and is frequently expressed by the males following the females;

(d) agonistic behaviour, in which males fight vigorously, particularly in the presence of an oestrous female. Females are less antagonistic, but display a weak and flexible social hierarchy.

Imprinting of neonatal guinea-pigs has been studied by Sluckin (1968). Young guinea-pigs aged 5–7 days were separated from their mother and exposed to a moving cube or ball for one hour per day for four days. They subsequently showed strong preferences for associating with the familiar rather than the unfamiliar object. Harper (1970) extended these studies to include objects of various shapes, textures and temperatures, with associated sounds. He found that all these factors had some influence on the imprinting, but the most important factor was the age at which the animals were exposed to the stimulus. In general, a low quality stimulus presented within the first 48 hours was preferred to a higher quality stimulus presented at a later age.

Vocalisation appears to play an important part in guinea-pig social behaviour. Berryman (1970) recorded at least eleven different vocalisations, some of which could be correlated with various aspects of social behaviour. Some of these vocalisations occurred at frequencies exceeding the human audible range.

Breeds, strains and genetics

Three main breeds of guinea-pigs are generally recognised: the short-haired (3–4 cm long) English; the long-haired (over 12 cm) Peruvian (Fig. 20.1); and the Abyssinian, which is characterised by a rosetted hair pattern (Fig. 20.2).

Fig. 20.1. A 'fancy' Peruvian guinea-pig. Some side hair has been wrapped in cloth to prevent it from becoming damaged.

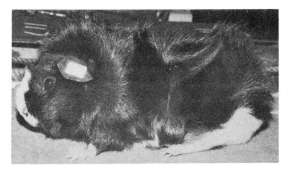

Fig. 20.2. A 'fancy' Abyssinian guinea-pig. The paper label on the ear serves for identification during show judging.

Coat colour depends on the genotype at six main (and several minor) coat-colour loci, some of which have multiple alleles. The genetics of coat colour in guinea-pigs have been intensively studied by Sewell Wright (see Searle (1968) for a review). These loci, and their effects on coat colour are:

Agouti locus. The non-agouti mutant produces a black coat colour.

Brown locus. The brown mutant alters areas of black pigment to brown.

Albino locus. There are at least four alleles at this locus, causing varying degrees of dilution of the coat and retinal pigments. The most extreme is the albino or himalayan mutant which has a white coat except at the extremities. The common albino laboratory guinea-pig has this allele together with other dilution factors, so that there is no pigment except on the ears.

Extension locus. There are three alleles at this locus. Partial extension (e^r) produces the tortoiseshell coat pattern, with streaks of black and yellow hairs, whereas the mutant *e* causes a yellow coat colour.

Pink-eye locus. There are three alleles at this locus, though the ruby-eye mutant may not be common. The pink-eye mutant causes a dilution of the pigment in the coat and the retina.

White-spotting locus. The recessive gene *s* causes white spots which may vary in size from just a few white hairs to almost a complete white coat. This variation depends on both environmental and genetic factors which modify the expression of the mutant gene. Spotting combined with partial extension produces the tricolour pattern, with patches of white, black and yellow.

In addition to the loci described above, there are a number of others which influence coat colour, though their effect may be only slight. A list of the known mutants in guinea-pigs is given in Table II. Further details of these mutants and a review of the genetics of guinea-pigs are given by Festing (1975).

A number of closed-colony, outbred stocks of guinea-pigs have been developed for use in the laboratory. The Dunkin-Hartley stock established in 1926 (Dunkin *et al.*, 1930) now has a world-wide distribution, and a number of sub-lines have been developed, such as the Pirbright and the Jap-Hartley stocks. Other genetically heterogenous colonies, such as the multicoloured NIH stock (Hansen *et al.*, 1973) are also in widespread use. Most of these stocks are of the short-haired English variety, though some of them may have rosettes and/or long hair.

Inbred strains of guinea-pigs were developed as early as 1915 (Wright, 1922). So far, only strains 2 and 13 are in widespread use, but recent studies of the genetics of the immune response has fostered interest in some of the other inbred strains currently under development. A

TABLE II
Summary of established mutants of the guinea-pig
(Reprinted, by permission, from Festing, 1975)

Gene effect*	Gene symbol	Name
1	A	Light or yellow bellied agouti
1	A^r	Ticked-bellied agouti
1	a	Non-agouti
1	b	Brown
9	C4	C4 complement deficiency
1	c^k	Dark dilution
1	c^d	Light dilution
1	c^r	Red-eyed dilution
1	c^a	Albinism or Himalayan
9	CA II	Carbonic anhydrose
9	Ca	Hypocatalasaemia
1	di	Dilution at base of hairs
1	dm	Diminished
1	E	Extension (+)
1	e^r	Partial extension
1	e	Yellow
10	GPIr-1	Immune response locus, includes PLL, GT, GA alleles
10	GPL-A	Main histocompatibility locus (several alleles)
2	Fz	Fuzzy
1	f	Fading yellow
2	gr	Grizzled
2	l	Long hair
2	m	Rough modifier
1	p^r	Ruby-eye
1	p	Pink-eye dilution
9	PGM	Phosphoglucomutase
3	Px	Polydactyly
2	R	Rough
2	Re	Rough eye (modifier of rough)
1	s	White spotting
10	S^hy	Serum factor—hypersensitivity
1	si	Silvering
1	sm	Salmon eye
2	St	Star
2	sth	Sticky haired
1	W	Whitish or white tipped
6	wtz	Recessive waltzing
6	Wz	Dominant waltzing

* *Gene effect code (as used in Mouse News Letter)*
1. Pigmentation
2. Hair direction and texture
3. Skeleton
4. (Tail)
5. Eye
6. Ear and circling behaviour
7. Neuromuscular
8. Blood, endocrine, internal defects, dwarfs, sterility
9. Enzymes
10. Antigens, disease resistance etc.
11. Miscellaneous

current list of inbred strains is published annually in *Guinea-pig News Letter*, obtainable from the Medical Research Council Laboratory Animals Centre, Woodmansterne Road, Carshalton, Surrey SM5 4EF, England.

Standard biological data

Data on the physiology of guinea-pigs is given in Table III, and on guinea-pig blood in Tables IV and V. Organ weights are recorded in Table VI. It should be noted that there are strain differences in relative organ weights. Some data on blood biochemistry during ageing is given by Rogers (1951). Nixon (1974) recorded the breathing pattern, and noted that the guinea-pig normally only breathes through the nose.

Post-natal growth rates are shown in Figs. 20.3 and 20.4. Again, it should be noted that growth is strain- and diet-dependent.

Uses in research

The guinea-pig is used in a wide range of disciplines, either because it is of a convenient size and is readily available, or because it has special features not found in other common laboratory species.

The species is widely used in immunology, particularly in studies of delayed hypersensitivity in which its immunological response resembles that of man (Gell & Benacerraf, 1961). The guinea-pig has also played a leading part, together with the mouse, in studies of the immune response and its genetic control. Kantor *et al.*, (1963) showed that the immune response to 2, 4-dinitrophenol poly-L-lysine (DNP-PLL) was under genetic control. All guinea-pigs of strain 2 were found to be responders and all strain-13 animals non-responders to this antigen; the ability to respond is controlled by an autosomal

TABLE III

Some physiological data (C. porcellus)

(from Spector, 1956)

	Mean	Range
Body temperature (°C) *Cavia* spp.	39·1	(38·4–39·8)
Cavia porcellus	37·9	(36·0–40·5)
Total body water (ml/kg) foetus, < 12 g	894	(863–966)
71–112 g	716	(684–754)
newborn	710	—
adult	635	(524–746)
Water balance (450 g body wt):		
water intake (g/100 g body wt/day)	14·5	
metabolic water produced (g/100 g body wt/day)	2·5	
Life span (yrs) average	> 2·0	
maximum	> 8·0	
Respiration rate (cc oxygen/kg/hr) (resting adult)	1250	
Respiration frequency (breaths/min)	90	(69–104)
Tidal volume (air inspired/breath, ml)	1·8	(1·0–3·9)
Minute volume (resp. freq. × tidal volume, litres)	0·16	(0·09–0·38)
Heart rate (NB. This may be very variable)	280	(150–400)
Blood pressure (mm Hg) systolic	77	
diastolic	47	
Rectal temperature (°C) normal	—	(38·5–39·9)
minimum	21	
Critical air temperature (temp. at which a change in rectal temp. first occurs) low (°C)	− 15·0	
high (°C)	29·5	
Thermoneutrality zone (range of air temp. at which metabolic rate is lowest and constant in the un-anaesthetized animal) (°C)	30–31	
Basal metabolic rate (animal of 0·76 kg, 0·07 m² surface) (Cal/m²/day)	700	
Heat production (Cal/m²/24 hrs) at various environmental temperatures: 15·1°C	1050	
20·0°C	913	
24·8°C	784	
29·8°C	601	
35·5°C	716	
Heat tolerance (27/38 survival for 7 hrs at) (°C)	44	

TABLE IV

Guinea-pig blood

(Abstracted from Altman & Dittmar, 1961)

	Mean	Range
Blood pH	7·35	(7·17–7·55)
Blood volume (ml/kg body wt)	75·3	(67·0–92·4)
Plasma volume ,,	39·4	(35·1–48·4)
BLOOD ELECTROLYTES		
Whole blood		
Calcium (mg/100 ml)	—	(8·60–11·29)
Iodine, total (μg/100 ml)	7·2	—
Phosphorus		
Inorganic P (mg/100 ml)	4·9	(3·5–6·1)
Total acid-sol P ,,	28·1	(23·6–32·2)
Organic-acid-sol P ,,	23·4	(20·0–27·8)
Nucleotide P ,,	2·0	(1·6–2·3)
Erythrocyte		
Phosphorus		
Organic-acid-soluble P		
(mg/100 ml)	64·6	(61·5–69·2)
Adenosine triphosphate P		
(mg/100 ml)	—	(10·4–11·2)
Diphosphoglycerate P		
(mg/100 ml)	—	(28·4–34·8)
Nucleotide P (mg/100 ml)	5·5	(5·2–5·7)
Plasma or Serum		
Calcium (mg/100 ml)	10·7	(7·4–13·6)
Magnesium ,,	2·3	—
Phosphate ,,	5·3	—
Potassium ,,	—	(23·7–27·3)
BLOOD NON-PROTEIN NITROGENOUS SUBSTANCES		
Whole blood		
Glutathione, total (mg/100 ml)	—	(80–175)
Plasma or serum		
Glycine ,,	2·5	—
Urea N ,,	19·0	(8·0–28·0)
Uric acid ,,	2·5	(1·3–5·6)
BLOOD LIPIDS		
Plasma		
Lipids ,,	169	(94–244)
Cholesterol ,,	32	(21–43)
Cholesterol, free (mg/100 ml)	11	(7–15)
Cholesterol, ester ,,	21	(11–30)
Fat, neutral ,,	73	—
Fatty acids, total ,,	116	(92–140)
Phospholipids, total ,,	51	(25–77)
BLOOD GLUCOSE		
Whole blood		
Glucose, fed (mg/100 ml)	96	(82–107)
Glucose, fasting ,,	95	(60–125)
Erythrocytes		
Glucose ,,	53	—
Serum		
Glucose ,,	155	—
BLOOD VITAMINS		
Whole blood		
Ascorbic acid, (μg/100 ml)	120	—
Plasma or serum		
Ascorbic acid, C ,,	300	—
Choline, free ,,	—	(2,000–12,000)
BLOOD PROTEINS		
Serum		
Protein, total (g/100 ml)	5·4	(5·0–5·6)
Albumin ,,	3·2	(2·8–3·9)
Globulin ,,	2·2	(1·7–2·6)

TABLE V

Haematological Values[1]

	Range[2]
Haemoglobin (g/100 ml)	12·4–15·4
Erythrocytes (millions/mm³)	4·4–5·4
Haematocrit %	39·0–47·6
Mean corpuscular vol (μ^3)	83·4–89·6
Mean corpuscular haemoglobin ($\gamma\gamma$)	26·1–28·6
Mean corpuscular haemoglobin content (%)	30·4–31·9
Plasma viscosity (centistokes)	1·45–1·61
Total leucocytes (millions/mm³)	4·46–10·79

Differential leucocyte counts (in 10³/mm³ of blood)	
Neutrophils	4·2
Eosinophils	0·4
Basophils	0·07
Lymphocytes	4·9
Monocytes	0·43

[1] First 8 values from Williamson & Festing (1971), rest from 'Data on Common Laboratory Animals', Teklad, Inc. U.S.A. (undated).
[2] Across both sexes and three strains.

TABLE VI

Some organ weights in Strain 2 and Strain 13 guinea-pigs

(10 animals of each sex, wts in g/kg body wt)

Organ	Strain 2		Strain 13	
	♂♂	♀♀	♂♂	♀♀
Liver	31·63	37·45	32·27	37·76
Lungs	6·50	6·64	6·92	7·91
Heart	2·64	2·65	2·31	2·40
Thyroid	0·076	0·074	0·075	0·079
Kidney	3·67	3·58	2·46	2·48
Adrenal	0·501	0·505	0·297	0·302
Spleen	0·97	1·32	0·699	0·99
Body wt	802 ± 65	780 ± 69	1044 ± 69	940 ± 99

Calculated from Strandskov, H. H. (1939).

dominant Mendelian gene. This and other immune response genes have been studied in detail (Green, 1970).

Other areas of immunology in which the guinea-pig has been used are in studies of anaphylactic shock (Stone *et al.*, 1964), and allergic encephalomyelitis (Takino *et al.*, 1971).

Biochemical, toxicological, physiological and pharmacological investigations are commonly carried out on guinea-pigs. In many cases detailed studies are made on the biochemistry and pharmacology of isolated organs rather than in the intact animal. The guinea-pig is apparently often chosen because it is a convenient size, but in other instances it is selected as a contrasting species; either because it is known to react with certain drugs, or because the investigation is part of a larger research programme in which other characteristics of the guinea-pig are useful.

The guinea-pig is a suitable host for a number of disease-causing organisms. This is particularly true of the Mycobacteria, and widespread use is made of guinea-pigs in the study of tuberculosis.

Finally, the species is widely used in studies of hearing (otology) and in ascorbic acid metabolism, the guinea-pig being the only common laboratory animal which resembles man in that it has a requirement for dietary vitamin C.

Husbandry

Housing and caging

Little work has been done on the optimum environmental conditions for guinea-pigs, but the species is relatively hardy, and is capable of living under a wide range of conditions. In the laboratory, the usual recommendations include a temperature of 18–22°C, a relative humidity of 50–70 per cent and a cycle of 12–16 hours light per day. Guinea-pigs can withstand cold conditions provided they have good shelter and plenty of bedding, and are housed in groups rather than singly. They are probably less able to withstand high temperatures, 11 out of 38 animals succumbing after 7 hours at a temperature of 44°C (Table IV). Heavily pregnant animals are particularly prone to heat prostration, and in tropical climates, and preferably even in temperate regions, air conditioning is recommended.

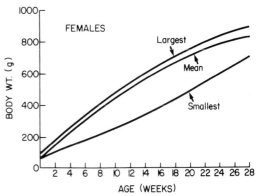

Fig. 20.4. Growth in female outbred guinea-pigs. (Smoothed curves calculated from Poiley, 1972).

A wide variety of caging is available, including floor pens, permanent shelving (often made of wood), mobile racks of metal cages and plastic cages suitable for holding small numbers of animals. Tables VII and VIII give details of recommended cage space requirements, and areas of animal room required to house stock of various types.

Floor pens. These may be constructed from a variety of materials ranging from permanent brick or concrete blocks to moveable aluminium partitions held in place by specially constructed corner posts. Walls need to be about 40 cm high. Whatever the method of construction, it is essential that floor pens should be easy to keep clean, with well insulated floors. The pens should be kept dry, and draughts avoided. The main disadvantage of floor pens is the relatively inefficient use that they make of the available space. However, they are cheap to construct and generally give good results.

Tiered compartments. Permanent wooden shelf cages measuring about 60 × 120 × 30 cm high in 4 or 5 tiers are commonly used by commercial guinea-pig breeders. Such pens accommodate a harem of about 1 male to 8 females. More rarely, concrete or other more permanent materials are used. Wooden shelves with a wire mesh front are cheap to construct but are difficult to keep clean, and could become a disease hazard.

Cages. Fig. 20.5 shows the type of cage used by the Medical Research Council Laboratory Animals Centre. Group breeding is conducted in anodised aluminium cages measuring 90 × 60 ×

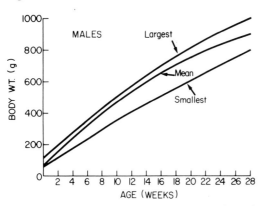

Fig. 20.3. Growth in male outbred guinea-pigs. (Smoothed curves calculated from Poiley, 1972).

Fig. 20.5. Solid-floor aluminium cage manufactured by Forth-Tech. Note provision for automatic watering on the right of the cage and external food hopper on the left.

23 cm high, in three tiers. The cages have solid floors, and provision is made for automatic watering using nipple drinkers. Each cage contains one male and four females with young. The advantage of this type of cage is that the whole unit can be changed when the litter becomes soiled, and can be washed regularly. Other common types of cage are shown in Fig. 20.6.

Wire-mesh floors are commonly used and reduce the amount of time spent in cleaning the cages. However, such floors may result in broken legs in a considerable proportion of young stock and it is essential that the correct mesh size is used. Hoyland (1963) recommended a mesh of 1.27 cm^2 for growing stock, and 1.6 cm^2 for breeding stock. In many cases, a generous ration of hay is also given which partially covers the wire mesh and can help to reduce leg injuries. Guinea-pigs which have been reared on solid floors should be provided with substantial amounts of hay for the first two weeks after transfer to mesh-floored cages in order to allow them gradually to become used to this form of flooring.

Solid-floored plastic tubs with a wire-mesh lid (Fig. 20.6) are satisfactory for housing small numbers of experimental animals or growing stock and monogamous pairs of inbred strains. They are easily transportable and can be handled by a single animal technician.

TABLE VII
Recommended cage-space requirements

ILAR-NRC[1]

Body weight (g)	Area cm^2	Height (cm)
up to 250	277	18
250–350	374	18
over 350	652	18

BRITISH HOME OFFICE[2]

Body weight (g)		
up to 350	475	25
350 to 550	563	25
550 to 750	700	25
over 750	940	25
♀ plus litter	1485	25

[1] Rodents. Standards and Guidelines for the Breeding, Care and Management of Laboratory Animals. National Acad. Sci., Washington, D.C. (1969).

[2] Guidance Notes on the Law Relating to Experiments on Animals in Great Britain. Compiled by the Research Defence Society with the collaboration of the Home Office Cruelty to Animals Act Inspectorate and available from the Research Defence Society, 11 Chandos Street, Cavendish Square, London WIM 9DE. These are recommended values and are not mandatory.

TABLE VIII
Recommended animal room areas
(Clough, 1975: pers. comm.)

Type of stock	Animals/m^2
Breeding ♀, allowing space for male and young	7
Stock animals	27
Experimental animals	24

Fig. 20.6A. Mesh-floored guinea-pig breeding unit in stainless or galvanized steel (All-Type Tools, Ltd.)

Food and water utensils

Pellets for guinea-pigs are usually both smaller and softer than mouse pellets and require hoppers designed to avoid wastage (Figure 20.7), while allowing the animal easy access to the food. Some hoppers have a fine-grid floor to prevent them from becoming clogged with dust from broken pellets. In any case, food hoppers should be regularly cleaned to prevent the accumulation of stale and contaminated food.

The provision of an adequate supply of water presents some problems. The simplest method is to use open earthenware bowls (designed for pets), but these soon become contaminated with faecal matter. Inverted glass bottles with a metal or pyrex glass cannula (not soft glass which the guinea-pigs will break) are widely used. Unfortunately, the animals spend much time playing with such bottles, and commonly waste large quantities of water, so that they may be without for long periods. If such bottles are used, they should be filled twice a day. An improved water bottle has been described by Wilson (1974).

Fig. 20.6B. Solid-floored galvanized-steel guinea-pig breeding unit. (All-Type Tools, Ltd.)

Automatic watering can be used, as guinea-pigs will readily learn how to operate a nipple valve, but, if solid-floored pens are used, provision should be made for water to fall outside the cage as the animals will spend time playing with the nipples. The nipples should also be strong, as it is surprising how much damage guinea-pigs can cause with their teeth.

Bedding

Softwood shavings or coarse sawdust are the most commonly used bedding materials, but peat-moss has also been used. Hay, which should primarily be used as a food, is usually fed loose and so can act as a bedding material, though it may be expensive. Paterson (1972) does not recommend straw.

Identification

There is no entirely satisfactory method of identifying individual guinea-pigs. Ear tattoos can be used with letters 0.5×0.25 cm in size, using black ink for albino animals and green ink for coloured ones, but such tattoos can be difficult to read and ears can become damaged by fighting. Ear tags may be used, although these tend to pull out.

Coloured animals may often be identified by sketches of the colour pattern. Stains may be used for temporary identification, but these must be renewed every 3–4 weeks. A range of colours may be obtained, such as yellow (saturated picric acid in alcohol), red (fuchsin), violet (methyl violet) or blue (trypan blue). Often the need for individual identification can be avoided by caging litter mates together in small groups or even by caging individuals separately.

Handling and sexing

Guinea-pigs are relatively easy to handle—at least they seldom bite. They should be picked up with both hands—one round the shoulders and the other supporting the hindquarters (Fig. 20.7). Great care should be taken in handling heavily pregnant females, which should not be allowed to struggle.

Sexing is relatively easy. The animal should be held as in Fig. 20.8, and gentle pressure applied round the genital area. In the male, the penis is easily extruded.

Fig. 20.7. Method of handling guinea-pig. (Photograph: UFAW).

Transport

Transport of guinea-pigs presents no special problems. The animals must be secure and have adequate ventilation, insulation and protection. Further details are given in Chapter 8.

Feeding

Nutrition

Guinea-pigs are naturally herbivorous and, in the wild, live on a range of green foods, seeds and roots. In the laboratory, minimum dietary requirements have been determined for some nutrients, but in other cases only levels which are assumed to be adequate can be stated. The nutrient requirements of the guinea-pig have been reviewed by Ford (1974) and Reid & Bieri (1972). Table IX gives the vitamin requirements, Table X the formulation of some common commercially available diets, and Table XI the calculated composition of these diets. Ford (1974) emphasised that the requirements for an individual nutrient cannot be considered in isolation from other nutrients. Protein requirement

also depends on the biological value of the proteins, so that diets of equal crude protein content may differ in their effects on growth and reproduction.

In addition to pellets, guinea-pigs appear to have a requirement for hay. The exact nature of this requirements is not known, but in the absence of hay there may be outbreaks of fur chewing, and there is a suspicion that breeding performance may decline.

Practical feeding

It is advisable to feed guinea-pigs on commercially formulated pellets and good quality hay and to ensure that they have an adequate supply of vitamin C. The latter can be supplied, either in the diet at the rate of 800 mg/kg diet, or by supplementing the water with 1 g/litre of ascorbic acid. In this case, the drinking bottles must have glass or stainless steel cannulas, as other metals may accelerate the decomposition of the vitamin. If the ascorbic acid is given in the drinking water, the solution must be made up

Fig. 20.8. Sexing a guinea-pig. The male's penis is easily extruded from the round genital opening using gentle pressure. (Photograph: UFAW).

daily. Dietary vitamin C should remain at an adequate level for 2–3 months, provided the diet is kept in a cool, dry place.

Green food can also be used as a source of vitamin C, but there may be a danger of bacterial contamination leading to a disease outbreak. As entirely satisfactory results have been achieved without the feeding of any green food, it should not normally be fed unless other sources of vitamin C are not available. In this case, care should be taken to ensure that the greens are fresh and clean. Paterson (1972) recommended any of the brassicas, chicory, lucerne (alfalfa), grass, carrots and swedes as being suitable green foods, although the white hearts of cabbages are not so satisfactory.

Care should be taken not to change the type of food too suddenly and in frosty weather greens should be thawed out before being fed. Animals which have been reared on diets free of green food may take some time to become accustomed to it.

Breeding

Selection of breeding stock

Selection of breeding stock will depend partly on genetic considerations and partly on general health and vigour. Chapter 2 provides details of breeding systems appropriate to the main-

tenance of inbred strains and outbred stocks. In any case, all animals chosen as breeding stock should be healthy and free of obvious defects and should not have suffered any setback during the growing period. If the aim of the breeding programme is to maintain the stock or strain characteristics without change, then no strong directional selection should be practised. Selection within an inbred strain for improved breeding performance may help to prevent a decline in performance caused by the accumulation of marginally deleterious mutants but there is little hope of improving the breeding performance, or altering body weight or any other character in an inbred strain by selection.

Selection can be used within an outbred stock to alter characteristics such as breeding performance or body weight. Characters with a low heritability may be improved by family selection (Falconer, 1960), while characters with a high heritability should be improved by mass selection. Wright (1960) gives a considerable amount of data on the genetics of various aspects of reproductive performance in guinea-pigs. Care should be taken to define the breeding objectives carefully. In guinea-pigs, selection for increased litter size may simply increase the incidence of stillbirths without altering the total productivity (Goy et al., 1957). Total number of young reared to weaning during a breeding life rather

TABLE IX

Vitamin requirements

(Reprinted, by permission, from Ford, 1974)

Vitamin	Adequate level/kg diet*	Criterion
A	1·7–9·9 mg	Optimum growth
D	1600 i.u.	None required if Ca:P ratio balanced
E	Not known—50 mg	—
K	2 mg	—
C	0·5–5 mg/100 g body wt/day	Maximum growth and wound healing
Thiamin	2 mg	Growth of young
Riboflavin	0·67–1·00 mg	Maximum growth
Pantothenic acid	15–20 mg	—
Nicotinic acid	10–20 mg	Growth on 30% casein diet
Pyridoxin	2–3 mg	Growth on a 30% casein diet
Folic acid	3–6 mg	Growth
Choline	1·0–1·5 g	
B12 Biotin Inositol PABA	no requirement found	

* unless otherwise stated

TABLE X

Diets commercially available in the U.K.

Formulations

Diet Name

Ingredient	RGP	SG1	18	GR3EK	TR2
Wheatfeed (Middlings)	15·0	18·0	—	1·3	+
Ground wheat	—	—	—	36·2	+
Bran	—	40·0	15·0	—	—
Barley meal	40·0	—	20·0	25·0	+
Sussex ground oats	12·5	12·0	—	—	+
Ground maize	—	—	—	12·5	+
Dried grass meal	15·0	20·0	30·0	—	+
Dried milk powder	—	—	—	—	+
Meat and bone meal	—	—	8·0	—	+
White fish meal	7·5	10·0	—	9·8	+
Linseed meal	10·0	—	10·0	—	—
Soya bean meal	—	—	—	10·7	—
Groundnut meal	—	—	15·0	—	—
Molasses	—	—	—	2·7	—
Vitamin/mineral supplement	+	—	2·0	1·8	+

References: RGP: Paterson (1972)
SG1: Short & Gamage (1959)
18: Bruce & Parkes (1947)
GR3EK: Ley, Bleby, Coates & Paterson (1969)
TR2: Pilsbury Ltd.*
* Pilsbury's Ltd., 21 Priory Road, Edgbaston, Birmingham B5 7UG.

TABLE XI

Diets commercially available in the U.K.

Calculated compositions

Ingredients	RGP	SG1	18	GR3EK	TR2	FD1
Crude protein %	17·9	19·4	23·6	19·2	21·5	18·7
Oil %	3·3	3·8	3·8	2·3	3·3	3·6
Crude fibre %	8·7	11·3	11·4	2·7	8·1	9·8
Metabolizable energy Cals/kg	2176	1700	1806	—	—	—
Calcium %	0·9	1·0	1·7	0·8	1·2	1·3
Phosphorus %	0·7	1·1	0·9	0·8	0·9	0·9
Sodium chloride %	0·9	0·4	1·5	1·1	1·0	—
Potassium %	0·9	1·3	1·5	0·7	1·2	1·4
Magnesium %	0·2	0·4	0·4	0·2	0·4	0·3
Vitamin A (iu/kg)	8000	—	—	5167	28,874	124,320
Carotene (mg/kg)	16·4	21·9	33·0	1·0	18·8	72·2
Vitamin D (iu/kg)	1000	—	—	2500	1000	2575
Vitamin E (mg/kg)	62·6	46·5	42·1	52·7	68·6	78·1
Thiamin (mg/kg)	8·4	9·0	5·2	4·1	6·4	9·1
Riboflavin (mg/kg)	13·1	5·2	7·5	8·1	10·4	17·2
Nicotinic acid (mg/kg)	73·0	123·0	94·1	47·7	97·0	75·8
Pantothenic acid (mg/kg)	17·8	22·9	24·1	14·2	26·3	34·2
Pyridoxin (mg/kg)	5·1	8·2	6·8	4·8	4·6	13·9
Ascorbic acid (mg/kg)	1500	0	0	0	1500	1000
	(added at manufacture)				(added at manufacture)	
Arginine (g/kg)	12·0	12·8	19·4	11·1	14·0	12·7
Methionine (g/kg)	3·8	3·8	3·3	3·9	3·0	4·1
Lysine (g/kg)	9·5	10·9	9·3	11·2	10·8	10·7
Tryptophan (g/kg)	1·9	2·5	2·9	2·5	2·9	3·0

References: RGP, SG1, 18: Ford (1974)
GR3EK: E. Dixon & Sons Ltd.*
TR2: Pilsbury's Ltd.
FD1: Cooper's Nutritional Products†
* E. Dixon & Sons Ltd., Crane Mead Mills, Ware, Herts.
† Cooper Nutritional Products, Stepfield, Witham, Essex.

than litter size might be the best criterion to select for if the aim is to improve productivity.

Breeding systems

The most common breeding system and the one which probably gives the highest output per unit area is to mate the animals in polygamous groups of one boar and 4–20 sows. Once the group has been set up, they are left together for the whole of their economic breeding life, which is usually about 18 months. Group size depends largely on the available facilities, but the optimum is probably about 12 sows to one boar (Rowlands, 1949). With this system, a high proportion of the females conceive at the post-partum oestrus.

Monogamous pairs can be used to propagate inbred strains, but, because of the large number of boars that have to be maintained, they are less efficient than polygamous groups in the utilization of space.

In cases where accurate dating of birth is essential, heavily pregnant females can be separated from the group just prior to parturition and allowed to litter down by themselves. Young are weaned at 18–21 days, or when they have reached about 180 g.

Breeding results

Table XII provides some data on reproduction in the guinea-pig and Table XIII data on the reproductive performance that can be expected under practical conditions. It should be noted that output is substantially lower in inbred than in outbred guinea-pigs.

Additional information on reproduction in this species is given by Phoenix (1970), including information on artificial insemination.

Extensive data on litter size, perinatal mortality, early growth rate and their inter-relationships are given by Wright (1960).

Laboratory procedures

Anaesthesia

The general principles of anaesthesia, together with details of various anaesthetic agents, are given in Chapter 13. Schaffer (1965) provides a general review of anaesthesia in laboratory

TABLE XII

Data on reproduction

Age at puberty (days)	55–70
Oestrus cycle duration (days)	16–19
Duration of oestrus (hrs)	6–11
Ovulation time (hrs from onset of oestrus)	10
Number of ova released	2–4
Sperm transit, vagina to fallopian tube (mins)	15
Gestation period (days)	68 (58–75)
Return to oestrus postpartum (hrs)	6–8
Chromosome number, diploid	64
Birth weight (g)	75–100
Age at weaning (days)	14–21
Eats solid food (days of age)	5
Lactation period (days)	21
Litters/♀/year	3·7
Young born/♀/year	16·6
Young weaned/♀/year	13·9
Av. litter size at birth	4·4
Av. litter size at weaning	3·7
Av. interval between litters (days)	96·3
% still births	8·5
% preweaning mortality of live born	9·6

From Spector, W. S. (1956) and Lovell *et al.* (1972).

animals and Graham-Jones (1964) considers anaesthesia in a wide range of small animals.

The choice of anaesthetic depends largely on the objective, which may be surgery, monitoring of physiological parameters or withdrawal of body fluids. Some anaesthetics may be unsuitable for a particular purpose as they may influence the physiological function under study.

Animals chosen for experimental work involving surgery and anaesthesia should be healthy and vigorous and should squeal complainingly when handled. If ether is to be used, food should be withheld for at least twelve hours, but fasting is not necessary with barbiturates. In the latter case, however, care should be taken to ensure that the animals have not recently consumed large quantities of food, as this may lead to an over-estimation of bodyweight and the administration of too high a dose.

Pre-medication is not usually given, though atropine sulphate (2·5 mg) administered subcutaneously at the point of the shoulder may be used to reduce salivation (Croft, 1958) and is recommended prior to the use of ether.

There may be strain differences in the susceptibility to anaesthetics, so care should be exercised in establishing dose rates for each strain.

Volatile anaesthetics. Ether may be used for

short or long procedures, or to potentiate the effects of barbiturates. Great care should be taken, as ether is highly flammable. Anaesthesia may be induced by placing the animal in a bell jar containing ether-soaked cotton wool under a grid floor, care being taken to avoid anoxia. The animal must be observed and removed when the desired level of anaesthesia is obtained. Alternatively (and preferably), a close fitting mask may be used ventilated with an ether/air mixture. Care should be taken to ensure that there is no damage to the cornea due to drying. Schaffer (1965) recommends the use of an ophthalmic ointment.

Apparatus suitable for halothane anaesthesia in rodents is described by Sebesteny (1971). In view of the risk of explosions with ether, the use of halothane is strongly recommended, even though it is more expensive and calls for slightly more complex apparatus.

Injectible anaesthetics. Pentobarbital sodium may be used to give prolonged anaesthesia, though it can damage young *in utero*. The recommended dose is approximately 30 mg/kg body weight administered intraperitoneally. A stock solution of a veterinary formulation containing 60 mg/ml is usually diluted with 5 parts of normal saline before use, providing a 10 mg/ml solution. Deep anaesthesia is reached in about 15 minutes, and will last for about 2–3 hours. It is essential that the animals should be kept warm whilst under the anaesthetic.

For short-term anaesthesia, a mixture of alpha-xalone and alphadolone* is recommended (Sparrow, personal communication). This can be given intramuscularly or subcutaneously, and is exceptionally safe. Anaesthesia lasts about 15 minutes but additional injections can be given to prolong the anaesthesia. The dose rate should conform to the manufacturer's recommendations.

Spinal anaesthesia using mepivacaine chloride has been described by Thomasson *et al.* (1974). They considered the method to be quick, inexpensive and safe; it gives excellent abdominal relaxation and is especially useful for foetal surgery.

Collection of specimens

Blood. Small samples of blood (usually less than 0·25 ml) may be obtained by simple venesection of the marginal ear vein. Larger quantities must be obtained by cardiac puncture from the metatarsal vein (if the animal is not to be sacrificed), or by section of the jugular vein in the anaesthetised or stunned animal if it is to be exsanguinated. The use of anaesthesia gives a higher yield than stunning.

Dolence & Jones (1975) describe the collection of small quantities of blood (3–20 ml) from the lateral metatarsal vein using a vacuum-assisted bleeding technique. No anaesthetic is necessary and repeated small samples may be obtained. Mortality was zero and 91 per cent of samples were sterile. The vein may also be used for intravenous injections.

*Saffan: Glaxo.

TABLE XIII
Summary of some published productivity statistics in guinea-pigs

Source	Strain	Young born/ sow/year	Young weaned/ sow/year	Litter size	Pre-weaning mort. %
Rowlands (1949)	Not named, probably Frant Dunkin Hartley	19·8	16·8	4·0	15
Wills & Sutherland (1970)	Frant Dunkin Hartley	20·5	16·9	4·2	14
Hoyland (1972)	Pirbright Hartley	19·3	18·9	4·2	—
	Strain 13 (inbred)			3·7	5
Festing (1971)	Strain B (inbred)	10·4	9·3	2·5	11
	Strain OM3 (inbred)	6·5	5·6	2·3	14
	Strain R9 (inbred)	8·3	7·0	2·5	16
Hansen & McEleney (1971)	Strain 13 (inbred)	—	7·1	2·5	—
	Strain 2 (inbred)	—	5·5	2·7	—
Lovell *et al.* (1972)	Lac: DHP (4-star)*	16·6	13·9	4·4	10

*MRC category 4-star, i.e. hysterectomy derived, barrier-maintained and free of a wide range of specified micro-organisms.

A convenient method of cardiac puncture is described by Brummerstedt (1971) as follows:

The animal is anaesthetised with ether and placed on its back between two small plastic bags nearly filled with sand.

The sternal region of the guinea-pig and the operator's left index finger are disinfected. The operator's left second finger is placed at the anterior end of the sternal bone and the thumb at the xiphoid process. The index finger is then placed exactly in the middle between thumb and second finger in an interval between two left ribs as near the sternal bone as possible. This indicates the insertion point of the needle. The index finger is moved a little to the left, and the needle (Gillette Disposable Needle No. 12) is introduced with the tip neither towards the animal's head nor its hind part (i.e. perpendicular to the animal). The needle is introduced about 1·5–2·0 cm, the plunger is withdrawn and the blood runs into the syringe. Small corrections in the position of the needle during the bleeding may be necessary. In this way up to 10 ml of blood may quite easily be taken, and only very few animals have died because of the bleeding. Death may happen, for example, to an animal suddenly disturbed by noise, which may awaken it from the anaesthesia with the needle in its heart.

The advantage of the method is that it is quick and may be carried out by a single person. The anaesthesia need not be deep as no time is used for fixation with strings and for looking for the insertion point estimated by pulsation of the heart.*

Urine. This may be collected either by manual manipulation, applying pressure over the posterior abdominal area, or by using a metabolism cage. Many cage manufacturers produce rodent metabolism cages that are suitable for guinea-pigs. Basically, the cage consists of a wire mesh floor over a funnel, which is designed to catch and separate the faeces and urine. Care should be taken to ensure that the food and water do not also contaminate the urine.

Milk. The collection of milk from guinea-pigs has been described by Gupta & Conner (1970).

Dosing and injection procedures

Intradermal. The injection site must first be clipped, shaved and swabbed with an antiseptic. A fine (23 s.w.g.) needle 19 mm long should be used. The whole length of the needle should be inserted into the tensed skin, and slowly with-

*Reprinted, by permission, from Brummerstedt, 1971.

drawn as the injection is made. Up to 0·5 ml may be administered at one site.

Subcutaneous. The skin is clipped, shaved and swabbed with an antiseptic. A 20 s.w.g. gauge needle should be used and inserted under the lifted loose skin, parallel with the underlying muscle. As there may be some leakage, the site should be swabbed with an antiseptic after the injection has been made if infectious agents are involved. As the skin of the guinea-pig is very loose, subcutaneous injections can be made in a wide range of sites.

Intraperitoneal. The guinea-pig should be held on its back with the head slightly lower than the hindquarters to allow the stomach and intestines to fall forwards. The needle (20 s.w.g. × 38 mm) should be inserted slightly to the right of the midline, about 2·5 cm in front of the pubis, directed anteriorly at an angle of about 45 degrees, and passed right through the skin and muscle wall of the abdomen.

Intramuscular. The thigh muscle is most suitable. A site is prepared on the posteriolateral aspect, 6–12 mm above the stifle. The needle should be passed through the skin and into the muscle mass, taking care to avoid hitting the femur.

Intravenous. Two methods can be used, though neither is entirely satisfactory. Large animals may be injected in the marginal ear vein, using a fine gauge (23 s.w.g. × 25 mm) needle. The operation must be done in a warm room, and local heat will help to distend the vein, which can be further enlarged by slight pressure at the base of the ear.

The second method is by injection into the saphenous vein at a point about half an inch above the hock. Unfortunately, the vessel is very mobile and, to ensure that the needle is accurately placed, the animal should be anaesthetised, and the skin cut in order to expose the vein. After the injection has been made, the incision in the skin must be sutured.

Dosing. Small volumes of liquid (up to 5 ml) can be given by pipette. One person holds the animal and forces the mouth open by exerting pressure on both sides of the lower jaw, while another person drops the liquid on the back of the tongue.

Larger volumes of liquid must be given by stomach tube. Various designs of tube can be used. Paterson (1972) recommended the use of a soft pliable rubber or plastic tube, 1·5 to 6 mm in diameter. The animal is lightly anaesthetised, and the tube passed down the oesophagus into the stomach. Care must be taken to ensure that the tube does not go into the trachea, which may be indicated by respiratory stress. Once the tube is in the stomach, the fluid may be injected. Moreland (1965), however, recommends the use of a 15 or 16 gauge hypodermic needle, 8–10 cm long. The end should be blunted and covered with a ball of solder and there should be a gentle 20–30° bend about 2 cm from the solder. The needle is introduced through the interdental space and advanced gently by rotating it over the tongue and into the oesophagus. As it passes into the oesophagus, a 'loose' feeling is detected and the fluid may then be discharged.

Small quantities of either solid or liquid may be given by admixture with food, but, if this involves a change in the diet, this should be introduced gradually.

Euthanasia

The use of carbon dioxide in a properly designed euthanasia cabinet is recommended (see Chapter 14 for details). Alternatively, guinea-pigs may be killed by stunning from a sharp blow with a blunt instrument on the back of the head, followed by section of the carotids. An overdose of anaesthetic may also be given, though this may be expensive if large numbers of animals are involved.

Disease control

The guinea-pig is subject to a number of infections and infestations which, at the least, can be troublesome and, at the worst, could invalidate experimental results. Wherever possible, colonies should be founded with disease-free stock. Specific-pathogen-free (SPF) colonies have been developed at a number of centres (Owen & Porter, 1967), and category three- and four-star stock is available in the United Kingdom through the Medical Research Council Laboratory Animals Centre's Accreditation Scheme. (Chapter 4).

Ideally, once a colony has been founded with SPF-quality animals it should be maintained under full SPF conditions (Chapter 10). How-

ever, if this is not practicable, all possible precautions should be taken to ensure that infectious agents do not gain entry into the colony. No outside stock should be introduced unless it comes from a colony of comparable quality. The feeding of green food should be avoided and a pasteurised or sterilized pelleted diet should be used. Staff should be encouraged to maintain high standards of hygiene and special clothing should be reserved exclusively for work in the animal rooms. Great care should be taken to exclude wild rodents from the buildings, as these might introduce infections.

In experimental units where it is essential to buy in stock, all such animals should be either of category three- or of category four-star status; alternatively, all purchases should be made from a single supplier.

If stocks from different conventional colonies have to be mixed, they should be quarantined separately for at least three weeks prior to mixing. This will give an opportunity to assess whether any of the batches are carrying infectious organisms which may cause clinical outbreaks of disease in animals suffering stress as a result of a change of environment.

All animals in a colony, whether SPF or conventional, should be inspected and handled regularly. Sick animals will lose weight and appear listless and their hair may become rough and staring. Such animals should be culled immediately as they will be useless for experiments and might be a danger to the rest of the colony.

Some of the more common diseases of guinea-pigs are given below and more details are provided by Wagner & Manning (1975).

Parasitic diseases

Some of the common ecto- and endo-parasites of guinea-pigs are listed in Table XIV, and further details are give by Owen (1971; 1972). Lice are common in conventional guinea-pigs: most colonies are infected with *Gliricola porcelli* and about one third carry *Gyropus ovalis* (Sparrow, pers. comm.). It is common to find both species on a single host. The large body louse, *Trimenopon jenningsii* is much rarer. The only other ectoparasite of any significance is *Trixacarus caviae* (Fain et al., 1972), a mite which can cause mange, leading to loss in condition.

TABLE XIV

Common parasites of the guinea-pig

(Owen 1971; 1972)

Ectoparasites	Comment
Lice	
Gliricola porcelli	Chewing lice found in fur. Eggs
Gyropus ovalis	cemented to hairs. Probably of
Trimenopon jenningsii	little pathogenic significance except when animal is stressed.
Mites	
Chirodiscoides caviae	Fur mite widely distributed over body surface. Little pathogenic significance.
Trixacarus caviae	Burrowing mites. Can cause mange in some circumstances.
Endoparasites	
Giardia muris	Caecal flagellates, probably of no
Trichomonas caviae	pathogenic significance.
Entamoeba cobaye	
Endolimax caviae	
Balantidium caviae	Caecal ciliates, probably of no
Cyathodinium piriforme	pathogenic significance.
Eimeria caviae	Coccidian that invades caecal epithelium and can be pathogenic in some circumstances.
Paraspidodera unicata	Caecal nematode. Probably of little pathogenic significance.

Although there are many endoparasites, most seem to have a low level of pathogenicity. However, the coccidian *Eimeria caviae* may cause outbreaks of disease in young guinea-pigs. It is an invasive organism which penetrates the epithelial cells of the caecum and colon. A gradual build-up of the organism can result in a disease outbreak, characterised by loss of condition and diarrhoea.

Viruses

Relatively little work has been carried out on the naturally occurring viruses of guinea-pigs, but it appears as though the species is susceptible to relatively few (Blackmore, 1971). Only four viral diseases of major importance have been recorded.

Lymphocytic choriomengitis. (LCM) In guinea-pigs this leads to a generalised infection with patchy pneumonia and hepatitis (Blackmore *et al.*, 1973). Persistent tolerant infections which are common in mice probably do not occur in guinea-pigs. Circulating antibodies can be detected by the complement fixation test.

Salivary gland virus of guinea-pigs. This is a herpes virus which can infect the bronchi and bronchioles and is associated with a pneumonic syndrome (Pappenheimer & Slawetz, 1942). The virus can also occur as an inapparent infection and may be relatively common. Detection of the infection depends on the demonstration of the typical intranuclear inclusions and the isolation of the agent in tissue culture.

Herpes virus. Infection by this virus, which is distinct from the salivary gland virus described above, may result in the formation in the lung and kidney cortex of multiple grey foci 1–2 mm in diameter (Bhatt *et al.*, 1971). The lung lesions are foci of lymphoid hyperplasia, while the kidney lesions take the form of tubular degeneration. Inapparent infections can also occur.

Leukaemia virus. The occurrence of this virus has been described by Opler (1967).

Bacterial diseases

The guinea-pig suffers from a wide range of bacterial diseases, some of which are acute and others chronic. The most important of these are salmonellosis and pseudotuberculosis.

Salmonellosis, one of the most acute and important of all guinea-pig diseases, is caused by *S. typhimurium, S. dublin, S. enteritidis* and other salmonellae. These organisms can also be transmitted to man.

Infection is by ingestion, and the outbreak begins with a few deaths followed shortly afterwards by an explosive epizootic which may result in the almost total destruction of the colony. In some cases, a less acute form is observed, with the animals surviving for two or three weeks. Survivors may become carriers, which can give rise to fresh outbreaks of the disease when non-immune animals come into contact with them (Paterson, 1972). Diagnosis involves the isolation and identification of the causal organism.

Once the disease has gained a firm foothold the only practical method of elimination is by killing out the whole colony, thoroughly disinfecting the premises and utensils and fumigating the building before restocking.

Pseudotuberculosis is caused by *Yersinia* (formerly *Pasteurella*) *pseudotuberculosis.* It is a chronic disease characterised by caseous nodules, particularly in the viscera and lymph nodes. Three types of this disease occur (Paterson, 1972). The first is an acute septicaemia with rapid death, the second a more chronic disease leading to emaciation and death in three to four weeks. In this form the young may become congenitally or neonatally infected. The third type of the disease is usually non-fatal with the lesions restricted to the lymph nodes of the head and neck. These lesions may burst externally and spread the infection.

Some success has been achieved in the eradication of the disease by palpation of the animals in order to detect enlarged mesenteric and colonic glands (Cook, 1954), followed by culling affected animals. However, the safest way of elimination is by killing the whole colony and re-stocking with disease-free animals. Genetically valuable strains may be saved by hysterectomy and fostering the young on to SPF foster mothers (Chapter 10).

Other bacterial infections. Bacteria of the Pasteurella group, including *P. pneumotropica*, *P. multocida* and *P. haemolytica*, may cause disease in certain circumstances, though they are not usually responsible for major outbreaks unless the colony is suffering from stresses and other infections.

Respiratory infections are sometimes associated with *Streptococcus pyogenes*, *S. pneumoniae*, *Bordetella bronchiseptica* (Wood & McLeod, 1967), and *Diplococcus pneumoniae* (Keyhani & Naghshineh, 1974). Local abscesses can be caused by streptococci (Fraunfelter *et al.*, 1971).

Diagnosis in each case rests on isolation and identification of the organism. Treatment is difficult (some antibiotics, e.g. penicillin, can be toxic to guinea-pigs) and reliance should be placed on prevention rather than cure. Animal quarters should be kept clean and free of draughts, there should be no overcrowding and the building should be well ventilated. Rigorous culling of unhealthy stock should also be practised.

Miscellaneous conditions. A wide range of other conditions has been reported in the literature. These include: soft tissue calcification (Wickham, 1959, Maynard *et al.*, 1958); a variety of tumours (Mosinger, 1961); dystocia and abortion (Edwards, 1966) and aflatoxin poisoning from feeding contaminated hay (Schoenbaum *et al.*, 1972).

REFERENCES

Altman, P. L., & Dittmar, D. S. (1961). *Blood and Other Body Fluids*. Washington, D. C.: Fedn. Amer. Soc. exp. Biol.

Berryman, J. C. (1970). Guinea-pig vocalizations. *Guinea-pig News Letter*, **2**, 9–18.

Bhatt, P. N., Percy, D. H., Croft, J. L. & Jonas, A. M. (1971). Isolation and characterisation of a Herpes-like (Hsuing-kaplow) virus from guinea-pigs. *J. infect. Dis.* **123**, 178–189.

Blackmore, D. K. (1971). Virus diseases of the guinea-pig. *Guinea-pig News Letter, 4,* 21–26.

Blackmore, D. K., Guilon, J. C. & Schwanzer, V. (1973). *The Viruses of Laboratory Rodents and Lagomorphs*. London: BVA Editorial Services.

Bruce, H. M. & Parkes, A. S. (1947). Feeding and breeding laboratory animals. *J. Hyg., Camb.*, **45**, 70–78.

Brummerstedt, E. (1971). A technique for bleeding guinea-pigs by heart puncture. *Guinea-Pig News Letter*, **3**, 5–6.

Cannell, H. (1972). Pentobarbitone sodium anaesthesia for oral and immunological procedures in the guinea-pig. *Lab. Anim.*, **6**, 55–60.

Cook, R. (1954). Diseases of laboratory animals. *J. anim. Techns. Ass.*, **4**, 71–74.

Croft, P. G. (1957). In *The UFAW Handbook on the Care and Management of Laboratory Animals*, 2nd edition. Edinburgh: Livingstone.

Dolence, D. & Jones, A. E. (1975). Percutaneous phlebotomy and intravenous injection in the guinea-pig. *Lab. anim. Sci.*, **25**, 106–107.

Dunkin, G. W., Hartley, P., Lewis-Faning, E., & Russell, W. T. (1930). Comparative biometric study of albino and coloured guinea-pigs from the point of view of their suitability for experimental use. *J. Hyg., Camb.*, **30**, 311–330.

Edwards, M. J. (1966). Prenatal loss of foetuses and abortion in guinea-pigs. *Nature*, **210**, 223–224.

Fain, A., Hovell, G. J. R., & Hyatt, K. H. (1972). A new sarcoptid mite producing mange in albino guinea-pigs. *Acta Zool. Pathol.*, Antwerp, **56**, 73–81.

Falconer, D. S. (1960). *Introduction to Quantitative Genetics*. Edinburgh & London: Oliver & Boyd.

Festing, M. F. W. (1971). A note on productivity in three part-inbred strains of guinea-pigs. *Guinea-pig News Letter*, **4**, 19–20.

Festing, M. F. W. (1975). Guinea-pig genetics. In *The Biology of the Guinea-Pig*, eds. Wagner, J. E. and Manning, P. J. New York: Academic Press.

Ford, D. (1974). Nutrition of the guinea-pig. *Guinea-pig News Letter*, **7**, 11–24.

Fraunfelter, F. C., Schmidt, R. E., Beattie, R. J. & Garner, F. M. (1971). Lancefield type C streptococcal infections in strain 2 guinea-pigs. *Lab. Anim.*, **5**, 1–13.

Gell, P. G. H. & Benacerraf, B. (1961). Delayed hypersensitivity to simple protein antigens. In *Advances in Immunology*, eds. Taliaferro, W. H. and Humphrey, J. H. New York: Academic Press.

Goy, R. W., Hoar, R. M. & Young, W. C. (1957). Length of gestation in the guinea-pig with data on the frequency and time of abortion and stillbirth. *Anat. Rec.*, **128**, 747–757.

Graham-Jones, O. (Ed.) (1964). *Small Animal Anaesthesia*. Oxford: Pergammon.

Green, I. (1970). Genetic control of immune responses in guinea-pigs. *Guinea-pig News Letter*, **2**, 40–47.

Gupta, B. N., Conner, G. H. & Langham, R. F. (1970). A device for collecting milk from guinea-pigs. *Am. J. vet. Res.*, **31**, 557–559.

Hansen, C. T., Judge, F. J. & Whitney, R. A. (1973). *Catalogue of N.I.H. rodents*. Pub. No. NIH 74-606. Department of Heath, Education and Welfare.

Hansen, C. T. & McEleney, W. J. (1971). Strain and season differences in the reproductive performance of inbred strains of mice, rats and guinea-pigs. Proc. IV ICLA Symposium. Washington, D. C.: Nat. Acad. Sci.

Harper, L. V. (1970). Role of contact and sound in eliciting filial responses and development of social attachments in domestic guinea-pigs. *J. Comp. Physiol. Psychol.*, **73**, 427–435.

Hoyland, F. (1963). Sizes of wire mesh in floors of guinea-pig breeding and holding cage units. *J. Anim. Techns Ass.*, **14**, 120.

Hoyland, F. (1972). Guinea-pig production at the Animal Virus Institute, Pirbright, Woking, Surrey. *Guinea-pig News Letter*, **5**, 13–17.

Huckinghaus, F. (1962). Vergleichende Untersuchungen über die formenmannig Faltigkeit der Unterfamilie-caviinae Murray 1886. *Z. wiss. Zool.*, **166**, 1–98.

Kantor, F. S., Ojeda, A. & Benacerraf, B. (1963). Studies on artificial antigens. I. Antigenicity of DNP-poly-lysine and DNP copolymers of lysine and glutamic acid in guinea-pigs. *J. exp. Med.*, **117**, 55–69.

Keyhani, M. & Naghshineh, R. (1974). Spontaneous epizootic of pneumococcus infection in guinea-pigs. *Lab. Anim.*, **8**, 47–49.

King, J. A. (1956). Social relations of the domestic guinea-pig living under semi-natural conditions. *Ecology*, **37**, 221–228.

Latimer, H. B. (1951). Weights, percentage weight and correlations of endocrine glands of the adult male guinea-pig. *Anat. Rec.*, **111**, 299–315.

Ley, F. J., Bleby, J., Coates, M. E. & Paterson, J. S. (1969). Sterilization of laboratory animal diets using gamma radiation. *Lab. Anim.*, **3**, 221–254.

Lovell, D., King, D. & Festing, M. F. W. (1972). Breeding performance of specific pathogen free guinea-pigs. *Guinea-pig News Letter*, **6**, 6–11.

Masson, J. H. (1940). The date of the first use of guinea-pigs and mice in biological research. *J. South Afr. vet. med. Ass.*, **11**, 22–25.

Maynard, L. A., Boggs, D., Fisk, G. & Segnin, D. (1958). Dietary mineral interrelations as a cause of soft tissue calcification in guinea-pigs. *J. Nutr.*, **64**, 85–97.

Medical Research Council (1974). *Survey of the Numbers and Types of Laboratory Animals Used in the United Kingdom in 1972*. Manual Series No. 3. Medical Research Council Laboratory Animals Centre, Woodmansterne Road, Carshalton, Surrey, SM5 4EF.

Moreland, A. F. (1965). Collection and withdrawal of body fluids and infusion techniques. In *Methods of Animal Experimentation*, ed. Gay, W. I. New York: Academic Press.

Mosinger, M. & Generalits, A. (1961). Sur la carcino-résistance du cobaye. Première partie. Les tumeurs spontanées du cobaye. *Bull. l'Ass. Française pour l'étude du Cancer*, **48**, 217–235.

Nakagawa, M., Muto, T., Nakano, T., Yoda, H., Amdo, K., Isobe, Y. & Imaizumi, K. (1969). Some observations on diagnosis of *Bordetella bronchiseptica* infection in guinea-pigs. *Exp. Anim. (Japan)*, **18**, 105–116.

Nixon, J. M. (1974). Breathing pattern in the guinea-pig. *Lab. Anim.*, **8**, 71–77.

Opler, S. R. (1967). Observations on a new virus associated with guinea-pig leukemia. Preliminary note. *J. nat. Cancer Inst.*, **38**, 797–800.

Owen, D. & Porter, G. (1967). The establishment of a colony of specified pathogen free guinea-pigs. *Lab. Anim.*, **1**, 151–156.

Owen, D. (1971). Some common parasites of guinea-pigs. *Guinea-pig News Letter*, **3**, 2–4.

Owen, D. (1972). *Common Parasites of Laboratory Rodents and Lagomorphs*. London: H.M.S.O.

Pappenheimer, R. M. & Slawetz, C. A. (1942). A generalised visceral disease of guinea-pigs associated with intranuclear inclusion. *J. exp. Med.*, **76**, 299–306.

Paterson, J. S. (1972). The Guinea-pig or Cavy. In *The UFAW Handbook on the Care and Management of Laboratory Animals*. 4th edn. Edinburgh: Churchill-Livingstone.

Phoenix, C. H. (1970). Guinea-pigs. In *Reproduction and Breeding Techniques for Laboratory Animals*. (ed.) Hafez, E.S.E. Philadelphia: Lea & Febiger.

Poiley, S. M. (1972). Growth tables for 66 strains and stocks of laboratory animals. *Lab. anim. Sci.*, **22**, 757–779.

Reid, M. E. & Bieri, J. G. (1972). Nutrient requirements of the guinea-pig. In *Nutritional Requirements of Laboratory Animals*. Washington: Nat. Acad. Sci.

Rogers, J. B. (1951). The ageing process in the guinea-pig. *J. Geront.*, **16**, 13–16.

Rothschild, Lord (1961). *A Classification of Living Things*. London: Longmans.

Rowlands, I. W. (1949). Post-partum breeding in the guinea-pig. *J. Hyg., Camb.*, **47**, 281–287.

Schaffer, A. (1965). Anaesthesia and sedation. In *Methods of Animal Experimentation*. ed. Gay, W. I. New York: Academic Press.

Schoenbaum, M., Klopfer, U. & Egyed, M. N. (1972). Spontaneous intussusception of the small intestine in guinea-pigs. *Lab. Anim.,* **6,** 327–330.

Searle, A. G. (1968). *Comparative Genetics of Coat Colour in Mammals*. London: Logos Press.

Sebesteny, A. (1971). Fire-risk from anaesthesia of rodents with halothane. *Lab. Anim.,* **5,** 225–231.

Short, D. J. & Gammage, L. (1959). A new pelleted diet for rabbits and guinea-pigs. *J. anim. Techns Ass.,* **9,** 62–69.

Sluckin, W. (1968). Imprinting in guinea-pigs. *Nature,* **220,** 1148.

Spector, W. S. (1956). *Handbook of Biological Data*. Philadelphia: Saunders.

Stevenson, D. E. (1964). General aspects of rodent anaesthesia. In *Small Animal Anaesthesia*, ed. Graham-Jones, O. Oxford: Pergamon.

Stone, S. H., Liacopoulos, P., Liacopoulos-Briot, M., Neven, T. & Halpern, B. N. (1964). Histamine differences in amount available for release in lungs of guinea-pigs, susceptible and resistant to acute anaphylaxis. *Science,* **146,** 1061–1062.

Strandskov, H. H. (1939). Inheritance of internal organ differences in guinea-pigs. *Genetics,* **24,** 722–727.

Takino, Y., Sugahara, K. & Horino, I. (1971). Two lines of guinea-pigs sensitive and non-sensitive to chemical mediators and anaphylaxis. *J. Allergy,* **47,** 247–261.

Thomasson, B., Ruuskanen, O. & Merikanto (1974). Spinal anaesthesia in the guinea-pig. *Lab. Anim.,* **8,** 241–244.

Wagner, J. E. & Manning, P. J. (1975) (Eds). *The Biology of the Guinea-pig*. New York: Academic Press.

Walker, E. P. (1964). *Mammals of the World*. Baltimore: Johns Hopkins.

Weir, B. J. (1972). Some notes on the history of the domestic guinea-pig. *Guinea-pig News Letter,* **5,** 2–5.

Weiss, C. (1958). Care of guinea-pigs used in clinical and research laboratories. *Amer. J. clin. Path.,* **29,** 49–53.

Wickham, N. (1959). Calcification of soft tissues associated with dietary magnesium deficiency in the guinea-pig. *Aust. vet. J.,* **34,** 244–248.

Williamson, J. & Festing, M. F. W. (1971). A note on some haematological parameters in three strains of guinea-pigs. *Guinea-pig News Letter,* **3,** 7–11.

Wills, J. E. & Sutherland, S. D. (1970). Increased productivity in a large guinea-pig colony. *J. Inst. Anim. Techns,* **21,** 134–147.

Wilson, J. E. (1974). An improved guinea-pig drinking bottle. *Lab. Anim.,* **8,** 45–46.

Woode, G. N. & McLeod, N. (1967). Control of acute *Bordetella bronchiseptica* pneumonia in a guinea-pig colony. *Lab. Anim.,* **1,** 91–94.

Wright, S. (1922). The effects of inbreeding and cross breeding guinea-pigs. *U.S. Dept. Agric. Bull.* No. 1090.

Wright, S. (1960). The genetics of vital characters in the guinea-pig. *J. cell. comp. Physiol.,* **56,** 123–151.

21 Hamsters

MICHAEL F. W. FESTING

Introduction

According to Walker (1964) the true hamsters are classified as one of the seven tribes of the family Cricetidae, order Rodentia. Six genera are known, the general description being that they are 'mouse-like animals with thick-set bodies, short tails and cheek pouches'. Distribution is throughout Eurasia, Africa and as far east as China.

Although little is known about the habits of of wild hamsters, most species are solitary, territorial animals, living in extensive burrows in which they store food consisting of seeds and vegetable matter. This food-hoarding habit leads some species to be classified as pests in their native areas.

Most hamsters appear to be active in the evening and early morning, and possibly throughout the night, and some species hibernate in the winter.

Breeding usually occurs in the spring and summer, when the males leave their winter burrows and seek out the females, with whom they stay for only a short period. Several litters may be born each season, each consisting of four to six pups.

Natural predators of hamsters are birds of prey, mink, foxes, dogs, cats, and man. The European hamster (*Cricetus cricetus*) is trapped for its fur, and is destroyed as a pest.

Taxonomy

There are at least 54 different species, varieties or subspecies of hamster, and the taxonomic relationship between these is currently in an unsettled state owing to recent cytological work on the various groups. A classification which takes account of this cytological evidence has been proposed by Yerganian (1968) and is given in Table I. The classification is incomplete in that it includes only those species on which cytological analyses have been performed. The table does, however, list the main laboratory

TABLE I

Present cytotaxonomy of old world hamsters of the subfamily Cricetinae[1]

Genus	No. Chromosomes
1. *Cricetulus* (Eastern Europe, Asia Minor, Siberia, Tibet, Mongolia)	
* *C. griseus* Milne–Edw. (Chinese hamster) Lat. 40°N, Long. 117°	22
* *C. migratorius* Pallas (Grey or Armenian hamster) Lat. 40°N, Long. 45°	22
C. barabensis Pallas (subspecies of *C. griseus*)	20
Subgenus *Tscherkia* (Korea, North China) *T. Triton*, De Winton	30
2. *Allocricetulus* (China, Mongolia, Siberia)	
A. eversmanni Brandt	26
A. curtatus	20
3. *Mesocricetus* (Eastern Europe, Asia Minor, Iran)	
* *M. auratus* Water. (Golden or Syrian hamster) Lat. 32°N, Long. 35°	44
* *M. brandti* Nehr. (Caucasian hamster) Lat. 40–45°N, Long. 45°	42
4. *Cricetus* (Central and Western Europe)	
C. cricetus	22
5. *Phodopus* (Siberia, Mongolia, China)	
* *P. sungorus* Pallas	28
P. raborovskii	?

* Currently bred in the laboratory.   [1] From Yerganian (1968).

species, including the Chinese hamster, *Cricetulus griseus*, the Golden or Syrian hamster, *Mesocricetus auratus*, and the common or European hamster, *Cricetus cricetus*. Other species which have been bred in the laboratory include the Armenian hamster, *Cricetulus migratorius*, the Caucasian hamster, *Mesocricetus brandti*, and the hairy-footed hamster, *Phodopus sungorus* (Pogosianz & Sokova, 1967). Three of these species are shown in Figure 21.1.

There is some controversy over whether the common name of *M. auratus* should be the 'Golden' or the 'Syrian' hamster. A recent monograph (Hoffman *et al.*, 1968) prefers 'Golden' hamster, on the grounds that more than one species of hamster live in Syria. However, the occurrence of coat colour mutations in laboratory populations also makes the term 'Golden' ambiguous, and as geographical descriptions are used for other common species of hamster, the term *Syrian* hamster will be used in this article.

Introduction as a laboratory animal

The domestication of the Syrian hamster and introduction to the laboratory is well documented, and has been described by Adler (1948). All laboratory strains of this species are believed to have originated from three litter-mates captured near Aleppo in Syria in 1930, progeny of which were used in experimental work on Kala-azar. These animals were later shipped to laboratories throughout the world, where extensive breeding and selection has resulted in the development of numerous strains, inbred lines

and mutant types. Hamsters were first introduced into the UK by Hindle in 1931.

In contrast, the introduction of the Chinese hamster into the laboratory is not so well documented. The earliest paper on this species, listed by Weihe & Isenbügel (1970) in their extensive bibliography, is by Hsieh (1919), but it is not clear whether present colonies of Chinese hamsters originate from one or several trappings from the wild. However, since Chinese hamsters were used as laboratory animals before it was possible to breed them in captivity (Fulton, 1968) it is likely that modern colonies are derived from several different wild groups imported at various times.

Behaviour of the hamster in captivity

There have been surprisingly few studies on the behaviour of the Syrian hamster, and virtually none on the Chinese hamster in captivity.

The main studies on the Syrian hamster have been those of Richards and co-workers on activity and maternal behaviour (Richards, 1964; Noirot & Richards, 1966; Richards, 1966), and an extensive study on social dominance by Lawlor (1963). Swanson (1966 and 1967) has used the hamster in psychological research, and Yaron *et al.* (1963) have studied the influence of handling on reproductive behaviour. In addition, Bowland & Waters (1955) compared the maze-learning ability of the hamster with the rat, Gumma *et al.* (1967) examined temperature preference, and Avcari *et al.* (1968) showed that physical restraint of the hamster could induce ulcers within 24 hours. Recently Murphy &

Fig. 21.1. Three species of hamster. L. to R. *Mesocricetus auratus*, *Phodopus sungorus* and *Cricetulus griseus*. (Photograph: Festing.)

Schneider (1970) have shown that the mating behaviour of the male hamster depends entirely on the sense of smell.

Hamsters are most active during the dark period, and in laboratory conditions hand-mating is usually most successful within about an hour of nightfall. Richards (1966) studied the home-cage activity of Syrian hamsters for a one hour period each day, starting half an hour after the start of the dark period, i.e. at the time of peak activity. There were detectable changes in activity according to the stage of pregnancy but during the first eight days approximately 25 per cent of the time was spent sleeping, 22 per cent in the nest, 18 per cent on bar chewing, 15 per cent on grooming, 7 per cent on nest building, 6 per cent on movement, and 5 per cent on hoarding.

Richards (1966) and Goodchild & Frankenberg (1962), examined the wheel-running activity of caged hamsters. Richards found that females averaged about 8,000 revolutions of a 100 cm activity-wheel, or a distance of about 8 km per day. This is a remarkable level of activity, and suggests that the forced inactivity of the normal laboratory hamster may be physiologically abnormal. The hamster has been domesticated for only a relatively short time, and may not yet be genetically adapted to captivity in the same way as the rat and mouse have become. Clearly there is scope for further ethological investigation of hamsters kept under laboratory conditions, particularly in relation to cage design and the physiological effects of an enforced low level of activity.

Inbred strains and mutants

According to Robinson (1968) there are now ten known mutant genes in the Syrian hamster, eight of which are autosomal and two sex-linked. All these mutations affect coat-colour, though some of them also have pleiotropic effects. In addition, Nixon & Connelly (1968) and Nixon et al. (1969) have recorded two more mutations, one affecting coat-colour and the other causing hind-leg paralysis, and three more probable mutations affecting the coat are known to the author though these have not yet been described in the scientific literature. Two of these cause hairlessness, and the other one affects the quality of the coat. A list of the named mutants (after Robinson) is given in Table II.

Yerganian (1967) lists four mutations in the Chinese hamster, though none of these appear to be inherited in a simple Mendelian way.

In the European hamster, *Cricetus cricetus*, there is a naturally occurring coat-colour polymorphism. According to Searle (1968) the usual wild type has a black belly, an agouti back, a band of chestnut round the head, with some white on the legs, side and nasal area. In parts of Russia and Germany, however, a melanic form has been common for at least 200 years. This melanic form, which is completely black except for some white hairs on the extremities, and occasional white spots on the throat and chest, is inherited as a dominant.

Jay (1963) lists five established inbred strains of *M. auratus*, and a further 27 strains in development, ranging from about F5 to F17. By now a substantial number of these strains are

---

TABLE II

Known genes of the Golden Hamster and probable homologues in other laboratory rodents[1]

Gene	Typical Phenotype	House Mouse	Rabbit	Rat	Guinea-pig	Deermouse
b	Brown eumelanism	b	b	b	b	b
Ba	White belt	..	..	..	..	..
c^d	Acromelanic albinism	c^h	c^h	..	c^d	..
e	Restriction of eumelanism	..	e	..	e	y
Mo	Sex-linked mottling	Mo	..	..	..	..
ru	Ruby-eye dilution	p	..	p	p	p
s	Irregular spotting	s	..	..	s	..
To	Sex-linked yellow	..	..	..	..	..
Wh	Anophthalmic white	mi^{wh}	..	..	..	..

[1] From R. Robinson (1968).

probably fully inbred (i.e. have been mated for more than 20 generations by brother × sister matings). A total of 19 inbred colonies are listed in Festing & Conn (1968), and eighteen inbred or partially inbred strains are listed in Yager & Sundborg (1968).

The uses of hamsters in research

Both the Syrian and the Chinese hamster are widely used in fundamental types of study which frequently do not require intact animals. Syrian hamsters are particularly favoured by virologists, but studies in the fields of cancer research (Hornburger, 1968), dental research, and biochemistry also make extensive use of this species. In contrast, few studies have been carried out on the behaviour, growth, reproduction and physiology of the Syrian hamster.

The Chinese hamster is widely used in pathology, largely because some strains have a high incidence of diabetes mellitus (Yerganian 1964; Butler 1967), thus making it a useful model of diabetes in man. The species is also used as a host in certain parasitological studies. The low chromosome number makes the Chinese hamster particularly useful for cytological investigations, so a large proportion of the research work using this species falls in the general categories of genetics, tissue culture and radiation.

Both species of hamster have unusual or unique features that make them particularly useful for certain studies, though the actual numbers used in these studies may be small. The Syrian hamster is the only common laboratory animal that hibernates, and Chaffee (1966) showed that the tendency to hibernate is genetically controlled. By selection he was able to develop lines with a high (74 per cent) and a low (25 per cent) proportion of animals which would hibernate under standard conditions of a low light intensity, an 8-hour day and a temperature of 5 to 6°C. Smith (1961) reported that hamsters have been frozen to −5°C for up to 60 minutes, with colonic temperatures of −1°C, and have subsequently been resuscitated. A proportion of these animals lived for periods up to 450 days in apparent good health. The short gestation period of the Syrian hamster has also

been considered an advantage in experimental teratology (Ferm, 1967).

The presence of a cheek pouch in both species has also proved to be exceedingly useful, especially as the cheek pouch of the Syrian hamster is an immunologically privileged site capable of accepting many tumour grafts from other species. Fulton *et al.* (1968) listed the advantages of using the cheek pouch of the Syrian hamster to study tumour growth, as follows:

1. The tumour transplants grow freely and symmetrically.
2. The same transplants may be measured and photographed at regular intervals without harm to the hamster.
3. The tumour is accessible for experimental procedures such as X-irradiation, perfusion or chemotherapy.
4. The early stages of growth and vascularization may be studied by microscopy *in vivo*.
5. The takes are nearly 100 per cent successful.
6. Serial transplantation will result in progressive reduction in host reaction with each passage, increase in the rate of growth and, finally, reproducible tumours.

The unusual response of the Syrian hamster to skin homografts has attracted a considerable amount of attention (e.g. Billingham *et al.*, 1960; Hildeman & Walford, 1960). In many cases skin grafts between members of a closed colony will survive indefinitely, though grafts between animals from different colonies may be rejected as vigorously as in other species. Billingham & Hildeman (1958) showed that the rate of rejection of homografts within a closed colony of *C. griseus* was comparable with other species, and suggested that the unusual situation in *M. auratus* was due to a paucity of transplantation antigens (genes). Since all Syrian hamsters kept in laboratories are derived from a single trio, the paucity of histocompatibility genes may be a reflection of the number of genes carried by the original stock rather than a reflection of the natural state of wild hamsters. More recently, however, Palm *et al.* (1967) studied skin transplantation in *M. brandti*, a subspecies of *M. auratus*, and found a similar paucity of histocompatibility antigens suggesting that '. . . the anomalous histocompatibility situation in domestic populations of *M. auratus* is a reflection of a paucity of segregating histocompatibility factors in wild hamster populations'.

Standard biological data

Breeding data for the Syrian and Chinese hamster are given in Table III. In many cases, however, the data are subject to large variations depending on the strain and the husbandry conditions, so even a relatively large departure from the figures given may still lie within the normal range.

The main differences between the two species are that the Chinese hamster matures later and has a longer gestation period.

Some organ weights in the two species are given in Table IV. The bodyweight of the mature male Chinese hamster is about half that of the Syrian hamster, and there are some striking differences in the relative sizes of some of the organs. In particular, the spleen, testes and brain of the Chinese hamster are all relatively larger than those of the Syrian hamster, while the adrenals appear to be relatively smaller.

Average normal haematological values for the two species are given in Table V. There appear to be no significant differences between the species except in the proportion of non-segmented polymorphonuclear neutrophils and monocytes. Such differences could be attributable to differences in haematological technique.

TABLE III
Breeding data for Syrian and Chinese hamsters

Character	Syrian (*M. auratus*)	Chinese (*C. griseus*)
Age at puberty	45–60 days	48–100 days
Min. breeding age	6–8 weeks	10–12 weeks
Breeding season	All year, may be a decrease in winter	All year in laboratory conditions
Oestrus cycle	Polyoestrus: all year	Polyoestrus: all year
Duration of oestrus cycle	4 days	4 days
Duration of oestrus	4–23 hours	6–8 hours
Gestation	15–18 days	21 days
Average litter size	5–7	4·5–5·2
Ovulation time	Early oestrus	Shortly before oestrus
Copulation	About 1 hour after night fall	2–4 hours after start of dark period
Implantation	5 or more days	5–6days
Birth weight	2 g	1·5–2·5 g
Weaned	20–25 days	21 days
Chromosome No.	44	22
Return to oestrus post partum	5–10 mins	? post partum mating does occur
No. of mammae	14–22	8

TABLE IV
Some organ weights of hamsters

(Weight±Standard Deviation)

Organ	Syrian (*M. auratus*)	Chinese (*C. griseus*)[2]
Number of animals	22[1]	12
Body weight (g)	97·2 ±11·7[1]	42·5 ± 3·2
Liver (g/100 g)	4·81± 0·75[1]	5·00± 0·28
Spleen (mg/100 g)	117·2 ±22·0[1]	653 ±170
Thymus (mg/100 g)[3]	141·2 ±38[1]	93·8 ± 25·2
Kidneys (mg/100 g)[3]	728 ±56[1]	982 ± 72
Adrenals (mg/100 g)[3]	23·5 ± 4·2[1]	15·6 ± 3·3
Testes (g/100 g)[3]	2·67± 1·65[1]	4·04± 0·54
Heart (mg/100 g)	470[4]	539 ± 62
Lungs (mg/100 g)	460[4]	660 ±115
Brain (g/100 g)	0·88[4]	1·67± 0·14

[1] Juszkiewicz & Stefaniak (1969)
[2] Original data, R. F. Parrott (1969).
[3] Total weight of both organs combined.
[4] Kaplan (date unspecified).

TABLE V

Average normal blood values in M. auratus *and* C. griseus

	Syrian (*M. auratus*)[1]	Chinese (*C. griseus*)[2]
RBC (Millions/mm³)	7·50± 2·40	8·52±1·07
Hemoglobin g (%)	16·8 ± 1·2	15·8 ±0·98
Hematocrit (%)	52·5 ± 2·3	51·8 ±1·0
Mean corpuscular vol. (μ^3)	71·2 ± 3·19	61·5 ±6·0
Mean corpuscular hemoglobin ($\mu\mu$g)	22·3 ± 1·27	17·5
Mean corpuscular hemoglobin concentrate (%)	32·0 ± 2·23	29·5
Reticulocytes (%)	2·5 ± 1·2	
Platelets (thousands/mm³)	670 ± 1·5 (indirect method) —	
	310 ±62·8 (direct method) —	
Total white blood count (thousands/mm³)	7·62± 1·3	6·05±1·66
Segmented polymorphonuclear neutrophils (%)	21·9 ± 5·5	18·5 ±5·9
Non-segmented polymorphonuclear neutrophils (%)	8·0 ± 2·5	0·64± ?
Eosinophils (%)	1·1 ± 0·02	1·3 ± ?
Lymphocytes (%)	73·5 ± 9·4	78·3 ±6·4
Monocytes (%)	2·5 ± 0·8	0·64

[1] Desai (1968).   [2] Moore (1966).

TABLE VI

Blood constituents of hamsters

Character	M. auratus	C. griseus
Serum glucose (mg/100 ml)	73—155[2]	111–119
Serum cholesterol (mg/100 ml)	153·9±8·9[3]	115–118
Total protein (g/100 ml)	4·5±0·73[2]	approx. 5·25
Blood pressure (mmHg)	111[4]	?

[1] Sirek & Sirek (1967) non-fasted animals (additional biochemical data also given).
[2] P. F. Robinson (1968).
[3] Lee *et al.* (1959)—Highest amongst monkey, dog, cat, rabbit, guinea-pig, rat and mouse. See also in Jaszkiewicz & Stefaniak (1969).
[4] Stroia (1954) Cannulation of carotid artery.

Some additional blood parameters are given in Table VI. According to Lee *et al.* (1959) the serum cholesterol level is the highest among common laboratory animals.

The life span, tumour incidence and occurrence of intercapillary glomerulosclerosis in the Chinese hamster has been studied by Kohn & Guttman (1964). The median age at death of animals surviving at 230 days was: male 1,045 days (99 per cent confidence interval (CI) 887–1,170); female 959 days (99 per cent CI 820–1,003). Of the males 81 per cent had died by 1,450 days—the time at which the experiment was terminated. Tumour incidence was found to be 16 per cent of all animals autopsied.

Grindeland *et al.* (1957) studied the influence of exposure to cold on the life span of the Syrian hamster. The mean life span of 72 control hamsters maintained under typical laboratory conditions was 601 days for males and 616 days for females, but exposure to a temperature of 6°C for four months killed about 50 per cent of the experimental animals.

The relatively short life span of Syrian hamsters suggests that they might be well suited to studies on ageing, though further longevity studies should first be made to confirm the work of Grindeland *et al.* Ortiz (1955) has used the hamster in a study of the effects of age on the female reproductive system.

Husbandry

Housing and caging

For all routine purposes, cages and equipment designed for rats are perfectly adequate for housing Syrian hamsters, and cages designed for mice are adequate for Chinese hamsters. The Institute of Laboratory Animal Resources (1960) has published standards for the area of accommodation that should be provided for housing Syrian hamsters. These standards are given in Table VII.

TABLE VII
NRC—ILAR space standards for the care of Syrian hamsters

Age	Recommended cage area	
	Sq. in/animal	Sq. cm/animal
Weaning—5 weeks	10·0	65
5 weeks—3 months	12·5	81
Over 3 months	15·0	97
Breeding ♀ and litter	150·0	970

Choice of a suitable breeding cage will depend on the breeding system that is to be used. For both species there are three well-established mating systems, namely, hand mating, harems, and permanently-mated groups. The choice of breeding system will depend on circumstances, but will dictate the type of breeding cage required.

It has sometimes been suggested that regular exercise might enhance reproductive performance, though there does not appear to be any experimental foundation for such a belief. It is not usual practice to provide exercise wheels, though in certain circumstances the provision of such facilities could be advantageous. A simple exercise disc for Syrian hamsters has been described by Magalhaes (1968), and Yerganian (1967) described a stainless steel exercise wheel used for Chinese hamsters. He stated that 'the exercise wheel has been found to encourage many social and physiological conditions desired for selective inbreeding of a naturally solitary species'.

Food and watering utensils

Most cages designed for rats and mice have built-in food hoppers, frequently in the lid. Such hoppers are usually satisfactory for hamsters, though if Syrian hamsters are housed in rat breeding-cages the hopper and water bottle spout may be too high for young animals.

Sometimes a wet mash or green food is given instead of water, in which case the food hopper has to be designed so that it can be cleaned easily. For normal laboratory work the use of a pelleted diet and fresh water is strongly recommended.

Water can be supplied in the same manner as is usual for rats and mice, fresh water being given at least once per week.

Bedding

The Syrian hamster is an active nest-builder, so breeding cages should be well supplied with suitable materials. Sterilized wood chips, shredded paper, hay, cellulose wadding and similar materials have all been found suitable, and sawdust or a similar material may also be provided to help to absorb moisture. As the young of the Syrian hamster are very immature when born, plenty of nesting materials should be available so that the mother can regulate their temperature. The bedding requirements of the Chinese hamster are similar, though it is not such an active nest-builder as the Syrian hamster.

Environmental conditions

Relatively little is known about the optimum environmental conditions for maintaining hamsters. In general, conditions suitable for rats and mice appear to be adequate, though it seems more difficult to avoid a seasonal pattern of reproduction in hamsters than in rats and mice.

The Syrian hamster is well known as a hibernating animal, and the conditions necessary for hibernation have been reported by several authors. These include low temperatures, short day lengths, quietness, ample nesting material, and enough food to enable a food store to be accumulated (Gumma *et al.*, 1967). These conditions could be approached if the animal house facilities were inadequate. Animal house temperatures should therefore be maintained at 20 to 21°C, and a day length of at least 12 hours is recommended. In windowless animal rooms the lighting schedule may be reversed, if necessary, to allow hand-mating during working hours. In animal rooms with windows the day length should be extended to 14–16 hours (controlled by a time clock) to avoid a seasonal pattern of daylight.

Identification

The marking of individual animals should be avoided wherever possible, as no completely satisfactory method has been reported. Short & Woodnott (1969) recommended the use of an ear tattoo with forceps and numerals 5 mm × 6

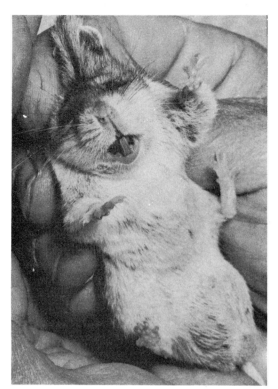

Fig. 21.2. Chinese hamster held by dorsal skin. (Photograph: UFAW.)

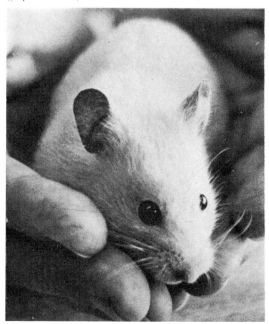

Fig. 21.3. Syrian hamster held in cupped hands. (Photograph: UFAW.)

mm. Animals may be temporarily marked with stains, and ear-punching may be carried out in the same manner as for mice.

Handling

Although relatively aggressive among themselves, the Chinese hamster seldom presents any handling problems, and bites are rare. Most individuals become extremely tame if handled frequently. Chinese hamsters are usually held by the loose dorsal skin (Fig. 21.2) or in the cupped hands.

In contrast, Syrian hamsters can be somewhat more troublesome since they tend to resent being handled. For routine handling (such as changing cages, etc.) the animals should be lifted in the cupped hands (Fig. 21.3) care being taken not to hold them tightly. If manipulation is necessary the dorsal skin should be held while the hamster is cupped in the palm of the hand; the little finger may then be placed over the hind leg, so that the legs are held between the fourth and fifth fingers (as in Figure 21.2, but with the fingers extended over the hind legs). For sexing, Syrian hamsters are usually held as shown in Figures 21.4 and 21.5.

Transport

The transportation of hamsters presents no serious difficulties. The only point worth noting is that cardboard containers should not be used,

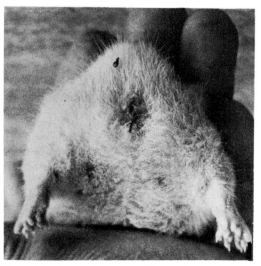

Fig. 21.4. Sexing hamsters. Female Syrian hamster. (Photograph: UFAW.)

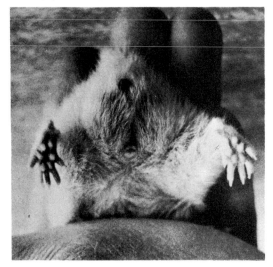

Fig. 21.5. Sexing hamsters. Male Syrian hamster. (Photograph: UFAW.)

since they present very little problem to a hamster determined to escape.

Feeding

Natural Foods

Little is known about the natural foods of the Syrian hamster, though examination of the food stores of other species of hamster suggests that a large proportion of the diet consists of seeds and berries. The extent to which insects and other animal matter also contribute to the diet is not known, though there is some evidence that, given the opportunity, hamsters will consume such material.

Feeding in the laboratory

According to Porter (1970) the nutritional requirements of the hamster have not been investigated in detail, and there is little information of the quantitative requirements of the nutrients that have been studied. Even basic information such as the protein requirements for optimum growth and reproduction has not yet been obtained (Granados, 1968). Experience has shown however, that hamsters will thrive on many of the commonly-used rat and mouse diets. Schweigert (1962), has prepared a table of the 'Satisfactory Levels of Nutrients to Meet the Nutritional Requirements of the Growing Hamster' (Table VIII), based largely on practical experience rather than critical experimentation, and he has also prepared a satisfactory synthetic diet (Table IX).

TABLE VIII
Satisfactory levels of nutrients to meet the nutritional requirements of the growing hamster [1]

Nutrient	Per 100 g diet
Total protein, g	24
Minerals	
Calcium, g	0·6
Phosphorus, g	0·35
Vitamins	
Vitamin A, IU	1,300
Vitamin D	not required[2]
α–Tocopherol, mg	2·5
Vitamin K	?[3]
Thiamine, mg	0·6
Riboflavin, mg	0·6
Vitamin B_6, mg	0·6
Niacin	?
Pantothenic acid, mg	4
Biotin	not required
Folic acid	not required
Choline	?
Vitamin B_{12}	not required
Inositol	?

[1] Quantitative values are estimated from various adequate rations, hence are probably in excess of the actual requirements.
[2] Vitamin D may be required in diets with an unsatisfactory calcium to phosphorus ratio.
[3] ? signifies that a qualitative requirement for the nutrient is in doubt.

Reprinted by permission from Nutrient Requirements of Laboratory Animals, Publication 990, Committee on Animal Nutrition, National Academy of Sciences—National Research Council, Washington, D.C., 1962.

In practice, Syrian hamsters may be fed simply on pelleted mouse or rat diet, with an adequate supply of water. Many breeders, however, add a supplement of grains, green food, meat, milk or vitamin concentrates. The benefits to be derived from such supplements will of course depend largely on the nature and quality of the basic diet.

A similar diet may be used for Chinese hamsters, though in this case it is usual to supplement with grains; if fresh green food is given daily it is possible to do without water. Moore (1965) found that Chinese hamsters preferred rolled oats to any of the pelleted diets and that lettuce was very palatable. Wheat germ was also fed to his colony, but this was discontinued owing to a very high level of wastage. The

TABLE IX

Example of the composition of a satisfactory purified diet for the hamster

Ingredient	Per cent	Ingredient	Mg per 100 g
Casein, purified	24	α–Tocopherol	2·5
Sucrose	62·4	Menadoine	0·6
Cellulose[1]	3	Thiamine hydrochloride	0·6
Salt mixture[2]	4	Riboflavin	0·6
Cod Liver Oil[3]	0·3	Pyridoxine hydro-	
Liver Extract[4]	1	chloride	0·6
Corn oil	5	Calcium pantothenate	4
Cystine	0·3	Niacin	2
		Choline chloride	100
		Folic acid	0·2
		Biotin	0·01
		Vitamin B_{12}	0·005
		Inositol	100
		p–Aminobenzoic acid	30

[1] Celluflour, Chicago Dietetic Supply Co., Chicago, Ill.
[2] Hegsted's Salts IV (9) which is composed of $CaCO_3$, 600; K_2HPO_4, 645; $CaHPO_4$, $2HO_4$, 150; $MgSO_4.7H_2O$, 204; NaCl, 335; ferric citrate. $6H_2O$, 55; Kl, 1·6; $MnSO_4.4H_2O$, 10·0; $ZnCl_2$, 0·5; $CuSO_4.5HO_2O$, 0·6 parts by weight.
[3] 6,000 i.u. of vitamin A and 850 i.u. of vitamin D per g.
[4] 1:20 Liver Extract, Wilson Laboratories, Chicago, Ill.

Reprinted by permission from Nutrient Requirements of Laboratory Animals, Publication 990, Committee on Animal Nutrition, National Academy of Sciences—National Research Council, Washington, D.C., 1962.

colony was eventually maintained on a diet of rolled oats and rat pellets, with lettuce once per week.

Breeding

Selection and discarding of breeding stock

As in all breeding colonies the animals selected should be healthy, vigorous and free from defects, but once these criteria have been met the choice should depend on the genetic aims of the colony.

Generally, the suitability of an animal for breeding will depend on its own or its parents' performance in respect of growth, reproduction and the presence of other desirable attributes, and the pedigree relationships with other animals. If the colony is in the early stages of inbreeding, for example, considerable attention should be paid to selection for high reproductive performance if the line is not to die out. Conversely, if a small outbred colony is to be maintained, a system giving maximum avoidance of

inbreeding should be used, and selection of any particular individual for the breeding colony should depend largely on its pedigree.

Full details of genetic aspects of maintaining inbred strains or outbred stocks are given in Chapter 2.

Breeding methods

Syrian hamsters

There are three established methods for breeding Syrian hamsters. These are:

(i) Hand-mating.
(ii) Harems.
(iii) Monogamous pairs.

Hand-mating is the commonest breeding system among commercial breeders, though it presents problems to the smallscale producer. The hamsters are mated soon after dark, at the time when they are naturally most active. The usual procedure is to place the female with the stud male, and to observe the animals to see whether mating occurs. If the female is on heat (and approximately a quarter of the females are likely to be on heat on any one night) mating will occur almost immediately, especially if experienced males are used. If not, the male and female may start to fight, and they should be separated as quickly as possible. Care must be taken in separating the combatants, as there is a danger of getting bitten. A substantial leather glove may be used as a safety precaution.

Where windowless animal rooms are available the lighting schedule may be partially reversed, so that the lights go off at, say, 1400 hours; hand-mating can then be carried out during the normal working day.

Hand-mating is not always a satisfactory breeding method owing to the labour involved and the absence of windowless animal rooms.

Harem-mating has been used as a breeding method, the usual procedure being to group one to four males with several females and to separate the females prior to parturition and between four and twelve days after pairing. The exact ratio of males to females is not critical. For example, Bacharach et al. (1958) used a ratio of five females to one male and obtained an output of about 520 weanling hamsters per m² of animal-room per year, a level of output that is only slightly lower than could be expected in a

TABLE X

*Reproductive performance of two inbred strains of Syrian hamster bred as monogamous pairs**

| Strain | No. | Age 1st litter | Numbers born alive | | Productivity (Y/♀/wk) |
			1st litter	2nd litter	
ALAC	41	74·1±8·0	4·7±1·2	5·7±1·8	0·50
CLAC	42	71·4±6·3	6·8±2·0	7·9±2·6	0·89

* Mean ± standard deviation.

breeding colony of rats. The main problem with harem-mating systems is that there may be a certain amount of fighting when the females are returned to the mating cage.

Monogamous pairs. Syrian hamsters may also be bred as monogamous pairs, though such a system has apparently not been previously reported in the literature. The colonies maintained at the MRC Laboratory Animals Centre have been bred in this way for several years with good results, considering that inbred strains are involved. Fighting does not occur provided that the animals are paired at weaning, and left together for their whole breeding life. Some data on breeding performance under there conditions are given in Tables X and XI. The output of weanling hamsters of inbred Strain CLAC, amounting to 0·89 young/female/week, or 46 young per year, compares very favourably with the output reported by Bacharach *et al.* (1958). (In the LAC colonies about 0·09 m² of animal-room is allowed per breeding female).

TABLE XI

Interval between litters in two strains of permanently mated Syrian hamsters

Strain	No. of Intervals	Interval between litters (days)
ALAC	27	42·1±5·4
CLAC	30	37·1±4·5

Chinese hamsters

Difficulties in the breeding of Chinese hamsters probably account for its relative unpopularity as a laboratory animal. The following methods have been used, with varying degrees of success.
  (i) Hand-mating.
  (ii) Harems.
  (iii) Polygamous groups using the 'collar' method.
  (iv) Monogamous pairs.

Hand-mating. Moore (1965) described three different hand-mating schedules, the most successful of which involved mating once at 11.00 hours and again at 15.00 and 16.00 hours. Yerganian (1958) found that a reversed lighting schedule providing 13 hours of light and 11 of darkness was necessary to obtain optimum reproduction, since mating could then be carried out during the early part of the dark period during a normal working day. According to Yerganian (1967) oestrus can be recognized by the superficial appearance of the vagina and by the heightened activity of the females, resulting in a flattening of their nesting material from increased running activity. Females judged to be on heat were placed with tester males, their behaviour confirming or refuting their presumed physiological state. With the reversed lighting schedule it was found possible to mate immediately all females on heat, or if a male was not available a mating could be scheduled for four days later.

Moore (1965) also describes a simpler breeding system whereby the males and females were placed together at weaning and the females were separated just before parturition. Little fighting was observed during adult life, and a generation could be obtained every 80 to 100 days. After the birth of the first litter a hand-mating schedule was used.

Harem systems in which the males and females are left together until just before parturition (as described by Moore, above) have been used for continued breeding, but there is always a danger that the male will be attacked by the female. This risk can be partially overcome by providing large boxes and ensuring that the males remain in the home cage, and placing the female in the male's cage.

Placing the sexes together in the ratio of two males to one female also helps. These methods have not, however, proved completely satisfactory.

The collar method. A collar method of breeding Chinese hamsters which overcomes the problem of the females attacking the male has been described by Belćić & Weihe (1967). The hamsters are placed in a series of cages connected by passageways so that the male can move freely between his own cage and those of the females. All the females are fitted with collars which have an internal diameter of 16 mm and an external diameter of 25 mm. The collars prevent them from entering the passage ways, while the male (without a collar) can move freely between cages. When a female becomes pregnant she is removed to another cage to litter, and another female is placed in the mating cage. This system can be recommended for breeding Chinese hamsters, the main disadvantage being the need for specially-modified mating cages and the necessity of collars for the females.

Monogamous pairs. Apparently certain strains of Chinese hamster are now becoming genetically adapted to laboratory conditions, so that they can be successfully bred as monogamous pairs. Porter & Lacey (1969) described such a colony in which replacement breeding stock was chosen each generation from the progeny of non-aggressive females. Although attacks by females were not entirely eliminated, the colony was successful, and the incidence of attacks will probably decline over a period of several generations. Some breeding data for Porter & Lacey's colony are given in Table XII.

TABLE XII
Breeding data on the LAC colony of Chinese hamsters (Bred as monogamous pairs)

Character	Value
No. of pairs studied	50
Average litter size at birth	4·5
Average litter size at weaning	3·6
Per cent post partum conceptions	40
Average age at first litter (days)	122
Productivity (young/♀/week)	0·42

Weaning and rearing

In both species weaning is usually carried out at 20–21 days. Young animals may be grouped, though the sexes should be separated if breeding is to be prevented. Fighting will not usually occur until the animals reach sexual maturity, but in many strains there will not be any fighting even then provided that there is no mixing of strange animals. The appearance of the external

genitalia in male and female Syrian hamsters is shown in Figures 21.4 and 21.5.

Laboratory procedures

Selection of animals for experiment

Hamsters to be used for experimental work should be healthy and free from obvious abnormalities. Groups of experimental animals should be as uniform as possible in weight, age and genotype. Where, however, there is a considerable variation in either weight or age, the randomized block principle of experimental design should be used to lessen the risk of bias owing to chance allocation of, say, heavy animals in one experimental group.

Where it is suspected that the response to a treatment may depend on genetic factors, several different strains should be used in a factorial type of experimental design. A statistically significant 'strain by treatment interaction' would indicate that strains reacted differently to the experimental treatments. The value of inbred strains, in which all animals are genetically identical, should not be overlooked. The advice of a statistician should be sought whenever an experiment is being planned as this may result in the use of considerably fewer animals.

Anaesthesia and euthanasia

Whitney (1963) has tabulated the doses of pentobarbitone sodium* needed for surgical anaesthesia in the hamster (Table XIII). Ether

TABLE XIII
Doses of pentobarbitone sodium suitable for surgical anaesthesia in Syrian hamsters*

Weight of Animal in grams	Dose of Nembutal in ml
40–50	0·07
51–57	0·08
58–65	0·09
66–69	0·10
70–74	0·11
75–78	0·12
79–85	0·13
86–97	0·14
98–106	0·15
107–115	0·16
116–130	0·17
131–150	0·18
To sacrifice	0·50

* Veterinary Nembutal (Abbott): 60 mg pentobarbitone sodium/ml solution.
Reprinted by permission from Whitney (1963).

has also been used, but chloroform is unsuitable as it causes respiratory complications. Carbon dioxide is recommended for chemical euthanasia, although an overdose of pentobarbitone sodium may also be used. Dislocation of the neck is also recommended for Chinese hamsters. (See Chap. 14).

Collection of specimens

Blood. The most convenient method of bleeding both species of hamsters is from the orbital venous plexus (Schermer, 1967) but it is desirable to anaesthetize the animals first. (For a description of this method see Chapter 17, page 207).

Hamsters can also be bled from the tail, though this is so short that it presents some problems if repeated samples are required. Large samples of blood must be obtained from the jugular vein, the femoral vein on the inside of the hind leg (dissected free) or by heart puncture. Anaesthesia should, of course, be used to avoid causing pain.

Urine and faeces. Metabolism cages suitable for the collection of urine and faeces in rats and mice can also be used for hamsters.

Dosing and injection procedures

These procedures should be carried out in the same way as for rats and mice.

Disease

Hamsters appear to be relatively free from serious infectious diseases, though endoparasites are frequently present. Probably the most serious disease is the so-called 'wet tail', a fatal diarrhoea that has been known in some cases to destroy whole colonies of Syrian hamsters. The aetiology of this disease is not known, though it has been suggested that *E. coli* may be implicated since this organism is frequently found in the gut content of affected hamsters, and neomycin sulphate appears to be effective as a prophylactic agent (Sheffield & Beveridge, 1962).

Handler & Chesterman (1968) list several spontaneous diseases, which appear to be due to nutritional deficiency, genetic factors and unknown causes. Chesterman (1970), in a study of the background pathology of a hamster colony over a 15 year period also noted a 20 per cent incidence of cirrhosis of the liver.

Attempts to produce germfree or specific-pathogen-free hamsters have so far been unsuccessful, owing to the difficulty of hand-rearing However, colonies maintained under hygienic conditions as closed colonies have achieved category 4 status under the MRC Laboratory Animals Centre classification scheme. This is equivalent to specific-pathogen-free status for rats and mice, and includes the complete absence of all ectoparasites, helminths, and pathogenic protozoa and of a comprehensive list of bacteria.

REFERENCES

Adler, S. (1948). Origin of the Golden Hamster, *Cricetus auratus*, as a laboratory animal. *Nature, Lond.*, **162**, 256–7.

Avcari, G., Gaetani, M., Glüsser, A. H. & Turolla, E. (1968). Restraint-induced gastric ulcers in the Golden Hamster. *J. Pharm. Pharmac.*, **20**, 73.

Bacharach, A. L., Cuthbertson, W. F. J. & Flynn, C. W. (1958). The economics of laboratory animal breeding—rats and mice. *Lab. Anim. Cent. coll. Pap.*, **7**, 31–44.

Belcic, I. & Weihe, W. A. (1967). Application of the collar method for breeding Chinese hamsters. *Lab. Anim.*, **1**, 157.

Billingham, R. E., Sawchuck, G. H. & Silvers, W. K. (1960). Studies on the histocompatibility genes of the Syrian hamster. *Proc. natn. Acad. Sci. U.S.A.*, **46**, 1079–90.

Billingham, R. E. & Hildemann, W. H. (1958). Studies on the immunological responses of hamsters to skin homografts. *Proc. R. Soc.* Ser. B., **149**, 216–33.

Bowland, J. A. & Waters, R. H. (1955). Maze learning by the Golden hamster and the albino rat. *Psychol. Rep.*, **1**, 437–40.

Butler, L. (1967). The inheritance of diabetes in the Chinese hamster. *Diabetologia*, **3**, 124–9.

Chaffee, R. R. J. (1966). On experimental selection for super-hibernating and non-hibernating lines of Syrian hamsters. *J. theor. Biol.*, **12**, 151–4.

Chesterman, F. C. (1971). Background pathology in a colony of golden hamsters. In *Symposium on Hamster Pathology*. Basel: Karger.

Desai, R. G. (1968). Haematology and microcirculation. In *The Golden Hamster*, ed. Hoffman, R. A., Robinson, P. F. & Magalhaes, H. Iowa State University Press.

Ferm, V. H. (1967). The use of the Golden hamster in experimental teratology. *Lab. Anim. Care*, **17**, 452–62.

Festing, M. & Conn, H. M. (1968). *International Index of Laboratory Animals*. Carshalton: MRC Laboratory Animal Centre.

Fulton, G. P. (1968). The Golden Hamster in bio-medical research. In *The Golden Hamster*, ed. Hoffman, R. A., Robinson, P. F. & Magalhaes, H. Iowa State University Press.

Fulton, G. P., Lutz, B. R., Patt, D. I. & Yerganian, E. (1954). The cheek pouch of the Chinese hamster for cinephotomicroscopy of blood circulation and tumour growth. *J. Lab. clin. Med.*, **44**, 144–8.

Goodchild, C. G. & Frankenberg, D. (1962). Voluntary running in the Golden hamster, *Mesocricetus auratus* (Waterhouse, 1839), infected with *Trichinella spiralis* (Owen, 1835). *Trans. Am. microsc. Soc.*, **81**, 292–8.

Granados, H. (1968). Nutrition. In *The Golden Hamster*, ed. Hoffman, R. A., Robinson, P. F. & Magalhaes, H. Iowa State University Press.

Grindeland, R. E., Folk, Q. E. & Farrand, R. L. (1957). Some factors influencing the life span of Golden hamsters. *Proc. Iowa Acad. Sci.*, **64**, 638–43.

Gumma, M. R., South, F. E. & Allen, J. N. (1967). Temperature preference in Golden hamsters. *Anim. Behav.*, **15**, 534–7.

Handler, A. H. & Chesterman, F. C. (1968). Spontaneous diseases. In *The Golden Hamster*, ed. Hoffman, R. A., Robinson, P. F. & Magalhaes, H. Iowa State University Press.

Hildeman, W. H. & Walford, R. L. (1960). Chronic skin homograft rejection in the Syrian hamster. *Ann. N.Y. Acad. Sci.*, **87**, 56–77.

Hoffman, R. A., Robinson, P. F. & Magalhaes, H. (1968). *The Golden Hamster*. Iowa State University Press.

Hornburger, F. (1969). Chemical carcinogenesis in the Syrian Golden hamster. A Review. *Cancer*, **23**, 313–38.

Hsieh, E. T. (1919). A new laboratory animal (*Cricetulus griseus*). *Natn. med. J. China*, **5**, 20–4.

Institute of Laboratory Animal Resources (1960). *Standards for the Breeding, Care and Management of Syrian Hamsters*. National Research Council. Washington: Nat. Acad. Sci.

Jay, G. E. (1963). Genetic strains and stocks. In *Methodology in Mammalian Genetics*, ed. Burdette, W. J. San Francisco: Holden Day.

Juszkiewicz, T. & Stefaniak, B. (1969). Some normal values for blood and organs in the Golden hamster. *Vet. Rec.*, **85**, 501.

Kaplan, H. M. (date unknown). *Data on Common Laboratory Animals*. Booklet prepared by Teklad Inc., Monmouth, Illinois, U.S.A.

Kohn, H. J. & Guttman, P. H. (1964). Life span tumour incidence and intercapillary glomerulo-sclerosis in the Chinese hamster (*Cricetulus griseus*) after whole body and partial body exposures to X-rays. *Radiat. Res.*, **21**, 622–43.

Lawlor, M. (1963). Social dominance in the Golden hamster. *Bull. Br. psychol. Soc.*, **16**, 25–38.

Lee, C. C., Herrman, R. G. & Froman, R. O. (1959). Serum, bile and liver total cholesterol of laboratory animals, toads and frogs. *Proc. Soc. exp. Biol. Med.*, **102**, 542–4.

Magalhaes, H. (1968). Housing, care and breeding. In *The Golden Hamster*, ed. Hoffman, R. A., Robinson, P. F. & Magalhaes, H. Iowa State University Press.

Moore, W. Jr. (1965). Observations on the breeding and care of the Chinese hamster, *Cricetulus griseus*. *Lab. Anim. Care*, **15**, 94–101.

Moore, W. Jr. (1966). Hemogram of the Chinese hamster. *Am. J. vet. Res.*, **27**, 608–10.

Murphy, M. & Schneider, G. E. (1970). Olfactory bulb removed eliminates mating behaviour in the male Golden hamster. *Science, N.Y.*, **167**, 302–3.

Nixon, C. W. & Connelly, M. E. (1968). Hind leg paralysis: a new sex-linked mutation in the Syrian hamster. *J. Hered.*, **59**, 276–8.

Nixon, C. W., Whitney, R., Beaumont, J. & Connelly, M. E. (1969). Dominant spotting: a new mutation in the Syrian hamster. *J. Hered.*, **60**, 299–300.

Noirot, E. & Richards, M. P. M. (1966). Maternal behaviour in virgin female Golden hamsters. Changes consequent upon initial contact with pups. *Anim. Behav.*, **14**, 7–10.

Ortiz, E. (1955). The relation of advancing age to reactivity of the reproductive system in the female hamster. *Anat. Rec.*, **122**, 517–37.

Palm, J., Silvers, W. K. & Billingham, R. E. (1967). The problem of histocompatibility in wild hamsters. *J. Hered.*, **58**, 40–4.

Parrott, R. E. (1969). Unpublished data.

Pogosianz, H. E. & Sokova, O. I. (1967). Maintaining and breeding of the Djungarian hamster under laboratory conditions. *Z. Versuchstierk.*, **9**, 292–7.

Porter, G. (1970). Personal communication.

Porter, G. & Lacey, A. (1969). Breeding the Chinese hamster (*Cricetulus griseus*) in monogamous pairs. *Lab. Anim.*, **3**, 65–8.

Richards, M. P. M. (1964). Cyclical behavioural activity in the pregnant Golden hamster. *Nature, Lond.*, **204**, 1327–8.

Richards, M. P. M. (1966). Activity measured by running wheels and observation during the oestrus cycle, pregnancy and pseudopregnancy in the Golden hamster. *Anim. Behav.*, **14**, 450–8.

Robinson, P. F. (1968). General aspects of physiology. In *The Golden Hamster*, ed. Hoffman, R. A., Robinson, P. F. & Magalhaes, H. Iowa State University Press.

Robinson, R. (1968). Genetics and karyology. In *The Golden Hamster*, ed. Hoffman, R. A., Robinson, P. F. & Magalhaes, H. Iowa State University Press.

Schermer, S. (1967). *The Blood Morphology of Laboratory Animals*, 3rd ed. Philadelphia: Davis.

Schweigert, B. S. (1962). Nutrient requirements of the hamster. In *Nutrient Requirements of Laboratory Animals*. N.R.C. Publ. No. 990. Washington: Nat. Acad. Sci.

Searle, A. G. (1968). *Comparative Genetics of Coat Colour in Mammals*, London: Logos Press.

Sheffield, F. W. & Beveridge, E. (1962). Prophylaxis of 'wet tail' in hamsters. *Nature, Lond.*, **196**, 294–5.

Short, D. J. & Woodnott, D. P. (1969). *The IAT Manual of Laboratory Animal Practice and Techniques*. London: Crosby Lockwood.

Sirek, O. V. & Sirek, A. (1967). The colony of Chinese hamsters of the C. H. Best Institute. A review of experimental work. *Diabetologia*, **3**, 65–73.

Smith, Audrey U. (1961). *Biological Effects of Freezing and Super-cooling*. London: Arnold.

Stroia, L. N., Bohr, D. F. & Vocke, L. (1954). Experimental hypertension in the hamster. *Am. J. Physiol.*, **179**, 154–8.

Swanson, H. H. (1966). Sex differences in behaviour of hamsters in open field and emergence tests: Effects of pre- and post-pubertal gonadectomy. *Anim. Behav.*, **14**, 522–9.

Swanson, H. H. (1967). Alteration of sex-typical behaviour of hamsters in open field and emergence tests by neo-natal administration of androgen or oestrogen. *Anim. Behav.*, **15**, 209–16.

Walker, E. P. (1964). *Mammals of the World*. Baltimore: Johns Hopkins Press.

Weihe, W. H. & Isenbügel, E. (1970). Bibliographie des Chinesischem Hamsters. *Z. Versuchstierk.*, **12**, 115–29.

Whitney, R. (1963). Hamsters. In *Animals for Research*, ed. Lane-Petter, W. London & New York: Academic Press.

Yager, R. H. & Sundborg, M. B. (1968). *Animals for Research*, 6th ed. Publ. No. 1678. Washington: Nat. Acad. Sci.

Yaron, E., Chovers, I., Locker, A. & Groen, J. J. (1963). Influence of handling on the reproductive behaviour of the Syrian hamster in captivity. *J. psychosom. Res.*, **7**, 69–82.

Yerganian, G. (1958). The striped-back or Chinese hamster (*Cricetulus griseus*). *J. natn. Cancer Inst.*, **20**, 705–27.

Yerganian, G. (1964). Spontaneous diabetes mellitus in the Chinese hamster (*Cricetulus griseus*). In *Ciba Foundation Colloquia on Endocrinology, Vol. 15 (Diabetes mellitus)*, ed. Cameron, M. P. & O'Connor, C. M. London: Churchill.

Yerganian, G. (1967). The Chinese hamster. In *The UFAW Handbook of the Care and Management of Laboratory Animals*, 3rd ed. Edinburgh: Livingstone.

Yerganian, G. (1968). Present cytotaxonomy of old world hamsters of sub-family Cricetinae. Quoted by G. P. Fulton in *The Golden Hamster*, ed. Hoffman, R. A., Robinson, P. F. & Magalhaes, H. Iowa State University Press.

22 The Mongolian Gerbil

J. H. MARSTON

Introduction

The Mongolian gerbil (*Meriones unguiculatus* Milne-Edwards, 1867) now has a well-established reputation as an extremely attractive and vigorous laboratory animal which breeds readily under ordinary systems of management.

The scientific usefulness of *M. unguiculatus* rests on the fact that it is adapted to desert life and shows almost entirely diurnal activity. The species is one of the many cricetid rodents now available as laboratory animals and this factor becomes important when comparative studies are undertaken. Most of the available observations on *M. unguiculatus* have been made with the intention of establishing the baseline values of its physiology, behaviour and susceptibilities. Schwentker (1969) has published a comprehensive bibliography which shows how widely the animals have been used, and how they have yet to establish a strong position as the animal of choice for particular types of work. Mongolian gerbils have recently been much used for behaviour studies but this may be due to the attraction of having a new species to study rather than any intrinsic value in *Meriones unguiculatus* as such. References to the most recent literature can be found in Swanson (1974) and Yahr & Kessler (1975).

Previous editions of this Handbook noted the introduction of the Mongolian gerbil, but tended to give more attention to other *Gerbillinae* which have since been found generally unsuitable as laboratory species. At the present time, *M. unguiculatus* is the only species of gerbil that is readily available in the UK and USA; it is also the only one being bred under controlled conditions in laboratory and commercial animal colonies. Various species of *Gerbillinae* are occasionally imported and made available by dealers.* It is not worth while to use such animals in the laboratory, as they must have

*The effect of the Rabies (Importation of Dogs, Cats and Other Mammals) Order, 1974 will be to stop all such importation into the U.K.

Fig. 22.1. Mongolian Gerbils; (male L, pregnant female R).

been trapped from the wild and therefore their origins and identities cannot be known. Management of such exotic gerbils requires special knowledge and considerable experience: the risk of introducing zoonoses to the laboratory must always be appreciated. Until very recently *M. unguiculatus* also ranked as an exotic gerbil, but its history as a domesticated species is surprisingly long. The Mongolian gerbil is therefore the only gerbil that can be recommended for general laboratory use.

History as a Laboratory Animal

In 1935 Kasuga captured twenty pairs of *M. unguiculatus* from the region of the Amur River basin in eastern Mongolia and Manchuria. The animals were taken to Japan and maintained as a closed, random-bred colony in the Kitasato Institute. A sub-colony from this stock was started by Nomura in 1949 at the Central Laboratories for Experimental Animals, and it was from there that Schwentker imported four breeding pairs into the USA in 1954 (cf. Schwentker, 1963; Rich, 1968). Schwentker's random-bred, closed colony at the West Foundation, Brant Lake, New York, U.S.A., then became the centre from which animals were distributed throughout the United States, to the United Kingdom and thence to Europe. The UK population of *M. unguiculatus* was established in 1964 from an importation of 12 pairs from the colony described by Marston & Chang (1965). The new colony was random-bred and stock was widely distributed to schools and laboratories in the UK and in Europe. Animals were sent to the Laboratory Animals Centre, Carshalton, but a successful sub-colony was not established there until 12 pairs were imported directly from Schwentker's foundation colony. This importation was made in April, 1966. Despite several attempts, a successful inbred line of Mongolian gerbil has yet to be established: there are no reports of genetic variations appearing in any laboratory colonies, and uniform, random-bred strains do not yet exist.

Taxonomy

Simpson (1945) classifies *Meriones unguiculatus* Milne-Edwards, 1867 as follows: Rodentia; Myomorpha; Muroidea; Cricetidae; Gerbillinae.

Origin

M. unguiculatus is the commonest of the five *Gerbillinae* listed for China and Mongolia, and may be the most abundant small mammal in that region (Allen, 1940). There is very little information about the species' ecology and natural history, and even its geographical distribution has not been precisely defined. It seems to inhabit the desert and semi-desert areas of the Mongolian Plateau and the Loess Plateaux of north-eastern China. The eastern limits of distribution extend into and perhaps beyond Manchuria; northwards, they reach Transbaikalia, and to the west, in the northern parts of Kansu, they approach the Tibetan Massif. The southern limit of distribution seems to be the province of Shansi. Nothing is known about the occurrence of geographical variation within the species, and its relationship to the other *Gerbillinae*, which are so widely distributed across Asia, the Middle East and North Africa, has not been adequately studied.

Père David collected the type specimens of the species, most probably whilst he was in the vicinity of Sa-La-Ch'i (40.35N–110.32E . . . previously called Saratsi). The gerbils captured by Kasuga in 1935, from which all laboratory colonies are descended, were taken much farther to the north-east in the basin of the Amur River: the exact location has not been recorded.

Habitat

Mongolian gerbils are distributed over a very wide area and so it is only possible to generalize about their habitat. The following details refer mostly to the area of northern Shansi where Père David was travelling when he first recorded his observations on *M. unguiculatus*.

The climate is markedly continental in type and is affected by the summer monsoon. Winter temperatures can go as low as $-10°C$, and the normal January range is from -10 to $+2°C$. Summer temperatures are high and range from $20–30°C$ during July. The annual rainfall is very low, usually less than 400 mm and always less than 500 mm. Severe droughts can occur, and the rainfall, being dependent on the summer monsoon, is unevenly distributed throughout the year. Over 80 per cent of the rainfall occurs in summer, and more than half falls in July and August. The relative humidity averages 50–60 per cent during the summer, but can become very low during winter when the only precipitation occurs as snow. The growing season ranges from 200–250 days, and spring planting is possible at the end of May. The whole area is one of sub-marginal agriculture, although prehistorically it was covered by deciduous forests. Present-day crops are winter or spring wheat and, principally, millet. Northern Shansi is part of the Loess Plateaux region and the terrain is characterised by uplifted block plateaux, with fault scarps, and loess filled basins. Agriculture is confined to the loess basins and, in these, erosion has destroyed much of the soil structure.

Natural history

Observations of the Mongolian gerbil's natural history are brief, incomplete and sometimes contradictory. The most reliable were made by Kasuga (cf. Schwentker, 1969) and apply to the Amur River basin. He found the animals living in sizeable colonies, perhaps as large family groups, in complex galleries of burrows which had usually been excavated in dry, sandy soil. Burrows were not found in wet or clayey areas, and were always well away from sites of human habitation. One gallery of the burrow complex could spread out for 3–4 metres, and the galleries were arranged in levels, with the lowest lying about 1·5 metres underground. A round nest chamber was usually located at the centre of each gallery, and had one or two large storage chambers close beside it. The nest was circular (ca. 18–25 cm in diameter) and lined with leaves. The storage chambers contained seeds (Graminaceae and Cyperaceae) from September through to March. The burrow system had several openings and the burrows themselves were ca. 4 cm in diameter. Gerbils were active during both day and night, which may suggest that their habits were largely crepuscular. Hibernation was not observed, and the animals fed on leaves during the summer and seeds in winter. Wheat and millet seeds were eaten when the gerbils were close to cultivated land.

Père David noted on observation (Fox, 1949) of large numbers of *M. unguiculatus* active during the day in sandy areas close to fertile land; they were 'gambolling in front of their holes'. Local peasants told him the animals were diurnal, as abundant in desert areas as in the cultivated lands, and tended to lay up stores of grain for the winter. The hear-say report also suggests that the animals emerged from their retreats from time to time during the winter. Allen (1940) interpreted this as evidence that *M. unguiculatus* hibernates, but there is nothing to support the conclusion (cf. Kasuga, 1935 in Schwentker, 1969).

Anderson (Allen, 1940) found at first-hand that *M. unguiculatus* could be trapped at all times of the day but were most often observed between sunset and dark sitting erect, squirrel-like, in front of their burrows. He could approach to within three yards of the sitting gerbil before it finally took cover. Within their burrow systems gerbils made a curious and characteristic 'drumming' sound, presumably with their hind feet.

Robinson (1959) has made some important experimental observations showing that Mongolian gerbils have a much greater capacity for heat regulation than most desert rodents: the latter usually avoid the extreme day-time stresses by remaining within their relatively cool, underground abodes. *M. unguiculatus* has a critical temperature of 30°C and its range of thermal neutrality extends to 40°C: individually caged animals can tolerate temperatures of 40°C for five hours without apparent discomfort and without attempting to drink. He concluded that, with its degree of heat tolerance, the Mongolian gerbil could spend much of its time in activity outside the burrow during the hottest parts of the day.

The field reports are limited but they do suggest that *M. unguiculatus* is naturally diurnal with a tendency toward the crepuscular habit. The gerbil's behaviour in the laboratory is very different, and animals can be active during the day, the twilight hours and at night. Thiessen *et al.* (1968) found that when gerbils were given a free choice, they preferred a darkened environment to a light one: they also showed much more activity in exercise wheels during the period of darkness, and this pattern was maintained when the lighting regimen was reversed. Epileptiform seizures can be more readily induced during the night, and other behavioural activities are also greater at this time (Thiessen, Lindzey & Friend, 1968). This evidence points to Mongolian gerbils being much more aroused during the period of darkness, and is therefore consistent with a nocturnal habit. It may be difficult to resolve the conflicting reports from the field and from the laboratory; but it is perhaps important to remember that gerbils are naturally fossorial and must spend much time in activity underground, in total darkness.

The evidence that *M. unguiculatus* hibernates is very uncertain, but it is essential to have accurate information on this point. The related species *M. meridiani* cannot be forced to hibernate (Kalabukhov, 1965). In the U.K., Mongolian gerbils show no tendency to hibernate when they are given ample food and nesting material, and kept exposed to the rigours of winter. It seems possible that *M. unguiculatus* does not hibernate, and by assembling stores of seeds can remain active through the winter. Winter is naturally the gerbil's most arid season, and this may be the time when the ability to do without water, other than metabolic water, has the greatest significance. Perhaps it may be wise to question the rather general assumption that the species is principally adapted to live in hot, arid conditions.

The natural history of *Meriones unguiculatus* is also described by Bannikov (1954) (cf. Yahr & Kessler, 1975).

Natural behaviour

Mongolian gerbils live in large groups, but nothing is known about social organization within a group. The community may have an extended family relationship, a situation similar to that in captivity when litters are not weaned away from their parents. If the accommodation is sufficiently large, several consecutive litters, and even generations, can live together as adults in harmony.

Although gerbils do not obviously vocalize, the behaviour of foot 'drumming' or 'stomping' may be a significant communicating or alerting mechanism: it may even function as a reward process (Routtenberg & Kramis, 1967). Various aspects of socialization and territory marking seem to depend on olfactory cues. Both sexes possess a large, sebaceous skin gland located mid-ventrally close to the umbilicus. The gland produces a yellowish-brown sebum of musky odour which is deposited when the gerbil deliberately rubs its belly on an object. Such scent-marking is seen more frequently in males than females, and the male's behaviour is also more severely depressed when he enters a territory already marked by another gerbil (Thiessen, Friend & Lindzey, 1968; Thiessen, Lindzey & Blum, 1969). Individual gerbils spend time grooming the skin gland, and much social inter-action between individuals involves examination of the gland. Urination and defaecation also appear to have a role in territory marking, particularly by the male gerbil.

When Mongolian gerbils are exposed to an unexpected stimulus or a new environment in the laboratory they can undergo epileptiform convulsions of disturbing severity. Both sexes can be affected, and sensitivity to the trait may be genetically determined (Thiessen, Lindzey & Friend, 1968). Quite minor procedures, such as handling or cage changing, will induce convulsions; and in their most extreme form these may last for a few minutes, during which time the gerbil is completely prostrate. If an adequate stimulus is given, most gerbils can be made to show convulsions. This rather extreme behaviour pattern may well obtain under natural conditions, and could just possibly have some survival value by acting to confuse and alarm an intending predator. (J. Mackintosh . . . personal communication).

Although there are no observations of aggression under natural conditions, such behaviour becomes a serious problem when gerbils are mis-managed in the laboratory: almost invariably, vicious fighting occurs when adults which have been isolated previously as groups or individuals, are caged together. Serious fighting can also occur when *adult* males and females are paired for the first time (e.g. after the death of a previous mate) or paired again after a period of separation: the male is more likely to be defeated and killed by the female. If pre-pubertal gerbils are caged together, there is no fighting: large groups numbering 50–100 individuals can be assembled at this time and reared, without hazard, into adulthood. It is advisable, but possibly not essential, to separate the sexes when such rearing groups are set up. Various procedures to minimise fighting when groups of adult gerbils are mixed have been devised, but none is really successful: it is best to observe the animals very closely, to house them in a completely fresh cage, and provide plenty of bolt-holes where individuals can escape from an aggressor. (Norris & Adams, 1972c).

Other behaviours of the gerbil follow the typical rodent pattern. Sexual behaviour is particularly obvious. The male actively pursues the female and shows a much repeated pattern of mounting, thrusting, ejaculating, dismounting and preputial grooming. The female briefly exhibits lordosis as the male mounts and copulates; she is not usually lordotic as an invitation to mounting. The male takes quite an active role in parental care. Both sexes will spend much time nest-building, shredding nesting materials, and burrowing in their bedding. If the caging is suitable, gerbils will excavate extensive tunnel systems and nest chambers.

Towards man *M. unguiculatus* is completely docile but not timid; it will immediately investigate any new object, any new hand, placed in the cage and make no resistance to being handled, stroked or generally pushed around. Rarely, individual animals can be vicious; and groups can become hostile if they are roughly or too frequently handled and generally mis-managed.

TABLE I
Post-natal development of the Mongolian gerbil in relation to age (days)

Opening of ears	3– 7
First appearance of hair	5– 7
Eruption of incisors	10–16
Opening of eyelids	16–20
Descent of testes	30–40
Opening of vagina	40–60
Sexual maturity	63–84

Standard Biological Data

Development

The young are born blind and naked, after a gestation period of 24–26 days, and are suckled

for 21–28 days. Post-natal development, as observed by Marston & Chang (1965), is summarised in Table I. The growth rate can be very variable: details of one study are given in Table II.

TABLE II
Body-weight in relation to age

Age	Weight (g) Male	Female
birth	2·5– 3·5	2·5– 3·5
3 weeks	11– 18	11– 18
8 weeks	36– 61	33– 47
6 months	75–105	65– 85
18 months	100–135	70–125

Puberty or functional maturity occurs in both males[*] and females at between 9 and 12 weeks of age. The fertile life of the female is very variable, with the birth of the last litter occurring at from 7 to 20 months of age, after a total production of from 3 to 10 litters. The male can produce fertile matings up to and possibly beyond 24 months of age. There are no accurate data regarding the total life span, but in both sexes it is greater than 24 months and may exceed 36 months.

Organ weights
Some data are given by Kramer (1964).

Haematology
The reader is referred to Ruhren (1965) and Mays (1969) for details. Table III gives a range of mean values which summarises some of Ruhren's results.

TABLE III
Range of mean haematological values

Packed cell volume	45–47 per cent
Haemoglobin	13·9–14·4 g/per cent
Red blood corpuscles	7·7–8·7 × 10⁶ mm³
Total white blood corpuscles	8·3–11·2 × 10³ mm³

A characteristic feature of the Mongolian gerbil is the presence of high levels of plasma lipid. These levels become excessive on certain diets, e.g. those with a high proportion of sunflower seeds. There is also a high concentration of reticulocytes and basophilic erythrocytes, indicating a short erythrocyte life span (Ruhren, 1965). (See also *Haemobartonella* infection, p. 272).

[*]Males may not sire young until they are more than 18 weeks old (Norris & Adams, 1972a).

X-ray sensitivity

Chang *et al.* (1964) studied the resistance of Mongolian gerbils to irradiation and found that they could survive 1500 rad whole body irradiation.

Karyology

The Mongolian gerbil shows normal nuclear sex dimorphism (Garner *et al.*, 1969). The diploid number of chromosomes is 44 (Pakes, 1969).

Husbandry

The husbandry of the Mongolian gerbil has been discussed by Schwentker (1963), Marston & Chang (1965), Aistrop (1968), Jansen (1968) and Rich (1968).

Housing and caging

Any standard laboratory cage with a solid bottom and which is suitable for rats or Syrian hamsters can be used. Cages should preferably have raised lids, so that the animals can stretch to their full height. Two suitable cages which are available commercially are illustrated in Figure 22.2: one of these (right) is actually a square washing-up bowl with rounded corners, made of soft plastic and fitted with a specially designed lid. Cages can be improvised by fitting simple wire-mesh lids to such bowls; the water bottle is placed on this lid and, instead of providing a hopper, pellets are also placed on the wire. Contrary to expectations, gerbils do not damage these bowls by gnawing.

A monogamous breeding pair with young requires a floor area of about 900 cm², while group cages should allow about 100 cm² for each animal. It is preferable to place gerbils in large groups as soon as they are weaned in order to prevent the fighting which will occur when older animals meet as strangers. Groups of up to 100 such animals can be reared in wooden boxes with a floor area of about 1 m². The box should be fitted with a wire-mesh lid.

Gerbils can also be kept in glass aquaria. They should be provided with ample bedding material in the form of earth and shavings, and possibly small pipes in which to burrow. In addition they should be given a small wooden nesting box. Aquaria should be placed in a shady part of the room away from direct sunlight and the animals should be disturbed as little as possible (see also Handling, p. 269).

Food and watering utensils

Water bottles and food hoppers should be provided. Hoppers should never be placed on the floor of the cage, as gerbils will burrow under them and, in doing so, may injure them-

Fig. 22.2. Two commercially available cages suitable for housing one breeding pair of Mongolian gerbils and their litter.

selves. Gerbils prefer to handle their food and will remove pellets from the hopper to eat in another part of the cage.

Bedding

Wood shavings are the best material for bedding; although sawdust or clean, dry sand can be used they are less satisfactory. The bedding should be about 2 inches deep. Plentiful nesting materials in the form of shredded paper or wood wool must also be given. This is particularly important for breeding pairs, who will make a nest in the corner of their cage.

Environmental conditions

With adequate food and bedding Mongolian gerbils will thrive in temperatures as low as 0°C. and even breed under such conditions. In the laboratory a temperature of 15–21°C is the most suitable. The lower level of 15°, coupled with ample nesting material, is probably better for adults, but weanlings need a higher temperature (20–24°C), particularly when taken from their mothers at 21 days of age. At 28 days, which is the latest age at which they should be weaned, they should be able to withstand lower temperatures.

Controlled lighting, giving a 12- to 14-hour day, is recommended. Where light and temperature are controlled, Mongolian gerbils should breed throughout the year, but in situations where such controls do not exist there may be an underlying pattern of seasonal breeding; this may be influenced by light or temperature, or both, but no experimental evidence concerning this is available.

Cage hygiene

As the Mongolian gerbil is comparatively odourless, and passes only hard, dry faeces and very little urine, cages need not be thoroughly cleaned oftener than once a month, though bedding should be changed every week. If there are young in the nest, it should not be disturbed, but should be moved in its entirety to a clean cage.

Handling

Even gerbils without experience of handling tend to be tame and are easily picked up with the cupped hands. Since they are extremely active animals they should always be handled over a table, with the head kept pointing towards the handler. For the same reason the base of the tail should always be held during handling in the laboratory; this should be done *gently*. Gerbils should not be picked up by the tail as this may cause the outer skin sheath to strip off, exposing the vertebrae. This may even happen when the base of the tail is gripped too firmly. Gerbils should not be restrained on their backs (see also p. 272).

It is perfectly safe to handle pregnant females, breeding pairs with suckling young, or even new-born young, provided the handling is done in moderation (but see stress behaviour, p. 266). Too much disturbance at parturition or before the young are fully furred may lead to cannibalism. *No gerbil should be handled extensively, nor should the cage be disturbed unnecessarily.*

Identification

The preferred method is by cage identification combined with a dye-marking code using indelible dyes. If permanent identification of individuals is essential, toe-clipping and ear-punching can be carried out, preferably under anaesthesia.

Transport

See Chapters 8, 17 and 18.

Feeding

A high-protein rodent-breeding diet is the most suitable: diet 41B is also an adequate diet. Seeds mixtures may be given, but animals tend to select those seeds which they prefer and may consume an unbalanced diet. This is most undesirable; for example, too high an intake of sunflower seeds will cause lipaemia. Seeds mixtures should be given only as a relish, not as the main diet. A limited amount of green food (lettuce, spinach or carrot) such as can be eaten in 10–15 minutes, should also be fed.

Water

Water should always be available, preferably in a water bottle. Studies by Winkelman & Getz (1962) and Arrington & Ammerman (1969) have shown voluntary daily water consumption to be *ca* 4 ml/100 g body-weight. Total intake, including metabolic water, was 7·5 ml/day/100 g. Body-weight was maintained by a consumption of 2 ml/day/100 g, while animals deprived of water lost 17 per cent of their body-weight, at which point their weight became stabilized. Animals can survive without water for more than 45 days provided they are not stressed (Boice & Witter, 1970). Mongolian gerbils do not breed if deprived of water, and water availability may function as the practical population controller under natural conditions (Yahr & Kessler, 1975).

Breeding

Reproductive physiology

Detailed studies have been carried out by Nakai *et al.* (1960), Schwentker (1965), Marston & Chang (1965, 1966) and Barfield & Beeman (1968).

The female is polyoestrous with a short oestrus cycle of about 4 to 6 days (Marston & Chang, 1965; Barfield & Beeman, 1968). Ovulation is spontaneous; a mean number of 6·6 eggs ovulated, with a range of 4 to 9, was recorded by Marston & Chang (1965). Attempts have been made to diagnose the presence of oestrus by the examination of vaginal smears, and Barfield & Beeman (1968) observed that smears taken on the morning of the day on which mating occurred were always devoid of leucocytes and usually had nucleated cells which were often cornifying. However, the examination of smears does not provide a certain means of diagnosis, and the behaviour of the two sexes is a better guide to the presence of oestrus (p. 267).

Mating occurs in the evening, usually after 17.00 hours: there are multiple ejaculations directly into the uterus. A copulation plug is formed during mating, but it is small and lies deep in the vagina, so that it cannot be readily detected. Definite evidence of mating is obtained from the presence of vaginal sperm the morning after oestrus.

An infertile mating may be followed by a pseudopregnancy lasting for 14–16 days (Marston & Chang, 1965; Barfield & Beeman, 1968). True pregnancy lasts for 24–26 days (Schwentker, 1963; Marston & Chang, 1965).

The mean litter size at birth is $4·5 \pm 0·04$; a 30 per cent prenatal mortality must therefore occur. The sex ratio at both birth and weaning is close to and probably not significantly different from unity (Marston & Chang, 1965).

Most females show a post-partum oestrus. In one study (Marston & Chang, 1965) mating was found to occur at this time in 59 per cent of females, and 86 per cent of such matings were fertile. Implantation is delayed when the female is suckling more than two young, and this can result in an interval of up to 42 days from the post-partum mating to the birth of the resulting litter (Norris & Adams, 1971) rather than the expected 26 days. The female will usually produce a litter every 30–40 days, and may produce a total of up to ten litters.

Selection of breeding stock

Animals should be selected just before puberty. They are selected on their appearance, the sleeker and more active, though not necessarily the larger, individuals being chosen. Any animals which are highly prone to epileptiform convulsions should be discarded.

Inbreeding and outbreeding

To date, inbreeding programmes have all failed at the ninth or tenth generation owing to the development of 'ovarian tumours'* (see Schwentker, 1963; Benitz & Kramer, 1965). Breeding should, therefore, be by random selection.

Breeding methods

Breeding should always be carried out with monogamous pairs which are *never* separated. When such pairs are set up with mature animals, close observation is essential so that the animals can be immediately parted, should fighting occur. It is best to assemble breeding pairs at puberty. Once a pair is established the animals will live in perfect harmony and breed continuously.

Polygamous breeding groups are not recommended and must, in any case, be set up before puberty, otherwise vicious fighting will occur, usually ending in the death of the male.

*These could either be cystic ovaries or true ovarian tumours. The former is more likely, but accurate information is required (see also p. 272).

Artificial insemination

This can be carried out surgically using epididymal sperm. As the female appears to need coital stimulus to initiate pregnancy, she should be allowed to mate with a vasectomised male.

Rearing

In most colonies about 70 per cent of the young born alive are successfully weaned. First and second litters are more likely to be lost than subsequent litters. This may be because breeding has occurred before the mother is physiologically or behaviourally capable of raising a litter. Norris & Adams (1972b) give data on neo-natal mortality. Certain strains appear to be addicted to cannibalism, but this may be due to bad management and excessive handling.

The young begin to nibble at solid food at about 16 days of age. Artificial rearing of orphans is not practicable, although in favourable circumstances it may be possible to foster them.

The young are usually weaned at 21 days, although they may be left with the mother until they are about 28 days old. If they are weaned at the earlier age great care must be taken to avoid deaths from chilling (see p. 269).

Sexing can be carried out from birth onwards by comparing the ano-genital distances; as in other rodents, this distance is much greater in the male. The dark-coloured scrotum of the young male is very obvious in recently weaned stock.

The best method of rearing is to place the newly weaned young in large, one-sex groups of 50–100 animals (see p. 266). Careful monitoring is necessary to ensure that no runts are included or that the very young animals do not suffer from competition. Further extensive data on growth rates are given by Norris & Adams (1972d).

Laboratory procedures

Selection of animals for experiment

Only well-grown, healthy animals should be used. Any animals prone to seizure should be discarded.

Anaesthesia

Satisfactory general anaesthesia lasting for 2 to 3 hours can be obtained with pentobarbitone sodium (60 mg/ml) at a dose rate of 0·01 ml/10 g body-weight, administered intraperitoneally, up to a total dose of not more than 0·10 ml. Mongolian gerbils are not good subjects for ether anaesthesia (Marston & Chang, 1965), although Herndon & Ringle (1969) have used pentobarbitone sodium as a basal anaesthetic at a dosage rate of 45 mg/kg, supplementing this with ether. With this method they frequently encountered post-surgical respiratory problems, but the situation was improved by the administration intraperitoneally of 1–2 mg of atropine.

Although the author has not had personal experience of the use of tribromoethanol* for anaesthesia of the Mongolian gerbil, he has used this successfully in rats and mice and suggests that it be used in gerbils at the dosage rate recommended for mice, i.e. 0·2 ml/10 g body-weight of a 1·25 per cent solution of the stock solution in warm normal saline (Jones & Krohn, 1960).**

Halothane is probably also a suitable anaesthetic for Mongolian gerbils, but, again, the author has had no personal experience of this method.

Euthanasia

See methods of euthanasia recommended for small rodents in Chapter 14. Cervical fracture can be carried out by the method described for mice, but using a rather more robust instrument than a pencil.

Collection of specimens

Blood may be taken from the tail by lateral vein incision or from the heart (Ruhren, 1965) or from the retro-orbital sinus at the medial canthus of the eye (Mays, 1969). The latter method is also described in Chapter 17, p. 207 According to Ruhren (1965) blood taken from the tail vein has a higher leucocyte count than that taken from other sites.

Faeces, which are passed as hard, dry pellets,

*Avertin: Winthrop Laboratories.

**Mandl & Zuckerman (1951) recommend the following dosage rates for rats: *adults*—0·8 ml/100 g of a 2·5 per cent solution in normal saline; *immature animals*—1·4 ml/100 g of a 1·25 per cent solution. If mice or immature rats are given the more concentrated solution recommended for adult rats peritonitis will result. It is advisable to observe the same precautions with gerbils.

may be sieved from the cage litter. Alternatively a metabolism cage can be used.

Urine. Mongolian gerbils pass very little urine, which they use as a territory marker. A method might be devised of training them to mark filter paper. They will also urinate on to the hand of the handler.

Dosing and injection procedures

The methods described for rats and mice (Chapter 17) can be used, with the exception of that for intraperitoneal injection. Mongolian gerbils object to being restrained on their backs and it is preferable for them to be held in the upright position by an assistant.

Disease

Mongolian gerbils usually remain healthy under laboratory conditions. In one colony which was studied over a period of 18 months (Marston & Chang, 1965) all the deaths which occurred after weaning could be ascribed either to the effects of an experiment or to fighting. Deaths before weaning appeared to result from failure of lactation and to general deficiencies of maternal care, and no obvious diseases of the new-born were recognised. However, Jansen (1968) has recorded a condition in suckling young from 10 days to weaning which resembles wet-tail in hamsters; it is associated with *Escherichia coli* but this may not be the primary cause. The author has seen a similar condition in weaned animals; the animals usually recovered if cared for. This condition could have been dietary in origin.

Some other spontaneously occurring disease conditions which have been reported are described below, together with some other conditions against which laboratory staff, etc. should be on their guard.

Tyzzer's disease. This is the only really serious disease hazard. It is usually contracted from mice via infected food or bedding and tends to run an acute and fatal course. According to White & Waldron (1969) there is a 70 per cent mortality, with death occurring suddenly. Diagnosis can be made post mortem from the presence on the liver of white, necrotic foci 1–3 mm in diameter. Carter *et al.* (1969) observed a 65–85 per cent mortality, with death occurring

one to three days after the appearance of clinical signs. Affected animals showed mild diarrhoea, lethargy and anorexia. Port *et al.* (1970) described a less acute form of the disease running a slower course and occurring in 10 per cent of the animals at risk. The incubation period was 10 days and clinical signs included lethargy and anorexia, but not diarrhoea. Liver lesions were present. All three reports state that *Bacillus piliformis* was isolated from affected animals.

Sore noses and eyes. This condition is of uncertain aetiology, but may be due to secondary infection by β-haemolytic streptococci of primary lesions incurred during burrowing. It is a persistent problem in animals kept in schools, where it is possibly due to cross-infection from the children or to bad management. The administration of antibiotics in the drinking water may lead to an improvement, but it is probably better to cull affected animals.

Haemobartonella infection. Najarian (1961) has reported that 94 per cent of healthy gerbils which he examined had erythrocytes infected with an organism which he tentatively identified as *haemobartonella* sp. He suggested that this infection is common in the gerbil and is probably transmitted from mother to offspring at birth or transplacentally. Ruhren (1965) has pointed out that the infected erythrocytes illustrated by Najarian are identical with the *stippled erythrocytes* which are normally present.

Ectoparasites. The author is not aware of any records of natural infection with ectoparasites, but individual laboratories should be alert to the possibility of cross-infection from other laboratory species.

Diseases of old age. Cystic ovaries are very common in females at the end of their breeding life, their incidence being strictly correlated with age (Adams, 1970). They can be unilateral or bilateral and need not cause infertility (Marston & Chang, 1965). Benitz & Kramer (1965) found par-ovarian and peri-ovarian cysts in 63 per cent of the aged females which they examined. These lesions were regressive with cystic changes occurring predominantly in atrophic ovaries.

Benitz & Kramer (1965) also found that 24 per cent of animals of about 2 years of age had spontaneous tumours, the most commonly observed being true ovarian tumours (in 7 of 56 females). They stated that this high rate of incidence 'renders this animal unsuitable for long-term toxicity studies as for instance necessary for the safety evaluation of drugs, food additives and cosmetics.'

Sensitivity to experimental diseases. The Mongolian gerbil has been used for a wide variety of endoparasitological, bacteriological and virological studies and these are referred to by Schwentker (1963, 1969) and Rich (1968). These animals are particularly sensitive to *Brucella abortus, Leptospira icterohaemorrhagiae* and *L. canicola* and the possibility of such diseases occurring spontaneously owing to contamination of food and bedding or contact with wild rodents should always be kept in mind.

REFERENCES

Adams, C. E. (1970). Ageing and reproduction in the female mammal with particular reference to the rabbit. *J. Reprod. Fert.*, Suppl., **12**, 1–16.

Aistrop, J. B. (1968). *The Mongolian Gerbil: a Guide for Teachers and Young Enthusiasts.* London: Dennis & Dobson.

Allen, G. M. (1940). The mammals of China and Mongolia. Part II, Volume XI. *Natural History of Central Asia.* (Publ. American Museum of Natural History; pp. 781–785).

Arrington, L. R. & Ammerman, C. B. (1969). Water requirements of gerbils. *Lab. Anim. Care*, **19**, 503–5.

Bannikov, A. G. (1954). The places inhabited and natural history of *Meriones unguiculatus*. In *Mammals of the Mongolian Peoples' Republic. USSR Acad. Sci.*, 410–415. (In Russian).

Barfield, M. A. & Beeman, E. A. (1968). The oestrous cycle in the Mongolian Gerbil, *Meriones unguiculatus. J. Reprod. Fert.*, **17**, 247–51.

Benitz, K. F. & Kramer, A. W. (1965). Spontaneous tumours in the Mongolian gerbil. *Lab. Anim. Care*, **15**, 281–94.

Boice, R. & Witter, J. A. (1970). Water deprivation and activity in *Dipodomys ordii* and *Meriones unguiculatus. J. Mammol.*, **51**, 615–8.

Chang, M. C., Hunt, D. M. & Turbyfill, C. T. (1964). High resistance of Mongolian gerbils to irradiation. *Nature, Lond.*, **203**, 536–7.

Carter, G. R., Whiteneck, D. L. & Julius, L. A. (1969). Natural Tyzzer's disease in Mongolian gerbils (*Meriones unguiculatus*). *Lab. Anim. Care*, **19**, 648–51.

Fox, M. (1949). *Abbe David's Diary.* Harvard University Press.

Garner, J. C., Meskill, V. P. & Hanan, F. (1969).

Nuclear sex dimorphism in the Mongolian gerbil. *Can. J. Genet. Cytol.*, **11**, 1004–7.

Herndon, B. L. & Ringle, D. A. (1969). Methods for neurological experimentation in the Mongolian gerbil, *Meriones unguiculatus. Lab. Anim. Care*, **19**, 240–3.

Jansen, V. (1968). The Mongolian Gerbil (*Meriones unguiculatus*) *J. Inst. Anim. Techn.*, **19**, 56–60.

Jones, E. C. & Krohn, P. L. (1960). Orthotopic ovarian transplantation in mice. *J. Endocr.*, **20**, 135–46.

Kalabukhov, N. I. (1965). Effects of vitamins E and C on hibernating rodents. *Fedn. Proc.* (Trans. Suppl.), **24**, 851–7.

Kramer, A. A. (1964). Body and organ weights and linear measurements of the adult Mongolian gerbil. *Anat. Rec.*, **150**, 343–8.

Mandl, A. M. & Zuckermann, S. (1951). Numbers of normal and atretic cocytes in unilaterally spayed rats. *J. Endocr.*, **7**, 112–9.

Marston, J. H. & Chang, M. C. (1965). The breeding, management and reproductive physiology of the Mongolian gerbil (*Meriones unguiculatus*). *Lab. Anim. Care*, **15**, 34–48.

Marston, J. H. & Chang, M. C. (1966). The morphology and timing of fertilization and early cleavage in the Mongolian gerbil and deer mouse. *J. Embryol. exp. Morph.*, **15**, 169–91.

Mays, A. (1969). Baseline hematological and blood biochemical parameters of the Mongolian gerbil (*Meriones unguiculatus*). *Lab. Anim. Care*, **19**, 838–42.

Milne-Edwards, A. (1867). Observations sur quelques mammifères du nord de la Chine. *Ann. Sci. Nat.* (*Zool.*), **7**, 375–7.

Najarian, H. H. (1961). Hemobartonella in the Mongolian gerbil. *Texas Rept. Biol. Med.*, **19**, 123–33.

Nakai, K., Nimura, M., Tamura, M., Shimizu, S. & Nishimura, H. (1960). Reproduction and postnatal development of the colony bred Meriones unguiculatus kurauchii (mori). *Bull. Exper. Animals* (Japan), **9**, 157–9.

Norris, M. L. & Adams, C. E. (1971). Delayed implantation in the Mongolian gerbil, *Meriones unguiculatus. J. Reprod. Fert.*, **27**, 486–487.

Norris, M. L. & Adams, C. E. (1972a). Aggressive behaviour and reproduction in the Mongolian gerbil, *Meriones unguiculatus*, relative to age and sexual experience at pairing. *J. Reprod. Fert.*, **31**, 447–450.

Norris, M. L. & Adams, C. E. (1972b). Mortality from birth to weaning in the Mongolian gerbil, *Meriones unguiculatus. Lab. Animals*, **6**, 49–53.

Norris, M. L. & Adams, C. E. (1972c). Suppression of aggressive behaviour in the Mongolian gerbil, *Meriones unguiculatus. Lab. Animals*, **6**, 295–299.

Norris, M. L. & Adams, C. E. (1972d). The growth of the Mongolian gerbil, *Meriones unguiculatus*, from birth to maturity. *J. Zool., Lond.*, **166,** 277–282.

Pakes, S. P. (1969). The somatic chromosomes of the Mongolian gerbil (*Meriones unguiculatus*). *Lab. Anim. Care*, **19,** 857–61.

Port, C. D., Richter, W. R. & Moise, S. M. (1970). Tyzzer's disease in the gerbil (*Meriones unguiculatus*). *Lab. Anim. Care*, **20,** 109–11.

Rich, S. T. (1968). The Mongolian gerbil (*Meriones unguiculatus*) in research. *Lab. Anim. Care*, **18,** 235–43.

Robinson, P. F. (1959). Metabolism of the gerbil, *Meriones unguiculatus*. *Science*, N.Y., **130,** 502–3.

Routtenberg, A. & Kramis, R. C. (1967). 'Foot Stomping' in the gerbil: rewarding brain stimulation, sexual behaviour and foot shock. *Nature, Lond.*, **214,** 173–4.

Ruhren, R. (1965). Normal values for haemoglobin concentration and cellular elements in the blood of Mongolian gerbils. *Lab. Anim. Care*, **15,** 313–20.

Schwentker, V. (1963). The gerbil, a new laboratory animal. *Illinois Veterinarian*, **6,** 5–9.

Schwentker, V. (1969). *The Gerbil: an Annotated Bibliography of the Gerbil as an Experimental Animal in Medical Research*. Privately published and distributed by Tumblebrook Farm Inc., Brant Lake, N.Y., USA.

Simpson, G. G. (1945). The principles of classification and a classification of mammals. *Bull. Amer. Mus. Nat. History*, **85,** 213–9.

Swanson, H. H. (1974). Sex differences in behaviour of the Mongolian gerbil (*Meriones unguiculatus*) in encounters between pairs of same or opposite sex. *Anim. Behav.*, **22,** 638–644.

Thiessen, D. D., Blum, S. L. & Lindzey, G. (1970). A scent marking response associated with the ventral sebaceous gland of the Mongolian gerbil (*Meriones unguiculatus*). *Anim. Behav.*, **18,** 26–30.

Thiessen, D. D., Friend, H. C. & Lindzey, G. (1968). Androgen control of territorial marking in the Mongolian gerbil. *Science, N.Y.*, **160,** 432–4.

Thiessen, D. D., Lindzey, G., Blum, S., Tucker, A. & Friend, H. C. (1968). Visual behaviour of the Mongolian gerbil (*Meriones unguiculatus*). *Psychon. Sci.*, **11,** 23–4.

Thiessen, D. D., Lindzey, G. & Friend, H. C. (1968). Spontaneous seizures in the Mongolian gerbil (*Meriones unguiculatus*). *Psychon. Sci.*, **11,** 227–8.

White, D. J. & Waldron, M. M. (1969). Naturally-occurring Tyzzer's disease in the gerbil. *Vet. Rec.*, **85,** 111–4.

Winkelman, J. R. & Getz, L. L. (1962). Water balance in the Mongolian gerbil. *J. Mammal.*, **43,** 150–4.

Yahr, P. & Kessler, S. (1975). Suppression of reproduction in water-deprived Mongolian gerbils (*Meriones unguiculatus*). *Biol. Reprod.*, **12,** 249–254.

23　The Chinchilla

BARBARA J. WEIR

Introduction

The chinchilla (*Chinchilla* Bennett, 1829) is an hystricomorph rodent of the family Chinchillidae and it is distantly related to the guinea-pig and the coypu. Like most hystricomorphs its native home is in South America where it lives at altitudes of up to 4,500 metres in the Andes of Peru, Bolivia, Chile and Argentina. But extensive hunting for their valuable pelts has severely reduced wild chinchilla numbers and few, if any, colonies are known to exist today. Although they are legally protected by the countries concerned, control is difficult to enforce because of the inaccessibility of their rocky habitat. It used to be considered that there were several types of chinchilla, varying according to the country and altitude of origin, but since this can no longer be checked on wild populations only two species are now recognised, *Chinchilla laniger* and *C. brevicaudata*. The latter is supposed to be heavier than *C. laniger* and also to have shorter ears and tail and a longer gestation period, but there are apparently very few brevicaudate chinchilla outside South America and the chinchilla which are commercially bred in the northern hemisphere are of the *C. laniger* type. The information contained in this chapter refers solely to *Chinchilla laniger*.

Commercial chinchilla farming for the fine pelts was begun in 1927 when Chapman took 13 animals from South America to California; most existing ranch animals have been derived from this stock, although some subsequent introductions of native stock have been claimed. The chinchilla fur industry is extensive and animals are bred both for their pelts and for live sales; several mutation strains have been developed, mainly towards colour and quality of the fur, but it is too early to pick out any consistent physiological features of the different strains.

In spite of the large numbers of farmed chinchilla they have not often been studied in the laboratory either for their own sake or as an alternative species for the elucidation of some particular problem. This is due partly to their high cost as research animals and also to their slow breeding rate (see below) which necessitates housing for considerable periods of time. Most laboratory studies have been related to commercial problems such as fur-chewing, nutrition, genetics and practicalities of breeding.

Standard biological data

Female chinchilla are heavier than males and a non-pregnant adult female averages about 450 g, although pregnant animals may weigh over 700 g. The males are rarely more than 500 g in weight and most tip the scales at 400 g. At birth the precocious young of both sexes weigh 35 g (range 30–50 g) but the weight difference of the sexes is obvious by 100 days of age. The gestation length in the chinchilla is 111 days (range 105–115 days; Weir, 1966) and parturition usually occurs before 08.00 hours. If the mating date is not known the imminence of parturition is difficult to detect since there is little change in the colour of the nipples and no appreciable relaxation of the pubic symphysis. Regular weekly weighing should indicate the rapid weight increase which occurs between days 50 and 60 of gestation. Litter-size ranges from 1–5 kits but 2 is the usual number born and many breeders actually prefer the smaller litters as lactation problems are less likely to occur. Lactation lasts for about 8 weeks, and the young are best weaned at 6–8 weeks although they can survive alone after 3 weeks.

In our laboratory colony chinchilla have shown a definite breeding season between November and May and, as expected, this is exactly 6 months different from that of wild chinchilla (Dennler, 1939) which in the southern hemisphere breed in May and give birth from September onwards. Within the breeding season the chinchilla is polyoestrous; the mean cycle length is 41 days and 80 per cent of cycles are between 30 and 50 days. Cycle lengths are rarely constant even within individual animals. However, when a male is present copulation usually takes place at the first oestrus of the breeding season (Fig. 23.1) and the female will probably mate again at the post-partum oestrus and so produce two litters a year. It is possible for a female to conceive at the second post-partum oestrus, but few produce three litters in a breeding season; as the long pregnancies and concurrent lactations are detrimental to health and fertility such conceptions should be prevented by removing the male. Many females do not mate at the first post-partum oestrus and instead may either have a particularly long summer anoestrous period, or they may mate again at the post-lactation oestrus which occurs about 55 days (range 35–84 days) after parturition.

As in the guinea-pig, oestrus in the chinchilla is detected by examination of the vaginal closure membrane which becomes perforate only at oestrus and parturition (Fig. 23.2). Vaginal smears taken during the period of vaginal opening at normal oestrus have not demonstrated any cell pattern which indicates the occurrence of ovulation and neither does this event take place at any particular time in relation to the time of vaginal opening, although ovulation is spontaneous. At the post-partum oestrus ovulation occurs on the second night, about 30 hours after parturition, although the

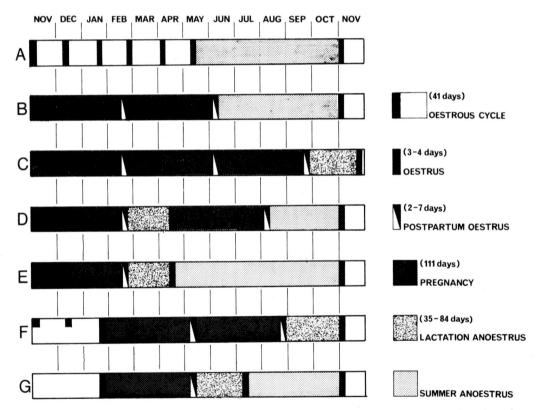

Fig. 23.1. Scheme of several possible breeding cycles experienced by chinchilla. (A) Oestrous cycles throughout the breeding season followed by the summer anoestrus. (B) Conception at the first oestrus of the season with impregnation at the post-partum oestrus and this pregnancy followed by the summer anoestrus; this is the ideal pattern. (C) Three consecutive pregnancies followed by a lactation anoestrus. (D) The first pregnancy of the season is followed by a lactation anoestrus if conception or mating does not take place at the post-partum oestrus. A second pregnancy may start at the post-lactation oestrus. (E) If conception does not occur at the post-lactation oestrus, the summer anoestrus usually follows. (F) A young female or an adult which has not conceived early in the breeding season may still have two consecutive pregnancies. (G) A late breeder will usually have only one pregnancy and the summer anoestrus follows the lactation anoestrus.

vagina stays open for 8–10 days. At normal oestrus the vagina is usually perforate (Fig. 23.2B) for 3–4 days although the range is $\frac{1}{2}$–10 days. As the chinchilla does not exhibit lordosis the duration of receptivity in intact animals has not been determined, but it is certainly the behaviour of the female which determines whether or not mating takes place (Bignami & Beach, 1968). Even when in full oestrus some females will refuse several males and then may suddenly accept a particular one. But as chinchilla are very aggressive and will fight to the death, random introductions are very risky. To minimise incompatibility much careful planning is needed in order to introduce animals to each other before the start of the breeding season.

Individual chinchilla may live for 15–20 years; two died in pregnancy at 18 years of age. The normal breeding life is probably about 10 years.

Spermatogenesis begins in males when they are 2–3 months old but copulatory maturity occurs at about 8 months of age (Fig. 23.2c). This delay is due to the fact that time is required for the development of the accessory glands, the vesicular and prostatic glands whose secretions form the copulatory plug that is often found after mating, and also because the male is behaviourally too immature to cope with the larger and aggressive female. Spermatogenesis continues throughout the year and male chinchilla will copulate at any time. In the female puberty is regarded as the age at first vaginal opening and usually occurs at $8\frac{1}{2}$ months, although the range is from 2–14 months depending on the time of year when birth occurs.

Chinchilla have one of the most disproportionate sex ratios known for mammals: $119\male\male:100\female\female$ (Galton, 1968) and this may be associated with the pronounced difference in size between the X- and Y-spermatozoa. The diploid chromosome number in *Chinchilla laniger* is 64 (Makino, 1953).

Husbandry

Caging

Chinchilla can be kept under a variety of conditions in many different types of cages. Mesh cages (constructed of $2\cdot5 \times 1\cdot3$ cm weldmesh) have been found to be preferable for laboratory purposes as cleaning is easier and copulatory plugs are less likely to be eaten.

Fig. 23.2 (A) External genitalia of a female chinchilla showing the closed condition of the vaginal membrane (v) when the animal is not in oestrus. (B) External genitalia of a female chinchilla in oestrus; note the perforated vaginal membrane (v). (C) External genitalia of an adult male chinchilla. The testes are inguinal in this species but sometimes the epididymides are descended into the post-anal sacs (s). (Photographs: UFAW.)

In small colonies or ranches pair-breeding is sometimes followed where one male and one female are kept together and the young are established elsewhere at weaning. This method is not suitable for either laboratory or ranch as it necessitates the housing of many unnecessary males and the spending of much time in re-establishing pairs in cases of incompatibility or infertility.

If it is not essential that the females are caught and checked daily for reproductive, or any other, condition, animals can be kept in open colonies where 1 male is housed in a large pen with about 6 females. The offspring may be established in a new group with other weanlings or left with the parent group. This system does have disadvantages, the most serious being that the young may be neglected and assaulted by other females while the mother is experiencing the post-partum oestrus. More nest-boxes than animals should always be provided so that each animal has somewhere to retreat to if it is attacked by others.

Polygamous colony breeding (Fig. 23.3A) is the system preferred by most ranchers and it is also very suitable for the laboratory (Weir, 1967). The adult females are housed in individual cages (Fig. 23.3B) measuring about $40 \times 45 \times 35$ cm and each cage is separated from the next by a hay rack. Up to 12 cages can be used in a row, depending on the virility of the male which lives in a tunnel (about 12×12 cm) running along the back of the cage row. The male can pass from the tunnel to the cage of a female via a door which can be shut if necessary, but the female is prevented from leaving her cage by a light metal or plastic collar placed round the neck. The females very quickly get used to these collars and can perform all their usual functions without apparent discomfort. The male feeds in the females' cages and is often to be seen with a preferred female; such atten-

5 cm

Fig. 23.3A. View of a chinchilla colony run as polygamous units. The male (♂), visible in the tunnel, can serve any of the four females in that row. Note the food hoppers, hay racks between cages and the sand baths. (Photograph: UFAW.)

tion can be a useful indication of incipient oestrus or parturition. At parturition the male is shut out if a post-partum mating is not required, or the young may be removed overnight and kept in a warm box. The male is also shut off when a new female is added, so that they may become fully acquainted through the mesh without fighting; after a few days they can usually be allowed together.

Environment

Chinchilla do well in a variety of conditions as long as they are kept away from draughts and direct sunlight. A normal laboratory temperature of 20°C and adequate light to each animal should be sufficient for most purposes; lower temperatures are often used on ranches for priming animals before pelting. Complete artificial lighting may be used but chinchilla do like to sun themselves provided they can retire when they get too hot. It has been reported that the use of blue lights changes the sex ratio so that a preponderance of females is born (Ott, 1964) but these results have not been substantiated. Optimum humidity conditions for chinchilla are not known.

The fur of chinchilla keeps in better condition if they are allowed to dust themselves in fine sand or Fuller's earth. In our laboratory, where the animals are checked daily for reproductive condition, the sand bath is put into the cage after this routine and acts as a reward to the animal. Each female has her own bath (Fig. 23.3A) to minimise the risk of spread of infection.

Identification

Chinchilla have large ears which can be utilised for identification purposes. A combination of nicks and holes in the ear can be used, although such markings may become erased if an animal gets chewed. A temporary mark with a biro can be made on the inside of the ear, or a permanent

Fig. 23.3B. Detailed view of a chinchilla in a polygamous unit. Round the neck of the female there is a light metal collar which prevents her from getting into the tunnel above where the male is resting. The male will eat and usually sleep in the cage of one of the females. (Photograph: UFAW.)

tattoo used. Tattooing is often used by ranchers and when animals come into the laboratory with tattooed numbers it is necessary to superimpose the laboratory number; this is best done by inserting an ear tag (Hauptner). Chinchilla take these tags very well; the numbers are nearly always visible and tag losses are rare.

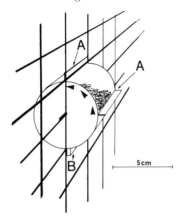

Fig. 23.4. Diagram to show a simple food hopper made from an evaporated milk tin. The counterweight (B) holds the hopper closed and the chinchilla pulls on the upper stop (A) to get the opening to the inside of the cage. (Photograph: UFAW.)

Handling and transport

Chinchilla are not difficult to handle, although it must always be remembered that their bite is very painful. Laboratory chinchilla can be picked up out of the cage by holding them round the body or by the tail. On ranches, where any risk of slipping fur must be avoided, the preferred method is to pick the animals up by the ears, but they should not be held by these longer than is necessary. For vaginal examinations (Fig. 23.5) the animal is held by

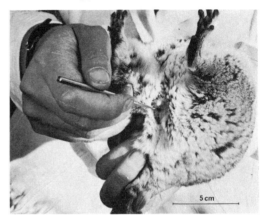

Fig. 23.5. The method for taking a vaginal smear of a chinchilla; the animal is held by the base of the tail with the forepaws resting on the observer's chest. (Photograph: UFAW.)

the tail and the forepaws rested on the observer's chest. For intraperitoneal injections the chinchilla is held on its back along the forearm; the base of the tail is held firmly between the second and third fingers while the thumb and little finger are used to keep the hind legs out of the way (Fig. 23.6). A method of restraint which is

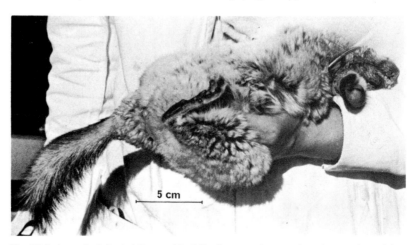

Fig. 23.6. A method for holding a chinchilla for procedures such as intraperitoneal injection. (Photograph: UFAW.).

useful when both hands must be free (e.g. for electro-ejaculation) is that described by Healey & Weir (1967) (Fig. 23.7): the chinchilla is restrained on its back on a box about 10 cm wide; a leather strap is then passed over the neck and attached at the other side over a carpet nail; canvas straps (2·5 cm wide) are then passed over the thorax and over the legs and fastened in the same way. A series of holes along the straps allows adaptation for all sizes of chinchilla.

For short distances the chinchilla is carried with the head tucked under the arm, the feet resting on the forearm and the base of the tail firmly held to prevent escape should this be attempted. For greater distances they can be placed in small mesh cages of about 20×20 cm. These are then placed inside a cardboard box and packed round with hay to exclude draughts.

Feeding

Little is known about the food of wild chinchilla, but many manufacturers produce a compounded diet pellet for the caged animals. Chinchilla will live and breed on standard guinea-pig diets, but they certainly prefer the chinchilla pellets probably because they are made as long, thin

pellets which the animals can easily hold in one forepaw. It is not necessary to feed greens to chinchilla and some ranchers claim that their animals do well without hay, but ours in the laboratory eat a lot of hay and it is always made available to them in hoppers between the cages. Hay is also placed in the cage of a female which is about to litter, as the chinchilla does not make a nest. The feeding of an enriched diet during pregnancy and lactation is recommended by some ranchers but this has never been practised in our laboratory.

Chinchilla easily become bored and will very quickly overturn food dishes left in the cages, or throw pellets out of the standard type of food hopper. Two ways of reducing this wastage have been found. One is to mix the chinchilla pellets with guinea-pig pellets; the animal then tends to throw away the cheaper guinea-pig pellet which it finds less easy to hold. The better method is to use home-made food hoppers (Fig. 23.4) from small evaporated-milk tins, the ends of which are partly opened. A third of the side of the tin is opened and the ends bent over. A wire is passed through the centre of the tin and a weight is fixed to the outside so that when the animal wants to feed it has to pull the hopper towards its cage so that the aperture lies within

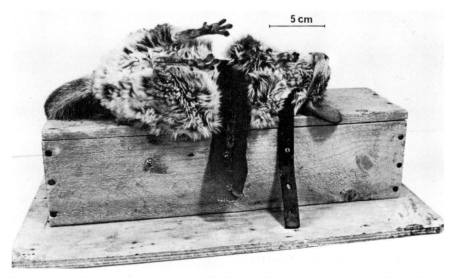

Fig. 23.7. A method for restraining a chinchilla for procedures such as intraperitoneal injection or electro-ejaculation. A leather strap is used at the neck and canvas ones for the thorax and legs. The different hole positions make the straps adjustable; these need to be firm round the animals but not tight. (Photograph: UFAW.)

the cage. No difficulty has been experienced in teaching chinchilla to use these hoppers and even baby chinchilla can reach the pellets.

Water

This is best provided in standard drinking bottles or by an automatic watering system; open dishes quickly get knocked over or fouled by faeces and pellets.

Breeding

Most chinchilla ranches use the polygamous system of breeding and both outbreeding and inbreeding is followed to produce animals which excel in fur qualities and reproductive performance (Houston, 1962). In the laboratory the problems of perpetuating specific characteristics have not yet arisen owing to the small size of laboratory colonies and the rapid turnover of animals.

Artificial insemination has not yet been shown to be feasible, although some commercial organisations in the United States do advertise the availability of frozen chinchilla semen. Semen can readily be obtained from chinchilla by electro-ejaculation (Healey & Weir, 1967). In their paper on artificial insemination in chinchilla, Hillemann, Gaynor & Dorsch (1963) do not actually state that they used the methods suggested, and no figures are given to indicate the proportion of successful insemination.

Some progress has been made towards a reliable technique for the induction of ovulation in the chinchilla with exogenous gonadotrophins (Weir, 1969). The gonadotrophins used were pregnant mares' serum (PMS) and human chorionic gonadotrophin (HCG) ('Gestyl' and 'Pregnyl', Organon) in various combinations. The animals were treated at all stages irrespective of their own reproductive cycles and at all times of year, and oestrus and ovulation were induced in about 50 per cent of the animals (Weir, 1973) and ovulation only in 86 per cent (Weir, 1969).

Laboratory procedures

Anaesthesia and euthanasia

Chinchilla respond well to most anaesthetics and will survive a wide variety of operative pro-

cedures. Surgical anaesthesia for periods of 30 minutes may be induced by an intraperitoneal injection of sodium pentobarbitone at a dose rate of 40 mg/kg. However, a few individuals are resistant to barbiturates and for these animals it is best to use a gaseous anaesthetic such as ether or halothane. If ether is used, mucous secretion and salivation should be controlled by a subcutaneous injection of 0·05 cc of a 0·5 per cent solution of atropine sulphate (i.e. 0·25 mg regardless of weight of the animals) before the start of the operation. Particular care is needed when chinchilla are recovering from anaesthesia. When the first signs of recovery are noted (e.g. response to touch inside the pinna or pulling of the vibrissae) the animal should be picked up and made to wake up fully, as some chinchilla, if not woken, will pass into a deep sleep which may become coma.

The simplest way to kill a chinchilla is to dislocate the neck. This technique is performed by holding the animal at the base of the tail in one hand and behind the ears with the other. A quick twist and pull results in instantaneous death. For some laboratory purposes this method is not suitable and then an overdose of barbiturates (400 mg/kg) injected intracardially or intraperitoneally is recommended.

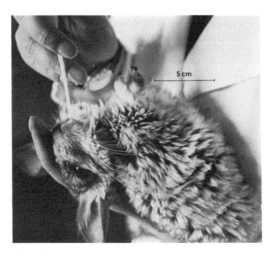

Fig. 23.8. The technique for oral administration of substances to a chinchilla; the head is held between the thumb and the first two fingers and the nozzle of the pipette is pushed through the diastema behind the incisors. (Photograph: UFAW.)

Injections

Injections in chinchilla can be given by all the usual routes; intravenous injections are best given via the ear veins but a very small gauge needle (26G) must be used.

Oral dosing

For oral administration (Fig. 23.8) the animal is held firmly against the operator's chest with the head held in the triangle formed by the thumb and first and second fingers. The third and little fingers are placed across the neck to prevent interference by the forefeet. A Pasteur pipette with rubber-tubing nozzle can then be inserted at the diastema and the substance expelled into the animal's mouth. A similar method is used to hand-rear baby chinchilla and the mixture given is diluted evaporated milk (50:50) with added glucose (25 per cent).

Diseases

The most important feature in preventing disease in chinchilla is good husbandry and maintenance of draught-free conditions. If subjected to draught, the animals will catch pneumonia very quickly and this can sometimes be treated successfully with injections of antibiotics. Other than this, chinchilla are more sturdy than is generally believed and, if they are a bit off-colour or sustain broken limbs or necks or tails, the best treatment is to leave them well alone.

The commonest disease is fur fungus (*Trichophyton mentagrophytes*) which starts on the nose and may spread to the eyes, ears and genitalia; in colony systems this will spread rapidly. In the laboratory, poor fur condition is not as important as it is to ranchers and, although the condition will clear up eventually even in the most severely affected animals, it is best treated orally with griseofulvin (15 mg daily) or by dusting with orthocaptan. This organic mercurial powder may also be put in the sand baths

which should be given to each animal at frequent intervals.

Chinchilla may sometimes suffer from dystokia which is recognised by restlessness of the female, crying and frequent attention to the genital region. Intramuscular injection of 0·1 i.u. oxytocin may be helpful in initiating birth. A similar dose may initiate milk let-down in cases of agalactia which is recognised when the kits fight amongst themselves and the mother is irritable towards the young.

REFERENCES

Bennett, E. T. (1829). The gardens and menagerie of the Zoological Society delineated. *Quadrupeds* vol. I.

Bignami, G. & Beach, F. (1968). Mating behaviour in the Chinchilla. *Anim. Behav.*, **16**, 45–53.

Dennler, G. (1939). Beitrage zur Kenntnis der Chinchilla. *Dte. Pelztierzüchter*, **14**, 388–90.

Galton, M. (1968). Chinchilla sex ratio. *J. Reprod. Fert.*, **16**, 211–6.

Healey, P. & Weir, B. J. (1967). A technique for electro-ejaculation in chinchillas. *J. Reprod. Fert.*, **13**, 585–8.

Hillemann, H. H., Gaynor, A. I. & Dorsch, A. (1963). Artificial insemination in chinchillas. *J. small Anim. Pract.*, **3**, 77–94.

Houston, J. W. (1962). *Chinchilla Care*, 3rd ed. Los Angeles: Borden.

Makino, S. (1953). Notes on the chromosomes of the porcupine and the chinchilla. *Experientia*, **9**, 213–4.

Ott, J. N. (1964). Some responses of plants and animals to variations in wavelengths of light energy. *Ann. N.Y. Acad. Sci.*, **117**, 624.

Weir, B. J. (1966). Aspects of reproduction in chinchilla. *J. Reprod. Fert.*, **12**, 410–1.

Weir, B. J. (1967). The care and management of laboratory hystricomorph rodents. *Lab. Anim.*, **1**, 94–105.

Weir, B. J. (1969). The induction of ovulation in the chinchilla. *J. Endocr.*, **43**, 55–60.

Weir, B. J. (1973). The induction of ovulation and oestrus in the chinchilla. *J. Reprod. Fert.*, **33**, 61–68.

24 Laboratory Hystricomorph Rodents other than the Guinea-pig and Chinchilla

Barbara J. Weir

Introduction

The order Rodentia is divided into 3 sub-orders: Sciuromorpha (squirrels, beavers), Myomorpha (rats, mice, hamsters) and Hystricomorpha (guinea-pigs, chinchilla) (see Rowlands & Weir, 1974).

The hystricomorphs differ from the other two groups in many aspects of their physiology and particularly in their reproductive patterns. For example, all except the coypu have been found to have a vaginal closure membrane which is perforate only at oestrus and parturition. The group is also characterised by its members having exceptionally long pregnancies; most gestation lengths are more than 100 days. For these reasons hystricomorphs are of great interest as laboratory animals. In addition to the domestic guinea-pig (*Cavia porcellus*) and the chinchilla (*Chinchilla laniger*), 8 other species have been kept by the author under laboratory conditions, namely:

Myocastor coypus	—	coypu
Dasyprocta aguti	—	agouti
Myoprocta pratti	—	acouchi
Cavia aperea	—	wild guinea-pig
Galea musteloides	—	cuis
Octodon degus	—	degu
Ctenomys talarum	—	tuco-tuco
Lagostomus maximus	—	plains viscacha

The last 5 species listed were established as colonies at the Wellcome Institute of Comparative Physiology after they had been collected in South America in 1967. *Dasyprocta*, *Myoprocta*, *Cavia*, *Galea* and *Lagostomus* have not previously been studied under laboratory conditions. There is, therefore, very little information, other than our own experiences, available on the breeding and husbandry of these animals. The following account may help as a general guide to those interested in starting and maintaining laboratory colonies of these and other new rodent species.

The coypu (*Myocastor coypus*)

Fam. Capromyidae; $2n = 42$ (Tsigalidou & Fasoulas, 1966).

The coypu is native to southern South America but, because of escapes and releases from fur farms, feral populations are now well established in many parts of the world, including Great Britain. In the wild, coypu spend much of the time in water, using the webbed hind feet for propulsion and the rounded tail as a rudder. They thrive better in captivity when they have access to water (Fig. 24.1A) and ideally they should be kept in groups in large pens (Flack, 1964; Capel-Edwards, 1967). Where this is not possible, we have found that double rabbit cages can be used for housing pairs of animals and a daily bath in the sink is sufficient to keep the fur in good condition. Rat pellets can be used as the main diet but fresh vegetables and a little fruit should be given daily, even if the animals have learnt to obtain their water from a standard drinking bottle. In our experience hay is not necessary unless the female is nesting, and wood chips make the best litter. A temperature higher than 15°C is reported to be necessary before breeding will occur (Illman, 1961).

Adult coypu weigh 3–7 kg. They are best handled by grasping the tail near the base and resting over the forearm; they may be carried in this manner for short distances (Fig. 24.1B). If the front claws are allowed to grip a mesh cage top, the animal will usually remain sufficiently still for an intraperitoneal or intramuscular injection or for vaginal smearing. Intravenous injections can be given by inserting a needle mid-ventrally into the tail, about 1 cm from its base, until it touches a vertebra; slight withdrawal of the needle usually leaves the point in the caudal vein. More detailed investigations necessitate restraint of the animal or light anaesthesia with ether or halothane. Butterworth *et al.* (1962) recommend anaesthesia by intraperitoneal injection of a mixture containing

Fig. 24.1A. Coypu (adult female and 6-week-old young) at the Wellcome Institute. The animals are housed in an outdoor pen which is partly sheltered and they have three sinks like the one shown in which to bath.

Fig. 24.1B, One way of carrying a coypu; the animal is picked up by the base of the tail and the forequarters rested on the observer's forearm. Note the webbed hind feet. (Photographs: UFAW.)

250 mg/ml chloralose and 250 mg/ml urethane at a dose rate of 2 ml/kg.

Since the coypu does not have a vaginal closure membrane, oestrous cycles have been investigated by means of vaginal smears and the results of various authors suggest that the cycle may vary from 4–30 days (Asdell, 1964). Coypu breed all the year round and under laboratory conditions a copulatory plug may be found as evidence of mating; blood may also appear in the vaginal smear at about 30 days of gestation. Pregnancy lasts for 127–138 days and averages 132 days (Newson, 1966); the 1–9 young are born fully furred and with their eyes open. The four pairs of mammae are situated dorso-laterally and the young will suckle for 6–10 weeks but have been successfully weaned at 5 days of age (Newson, 1966). The coypu has a post-partum oestrus and Newson found that 78 per cent of wild lactating females were also pregnant. The coypu in captivity may breed for 8–9 years (Federspiel, 1941).

The agouti (*Dasyprocta aguti*)

Fam. Dasyproctidae; 2n = 64 (Fredga, 1966).

The acouchi (*Myoprocta pratti*)

Fam. Dasyproctidae; 2n = 62 (Fredga, 1966). These two species can be considered together as they are very similar in habitat and behaviour. Both live solitarily in the undergrowth of forest regions of the West Indies and northern South America. The two genera are distinguishable by weight (the agouti being about 3 kg and the acouchi (Fig. 24.2A) about 1 kg), and by the fact that the acouchi has a small tail which is used as a signalling device (Morris, 1964).

Rat pellets and water are provided *ad libitum* and fresh fruit and vegetables given daily. Hay is given but is not much eaten.

These animals may be obtained from animal dealers and kept in fairly small cages, but they do better if housed in larger enclosures in pairs rather than in groups (Kleiman, 1970). A solid floor must be used to avoid broken legs and toes, and in large pens these animals can be very difficult to catch as they are very quick and can jump well.

Both agoutis and acouchis can be tamed, especially if obtained young, to permit handling without gloves or nets for simple procedures such as vaginal examination or injection. The acouchi is the easier to handle (Fig. 24.2B) as it is smaller but even the fiercest animals of both species can be caught and restrained without anaesthesia in a net specially devised by Lemmon & Weir (1968). As long as fingers are not stuck carelessly through the meshes the net can be used without mishap to animal or operator.

Fig. 24.2A. An adult acouchi male; note the short tail which distinguishes this species from the related, but much larger, agouti, which has no tail. (Photograph: UFAW.)

Daily examinations of the vaginal closure membrane indicate that the oestrous cycle is about 40 days (agouti) and 35 days (acouchi) and both species may show a period of anoestrus during the summer months in temperate latitudes.

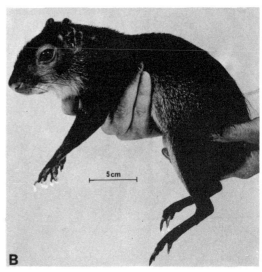

Fig. 23.2B. The acouchi can be picked up by holding it round the chest and supporting the rump with the other hand. (Photograph: UFAW.)

Agoutis have not bred under our laboratory conditions, even when extra space was provided for courtship activities (Weir, 1967). Acouchis do breed in captivity but are inclined to dystokia when the young are particularly large. Gestation length in this species is about 100 days and the 1–2 young are born fully precocious; they can be weaned before 2 weeks of age but do better if left until 4–6 weeks old (Kleiman, 1970).

The wild guinea-pig (*Cavia aperea*)

Fam. Caviidae; 2n = 64, George, Weir & Bedford (1972).

This cavy abounds on the pampas in Argentina and on grasslands in Brazil. In the laboratory it may be housed in the same way as the domestic guinea-pig (Weir, 1970). Although cavies do well in small cages on grids, a solid floor with wood-chips reduces the overgrowth of the claws. Mesh covers must be placed over floor pens to prevent the animals from jumping out. Standard guinea-pig pellets and water should always be available and cabbage should be fed 3 times a week. Hay, of which they eat a lot, should be provided as bedding and food.

These wild guinea-pigs have very dark fur (Fig. 24.3) and (as they will not tolerate ear clips or leg bands) identification of individuals is accomplished by spot-marking with carbol fuchsin solution which stays visible for about 2 months. Adult *C. aperea* are smaller than domestic guinea-pigs and rarely weigh more than 1 kg. They breed throughout the year and, as they have a post-partum oestrus, mature females are nearly always pregnant. The oestrous cycle length in isolated females varies from 13–30 days with a mean at about 21 days. The length of pregnancy varies from 59 to 74 days but the average is 61 days; the duration of gestation is inversely affected by litter-size (Rood & Weir, 1970), as in domestic guinea-pigs (Rowlands, 1949). The 1–6 young look like miniature adults at birth and can be weaned after a few days, although they are better left for about 2 weeks. Puberty occurs at about 2 months of age but may be as early as 27 days or as late as 6 months. Adult males will not usually tolerate the presence of another male.

Fig. 24.3. A wild guinea-pig (*Cavia aperea*) which is agouti-coloured and has a distinctive crest on the neck. (Photograph: UFAW.)

The cuis (*Galea musteloides*)

Fam. Caviidae; $2n = 68$, George, Weir & Bedford (1972).

This cavy occurs on the altiplano of Peru and also in the Coronel Pringles district in Argentina where our original animals were trapped. They

Fig. 24.4. A group of cuis (*Galea musteloides*) in a floor pen. Note the compact body and the eye rimmed with pale hairs. The galvanized-iron drainpipe elbow is very useful for catching up these animals; a mesh lid is needed to prevent their jumping out of the pens. (Photograph: UFAW.)

are quite different in temperament and physiology from *Cavia* species (Weir, 1970).

Like *Cavia* they are best housed on solid floors (Fig. 24.4) and should be fed as described above for *C. aperea*, although they do not eat as much hay. They are smaller animals, weighing between 300–700 g, but are more compact and sturdy than *C. aperea*. Ear-tags are not tolerated and carbol fuchsin is used to spot the agouti-coloured hair for individual identification. Sex difference in this species is very obvious because there is a wide separation of urinary and anal apertures in the male and the genital skin is heavily pigmented. When females were isolated to record oestrous cycle length no vaginal openings occurred; neither was there any response when they were separated from males by a mesh partition (Weir, 1971, 1973). However, in the presence of a vasectomised male the females cycled regularly at intervals of about 22 days. The secretions of a chin gland and a behaviour pattern are involved. Cuis nearly always mate at the post-partum oestrus, which occurs immediately after giving birth, and once she is mature a female is almost always pregnant; one female *Galea* had 15 immediately consecutive pregnancies. Gestation length is very constant at 52 or 53 days and the 1–5 young are born fully developed. Under our conditions of captivity, the aggressive nature of *Galea* adults makes it necessary to plan carefully the setting up of new groups of animals in order to prevent deaths from fighting. It is best to start a group by weaning several young together at about 3 weeks of age. Conceptions at 11 to 20 days of age are not uncommon, but in the laboratory puberty usually occurs at about 8 weeks of age. In Argentina the age at which puberty is reached is dependent upon the time of year of birth (Rood & Weir, 1970).

The degu (*Octodon degus*)

Fam. Octodontidae; 2n = 58 (Fernandez, 1968). This rat-like hystricomorph, which weighs 200–300 g, lives on the western slopes of the Andes at about 300 metres. Our original degu were obtained from a laboratory colony in Santiago. In our experience this species lives and breeds well in captivity (Weir, 1970) but the animals should be handled with great care as they have a

particularly painful bite. Handling is best achieved by scooping from one hand to the other (Fig. 24.5) as in the method used for hamsters (p. 255). They should not be held by the tail, as it skins readily. If any animal is too aggressive it can be run into a 10 cm diameter galvanised-iron drainpipe elbow, from which it can be gently pushed on to a cloth surface without further trouble. Once on a cloth surface, such as a sleeve, a degu will cling tightly and can quite safely be carried for short distances.

In our colony plastic, rat-breeding cages with grids are used to house either pairs or trios of 1 male and 2 females. Ear tags have been used with reasonable success as a means of identification.

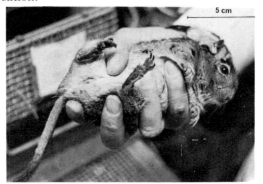

Fig. 24.5. How to hold a degu for examination or injection; note the brush on the tail and the ear-tag for identification. (Photograph: UFAW.)

We have found that this species is unusual in that animals can be moved from one cage to another without any preliminary introductions or subsequent fighting, although when littering a female may be disturbed by the presence of another female and neglect her young; she will not, however, resent a male.

Gestation length is 90 days and the 1–8 young (usually 5 or 6) are born rather less precocious than those of other hystricomorphs. The eyes are closed at birth and the hair covering is not dense, but growth is quick and weaning takes place at 6–8 weeks of age. Breeding takes place only once a year, in December in the northern hemisphere, and it is likely that the degu is an induced ovulator because the vagina remains open at oestrus for a period of 3–21 days. Closure occurs very soon after mating, which is indicated by blood at the vagina and sometimes by the recovery of a copulatory plug.

Very few spontaneous deaths have occurred in our colony although the animals tend to become obese. For this reason they are fed rat pellets and vegetables on alternate days only, although hay and water are always available.

Most of the degu in our colony exhibit lens cataracts and a preliminary biochemical assay indicated that these are glucose cataracts, although there is no associated hyperglycaemia, glycosuria or impaired tolerance to a glucose load. The cataracts are morphologically similar to those found in rats fed on high galactose diets, but since the animals are breeding and the colony is building up satisfactorily correction for this condition is not yet being sought.

The tuco-tuco (*Ctenomys talarum*)

Fam. Ctenomyidae; 2n = 48 (Kiblisky & Reig, 1966).
This genus occupies the same ecological niche in southern South America as does the gopher in North America. It is a burrower and is rarely seen above ground, although the mounds of soil it throws up and its characteristic vocalisation (a 'tuc-tuc-tuc' uttered at increasing tempo) are well known in Argentina where this small (100–200 g) species (Fig. 24.6) was caught in Longworth traps (Weir, 1971c).

They can be kept on grids but solid-floored cages are better, although the claws still have to be clipped every 3–4 months. Sawdust is used as litter but the animals spend much of their time

Fig. 24.6. A tuco-tuco female. Note the small pinna typical of a burrowing species and the nearly naked tail which is used as a sensory organ when the tuco-tuco is running backwards along its tunnel. The comb (Greek *ctenus*) of stiff hairs from which the genus gets its name, can be seen on the toes of the hind foot. (Photograph: UFAW.)

frantically trying to dig it into a corner of the cage, or even out of it. Hay is provided as bedding; some is eaten but most of it is cut up into 2–3 cm lengths. Rat pellets should always be available and fresh vegetables fed 3 times a week, as these animals have not learnt to drink from water bottles or bowls.

In spite of their large orange incisors, tuco-tucos rarely bite and are best picked up round the body and prevented from wriggling by a light grip behind the bullae. They object to being picked up by the tail. Being burrowers their pinnae are very small and the only way of identification is by staining with carbol fuchsin; this needs renewing about every 2 months. One male is housed with 3–4 females, but often serious fighting develops if groups have to be re-organised for any reason. As with *Galea* (p. 282) we have found it best to wean young tuco-tucos together as a group, although as they have not bred well in captivity this has rarely been possible.

In the wild the breeding season is from March–April, but in captivity we have found the occurrence of oestrus to be erratic and few males have been sexually active. The tuco-tuco is certainly an induced ovulator as the vagina remains open for a variable time at oestrus, sometimes for several weeks, and corpora lutea have been found only after mating. The gestation length is more than 100 days and is probably between 120–140 days, but the 1–7 young are more immature than degus at birth; only the guard hairs are visible and the eyes stay shut for longer. Usually, they are left with the mother for 4–5 weeks, but they can manage alone after 2 weeks. Total litter resorption is common in this species in our conditions of captivity, but the poor breeding record is thought to be due to the fact that many of the original animals and their F_1 offspring developed signs of diabetes (Wise et al., 1968, 1972). This was characterised by lens lesions, hyperglycaemia, glycosuria, pancreatic islet hyperplasia, β-cell degranulation and basement membrane thickening as well as the reproductive failure of both sexes. Tuco-tucos are resistant to bovine insulin (Weir et al., 1969) and it is possible that successful breeding may be obtained only if a new group of animals is caught from the wild and maintained under different conditions of diet and activity.

The plains viscacha (*Lagostomus maximus*)

Fam. Chinchillidae; 2n = 56 (George & Weir, unpublished)

This species (Fig. 24.7A) is the largest of the three chinchillid species and in its native habitat on the pampas in Argentina it digs extensive colonial burrow systems. Because of the damage they do to the ground and crops, they are classified as pests and efforts are being made to reduce their numbers. Only one extensive study, mainly ecological, has been reported (Llanos & Crespo, 1952), but details of laboratory husbandry have been given by Weir (1970) and reproduction in captivity has been investigated (Weir, 1970a and b; Roberts & Weir, 1973).

Viscacha weigh 3–8 kg and their size and burrowing proclivities present problems in their husbandry in captivity. Some have been kept as pairs in double rabbit cages but the majority are housed in groups of not more than 15 animals each, in large pens (about 3 metres square) which are partly exposed to the weather. Shelter is provided by one or two concrete bunkers in each pen. A series of metal and concrete indoor pens with outdoor runs, such as is used to accommodate dogs, would be an ideal arrangement for keeping viscacha. Smaller groups of animals could then be housed and thus reduce the problems of incompatibility. Some groups of viscacha tolerate the introduction of other animals, including young ones, but other groups may kill an introduced animal within 24 hours. Change is rarely required for established adult groups but weaning of young animals is a big problem when space is limited.

In our colony pregnant viscacha are removed to a small cage just before littering, as the young otherwise get accidentally kicked out of the bunker and die of inattention or trampling. At weaning, which is effected at 6–10 weeks after parturition according to other conditions of husbandry, the mother is returned to her original group, usually without much trouble, and if possible the young are also put with this group or grouped with other weanlings.

Wild viscacha have a breeding season in March and sometimes also in September but, in captivity, the females have become poly-oestrous and births now occur throughout the

Fig. 24.7A. A male and female plains viscacha in their outdoor pen. Note the larger size and blacker moustache of the male. The pad of stiff bristles (B) on the hind foot for grooming can be seen; the front feet are usually used for digging and the soil is pushed away with the nose. (Photograph: UFAW.)

year. Oestrus is detected by the condition of the vaginal membrane; oestrous cycles are very erratic and vary from 20–70 days in length (average 41 days). A large gelatinous copulatory plug is found after mating, usually towards the end of the period of vaginal opening, which lasts from 2–24 days although ovulation is spontaneous. There is an immediate post-partum oestrus but if the young are lost at birth the next oestrus is at 17 days instead of the usual post-lactation oestrus at 54 days. Two young (but sometimes singles and rarely triplets) are born after a pregnancy of 155 days, and they are fully developed at birth. For identification ears are notched or punched at birth and the animals are also marked with carbol fuchsin so that they can easily be identified in the pens.

Viscacha are not difficult to handle, although considerable care must always be exercised as they have large incisors and strong claws on the hind feet. Most of our animals can be picked up for vaginal examination or short transportation (Fig. 24.7B) by holding firmly at the base of the tail. A grip taken any further along the tail

results in autotomy at a predetermined plane. Difficult animals can be handled in the agouti-net (see p. 286; Lemmon & Weir, 1968). Further restraint for injection or anaesthesia is effected by quickly pinning the animal down with one hand behind the ears and the other at the base of the tail and round the hind feet. Viscacha respond well to surgery and have been suitably anaesthetised by halothane or by thiopentone administered intravenously.

As the animals tend to become obese, pellets and fresh vegetables are fed on alternate days, although hay and water are always provided.

The ovaries of viscacha are quite unlike those observed for any other mammal and it appears that at oestrus up to 800 eggs are ovulated, although only 2 young are brought to term (Weir, 1971b).

Disease

Captive hystricomorphs are remarkably free from external parasites although lice, ticks and fleas were abundant on the viscacha, tuco-tucos

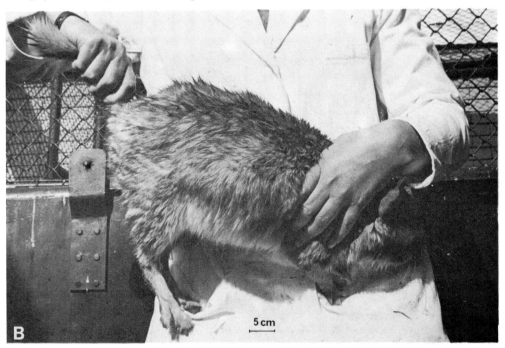

Fig. 24.7B. A method for carrying viscacha. The animal is picked up by the base of the tail and the shoulder is held tightly against the body, usually at the left side of the observer so that the head is under the elbow and the hind legs are in mid-air and have nothing to kick against. This method can only be used with placid viscacha; the one illustrated is a large female. (Photograph: UFAW.)

and cavies when they were caught by the author in Argentina in 1967. Coypu may carry a louse species which is indigenous to South America. Internal parasites, such as nematodes, are common but have not been known to reach such concentrations as to be the obvious cause of death of any animal. The commonest infection we have encountered in captivity is *Pasteurella pseudotuberculosis* and all species except the degu have suffered deaths from this organism. Middle-ear disease has occurred in the tuco-tuco and plains viscacha and *Salmonella* has also been identified in the latter. But apart from an unidentified epidemic which swept through the cuis and wild guinea-pig colonies, infection has been infrequent and most deaths have resulted from fighting.

REFERENCES

Asdell, S. A. (1965). *Patterns of Mammalian Reproduction*, 2nd ed. Ithaca, New York: Cornell Univ. Press.

Butterworth, K. R., Shillito, E. E., Spencer, K. E. V. & Stewart, H. C. (1962). The laboratory use of *Myocastor coypus* (the coypu). *J. Physiol., Lond.*, **162**, 36–7P.

Capel-Edwards, M. (1967). Management of the coypu. *J. Inst. Anim. Techns*, **18**, 60–5.

Federspiel, M. N. (1941). Nutria farming. *Am. Fur Breed.*, **13**, 12–13.

Fernandez, R. (1968). El cariotipo del *Octodon degus* (Rodentia-Octodontidae) (Molina 1782). *Archos. Biol. Med. exper.*, **5**, 33–7.

Flack, M. B. (1964). Some aspects of husbandry of the coypu (*Myocastor coypus*) in the laboratory. *J. Anim. Techns Ass.*, **14**, 145–56.

Fredga, K. (1966). Chromosome studies in five species of South American rodents (Sub-order Hystricomorpha). *Mammalian Chromosome Newsletter*, **20**, 45–6.

George, W., Weir, B. J. & Bedford, J. Chromosome studies in some members of the family Caviidae (Mammalia: Rodentia). *J. Zool., Lond.*, **168**, 81–99.

Illman, O. (1961). The coypu (*Myocastor coypus* Molina) as a laboratory animal. *J. Anim. Techns Ass.*, **12**, 8–10.

Kiblisky, P. & Reig, O. A. (1966). Variation in chromosome number within the Genus *Ctenomys* and description of the male karyotype of *Ctenomys talarum talarum* Thomas. *Nature, Lond.*, **212**, 436–8.

Kleiman, D. G. (1970). Reproduction in the female green acouchi, *Myoprocta pratti* Pocock. *J. Reprod. Fert.*, **23**, 55–65.

Lemmon, F. R. & Weir, B. J. (1968). A method for handling agoutis and acouchis. *J. Inst. Anim. Techns*, **19**, 49–51.

Llanos, A. C. & Crespo, J. A. (1952). Ecologia de la vizcacha (*Lagostomus maximus maximus* Blainv.) en el nordeste de la provincia de Entre Rios. *Revta Invest. agric., B. Aires*, **6**, 289–378.

Morris, D. (1962). The behaviour of the green acouchi (*Myoprocta pratti*) with special reference to scatter hoarding. *Proc. zool. Soc. Lond.*, **139**, 701–32.

Newson, R. M. (1966). Reproduction in the feral coypu (*Myocastor coypus*) *Symp. zool. Soc. Lond.*, **15**, 323–34.

Roberts, C. M. & Weir, B. J. (1973). Implantation in the plains viscacha, *Lagostomus maximus*. *J. Reprod. Fert.*, **33**, 299–307.

Rood, J. P. & Weir, B. J. (1970). Reproduction in female wild guinea-pigs. *J. Reprod. Fert.*, **23**, 393–406.

Rowlands, I. W. (1949). Post-partum breeding in the guinea-pig. *J. Hyg., Camb.*, **47**, 281–7.

Rowlands, I. W. & Weir, B. J. (Eds.) (1974). The biology of hystricomorph rodents. *Symp. zool. Soc. Lond.*, **33**, 1–472.

Tsigalidou, V., Simotas, A. G. & Fasoulas, A. (1966). Chromosomes of the coypu. *Nature, Lond.*, **211**, 994–5.

Weir, B. J. (1967). The care and management of laboratory hystricomorph rodents. *Lab. Anim.*, **1**, 95–104.

Weir, B. J. (1970). The management and breeding of some more hystricomorph rodents. *Lab. Anim.*, **4**, 83–97.

Weir, B. J. (1971a). The reproductive physiology of the plains viscacha, *Lagostomus maximus*. *J. Reprod. Fert.*,

Weir, B. J. (1971b). The reproductive organs of the female plains viscacha, *Lagostomus maximus*. *J. Reprod. Fert.*,

Weir, B. J. (1971c). A trapping technique for tuco-tucos (*Ctenomys talarium*). *J. Mammal.*, **53**, 836–9.

Weir, B. J. (1971d). The evocation of oestrus in the cuis, *Galea musteloides*. *J. Reprod. Fert.*, **26**, 405–408.

Weir, B. J. (1973). The role of the male in the evocation of oestrus in the cuis, *Galea musteloides*. *J. Reprod. Fert., Suppl.*, **19**, 421–432.

Weir, B. J., Wise, P. H., Hime, J. M. & Forrest, E. (1969). Hyperglycaemia and cataract in the tuco-tuco. *J. Endocr.*, **43**, vii–viii.

Wise, P. H., Weir, B. J., Hime, J. M. & Forrest, E. (1968). Implications of hyperglycaemia and cataract in a colony of tuco-tucos (*Ctenomys talarum*). *Nature, Lond.*, **219**, 1374–6.

Wise, P. H., Weir, B. J., Hime, J. M. & Forrest, E. (1972). The diabetic syndrome in the tuco-tuco (*Ctenomys talarum*). *Diabetologia*, **8**, 165–172.

25 The Heteromyid Rodents

John F. Eisenberg

Introduction

The family Heteromyidae is confined to the New World and is closely related to a second New World family of rodents, the Geomyidae or pocket gophers. It includes five genera: *Heteromys*, *Liomys*, *Perognathus*, *Microdipodops* and *Dipodomys*, all of which are nocturnal surface foragers which inhabit rather complicated burrows. In the wild, seeds and herbaceous material constitute the bulk of their diet, although insects may be eaten by certain species. All members of the family store quantities of seeds either in their burrow systems or in the soil immediately adjacent to the burrows. Seeds are transported in the fur-lined cheek pouches, which open externally.

Heteromys and *Liomys* are included in the sub-family Heteromyinae. They resemble small rats in size and appearance, and many species have spines scattered among their dorsal pelage. The genus *Heteromys* ranges from northern South America to the state of Veracruz in Mexico. It appears to occupy moist, forested habitats and may occur in the cloud forests at high altitudes in the northern limits of its range. The genus *Liomys* is widespread in Mexico and Central America. It includes three major groups of species and occupies a variety of habitats. In Mexico, *L. pictus* and *L. inornatus* occupy seasonally arid areas, whereas the *L. crispus* group is found in the more moist, semi-tropical and tropical lowland areas of southern Mexico and Central America (Fleming, 1974).

The three remaining genera are adapted to arid and semi-arid areas in the western United States and northern Mexico. *Perognathus* and *Microdipodops* are included in the sub-family Perognathinae. The genus *Perognathus* is divided into two sub-genera, *Chaetodipus* and *Perognathus*. The former are often characterized by the presence of spines mixed with the pelage of the dorsum. The members of the sub-genus *Perognathus* all possess a short, soft pelage. There is considerable variation in size among the

species of this genus. *P. hispidus* may have an average body-length of 104 mm, while that of *P. longimembris* averages 61 mm.

The genus *Microdipodops* is confined to the Great Basin in western North America. It is capable of bipedal locomotion and shows a lengthening of the hind foot and shortening of the forelimbs which is characteristic of all saltators. The tympanic bullae are tremendously enlarged. The animals are quite small with an average body-length of 71 mm.

The genus *Dipodomys* is the sole member of the third sub-family, the Dipodomyinae. All species are bipedal ricocheters, exploiting arid and semi-arid habitats in western North America and northern Mexico. They are quite distinctive in appearance, exhibiting lengthened hind feet, shortened forelimbs, enlarged tympanic bullae, and a long tufted tail. In size the species range from *D. deserti*, with an average weight of 103·3 g and a body-length of 143 mm, to *D. nitratoides* with an average weight of 36·6 g and body-length of 101 mm.

Species of all five genera can be taken in live-traps of a suitable size baited with oats and sunflower seeds and set near a burrow entrance. Since the animals all tend to exhibit high intraspecific aggressiveness, it is best to house them separately during transport and in the laboratory.

Caging

Liomys and *Heteromys* may be kept in cages with sawdust as a floor litter; the other three genera should be kept in cages with earth or unwashed sand as a floor covering. The pelage of *Dipodomys*, *Microdipodops*, and many species of *Perognathus* will become matted unless these animals have access to dry sand or earth; they exhibit a definite sand-bathing habit and by this means dress their pelage (Eisenberg, 1967).

Nest-boxes or cans should also be provided. The animals will spend the day inside the nest-box and will plug the entrance with earth or sawdust. Burrow-plugging is a characteristic behaviour pattern for the whole family. Nesting-material such as straw or paper strips is readily utilized by *Heteromys*, *Liomys*, and some species of *Perognathus*. The nest-building habit is reduced in *Microdipodops* and in some species of *Dipodomys*.

The cages should be large if breeding is to be attempted (see below), but holding-cages with a floor space of 40×30 cm are quite adequate for individual specimens of all the genera.

Handling

Kangaroo rats and pocket mice have extremely fragile tails and, if grasped by the terminal portion, the skin will often break and slip off from the distal portion of the tail. The muscular and osseous portion of the tail, deprived of its blood supply, will undergo necrosis and drop off. For this reason it is recommended that the animals not be handled by the tail but rather be caught in a small container or glass tube. The animals may be restrained by grasping the skin at the nape of the neck with one hand while the other hand is used to restrain the hind limbs. Restraint of the hind limbs is especially important in a species of kangaroo rat which will kick violently when seized by the nape.

Feeding

Feeding requirements vary from species to species, but the following diets have proved adequate. All genera are given a seed mixture consisting of 50 per cent of sunflower seeds, the remaining 50 per cent containing equal amounts of whole oats, millet and canary seed. Small amounts of dry dog-food are added from time to time as a protein and salt supplement. In addition, *Heteromys lepturus* and *H. anomalus* are given fruit (such as sliced apples or blackberries) once a week. Most species of *Dipodomys* will eat carrots, and these may be given as a moisture supplement in lieu of lettuce. *Liomys pictus* and *Heteromys* will eat raw potato and this is generally offered once a week. Meal-worms are readily taken by *M. pallidus*, *D. deserti*, *P. inornatus* and *H. lepturus*. Other species may only nibble at mealworms or ignore them.

Pregnant and lactating females of *Liomys*, *Perognathus* and *Dipodomys* receive lettuce every day. In addition, the *Liomys* females are allowed access to drinking-water. During lactation the females of *Perognathus californicus* and *Liomys pictus* often eat canned dog-food if it is offered them. Lactating females of *Dipodomys* eat very little meat, but it would be wise to offer it as species preferences seem to vary. Since *Microdipodops* and *Heteromys* have not been bred in the laboratory no special comments on feeding during lactation can be presented, but attention should certainly be given to protein and water requirements.

Water

Water requirements vary greatly with the species. *Heteromys* needs drinking-water at all times. *Liomys pictus* can be kept without free water if some succulent vegetable matter (lettuce or potato) is supplied to provide it with moisture. *Microdipodops pallidus* can survive for several weeks on dry food but, in order to maintain it in good health, a small amount of lettuce should be given every second day.

Many species of *Dipodomys* and *Perognathus* need no free water for drinking (Schmidt-Nielson, 1952) and *D. merriami*, *P. californicus*, *P. penicillatus* and *P. flavus* have been kept in good health for several months with only dry seeds for food. Although some species of these genera can subsist for long periods on metabolic water alone, other species cannot. *Dipodomys venustus* and *D. microps* will persistently lose weight on a dry diet, even though they can be kept for several weeks without a source of moisture (Church, 1961; Kenagy, 1973). There is evidence that the same is true for coastal races of the genus *Perognathus* Because of this variability in response to desiccation it is recommended that lettuce be fed every other day to all specimens. Free water should not be provided for *Perognathus*, *Microdipodops* and *Dipodomys* for two reasons: first, it is often ignored; and secondly, some individuals occasionally become addicted to drinking-water. This addiction has occurred with *M. pallidus*, *D. panamintinus* and *D. microps* and the animals develop symptoms similar to those encountered with the *diabetes insipidus* syndrome. They urinate copiously, drink often, and gradually waste away, becoming very thin and weak.

Breeding

With the possible exception of *Heteromys*, the heteromyid rodents are characterized by a high intraspecific aggressiveness (Wagner, 1961; Reynolds, 1960; Tappe, 1941; Bailey, 1931; Fleming, 1974), although this varies from species to species. *Dipodomys deserti* is intolerant of conspecifics while *D. merriami* and *D. nitratoides* are often more docile. *Heteromys desmarestianus, Perognathus californicus, P. penicillatus* and *P. flavus* are also highly aggressive, while *P. formosus* and *P. inornatus* have proved to be slightly more tolerant. *Liomys pictus* and *Microdipodops pallidus* are very intolerant of conspecifics, and *Heteromys anomalus* exhibits the least aggressiveness of any species studied. Even so, *Heteromys* pairs will fight and chase each other during an encounter (Eisenberg, 1963 and 1967).

In breeding these animals two factors must be borne in mind. First, the female is least likely to be aggressive if she is in oestrus; and secondly, an individual animal is always very antagonistic when a stranger is introduced into its cage. For maximum reproductive success three general methods have been employed. A pair may be simultaneously introduced into a large cage provided with two or more nest boxes and allowed to live there until mating has taken place, or they may be introduced simultaneously for half-an-hour to an hour on successive nights in a neutral arena until the female comes into oestrus, when mating will be accomplished. The third technique involves keeping the pair in adjoining cages with a removable partition and allowing them to meet on successive nights until mating takes place (see also Wilken & Ostwald, 1968).

The first method has been successfully employed with *D. nitratoides, D. merriami* and *L. pictus* placed in a cage measuring $150 \times 210 \times 60$ cm. Using a larger floor area of 6×8 m Butterworth (1961b) was able to breed *D. deserti*; and Day *et al.* (1956) were able to breed *D. ordii* by simultaneous introduction into a cage with a floor area of 76×76 cm. Chew (1958) bred *D. merriami* by introducing females to males (but only when the female was in oestrus) in a cage measuring $25 \times 25 \times 27.5$ cm.

The second method, involving brief encounters on successive nights, has been used in a cage with a floor area of 0.56 m^2. *Perognathus californicus, D. nitratoides* and *D. panamintinus* have been bred in this way. The technique has several advantages: the observer can separate the pair if sustained fighting or chasing results, and the successive introductions allow the animals to gain familiarity with the cage, so that the oestrous condition of the female is not likely to be upset by her sudden transfer to a strange situation.

The third method, which employs adjoining cages, has had wide success. *D. nitratoides, D. merriami, D. panamintinus* and *L. pictus* have been bred in cages with a floor space of 196×199 cm divided into two equal areas, and *P. flavus* in similar cages with a combined floor area of about $1,400$ cm^2.

If a pair are established in the same cage, the animals generally nest separately. Occasionally the male establishes a dominance relationship over the female, when they may nest together for several weeks. It seems that the enforced proximity of the male induces a prolonged anoestrous condition in the female and if she is not already pregnant no litters will be produced. Often the male indulges in the aberrant behaviour pattern of nibbling the fur of the female's back, almost denuding it. This condition has been observed in *L. pictus, D. merriami, D. nitratoides* and *D. panamintinus*. It seems that successful breeding is best accomplished where the male and female pair together for brief intervals.

The breeding behaviour has been described in Butterworth (1961b), Allan (1944) and Eisenberg (1963).

The females become quite intolerant at parturition, and if the male is still in the same cage he should be removed for the female will often attack and chase him repeatedly. Males of *L. pictus, D. merriami* and *P. flavus* have been severely injured or killed by pregnant females.

Chew (1958) stressed that fluctuations in the environmental temperature could be important for the successful breeding of *D. merriami*. Although no special attention was paid to this in my own breeding experiments, it may be an important factor. Three genera have been bred indoors, but the environmental temperature often varied in the laboratory by as much as $8.5°$C in a twenty-four hour period.

Data on reproduction

The heteromyids appear to be seasonally poly-oestrous. The oestrous cycle for *D. ordii* (Pfeiffer, 1960), and *P. inornatus* is 5 to 6 days in length. *D. nitratoides* females have shown a post-partum heat with conception taking place.

The gestation periods for species of *Dipodomys* range from 29 to 33 days, and these have been determined for *D. merriami* (Chew, 1958), *D. deserti* (Butterworth, 1961b), *D. nitratoides*, *D. ordii* (Day *et al.*, 1956) and *D. panamintinus*. The period for *Liomys* and *Perognathus* appears to be somewhat shorter—records indicate 24 to 26 days for *L. pictus*, 25 days for *P. californicus* and 25 to 26 days for *P. flavus* (Eisenberg & Isaac, 1963).

In general the species of *Dipodomys* show a tendency to have small litters (1 to 3 for *D. merriami* and *D. nitratoides*). *Liomys pictus* in captivity has an average litter-size of 3–5 with a range of from 2 to 5. The various species of *Perognathus* average about 4 young per litter.

Lactation persists for 24 to 28 days in *Liomys pictus*, and 22 to 24 days in *P. californicus*. Young *D. nitratoides* may be weaned at 21 to 24 days; *D. panamintinus* may nurse its young from 27 to 29 days.

The earliest age of conception for *D. nitratoides* is 84 days, while *Liomys pictus* has bred at 98 days of age. For further data on growth and maturation see Chew & Butterworth (1959), Butterworth (1961a) and Eisenberg (1963).

Diseases

The heteromyid rodents have proved very resistant to diseases in the laboratory. During a four-year period when as many as seventy animals were kept in the same laboratory, no epidemics occurred.

Ectoparasites

The animals are often infested with mites, ticks, or fleas when trapped in the wild, but these are removed by the animals themselves during grooming. During the first three weeks of captivity their nesting-material and cage-litter should be replaced every five days, so that the eggs of the parasites are destroyed. Once the reproductive cycle has been broken the animals remain uninfested unless new parasites are introduced with newly-captured animals. Heavy infestations can be treated with a commercial flea powder such as is recommended for cats.

Anaesthesia and euthanasia

Methods suitable for mice can be used.

REFERENCES

Allan, P. F. (1944). The mating behaviour of *Dipodomys ordii richardsoni*. *J. Mammal.*, **25**, 403–4.

Bailey, V. (1931). Mammals of New Mexico. *N. Am. Fauna*, **53**, 1–412.

Butterworth, B. B. (1961a). A comparative study of growth and development of the kangaroo rats *D. deserti* Stephens and *D. merriami* Mearns. *Growth*, **25**, 127–38.

Butterworth, B. B. (1961b). The breeding of *Dipodomys deserti* in the laboratory. *J. Mammal.*, **42**, 413–4.

Chew, R. M. (1958). Reproduction by *Dipodomys merriami* in captivity. *J. Mammal.*, **39**, 597–98.

Chew, R. M. & Butterworth, B. B. (1959). Growth and development of Merriam's kangaroo rat. *Growth*, **23**, 75–95.

Church, R. L. (1961). *A study of evaporative water loss in the kangaroo rat Dipodomys venustus.* M.A. thesis, University of California, Berkeley.

Day, B. N., Egosque, H. J. & Woodbury, A. M. (1956). Ord kangaroo rat in captivity. *Science, N.Y.*, **124**, 485–6.

Eisenberg, J. F. (1963). The vehaviour of heteromyid rodents. *Univ. Calif. Publs Zool.*, **67**, 1–100.

Eisenberg, J. F. (1967). A comparative study in rodent ethology with emphasis on evolution of social behaviour, I. *Proc. U.S. natn. Mus.*, **122**, 1–51.

Eisenberg, J. F. & Isaac, D. E. (1963). The reproduction of heteromyid rodents in captivity. *J. Mammal.*, **44**, 61–7.

Fleming, T. H. (1974). Social organization in two species of Costa Rican heteromyid rodents. *J. Mammal.*, **55**, 543–561.

Hayden, P., Gambino, J. J. & Lindberg, R. G. (1966). Laboratory breeding of the little pocket mouse, *Perognathus longimembris*. *J. Mammal.*, **47**, 412–23.

Kenagy, G. J. (1973). Adaptations for leaf eating in the Great Basin Kangaroo Rat, *Dipodomys microps*. *Oecologia*, **12**, 383–412.

Pfeiffer, E. W. (1960). Cyclic changes in the morphology and clitoris of *Dipodomys*. *J. Mammal.*, **41**, 43–8.

Reynolds, H. G. (1960). Life history notes on Merriam's kangaroo rat in southern Arizona. *J. Mammal.*, **41**, 48–58.

Schmidt-Nielson, K. & Schmidt-Nielson, B. (1952). The water metabolism of desert animals. *Physiol. Rev.*, **32**, 135–66.

Tappe, D. T. (1941). The natural history of the Tulare kangaroo rat. *J. Mammal.*, **22**, 117–48.

Wagner, H. O. (1961). Die Nagetiere einer Gebirgsabdachung in Sudmexico und ihre Beziehungen zur Umwelt. *Zool. Jb.*, **89**, 177–242.

Wilken, K. K. & Ostwald, R. (1968). Partial contact as a stimulus to laboratory mating in the desert pocket mouse, *Perognathus penicillatus*. *J. Mammal.*, **49**, 570–2.

26 Bats

P. A. Racey

Introduction

The order Chiroptera is the second largest order of mammals with 981 recorded species (Morris, 1965) showing extensive adaptive radiation of size, habit and diet (Figs. 26.1–26.4). Accounts of its unique biology are given by Allen (1939), Brosset (1966), Slaughter & Walton (1970) and Wimsatt (1970). Bats are the only mammals with the capacity for powered flight, and the sub-order Microchiroptera, together with the genus *Rousettus* of the sub-order Megachiroptera, have evolved a system of echo-location, by means of which they find their food (Griffin, 1958). In addition, those species living in temperate latitudes are heterothermic and are consequently able to adapt to food shortage during winter by hibernating, and their reproductive biology has become modified as a result (Wimsatt, 1969). The most notable reproductive adaptation is the delay in fertilization: mating generally takes place in autumn and spermatozoa are stored in the female reproductive tract until ovulation occurs in the following spring (Racey, 1975).

Many species of bats have been maintained in captivity for studies of echo-location (Möhres, 1951), thermo-regulation (Stones & Wiebers, 1965a), menstruation (Rasweiler, 1973) and the epidemiology of rabies and other diseases (Tesh & Arata, 1967). In addition, the wing membrane is ideal for studies of blood circulation (Wiedeman, 1957) and wound healing (Church & Warren, 1968). Selection of a species for research investigations now requires great care, as several European species are becoming increasingly rare (Racey & Stebbings, 1972).

As it is impossible to deal individually with all species maintained in captivity, these are listed in Table I.

Husbandry

General environmental conditions

Megachiroptera are tropical and when transported to temperate latitudes must be kept at

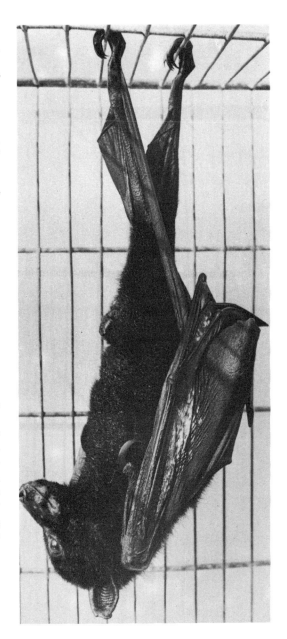

Fig. 26.1. The Indian fruit bat *Pteropus gigantens* (700 g.) is the largest species to be maintained in captivity. (Photograph: UFAW.)

TABLE I

Bat species maintained in captivity

Natural diet	Taxonomic name	Author	Date
Fruit	Cynopterus brachyotis	Lucas (Ed.)	1969
		Marshall & Lim	1968
	C. horsfeldi	,, ,,	1968
	C. sphinx	A. Novick	pers. comm.
	Ptenochirus jagori	,, ,,	,, ,,
	Rousettus amplexicaudatus	Marshall & Lim	1968
	R. aegyptiacus	Kulzer	1958, 1969a
	R. seminudus	A. Novick	pers. comm.
	Myonycteris torquata	J. D. Pye & A. Pye	,, ,,
	Pteropus hypomelanus	Marshall & Lim	1968
	P. vampyrus	,, ,, ,,	,,
	P. melanotus	Jarvis (Ed.)	1966
	P. giganteus	Kulzer	1963
		Lucas (Ed.)	1969
	Epomophorus wahlbergi	Kulzer	1959
	Epomops buettikoferi	Jarvis & Morris (Eds.)	1963
	E. franqueti	Lucas (Ed.)	1969
	Hypsignathus monstrosus	J. D. Pye & A. Pye	pers. comm.
	Eidolon helvum	Kulzer	1969b
	E. anurus	J. D. Pye & A. Pye	pers. comm.
	E. gambiana	,, ,, ,,	,, ,,
	E. wahlbergi	,, ,, ,,	,, ,,
		Rasweiler & Ishiyama	1973
	Phyllostomum discolor	J. D. Pye & A. Pye	pers. comm.
	Monophyllus robusta	A. Novick	,, ,,
	Leptonycteris nivalis	,, ,,	,, ,,
	L. sanborni	,, ,,	,, ,,
	Carollia perspicillata	Arata & Jones	1967
		Rasweiler & DeBonilla	1972
	Sturnina lilium	Rasweiler & Ishiyama	1973
	Artibeus lituratus	Rasweiler & Ishiyama	,,
	A. jamaicensis	Jarvis (Ed.)	1968
		A. Novick	1960, 1963
Pollen, nectar and fruit juices	Nanonycteris veldkampi	Buckland-Wright & Pye	1973
	Micropteropus pusillus	,, ,, ,, ,,	,, ,,
	Eonycteris spelaea	Marshall & Lim	1968
	E. robusta	A. Novick	pers. comm.
	Megaloglossus woermanni	J. D. Pye & A. Pye	,, ,,
	Glossophaga soricina	Rasweiler	1973
	Lonchophylla robusta	A. Novick	pers. comm.
	Anoura caudifera	Rasweiler & DeBonilla	1972
Blood	Desmodus rotundus	Dickson & Green	1970
		Ditmars & Greenhall	1934
		Greenhall	1965
		Trapido	1946
		Wimsatt & Guerrière	1961
	Diaemus youngi	J. M. Dickson & D. G. Green	pers. comm.
Fish	Noctilio leporinus	H. B. House	pers. comm.
		Lucas (Ed.)	1969
	Pisonyx vivesi	Carpenter	1968
		Orr	1954
Flesh	Megaderma lyra	E. Kulzer	pers. comm.
	Macroderma gigas		,, ,,
	Lavia frons	J. D. Pye & A. Pye	,, ,,
	Cardioderma (= Megaderma) cor	E. Kulzer	,, ,,
	Phylloderma stenops	Pye	1967
	Vampyrum spectrum	J. D. Pye & A. Pye	pers. comm.
Insects	Rhinopoma hardwickei	Kulzer	1966b
		Racey	unpublished
		J. D. Pye & A. Pye	pers. comm.
	R. microphyllum	Kulzer	1966b
	Noctilio labialis	A. Novick	pers. comm.
	Rhinolophus ferrumequinum	De Coursey & De Coursey	1964

Table I
(continued)

	Möhres	1951
R. hipposideros		
	J. D. Pye & A. Pye	pers. comm.
Hipposideros commersoni	A. Novick	,, ,,
Chilonycteris (5 species)	,, ,,	,, ,,
Pteronotus (both species)	,, ,,	,, ,,
Macrotus (all 3 species)	,, ,,	,, ,,
Myotis myotis	Möhres	1951
	J. D. Pye & A. Pye	pers. comm.
M. lucifugus	Gates	1938
	Mohos	1961
	Orr	1958
M. mystacinus	J. D. Pye & A. Pye	pers. comm.
M. nigricans	,, ,, ,, ,,	,, ,,
M. albescens	,, ,, ,, ,,	,, ,,
M. bechsteini	Möhres	1951
M. grisescens	Mohos	1961
M. austroriparius	,,	,,
M. yumanensis	Orr	1958
	Ramage	1947
M. thysanodes	Orr	1958
	Ramage	1947
M. sodalis	Gates	1936
	Orr	1958
	Ramage	1947
M. californicus	Orr	1958
M. subulatus leibii	Gates	1936
M. nattereri	Racey	1970
Pipistrellus pipistrellus	Möhres	1951
	Racey	1970
	Racey & Kleiman	1970
P. subflavus	Gates	1936
	Mohos	1961
Nyctalus noctula	Möhres	1951
	Racey	1970
	Skreb & Gjulic	1955
N. leisleri	J. D. Pye & A. Pye	pers. comm.
	R. E. Stebbings	,, ,,
Eptesicus fuscus	Davis & Luckens	1966
	Gates	1938
	Mohos	1961
	Orr	1958
	Ramage	1947
E. serotinus	Racey	1970
Tylonycteris pachypus	Marshall & Lim	1968
T. robustula	,, ,, ,,	,,
Scotophilus nigrita	Kulzer	1959
	J. D. Pye & A. Pye	pers. comm.
S. leucogaster	,, ,, ,, ,,	,, ,,
Lasiurus borealis	Gates	1936
L. seminola	,,	,,
L. cinereus	,,	1938
	Orr	1958
Barbastella barbastellus	Möhres	1951
Plecotus auritus	Racey & Kleiman	1970
	Racey	1970
Corynorhinus macrotus	Gates	1936
C. rafinesquei	Orr	1958
	Pearson, Koford & Pearson	1952
Euderma maculata	A. Novick	pers. comm.
Miniopterus medius	Marshall & Lim	1968
M. schreibersii	,, ,, ,,	,,
Antrozous pallidus	Orr	1958
		1954
Cheiromeles torquatus	A. Watson	pers. comm.
	A. Novick	,, ,,
Tadarida brasiliensis mexicana	Constantine	1952

TABLE I
(*continued*)

		Krutzsch & Sulkin	1958
		Tesh & Arata	1967
	T. limbata	Kulzer	1959
	T. angolensis	,,	,,
	T. condylura	,,	1962
	T. molossa	A. Novick	pers. comm.
	Molossus nigricans	,, ,,	,, ,,
	M. fortis	,, ,,	,, ,,
	M. milleri	,, ,,	,, ,,
Omnivorous	*Phyllostomus hastatus*	Rasweiler	1975

Arranged within each dietary grouping according to Morris, 1965. Records from the International Zoo Yearbook (Eds. Jarvis & Morris; Jarvis; and Lucas) refer to breeding in captivity.

temperatures above 22°C. They are homoeothermic and shiver when cold (Stones & Wiebers, 1965a). Microchiroptera of temperate latitudes are heterothermic and, during both daily sleep and hibernation, peripheral body temperature falls. Torpid bats can, however, be awakened by handling and exposure to warmth, but require several minutes of thermogenesis before they are fully active. They can be kept in a wide range of environments from thermoneutral temperatures of 33°C (Stones & Wiebers, 1965b) to those of a domestic refrigerator. For reproductive studies of hibernating bats, especially if these encompass all stages of the annual cycle, it is essential that the natural environment is simulated as closely as possible (Racey, 1974). Controlled environment rooms are ideal for this purpose but are expensive to construct and maintain and are not generally available. Any unheated, poorly-lit room will serve provided that the winter temperature is kept low enough, and this may be

Fig. 26.3. Vampire bats in cage, seen from below. (Photograph: UFAW.)

Fig. 26.2. The insectivorous pipistrelle *Pipistrellus pipistrellus* (5g.) is the commonest British bat. (Photograph: Racey.)

achieved by means of a ventilation fan set in an exterior wall (Racey, 1970). Increased environmental temperature during the second half of winter results in premature ovulation and pregnancy.

The effect of light on bats has been little investigated, although De Coursey & De Coursey (1964) showed that periodicity of light is an important factor in regulating daily cycles of activity. Skreb (1954) suggested that light was

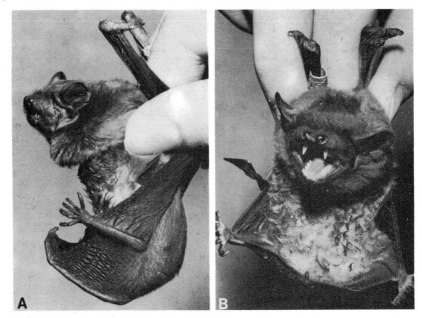

Fig. 26.4. The noctule *Nyctalus noctula* (25g.) is the only hibernating bat to breed regularly in captivity. Note method of handling and the forearm ring used for identification. A Side-on view. B Head-on view. (Photographs: Racey.)

a stimulus for ovulation in *Nyctalus noctula*, and Erkert (1970) examined the effect of light on the activity of *Rousettus* and *Eidolon*.

Little is known about the optimum humidity for bats, but they should not be kept in dry atmospheres as this appears to be deleterious to the wing membranes. A relative humidity of *ca.* 70 per cent appears to be satisfactory for many species. Many temperate Microchiroptera hibernate in caves where the relative humidity is often 100 per cent, and Pirlot (1946) showed that relative humidity below *ca.* 75 per cent was fatal to hibernating *Myotis* and *Rhinolophus*. A saturated atmosphere should, therefore, be provided when these bats are kept in artificial hibernacula.

Caging

Galvanized iron mesh cages with sawdust trays

5cms.

Fig. 26.5. Cage lined with grooved plywood used by the author for housing hibernating bats. The perspex fronts have been removed. (Photograph: Racey, reproduced by permission of *Laboratory Animals*).

have been used for the larger Megachiroptera. For Microchiroptera, several different designs, consisting basically of wooden boxes with perspex fronts, have been tried, but these are difficult to clean and cannot be recommended for long-term husbandry. In our laboratory, cages adapted from Jewell's (1964) design for a small mammal cage are used for hibernating vespertilionids (Racey, 1970) (Fig. 26.5). Wimsatt, Guerriere & Horst (1973) designed an aluminium cage for vampire bats (*Desmodus*) and Krutzsch & Sulkin (1958) also describe a cage to house infected bats. A. M. Greenhall (personal communication) housed vampires in large glass jars and Dickson & Green (1970) maintained their colony of these bats in polypropylene rat cages arranged in a battery with great economy of space.

So long as it is appreciated that small bats, like other small mammals, have a remarkable ability to squeeze through cracks, the type of cage used will depend on the work involved. Thus, for a study of ectoparasites, Marshall & Lim (1968) maintained *Tylonycteris* in glass jars of about the same size and shape as the bamboo internodes in which these bats normally live, and Tesh & Arata (1967) used disposable cardboard containers when studying virus pathogenicity in *Tadarida braziliensis*.

Handling

Several viruses pathogenic to man and other mammals have been isolated from bats (Sulkin, 1962), especially from those living in the tropics, and Garnham (1966) has reviewed the Haemosporidia found in bats. Great care should be taken when handling any bats until appropriate clearance tests have been carried out. Tissues from many vespertilionids which have died or have been killed for experimental purposes in the author's laboratory have been tested by intracerebral inoculation of baby mice. No virus has yet been detected (J. Boorman, personal communication).

Bite-proof gloves should be used when handling bats which may be infected and long-handled forceps may also be used to pick up small and medium-sized species. Some procedures do, however, require manipulation of the bats with bare hands and a detailed description of a suitable method is given by Racey (1970) (Fig. 26.4).

Marking

Numbered forearm rings are now available in sizes of 2·5, 3 and 4 mm closed internal diameter from the Mammal Society, Harvest House, 62, London Road, Reading, RG1 5AS. By means of specially shaped pliers or digital pressure a ring is applied to the forearm of the bat, loosely enough to allow it to slip along the distal half (Fig. 26.4). These rings are used routinely for identifying individual bats and, if applied with care, cause little irritation to the wing membrane. Neonates, whose arms are too small to take rings, can be punch-marked in the wing membrane (Racey, 1970) and this method has also been applied to adults (Bonaccorso & Smythe, 1972).

Exercise

In our laboratory, where large numbers of vespertilionids have been maintained for long periods, it has not been possible to provide facilities for flight. The consequent lack of exercise may have been responsible for the occasional incidence of sore and chronically infected wrist joints, and wing lesions associated with the presence of *Pseudomonas aeruginosa*, together with signs of pectoral muscle degeneration at *post-mortem* examination. One of the largest bat species *Pteropus giganteus*, which has a wing span in excess of 1 metre, can be kept healthy in a cage of about 1 m³. Nectar-feeding bats, such as *Glossophaga*, can exercise adequately within quite small cages, as they are able to hover. All bats should, however, be given enough space to spread their wings to ensure that they can perform all grooming operations adequately.

Rhinolophids deprived of exercise develop swollen wrist joints in a few days and make frantic attempts to escape from small cages, often causing themselves injury. Möhres (1951) used a cage 70 × 60 × 50 cm for greater horseshoe bats (*Rhinolophus ferrumequinum*) and several species of vespertilionids. He considers that it is not possible to keep these temperate Microchiroptera permanently confined and allows them free flight each night. When this is done it is important, in order that the bats can return to their cages, that the topography of the environment is not changed. In familiar surroundings they seem to rely on memory rather than on echolocation and will collide with any new objects in their path.

Feeding

Insects and substitutes

All temperate and many tropical Microchiroptera are insectivorous and the majority of species kept in captivity are included in this category (see Table I). There have been some comprehensive studies of the insect species comprising the natural diet (Chapman, 1958; Gould, 1955; Poulton, 1929; and Ross, 1967), but, in captivity, most species will accept a diet of mealworms* (*Tenebrio molitor* larvae). Pupae and adults of this insect are also taken, although attention has been drawn to the level of toxic quinones in mealworm imagos (Ladisch *et al.*, 1967).

Bats will also eat the larvae (gentles) and pupae of Calliphoridae* and these are a cheaper, if less preferred, source of food. The larger species will also take the early instars of many Orthoptera—the most commonly cultured of which are locusts* and cockroaches*. In the tropics, grasshoppers and locusts are often so plentiful that they can easily be caught in sufficient numbers to feed to bats (J. D. Pye & A. Pye, personal communication).

In the wild, bats usually catch their insect prey in free flight. In captivity, where food is provided in dishes, they require training before they will feed. Weanling bats, born in captivity, also require such training. The procedure has been explained several times in the literature (see, for example, Racey, 1970) but it should be emphasized that this is the most crucial and time-consuming stage in the adaptation of insectivorous bats to the laboratory. The speed with which it is achieved can be increased by introducing a bat, head-first, through the open end of a 50 ml plastic syringe. Mealworms are then offered through a hole cut in the delivery end. Several such syringes can be used simultaneously (Tesh & Arata, 1967). Nellis (1969) found that placing bats in a container of squirming mealworms was sufficient to cause them to snap, and this procedure is followed in

* For culture methods see Chapters 44, 47, 48 and 50 of this Handbook—Ed.

our laboratory after feeding has been initiated by hand. The success of training depends largely on the skill and efforts of the trainer: occasional individuals require little training but others require several sessions. Very few individuals of the five vespertilionid species maintained in our laboratory have been refractive to training. Rhinolophids and other groups which hang free (for example, the Megachiroptera and the Hipposideridae of the Microchiroptera) will not descend to a dish on the cage floor and cannot, therefore, be trained so easily. Many of them must be fed by hand, although some will learn to eat from an elevated dish.

Although mealworms are the most convenient food for insectivorous bats they are expensive and several authors (Gates, 1936 and 1938; Krutzsch & Sulkin, 1958; Mohos, 1961; Racey, 1970) have used compound diets. These consist basically of cottage cheese, banana, hard-boiled egg and vitamins mixed in a blender to a firm crumbly consistency. Mealworms can be added, as chitin appears to be necessary for alimentary health. Davis & Luckens (1966) incorporated canned dog-food into their mixture.

Quantitative measurements of food requirements have been obtained by Gates (1936), Mathias & Seguela (1940), Ramage (1947), Orr (1954) and Coutts *et al.* (1973). Stones & Wiebers (1965b) report that *Myotis lucifugus* maintained at the thermoneutral temperature of 33°C require only about a third of the food of those maintained at room temperature. In captivity, there is a tendency for bats to overeat and this, combined with lack of exercise, can result in obesity unless intake is controlled. Noctules and serotines in our laboratory have been weighed weekly and their food regulated in order to follow the annual cycle of loss and gain (Kleiman & Racey, 1970). If laboratory colonies are fed at a regular time they will modify their cycle of daily sleep accordingly, although individuals occasionally arouse after their cage mates have eaten all the food. With careful grouping, however, this can be avoided.

Fruit

Most Megachiroptera and some tropical Microchiroptera are frugivorous. Other Microchiroptera normally eat both insects and fruit, and some are partly carnivorous or nectivorous. The laboratory maintenance of frugivorous bats has been thoroughly reviewed by Rasweiler (1975 and in press). Greenhall (1957) lists the natural food preference of some neotropical fruit-eating bats. In captivity, soft sweet fruits such as bananas, mangoes, peaches, plums, melons, grapes and papayas are preferred. Apples are also eaten and sweet oranges may be accepted, although citrus fruits are not usually favoured. In general, if a variety of fruit is fed, it should be diced and served well mixed—but, even then, some species will sort out their preferences. Buckland-Wright & Pye (1973) revealed the deficiencies of a diet of fruit alone and illustrated the importance of additives such as Complan.*

Nectar, Pollen and Fruit Juices

Members of the Glossophaginae (a Central American subfamily of the Phyllostomatidae) and some Macroglossinae (a subfamily of Megachiroptera) feed on the nectar and pollen of night-flowering trees and shrubs and on the juices of overripe fruit although they will also take insects. They have a hovering flight and very long protrusible tongues, and are the nocturnal equivalent of humming birds. Several species of nectarivorous bats have been successfully maintained and bred in captivity by Rasweiler (Table I) who has also reviewed the methods and problems involved (Rasweiler, 1975; in press). Very small Megachiroptera, such as *Megaloglossus* and *Nanonycteris*, require a fluid pulpy diet (Pye, 1967). Although *Eonycteris* can be maintained on a diet of fruit, it has also been kept entirely on a liquid consisting of Complan, sugar solution and water (Marshall & Lim, 1968).

Blood

Ditmars & Greenhall (1934) were the first to draw attention to the adaptability of the common vampire *Desmodus rotundus* to the laboratory, and Greenhall (1965) has subsequently given an account of its behaviour in captivity. Trapido (1946) reported that this species had been maintained for up to twelve years in captivity, and laboratory husbandry has been described by Wimsatt & Guerrière (1961). Dickson & Green (1970) have recently described improved methods of care and handling of a large colony of vampires for studies on the fibrinolytic activity

* Glaxo Laboratories.

of their saliva (Hawkey, 1966). They are fed on citrated bovine blood which is stored in a deep freeze until needed and dispensed from drinking tubes. Defibrination has commonly been used to preserve blood in the past, but this method results in loss of volume. *Diaemus* (a vampire which feeds naturally on avian blood) has also been maintained by these authors (personal communication) on the same diet as *Desmodus*, although a weekly supplement of chicken blood has been supplied.

Flesh

Some Microchiroptera (such as *Vampyrum spectrum* and *Megaderma lyra*) are wholly carnivorous and others (such as *Phyllostomus hastatus*) have catholic tastes which include rodents and small bats, although this species can survive in captivity on a diet of bananas (J. D. Pye, personal communication). *Trachops* is reported to hunt and capture geckos and the closely related *Phylloderma* will accept these reptiles in captivity (Pye, 1969).

E. Kulzer & G. Neuweiler (personal communication) have kept most species of Megadermatidae in captivity. Of these, *M. lyra* has proved to be particularly adaptable to laboratory conditions.

Fish

The neotropical *Noctilio*, *Pisonyx* of Baja California, and the asiatic *Myotis macrotarsus*, fly low over water and catch fish with their enlarged hind feet. Griffin has maintained *Noctilio leporinus* in captivity (Pye, 1967) and this species has bred in New York's Bronx Zoo (H. B. House, personal communication). Orr (1958) maintained *Pisonyx vivesi* in captivity for over 3 years, and Carpenter (1968) has recently reported that a diet of frozen shrimps maintained this species in better health than did tinned fish (cat-food). *M. macrotarsus* has been kept in captivity on a diet of crayfish, fish and insects by Novick (1968) who considered that it was potentially a good laboratory animal.

Water

All bats, including nectar feeders, should have water available *ad libitum* although some species, like the tropical *Rhinopoma hardwickei* are able to live without drinking (Vogel, 1969). In general, bats will use drinking tubes and some will drink from the nozzles of inverted water bottles. Vespertilionids will also lap water from saturated cotton wool and this is a useful way of providing water when it is liable to spill, as during transport.

Vitamins and Minerals

It is impossible to provide bats in captivity with all the foods they eat in the wild and attempts should be made to replace the natural sources of vitamins and minerals with substitutes. The health of bats whose diets are so supplemented is generally better than that of those which do not receive them. Workers in USA, Britain and Germany have followed Orr (1958) in using Stuart Formula liquid.* Virol† is a similar preparation and both can be dissolved in drinking water. Alternatively, Vionate powder ‡ can be sprinkled over the diet. Mealworm diet (usually bran) can itself be supplemented with vegetables such as cabbage, with white flour or bread (which, in Britain, contains added calcium), or even with vitamin powder, in order to increase the nutrient value of the mealworms. The importance of supplementing the diet of fruit bats with calcium has been illustrated by Buckland-Wright & Pye (1973).

Breeding

The different types of reproductive cycles are concisely reviewed by Ramaswamy (1961), Anciaux de Faveaux (1973) and Wilson (1973). The majority of species are monoestrous and have a single young so that large laboratory colonies cannot be bred from a few founders. Field studies have revealed that males of some species have harems, e.g. *Nyctalus noctula* (Sluiter & Van Heerdt, 1966), *Phyllostomus hastatus, P. discolor*, and *Saccopteryx bilineata* (Bradbury, in press). This fact should be taken into account when mating groups are constituted.

Records published in the International Zoo Yearbook show that several species of fruit-eating *Megachiroptera* have bred successfully in captivity, notably *Pteropus giganteus, Cynopterus brachyotis* and *Rousettus aegyptiacus*. Kulzer (1958 and 1969a) has given a well-illustrated account of *R. aegyptiacus* breeding in his laboratory. Rasweiler (1973; 1975) reports the breeding in cap-

* Stuart Co., Pasadena, California.
† Virol Ltd., Enfield, Middlesex.
‡ E. R. Squibbs, Twickenham, Middlesex.

tivity of several species of nectarivorous and frugivorous phyllostomatids and Buckland-Wright & Pye (1973) report breeding the nectivore *Nanonycteris veldkampi*.

There are many accounts of bats giving birth in captivity (Kulzer, 1962; 1966a; Wimsatt, 1960), and Kleiman (1969) has described the maternal care, growth-rate and development of three such species of vespertilionids. Kleiman & Racey (1970) reported on the first colony of hibernating bats (*Nyctalus noctula*) to breed in captivity and these authors have also reviewed the reproductive activity of four vespertilionid species maintained in their laboratory (Racey & Kleiman, 1970). These bats normally mate in autumn and sperm is stored in the female reproductive tract until ovulation and fertilization in spring (Racey, 1975). Ovulation can be induced by the administration of gonadotrophins to animals aroused in the first half of winter, or simply by premature arousal from hibernation in the second half of winter (Racey, 1976).

Dickson & Green (1970), Greenhall (1965) and Trapido (1946) have all reported that the common vampire *Desmodus rotundus* will breed in captivity. The carnivorous *Megaderma lyra* has bred in laboratory conditions (G. Neuweiler, personal communication), as has the piscivorous *Noctilio* in New York's Bronx Zoo (H. B. House, personal communication). Both workers, however, report difficulties in weaning the young. In addition, Buckland-Wright & Pye (1973) report the nectivore *Nanonycteris veldkampi* breeding in captivity. Gestation periods of several species are recorded by Racey (1973), who, however, as a result of increasing the period of gestation by inducing torpor in pregnant female pipistrelles, doubts whether the concept of a standard gestation period for heterothermic bats is a valid one. Huibregtse (1966) has analysed the milk of *Leptonycteris sanborni* and *Tadarida brasiliensis* and Adams & Baer (1966) have described an artificial feeding device for suckling large numbers of bats delivered by caesarian section, although this procedure met with limited success. Better results were achieved by Taylor *et al.* (1974) in rearing one-week-old bats using a stomach catheter.

Adaptation to captivity

Comparison of the adaptability of different

groups of bats to captivity is difficult, because success often depends on the efforts of the investigators rather than on the characteristics of the species concerned. Nevertheless, some generalizations can be made. More difficulty seems to be encountered with the insectivorous than with the frugivorous species. Novick (1968) found that the Molossidae refused food entirely and behaved lethargically in captivity and J. D. Pye & A. Pye (personal communication) were likewise unable to coax any of several hundred specimens of nine species to feed. *Tadarida* is, however, an exception, for several authors have reported some success with bats of this genus (Table I). The Rhinopomatidae are also difficult and, in the author's laboratory, *Rhinopoma hardwickei* has required hand-feeding. Emballonuridae, Nycteridae and Rhinolophidae are difficult to feed. The latter do not thrive under the restrictions of captivity and are now a conservation problem in many parts of Europe (Racey & Stebbings, 1972). Many of the Hipposideridae are even more delicate, an exception being the large and hardy *Hipposideros commersoni* maintained by A. Novick (personal communication). The chilonycterids may live with patient feeding and several species have been kept by Novick for up to a year. Novick (1963) found the phyllostomatids difficult, mainly because they succumb so easily to chilling.

Many species of vespertilionids have been successfully maintained in the laboratory and Orr (1958) and Racey (1970) give records of longevity in captivity—the former reporting that *Antrozous pallidus* lived for over eight years in his care. The common vampire has also adapted very well to captivity, living for as long as twelve years (Trapido, 1946). J. Simmons (personal communication) found that for behavioural work involving rigorous conditioning methods *Eptesicus* and *Phyllostomus* were most responsive.

Laboratory procedures

In Pye's laboratory, pentobarbitone sodium* is used (1 vol. of commercial solution to 9 vols. of 10 per cent ethanol) intraperitoneally to induce surgical anaesthesia at a dose of 30–50 mg/kg body-weight. It is important that the bat should be fully aroused before injection and a constant

* Nembutal: Abbott Laboratories Ltd.

body temperature (37–40°C) must be maintained afterwards to ensure successful anaesthesia and recovery. Wiedeman (1957) and Huibregtse (1966) also used pentobarbitone sodium, the latter following it with oxytocin injections to aid the manual expression of milk. However, Neuweiler (1970) reports that it lowers the sensitivity of the colliculus inferior to ultrasound, and Mohos (1961) prefers ether to pentobarbitone sodium for anaesthesia.

Dickson & Green (1970) introduce a mixture of halothane, nitrous oxide and oxygen into their vampire cages to induce anaesthesia. The bats are then placed in perspex stalls which support and restrain them, so that pilocarpine can be administered to the buccal mucosa and the resulting flow of saliva collected. The present author prefers a halothane/oxygen mixture for the induction and maintenance of anaesthesia in vespertilionids.

Kleiman & Racey (1970) routinely took vaginal smears from noctules and serotines with a blunt probe (dissecting seeker), and Kanthor (1965) has described the use of an indwelling catheter for the collection of urine from small bats. Considerable mortality resulted when blood was withdrawn by cardiac puncture from *Myotis lucifugus* and *Pipistrellus subflavus* although 0·1 ml could be withdrawn successfully from the larger *Eptesicus fuscus* (La Motte, 1958). Baer (1966) recommends that for the best results blood should be taken from the orbital sinus. (For a description of this method in mice see Chap. 17, page 207).

Diseases and parasites

Hazards to man and other animals

Tesh & Arata (1967) list some of the microorganisms pathogenic for man and domestic animals which have been isolated from or identified serologically in bats, and it is obvious that until appropriate screening has taken place great care must be exercised in the protection of all personnel coming into contact with any species. One of the most serious of these diseases is rabies, which is carried not only by vampires but by several other species of New and Old World Microchiroptera and can also be transmitted by air in bat caves (Constantine, 1967). Pre-exposure immunization with duck embryo vaccine is, however, now available.

Ectoparasites

Bats carry many different ectoparasites: fleas, ticks, mites, bugs and both streblid and wingless nycteribiid flies. These can be removed by dusting with pyrethrum powder. There is evidence that DDT is very toxic to bats (Luckens & Davis, 1964; Jefferies, 1972) and this compound should not, therefore, be used for dusting. In general, ectoparasites decrease in numbers with the length of time that the bats spend in captivity, and often disappear altogether. For this reason dusting has not been adopted as routine practice in this laboratory. Ectoparasites seem to have caused little harm, the exception being an infection of the mite *Notoedres chiropteralis* (Trouessart) (identified by Professor A. Fain, Antwerp) which caused debility in a noctule. Some investigators have, however, preferred to eliminate any possibility of the transmission of arthropod-borne disease by routine dusting. Little attention has been paid to the damage caused by bat ectoparasites to their hosts and the work of Lavoipierre & Rajamanickam (1968) is of interest in this respect.

Pathology

There is little published information on pathology relevant to the husbandry of bats. Kleiman & Racey (1970) note that mortality in their noctule colony was highest in autumn and that relatively few bats died during the hibernation period. Racey (1970) has subsequently reported that 35 per cent of noctules dying in captivity succumbed to pneumonia of unknown aetiology. The cause is however speculative—lack of exercise being a possibility.

REFERENCES

Adams, D. B. & Baer, G. M. (1966). Caesarian section and artificial feeding device for suckling bats. *J. Mammal.,* **47,** 524–5.

Allen, G. M. (1939). *Bats.* Cambridge, Mass.: Harvard. Republished 1962. New York: Dover.

Anciaux de Faveaux, M. (1973). Essai de synthèse sur la réproduction de chiroptères d'Afrique (Region faunistique ethiopienne). *Period. biol.,* **75,** 195–199.

Arata, A. A. & Jones, C. (1967). Homeothermy in *Carollia* (Phyllostomatidae: Chiroptera) and the adaptation of poikilothermy in insectivorous northern bats. *Lozania,* **14,** 1–11.

Baer, G. M. (1966). A method for bleeding small bats. *J. Mammal.* **47,** 340–1.

Bonaccorso, F. J. & Smythe, N. (1972). Punch-marking bats: an alternative to banding. *J. Mammal.,* **53,** 389–390.

Bradbury, J. W. & Emmons, L. H. (1974). Social organization of some Trinidad bats 1. Emballonuridae. *Z. Tierpsychol.,* **36,** 137–183.

Brosset, A. (1960). *La Biologie des Chiroptères.* Paris: Masson.

Buckland-Wright, J. C. & Pye, J. D. (1973). Dietary deficiency in fruit bats. *Int. Zoo Yb.,* **13,** 271–277.

Carpenter, R. E. (1968). Salt and water metabolism in the marine fish eating bat, *Pizonyx vivesi. Comp. Biochem. Physiol.,* **24,** 951–64.

Chapman, R. F. (1958). Some observations on the food of bats. *Ann. Mag. nat. Hist.,* **13,** 188–92.

Church, J. C. T. & Warren, D. J. (1968). Wound healing in the web membrane of the fruit bat. *Br. J. Surg.,* **55,** 26–31.

Constantine, D. G. (1952). A program for maintaining the freetail bat in captivity. *J. Mammal.,* **33,** 395–7.

Constantine, D. G. (1967). Rabies transmission by air in bat caves. *Publ. Hlth Serv. Publs, Wash.* No. 1617.

Coutts, R. A., Fenton, M. B. & Glen, E. (1973). Food intake by captive *Myotis lucifugus* and *Eptesicus fuscus* (Chiroptera: Vespertilionidae). *J. Mammal.,* **54,** 985–990.

Davis, W. H. & Luckens, M. M. (1966). Use of big brown bats (*Eptesicus fuscus*) in biomedical research. *Lab. Anim. Care,* **16,** 224–7.

De Coursey, G. & De Coursey, P. G. (1964). Adaptive aspects of activity rhythms in bats. *Biol. Bull.,* **126,** 14–27.

Dickson, Janet M. & Green, D. G. (1970). The vampire bat (*Desmodus rotundus*). Improved methods of laboratory care and handling. *Lab. Anim.,* **4,** 37–44.

Ditmars, R. L. & Greenhall, A. M. (1934). The vampire bat: a presentation of undescribed habits and review of its history. *Zoologica, N.Y.,* **19,** 53–76.

Erkert, S. (1970). Der Einfluss des Lichtes auf die Aktivität von Flughunden (Megachiroptera). *Z. vergl. Physiol.,* **67,** 243–72.

Garnham, P. C. C. (1966). *Malaria Parasites and other Haemosporidia.* Oxford: Blackwell.

Gates, W. H. (1936). Keeping bats in captivity. *J. Mammal.,* **17,** 268–73.

Gates, W. H. (1938). Raising the young of red bats on an artificial diet. *J. Mammal.,* **19,** 461–4.

Gould, E. (1955). The feeding efficiency of insectivorous bats. *J. Mammal.,* **36,** 399–407.

Greenhall, A. M. (1957). Food preference of Trinidad fruit bats. *J. Mammal.,* **38,** 409–10.

Greenhall, A. M. (1965). Notes on behaviour of captive vampire bats. *Mammalia*, **29,** 441–51.

Griffin, D. R. (1958). *Listening in the dark.* New Haven: Yale University Press.

Hawkey, C. (1966). Plasminogen activator in saliva of the vampire bat *Desmodus rotundus. Nature, Lond.,* **211,** 434–5.

Howell, D. J. (1974). Bats and pollen: physiological aspects of the syndrome of Chiropterophily. *Comp. Biochem. Physiol.,* **48A,** 263–276.

Huibregtse, W. H. (1966). Some chemical and physical properties of bat milk. *J. Mammal.,* **47,** 551–4.

Jefferies, D. J. (1972). Organochlorine insecticide residues in British bats and their significance. *J. Zool., Lond.,* **166,** 245–263.

Jewell, P. A. (1964). An observation and breeding cage for small mammals. *Proc. zool. Soc. Lond.,* **143,** 363–4.

Jarvis, C. (1965–8). *International Zoo Yearbook.* vols 5–8.

Jarvis, C. & Morris, D. J. (1960–3). *International Zoo Yearbook,* vols 1–4.

Kanthor, H. A. (1965). Indwelling catheters for small bats. *J. appl. Physiol.,* **20,** 326–7.

Kleiman, Devra G. (1969). Maternal care, growth rate and development in the noctule (*Nyctalus noctula*), pipistrelle (*Pipistrellus pipistrellus*) and serotine (*Eptesicus serotinus*) bats. *J. zool., Lond.,* **157,** 187–211.

Kleiman, Devra G. & Racey, P. A. (1969). Observations on noctule bats (*Nyctalus noctula*) breeding in captivity. *Lynx,* **10,** 65–77.

Krutzsch, P. H. & Sulkin, S. E. (1958). The laboratory care of the Mexican free tailed bat. *J. Mammal.,* **49,** 262–5.

Kulzer, E. (1958). Untersuchungen über die Biologie von Flughunden der Gattung *Rousettus* Gray. *Z. Morph. Okol. Tiere,* **47,** 374–402.

Kulzer, E. (1959). Fledermäuse aus Ostafrika. *Zool. Jb. (Syst.),* **87,** 13–42.

Kulzer, E. (1962). Uber die Jugendentwicklung der Angola-Bulldog-fledermaus *Tadarida* (*Mops*) *condylura* (A. Smith, 1833) (Molossidae). *Saug. Mitt.,* **10,** 116–24.

Kulzer, E. (1963). Die Regelung der Körpertemperatur beim Indischen Riesenflughund. *Natur. Mus., Frankf.,* **93,** 1–11.

Kulzer, E. (1966a). Die Geburt bei Flughunden der Gattung *Rousettus* Gray (Megachiroptera). *Z. Saugetierk.,* **31,** 226–33.

Kulzer, E. (1966b). Thermoregulation bei Wüstenfledermäusen. *Natur. Mus., Frankf.,* **6,** 242–53.

Kulzer, E. (1969a). African fruit-eating cave bats. *Afr. wild Life,* **23,** 39–46; 129–37.

Kulzer, E. (1969b). Das Verhalten von *Eidolon helvum* (Kerr) in Gefangenschaft. *Z. Saugetierk.,* **34,** 129–48.

Ladisch, R. K., Ladisch, S. K. & Howe, P. M. (1967). Quinoid secretions in grain and flour beetles. *Nature, Lond.,* **215,** 939–40.

La Motte, L. C. (1958). Japanese B. encephalitis in bats during simulated hibernation. *Am. J. Hyg.,* **67,** 101–8.

Lavoipierre, M. M. J. & Rajamanickam, C. (1968). The skin reactions of two species of South-east Asia Chiroptera to notoedrid and teinocoptid mites. *Parasitology,* **58,** 515–30.

Lucas, J. (1969). *International Zoo Yearbook,* vol. 9.

Luckens, M. M. & Davis, W. H. (1964). Bats: sensitivity to DDT. *Science,* **146,** 948.

Marshall, A. G. & Lim, B. L. (1968). Observations on keeping some Malaysian bats in captivity. *Malayan Nature J.,* **21,** 165–70.

Mathias, P. & Seguela, J. (1940). Contribution à la connaissance de la biologie des chauve-souris. *Mammalia,* **4,** 15–19.

Mohos, S. C. (1961). Bats as laboratory animals. *Anat. Rec.,* **139,** 369–78.

Möhres, F. P. (1951). Uber Haltung und Pflege von Fledermäusen. *Zool. Gart. Lpz.,* **18,** 217–27.

Morris, D. (1965). *The Mammals.* London: Hodder & Stoughton.

Nellis, D. W. (1969). Acclimating bats to captivity. *J. Mammal.,* **50,** 370.

Neuweiler, G. (1970). Neurophysiologische Untersuchungen zum Echoortungssystem der Grossen Hufeisennase *Rhinolophus ferrumequinum* Schreber, 1774. *Z. vergl. Physiol.,* **67,** 273–306.

Novick, A. (1960). Successful breeding in captive *Artibeus. J. Mammal.,* **41,** 508–9.

Novick, A. (1963). Orientation in neotropical bats. II. Phyllostomatidae and Desmondontidae. *J. Mammal.,* **44,** 44–56.

Novick, A. (1968). Orientation in palaeotropical bats. I. Microchiroptera. *J. exp. Zool.,* **138,** 81–153.

Orr, R. T. (1954). Natural history of the pallid bat *Antrozous pallidus. Proc. Calif. Acad. Sci.,* **28,** 168–246.

Orr, R. T. (1958). Keeping bats in captivity. *J. Mammal.,* **39,** 339–44.

Pearson, O. P., Koford, M. P. & Pearson, A. K. (1952). Reproduction of the lump-nosed bat (*Corynorhinus rafinesquei*) in California. *J. Mammal.,* **33,** 273–320.

Pirlot, P. (1946). *Hibernation of bats: resistance to dessication.* Thesis. Univ. Louvain.

Poulton, E. B. (1929). British insectivorous bats and their prey. *Proc. zool. Soc. Lond.,* **1,** 277–303.

Pye, J. D. (1967). Bats. In *The UFAW handbook on the Care and Management of Laboratory Animals,* 3rd ed. Edinburgh: Livingstone.

Pye, J. D. (1969). The diversity of bats. *Sci. J.,* **5,** 47–52.

Racey, P. A. (1969). Diagnosis of pregnancy and experimental extension of gestation in the pipistrelle bat *Pipistrellus pipistrellus*. *J. Reprod. Fert.*, **19**, 465–74.

Racey, P. A. (1970). The breeding, care and management of vespertilionid bats in the laboratory. *Lab. Anim.*, **4**, 171–83.

Racey, P. A. (1973). Environmental factors affecting the length of gestation in heterothermic bats. *J. Reprod. Fertil., Suppl.*, **19**, 175–189.

Racey, P. A. (1974). The temperature of a pipistrelle hibernaculum. *J. Zool., Lond.*, **173**, 260–262.

Racey, P. A. (1975). The prolonged survival of spermatozoa in bats. In *The Biology of the Male Gamete*, eds. Duckett, J. G. & Racey, P. A. *Biol. J. Linn. Soc.* **7**, *Suppl.* **1**, 385–416.

Racey, P. A. (1976). Induction of ovulation in the pipistrelle bat, *Pipistrellus pipistrellus*. *J. Reprod. Fert.*, **46**.

Racey, P. A. & Kleiman, Devra G. (1970). Maintenance and breeding in captivity of some vespertilionid bats, with special reference to the noctule *Nyctalus noctula. Int. Zoo Yb.*, **10**, 65–70.

Racey, P. A. & Stebbings, R. E. (1972). Bats in Britain—a status report. *Oryx*, **11**, 319–327.

Ramage, M. C. (1947). Notes on keeping bats in captivity. *J. Mammal.*, **28**, 60–2.

Ramaswamy, K. R. (1961). Studies on the sex-cycle of the Indian vampire bat *Megaderma (Lyroderma) lyra lyra* (Geoffroy). *Proc. natn. Inst. Sci. India*, **27**, 286–301.

Rasweiler, J. J. IV. (1973). Care and management of the long-tongued bat, *Glossophaga soricina* (Chiroptera: Phyllostomatidae) in the laboratory, with observations on estivation induced by food deprivation. *J. Mammal.*, **54**, 391–404.

Rasweiler, J. J. IV. (1975). Maintaining and breeding neotropical frugivorous, nectarivorous and pollenivorous bats. *Int. Zoo Yb.*, **15**, 18–30.

Rasweiler, J. J. IV. (in press). The care and management of bats as laboratory animals. In *Biology of Bats* Vol. III, ed. Wimsatt, W. A. New York: Academic Press.

Rasweiler, J. J. IV & DeBonnila, H. (1972). Laboratory maintenance methods for some nectivorous and frugivorous phyllostomatid bats. *Lab. anim. Sci.*, **22**, 658–663.

Rasweiler, J. J. IV & Ishiyama, V. (1973). Maintaining frugivorous phyllostomatid bats in the laboratory: *Phyllostomus, Artibeus* and *Sturnina. Lab. anim. Sci.*, **23**, 56–61.

Ross, A. (1967). Ecological aspects of the food habits of insectivorous bats. *Proc. West. Fdn. Vert. Zool.*, **1**, 205–63.

Skreb, N. (1954). Experimentelle Untersuchungen über die äusseren Ovulationsfaktoren bei der Fledermaus *Nyctalus noctula. Naturwissenschaften*, **41**, 484.

Skreb, N. & Gjulic, B. (1955). Contribution à l'étude des noctules (*Nyctalus noctula* Schreb) en liberté et en captivité. *Mammalia*, **19**, 335–43.

Slaughter, B. H. & Walton, D. W. (Eds.) (1970). *About Bats—a Chiropteran Symposium.* Dallas; Southern Methodist Univ. Press.

Sluiter, J. W. & Van Heerdt, P. F. (1966). Seasonal habits of the noctule bat (*Nyctalus noctula*). *Archs. néerl. Zool.*, **16**, 423–39.

Stones, R. C. & Wiebers, J. E. (1965a). A review of temperature regulation in bats (Chiroptera). *Am. Midl. Nat.*, **74**, 155–67.

Stones, R. C. & Wiebers, J. E. (1965b). Laboratory care of little brown bats at thermal neutrality. *J. Mammal.*, **46**, 681–2.

Sulkin, S. E. (1962). The bat as a reservoir of viruses in nature. *Prog. med. Virol.*, **4**, 157–207.

Taylor, H., Gould, E., Allan, F. & Woolf, N. (1974). Successful hand-raising of one-week-old bats, *Eptesicus* and *Antrozous*, by stomach catheter. *J. Mammal.*, **55**, 228–231.

Tesh, R. B. & Arata, A. A. (1967). Bats as laboratory animals. *Hlth Lab. Sci.*, **4**, 106–12.

Trapido, H. (1946). Observations on the vampire bat with special reference to longevity in captivity. *J. Mammal.*, **27**, 217–9.

Vogel, V. B. (1969). Vergleichende Untersuchungen über den Wasserhaushalt von Fledermäusen (*Rhinopoma, Rhinolophus* und *Myotis*). *Z. vergl. Physiol.*, **64**, 324–45.

Wiedeman, M. P. (1957). Effect of venous flow on frequency of venous vasomotion in the bat wing. *Circulation Res.*, **5**, 641–4.

Wilson, D. E. (1973). Reproduction in neotropical bats. *Period. biol.*, **75**, 215–217.

Wimsatt, W. A. (1960). An analysis of parturition in Chiroptera, including new observations on *Myotis l. lucifugus. J. Mammal.*, **41**, 183–200.

Wimsatt, W. A. (1969). Some interrelations of reproduction and hibernation in mammals. *Symp. Soc. exp. Biol.*, **23**, 511–58.

Wimsatt, W. A. (Ed.) (1970). *The Biology of Bats* Vols. I & II. New York: Academic Press.

Wimsatt, W. A. & Guerriere, A. (1961). Care and maintenance of the common vampire in captivity. *J. Mammal.*, **42**, 449–55.

Wimsatt, W. A., Guerriere, A. & Hirst, R. (1973). An improved cage design for maintaining vampires (*Desmodus*) and other bats for experimental purposes. *J. Mammal.*, **54**, 251–254.

27   The Dog

J. MALCOLM HIME

There are more than 100 breeds of domestic dog (*Canis familiaris*) and the weight range varies between a few kilograms for the toy breeds to 70 kilograms for St. Bernards and Wolfhounds. Temperament and therefore suitability as an experimental animal also varies considerably between individuals, but those from a kennel environment are probably more predictable than cross-breds or mongrels. It should be noted that some strains of highly-bred domestic dog may be of unpredictable temperament. In general, medium to large breeds are often more placid than toy breeds but food consumption is considerably greater. The breeds most commonly used in nutritional and toxicological work where standard body proportions are required, are beagles and corgis. Cross-bred dogs and mongrels are still used largely in experimental surgery.

Uses

Dogs are used experimentally in pharmacological, toxicological, nutritional and surgical investigations. There are, however, some well documented differences in metabolic pathways in the dog compared with those in man and the reader is advised to refer to the published literature. The dog is used in veterinary research for the investigation of diseases and conditions affecting the dog. It is a satisfactory surgical subject and there is extensive veterinary literature available concerning clinical anaesthesia, surgical techniques, normal physiology, anatomy and disease. (See Suggested Reading, p. 329).

When breeding is not intended male dogs may be preferable to females since behaviour will not be disturbed by oestrus.

Persons intending to use dogs in experimental work would be well advised to arrange for veterinary advice and assistance to be available when required, since such advice may reduce suffering and save time and money.

Standard biological data

The rectal temperature of the healthy dog varies between 38°C and 38·7°C, the pulse rates are 62–80 in large breeds and 90–130 in small breeds. Respiratory rate is 16–30 per minute. In interpreting the recordings allowance must be made for fear, excitability and ambient room temperature, for otherwise erroneous conclusions may be drawn. An intermittent pulse is found in many normal dogs.

Further data are given in Tables I–IV.

Husbandry

Housing

The decision whether to house dogs individually, in pairs, or in groups must to some extent be dictated by the space available and the type of experiment for which the animals are being used. There is no doubt that dogs are happier, less noisy, and less destructive if they are housed in pairs or groups. Dogs undergoing surgical investigations should be housed separately, at least during post-operative recovery and wound-healing; this eliminates wound interference by other dogs.

The basic requirements of any dog accommodation are comfortable quarters with adequate ventilation and drainage and reasonable temperature-control within the range 5–15°C. Sound insulation may be necessary.

Ideally the quarters for each dog should be divided into an inside sleeping area and an outside run. The inside area should contain as a bed a place insulated from the ground (see p. 313). If outside runs are provided the dimensions of the inside area should be 1×0.75 m for dogs up to 14 kg body-weight and 1.3×1.0 m for dogs weighing between 14 and 23 kg.

Ottewell (1968) describes in detail a design of accommodation for dogs. Bowden (1970) suggests that granolithic cement over a polystyrene insulation provides an efficient and economical

TABLE I

Normal blood values

		Average
Erythrocytes (millions/cu mm)	5·5–8·5	6·8
Hemoglobin (gm/100 ml)	12·0–18·0	14·9
PCV (cc/100 cc)	37·0–55·0	45·5
MCV cu μ	66–77·0	69·8
MCH $\mu\mu$ gm	19·5–24·5	22·8
MCHC	31·0–34·0	33·0
Thrombocytes (cu mm)	$2\cdot0–9\cdot0 \times 10^5$	4·7
Icterus Index	2–5	Less than 5·0
Specific Gravity	1·054–1·062	1·057
Osmotic Pressure Serum Colloids	240–330 mm H_2O	300
Reticulocytes (% of total RBC)	0·0–1·5	0·4
RBC Diameter μ	6·7–7·2	7·0
Resistance to hypotonic saline Min.	0·40–0·50	0·46
Max.	0·32–0·42	0·33
M:E ratio	0·75–2·5 :	1·0
Leuckocytes/cu mm	6–18,000	11,000
Band Neutrophil (%)	0–3	0·8
Neutrophil (%)	60–77	70·0
Lymphocyte (%)	12–30	20·0
Monocyte (%)	3–10	5·2
Eosinophil (%)	2–10	4·0
Basophil	rare	rare

Absolute number of each leukocyte type in the dog per cu mm

(Mean and Standard Deviation)

Age	Number of Dogs	Total Leukocyte Count	Band Neutrophils	Mature Neutrophils	Lymphocytes	Monocytes	Eosinophils
1–6 mo.	14	11,000±2,300	50± 70	6,400±1,200	3,400±1,200	650±350	450±300
6–12 mo.	21	12,000±3,200	70± 80	7,700±2,250	2,700±1,200	750±300	750±400
1–2 yrs.	17	11,300±2,800	50± 90	7,200±2,400	2,800± 950	750±250	500±300
2 yrs.	18	11,000±2,250	100±150	7,000±2,000	2,500± 850	800±300	500±300
All ages	76*	11,500±2,800	70±100	7,300±2,200	2,800±1,000	750±300	550±350

* Included are six animals of unstated ages.

Blood or plasma chemical composition

(mg/100 ml)

(B)	=	Blood
(P)	=	Plasma
(S)	=	Serum

Nitrogenous Constituents, Non-protein

Non-protein Nitrogen   20–36 (B)

Urea Nitrogen   10–20 (B)

Urea = 2·14 × Urea Nitrogen

Creatinine   1–2 (B)

Uric Acid   0–1·0 (B) Values greater than 1·0 in the Dalmatian

Allantoin   0·8–1·35 (B) 0·57–0·64 (P) Dalmatian

Amino Acids   7–8 mg (B) Amino Acid Nitrogen 4·2–7·6 (B)

Protein Constituents

Total Proteins   5·3–7·5 gm/100 ml (P)

Albumin   3·0–4·8 gm/100 ml (P)

Globulin   1·3–3·2 gm/100 ml (P)

Albumin Globulin Ratio:

A/G Ratio   1·5–2·3

Fibrinogen   0·3–0·5 gm/100 ml (P)

TABLE I (*Continued*)

*Electrophoretic analysis of plasma proteins in
veronal citrate buffer*

(pH 8·6; ionic strength 0·1%)

Albumin	39·6
Alpha$_1$ Globulin	16·9
Alpha$_2$ Globulin	8·0
Alpha$_3$ Globulin	
Beta Globulin	13·0
Fibrinogen	13·3
Globulin Gamma	9·3

Carbohydrates
  Glucose   60–100 mg/100 ml (**B**)

Organic Acids
  Lactic   2–13 (**B**) 12·6–36 (**P**) mg/100 ml
  Citric   1·7–3·9 (**P**) mg/100 ml
  Pyruvate   0·1–0·2 (**B**) mEq/L

(mg/100 ml)

---

*Blood, plasma, serum constituents
chemical composition*

(mg/100 ml)

Lipids
Total Cholesterol   140–210 (**S**)
Cholesterol Esters   84–168 (**S**)
Free Cholesterol   28–84 (**S**)
Total Lipids   47–725 (**P**)

Electrolytes
Calcium   9–11·5 (**S**)
Inorganic Phosphorus   2·5–5 (**S**)
Sodium   137–149 mEq/L (**S**)
Potassium   3·7–5·8 mEq/L (**P**)
Magnesium   1·4–2·4 mEq/L (**S**)
Chloride   99–100 mEq/L (**P**)
Sulfate   2·0 mEq/L (**P**)
Carbon Dioxide   18–24 mEq/L (**P**)
   Combining Power
pH   7·31–7·42

Enzymes
Alkaline Phosphatase   3–6 Bodansky Units/100 ml (**S**)
Lipase   Less than 1 Sigma Tietz Units/ml (**S**)
Amylase   423–562 Somogyi units/100 ml (**S**)
                      Occasionally 1,000 Somogyi units/100 ml (**S**)

Transaminase
  SGOT   Below 23 Sigma Frankel Units
  SGPT   Below 22 Sigma Frankel Units

Miscellaneous
Bilirubin   0·1–1·0 (**S**)
17 Hydroxycorticosteroids   3–10μ g/100 ml (**P**)

Protein-bound iodine   2·5–7·0μ g/100 ml (**S**)
                              lower limit 2·6μ g/100 ml (**S**) (Kaneko)
Thyroid uptake of radioiodine I^{131} 11–40% 72 hours postinjection
Bromsulfalein Retention test   5 mg/kg dose = 0–10% after 30 minutes

TABLE I (*Continued*)

Bleeding Time

Dorsum of nose	2–4 min.
Lip	85–110 sec.
Ear	2·5–3 min.
Abdomen	1–2 min.
Coagulation Time	
Glass	6–7·5 min.
Silicone	28–40 min.
Prothrombin Time	6–9 sec. (puppies show prolonged values up to 55 sec. until the age of 2 days)
Prothrombin Levels	350 units
Factor V	158–203 units

Reproduced with permission from Schalm (1965).

floor. Under-floor heating has been used and has proved satisfactory in many premises but has the disadvantage of increasing the smell and making the surface difficult to clean as a result of the drying of faeces and urine on the warm surface. A common fault in animal building is poor drainage caused by inadequate fall on the floor, and it seems that the standard allowance is inadequate where small volumes of water need to run off. A suitable slope is 1 in 20.

TABLE II
Normal urine values
(mg/kg of body weight per day except where indicated)

Specific Gravity	1·018–1·060
Volume	24–41 ml/kg body wt/day
Calcium	1–3
Magnesium	1·7–3·0
Phosphorus	20–30
Phosphate	
Potassium	40–100
Sulfate Total	30–50
Sulfur Total S	25–40
Ethereal S	1·3–3·5
Neutral S	5–10
Allantoin	35–45
Creatine	10–50
Creatinine	30–80
Urea	800–4,000
Uric Acid	0·2–13·0
Nitrogen	
Total N	500–1,100
Ammonia	60
	(0·2–3·7 mEq/kg body wt/day) proportional to pH
17 Ketosteroids	0·040–0·100
Bicarbonate	0·05–3·2 mEq/kg body wt/day
Chloride	0–10·3 mEq/kg body wt/day
Sodium	0·04–13 mEq/kg body wt/day
Phosphate	0–1·04 mEq/kg body wt/day
Indican	Absent to a trace
Bilirubin	No reliable quantitative data
	1–2 + reaction Harrison spot test
Urobilinogen	Less than 1:32 Wallace & Diamond test
pH	5·0–7·0

Reproduced with permission from Bentinck-Smith (1971).

Heating may be achieved by simple warm pipes, ducted hot air or electric panels in the ceilings. Experience suggests that the simplest methods are still the most easily-controlled and economical. Air-extraction systems should be capable of causing at least 10 air-changes per hour, but they will considerably increase heating costs.

Good lighting is essential and will help to create a congenial working environment for staff.

Electrical fittings must be waterproof and out of reach of dogs. If infra-red lamps are used in breeding kennels they must be placed not less than 1 metre from the ground and cables should be run out of reach of adult dogs standing on their hind legs.

Feeding and watering

Individual feed bowls are satisfactory and enable more accurate observation of each dog

TABLE III
Water balance data
Water balance resting state
(Body weight value in grams. Others are g/100g of body weight per day)

Body weight	18,600
Water Turnover	6·0
Water Intake	
Food and H_2O	4·6
Metabolic Water	1·4
Urine Output	1·9
Other, sweat, lungs and incorporation into new protein	4·1

Body Water and Plasma Volume	Dog, adult lean
Total Body Water ml/kg	700 (619–756)
Extracellular Body Water ml/kg	320 (239–408)
Plasma Volume ml/kg	52·7 (35·0–70·4)

Reproduced with permission from Bentinck-Smith (1971).

TABLE IV

Cerebrospinal fluid

(mg/100 except where indicated)

Quantity	0·9–16 ml
Aspect	clear- colourless
Pressure	24–172 mm H_2O
Specific Gravity	1·003–1·012
Freezing Point Depression	−0·61°C to 0·63°C
Alkali Reserve	48·5–68·6 vol %
Cells/cu mm	1–8 lymphocytes occasional endothelial cell
Calcium	
Chloride	761–883 as NaCl
Sugar	61–116
Phosphorus	2·8–3·5
Magnesium	2·6–3·8
Nonprotein N	below 40
Total Protein	11–55
Albumin	16·5–37·5
Globulin	5·5–16·5
Pandy Test	negative

Reproduced with permission from Bentinck-Smith (1971).

to be made. Water bowls may be either un-attached or of the automatic-drinker type. It should be borne in mind that many dogs (especially if bored) are very destructive, and may destroy plastic or aluminium bowls; stainless steel is probably the best material. The bowls should be non-tippable.

Water should be available *ad lib* at all times unless the experiment requires special restriction, in which case Home Office authority may be required.

Dogs easily form habits, especially with regard to time of feeding. In general, adult dogs, non-pregnant and non-lactating dogs require feeding only once daily, and the late afternoon is the best time. The dog usually sleeps after food. Food should be withheld for 12 hours and fluids for 4 hours prior to general anaesthesia.

Bedding

Dogs should have wooden or stainless steel beds raised 10–12 cm from the floor; these may take the form of solid platforms separated from the ground by air bricks. It may be necessary to clad with metal the edges of wooden beds, to prevent dogs from chewing them. Many dogs remain healthy without any additional bedding, provided the quarters are heated in winter. Wood-wool, polythene bags filled with hay or straw, or rugs make satisfactory bedding. Bare straw is known to cause dermatitis in some individuals,

and some dogs will destroy and eat polythene bags. A variety of types of bedding should be available so that provision can be made for special cases, especially where there is a large throughput of dogs. Bedding should be of easily disposable material, although even dogs that are not house-trained seldom foul the bed area in which they sleep, and may be trained to defaecate only in an outside run; this will reduce labour demands.

Schumacher & Strasser (1968) describe in some detail an installation for breeding experimental dogs which includes points of special design. In their premises dogs share outside runs and they have found that there are special advantages in using gravel as a surface covering for these. It allows good drainage and easy removal of faecal matter under varied climatic conditions, and reduces to a minimum endo-parasitic reinfection.

Environmental conditions

The dog is susceptible to heat stroke; this can be brought about by air stagnation as well as by high ambient temperature. It can withstand low temperatures, but high temperatures may cause respiratory distress. At no time should a close-fitting muzzle be applied, since efficient panting is the main method of heat loss in the dog. It may be necessary to provide insulation in the roof of the building in regions where hot sun is common.

Identification

The most reliable method of identification is by tattooing of an ear or of the abdominal wall. It is inhumane to carry out either method except under general anaesthesia. Objections have been raised to collars and tags, but properly adjusted chain collars with steel labels can be satisfactory. Temporary identification marks may be cut into the fur of the coat with scissors or electric clippers. Such marks would need renewal at 3–4 week intervals.

Handling

Vicious and intractable dogs are uncommon and in any case are unsuitable animals for experimental work, except for acute non-recovery experiments. Most dogs respond to a confident approach but are quick to detect fear in an attendant. On all occasions a dog should be approached from the front, quietly, and the handler should talk to it. The dog's fear (probably the main cause of intractability) may be overcome if the attendant stands some distance away and crouches so that his height more nearly approximates that of the dog. The dog responds well to reward.

An intractable animal may be snared in a dog-noose (Fig. 27.1) whilst an assistant places a clove hitch of tape or 7·5 cm bandage (Fig. 27.2A) around the nose and jaw, which is tied behind the head. If this is to be effective it must be applied firmly. The dog may be restrained by holding the scruff (Fig. 27.2B) of the neck. Leads attached to collars are insecure and the safest

way of leading a dog is by a slip noose of soft rope around the neck. Dogs of up to 15 kg body-weight are most conveniently carried from one working area to another; again a slip noose around the neck should always be used to prevent escape. There are well-established restraint methods for manipulating the conscious dog and familiarity with these will prevent accident (see Figs. 27.3 and 27.4).

Transport

Dogs may be transported loose in vans or in crates. In either case special attention should be paid to ventilation, since in warm weather the death of dogs in improperly ventilated vehicles is all too common. It is inadvisable to transport

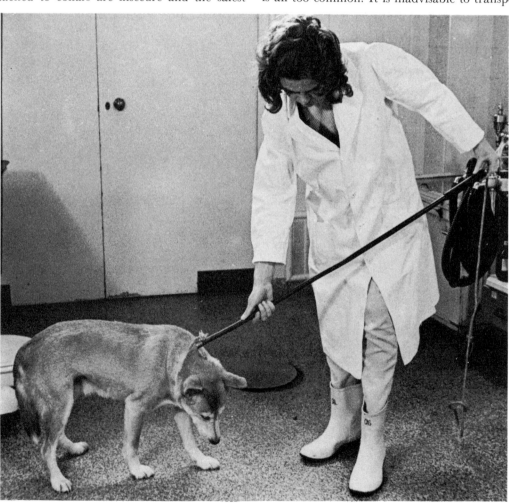

Fig. 27.1. Restraining a dog with dog catcher; note position of technician's right hand. (Photograph: UFAW.)

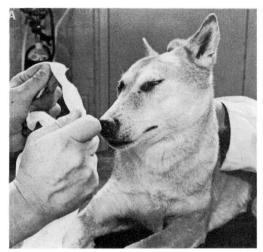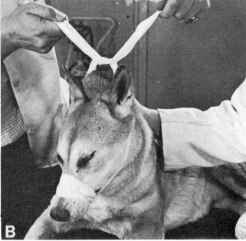

Fig. 27.2. Stages in the application of a tape muzzle. (Photograph: UFAW.)

experimental dogs in cars unless they are properly tethered or separated by wire mesh from the driver.

Dimensions and methods of construction of crates for transporting dogs by air are given in the International Air Transport Association Live Animal Regulations (1975). The same standards are equally appropriate to the construction of crates for rail or road transport and it is strongly advised that these recommendations be implemented. The dimensions are based on standard measurements made on the dog in-

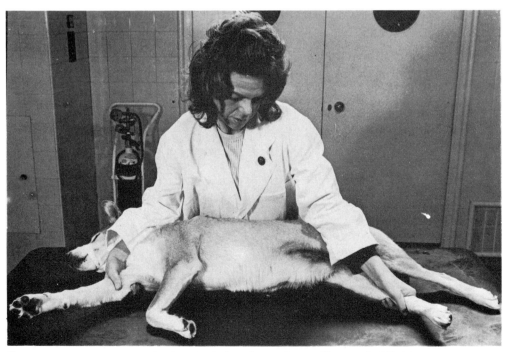

Fig. 27.3. The correct way to maintain a dog in lateral recumbency. (Photograph: UFAW.)

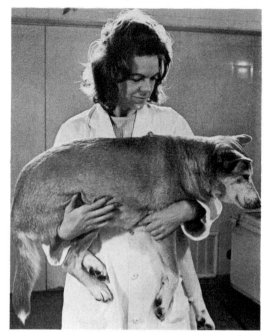

Fig. 27.4. Safe way of lifting a dog. (Photograph: UFAW.)

tended for transportation and therefore vary with each breed; the Manual should therefore be consulted for the formula.

The Home Office must be informed if it is intended to move animals controlled by the Cruelty to Animals Act from one licensed premises to another. Dogs may be sent by rail provided they are muzzled. Careful consideration should be given to the provision of water for dogs in transit. In warm weather it may be necessary to provide drinking water at least every three hours. Adult dogs need feeding at 24-hour intervals and pups between 3 and 6 months of age every five hours. Pups younger than 9 weeks old should be transported in properly designed boxes and should preferably be accompanied.

Feeding

The dog is naturally carnivorous, consuming whole carcases as opposed to isolated lumps of meat. This fact is important since nutritional deficiency may develop in dogs fed on a narrow exclusive diet. The dog has been used for many years in nutritional research and considerable literature exists concerning specific requirements. However, there is still some disagreement. The nutritional requirements of the dog are discussed by Abrams (1962; 1964), National Research Council (1962) and Worden (1964) and the reader is referred to these works for details and experimental results.

In practice it is most satisfactory to feed dogs to maintain body-weight and condition. The diet may consist of raw or cooked meat (fit for human consumption) supplemented with twice its quantity of cereal dog biscuit. A proportion of protein of high biological value is essential. It is also necessary to include 6–8 per cent fat in the diet; about 2 per cent of this should be vegetable fat to provide essential fatty acids. Deficiency of these will lead to skin diseases and hair defects. This mixture may be fed at the rate of 0·5 kg per 10 kg body-weight to an adult, exercised dog. It is also advisable to add a proprietary vitamin and mineral mixture. Care should be taken not to overdose with fat-soluble vitamins. Chronic overdosage with Vitamins A or D will lead to bone changes in the axial and appendicular skeletons respectively. Metastatic calcification in blood vessels and kidneys may occur in chronic hypervitaminosis D. The requirement of the dog for vitamin D may be as low as 13 i.u. per kg body-weight provided that the calcium to phosphorus ratio is maintained in the region 1·2 : 1. A number of satisfactory commercial diets are now available and it is probably more economical to feed one of those which has been shown to sustain health and fertility in successive generations of dogs (Walker, 1965).

The food intake of the bitch should be increased by about one third by weight from the middle pregnancy. However, provided that the formulation of the diet is correct, the general condition of the dog is a good guide to the total quantity which should be fed. Each pup will require 200 calories per kg body-weight and this increased requirement will be reflected in the bitch's diet; as lactation proceeds she will require 2·5–3·0 times her maintenance intake.

Breeding

The bitch is monoestrous and 2 oestral periods occur each year. The cycle of reproductive activity is divided into proestrus, oestrus, metoestrus and anoestrus. Proestrus, which lasts for 7–10 days, is characterized by vulval swelling and haemorrhagic discharge. Frequency of urination may be observed before physical signs become apparent. At this time mating is not permitted by the bitch although males are attracted. In true oestrus, which lasts from 4–10 days, the colour of the discharge changes to that of yellowish straw. Oestrus is the period of acceptance, during which conception may be expected to occur; this is most likely to result from a mating 11–17 days after the start of proestrus.

Oestrus may be detected by examination of a vaginal smear: the technique is clearly described by Harrop (1960). Artificial insemination, breeding and rearing of dogs is described by the same author and more recently by Kirk (1970).

If pregnancy does not occur oestrus is followed by metoestrus which lasts about 2 months. During this period the condition of pseudo-pregnancy may arise. A bitch in this condition will behave as if pregnant and towards the end of the period make a bed out of whatever materials may be available, and may also lactate. Occasionally cystic endometritis may affect the uterus, resulting in thirst, vomiting and an evil discharge from the vulva, although the latter is not constant; the bitch may become extremely ill and veterinary assistance should be obtained without delay since the condition may run a rapidly fatal course.

Pregnancy diagnosis

With practice pregnancy may be diagnosed by palpation at 21–30 days with the bitch in the standing position. Concepti may be detected with the fingers through the abdominal wall as a chain of discrete spherical swellings, up to 2·0 cm in diameter, occurring in the mid-abdomen.

Later in pregnancy, diagnosis by palpation becomes increasingly difficult since the concepti becomes less discrete and therefore less easily identifiable from other abdominal contents.

Suppression of oestrus and pseudo-pregnancy

It may be desirable to suppress oestrus that has begun or to prevent oestrus. Treatment with a preparation containing megestrol is effective for these purposes. David et al. (1963) reported the initial trials of megestrol acetate in dogs and Evans et al. (1969) describe the clinical use of megestrol acetate (Ovarid: Glaxo) to suppress oestrus in bitches. Reference should be made to the manufacturer's literature for dose rates, etc.

Care during pregnancy and parturition

The average gestation period of the bitch is 63 days but this may vary plus or minus 3 days. Litter size varies with the breed, from 2–3 in small breeds to 12 or more in medium and large breeds.

No special care is required apart from feeding a high protein, high calorie diet during the latter half of pregnancy. By the third week in lactation the nutrient intake may need to be increased three-fold over the normal maintenance diet.

During the last week of pregnancy the bitch should be housed separately from other dogs and provided with material such as newspaper with which to make a bed. The mammary tissue should be examined to ensure that there is no obstacle to suckling and that the glands are not painful.

Impending parturition is indicated by a change in the temperament of the bitch. She may become listless or excitable and will use any available material to make a bed. Many suffer a loss of appetite 24 hours before parturition and in most cases the rectal temperature becomes subnormal up to 24 hours before delivery. In most bitches parturition proceeds without assistance. It is normal for pups to be presented either head or tail first, and for the bitch to rest for up to 30 minutes between delivering pups. Should a bitch strain consistently for more than 30 minutes without delivering a pup, or alternatively if a prolonged period of straining occurs and then stops, qualified advice should be sought. It is normal for a bitch to consume the placenta. Following parturition the bitch should be left in peace, under observation to ensure that the pups are suckling.

Hereditary defects

The dog is heir to a wide range of inherited defects, some of which are associated with certain breeds. Where inbreeding is practised it is advisable to subject foundation stock to veterinary examination before final selection; any pups showing defects should also be examined by a veterinary surgeon. Of special importance in the experimental dog are those conditions affecting the eye, brain, heart and locomotor apparatus. At birth the most commonly-seen defect is cleft palate, and all pups should be examined routinely soon after birth for this defect, which may prevent suckling.

Rearing

Chemical regulation of body temperature in the newborn pup is inefficient and the pup must rely upon external sources of heat to maintain body temperature; if these are not provided hypothermia will develop and may be fatal (Crighton, 1968).

The rectal temperature of the newborn pup falls rapidly after birth from that of the bitch (approx. 38°C to approx. 32°C). It is therefore important to maintain an ambient temperature of about 29°C. Although the bitch may provide this, she will not settle or relax into lateral recumbency for the benefit of the pups if the ambient temperature is too low, or if she is disturbed. Pups which show signs of hypothermia (indicated by low rectal temperature, bradycardia, diminished reflexes and poor muscle tone) may be revived by gentle warmth in an ambient temperature of 29°C. A bath of warm water is suitable, and the pup should be partially immersed with the head above the surface of the water for 5–10 minutes and then carefully dried.

Fox (1965) emphasizes the importance of human contact if pups are not to become unbalanced and intractable as adults. Such contact must occur during the critical period of 7–16 weeks. Awareness of this fact will clearly improve the usefulness of the dog in an experimental environment.

Regimes for feeding pups are based upon requirements for 3 main age periods: from weaning to 3 months old; 3–6 months old; and 6–10 months old. For the first 7 weeks of life pups will be receiving bitch's milk (for composition see Table V) latterly supplemented with cow's milk, cereal baby-food and scrapings of meat. Weaning on to prepared rations should be complete by 8–9 weeks, at which time they should be fed 4 times daily with food rich in protein. Very satisfactory commercial puppy biscuit is available, and it is usually most efficient to feed this together with the supplements recommended by the manufacturers.

Kirk (1970) describes a successful technique for feeding neonatal pups through a 3–4 mm diameter stomach tube down which a measured feed is injected every 8 hours.

TABLE V
Table of composition of bitch milk

Protein	(g/100 g)	7·1
Fat	(g/100 g)	8·3
Lactose	(g/100 g)	3·8
Calories	(per 100g)	121
Calcium	(mg/100 g)	230
Phosphorus	(mg/100 g)	160

Reproduced with permission from Widdowson (1964).

Table VI gives the formula of a satisfactory artificial substitute for a bitch's milk.

TABLE VI
Preparation of a milk substitute for hand-rearing puppies

Components

Whole milk	800 ml
Cream, single, 12% fat	200 ml
Egg yolks	One
Bone flour, sterilized	6 g
Citric acid	4 g

Reproduced with permission from Abrams (1962).

The cream is stirred into the milk. Following this the egg yolk, the bone flour and the citric acid are added in that order, each being stirred in thoroughly.

Laboratory procedures

Home Office regulations require licencees under the Cruelty to Animals Act to possess special certificates (E and EE) before dogs may be used for experimental investigations; and dogs may be used only where other small vertebrates are unsuitable.

Selection of animals for experiment

The evidence suggests that dogs bred specifically for experimental work provide the most satisfactory and economical material. Buying them cheaply from an unspecified environment encourages the stealing of pets off the streets, as well as entailing the risk of introducing transmissible disease into the unit or deriving information from dogs which were not normal at the start of the experiment.

The Medical Research Council's Laboratory Animals Centre, Carshalton, operates a scheme whereby a number of species of laboratory animals, including dogs, may be obtained from breeders accredited by them (see Chap. 4).

Where dogs are obtained from non-accredited breeders they should be quarantined for 3 weeks away from existing stock and during this period subjected to clinical and laboratory examinations to ensure their fitness for use as experimental animals. The standard required will, of course, depend on the type and duration of the proposed experimental programme, but the existence of certain conditions should be eliminated (see pp. 326–327).

Anaesthesia

Food and water should be withheld for 12 hours and 4 hours respectively prior to general anaesthesia. General anaesthesia in the dog is a routine procedure in clinical veterinary surgery and the discipline is covered by an extensive literature. However, special conditions may prevail, and it is strongly advised that those responsible for anaesthetizing dogs should work under supervision in the first instance in order to obtain experience (Hall, 1964).

General anaesthesia may be induced by intravenous injection of barbiturate salts or by gaseous anaesthetics such as halothane with oxygen alone, or with oxygen and nitrous oxide, given via a dog mask. The choice of anaesthetic employed must depend upon the experimental procedure which will follow. Anaesthesia may be maintained by giving incremental doses of the intravenous agent or by intubation of the dog with a cuffed endotracheal tube which is then connected to an apparatus delivering measured quantities of anaesthetic and oxygen.

The simplest method depends upon the intravenous injection of pentobarbitone sodium, which will produce surgical anaesthesia for up to 40 minutes. Recovery is prolonged and may be noisy. Premedication with phenothiazine derivatives produces a quieter recovery and reduces the induction dose of barbiturate by up to 30 per cent.

For prolonged surgical interferences an efficient method consists of induction with thiopentone sodium given intravenously (perivascular injection may cause tissue reaction), intubation with a cuffed endotracheal tube and connection to any standard anaesthetic apparatus either with a reservoir bag and carbon dioxide absorption facility or incorporating an Ayre's T-piece system. In both systems oxygen is administered with halothane and/or nitrous oxide but, in the former, rebreathing is allowed to occur with consequent saving of halothane, which is expensive. However, the Ayre's T-piece system is physiologically more acceptable.

Brachycephalic breeds should always be intubated when anaesthetized, in order to maintain adequate airway in the presence of the abnormal oral and pharyngeal anatomy in such breeds.

The reader is referred to Hall (1974) and Lumb & Jones (1973) for detailed actions and uses of ataractic and anaesthetic agents in the dog.

In Great Britain the use of muscle relaxants in experimental surgery is controlled under Home Office regulations. In any case, the actions of some of these drugs on the dog are markedly different from those in the human subject and the experimental anaesthetist should refer to veterinary anaesthetic textbooks before contemplating their use. For example d-tubocurarine causes a severe fall in blood pressure and the release of histamine in the dog. The dog is also very sensitive to small doses of the suxamethonium and other quaternary ammonium compounds.

Euthanasia

Three methods of euthanasia are acceptable in the dog in laboratory conditions:

(1) Destruction of the brain with a captive-bolt pistol. This is no doubt humane if carefully used, but it is aesthetically unpleasant.

(2) The intravenous or intraperitoneal injection of a pentobarbitone sodium solution containing 180 mg/ml (3 times the concentration of pentobarbitone sodium solution BP). This method is efficient, humane and inexpensive and is to be preferred to other methods. The diagnosis of death should not be made until bulbar paralysis

has occurred as indicated by complete pupillary dilation.

(3) Exsanguination and perfusion with fixative under general anaesthesia may be used where intravital fixation is required.

Experimental feeding procedures

Oral feeding or dosing of dogs artificially is usually easy. Fluids should be poured into a pouch of the lower lip at the posterior labial angle with the nose held slightly higher than the occiput, pausing frequently to allow the dog to swallow. The stomach tube is usually easily passed in the trained amenable dog, but it may be necessary to sedate some animals first. The first few inches of the tube should be lubricated with an analgesic lubricant such as Xylocaine Gel (lignocaine 2 per cent: Astra Chemicals). Suitable tubes for the purpose are obtainable from Portex Plastics Ltd, Hythe, Kent (Fig. 27.5). An approximate measurement of the distance from last rib to tip of nose of the dog should be recorded on the tube before introduction so that the distal end of the tube may be more accurately placed within the gastric lumen.

The passage of the stomach tube is carried out by passing the tube through the mouth in the exact mid-line; otherwise the sectorial teeth will damage the tube. Alternatively, a gag may be inserted through which the tube may be passed.

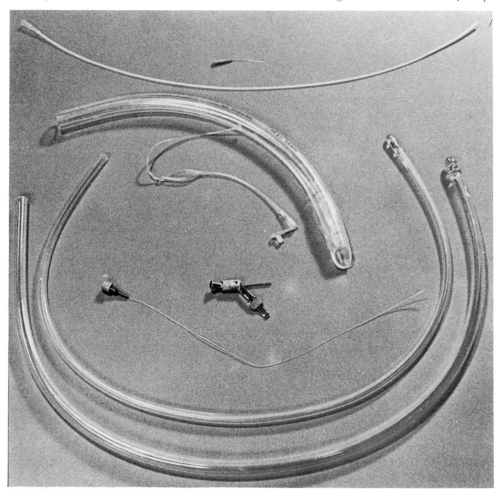

Fig. 27.5. Urethral and venous catheters, two sizes of stomach tube and cuffed endotracheal tube, marketed for clinical veterinary use in the dog by Portex Plastics, Mitchell indwelling needle and small irrigating cannula. (Photograph: UFAW.)

The end of the tube will travel naturally lateral to the larynx to the right side of the fauces and into the oesophagus.

Tablets may be administered by hand by dropping over the back of the tongue with the jaws open. Stroking the throat after closing the jaws will then induce the dog to swallow.

The dog has a well developed facility for rejecting unwanted swallowed material and observation should be kept to ensure that the dose has not been regurgitated.

Some preparations (e.g. insoluble powders) may be given mixed in gravy or minced meat.

Collection of specimens

Blood sampling: Three convenient sites are available in the dog:

(1) cephalic vein.
(2) saphena vein.
(3) jugular vein.

All these sites may be used in fully-conscious dogs and with minimal restraint.

Bleeding from the cephalic vein. The dog is restrained as shown in Figure 27.6. The assistant raises the vein by applying gentle pressure with the mid-part of the thumb applied to the anterior aspect of the forelimb just distal to the elbow. The vein may then be seen running subcutaneously along the anterior aspect of the limb. The hair over the vein is clipped and the skin cleaned with antiseptic. A short-bevel 19-gauge needle attached to a syringe is inserted through skin and vein wall, the shaft of the needle being aligned parallel to the lumen of the vein as the point is fully advanced up into the lumen. The grip is then changed so that the hand encircles the limb from below and the fingers grip the syringe above. Should the dog now snatch the leg away syringe and leg will move together.

To avoid vein wall collapse aspiration of blood should be performed with minimal negative pressure. It may be necessary to apply temporary strapping over the site after removal of the needle to prevent the formation of a haematoma.

Bleeding from the saphena vein. The manipulation of the syringe and needle is the same as for bleeding from the cephalic vein. To raise the vein, however, the assistant's hand should encircle the leg immediately above the tarsus (see Fig. 27.6).

Bleeding from the jugular vein. The dog is either restrained manually on its back or left in the standing position. The vein is occluded at the base of the neck in the jugular furrow and the hair removed and skin cleaned. It is possible to use a larger-bore needle and remove a greater volume of blood from this site. In this case it is reasonable to inject a small volume of local anaesthetic over the vein two minutes before venepuncture.

Cerebro-spinal fluid: Collection of cerebro-spinal fluid is made from the cisterna magna and

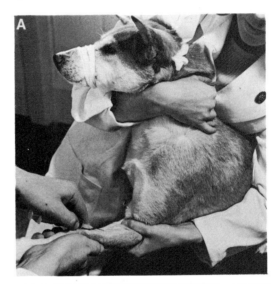

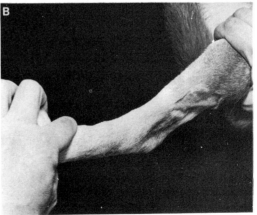

Fig. 27.6. Venepuncture. (A) Method of restraint when taking blood from cephalic vein. (B) Raising the saphena vein for venepuncture. (Photographs: UFAW.)

must be carried out under general anaesthesia, with tracheal intubation and proper surgical asepsis. Hoerlein (1965) describes the technique clearly but it should not be attempted by inexperienced workers.

Saliva: A mixed salivary sample may be obtained by collection from the mouth by aspiration after exciting the dog with strong smelling food.

Milk: This is collected by manual expression from the teat in the normally lactating or pseudo-pregnant bitch.

Urine: The technique adopted will depend upon whether aliquots are to be examined or whether it is intended to derive results based on total 24-hour excretion. The latter is usually done in a metabolism cage. Samples for bacteriological examination are usually only valuable if taken by catheter or are 'midstream'. Dog and bitch plastic catheters are available from Portex* (Fig. 27.5). As far as possible sterile technique should be used to avoid introduction of pathogens into the urinary tract. In certain bitches, where repeated catheterization is necessary, episiotomy may be carried out and usually results in less distress to the bitch.

Faeces: Samples of faeces may be collected from the floor or immediately after defaecation in dogs that are exercised in small groups, or by removal from the rectum with a spatula. Total, uncontaminated faecal collections can be made only from metabolism cages.

Dosing and injection procedures

Oral administration. The technique for this and for administration by stomach tube are described under experimental feeding, p. 322.

Intradermal inoculation. No anaesthetic is necessary; the area should be free of hair and cleaned. The most usual sites are on the ventral abdomen and inside the thigh. A short needle firmly attached or locked to the syringe is usually required to allow adequate pressure to be applied during injection into the dermis.

Subcutaneous injection. Subcutaneous injections of small volumes are usually made in a fold of skin on the back of the neck. Large volumes should be injected over the shoulder blade so that the action of walking assists distribution and absorption.

Intramuscular injection. Intramuscular injec-

* Portex Plastics Ltd, Hythe, Kent.

tion is usually made into the lateral aspect of the thigh midway down the leg, one quarter of the width from the posterior edge of the thigh. A slight negative pressure should always be created in the syringe initially after introduction of the needle into the muscle to ensure that no blood is withdrawn, as a safeguard against inadvertent intravenous injection. Injection should be made slowly and the point of the needle should be repositioned if undue pain or foot movement is observed, since these signs may indicate sciatic nerve involvement.

Intramuscular injection may also be made into the triceps group of muscles on the posterior aspect of the humerus, but this site is less satisfactory.

Intravenous injection. For intravenous injection the technique of entry into the vein is as described above for blood sampling (p. 319). Air should be expelled from the syringe and needle before venepuncture and all injections should be made slowly. Careful observation of the injection site should be made so that inadvertent perivascular injection will be detected and the needle either repositioned or fresh venepuncture carried out. It may be necessary to apply strapping over the site for a few hours to prevent haematoma formation.

Intraperitoneal injection. Aseptic technique should be used and the skin should be washed and shaved prior to disinfection. The site of injection is about 2·0 cm lateral to the umbilicus. The dog is restrained on its back and an approx. 18 gauge by 4·0 cm needle is passed through the skin at an acute angle and then introduced through the abdominal wall. Confirmation of the position of the needle point is made by attempted aspiration into the syringe. No material or fluid should be aspirated. No resistance to injection should be detected. The risk is considerably increased in pregnant bitches or dogs with full bladders, and it must be remembered that in dogs under barbiturate anaesthesia the spleen is dilated and may cover the anterior abdominal viscera on the ventral aspect.

Epidural injection. The technique is described by Hall (1974) and this account should be studied for details of the technique and necessary precautions.

Canine red cell antigens

In view of the increasing use of dogs in surgical procedures where large volumes of canine blood may be used it is appropriate to mention the red cell antigen/antibody systems in the dog. Swisher & Young (1961) describe seven blood group systems in the dog. Of these the A group is potentially of the greatest clinical importance. Transfusion with blood from random donors will result in 25 per cent incidence of incompatibility for canine group A. Destruction of transfused cells with typical clinical signs of transfusion reaction may occur in subsequent transfusions of A-positive cells to A-negative dogs. Haemolytic disease of the newborn may occur in certain pups born to immunized A-negative dams (Hime, 1963). In suspected cases pups should be artificially fed for the first 24 hours of life, since colostrum may contain a high titre of antibody to the A antigen. Christian *et al.* (1949) found that pups allowed to suckle only after the first 24 hours did not develop haemolytic anaemia. Dudok de Wit *et al.* (1967) draw attention to this problem in experimental breeding situations. Hall (1970) reports a naturally occurring red cell antigen/antibody system in the beagle, in which an agglutinogen has been identified which does not fall within the canine A system. Hall suggests that there appears to be a higher incidence of this in the beagle than in other breeds of dog.

Experimental surgery

The success and validity of the results of experimental surgical interventions in the dog, followed by recovery, depend to a great extent on the selection, preoperative preparation and postoperative care of the dog.

It is inadvisable to use dogs for surgical procedures until they have been held in a disease-free environment (unless obtained through the MRC accreditation scheme) and are apparently disease-free for at least three weeks. Diseases in the incubation stage will be exacerbated by the stress of operation; at the same time a holding period will allow the dog to become accustomed to its environment and attendants. Dogs which have travelled on the day of operation may prove difficult to anaesthetize smoothly; ideally they should be allowed to settle overnight on the surgical premises.

Preoperative care. The dog should have been allowed to settle down in the environment for at least 24 hours prior to surgery. Food should be withheld for 12 hours and water for 4 hours but enemata or purgation are unnecessary and unsound unless surgery of the gut is involved. The dog should be allowed to exercise for defaecation and urination half an hour prior to premedication.

Postoperative care. The experimental surgical case is entitled to at least as good postoperative care as it would receive in a Veterinary Hospital. It appears that this is not always properly appreciated by non-veterinarians. Following surgery the patient should be placed on a receptive surface, such as enclosed 2·0 cm foam rubber, and all except the head should be covered with a clean lightweight blanket. It is inadvisable to apply excessive heating, but body-heat loss should be prevented. There should be adequate facilities for observation and resuscitation. No protruding fixtures or full water-bowls should be present in the recovery area, since these are hazardous to the animal. It is wise to allow the dog to sleep off the anaesthetic rather than stimulate it into tiring and useless activity. The dog should be turned every two hours to prevent hypostatic congestion of the lungs. Bowden (1970) emphasizes that where advanced surgery is performed the unit should be administered in the same way as a first-class modern hospital, and describes a mobile intensive care recovery cage with monitoring facilities developed at Guy's Hospital, London (see Fig. 27.7). Routine dressing of simple surgical wounds in the dog is usually unnecessary, and may be contraindicated since most dressings act as a stimulus to over-attention by the dog. It is rarely necessary to apply surgical dressings and bandages after abdominal or thoracic surgery. Skin sutures should be removed after eight days.

The type of postoperative feeding must depend upon the nature of the surgical interference, but unless this has involved the alimentary tract a normal light diet and fluids may be given within a few hours of surgery. If food is not given by mouth parenteral nutrients must be administered; there is no necessity for severe food restriction, and neither is it humane. It should at least be possible for a nurse or a technician to spoon-feed the dog over a critical

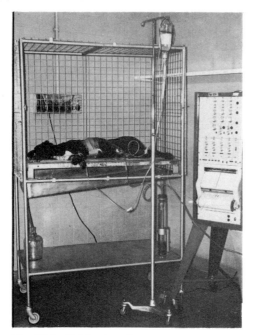

Fig. 27.7. Mobile intensive care recovery cage. (Photograph: N. L. R. Bowden. Reproduced by courtesy of Guy's Hospital Medical School.)

period. Markowitz *et al.* (1964) lay down standards which should be reached in any well-administered experimental surgical department.

Disease control

It is strongly recommended that only stock from accredited suppliers be used, in which case certain standards are fulfilled and the probability of specified diseases reduced to a minimum (see p. 321). All new stock should be housed for three weeks in isolation from existing stock and subjected to a competent clinical examination with minimum delay.

There is considerable literature on the diseases of dogs and brief mention only of the signs of health or ill-health will be made here. The diagnosis and treatment of disease in dogs is a matter for a veterinary surgeon whose advice should be sought in these matters. From the point of view of both dog and research worker it is strongly advised that a veterinary surgeon should be appointed to attend and advise at any establishment where dogs are kept. Such action is likely to reduce suffering caused through ignorance and expense through the use of unsatisfactory experimental material. The source

from which the dog is obtained plays a major part in determining its disease status.

Internal parasites

Helminthiasis. The important helminth parasites of the dog are hookworms of the genera *Ancylostoma* and *Uncinaria*, large roundworms of the genera *Toxocara* and *Toxascaris*, and tapeworms of many species, the commonest being *Dipylidium caninum*.

Severe hookworm infestation may lead to anaemia, haemorrhagic diarrhoea and death. Young dogs are at greatest risk.

Of the large roundworms *Toxocara canis* undergoes a migratory cycle through the tissues of the host and pre-natal infection of pups is common after the 42nd day of pregnancy. This worm is of public health importance since similar infection may occur in man causing, in some cases, serious disease. In the pup unthriftiness and diarrhoea are common signs.

It is advisable to treat breeding bitches in the last third of pregnancy, and pups at 14 days of age and subsequently at 1, 2 and 3 months of age. Piperazine preparations are effective and safe ascarifuges.

Coccidiosis. Four species of coccidia are reported in dogs, and pups kept under poor conditions are especially susceptible. The condition is characterized by acute or chronic dysentery.

Where infection with endoparasites is suspected examination of faecal samples and identification of pathogens is essential to enable the correct specific therapy to be carried out.

Ectoparasites

The following description is confined to the common ectoparasites of the dog.

Fleas. The dog flea (*Ctenocephalides canis*) is common and may promote an allergic skin reaction; since it is a blood sucker, a heavy infestation may cause anaemia. The flea is also the intermediate host for the tapeworm *Dipylidium caninum*. The live flea itself may be difficult to see but diagnosis is established by finding black coal-dust-like particles at the base of the hair: these are flea faeces. Since part of the life-cycle of the flea takes place in the bedding it is necessary to treat or destroy this to prevent reinfection.

Lice. There are three species of lice which affect

the dog. As far as is known dog lice of the genera *Heterodoxus* and *Linognathus* are not important disease vectors, but infestation may produce discomfort and pruritus, which may lead to self-inflicted skin lesions. The larva of the biting louse, *Trichodectes canis*, is a vector for the dog-tapeworm *D. caninum*.

Diagnosis requires careful examination of the skin for moving objects resembling sand particles.

Ticks. Pediculosis is most commonly due to the brown dog-tick, *Rhipicephalus sanguineus* but may also be due to species of *Dermacentor*, *Ixodes*, or *Amblyomma* and in certain countries these ticks act as vectors of disease. The brown dog-tick is well adapted to life in canine and human habitation, especially in premises housing a number of dogs.

Treatment of affected dogs usually consists of dusting into the coat, or bathing with, gamma BHC, powdered derris, or pyrethrum. Recently a preparation 5-bromomethyl-1,2,3,4,7,7,-hexachlorobycyclo(2,2,1)-heptene(2) (Alugan: Hoechst) has proved both safe and effective as an ectoparasiticide in a range of species including the dog and is available as a concentrate, dusting powder or aerosol spray.

Mange. Canine skin and ear diseases (mange) may be caused by the mites *Sarcoptes scabiei*, *Demodex follicularum* and *Otodectes cynotis*.

Sarcoptic mange is characterized by intense scratching and dermatitis with multipapular skin eruptions; these are most commonly seen on the abdomen and in the axillae and groins. It is transmissible to other dogs, particularly young stock or stock which is in poor health. Diagnosis is based upon demonstration of the mite in skin scrapings, but repeated examinations may be necessary.

Treatment with benzyl benzoate BP under veterinary supervision is usually effective. Affected dogs should be isolated and the premises carefully cleaned and treated with insecticide.

Demodectic mange (caused by *Demodex follicularum*) is apparently not easily transmissible by contact but may be passed to pups by contact during nursing and suckling. Lesions usually begin on the head as characteristic thickened hairless skin which may develop pustules. Treatment is protracted and should be carried out under qualified supervision.

Ear-mite infestation with *Otodectes cynotis* is less serious but nevertheless irritating to the dog. The mites live in the ear canal and cause repeated head shaking. Neglected cases may progress to intractable ear disease. The mite is easily killed with proprietary ear drops or a few drops of medicinal liquid paraffin instilled into and massaged along the ear canal.

Ringworm may be caused by fungi of the genera *Microsporum* or *Trichophyton*. Typically, circular, encrusted lesions with broken hairs and a spreading edge are seen. These may affect the head and feet (by direct contact) and claws. The disease is transmissible to other dogs and man.

Diagnosis is based upon demonstration of fungal spores or hyphae in material (which should include *broken* hairs) scraped, or stuck to adhesive tape. Examination of material by Wood's light has been advocated and may be useful to screen a number of suspected animals, but false positives occur with certain substances and absence of fluorescence does not establish the case as free from ringworm.

Treatment should be directed at the premises and equipment as well as the dog. It is essential that brushes and electric clippers are sterilized immediately after use, and before using on another dog.

Treatment of localized lesions with Whitfield's ointment (Ung. Acid Benz Co. BNF) or proprietary fungicidal preparations is effective but in generalized cases systemic therapy with griseofulvin is usually necessary. All cases should be maintained in strict isolation.

Transmissible diseases

The important transmissible diseases of the dog caused by viruses are distemper, infectious canine hepatitis, canine herpes and rabies. *Leptospira icterohaemorrhagiae* and *Leptospira canicola* cause important bacterial diseases in the dog. Fortunately there are combined vaccines available which will give effective protection against distemper, infectious canine hepatitis and the leptospiral diseases, and vaccination should always be carried out routinely where dogs are housed in numbers.

Distemper. Distemper is spread by droplets and by direct contamination. The average incubation period is 6–9 days. A high proportion of stray dogs are incubating or showing signs of distemper when removed from the streets and any dog which on arrival is febrile should either be rejected or kept in a barrier nursing environment for 2–3 weeks. Veterinary advice should be sought if signs of this disease occur. The following signs may indicate the presence of distemper: pyrexia, mucopurulent nasal or ocular discharge, anorexia and diarrhoea, occasionally a dry cough. Pneumonia caused by secondary bacterial infection is not uncommon. Recovery after treatment is rarely complete and varying degrees of central nervous damage may remain with choreiform twitching or convulsions.

Effective vaccines are available from a number of manufacturers and their recommendations as to age of initial vaccination and revaccination should be followed. Primary vaccination is usually carried out at 12 weeks or over; if it is done in younger animals maternally derived antibody may interfere with the development of active immunity and revaccination at 12 weeks is usually advised.

Infectious canine hepatitis (ICH).   This   is caused by an adenovirus which is transmitted in the dog's secretions during acute illness. The incubation period is 6–9 days. The signs range from slight fever to severe depression and leukopenia. Following the acute disease the virus localizes in the kidney and may be shed in the urine for many months. Prolongation of bleeding time is a cardinal feature of the acute disease.

Canine herpes.   Recent work (Cornwell, 1969) has incriminated a virus of the herpes group as a cause of death in pups up to 3–4 weeks old. Infection is transmitted to pups *in utero*.

Leptospirosis.   *L. icterohaemorrhagiae* produces a serious disease, characterized by jaundice, which may be fatal unless diagnosed in the early stage. The rat acts as a reservoir of infection and the disease is most common in dogs living in buildings away from human habitation, such as ranges of kennels where contamination of food and feeding utensils with infective rat urine occurs.

L. canicola is not rat-borne but is spread between dogs by infective urine from acute or apparently recovered cases. The organism gains entry via the oral and nasal mucosa. The disease is characterized by fever, anorexia, and acute nephritis. The mortality is low but chronic progressive renal failure over a period of years

is a common sequel so that many stray dogs must inevitably be suffering varying degrees of renal incompetence, which would invalidate some experimental data derived from them. Leptospirosis can be transmitted to man, and may cause severe or fatal disease.

Rabies   In some countries the virus disease rabies is endemic in the wild animal population. Where stray dogs are acquired from an uncontrolled environment the possible presence of this disease should be considered. Great care should be taken when handling any dog that appears to have a fit or whose temperament becomes noticeably changed to the moroseness seen in dumb-rabies. The medical advisers to an institution handling a large number of dogs not bred for experimental purposes should consider routine vaccination against rabies for the protection of the staff.

Effective vaccination is clearly the most satisfactory method of controlling these diseases.

REFERENCES

Abrams, J. T. (1962). *The Feeding of Dogs*. Edinburgh: Green.

Abrams, J. T. (1964). A review of specialised canine diets. In *Canine and Feline Nutritional Requirements*, ed. Graham-Jones, O. Oxford: Pergamon Press.

Bentinck-Smith, J. (1971). A roster of normal values. In *Current Veterinary Therapy*, 4tlr edn. Philadelphia: Saunders.

Bowden, N. L. R. (1970). Care of the experimental dog. In *Symposium of Nutrition and Disease in Experimental Animals*, ed. Tavernor, W. D. London: Baillière, Tindall & Cassell.

Christian, R. M., Ervin, D. M., Swisher, S. N., O'Brien, W. A. & Young, L. E. (1949). Haemolytic anaemia in new born dogs due to absorption of isoantibody from breast milk during the first day of life. *Science, N.Y.*, **110**, 443.

Cornwell, H. J. C. (1969). Neonatal canine herpes virus infection. A review of present knowledge. *Vet. Rec.*, **84**, 2–6.

Crighton, G. W. (1968). Thermal regulation in new born dogs. *J. small anim. Pract.*, **9**, 463–72.

David, A., Edwards, K., Fellews, K. T. & Plummer, J. M. (1963). Anti-ovulatory and other biological properties of megestrol acetate. *J. Reprod. Fert.*, **5**, 331–46.

Dudok de Wit, C., Coenegracht, N. A. C. J., Poll, P. H. A. & v.d. Linde, J. D. (1967). The practical importance of blood groups in dogs. *J. small Anim. Pract.*, **8**, 285–9.

Evans, J. M. (1969). Hormonal control of the oestrus cycle in the bitch. *Vet. Rec.*, **85**, 233–4.

Fox, M. W. (1965). *Canine Behaviour.* Springfield, Illinois: Thomas.

Hall, D. E. (1970). A naturally occurring red-cell antigen antibody system in beagle dogs. *J. small Anim. Pract.*, **11**, 543–51.

Hall, L. W. (1964). Current methods of anaesthesia for dogs. In *Symposium of Small Animal Anaesthesia*, ed. Graham-Jones, O. Oxford: Pergamon Press.

Hall, L. W. (1966). *Wright's Veterinary Anaesthesia and Analgesia.* London: Baillière, Tindall & Cassell.

Harrop, A. E. (1960). *Reproduction in the Dog.* London: Baillière, Tindall & Cox.

Hime, J. M. (1963). An attempt to simulate the fading syndrome in puppies by means of an experimentally produced haemolytic disease of the new born. *Vet. Rec.*, **75**, 692–4.

Hoerlein, B. F. (1965). *Canine Neurology.* London: Saunders.

International Air Transport Association (1975). *IATA Live Animal Regulations.* Fourth Edn. Geneva: IATA.

Kirk, R. W. (1970). Dogs. In *Reproduction and Breeding Techniques in Laboratory Animals*, ed. Hafez, E. S. E. Philadelphia: Lea & Febiger.

Lumb, W. V. & Jones, W. E. (1973). *Veterinary Anaesthesia.* Philadelphia: Lea & Febiger.

Markowitz, J., Archibald, J. & Downie, H. G. (1964). *Experimental Surgery, including Surgical Physiology.* Baltimore: Williams & Wilkins.

Medical Research Council (1969). *The Accreditation and Recognition Schemes for Suppliers of Laboratory Animals.* Carshalton: M.R.C.

National Research Council (1962). *Nutrient Requirements of Dogs.* Washington: National Academy of Sciences.

Ottewell, D. (1968). Planning and design of accommodation for experimental dogs and cats. In *The Design and Function of Laboratory Animal Houses*, ed. Hare, R. & O'Donoghue, P. N. Lab. Anim. Symp., **1**, 97–112.

Schalm, O. W. (1965). *Veterinary Hematology*, 2nd Edition, Philadelphia: Lea & Febiger.

Schumacher, H. & Strasser, W. (1968). Breeding dogs for experimental purposes. *J. small Anim. Pract.*, **9**, 597–602.

Swisher, S. N. & Young, L. E. (1961). The blood grouping systems of dogs. *Physiol. Rev.*, **41**, 495–520.

Walker, A. D. (1965). Rearing beagles on controlled diets. In *Canine and Feline Nutritional Requirements*, ed. Graham-Jones, O. Oxford: Pergamon Press.

Widdowson, E. M. (1964). Food, growth and development in the suckling period. In *Canine and Feline Nutritional Requirements*, ed. Graham-Jones, O. Oxford: Pergamon Press.

Worden, A. N. (1964). Feeding laboratory dogs and cats. In *Canine and Feline Nutritional Requirements*, ed. Graham-Jones, O. Oxford: Pergamon Press.

Suggested reading

Anderson, A. C. (1970). *The Beagle as an Experimental Dog.* Iowa State University, Ames, Iowa.

Archibald, J. (1965). *Canine Surgery.* Illinois: American Veterinary Publications.

British Veterinary Codex (1965). London: Pharmaceutical Press.

Catcott, E. J. (1968). *Canine Medicine: Text and Reference Work.* Illinois: American Veterinary Publications.

Felson, B. (1968). *Roentgen Techniques in Laboratory Animals.* Philadelphia: Saunders.

Miller, M. A. (1964). *Anatomy of the Dog.* Philadelphia: Saunders.

28 The Cat

PATRICIA P. SCOTT

General biology

Taxonomic position and origin

The domestic cat is one of the smallest members of the genus *Felis*, an ancient group of predatory carnivorous mammals including lions and tigers, whose members vary mainly in size. They have characteristic rounded skulls with small incisors, prominent sharp canines and large carnassial teeth. They can climb, leap and move rapidly for short periods (cheetahs are the fastest living animals), but are otherwise rather lazy and do not benefit by regular prolonged exercise as does a dog. In Ewer's excellent monograph on the Carnivora (1973) the anatomy, physiology and behaviour of wild Felidae are related to those of the domestic cat. The European domestic cat originated, according to Pocock (1951), through the crossing of two wild species, *F. sylvestris* and *F. lybica*, and has been domesticated for thousands of years. The tabby markings of *F. catus* appear in two distinct forms, viz. *torquata*, with narrow vertical stripes on the body, and *catus* with three dorsal stripes running to the root of the tail and a looped or spiral arrangement of blotched stripes on the sides. The *catus* form may have arisen as a mutation from the *torquata*. The normal diploid complement of chromosomes for the cat is 38, the same number as in the lion (Makino & Tateishi, 1952); karyotyping has been carried out by O'Reilly & Whittaker (1969).

Size range and litter size

According to H. B. Latimer (personal communication and 1967) common domestic cats are rather stable compared with rabbits and dogs, and show surprisingly small inherent variability in ponderable and linear dimensions, or in growth-rate under normal conditions (Latimer, 1947). The average adult female (queen) weighs 2·25 to 3·0 kg, while fertile males (toms) weigh between 3·5 and 5·9 kg. Neutered cats are often heavier owing to deposition of fat.

Cats may have 1–8 kittens in a litter. Normally 3–5 kittens are born in spring and late summer, each breeding cycle taking a minimum of $4\frac{1}{2}$ months. Normal full-term kittens weigh 110 ± 20 g at birth; average growth curves are shown in Figure 28.1.

Genetics, breeds and varieties of domestic cat

Robinson has published a review (1959) and a book (1971) on the genetics of domestic cats. Further information can be found in 'Carnivore Genetics Newsletter' obtainable from P.O. Box 5, Newtonville, Mass. 02160, U.S.A., and 12 The Crossway, Ealing, London, England. The nomenclature of monogenic colour variants, polygenic colour variants (eye-colour, rufinism, ticking) and coat variants (long-hair, short-hair, rex) has been standardized and their modes of inheritance studied (Searle, 1968). It is often difficult to determine whether congenital defects

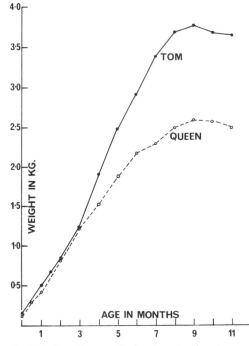

Fig. 28.1. Growth curves for female and male cats.

330

are of genetic or environmental origin, but the following conditions are among those found to be heritable in the cat: folded ears with multiple epiphyseal dysplasia (Dyte & Turner, 1973; Jackson, 1974); brachyury and anury as in the Manx breed (Darwin, 1968; James *et al.*, 1969); ectrosyndactylia (Searle, 1953); polydactyly (Danforth, 1947 a and b); strabismus convergens (Moutchen, 1950); porphrynuria (Glenn *et al.*, 1968). Breeds have been obtained by selective inbreeding of mutants, for example 'Siamese' are partial albino recessives, whose coat colour is affected by age and environment (temperature, light, etc.). Holmes (1953) has demonstrated the existence of at least three blood groups in cats.

Behaviour

Cats are paired territorial predators, although a harem occasionally develops. In the wild the territory is of sufficient area to feed the pair and their offspring. It contains a place of maximum security, a treehole or cave forming a den, where the cat can sleep in a relaxed state; here kittens are born and reared. A territory also contains subsidiary refuges, a drinking place and defined points for urination and defaecation. Scratch posts and demarcation points, regularly sprayed from the anal glands, mark the territorial boundary, which is defended by the male (tom) and by the female (queen) especially when she is rearing kittens. Domestication only slightly modifies the territorial habit. The territory now coincides with a human habitation and its surroundings, a chair or cupboard replaces the den, and although the feeding area is chosen by the housewife, the cat still selects urination and defaecation points where it can dig in soft ground. Domestic territories are vigorously defended against intruders. Laboratory conditions impose a further reduction in territory, but the territorial habit must be respected for success with breeding toms and queens. At other times queens, young stock and castrates will adopt a colonial habit, but this relatively unnatural situation may result in fighting, subjugation and the establishment of a peck order. In extreme overcrowding an outcast may be prevented from feeding or resting, and ultimately dies of stress syndrome and endemic disease.

Daily cycles of activity in the wild consist of hunting or scavenging, visiting watering, urination and defaecation spots each once daily, and maintaining the territory against other cats. Intervals between hunting expeditions may be greater than one day, depending on success; hunting takes place at dusk and dawn, rather than at mid-day or midnight. After feeding and drinking followed by urination and defaecation, the cat will sleep for long periods; it is only active in short bursts and incapable of running long distances like the dog. However, the territory of a tom may cover a considerable area e.g. a number of farms or gardens each containing a female territory. He will move from one female territory to another as the queens come into season. These habits have been confirmed by observations in the colours of offspring, and by the spread of epidemic disease, in which the tom acts as a carrier. In cat breeding it is usual to establish a territory for the tom, and to take the oestrous females to him. In other instances a group of females may be run with the tom in his territory, removing the queens just before they litter.

Uses

The cat began to be used as an experimental animal about 1881, when the well-known monograph by Mivart was published. Acute experiments in which the cat is maintained under anaesthetic throughout the experiment and not allowed to recover consciousness have resulted in many important discoveries concerning reflex action, synaptic transmission, the perception of light and sound, the secretions of digestive glands, and the behaviour of the cardiovascular, respiratory, excretory and nervous system towards natural and synthetic drugs. Cats are important subjects for physiological experiments because they are of a convenient size enabling the investigator to carry out dissection and to set up recording apparatus in an exact manner. Cats are easily anaesthetised for long periods with a good blood-pressure. Where results are likely to be of medical significance, the behaviour of the system under investigation should correspond as closely as possible to that of man. In this respect the circulatory, digestive and neuro-muscular systems of the cat appear to be more satisfactory than those of rodents. Cats are employed in recovery experiments to a less extent than other common laboratory species but are of particular value in studying the effects of drugs on the nervous system and in behavioural studies, especially in the study of sensory perception and motor function (see review by Kling *et al.*, 1969).

In Britain the law requires that, before using cats for survival experiments, a licensee must furnish Certificate E or EE and await allowance of this by the Home Office.

Standard biological data

These will be found in Tables I, II and III and Figures 28.1 and 28.2.

Husbandry

Housing and caging

Cats kept indoors may be confined in cages, or allowed to move freely in a room or part of a room forming a pen. The number of rooms required and their actual dimensions depend on the size and type of cat colony contemplated— whether continually receiving animals into a holding unit, or maintained as a closed breeding colony. In either case, a small room kept at a steady temperature of 22 to 24°C should be set aside for the isolation of suspects and the nursing of infectious complaints. A separate intake room is a great asset in running a holding colony; incoming cats should be kept in individual cages in this room for not less than two (preferably three) weeks before introducing them into the main colony, the individuals of which can be allowed a greater measure of free-

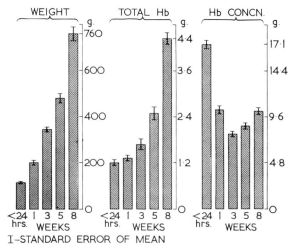

Fig. 28.2. Weight gain and total haemoglobin of kitten.

dom in rooms or pens. The work of cleaning and feeding is reduced when cats or kittens are run together in groups, although the risk of contact infection is greatly increased. A room should be provided for mating and running resting or pregnant queens in a breeding colony, a second for littering and rearing litters to weaning, and a third for rearing growing kittens after separation from their mothers at 8 weeks.

Littering queens should be separated from the colony a week before they are due and placed in individual cages in the breeding room, either the folding cage shown in Figure 28.3, or special breeding cages such as that designed by Jackson & Scott (1970) (Fig. 28.4).

Pens and cages must be provided with adequately sized dirt trays or floor units filled with sawdust, peat or other suitable loose

TABLE I
*Blood values for healthy adult non-parous cats**

	Mean ± s.d.	Range
Blood volume ml/100 g body-weight	5·1±0·5	4·2–5·9
Packed cell volume %	36·9±4·4	
Red cell count 10^6/cu mm	7·14±1·45	
Haemoglobin concentration g/100 ml blood ♂	14·0±1·0	
" " " " " ♀	12·7±1·2	
Total haemoglobin g/100 g body-weight		0·56–0·78
Mean corpuscular haemoglobin concentration %		33·7–35·1
White cell count cu mm	15,000	5,600–28,900
Neutrophils %	59	35–79
Eosinophils %	8	2–31
Lymphocytes %	32	11–52

* Values from cats in the author's laboratory described by S. Gurubatham, Ph.D. Thesis 1961, University of London 'Studies in the haematology of healthy cats throughout life: the changes in growth and haematology associated with infection and with various experimental procedures'.

TABLE II
Values for some constituents of urine from normal cats (O. F. Jackson (1971))

		Mean	SD	Range	No. of samples
Daily urine volume ml		68·6	25·1	0–174	510
pH		6·1	0·6	5·2–7·5	409
SpG		1·046	0·011	1·022–1·069	510
Osmolarity (mixed diet) mOsm		1425	92	1322–1688	13
Osmolarity (all meat diet) mOsm		2855	—	2720–3080	—
Calcium	mg/100 ml	3·8	1·4	1–6·6	21
Chloride	mEq/l	135	73	31–360	167
Creatinine	mg/100 ml	272	97	138–478	107
Magnesium	mg/100 ml	13·1	6·5	4–37	36
Uric acid	mg/100 ml	14·8	2·7	10·2–17·6	6

All these figures are (to a greater or lesser extent) dependent on diet, and refer principally to mixed feeding.

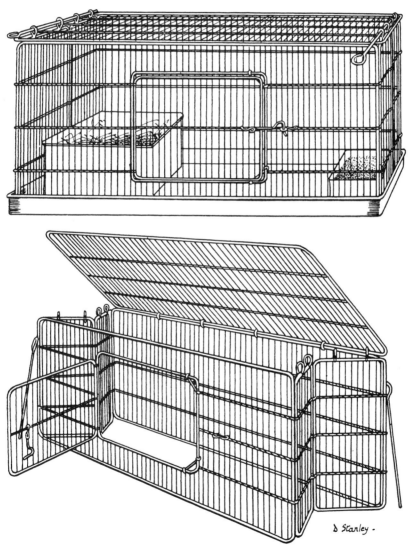

Fig. 28.3. Folding cat cage.

TABLE III
Length and weight of cat foetuses in relation to age

Age (days after mating)	C-R length range (mm) No. of foetuses in parentheses	C-anus length range (mm)	Weight range (g)
19 to 25	7–20 (27)	—	—
26 to 30	21–30 (18)	20–40	1 to 4
31 to 35	35–45 (26)	44–58	4 to 6
36 to 40	50–60 (16)	60–80	6 to 16
41 to 45	75–80 (17)	82–94	16 to 30
46 to 50	90–103 (17)	94–108	30 to 54
51 to full term	110–140 (12)	112–152	54 to 120
At term	125–150	—	—

Second column: Figures from Windle & Griffin (1951).
Third and fourth columns: Figures from Coronois (1933).

absorbent material, in which the cat can scratch.

Rooms should be well lighted, facing south if possible, with windows which can be thrown open in the summer, guarded by wire-netting. Cold water and electricity should be supplied to

Fig. 28.4. Breeding cage unit for cats designed by O. F. Jackson and P. P. Scott (Lab. Animals 4, 135, 1970). The cage 'a' is suspended on nylon runners in a rack, 'b' is a support carrying the detachable door; 'c' and 'g' mark the welded polypropylene kittening box, 'd' is a sliding tray accepting dirt-tray, food and water receptacles, 'f' is a strip of aluminium 2 cm wide rivetted inside the cage at an angle of 60° to overhang the dirt-tray and prevent the cat from scattering sawdust. The cage is made in aluminium alloy.

each room. Walls and floors should be non-porous and washable, with adequate slope and drainage to the floor to permit hosing down. Asphalt or plastic sheeting-covered floors are preferable to concrete. Ledges, skirtings and dust-collecting corners should be avoided since scrupulous cleanliness is essential for satisfactory maintenance of the colony. Attention to design also enables cleaning to be carried out efficiently by unskilled assistants. Ideally all the structures with which the cat comes into contact should be made of metal, plastic or fibreglass and be removable for sterilization. This includes racks, cages, boxes, dirt trays etc., although disposable cardboard boxes may also be used.

Cats must be provided with soft-wood posts against which they can stretch and scratch. The significance of this behaviour is uncertain; it may be a form of territorial demarcation, or for cleaning and trimming of claws (Hediger, 1950). A small, hard-topped table with wooden legs provides scratching posts and a convenient place for weighing, examination and a temporary rest for trays of food, etc.

A system of cat housing in the open which has proved satisfactory has been adapted to laboratory conditions by Weipers (1953) and modified by Quinn & Pearson at Tallents Farm, Kimpton, Herts (Smith, Kline & French Ltd.) for littering and rearing (Fig. 28.5). Housing for rearing kittens from weaning onwards, with internal heated pens and external runs, is very useful when the units are separated from one another by solid walls (Fig. 28.6). Cats which have been housed indoors will maintain good health when transferred to an open-air sanatorium system even under winter conditions. There are two

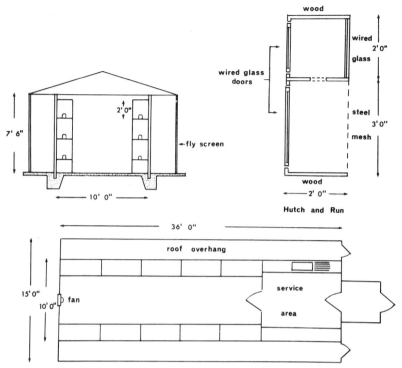

Fig. 28.5. Pagoda for housing pregnant and lactating cats and their offspring. Mating is carried out in a separate breeding unit to which the females are returned after the kittens are weaned. (Reproduced by courtesy of Smith Kline and French Laboratories.)

exceptions: firstly, cats recovering from an anaesthetic, particularly a barbiturate, require artificial heat in cold weather and, secondly, cats showing shock require some additional source of heat. Young kittens thrive poorly in the cold, and benefit from some artificial heating. Many workers who considered it important to keep cats warm sacrificed satisfactory ventilation to achieve a high temperature and have had serious trouble with infectious respiratory diseases, with catastrophic results. While the outdoor system has the advantage of relative freedom from infectious respiratory disease, a disadvantage is the slightly increased work due to the dispersal of the animals, and some discomfort for staff attending animals in bad weather. Cats kept in the open in rabbit-hutch types of cage have more opportunities of escaping, but this risk can be minimized by keeping the hutches in completely covered-in enclosed runs.

To overcome the disadvantages of dispersal, and retain the advantages of a colony of houses, these can be arranged under one roof, either

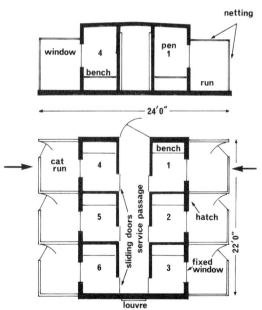

Fig. 28.6. Unit for rearing kittens at the cattery of the Royal Free Hospital School of Medicine and School of Pharmacy, London.

back-to-back or with a central passage. This latter type has many advantages; the house can be made escape-proof, heated and an annexe provided for handling the animals' food etc. Post-operative cases do well in this type of unit, fitted with a small tubular heater in the sleeping-quarters. The most serious objection to wooden construction is difficulty in cleansing and sterilizing but, in actual practice, this has not proved a serious disadvantage. As animals are removed, the cages are treated with a high Rideal-Walker coefficient disinfectant spray (Short, 1969). Care must be taken to spray the entire cage including the under surface of the roof.

Kennels, built of brick or concrete, well insulated and surfaced with a smooth durable paint, with 'pop holes' 0·6 m from the ground to outside runs are convenient for rearing weaned kittens. Six can be reared in a kennel 1·35 × 1·5 × 2·25 m with a wired run 1·5 × 1·5 × 1·8 m. The kennel should be provided with a hard-wood bench 0·6 m off the ground, warmed in cold weather by a tubular electric heater beneath it. Each kennel is fitted with a sliding door opening on to a central corridor, and external access is also provided to the run for cleaning and maintenance.

Food and watering utensils

The problem of receptacles for food should be considered carefully because transfer of un-sterilized dishes from one cat to another is a major factor in spreading infection. For in-dividual feeding disposable cardboard recept-acles about 14 cm diameter, with a smooth impervious internal finish are useful e.g. Uni-bowls made for the food industry by Bowater Scott Corporation. Stainless steel or plastic bowls must either be sterilized on each occasion of their use, or remain permanently with a particular cat or group of cats. Dishwashing fluid must be non-toxic to the cat and not repel it so that food intake is affected; fluids designed for human use are usually satisfactory.

Cats must always have access to clean, un-contaminated drinking water in a non-spill bowl, chick font or special trough fitted with an inverted blood bottle. As they lap, it is difficult to train them to use automatic watering devices. As in the case of feeding bowls, special efforts should be made to prevent cross-infection by the accidental transfer of unsterile drinking vessels. Liquid milk is an expensive and not essential luxury, but if offered particular care should be taken to see that only sufficient is given for immediate consumption and extra care taken over cleansing bowls.

Bedding

Cats maintained at over 15°C environmental temperature do not require bedding, and should be given hard-wood benches 0·3 to 0·45 m wide and 0·3 to 1·0 m off the ground for sleeping, arranged singly or in tiers, with electric tubular heaters beneath them for cold weather, thermo-statically controlled if possible (place thermo-stats out of reach of cats). For plastic or card-board disposable littering boxes, bedding is un-necessary but it is essential for metal boxes. Sheets or off-cuts of clean newsprint or other non-surfaced paper are suitable for littering; wood wool and shavings must be avoided as they contribute to the loss of kittens (dust, strangula-tion).

Environmental conditions

Heating and ventilation should be controlled particularly when the cats are confined to cages and cannot take enough exercise to keep them-selves warm. The temperature should be between 15 and 22°C. Thermostatically-con-trolled electrical heating is very useful, but central heating can also be used. An efficient mechanical means of ventilation should be installed in cat rooms, either an extraction or an input system, using fans capable of providing 10 to 18 changes of air per hour. Stagnation of the air, or any environmental factor raising the bacterial count in air samples, will result in a high incidence of respiratory disease. Air changing is particularly important where breed-ing is contemplated. Humidity requirements are not known, but can be assumed to lie between 45 and 60 per cent.

Efficient artificial lighting, continuously ad-justed to provide 14 hours of light in 24, eliminates anoestrus and allows continuous breeding throughout the autumn and winter months (Scott & Lloyd-Jacob, 1959; Sadleir, 1969).

Identification

Cats are usually identified by their sex, colour and individual markings. When a cat is received or born into the animal house particulars should be entered into a book or on to a file card. Temporary marking of kittens of similar colour can be achieved by clipping small areas of fur on the head or back. In a breeding colony it is useful to have a serial number or letter to identify a queen, and her offspring are numbered sequentially following this serial. Metal discs or thin brass tubing stamped with numbers or experimental details can be attached to leather collars. Adjustable plastic tubular bands (Identi-band by Hollister distributed by Thackray, 10 Park Street, Leeds) designed for a human infant's ankle, the ends joined by a metal clip, contain a card with space for experimental details and make useful collars. As kittens grow collars must be let out. Ears may be tattooed with letters and numbers provided that the skin is not densely pigmented, but the method is not very satisfactory. Small brass numbered ear-tags by Hauptner (distributed by Brookwick & Ward, 8 Shepherds Bush Road, W.6.) are better and cause little trouble provided they are firmly inserted into the *base* of the external ear. When cats are running in a group, or when there is more than one in a cage, the Home Office Inspector will insist that each individual cat under experiment is properly marked and identified on a cage label. Rigid sheet-plastic makes indestructible labels on which particulars can be written with a china pencil or felt-tipped pen. Card labels are rapidly destroyed by cats.

Handling

Cats are highly intelligent but nervous animals, apt to regard strangers with suspicion, so that it is worthwhile spending time making friends with them. Persons who dislike cats should avoid handling them, and should not be responsible for their care. An ordinary domestic cat appreciates being handled in a firm, confident, but gentle way, and seldom scratches unless held so

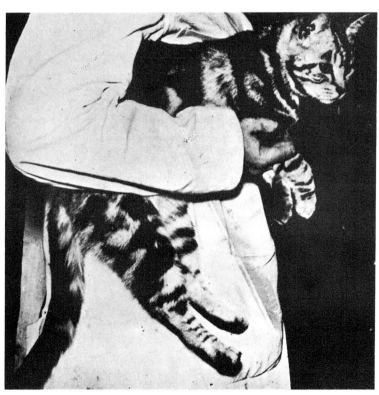

Fig. 28.7. Method of holding a cat. (Photograph: Scott.)

Fig. 28.8. Method of holding a cat for examination. (Photograph: Scott.)

gingerly that it feels unsafe and stretches out its claws to grasp for support. Approach a cat quietly, speaking to it and observing its re-actions. Stroke the head behind the ears and along the length of the back. Take care when you pass your hand over the lumbar region as an oestrous female or a cat suffering from calcium deficiency may bite when touched in this region. Pick a friendly animal up by placing an arm across its back and passing the hand forward under the chest to support its weight, controlling the uppermost parts of the forelegs with the fingers; the body and hind legs of the cat should then be tucked firmly under the handler's arm, so that the other hand remains free (Fig. 28.7). Cats may be examined on a smooth-topped table, the scruff of the neck being held firmly, the arm being kept parallel to the spine and the cat's feet pressed on to the table (Fig. 28.8). Its eyes, nose, mouth and ears can then be examined and cleaned with the attendant's other hand.

An intractable cat when newly received into the laboratory can be laid on its side on a table, one assistant holding the scruff of the neck in one hand and the crossed forepaws in the other, while a second assistant holds the upper part of a hind leg in each hand. A restless cat can be handled by a single person using the method described by Short (1969) (Fig. 28.9). On replacing the animal in its cage or basket care should be taken that both assistants let go at the same moment. In dealing with wild cats strong gloves should be worn, and in cases where the animal is intractable or in pain through accident or sickness, a tranquillizer or suitable short-acting anaesthetic should be given by parenteral injec-tion without hesitation. Greater restraint of a wild cat can be obtained by using a special cage (Graham-Jones, 1964) with crush bars in two planes, operated from outside. By moving the bars inwards the cage space is so reduced that the animal cannot struggle, and injections can be given between the bars, with little chance of injury to either the animal or the handler.

Restraint can be imposed by wrapping a cat in a large strong cloth. This is spread over the back and then the sides, and the end is folded in to enclose the hind legs. The front ends are crossed under the chin, enclosing the forepaws, and are brought back and tied behind the head

Fig. 28.9. Method of holding a cat in order to immobilize the four sets of claws. (A.T.A. Manual, 1963.)

or fixed with a large safety-pin. Alternatively, the cat may be placed in a strong bag with a drawstring around the neck, and a hole in one corner through which a limb can be withdrawn for venepuncture. Another method is to intro-duce the cat into a strong, rigid, transparent plastic tube about 9 cm in diameter; this is use-ful for electro-ejaculation or stomach-tubing.

Cat bites and scratches must be regarded as potentially dangerous. Scratches should be washed immediately with a suitable antiseptic solution (e.g. Cetavlon, ICI) and covered with a dry dressing; later they should be carefully inspected for signs of inflammation. Cat-bites are usually of the puncture type and difficult to wash clean, however much care is taken. It should be an absolute rule, for the safeguarding of the individual and of the Institution, that bites or inflamed scratches should be immediately reported to medical authority. Injuries from cats carry special risks due to *Pasteurella septica* and cat-scratch fever.

Transport

Cats may be transported in wicker baskets made for the purpose by the Association for the Welfare of the Blind, 257 Tottenham Court Road, London, W.1. These are useful in the laboratory and for sending cats on journeys but are not easily cleaned and do not withstand steam sterilization. Fibre-glass or plastic carrying baskets are easily kept clean, but the ventilation provided is often insufficient in hot weather, or for a litter of kittens. A box which can be used in the laboratory for induction of anaesthesia as well as transport is described on p. 346. The RSPCA recommends crates 90×42.5 cm, height 37.5 cm for sending cats on journeys. They can be divided into three by movable wooden partitions. The floor should be solid and the top and sides of 5 cm slats, 1.25 cm thick and 1.25 cm apart. Sheets and torn strips of newspaper make excellent bedding which should be burnt and renewed each time the basket or crate is used, since disease can be transmitted from one cat to another unless the container is kept scrupulously clean, preferably sterilized.

Generally speaking, it is unwise to leave food and water with the cat in its basket; it is liable to result in the cat being sick or becoming wet and dirty from upset fluid. It is better to reduce journeys to the shortest possible time and to feed and water immediately on arrival. Nervous cats can be given a tranquiliser before a journey. Whenever possible, cats should be transported in private vehicles (this is obligatory for any cat under experiment that is moved from one laboratory to another). Cats sent by rail should be delivered to railway stations as near the time of departure as possible, and consigned by passenger train. Efficient arrangements must be made for the consignee to collect them on arrival. Cats must never be left on cold and draughty platforms or in sheds, but kept in a warm room while awaiting transport or collection. Cats exposed to sudden chilling are liable to develop clinically apparent respiratory infections.

Cats can be exported to many countries by air without difficulty but Great Britain requires six or more months' quarantine for imported cats as an anti-rabies precaution. British Standard No. 3149, part 4, 1961 recommends that a specially designed container with vessels and trays for food, water and toilet purposes should be used for unaccompanied air journeys of any length so that the cat can be fed and watered and kept clean en route without being released, and all danger of escape avoided. It also states that cats (and dogs) are sensitive to noise on journeys, but advises against the use of narcotics on unaccompanied air journeys.

The reader is also referred to the latest (1975) IATA regulations for the transport of live animals by air. These are, however, less detailed than BSI No. 3149, merely specifying the size of container and ventilation holes and the recommended construction.

Feeding

Natural foods and feeding habits

Cats are true carnivores; wild or feral cats eat rodents, frogs, insects, and occasional birds, supplementing these by raids on human food-stores and waste-bins (Eberhard, 1954; Petterson, 1968). Both large and small cats consume all their prey, skin, bones and viscera as well as muscle, and thus obtain adequate supplies of all essential foods.

A well-nourished adult cat is an occasional feeder (Hediger, 1950) and has considerable reserves of fat and protein. It can withstand periods of starvation remarkably well, as evidenced by the survival of cats accidentally shut into outhouses. Adult cats can be fed once daily and trained to consume all their day's food within half an hour of offering it, but this feeding technique will not produce growth in kittens.

Acceptability

Cats differ from rodents in that they will starve to death rather than eat a diet they dislike, no matter how desirable, theoretically, the diet offered to them. The first criterion of a satisfactory diet must be that the animal will eat sufficient to supply enough calories for maintenance, growth, and reproduction. In spite of many efforts to determine factors influencing food-intake these are but poorly understood. Any environmental change which disturbs or frightens a cat, or any factor which affects its general health, will be immediately reflected in its food-intake, even to changes in personnel offering food to the animal. Continuous records of food-intake suggest that there are also rhythmic fluctuations in the amount of food consumed even by growing kittens.

It is clear that the taste of food is very important. Meat flavours (as meat extracts, but more particularly fresh raw meat), sweet-tasting foods such as milk powder, condensed milk or chocolate and fats (as butter and margarine), are all appreciated by cats. Foods, especially fish, must be fresh and free from bacterial decay for good acceptance. Overheating in processing or overcooking food (above 120°C) are likely to result in its rejection by the cat.

At weaning a kitten can be trained to like one particular food in preference to others, or to eat a variety of foods, but once the cat's tastes have become fixed it may prove difficult to persuade it to eat a different diet. The best method is to wean the kitten on to the complete form of any diet under investigation.

Fluid requirements

According to Carver & Waterhouse's experiments (1962) a six-month-old kitten weighing 2 kg will need about 7 g of water daily when given a diet containing 74 per cent of water; on a dry diet ten times as much water would be needed. An adult weighing 3 to 4 kg would drink about 30 g daily on a wet diet and seven times as much on a dry diet. Thus the fluid requirements of kittens and adults can be met quite easily by giving liquid milk, although access to a clean water supply is always desirable. Fluid requirements of lactating mothers are much greater, but have not been measured.

Special requirements

The nutritional requirements of carnivores have been reviewed by Scott & Scott (1967). The cat requires a greater proportion of protein to ensure satisfactory maintenance and growth, than does the dog (Miller & Allison, 1958). Greaves (1959, 1965) has shown that the ratio should be 11 to 13 mg N/kcal for growth and 7·5 to 9·0 mg N/kcal for maintenance; that is, kittens should have at least 32 per cent of the dry constituents of an average mixed diet as protein, and an adult cat not less than 21 per cent (Dickinson & Scott, 1956 a and b; Greaves & Scott, 1960; Table IV). In a diet containing about 75 per cent of water these minimal figures would be equivalent to 10 per cent of protein for growth and 6 per cent for maintenance, but good diets for cats usually contain 12 to 14 per cent of protein on a wet basis.

Fat should form 15 to 40 per cent of the dry weight of the diet, more than 60 per cent of fat has been fed without inducing vascular changes (Humphreys & Scott, 1962a). Oxidized unsaturated fatty acids are toxic and ingestion leads to steatitis. Carbohydrates are a cheap source of calories, but are not an essential part of the diet.

Vitamin and mineral requirements have been discussed by Scott (1965). Deficiency of thiamine is liable to occur on canned or dried diets (Loew et al., 1970); and deficient intake of vitamin A is not uncommon since it is normally only present when liver, cod liver oil or synthetic vitamin A are added to the diet. Vitamin C is not required by cats and, unlike the dog, their requirement for vitamin D is small. Calcium deficiency is the commonest mineral inadequacy encountered during growth and lactation, especially on meat diets free from bones; iodine intake is frequently below acceptable levels (see Table V).

There is good agreement in the literature on the calorie requirements of the cat (Table IV, Scott & Scott, 1967). A cat feeding four kittens will eat 500 g of meat a day; weaned kittens need about 100 g of wet diet daily at first, rising to 200 g a day at 10 weeks.

Recommended diets, including compounded foods

Cats can be fed fresh foods, or foods may be purchased ready mixed as in tinned or dried diets. The success of the diet may be measured by weighing weaned kittens weekly to record growth (Fig. 28.1) and by recording numbers and weights of kittens successfully reared by each queen in a colony.

Diets compounded in the laboratory are based on fresh meat offals and fish, cooked in the laboratory to destroy possible sources of infection. The protein source can be diluted with relatively less expensive biscuit meal, boiled rice or cooked potatoes up to one half of the dry weight of the ration. Such a diet should be supplemented with a vitamin and mineral mix. Well-balanced commercial tinned diets produced for pets, containing adequate levels of vitamins and minerals, are a convenient (but expensive) method of feeding, provided the contents of the can is acceptable to the cat. Expanded dried cat foods are very convenient and, in spite of earlier difficulties, are now very acceptable to cats. However, some laboratory colonies receiving this type of diet have shown an increase in the incidence of feline urological syndrome (FUS), formerly referred to as urolithiasis (Jackson, 1972). In addition, a reduction in the biological value of protein and a loss of vitamins tend to occur during drying at a high temperature in the presence of oxygen. Compared with tinned foods, shelf-life may be short, especially where storage conditions are not ideal. Research and development is being actively pursued in this field and more satisfactory products should shortly become available.

In the laboratory an inadequate food-intake is more likely to be due to anorexia than to the provision of insufficient food. Anorexia may result from an incorrect diet, deficient in protein and fat, and possibly low in vitamin A, thiamine,

or mineral elements; but the most common cause is infectious disease. In respiratory infections anorexia results from obstructed breathing and the kitten will eat its food once this symptom is alleviated; in alimentary disorders, on the other hand, it is usually best to discourage the kitten from eating for a time and, where necessary, to give 1 per cent glucose-saline parenterally.

Breeding

Reproduction

Aspects of reproduction in the cat have been reviewed by the author (Scott, 1970). A fertile queen is polyoestrous during the breeding season January to October, which can be prolonged over the winter months by providing 12–14 hours of artificial daylight. Cycles occupy about 14 days, full oestrus with acceptance of the tom lasting 3–6 days; cycles can be detected by oestrogen-dependent changes in behaviour and by vaginal smears (Figs. 28.10–28.13). In the absence of mating, cycles are anovular. Ovulation is a neuro-hormonal process induced by trains of afferent stimuli from the vagina and cervix acting on the pituitary via the hypothalamus. Several matings are normally necessary to bring about ovulation, but ovulation can also be induced in an oestrous queen by mechanical stimulation of the vagina or an injection of luteinising hormone. Ovulation occurs 25–27 hours after successful mating or injection of LH, and implantation of the fertilised blastula on about the 14th day. Corpora lutea are essential for the maintenance of pregnancy up to the 45th day, after which the ovaries can be removed. Delivery occurs on the 65th day ±4. Kittens can

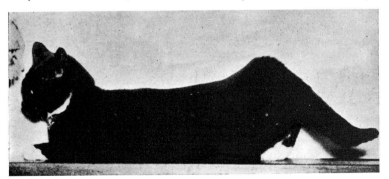

Fig. 28.10. Pre-coitus display by female in presence of male. (Photograph: Scott.)

Fig. 28.11. Coitus. (Photograph: Scott.)

be successfully removed seriatim from the uterus during pregnancy (McCance *et al.*, 1966). Data is available concerning cat embryos (Table III) and the growth, development and physiology of foetal and neonatal kittens which have been used in a variety of researches.

Selection and discard of breeding stock

Breeding toms should be between one and six years old, selected on the basis of weight (up-wards of 3·5 kg), health, vigour and general liveliness. Both testes must be fully descended into the scrotal sacs, firm and not excessively large. The ejaculate of about 0·3 ml, should contain at least 12×10^8 spermatozoa per ml, 80 per cent actively swimming, with a minimum of abnormally shaped sperm (Sojka, 1968). The tom should be provided with his own territory which he can mark. His behaviour towards an oestrous female should be carefully observed;

Fig. 28.12. Post-coital rolling. (Photograph: Scott.)

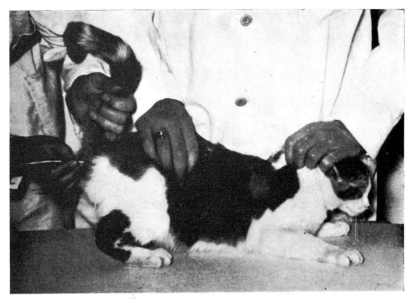

Fig. 28.13. Method of taking vaginal smears. (Photograph: Scott.)

about one in three randomly collected domestic toms do not perform successfully in laboratory surroundings.

Breeding queens should be between eight months and eight years or more. They should weigh upwards of 2·5 kg, be healthy, alert and have a good appetite. They must show regular oestrous cycles in the breeding period.

Breeding cats should be checked for intestinal and ectoparasites and must be vaccinated against panleucopenia. As far as possible carriers of bacterial, fungal and viral pathogens should be excluded. The establishment of a specific pathogen free colony of cats has been described by Bleby & Lacey (1969).

First litters seldom indicate the breeding potential of a queen but if she fails to rear a second litter or shows definite signs of cannibalism she should be discarded. Four or more kittens per litter should be aimed at; five litters can be successfully reared in the course of two years using artificial daylight during the winter months. Robinson & Cox (1970) have described the reproductive performance of a closed cat colony over a period of ten years.

Inbreeding and outbreeding

Common domestic or farm cats are out or random-bred, but pedigree varieties produced by breeders in the Cat Fancy are inbred to a greater or lesser extent. Homozygous, and heterozygous cats carrying known recessives, are available, mainly for coat and eye colour genes but also for certain other physical characteristics such as tailessness in Manx and strabismus in Siamese. Most laboratories practice more or less random breeding (since few are concerned with genetics) and produce numbers of half-sibs on the harem system.

Monogamous and polygamous units

Monogamous groups are not used in cat breeding since the extra males required occupy a great deal of space and are expensive to feed. Polygamous groups consist of 5–15 queens to one tom. The tom may be confined in a special mating area or cage which becomes his territory, oestrous queens being brought to him for mating. Alternatively he may be run with his harem. Some colonies run more than one tom with a group of queens, in such cases the dominant male will carry out 90–95 per cent of the matings.

Artificial insemination

Sojka (1968) has developed a technique for the successfully electro-ejaculation of tom cats; some toms can be induced to use an artificial vagina when 'teased' with an ovariectomised queen maintained in artificial oestrus. The technique is described by Scott (1970). Saline diluted ejaculate is used to inseminate queens at the height of oestrus with 1.2 to 3.0×10^8 sperm, via the cervix. To induce ovulation the queen is given 20–50 I.U. of luteinising hormone at the time of insemination. Fertilised and cleaving eggs may be recovered from the uterine (fallopian) tubes over the next six days.

To produce pregnant cats at any season of the year for teaching and research, Colby (1970) used pregnant mares' serum gonadotrophin to induce artificial oestrus in sexually mature random-source cats. The intramuscular dose rate was varied with the season; successful natural mating usually occurred (over 80 per cent of cases) on the 7th or 8th day. Immature females treated in this way tended to produce cystic follicles, superovulation, or over-production of follicles without ovulation.

Weaning and rearing

Depending on the number in the litter and the breed (strain) of cat, normal kittens weigh between 90 and 140 g at birth; underweight kittens are a poor risk and are often best discarded. A queen will not clean a still-born kitten, which sometimes gets eaten with the afterbirth and so overlooked by attendants.

Kittens can be very successfully transferred from their own mother to a foster-mother; by this means a female can be made to lactate continuously and rear at least three litters in succession. New-born kittens usually commence suckling immediately they have been washed but if not fed within 12 hours of birth they suffer fatal hypoglycaemia and hypothermia (Widdowson, 1965). Immediate post-partum period feeding occurs about every two hours but falls to four hourly and longer intervals very rapidly.

Kittens do not lose weight post-partum unless they are unhealthy or have been accidentally starved; gains of 2–5 g per feed, or 10 g in 24 hr are satisfactory. Average weight-gains (Fig. 28.1) should be 80 to 100 g per week, although some fluctuations may occur; after weaning, for example, a kitten may not gain for a week or two, but then put on 300 g in the third week. Failure to gain weight indicates that the food-consumption, i.e. the calorie intake, is below that appropriate to the kittens' age and weight and the cause of growth-arrest (insufficient food, over-competition, infection etc.) should be determined.

When a mother cat's milk-supply is deficient

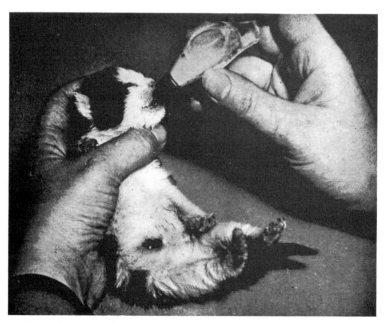

Fig. 28.14. Supplemental feeding of kitten using a doll's feeding bottle. (Photograph: Scott.)

the kittens, besides failing to gain weight, continually cry and nose around for supplies; the maternal milk supply should then be supplemented with a synthetic milk mixture using a doll's or special cat's feeding bottle (Fig. 28.14). Milk substitutes can be based on the composition of cat's milk, which contains 9·5 g protein, 6·8 g fat, 10·0 g lactose, 35 mg calcium and 70 mg phosphorus and provides 142 calories per 100 g liquid milk (Widdowson, 1965). Skimmed dried cow's milk reconstituted by adding 20 g to 90 ml water, then warming to 37°C and adding 10 ml vegetable oil, is a suitable substitute.

Caffyn (1965) recommends 2 parts by weight

TABLE IV
Daily food requirements of cats according to age
(modified from Scott & Scott, 1967)

Age	Expected weight kg	Expected weight lb	Daily calorie requirements kcal/kg body-weight	Daily calorie requirements kcal/lb body-weight	Daily ration g	Daily ration oz
Newborn	0·12	¼	380	190	30	1
5 weeks	0·5	1	250	125	85	3
10 weeks	1·0	2	200	100	140	5
20 weeks	2·0	4½	130	65	175	6
30 weeks	3·0	6½	100	50	200	7
Adult ♂	4·5	10	80	40	240	8½
Adult ♀ (pregnant)	3·5	7½	100	50	240	8½
Adult ♀ (lactating)	2·5	5½	250	125	415	14
Neuter ♂	4·0	9	80	40	200	7
Neuter ♀	2·5	5½	80	40	140	5

Ratio for newborn: 72% water, 9·5% protein, 6·8% fat, 10·0% lactose giving 142 kcal/100 ml milk mixture (Widdowson, 1965).
Ration for kittens and adults: 70% water, 14% protein, 10% fat, 5% carbohydrate giving about 150 kcal/100 g diet.

TABLE V
Recommended vitamin and mineral allowances for the cat

Vitamin	Daily dietary allowance	Comment
A (alcohol or ester)	1000–2000 i.u. (300–600 μg)	Cannot utilize carotene
D (cholecalciferol)	50–100 i.u.	May synthesize in skin
Essential fatty acids	1 per cent total FA	Tolerates high saturated FA intake
K (menadione)	Negligible	?Intestinal synthesis
E (α-tocopherol)	0·4–4·0 mg	Proportional to poly-unsaturated FA content
B_1 (thiamine)	0·2–1·0 mg (or 0·1 mg/50 cal diet)	Increase in lactation or fever
B_2 (riboflavin)	0·15–0·2 mg	Increase in lactation or fever and on high fat diet
Niacin (nicotinic acid)	2·6–4·0 mg	Increase in lactation or fever; cannot synthesize
B_6 (pyridoxine)	0·2–0·3 mg	Increase in lactation or fever
Pantothenic acid	0·25–1·0 mg	
Biotin	0·1 mg	
Choline	100 mg	
Inositol	10 mg	Essential
B_{12} (cobalamines)	?	Intestinal synthesis (Co present)
Folic acid	?	
C (ascorbic acid)	Negligible	Metabolic synthesis
Major Mineral Elements		
Na	20–30 mg	this is minimum intake
NaCl	1·5 g	common salt requirement
K	80–200 mg	
Ca	200–400 mg	increase in growth and lactation
P	150–400 mg	Ca/P ratio 0·9 to 1·1
Minor and Trace Elements		
Mg	8–10 mg	
Fe	5 mg	
I	100–200 μg	
Mn	200 μg	
Zn	250–300 μg	
Co	100–200 μg	

of full-cream dried cow's milk and 1 part of casein. The mixture has a protein/fat ratio of 1:0·65 and a fat/carbohydrate ratio of 1:1. Since queen's milk contains 1 per cent of minerals, these can, with advantage, be added to this mixture. Figure 28.14 shows a method for hand-feeding a kitten; frequency and amounts are indicated in Table IV, based on the work of Bleby & Lacey (1969) in hand-rearing caesarian-derived kittens.

The average length of lactation is seven weeks. Weaning commences during the fourth and fifth week post-partum. It is essential to supplement the mother's milk supply from the fourth week onwards and to encourage the kittens to take their first meal of finely minced beef, cooked fish or tinned cat food in a positive way. Weaned kittens need three meals a day, reduced to two at 10 to 12 weeks. Alternatively, enough food can be given to cover the whole 24 hour period, the kitten being allowed access at all times, but this is poor feeding technique.

Laboratory procedures

Selection and care of cats for experiment

A routine for the examination of cats, described by Custer (1964), can be applied before acute and during long-term experiments. Cats selected for experiment should be in perfect health; alert, with shining eyes and coat, neither fat nor thin, eating well and entirely free of discharges from the eyes, nose or ears. Cats for long-term experiments must have been vaccinated against panleucopenia and should be held for 2–3 weeks to ensure that they are free from infection. The use of specific-pathogen-free (SPF) cats, obtainable from a small number of accredited breeders, will prevent loss and disease from interfering with long-term experiments. At present category 3 cats, as specified in the LAC Accreditation and Recognition Schemes for Suppliers of Laboratory Animals (see Chapter 4, p. 54) are available for purchase in the UK and it is hoped that category 4 animals will be available soon. It is often asked whether SPF cats will maintain their superiority when brought into contact with apparently clean conventional cats, as may occur in experimental laboratories. It is most inadvisable to allow newly-weaned SPF kittens to have

contact with cats carrying respiratory viruses, but, by the time they are one year old, their reticulo-endothelial system is well established and, if they should become infected, the illness will be no more severe than in any conventional cat of the same age.

Cats are excellent subjects for experimentation whether involving surgical procedures or not, but success will depend on the provision of a suitable environment. Pre-operative care of conventional cats consists of treating for worms and ectoparasites, and building up resistance by feeding a good protein-containing diet with an adequate supply of vitamins and constant access to drinking-water. Cages should be sufficiently large for the animals to move freely. While they are adapting themselves to cage life, the cats' confidence can be won by their attendant; those which remain savage and intractable after a few days should be rejected because they are not suitable for long-term experiments. Personal relations between the cat, the experimentalist, and the technician who carries out daily routines of feeding and cleaning are most important. Gentle, capable handling, avoidance of sudden movements and loud noises are as important as the maintenance of a constant temperature in the comfort zone of the cat (20 to 22°C). Daily exercise for cats confined to cages and the provision of play objects such as ping-pong balls, or paper tied to a suspended string, keep the animal alert and improve food-intake. Grooming will also help to stimulate interest in a listless cat. Many cats do better when kept in pairs or groups rather than singly, provided that territorial fighting between entire males is avoided. Since cats become habituated to a particular environment and routine, moving them from one laboratory to another, or changing their companions or human attendants, may temporarily affect their behaviour and therefore the course of an experiment. Before starting an experiment cats should be given one or two weeks to adapt to new circumstances—for example, to a metabolism cage or to changes in personnel. Attention should be given to the problem of the accidental transfer of infection from one area to another when moving animals, and personnel, about a building.

Experimentalists who contemplate maintaining a colony of cats for experimental purposes, especially where breeding is desired, should make arrangements for obtaining veterinary advice. Before setting up their colony they should visit other active centres and are strongly advised to begin with SPF stock produced by recognised procedures (see Chap. 10) or obtained from an accredited breeder. Post-weaning losses are markedly reduced in SPF kittens; they grow more quickly and therefore have consumed less food by the time they attain a suitable weight for experimental use (usually 2 kg) (Festing & Bleby, 1970).

Details of experimental techniques and the methods developed by veterinarians for the medical and surgical care of the cat are excluded by considerations of space but appropriate texts can be consulted, e.g. Catcott, 1964. An experimentalist should try to obtain practical guidance from research personnel who are highly skilled in the particular procedure he wishes to adopt, this reduces the number of animals needed and saves times. Before undertaking recovery experiments, for which certificates B and EE are required in Britain, novel surgical procedures should be carefully worked out on carcases, followed by terminal experiments in fully anaesthetised cats before proceeding to recovery experiments. Operating-theatre techniques are well described by Brigden (1974).

Anaesthesia and euthanasia

General remarks. Cats and kittens are good subjects for anaesthetisation provided that they are in a healthy condition; general debilitation, dehydration, liver disease, and respiratory infections are contra-indications. Clinically apparent respiratory disease is a common cause of failure in acute experiments, and can be recognized by listening to the chest; waste of valuable time and material is avoided by refusing cats with dyspnoea, noisy breathing, or discharging eyes, nose or ears. The cat can be given a light meal the afternoon before the operation, but no food on the same day. Suitable anaesthetisation is obligatory in all experimental interferences likely to cause pain. Curare and other muscle relaxants are not anaesthetics, while morphine should be avoided as a narcotic, since it produces acute excitement in the cat, which becomes difficult to handle. Modern methods of anaesthetisation for cats are described by Hall (1971) and Stock

(1973). There is no one method of anaesthesia that is suitable for all occasions and for all experiments. The object of any anaesthetic procedure should be to produce as undisturbed an environment as possible for the experiment contemplated. Proper equipment is essential, and will include gas cylinders, flowmeters, mixing chambers, rebreathing bags and suitably sized masks, all of which can be obtained from suppliers to small animal veterinary practitioners.

Local anaesthesia. For the production of local anaesthesia by general infiltration of the skin or gums, a 2% aqueous solution of procaine (Novocain) is recommended, with the addition of 1/100,000 parts of adrenaline hydrochloride. At this strength procaine may be used almost *ad libitum* without danger of toxic reaction. Hall (1971) describes other local anaesthetics and gives a helpful account of the technique of infiltration.

Stages and signs of general anaesthesia. An excellent account of the signs and stages of anaesthesia in domestic animals has been given by Campbell and Lawson (1958). This should be studied by anyone wishing to undertake experiments involving surgery in the cat. The classification is based on the stages described by Guedel (1937).

Stage I: Analgesia. This begins from the commencement of induction and lasts until loss of consciousness is complete; there is a progressive loss of reactivity to painful stimuli. The inhalation of irritant vapours (e.g. ether) frequently gives rise to excitement and active resentment by the cat, making it difficult to handle.

Stage II: Delirium. The control of respiratory movements by the higher centres of the brain diminishes with loss of consciousness and is replaced by reflex (automatic) breathing regulated by the brain stem and mid-brain centres. During this change (which can occur during both induction and recovery), animals may struggle, hold their breath or vomit. Intravenous anaesthetics cause such rapid transit through Stages I and II that they may be missed, but in recovery from pentobarbitone (Nembutal) anaesthesia Stage II lasts for several hours.

Stage III: Surgical Anaesthesia. This lasts from the onset of automatic breathing to apnoea from respiratory paralysis; there is progressive muscular relaxation as anaesthesia deepens. Four planes can be identified; the important signs which distinguish these are described below under *Signs*.

Stage IV: Respiratory Arrest. As anaesthesia deepens, respiratory muscles are paralysed and death occurs from cardiac failure, due to anoxia of heart muscle. Artificial respiration at this stage will prevent cardiac arrest provided the plane of anaesthesia can be lightened.

Signs. Respiratory rate and depth must be carefully monitored, since changes in breathing indicate the stage and plane of anaesthesia. Inspiratory volume, resulting from the combined action of intercostals and diaphragm, may be reduced by lack of co-ordination between these very separate groups of muscles. Some anaesthetics stimulate the respiratory centres initially, resulting in an increase in respiratory volume in Stage I. In Stage II, breathing often becomes irregular and jerky, in contrast to Planes 1 and 2 of Stage III, when breathing settles down to a steady, regular, automatic rhythm. In Plane 3 (deep surgical anaesthesia), the intercostal muscles become progressively paralysed, chest movements occurring slightly later than diaphragmatic movements. At this stage, respiratory rate may increase to compensate for diminished depth; this change must not be mistaken for a lightening of anaesthesia. At the lower border of Plane 3, respiration becomes abdominal and gasping, which may interfere with surgery, observations or recording procedures. This passes into Plane 4, when the thorax becomes passive, so that the normal movements are reversed and the chest wall collapses on inspiration. When the cat is lying on its side, 'see-saw' movements of the chest and abdomen are obvious.

Information about the stage of anaesthesia can be gained from studying pupil size and eye reflexes to light. When the animal enters Stage III, the pupil constricts, but as anaesthesia deepens, it begins to dilate until, by Plane 4, it is fully dilated. Moreover, at this plane, no light reflex can be elicited, indicating that anaesthesia should be lightened. The eyeball remains moist until Plane 4 is reached, when lachrymation ceases, at which point the cornea loses its sheen and appears glazed. Since the eye may be damaged by repeatedly testing for corneal or conjunctival reflexes, it is preferable to rely on the palpebral reflex, a twitch elicited by touching the inner canthus. This reflex shows some variation, but is usually abolished by the end of Plane 1.

With any anaesthetic, heat regulating mechanisms are depressed very early in Stage III so that the cat loses heat. Thus, during prolonged manipulations, some form of external heating must be provided to prevent hypothermia, particularly when the body cavities are open, and body temperatures should be monitored with a rectal thermometer.

Anaesthetic Agents. For rapid induction and experimental interferences of short duration, there is a choice of routes and anaesthetics. In the past, inhalation anaesthesia has been used for both induction and maintenance, commonly by anaesthetic ether applied on a mask. However, the hazards of using this highly inflammable volatile substance, and the severe irritation caused to the lungs and other mucous membranes of the cat and resulting in postoperative oedema, suggests that the use of ether and related volatiles should be discontinued.

Halothane (Fluothane: I.C.I.), administered by mask at 3 to 4 per cent with oxygen can be safely used for the induction of cats and kittens, even newborn ones. If necessary, anaesthesia can be maintained on 2 per cent halothane and oxygen. No obvious damage to lungs or liver has been observed by the author in young kittens anaesthetised in this way on successive occasions, for blood collection by cardiac puncture. Methoxyflurane (Penthrane: Abbott) with oxygen, replaces halothane in the USA, probably because it is more suitable for warmer environments.

Both thiopentone and pentobarbitone have been widely used as anaesthetics, induction and maintenance being by the intravenous route. Intraperitoneal pentobarbitone was formerly popular but has now been superseded. These barbiturates have a narrow margin of safety in the cat, over-dosage leading to respiratory arrest and cardiovascular collapse. Recovery from pentobarbitone is very slow, and during it the cat may undergo a period of excitement which is unnecessary and sometimes destructive of experimental work.

Thiopentone is a useful induction agent, a 2·5 per cent solution being given intravenously at the rate of 20 mg/kg; half the dose is administered rapidly, followed by as much of the remainder as is required to produce relaxation. If given extravascularly, thiopentone will produce a tissue reaction resulting in a large sterile abscess.

Saffan (Glaxo) is a recently introduced steroid (alphaxalone and alphadolone acetate) anaesthetic having many advantages over barbiturates. It has a greater therapeutic ratio, is non-irritant to tissues, and is rapidly metabolised, so that repeated doses can be given without prolonging the anaesthetic recovery time which is brief. It produces only slight respiratory depression and transient hypotension, and its effects can be regulated. Two mg/kg, given intravenously, will result in light anaesthesia for 20 minutes; at 15 mg/kg, intramuscularly, light anaesthesia occurs after about 10 minutes, followed by a slow, deep sleep, which is useful for procedures requiring a high degree of restraint and relaxation, rather than complete surgical anaesthesia. Saffan can safely be supplemented by inhalation anaesthetics. Recovery is rapid, with the cat immediately becoming alert and eating well. Saffan has not shown any evidence of mineralocorticoid, glucogenic or progestational activity, but it does possess very weak anti-oestrogenic activity.

Ketamine hydrochloride (Vetalar: Parke Davis) is closely related to phencyclidine, but acts for a much shorter time. At doses of 10–15 mg/kg, given intramuscularly, it produces immobility and catatonia within 5 minutes. Larger doses (22–44 mg/kg) are profoundly analgesic and can be used for surgical procedures. Compared with adults, kittens require relatively higher doses in relation to body weight. Vetalar can also be supplemented with inhalation anaesthetics. Its chief disadvantage is a rather prolonged recovery time.

Alpha-chloralose, a compound of chloral and glucose ($C_8H_{11}Cl_3O_6$) may be used to produce light anaesthesia for acute experiments in cats. It can be given orally, intravenously or intraperitoneally in doses of 80 to 100mg per kg according to the batch of the drug and the depth of anaesthesia required. Alpha-chloralose can be dissolved in saline (0·9-per-cent NaCl) with the aid of heat (do not boil) to form a 1 per cent solution. When given intravenously, induction with a respiratory anaesthetic is desirable;

the injection should be made rapidly through a cannula or large-bore needle inserted into the saphenous or other suitable vein. When given intraperitoneally, a 2-per-cent suspension given through a large-bore needle will produce light anaesthesia in 30 minutes. In chloralose anaesthesia respiration and circulation are not depressed, blood-pressure tends to be high, and water-excretion is said to be normal. The animal sometimes exhibits periodic muscular spasms which can usually be abolished by intravenous injections of small doses of pentobarbitone sodium (10 to 20 mg) or by allowing the animal to inspire ether, for up to one minute, from a pad lightly applied to the nose or tracheal cannula.

Procedures. Inhalation anaesthetics are usually administered by veterinary masks of a suitable size and designed for cats or kittens. However, where surgery involves the face, eye, mouth or ear it may be necessary to pass an endotracheal tube. Intubation is more difficult in the cat than in the dog and it is necessary to desensitize the larynx with a local anaesthetic (Xylocaine: Astra). A 3-mm or 5-mm uncuffed tube is a suitable size for a cat. In acute experiments, a tracheotomy tube should be inserted.

The intravenous route is highly satisfactory provided the operator is skilled; induction is rapid and the depth of anaesthesia can be gauged as injection proceeds. Dissection and practice of the technique should be carried out on a dead cat, and then at the termination of an acute experiment. A short bevel needle should be slid into either the saphenous or the cephalic vein pointing towards the heart (Fig. 28.15), the area over the vein having been cleaned of hair by clipping or shaving and the vein raised by digital pressure applied proximally. If the cat is difficult and inclined to struggle it may be easier to induce by inhalation or intramuscular injection.

If the cat's respiratory movements cease during induction or while it is on the operating table, respiratory anaesthetics should be lightened immediately, and artificial respiration started by gently, but firmly and rhythmically, compressing the chest wall at about 30 times a minute until breathing is re-established; 7 per cent carbon dioxide in oxygen can be used as a respiratory stimulant. If the heart stops beating, massage and cardiac stimulants (adrenaline, coramine, picrotoxin) may be tried, but it is much more difficult to overcome syncope than respiratory arrest. It is often necessary to pump air into the lungs by a tracheal cannula; in acute experiments this should always be inserted and tied in immediately after induction; the pump must be carefully adjusted to give the correct rate and stroke volume.

Euthanasia. The recommended dose of pentobarbitone sodium (Nembutal) should be increased by twice or three times, and given intraperitoneally or intracardially. The animal is left in a comfortable position until it is stiff and cold. Where rapid death is essential, after deep anaesthesia has been induced by basal or respiratory anaesthetic, the chest wall can be opened, or a peripheral artery or vein can be cannulated to collect blood, severing the vessel when collection is complete. A high concentration of carbon dioxide in the bottom of an air-tight box will kill a cat or kitten in less than one minute without signs of distress. Etherised animals are dangerous; placed in a plastic bag they explode in an incinerator, and have caused explosions and fires in refrigerators. Anaesthetised cats have also been known to revive after refrigeration.

Post-operative care. A healthy cat recovers rapidly from operative procedures and attention to a few simple points will ensure absolute success. Firstly, food must be withheld for at least 18 hours before anaesthetisation, especially if abdominal surgery is contemplated; otherwise there is a considerable risk of the cat suffocating as a result of vomiting. Next, even with the best aseptic technique it is wise to give antibiotic cover post-operatively, e.g. 150,000 units procaine penicillin by subcutaneous injection at operation, and should the interference be extensive, the same dose on three to five succeeding days. Wound infections, though rare, give rise to pain, vitiate the experiment, and may result in the necessity of destroying the animal under the Pain Condition of Certificate B.

For useful details of the procedures undertaken in a well-run operating theatre, methods of stitching, etc., reference should be made to Brigden (1974).

After minor operative procedures involving the use of local or short-acting anaesthetics, a cat can be returned to its ordinary cage as soon as it recovers consciousness and resumes its normal routine. After major operations it must be allowed to rest quietly for 18 to 24 hours, away from its fellows, in a warm place (24°C) under subdued lighting, which will encourage sleep. When pentobarbitone is used it is sufficient to keep the animal quiet, but otherwise a post-operative tranquillizer may be advisable, especially if the cat has a tendency to scratch. Scratching can be discouraged by lightly clipping claws with bone-forceps (but not down to the quick), and by the use of a cardboard ruff padded with bandage and fastened round the neck. Properly stitched wounds should not be covered or bandaged. If the shaved skin is correctly everted and stitched with interrupted mattress sutures leaving long ends (preferably with silk), perfect healing will result; the cat will pull out the sutures in time, or they can be removed on about the fifth day. Alternatively Michel skin clips can be used and removed with the aid of special forceps on about the fifth day. After major surgery a cat should be placed in a relatively small cage or basket from which it cannot escape, and in which it can be readily observed without being disturbed. Sterilizable metal or plastic recovery boxes, with a floor area of 38×46 cm, and a depth of 20 cm, and clip-on wire-mesh covers are useful. The floor is covered with a thick layer of newspaper strips or cotton waste on which the cat is placed, loosely wrapped in one of the towels from the operating table. This prevents the bedding from sticking to the wound or obstructing breathing. The cat is laid on its side with its head slightly elevated. Water and food are withheld until the cat is fully conscious; an attempt to feed a partially anaesthetised cat usually ends in asphyxiation. After recovery from the anaesthetic cats usually eat well; anorexia signals trouble, perhaps an infection. According to Clark (1964) a diet made up of 1–2 g of fat, 2–3 g of carbohydrate and 1 g of amino acids per 454 g of body-weight is excellent for treating anorectic cats. Dissolved in water it can be given by stomach tube (see below) in two feeds during the day, (25 g dissolved in 50 ml water on each occasion) to a 3 to 4·5 kg cat. Other useful mixtures for post-operative care are that recommended as a cat's milk supplement (p. 345) or two parts by weight of full-cream dried milk powder to one part of casein, given mixed with enough water to allow it to pass along the stomach tube—about one part solid to three parts water; 30 to 50 g of solids should be given in this way each day.

Experimental feeding procedures

Oral administration of drugs, etc. A few taste-less substances can be given with milk or food, but most have to be given separately. Many tablets, powders, watery and oily solutions placed in a cat's mouth cause intense salivation and discomfort, followed by vomiting. The most practical method is to put all materials, solids or liquids, into gelatine capsules which can be obtained ready for filling as required; this must always be done for substances liable to cause salivation. Capsules are readily swallowed by cats and the coating of gelatine dissolves in the stomach; capsules are particularly useful for materials containing radioactive isotopes, for example, since the whole dose is swallowed cleanly.

To give capsules or pills single-handed, hold the cat under the left arm, pressing its body against your own to prevent struggling. (Alternatively the cat may be held by an assistant). Open the mouth by pulling the head back with the left hand, holding the upper jaw with the lips folded inwards around the teeth. Place the capsule or pill on the back of the tongue with the right hand, using small artery-forceps to hold the capsule for kittens with small mouths. Close the mouth firmly, and massage the throat with the right hand until swallowing is observed. A spoonful of milk or water, as a chaser, is a help in preventing regurgitation of the capsule.

Soluble medicines can be mixed with milk, which cats will often drink without compulsion, while sulphathiazole or other drugs can be added to drinking-water, care being taken to see that the fluid intake is not reduced because the cat objects to the additive.

Passing stomach and nasal tubes. This technique is useful in anorexia, and also for giving test meals or drugs directly into the stomach. A small gag of hard-wood or plastic, about 40×10 mm is placed in the mouth behind the canine teeth, and a Ryle rubber tube passed down a central hole in the gag. Measure the length necessary to reach to the stomach and mark the tube before passing it. Clark (1964) recommends the use of a flexible human urethral catheter as a nasal tube. After a local anaesthetic has been applied, the catheter is passed slowly and easily, with the tip curved downward, into the lower part of the nasal passage, through the nasal cavity and down the oesophagus and thus into the stomach.

Collection of specimens

Blood. A method described by Loeb (1964) for collecting small samples from the external ear into a glass capillary tube, consists of applying liquid paraffin to a clean ear, and pricking a vein with a sharp blade under the oil, thus preventing dispersion of the small amount of blood produced. Samples for smears may be collected directly on to a glass slide from a claw which has been clipped to the quick.

Techniques for giving intravenous injections or withdrawing larger samples of blood are described by Loeb (1964) and by Wright & Hall (1966). Well supported veins are used, the cephalic in the forelimb, the saphenous in the hindlimb and the external jugular vein in the neck. The novice is recommended to dissect these veins in a carcass so that he may become acquainted with their situation and course. The chosen area is clipped and shaved, but electric clippers are to be avoided since vibrations upset a conscious cat and blood values alter in a frightened animal. The area is cleaned either with soap and water or 1 per cent Cetavlon* (cetyltrimethyl-amonium bromide) in water, or 0·1 per cent Cetavlon in 70 per cent alcohol. Plastic disposable 2 or 5 ml syringes are useful for collecting blood, but glass-in-glass sterile syringes are better for injecting accurately measured doses. Tuberculin syringes are useful for volumes less than 1 ml; the amount delivered can be accurately determined by weighing the syringe before and after injection. The syringe must be clean, dry and sterile, and should run easily and efficiently. Needles (preferably sterile disposable) should be of 14 to 20 gauge depending on the size of the vein, short-bevelled, and sharp. Needle and syringes can be coated with silicone, citrate, EDTA or heparin to prevent clotting. Adopting the handling technique shown by Loeb (1964) for the external jugular, or as illustrated in Figure 28.15, see that the syringe is filled to the correct level, avoiding air

* ICI.

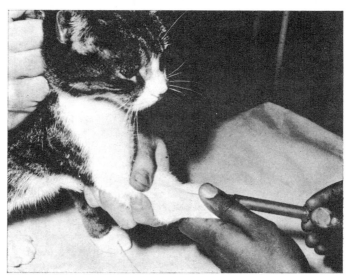

Fig. 28.15. Venepuncture. Method of holding cat applying pressure and inserting needle into cephalic vein. (Photograph: Scott.)

bubbles, and insert the needle into the vein, which has been raised by the application of pressure by an assistant. Direct the needle towards the heart, and insert it nearly parallel to the surface; withdraw the plunger slightly, and if blood shows in the syringe, proceed with the injection or collection. If no blood appears try again, without withdrawing through the skin, to direct the point laterally into the vein.

Fig. 28.16. General view of metabolism cage in rack. (Photograph: Scott.)

After withdrawing the needle apply a piece of gauze with digital pressure until bleeding ceases.

A short-acting respiratory anaesthetic, such as halothane and oxygen, should be given to cats which are difficult to handle or when the experimenter is single-handed. Indwelling Mitchell needles (Medical and Industrial Equipment Ltd.) or Guest Cannulae, secured to the cat's leg with adhesive tape, are useful for giving intravenous infusions of anaesthetic from time to time, or for taking serial samples of blood. Blood can be obtained by cardiac puncture from anaesthetised kittens of all ages and provided the amount is small (0·25–0·5 ml) collections can be made daily or weekly.

Urine and faeces. A single urine specimen may be collected by careful manual expression of the urinary bladder. For continuous collection of urine and faeces the cat must be kept in a metabolism cage (Figs. 28.16 and 28.17) with a grid floor through which the urine passes into a sloping tray and thus into a collecting-bottle (Humphreys & Scott, 1962b). The collecting-bottle may contain acid or toluene and the tray washings (with distilled water) be collected and combined, so that the total 24-hour collection is contained in a known volume of fluid. Faeces are usually well formed, and are collected as soon as possible with forceps and either placed in screw-top containers in the refrigerator or in

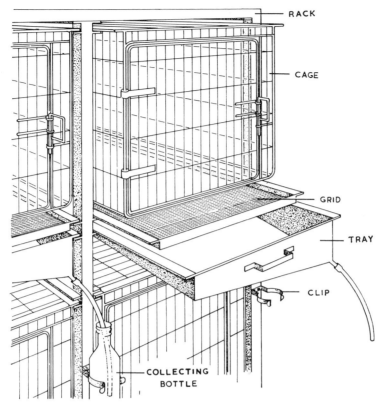

RACK

CAGE

GRID

TRAY

CLIP

COLLECTING
BOTTLE

Fig. 28.17. Details of construction of metabolism cage.

sulphuric acid. The minimum significant period for collecting urine and faeces for metabolic experiments is five days, while seven or eight day periods are more satisfactory. At least four days should be allowed for the animal to adapt to the cage, or to any change of diet, before the balance experiment is begun.

Dosing and injection procedures

The amount of restraint necessary for giving injections (see p. 351) depends on the volume and nature of the fluid to be injected; 1 or 2 ml of bland fluids can be given single-handed, either subcutaneously, intramuscularly, or intra-peritoneally, to a good-tempered cat. To give a subcutaneous injection pick up the loose skin over the shoulders and slide a sharp hypodermic needle (14 to 19 gauge, short bevel) into the subcutaneous space, parallel to the surface. Give intramuscular injections into the fleshy region of the thigh or the lumbar bundle, avoiding the thorax and the abdomen, and the sciatic nerve.

Push the hypodermic needle through the skin over the region, then vertically down into the muscle bundle for about 6 mm.

For intraperitoneal injection it is best to have the cat standing on all four legs with the viscera dependant and the abdominal muscles relaxed. Hold the loose skin over the back of the neck with the left hand to restrain the cat, place the right arm around the body, feel the border of the ribs and introduce a sharp hypodermic needle through the skin in the midline of the abdomen about 5 cm tailwards from the caudal end of the sternum; then push upwards, until you feel you have passed through the resistance of the muscle coats. The syringe can now be gently rocked to and fro to test that the point is lying free in the peritoneal cavity. Avoid going too deep, and see that the bladder is empty, if necessary emptying it by manual pressure before commencing. If little effect is apparent twenty minutes after injecting an anaesthetic such as pentobarbitone sodium, it can be assumed that the injection has accidentally been made into the stomach, intestinal lumen, or bladder. Re-anaesthetisation must then be carried out with the greatest care, as over-dosing can easily result.

The technique for intravenous injections is similar to that for collecting blood from a vein described above.

Disease control

The cat colony should be examined twice or thrice a week by a responsible senior person. On other days, including weekends, technicians undertaking feeding and cleaning must report changes, however slight, in the condition of the animals. A daily log should be kept in which all events, important or trivial, affecting the colony are recorded; such a record often proves valuable at a later date. The maintenance of a healthy colony will depend on the rapidity with which steps are taken to deal with trouble; the most fatal policy is that of waiting another twenty-four hours to see what happens. If the cause cannot be diagnosed it is best to isolate the cat or cats under suspicion and send for the veterinarian.

The condition of the cats is best assessed at feeding-time. A healthy cat will move to the front of its cage, or walk towards the attendant with tail erect, and consume food offered to it with avidity. A normal cat, which is unafraid, has an alert bearing and moves about readily with head held straight, showing interest in its surroundings. It will spring to the floor from a table with an easy controlled movement. Its skin is clean, free from blemishes, with a thick bushy undercoat and short or long silky guard hairs according to variety. Its eyes are shining, its teeth white and free from tartar, its tongue and gums clean and pinkish, and the auditory meatus is clean. The healthy cat grooms itself frequently and has a sleek appearance. The lumbar muscles form a firm bundle on either side of the back of a well-fed cat, and the vertebrae should not be very easily felt in the adult.

If the cat remains in its box at a meal-time, displaying little interest in food, it should be carefully examined and, if possible, the cause ascertained. Pregnant cats often sit or sleep in their boxes for long periods, but they leave these quite briskly at feeding-time. Watery lacrimation and purulent discharges from the eyes, nose or ears, excessive salivation, vomiting or diarrhoea, are all indications for immediate isolation, at least until it has been ascertained that the cause of the trouble is not an infection—i.e. due to accident or fighting.

Two reference books, *Feline Medicine and Surgery* edited by Catcott (1964) and *Diseases of the Cat* by Wilkinson (1966) will be found very helpful in the diagnosis and treatment of disorders. Particular attention should be paid to the elimination of protozoal, helminth, arthropod and fungal parasites, both external and internal, and the avoidance of reinfection or reintroduction of these pests into the animal-quarters. Bacterial infections are usually controllable, by disinfection and appropriate antibiotic treatment. Viral diseases pose a real problem in maintaining cats under colonial conditions. These diseases may be lethal and depress growth and reduce reproductive efficiency in survivors. Panleucopenia, that is feline infectious enteritis (FIE) can be controlled by a positive vaccination policy, but the viruses attacking the respiratory system, which include feline viral rhinotracheitis (FVR), picorne and pneumonitis viruses, are more difficult to control. The development of a high level of immunity is desirable in some cases, in other instances caesarian-derived, barrier-maintained cats may be the best method of eliminating 'carriers' from the colony.

Acknowledgement. O. F. Jackson has given valuable assistance with the section on *Anaesthesia.*

REFERENCES

Anderson, L., Wilson, R. and Hay, D. (1971). Haematological values in normal cats from four weeks to one year of age. *Res. vet. Sci.*, **12**, 579–583.

Bleby, J. & Lacey, A. (1969). The establishment of a specific pathogen free cat (*Felis catus*) colony. *J. Small Anim. Pract.*, **10**, 237–48.

Brigden, R. J. (1974). *Operating Theatre Technique* 3rd edn. Edinburgh: Churchill-Livingstone.

Caffyn, Z. E. Y. (1965). Artificial rearing of kittens. *J. Anim. Techns Ass.*, **16**, (1) 1.

Carver, D. S. & Waterhouse, H. N. (1962). The variation in the water consumption of cats. *Proc. Anim. Care Panel*, **12**, 267–70.

Catcott, E. J. (1964). *Feline Medicine and Surgery*. Illinois: American Veterinary Publications.

Clark, C. H. (1964). Management and treatment of hospitalized cats. In *Feline Medicine and Surgery*, ed. Catcott, E. J. Illinois: American Veterinary Publications.

Clifford, D. H. & Soma, L. R. (1964). Anesthiology. In *Feline Medicine and Surgery*, ed. Catcott, E. J. Illinois: American Veterinary Publications.

Colby, E. D. (1970). Induced estrus and timed pregnancies in cats. *Lab. Anim. Care*, **20,** 1075–80.

Custer, M. A. (1964). Physical examination. In *Feline Medicine and Surgery*, ed. Catcott, E. J. Illinois: American Veterinary Publications.

Danforth, C. H. (1947a). Morphology of the feet in polydactyly cats. *Am. J. Anat.*, **80,** 143–71.

Danforth, C. H. (1947b). Heredity of polydactyly in the cat. *J. Hered.*, **38,** 107–12.

Darwin, C. (1868). *The Variation of Animals and Plants under Domestication*. London: Murray.

Dickinson, C. D. & Scott, P. P. (1956a). Nutrition of the cat. 1. A practical stock diet supporting growth and reproduction. *Br. J. Nutr.*, **10,** 304–10.

Dickinson, C. D. & Scott, P. P. (1956b). Nutrition of the cat. 2. Protein requirements for growth of weanling kittens and young cats maintained on a mixed diet. *Br. J. Nutr.*, **10,** 311–6.

Dyte, C. E. & Turner, P. (1973). Further data on folded-ear cats. *Carnivore Genetics Newsletter*, **2,** (5), 122.

Eberhard, T. (1954). Food habits of Pennsylvania house cats. *J. Wildl. Mgmt.*, **18,** 284–6.

Ewer, R. S. (1973). *Carnivores*. London: Weidenfield & Nicolson.

Glenn, B. L., Glenn, H. G. & Omtvedt, I. P. (1968). Congenital porphyria in the domestic cat (*Felis catus*). *Am. J. vet. Res.*, **29,** 1653–7.

Graham-Jones, O. (1964). Restraint and anaesthesia of some captive wild mammals. *Vet. Rec.*, **76,** 1216–43. Discussion, pp. 1243–8.

Greaves, J. P. (1959). *The nutrition of the cat: protein requirements and other studies*. Ph.D. thesis, University of London.

Greaves, J. P. (1965). Protein and calorie requirements of the feline. In *Canine and Feline Nutrition*, ed. Graham-Jones, O. Oxford: Pergamon.

Greaves, J. P. & Scott, P. P. (1960). Nutrition of the cat. 3. Protein requirement for nitrogen equilibrium in a mixed diet. *Br. J. Nutr.*, **14,** 361–9.

Hall, L. W. (1971). In *Hall/Wrights Veterinary Anaesthesia and Analgesia* 7th edn. London: Bailliere Tindall.

Hediger, H. (1950). *Wild Animals in Captivity*. London: Butterworth.

Holmes, R. (1953). The occurrence of blood groups in cats. *J. exp. Biol.*, **30,** 350–1.

Humphreys, E. R. & Scott, P. P. (1962a). The addition of herring and vegetable oils to the diet of cats. *Proc. Nutr. Soc.*, **21,** xiii.

Humphreys, E. R. & Scott, P. P. (1962b). A new metabolism unit for cats. *J. Physiol.*, **162,** 6P.

International Air Transport Association (1975). *IATA Live Animal Regulations*. Fourth Edn. Geneva: IATA.

Jackson, O. F. (1971). *An Experimental Study of the Factors promoting the Formation of Uroliths in the Cat*. Unpublished thesis for Ph.D., University of London.

Jackson, O. F. (1972). The dry cat food controversy. *Vet. Rec.*, **91,** 292–293.

Jackson, O. F. (1974). A heritable osteodystrophy of the extremities of the cat. *Proc. Netherlands small Anim. Vet. Ass.*, p. 21.

Jackson, O. F. & Scott, P. P. (1970). A breeding cage unit for cats. *Lab. Anim.*, **4,** 135–7.

James, C. C. H., Lassman, L. P. & Tomlinson, B. E. (1969). Congenital anomalies of the lower spine and spinal cord in Manx cats. *J. Path.*, **97,** 269–76.

Kling, A., Kovach, J. & Tucker, T. (1969). The behaviour of cats. In *Behaviour of Domestic Animals*, ed. Hafez, E. S. E. London: Baillière, Tindall & Cassell.

Latimer, H. B. (1939a). The prenatal growth of the cat. VII. Weights of kidneys, bladder, gonads and uterus with weights of adult organs. *Growth*, **3,** 89–108.

Latimer, H. B. (1939b). Weights of hypophysis, thyroid and suprarenals in the adult cat. *Growth*, **3,** 435–45.

Latimer, H. B. (1947). Correlations of organ weights with body weight, body length, and with other weights in the adult cat. *Growth*, **11,** 61–75.

Latimer, H. B. (1967). Variability in body weight and organ weights in the newborn dog and cat compared with that in the adult. *Anat. Rec.*, **157,** 449–56.

Loeb, W. F. (1964). Laboratory procedures. In *Feline Medicine and Surgery*, ed. Catcott, E. J. Illinois: American Veterinary Publications.

Loew, F. M., Martin, C. L., Dunlop, R. H., Mapletoft, R. J. & Smith, S. I. (1970). Naturally-occurring and experimental thiamine deficiency in cats receiving commercial cat food. *Can. vet. J.*, **11,** 109–13.

Makino, S. & Tateishi, S. (1952). A comparison of chromosomes in the lion, leopard cat and house cat. *J. Morph.*, **90,** 93–102.

McCance, I., Phillis, J. W. & Wright, B. (1966). The seriatim removal of foetal kittens. *J. Reprod. Fert.*, **12,** 229–32.

Miller, S. A. & Allison, J. B. (1958). The dietary nitrogen requirements of the cat. *J. Nutr.*, **64,** 493–501.

Mivart, St. G. (1881). *The Cat—an Introduction to the*

Study of Boned Animals, especially Mammals. London: Murray.

Moutchen, J. (1950). Quelques particularités héréditaires du chat siamois. *Naturalistes belg.*, **31**, 200–3.

O'Reilly, K. J. & Whitaker, A. M. (1969). The development of feline cell lines for the growth of feline infectious enteritis (panleucopenia) virus. *J. Hyg., Camb.*, **67**, 115–24.

Petterson, R. (1968). Metodik vid studier av varingeval vos katter. *Sartrych ur Zoologisk Revy årg.*, **29**, 1–9.

Pocock, R. I. (1951). *Catalogue of the genus Felis*. London: British Museum (Nat. Hist.).

Robinson, R. (1959). Genetics of the domestic cat. *Biblphia genet.*, **18**, 273–362.

Robinson, R. (1971). *Genetics for Cat Breeders*. Oxford: Pergamon Press.

Robinson, R. & Cox, H. W. (1970). Reproductive performance in a cat colony over a 10-year period. *Lab. Anim.*, **4**, 99–112.

Sadleir, R. M. F. S. (1969). *Ecology of Reproduction in Wild and Domestic Animals*. London: Methuen.

Scott, P. P. (1965). Minerals and vitamins in feline nutrition. In *Canine and Feline Nutritional Requirements*, ed. Graham-Jones, O. Oxford: Pergamon.

Scott, P. P. (1970). Cats. In *Reproduction and Breeding Techniques for Laboratory Animals*, ed. Hafez, E. S. E. Philadelphia: Lea & Febiger.

Scott, P. P. (1974). The nutritional requirements of cats. In *Basic Guide to Canine Nutrition*, 3rd edn. Gaines Dog Research Centre, White Plains, N.Y. 10625.

Scott, P. P. & Lloyd-Jacob, M. A. (1959). Reduction in the anoestrous period of laboratory cats by increased illumination. *Nature, Lond.*, **184**, 2022.

Scott, P. P. & Scott, M. G. (1967). Nutritive requirements for carnivores. In *Husbandry of Laboratory Animals*, ed. Conalty, M. L. London & New York: Academic Press.

Searle, A. G. (1953). Hereditary 'split-hand' in domestic cats. *Ann. Eugen.*, **17**, 279–82.

Searle, A. G. (1968). *Comparative Genetics of Coat Colour in Mammals*. London: Logos Press & Academic Press.

Short, D. J. (1969). Transport of Laboratory Animals. In *The A.T.A. Manual of Laboratory Animal Practice and Techniques*, 2nd ed., ed. Short, D. J. & Woodnott, D. P. London: Crosby Lockwood.

Sojka, N. J. (1968). Personal communication.

Standardized Genetic Nomenclature for the Domestic Cat. *J. Hered.*, **59**, 39–40.

Stock, J. E. (1973). Advances in small animal anaesthesia. *Vet. Rec.*, **92**, 351–354.

Weaver, B. M. Q. (1964). Current methods of anaesthesia for cats. In *Small Animal Anaesthesia*, ed. Graham-Jones. O. Oxford: Pergamon.

Weipers, W. L. (1958). Personal communication.

Widdowson, E. M. (1965). Food, growth and development in the suckling period. In *Canine and Feline Nutritional Requirements*, ed. Graham-Jones, O. Oxford: Pergamon.

Wilkinson, G. T. (1966). *Diseases of the Cat*. Oxford: Pergamon.

Wright, J. G. & Hall, L. W. (1966). *Veterinary Anaesthesia and Analgesia*, 6th ed. London: Baillière, Tindall & Cox.

Veterinary Library Lists 1–6, containing references to work on cats, may be obtained from the Feline Advisory Bureau Librarian, Mrs Ruth Goodwin, M.Sc., 92 Church Road, Horley, Surrey.

Carnivore Genetics Newsletters are obtainable from the editor, Roy Robinson, St. Stephen's Nursery, London, W13 8HB, England, and contain many papers on cats.

29   The Ferret

J. HAMMOND JR. AND F. C. CHESTERMAN

Introduction

The ferret (*Mustela putorius furo* L.) is a mustelid carnivore: that is, it belongs to the same general group as otters, weasels and mink. It has been a domesticated animal for at least 2,000 years, and is generally held to be a form of the European polecat (with which it will interbreed), though some believe it is a hybrid from this and other species. As a domestic animal it has been used principally for hunting rabbits and, to some extent, for destroying rats. It is perhaps because of this that the ferret has a reputation for ferocity, which is quite undeserved.

There appear to be no standard breeds or strains although there are minor variants of wild-type markings (generally known as polecat) and also albinos. Albinos tend to have a rather yellowish hue, particularly when the coat is old; the colour derives from sebaceous secretion. Albino is recessive to polecat.

The ferret has been relatively little used as a laboratory animal. The reasons for this neglect, apart from unfamiliarity, are perhaps the restricted breeding season and the implications of the animal's naturally carnivorous diet. Both these difficulties are in fact easily overcome. It is possible to maintain ferrets on a pelleted diet and, by simple photoperiod control, to breed from them at any time of the year. Ferrets have been used for work on the pathology and immunology of virus diseases (e.g. distemper, influenza, vesicular stomatitis) and in experiments on the viral aetiology of spontaneous ataxia of cats (Kilham & Margolis, 1966). They are susceptible to the oncogenic action of polyoma virus (Harris *et al.*, 1961) but not to Rous sarcoma virus (Harris & Chesterman, 1964). They have been used for work on reproduction, control of breeding season by day-length change, growth, and pelt cycles. They might prove useful additional animals for the safety testing of drugs, and a satisfactory alternative to the cat for pharmacological studies.

This account of their management is largely based on personal experience in these fields, with particular reference to the colony at Cambridge (J.H.).

Standard biological data

Longevity, weight and breeding cycle etc. will be considered later. The structure, development and cytology of the spleen and bone marrow together with a haemogram of marrow cells are described by Kohler (1958). Haematological values on blood obtained by cardiac puncture of normal ferrets 15 months of age are given by Pyle (1940). Total red cell count is 8·5 million per mm³ and white cell count 8,900 per mm³. Polymorphonuclear leucocytes form 65 per cent of cells of the stained film. The differential white blood cell count for the polecat (*Mustela putorius*) is: polymorphonuclear leucocytes 42 per cent; lymphocytes 55 per cent; large mononuclears 2 per cent; and eosinophils 1 per cent (Fox, 1923). Macromolecular uptake by the gut epithelium continues as late as 40 days postnatally (Clarke & Hardy, 1970).

The average rectal temperature is 38·8°C but may vary from 37·8°C to 40·0°C (Woodnott, 1969). Respiration rate ranges between 33 and 36 per minute (Pyle, 1940).

Husbandry

Housing

Although ferrets can be caged in the same way as mink* such methods are not generally what are required for laboratory animals. Mink are ferocious animals and so are handled as little as possible; they are often housed outdoors so as to minimise the labour of dung disposal and to avoid the need for artificial ventilation. Ferrets, however, are very tame when frequently handled and have to be housed indoors, where they require little space, if there is to be artificial

---

* For an account of mink husbandry see *UFAW Handbook on the Care and Management of Farm Animals* (1971). Edinburgh: Livingstone.

control of day-length exposure. They will play together, particularly when young, but they seem to spend most of their time asleep, and they do not need to be provided with a run for exercise.

They may be satisfactorily kept in wire cages, but we prefer a solid floor covered with wood shavings. A solid floor is essential if pellets are fed. Three females or two males can be comfortably housed in a 60 cm square compartment of a wooden rabbit hutch (Fig. 29.2), or a group of up to 35 females can be kept on the floor of a room measuring 2·5 by 3 metres. One or more corners of the cage, depending on the size of the group, will be used for droppings. Cleaning out necessitates the daily removal of shavings from the defaecation area, with an occasional addition of more shavings to the floor-covering. An alternative method of cleaning out pens or cages is to have removable sawdust-filled receptacles with low fronts and high backs in the defaecation corners. It is necessary to devise a simple method of anchoring such receptacles securely since it is the nature of ferrets to investigate and if possible enlarge any hole which may be present.

If fed a diet of meat or of bread and milk another corner of the cage will be used as a larder, and will also need cleaning; pellets are not transferred to the larder. If a mash is fed with pellets it may be placed by the ferret in a larder corner, or it may be stored in the pellet dish.

If caging of the rabbit-hutch type is used the lower part of the opening, inside the door, must be closed with a strip of metal sheet so as to retain the bedding material. To retain droppings this strip needs to be 10 to 12 cm high at the corners (Fig. 29.2).

Dishes for pellets should be heavy and have high sides, so that they are not easily overturned nor the pellets easily scraped out. No doubt a suitable hopper for pellet feeding can be devised, but with the ordinary type used for rabbits all the pellets are liable to be soon scraped out, and the ferret (which is adept at squeezing through small apertures) is likely to escape from the cage through the hopper.

One of us (J.H.) has kept ferrets in unheated (but moderately insulated) buildings in England and in Massachusetts, and at constant temperatures of 7–10°C and 27–32°C, and has also kept adults in a refrigerator room at near freezing point. The animals thrived under all these conditions. Litters were reared successfully at 7°C, but it is probably advisable that unweaned young should be kept at room temperatures of at least 15°C.

On the neck dorsum of newborn young there are extremely well developed apocrine glands, but otherwise the ferret has little in the way of sweat glands and cannot endure very high environmental temperatures. Gentle air movement through the animal-house is desirable to keep down humidity and remove smell from stale droppings. Like the mink and other carnivores, the ferret has anal scent glands which produce a potent and persistent odour. They are discharged only when the animals are frightened, and this should rarely happen under laboratory conditions.

Because photoperiod affects reproduction, the room should either have adequate natural lighting to allow proper supervision, and no artificial lighting, or there should be automatically-controlled, exclusively artificial, lighting. Suitable lighting schedules are described below. A natural pattern of day-length change, using artificial light, can be obtained by modification of a time-switch such as is used for control of street or shop-window lighting.

Identification

The ears of the ferret are small and do not lend themselves to tattooing or notching, and ear tags are very liable to get torn out. A collar (like those worn by cats) can be used to carry an identification code. If albino animals are used they can be readily and easily identified by the use of hair dyes (ICI Durafur dyes or proprietary hair dyes are very satisfactory). The markings have to be renewed when the animals moult. Details of moult cycles are given below.

Handling

The use of gloves may be of some comfort to the novice handler, but none to the ferret. If gloves are used (that are tough enough to give full protection from a ferret's canines) the animal cannot easily be held firmly without being hurt. However, handled confidently and treated kindly, only a sick animal, or one with a litter of young,

Fig. 29.1. Holding a ferret. The thumb and forefinger is around the neck and the other fingers around the chest behind the fore limb. (Photograph: D. M. Auger.)

is likely to attack. The ferret should be picked up by placing the hand across the animal's shoulders—the thumb and forefinger around the neck, the other fingers around the chest behind the forelimb (Figs. 29.1 and 29.2).

If it appears possible that the animal may resent interference (when nursing, for example) then the back of one hand (which is too big to be bitten) should be presented for the ferret to smell at and the animal seized with a rapid movement of the other hand. To introduce one hand into the cage and then withdraw it nervously as the animal approaches is to imitate the behaviour of its natural prey and to invite attack. Rarely an animal may bite and, although free to escape, retain its hold. If this happens, do not try to force the mouth open, but hold the animal's head under the cold running water tap.

Transport

Ferrets will not be accepted for transport in Britain if the journey is to exceed 12 hours (Short & Woodnott, 1969). If there is any danger of warm travelling conditions, cages for transport should be constructed to allow good air movement, because the ferret relies for cooling on evaporative heat loss in respiration. At least two sides of the cage should then be of strong wire mesh, with apertures too small to allow the muzzle to be pushed through. The cage

floor should preferably be solid. If the cage is of all wire construction a loop handle which stands clear of the mesh should be provided, to allow lifting without directly touching the mesh. Food and water should not be needed for a 12 hr transit period.

Feeding

It is hardly proper to speak of a natural diet for the ferret since, except in New Zealand where it has successfully gone feral, it does not live apart from man. In the wild its diet, besides rabbit and other small mammals, probably extends to molluscs, arthropods and birds' eggs. The large intestine consists only of colon and rectum, so fibre digestion must be negligible. There is a close similarity in body size to mink, for which the nutritional needs have been defined (National Academy of Sciences, 1968); the requirements of the ferret are probably much the same.

It is possible to keep ferrets on bread-and-milk only. Addition of margarine is probably bene-

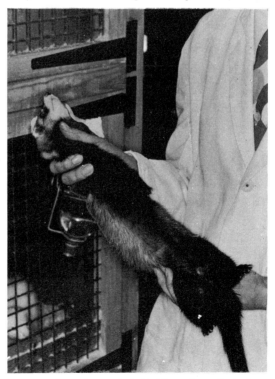

Fig. 29.2. Alternative method of holding a ferret. In the front of the wooden hutch is a metal strip to retain droppings when the door is opened. (Photograph: D. M. Auger.)

ficial, but in either case the droppings are very wet and unpleasant to deal with. In some laboratories it may be practicable to feed exclusively with the flesh of other experimental species. On a bread and milk diet or meat diet King & Glover (1945) found that the addition of short lengths of bone to the diet prevented the development of paradontal disease. Generally it will be desirable to provide a less expensive and more convenient diet. On mink farms* it is usual to feed ground fish or poultry waste and horsemeat, extended by the addition of cooked starch and fat, or by a commercial meal. This is made into a mash, deep frozen for use as required, and fed once daily, drinking water being also available. Ferrets do very well on such a regime. Complete powdered diets, designed to be fed as a mash and mixed as required, are now commercially available for mink.

A pelleted diet eliminates the labour of food preparation and the daily cleaning of food dishes, and much simplifies food hygiene. To find something suitable and readily available, one of us (J.H.) has tried a variety of commercial pellets. Pellet diets for early-weaned piglets have been generally satisfactory, and over a period of 10 years well over a thousand ferrets have been reared on such diets alone.† Preference has been given to diets which are not supplemented with copper, which is frequently added to pig diets as a growth promoter.

It is not suggested that such a diet is ideal, but with one modification during lactation it has been generally satisfactory. For experimental reasons litter size has been limited to 6 young, but even so, to secure good growth, it has been found advantageous to provide some wet feed from 3 weeks after birth until weaning. A mash made by soaking pellets in hot water is given together with dry pellets. The young at weaning are found to prefer the pellets, but they appear to lack the strength of jaw (and perhaps also sufficient saliva) to deal with the dry pellets any earlier than this. The wet mash also seems to allow greater maternal food intake.

On a meat diet the mother brings food to the

young in the nest, but does not appear to take pellets. When the young are 25–28 days old— well before the eyes are open—they wander from the nest and will readily take the warm mash, which is best provided in a rather fluid state at this stage. Water can be given either in a drinking cup or from a drinking bottle, but it is difficult to get the bottle spout sufficiently low, without touching the bedding, for the young to reach it.

The ferret's milk has a high fat content; if, in emergency, young have to be hand-reared, cream-enriched cows' milk, or a milk and egg-yolk mixture, can be fed by medicine dropper until they can drink from a saucer.

Breeding

Under natural lighting conditions the testes of the male begin to enlarge in midwinter and are fully developed after about 8 weeks. They begin to regress in July or August, on average towards the end of July. The female does not ovulate spontaneously, and the vulva provides an excellent indicator of ovarian activity, swelling under the influence of oestrogen—an effect which is antagonised by progesterone (Marshall & Hammond, 1945). About a month after the vulval swelling first becomes apparent it reaches full size and the animal becomes oestrous (Fig. 29.3). The season for the female begins and ends later than that for the male. Generally they

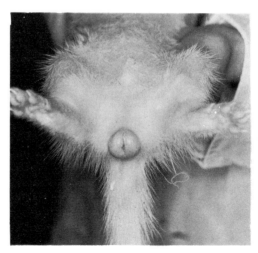

Fig. 29.3. Vulval swelling during oestrus. (Photograph: D. M. Auger.)

---

* See *UFAW Handbook on the Care and Management of Farm Animals* (1971). Edinburgh: Livingstone.

† Mainly Amvilac No. 2 with antibiotic (Glaxo); no longer commercially available.

become oestrous early in April, and a few may still be in breeding condition in September—or even later, though some become anoestrous in July; occasional females may be oestrous as early as the end of February.

Three or four days after mating the vulva becomes dry. It becomes wrinkled and subsides almost completely in the following four weeks although if oestrus has been prolonged subsidence is slower and less complete. Pseudopregnancy and pregnancy are of the same duration, nearly always 41–43 days. Should there be only one or two young they usually die *in utero* at 45 days and are delivered a few days later.

Within about a fortnight of the end of pseudopregnancy, or of pregnancy with the loss of the litter, or after weaning, the vulva again begins to swell, reaching full size in about ten days. When only a few young are nursed the female may come on heat for a short time soon after parturition, but becomes anoestrous again as the strain of lactation increases. It is doubtful whether concurrent pregnancy and lactation are practicable. If mating is not delayed it should be possible for a female to rear two litters a year; the litter size may vary from 3 to 15. It is not wise to breed from a female for more than 3 seasons. The normal life expectancy is probably 5 or 6 years, although Kirk (1966) states that it may reach 10–11 years.

If females reared on natural day-length are given long days in late autumn or winter their next breeding season begins prematurely. There is an optimum stimulating day-length (c. 14–15 hours) with a slower response to photoperiods longer than the optimum.

If it is desired to breed and rear ferrets at all times of the year half the colony can be kept on natural lighting and half on a reversed natural lighting pattern. But probably it is best to keep the whole colony on constant short photoperiod (say 6 to 8 hours light daily) but to have available also a long-day room (say 14–15 hours light/day).

Ferrets have been kept from birth (by J.H.) on 2-hour, 6-hour and 9-hour constant short-days, and on 14-hour and 24-hour long daily photoperiods. Males on constant long-day have alternating periods of sexual activity and rest; on short-day they remain in breeding condition

almost indefinitely, for periods of a year or more. Females on constant day-length follow in general an annual rhythm of breeding and rest periods, much as they do under natural lighting. However, on short-day the age at first oestrus is 7–8 months, while on 24-hour (continuous light) it is about a year. On 14-hour days many females fail to come on heat at all.

Thus on constant short-days males will nearly always be available for breeding, and females will be breeding at times determined by their dates of birth (and inherent characters). Females that fail to breed on constant 14-hour days come on heat about 6 months after transfer to short-day; but they will respond to long-day stimulation after only 8 weeks on short-days. Females that have completed a breeding season on short-day will comes into oestrus again if moved to long-day, and young animals reared on short-day will be breeding earlier (at $4\frac{1}{2}$–5 months) if transferred to long-day. However, if transfer is made too early (before about 90 days) they become refractory before full breeding condition is attained, and relapse into anoestrus. If long-day conditions are available there is the possibility of arranging matings at fairly short notice, should sufficient naturally oestrous females not be available when needed.

Moulting

Hair shedding and replacement seem to be determined by an ageing process in the hair follicle, modified by adrenal corticoids acting synergistically with oestrogen. Glucocorticoid inhibits hair shedding and replacement. Under natural lighting conditions there is an autumn moult in October-November. The coat thins out by partial shedding and in males is not usually replaced until just at the end of the breeding season. In females a moult follows the first ovulation of the season, and usually also subsequent ones, but it may be delayed during lactation.

If a small patch of hair is plucked from a female in pro-oestrus the skin remains bare until regrowth is initiated after ovulation (or at the end of oestrus). On the loin dorsum the regrowth reaches the skin surface 21–24 days after mating—which happens also to be the stage at which uterine swellings can most conveniently be palpated. If females are allowed to run with a male it is then possible by periodical checks on a plucked area to predict date of parturition within a day or two, even though mating has not been observed.

Mating

Generally dated matings are required. Small groups of males may be caged together if they have been reared together or have been put together in the non-breeding season, but it is best to separate them at the time of mating. If a female which is not fully oestrous is placed in the male's cage they will generally be found in different corners within half an hour. Mating itself is prolonged, particularly at the start of the season, and takes from half to several hours. Part of this time is spent in achieving intromission—at which some males are particularly clumsy. Induction of ovulation does not require intromission, and it is advisable to check that males are mating effectively.

Unless heavy use has to be made of the males, it is convenient to leave the female with the male for 24 hours, and if pedigree is unimportant she can then be placed with another male. Ovulation occurs about 30 hours after mating, and fertilization can result from insemination 12 hours or more after ovulation. Recently-mated females should not be placed with other oestrous females, as the latter would then be very likely to become pseudopregnant. However small groups of oestrous females can, unlike rabbits, generally be caged together without ovulation occurring.

Mated females can be caged together (or with oestrous females, provided they are not put together until 4 or 5 days after mating). They should, however, be isolated a few days before parturition is due. It may also be necessary to separate pseudopregnant females at this stage, as they develop maternal instincts and tend to try to drag other females into their nest and this can lead to fighting. Such behaviour seems to delay a return to oestrus; if it develops it can be soon stopped by caging the female for a few days with a male.

When wood shavings are provided for bedding, the nest made by the pregnant female is simply a depression in a pile of shavings. Shedding of the hair on the belly may be accelerated by licking of the nipples, but shed hair is not collected to make a nest. Although there is no need to provide special nesting material, it is desirable that the cage should have an enclosed nest-box, or separate nesting compartment. An appreciable loss of whole litters may occur within the first 3 days. This loss occurs in several ways: the young may be eaten with the placenta, they may be collected into the nest but apparently not nursed, or they may (rarely) become tangled together in the nest because the placentae have not been eaten. Agalactia is, according to Rowlands (1967), encountered only during the early part of the breeding season. Parturition can sometimes be prolonged, and it seems likely that loss of whole litters arises largely from disturbance of the female at about the time of birth. In some colonies preweaning losses can be as high as 70 per cent (Seamer & Chesterman, 1967). Grinham (1952) finds that a greater proportion of young survive when rearing takes place in the more open environment of a wire-mesh cage than in wooden hutches.

If the litter survives for the first three days there is generally no further trouble (except occasionally from mastitis) though there may be loss of an odd young one. The young are nursed frequently, and a female can rear more young than she has nipples. For experimental reasons in the Cambridge colony the number reared is usually restricted to 6. It is usually possible to foster young from a large litter on to another female with a small litter, provided the two litters are of approximately the same age. The transfer should be made fairly early, as in females with a small litter not all the mammae remain functional. Pseudopregnant females will often foster young at about day 44 of their pseudopregnancy. Nursing females caged together may nurse young communally, but they may also fight for possession of individuals.

Size range

Adult body weights range from 400 to 3,500 g. This wide range is partly due to sex differences (the male is nearly twice the size of the female), partly to individual differences, and partly to large seasonal changes in weight. A male weighing 2,800 g in January may weigh only 1,800 g at the end of July. Females lose weight more rapidly after they become oestrous. This, together with a susceptibility to uterine infections after long periods of oestrus, probably accounts for the belief that an unmated ferret will pine away and die. After mating—whether pregnancy or pseudopregnancy ensues—there is considerable gain in weight. This, like the seasonal gain on decreasing photoperiod, is largely accounted for by deposition of a thick layer of subcutaneous fat. In the Cambridge colony, which is selected for large (and, hopefully, more uniform) body-size, the young weigh 6–12 g at birth and about 300–450 g when weaned at 7 weeks old. On natural lighting males born late do not, like those born earlier in the year, reach their maximum weight by January.

Laboratory procedures

Selection of animals

Because ferrets are relatively seldom used for experimental purposes, purchased animals are likely to have derived from breeders who use their ferrets for rabbiting, and are likely to be a mixture of young animals and discarded breeding stock.

Except in very young animals, size is not a useful indicator of age, nor is the replacement of temporary by permanent dentition. An old animal is likely to show wear of the canine and molar teeth, but the degree of wear and discolouration may depend much on diet. Outside the breeding season the testes of immature males are small and firm, while in older males they are larger and a little flabby. The vulva of a female which has been oestrous usually retains some trace of having been swollen, and projects a little and has a wrinkled appearance. A very large vulva in an oestrous female suggests a prolonged period of oestrus and the likelihood of her being a poor prospect for breeding.

Anaesthesia

Ether is very satisfactory for brief anaesthesia. It is administered by placing the otherwise unrestrained animal, with a pad moistened with anaesthetic, in a bell jar or glass-lidded box; it is removed for surgery as respiration slows. For more prolonged anaesthesia pentobarbitone sodium* can be used. The dose is 0·25 ml/kg

* Veterinary Nembutal: Abbott Laboratories. This contains 60 mg pentobarbitone sodium per ml.

body-weight given intraperitoneally and supplemented by ether, administration of which should be started not less than ten minutes after the injection. If the dose of pentobarbitone sodium is 0·6 ml/kg supplementary ether may be unnecessary (Donovan, 1964) but recovery from anaesthesia takes much longer. Chloroform should not be used in the animal house: low concentrations of vapour are lethal to the males of certain strains of mice and cause nursing ferrets to scatter young all over the cage.

Oral administration

Oil or water solutions or suspensions of drugs are very easily administered by mouth. If the animal is held vertically by the neck in one hand, a hard glass pipette may be introduced into the side of the mouth and between the teeth with the other (Fig. 29.4). On gently emptying the syringe on to the upper surface of the tongue, its contents are swallowed. A stomach tube can be passed through a central hole in a piece of dowelling on which the ferret bites (Pyle, 1940).

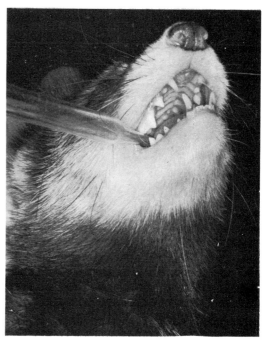

Fig. 29.4. A hard glass dropper may be inserted into the mouth behind the teeth on top of the tongue. With slight pressure on the rubber teat on the end of the pipette fluids etc. can then be given by mouth. (Photograph: D. M. Auger.)

Injection procedures

Subcutaneous injections forward into the neck dorsum may be made without assistance from a second person, if the hand not holding the syringe is placed on the ventral surface of the neck and the animal is held by the neck, with the weight of its body supported on the forearm. Intraperitoneal injection may be made by placing the ferret on a bench and restraining it during the injection by holding it against the operator's body, at the edge of the bench. One hand is placed across the shoulders with the thumb and forefinger behind the forelimb and the other fingers holding the neck firmly against the operator's body. During the injection the hind limbs are held between the bench, the operator's body and the digital border of the hand holding the syringe. Because of the small size of the ears and the presence of a thick layer of subcutaneous fat, there is no large superficial vein suitable for intravenous injections. Heart puncture under anaesthesia is described by Baker & Gorham (1951). The technique of intranasal inoculation is described by Hoskins (1967).

Diseases

Ferrets acquired from dealers or other laboratories should be isolated for 40 days before introduction to a colony in order to minimise disease risks.

In a survey of diseases of laboratory animals which included six laboratories using ferrets, Seamer & Chesterman (1967) reported influenza, otitis, pyometra and adrenal haemorrhage from these animals.

A miscellaneous group of pathological disorders seen in the Imperial Cancer Research Fund's Mill Hill colony includes pneumonia, pyelonephritis, gastric ulcer, middle-ear disease, and enteritis. Small yellow spots are sometimes seen on the surface of the lungs at autopsy: these are subpleural emphysematous lesions with macrophages in the alveoli. Fatty degeneration of the liver similar to that described in mink by Chaddock (1947) has been seen.

Diseases occurring in the Cambridge colony have been described by Hammond (1961 and 1969). These include uterine infection, urinary calculi, ovarian tumours and mastitis.

Viral diseases

Ferrets are exceedingly susceptible to the virus of canine distemper. Mortality may reach 100 per cent in natural outbreaks (Dunkin & Laidlaw, 1926). Transmission is by contact or droplet infection and the incubation period is about ten days. There is first a watery and then a purulent discharge from the eyes and nose (Pockson, 1956). Later, muscular tremors, twitching and convulsions may occur. Diagnosis is suspected on the clinical signs and the finding of inclusion bodies in tissue smears or sections, and confirmed by isolation of the virus (Hogan, 1966). All ferrets showing signs of the disease should be killed humanely, but immediate vaccination of others may give protection. Cages and equipment must be disinfected and sterilized. To help in preventing future outbreaks the source of infection must be found. This could be from a technician who is working with dogs or is keeping unvaccinated pets.

Cross-infection with influenza virus between staff and ferrets may occur (Bell & Dudgeon, 1948; Wilson Smith & Stuart-Harris, 1936). An epizootic disease of ferrets caused by a virus was described by Slanetz & Smetana (1937). There was fever, lethargy, and anorexia with pneumonia and death of the animals within 2 months, but there was no immunological relation between the virus of this disease and that of canine distemper or human influenza.

There is some evidence that Aleutian disease has been transmitted from mink to ferret (Kenyon et al., 1966) and that the disease may also occur spontaneously in the ferret (Kenyon et al., 1967).

Bacterial diseases

Abscesses around the neck or jaw may arise from perforation by fish bones in the diet. Abscesses may also occur in the groin, and the anal scent glands occasionally become infected. If not too extensive, abscesses may respond to treatment with systemic antibiotics and local surgery.

Group C streptococci have been isolated from the uterus in sporadic outbreaks of metritis and also from the lungs of animals dying of pneumonia (Hughes, 1947). Ferrets are relatively resistant to salmonella infection (Gorham et al., 1949). They are susceptible to the human, bovine and avian strains of *Mycobacterium tuberculosis*; the infection in ferrets is primarily a disease of the alimentary tract and abdominal lymph nodes. Animals may appear well until a late stage of the disease and there may be a risk of transmitting the infection unwittingly by experimental procedures or in vaccines made from ferret tissue (Symmers *et al.*, 1953). There do not appear to be any reports on chemotherapy of tuberculosis in the ferret but Grinham (1952) describes a method for eradicating the disease in a breeding colony.

Ectoparasites

The ear mite *Otodectes cyanotis* may give rise to inflammation and irritation and resultant scratching of the ears (Lapage, 1962). Foot-rot is also caused by a mange mite, the feet becoming swollen, scabby and eventually clawless.

The claws should be cut back and the scabs removed after soaking in soapy water; sulphur ointment or benzyl benzoate is then applied. Rapid immersion of the feet in kerosene or treatment with 10 per cent chloramphenicol (Sellers, 1955) have also been effective. Infected animals should be isolated and their companions examined carefully and, if unaffected, removed to clean quarters. The bedding should be burned *in situ* with a blow-lamp and the floors and walls washed with a hot, strong solution of washing soda. Particular attention should be given to females early in pregnancy as treatment is not possible during late pregnancy or lactation.

Mite eradication in general is described by Tuffery (1962).

Endoparasites

Hoare (1935) describes the reaction of the intestine to three distinct species of coccidia and Petrov (1928) the parasitic nematode *Filaroides* in the lungs of *Putorius putorius* and the parasite *Spiroptera nasicola* in the frontal sinus.

Zoonoses

We have already noted some of the diseases which may be transmitted from ferrets to man or vice versa: for example, tuberculosis, influenza. Other diseases of which the ferret may act as a reservoir of infection include leptospirosis, glanders, listeriosis (Morris & Norman, 1950) *Ascaris devosi* (visceral larva migrans) and in the

USA tularaemia and Chaga's disease (American trypanosomiasis) (see Hull, 1963, Van der Hoeden, 1964 and Bisseru, 1967).

Tumours

The clinical and pathological data of published cases of spontaneous tumours in ferrets and polecats have been summarised by Chesterman & Pomerance (1965). Single or multiple cutaneous basi-squamo-sebaceous carcinomata are the most frequently reported tumours. Other tumours include a squamous carcinoma of the skin (Engelbart & Strasser, 1966).

In the Cambridge colony ovarian fibromyomas are the commonest tumours observed, although haemangioma of the liver and lymphocytic leukaemia have also been seen.

REFERENCES

Baker, G. S. & Gorham, J. R. (1951). The technique for bleeding ferrets and mink. *Cornell Vet.*, **41**, 235–6.

Bell, F. R. & Dudgeon, J. A. (1948). An epizootic of influenza in a ferret colony. *J. comp. Path.*, **58**, 167–71.

Bisseru, B. (1967). *Diseases of Man Acquired from his Pets*. London: Heinemann.

Chaddock, T. T. (1947). Ten year autopsy study of mink. *Vet. Med.*, **42**, 409.

Chesterman, F. C. & Pomerance, A. (1965). Spontaneous neoplasms in ferrets and polecats. *J. Path. Bact.*, **89**, 529–33.

Clarke, R. M. & Hardy, R. N. (1970). Structural changes in the small intestine associated with the uptake of polyvinyl pyrrolidone by the young ferret, rabbit, guinea-pig, cat and chicken. *J. Physiol.*, **209**, 669–687.

Donovan, B. T. (1964). Anaesthesia of the ferret. In *Small Animal Anaesthesia*, ed. Graham-Jones, O. Oxford: Pergamon.

Dunkin, G. W. & Laidlaw, P. P. (1926). Studies in dog distemper. I. Dog distemper in the ferret. *J. comp. Path.*, **39**, 201–4.

Engelbart, K. & Strasser, H. (1966). Squamous carcinoma in the ferret. *The Blue Book for the Veterinary Profession*, **11**, 28–30.

Fox, H. (1923). *Diseases in captive wild mammals and birds*. Philadelphia: Lippincott.

Grinham, W. E. (1952). The management of a breeding colony of ferrets. *J. Anim. Techns Ass.*, **2**, (4). 3–6.

Gorham, J. R., Cordy, D. R. & Quortrup, E. R. (1949). Salmonella infections in mink and ferrets. *Am. J. vet. Res.*, **10**, 183–92.

Hammond, J. Jr. (1961). Ferret mortality on a pellet diet. *J. Anim. Techns Ass.*, **12**, 35–6.

Hammond, J. Jr. (1969). The ferret as a research animal. *Carnivore Genetics Newsletter*, **6**, 126–7.

Harris, R. J. C. & Chesterman, F. C. (1964). Growth of Rous sarcoma in rats, ferrets and hamsters. *Natn. Cancer Inst. Monogr.*, **17**, 321–35.

Harris, R. J. C., Chesterman, F. C. & Negroni, G. (1961). Induction of tumours in newborn ferrets with Mill Hill polyoma virus. *Lancet*, **1**, 788–91.

Hoare, C. A. (1935). The endogenous development of the coccidia of the ferret and the histopathological reaction of the infected intestinal villi. *Ann. trop. Med. Parasit.*, **29**, 111–2.

van der Hoeden, J. (1964). *Zoonoses*. Amsterdam: Elsevier.

Hogan, P. A. (1966). In *Infectious Diseases of Domestic Animals*, ed. Brauer, D. W. & Gillespie, J. H. London: Baillière, Tindall & Cassell.

Hoskins, J. M. (1967). *Virological Procedures*. London: Butterworth.

Hughes, D. L. (1947). *UFAW Handbook on the Care and Management of Laboratory Animals*, 1st ed. London: Baillière, Tindall & Cox.

Hull, Thomas G. (1963). *Diseases Transmitted from Animals to Man*, 5th ed. Springfield, Illinois: Thomas.

Kenyon, A. J., Howard, E. & Buko, L. (1967). Hypergammaglobulinemia in ferrets with lymphoproliferative lesions (Aleutian disease). *Am. J. vet. Res.*, **28**, 1167–72.

Kenyon, A. J., Magnano, T., Helmboldt, C. F. & Buko, L. (1966). Aleutian disease in the ferret. *J. Am. vet. med. Ass.*, **179**, 920–3.

Kilham, L. & Margolis, G. (1966). Viral aetiology of spontaneous ataxia of cats. *Am. J. Path.*, **48**, 991–1004.

King, J. D. & Glover, R. E. (1945). Effect of diet on calculus formation and gingival disease in the ferret. *J. Path. Bact.*, **57**, 353.

Kirk, R. W. (1966). *Current Veterinary Therapy—Small Animal Practice*. Philadelphia: Saunders.

Kohler, H. (1958). In *Pathologie der Laboratoriumstiere*, ed. Cohrs, P., Jaffe, R. & Meesen, H. Berlin: Springer.

Lapage, G. (1962). *Monnig's Veterinary Helminthology and Entomology*. Baltimore: Williams & Wilkins.

Marshall, F. H. A. & Hammond, J. Jr. (1945). Experimental control by hormone action of the oestrous cycle in the ferret. *J. Endocr.*, **4**, 159–68.

Morris, J. A. & Norman, M. C. (1950). The isolation of *Listeria monocytogenes* from ferrets. *J. Bact.*, **59**, 313.

National Academy of Science (1968). *Nutrient Requirements for Domestic Animals No. VII*. Publ. 1676. Washington: Nat. Acad. Sci.

Petrov, A. M. (1928). Addition to the explanation of systematics of nematode parasites in the frontal sinus and lungs of Mustelidae. *Ann. trop. Med. Parasit.*, **22**, 259–64.

Pockson, A. P. (1956). The breeding and management of a small colony of ferrets. *J. Anim. Techns Ass.*, **7**, 7–9.

Pyle, N. J. (1940). The use of ferrets in laboratory and research investigations. *Am. J. publ. Hlth*, **30**, 787–96.

Rowlands, I. W. (1967). In *UFAW Handbook on the Care and Management of Laboratory Animals*, 3rd ed. London: Baillière, Tindall & Cox.

Seamer, J. & Chesterman, F. C. (1967). A survey of disease in laboratory animals. *Lab. Anim.*, **1**, 117–39.

Sellers, K. C. (1955). Personal communication quoted in 3rd edition of this Handbook.

Short, D. J. & Woodnott, D. P. (1969). *The I.A.T. Manual of Laboratory Animal Practice and Techniques*, 2nd ed. London: Crosby Lockwood.

Slanetz, C. A. & Smetana, H. (1937). An epizootic disease of ferrets caused by a filterable virus. *J. exp. Med.*, **66**, 653–66.

Symmers, W. St. C., Thomson, A. P. D. & Iland, C. N. (1953). Observations on tuberculosis in the ferret. *J. comp. Path.*, **63**, 20–30.

Tuffery, A. A. (1962). In *Notes for Breeders of Common Laboratory Animals*. ed. Porter, G. & Lane-Petter, W. London & New York: Academic Press.

Wilson Smith & Stuart-Harris, C. H. (1936). Influenza infection of man from the ferret. *Lancet*, **2**, 121–3.

Woodnott, D. P. (1969). In *The I.A.T. Manual of Laboratory Animal Practice and Techniques*, 2nd ed., ed. Short, D. J. & Woodnott, D. P. London: Crosby Lockwood.

30 Ungulates

F. A. Harrison

In the classification of mammals, ungulates are represented by two orders; the Artiodactyla or even-toed ungulates such as cattle, sheep, goats, pigs, deer and camels; and the Perissodactyla or odd-toed ungulates which includes horses and donkeys. All of the species mentioned have been and are being used for laboratory experiments of varying kinds, some of which will directly benefit the species concerned by increasing the understanding of, for example, their special digestive functions or control of body temperature. Pigs, sheep and cattle are also being used in human medical research, for example, in the study of the immune response to organ homotransplantation.

For detailed consideration of each species and the general housing, feeding and maintenance reference should be made to the *UFAW Handbook on the Care and Management of Farm Animals*†.

† Published by E. & S. Livingstone, Edinburgh (1971).

Table I
Clinical data

	Temperature ±0·5°C	Pulse min⁻¹	Respiration min⁻¹
Horse	38·0	36–42	8–12
Cattle	38·5	50–60	12–16
Sheep	39·5	70–80	15–25
Goat	39·5	70–80	10–16
Pig	39·0	70–80	12–30

Adapted from Miller & Robertson (1952).

Biological Data

Some relevant data are given in Tables I, II and III.

Table III
Female reproductive data

Species	Oestrus (days)	Gestation (days)	
Sow	21 (19–23)	115 (111–117)	
Goat	21	(144–157)	S
Cow	21 (14–23)	281 (210–315)	
Donkey	22	365 (340–395)	
Mare*	21 (19–26)	336 (325–341)	
Deer*	—	234 (225–246)	S
Sheep	16·5	147 (142–150)	S
Camel	10–20	315–410	

S = seasonally polyoestrus ( ) = range.

* Data from Spector (1956).

Other data adapted from UFAW (1971).

Husbandry

Housing and caging

Horses, donkeys, cows, calves, sheep and goats are gregarious animals and ordinarily do best housed in groups as described in *The UFAW Handbook on the Care and Management of Farm Animals*. Animals that have been surgically prepared for experimentation may have to be housed individually in pens or loose boxes and these animals should be allowed to see others of their own species.

Table II
Blood data

	Cattle	Goat*	Horse m-equiv. l⁻¹	Sheep	Pig
Haematocrit	30–40	25	30–40	27–37	35–45
Serum Ca	5·4	4·4	6·1	5·7	5·6
,, Cl	104	109	102	116	103
,, Mg	2·3	2·3	1·5–2·1	1·8	2·2
,, K	4·8	4·2	3·3	4·8	5·9
.. Na	142	153	149	160	155
Protein g%	6·9	7·5	7·6	5·7	8·7
Blood Hb g%	11·1	10·0	10·1	10·3	14
Blood glucose mg%	46	61	73	30–57	45–75

Adapted from Spector (1956).

* Published and unpublished data kindly supplied by Dr J. L. Linzell.

Sheep, goats and calves. Pens should be approximately 2·1 m × 1·05 m and should have walls made of wood, sheet metal or smooth cement, to prevent injury. The most suitable arrangement is for animal rooms or houses to hold individual pens for 6 to 12 animals. A low-temperature (65°C) underfloor system is most satisfactory for heating since it serves not only to heat the room but also to keep the bedding and floor dry. Sawdust may be provided for bedding; alternatively the pens may have floors with wooden slats which are kept clean by mechanical scraping or hosing down with water. Adequate ventilation with suitable extract fans must be provided.

Food and water containers can be fastened on the outside of the pen door, which must then be provided with suitable openings to allow the animal access to the supply (Fig. 30.1). A disadvantage of this system is the potential hazard to animals which have surgically exteriorized vessels or cannulae in the neck region. Alternatively a galvanized or plastic container (15 cm × 20 cm × 20 cm, holding 4–6 litres of water), may be hung on one wall (Fig. 30.2) and the food supplied in a deep plastic box (32 cm × 46 cm × 18 cm) placed on the floor.

Fig. 30.2. Sheep pen with plastic food container on the floor and water container fastened to one wall of the pen. (Photograph: ARC Institute of Animal Physiology.)

Pigs. Small and medium pigs can be kept in individual pens similar to those described for sheep, but with some modifications to the food and water containers (see *UFAW Handbook on the Care and Management of Farm Animals* for details). Large pigs require bigger pens with at least twice the floor area specified for sheep.

Horses and cattle. Individual housing for the larger species will be in loose-boxes or the appropriate stalls (see *UFAW Handbook on the Care and Management of Farm Animals* for details).

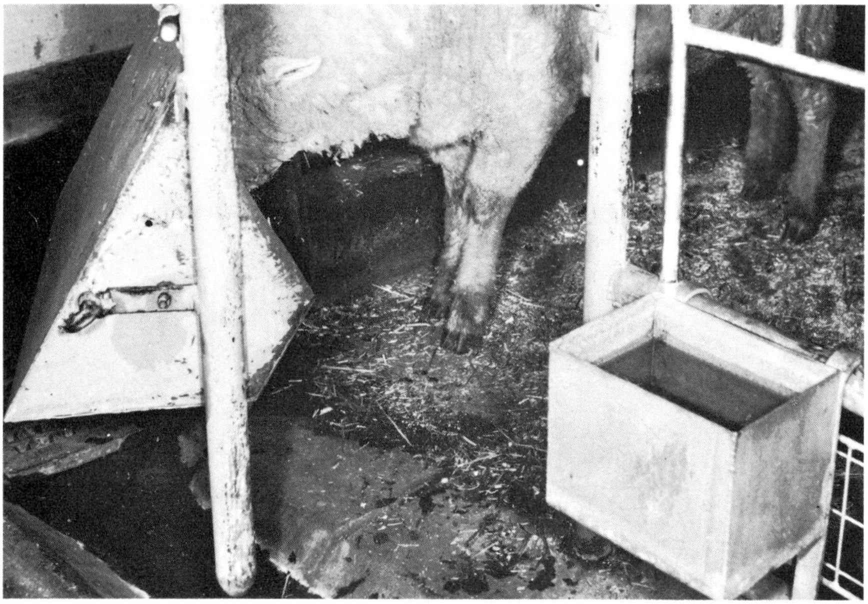

Fig. 30.1. Food hopper built into door of sheep pen. The water container is attached to the outside of the pen. (Photograph: ARC Institute of Animal Physiology).

Metabolism cages

For many experiments, additional restraint is required and animals may have to be kept for varying periods of time in metabolism cages. Such cages usually restrict the overall movement of the animal and prevent it turning round, thus permitting the separate collection of urine and faeces.

Sheep. At the Agricultural Research Council's Institute of Animal Physiology at Babraham cages of the type illustrated in Figure 30.3 have been used successfully for more than ten years (Harrison, 1974). They are constructed of tubular steel. The animal enters by a ramp attached to the back (Fig. 30.3A) and at the end of the experiment leaves by the same ramp after this has been transferred to the front (Fig. 30.3B). Sheep soon learn the routine handling involved and suffer very little distress. A urine and faeces separator under the animal compartment allows adequate separation and collection of excreta from both male and female sheep. The elevated

position enables easy access to the animal and greatly facilitates experimental procedures.

Pigs. The type of cage described above can be easily modified for use with pigs up to medium size. One side of the cage must be adjustable so that smaller animals can be prevented from turning round. Sows and boars can be restrained for short-term experiments in standard farrowing crates firmly anchored to the floor.

Cattle. A suitable arrangement for the individual confinement of cows has been developed by Balch and co-workers. The method enables the separate collection of urine and faeces as well as individual food and water recordings. A detailed description of the housing and equipment is given by Balch, Johnson & Machin (1962).

Identification

It is most important to identify experimental animals and reference should be made to the *UFAW Handbook on the Care and Management of*

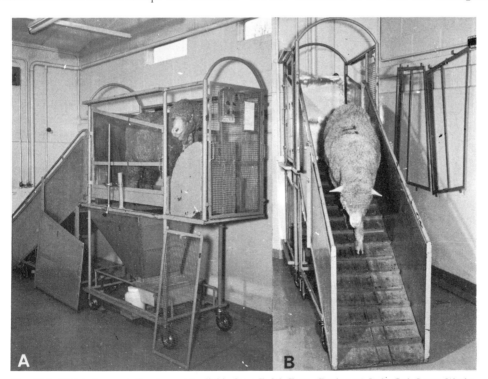

Fig. 30.3. Metabolism cage for sheep. (Available from F. W. Frost (Engineers) Ltd., Oak Street Works, Norwich, NOR 83K.) A. With ramp attached to back for entry and one side open to show sheep in animal compartment. B. With ramp attached to front, allowing sheep to leave cage. (Photographs: ARC Institute of Animal Physiology.)

Farm Animals for details of the most suitable methods for different species.

Handling

Most if not all domesticated animals respond to gentle, though firm and careful, handling and this consideration is of paramount importance in the care and maintenance of experimental animals generally. Handling of ungulates should ideally commence at birth, but obviously such perfection cannot always be achieved. Thus most experimental animals require a period of adaptation to human contact and handling as well as to laboratory accommodation and diet. In all except the newborn and very young, a period of three weeks should be allowed for initial acclimatization to the laboratory practice and handling routine before any surgical preparation or experimental work is commenced.

Again, in many long-term physiological experiments on surgically prepared animals up to three weeks should usually be allowed after operation before the animals are used for

Fig. 30.4. Trolley for transporting sheep and goats. Note ramps at each end which are used to close the trolley. A flap of sheet rubber covers the gap between the ramp and trolley floor, which is also rubber-covered. (Photograph: ARC Institute of Animal Physiology.)

experimental observations, thereby enabling normal surgical healing to occur.

Transport

Horses, cattle and goats can usually be trained to walk on a halter and so can be led about the

Fig. 30.5. Trolley for pigs (based on a design by Dr K. F. Hosie). The hinged grids on top can be opened, and the sides can be removed for easy access to the sedated or anaesthetized animal. (Photograph: ARC Institute of Animal Physiology.)

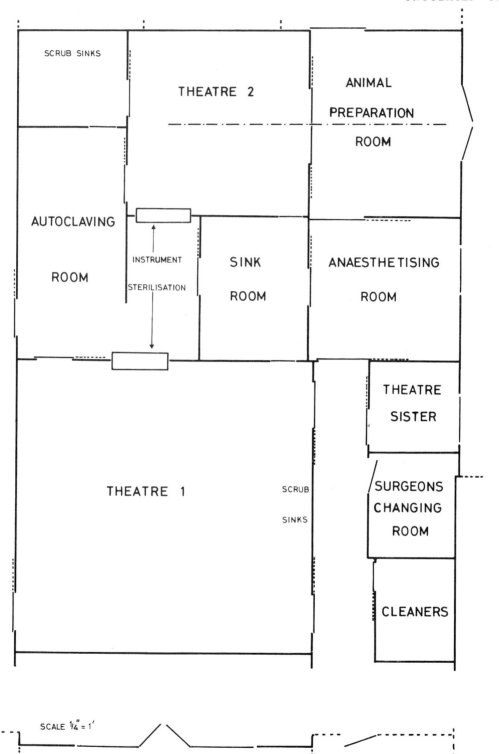

Fig. 30.6. Layout of the operating theatre suite at the Agricultural Research Council Institute of Animal Physiology. Theatre 2 is equipped with an overhead hoist for moving large animals.

experimental laboratory area. Sheep and pigs are not generally so easy to manage individually in this way and a purpose-made 4-wheeled truck is most suitable for transporting them (Figs. 30.4 and 30.5). Important points in construction are the avoidance of slippery ramps and an unnecessarily large gap between the ramp and truck where an animal can easily get a foot trapped. A flap of sheet rubber can be used to cover the small gap which is inevitable at the hinged joint. With all animals, but especially pigs, patience is required in coaxing them into a truck. Pigs can usually be enticed with a little food.

Feeding*

The food of experimental animals may differ from that of their commercial counterparts and they should be allowed time to adjust to any changes. This is particularly important for ruminants as the rumen flora need two to three weeks to become adapted to changes in diet. It is important to provide adequate supplies of fresh water to experimental animals.

Mineral supplements should be provided for ruminant animals when housed indoors for prolonged periods. It should, however, be noted that in several laboratories a condition of chronic copper poisoning has been encountered in sheep which have been kept for long periods with access to salt licks containing high levels of copper. Housed sheep should therefore have access to copper-free salt licks.

Laboratory Procedures

Selection of animals for experiment

The scientific ideal would be a healthy normal animal of known age and breeding and this can most easily be obtained by maintaining breeding stock of the species concerned. Not every research institute or laboratory can achieve this objective, and stock will then have to be purchased when required. Where possible the animals should come from established disease-free herds or flocks and should be clinically healthy animals of normal standard conformation for the particular breed. All newly-purchased ungulates should be isolated in quarantine for up to 4 weeks before coming into direct or indirect contact with existing stocks of the same species.

Surgery

All surgical operations should be conducted in a purpose-built operating theatre with facilities for high standards of surgical cleanliness. Aseptic techniques should always be used and

Fig. 30.7. Operating table base (made by F. W. Frost (Engineers) Ltd., Oak Street Works, Norwich, NOR 83K) used with one of three table tops.
  (i) for sheep and goats, 142 × 56 cm with headpiece, 46 × 51 cm;
 (ii) for medium and large pigs, 156 × 76 cm with headpiece, 46 × 51 cm;
(iii) for cattle, 193 × 127 cm with headpiece, 61 × 61 cm.
Headpiece is shown detached from table top (ii) which is suspended on the hoist over the table base. The table-tops self-centre on to the base and lock with four pins. When used for large pigs and cattle, the base is screwed to two floor sockets. (Photograph: ARC Institute of Animal Physiology.)

* See *UFAW Handbook on the Care and Management of Farm Animals* for detailed recommendations.

provision made for sterilizing drapes, gowns and instruments. The layout of the Babraham operating theatres is illustrated in Figure 30.6.

Flexibility in the operating theatre can be achieved with the type of operating table designed and used at Babraham. This has one table base and three separate table tops, for small, medium and large farm animals (Fig. 30.7). For the large pigs and cattle, use is made of an overhead compressed-air hoist which can elevate from the floor the table top bearing the anaesthetised animal and transfer it to the table base in the operating theatre (Fig. 30.8). After operation the still-anaesthetized animal can be transferred on a low trolley to a suitable recovery area (Fig. 30.9). Because anaesthetized animals are unable to regulate their body temperature, they should be covered with a blanket to avoid unnecessary thermal shock during transfer.

Anaesthesia

Equine species For anaesthesia of these reference should be made to Hall (1966).

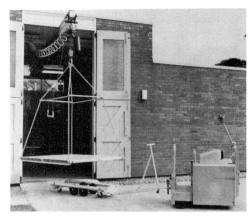

Fig. 30.9. Low-transporter trolley attached to an electric tug above which is the large table-top, suspended from the compressed-air hoist (Broom & Wade Ltd., High Wycombe, Bucks.). For transporting anaesthetized animals, the table-top is locked on to the trolley. After transfer to the recovery loose-box, the top can be tilted to facilitate removal of the animal on to a bed of straw. (Photograph: ARC Institute of Animal Physiology.)

Ruminants. General anaesthesia may be induced satisfactorily in cattle weighing 400–500 kg by the rapid intravenous injection of 50–70 ml of a 2·5 per cent solution of methohexitone sodium* 30–40 minutes after prior sedation with acepromazine maleate BPC† given intramuscularly at a dosage rate of 5 mg/100 kg bodyweight. Intravenous injection is greatly facilitated by the introduction percutaneously into a jugular vein of a No. 2 Braunula‡ or similar catheter, closed with a tap and secured with a suture. After induction, the animal should be intubated using some form of gag to immobilize and open the mouth (a Drinkwater's gag is most suitable). Additional injections of barbiturate may be required for adequate relaxation of the jaw and larynx. The animal can then be connected to an inhalation anaesthetic apparatus such as the Weaver circle absorber apparatus and anaesthesia maintained with mixtures of oxygen and halothane§ or cyclopropane in closed-circuit.

In sheep and goats, general anaesthesia is most easily induced by intravenous injection of

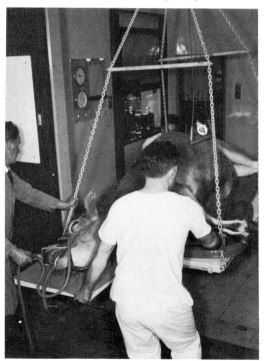

Fig. 30.8. Anaesthetized cow being transferred from animal preparation room (see Fig. 30.6) to Theatre 2. (Photograph: ARC Institute of Animal Physiology.)

* Brietal Sodium: Eli Lilly.
† Acetyl promazine: Boots.
‡ Braun, Melsungen, supplied by Armour Pharmaceutical Co.
§ Fluothane: ICI Ltd.

pentobarbitone sodium and, after endotracheal intubation with a cuffed Magill tube, maintained in closed-circuit with mixtures of oxygen and halothane or cyclopropane, with a sodalime absorber for the removal of carbon dioxide (see Harrison, 1964; Linzell, 1964). When studies on blood or urine are intended in the immediate post-operative period, freshly made solutions of pentobarbitone must be used since formulations using propylene glycol as a preservative cause haemolysis of sheep and goat red blood cells (Potter, 1958; Linzell, 1964).

In the ruminant animal undergoing surgical preparation an important consideration is the large volume of the contents of the reticulo-rumen with the continuing production of gas by the bacterial flora. Food should be withheld for 1 to 2 days and if prolonged and difficult abdominal surgery is contemplated it is often helpful to empty the reticulo-rumen of its contents preoperatively via a rumen fistula and store them in in an incubator at 39°C until the operation is completed. The contents are then poured back into the rumen.

Pigs. In small pigs, anaesthesia can be induced with a face mask and an oxygen-halothane mixture delivered in open-circuit. After intubation, closed-circuit anaesthesia can be maintained with halothane/oxygen mixtures. It is important to realise that the glottis of most animals is extremely sensitive to mechanical stimulation and in the small pig attempts to intubate will be severely frustrated due to spasm of the arytenoid cartilages if the animal is only lightly anaesthetized; this can even lead to death through asphyxia. An anaesthetic aerosol spray can be usefully applied to the epiglottis and arytenoid cartilages before introducing the endotracheal tube. For the larger pig, some form of pre-anaesthetic sedation is necessary and the use of chlorpromazine as described by Hill & Perry (1959) still has many advantages for the induction of anaesthesia and is to be preferred if recovery is intended. Perry (1964) and Tavernor (1964) have discussed the use of phencyclidine* for sedation in the large pig. The author has used this drug in both recovery surgery and for acute experiments terminated by death. The major disadvantage is that recovery from anaesthesia and surgery after the use of phen-

* Sernylan: Parke, Davis.

cyclidine can be complicated by the apparently hallucinatory behaviour of the pig and measures such as bandaging the jaws to prevent laceration and chewing of the tongue have to be used to protect the animal from self-inflicted injury; this drug is not, therefore ideal. After anaesthesia and surgery, all species should be allowed to recover in a well-ventilated and warm room with adequate protection against self-inflicted injury.

Euthanasia

Euthanasia is required by law (Cruelty to Animals Act 1876) at the end of the experimental life of each animal and this may be carried out either by administration of an overdose of the appropriate anaesthetic agent or by the use of a humane killer (see Chap. 14).

Collection of specimens

Blood. When single samples of blood are required on isolated occasions, simple needle puncture of a superficial vessel such as the external jugular vein may be satisfactorily performed with the minimum restraint provided that the needle is sharp and sterile. A small area of skin over the vessel should be exposed, usually by clipping the hair or fleece, and cleansed with 70 per cent alcohol prior to venepuncture. If large samples are required, it is advisable to use a large bore needle, and in this case local anaesthesia should be applied to the skin at the site of venepuncture. For repeated sampling of peripheral blood over periods of 12 to 24 hours, temporary catheterization of the external jugular vein should be performed using the percutaneous technique of Seldinger (1953). A small area of skin is locally anaesthetized and the skin is cut down to the level of the vessel. A sharp, sterile needle is then introduced into the vein and used to pass a wire or nylon stilette into the vessel. After withdrawal of the needle, a nylon catheter (OD 2·1 mm, FG6, 30 cm long; Portex Ltd, Hythe, Kent) can be passed over the stilette and positioned in the vein. The catheter is flushed through with heparinized saline (100 i.u. ml^{-1}) and closed with a suitable sterile tap. The skin incision is closed and the catheter anchored with one or two simple sutures (silk). The tap and free end of the catheter can be fixed to the skin with adhesive tape or, in sheep, tied to the fleece. This technique avoids repeated venepuncture with the associated hazards of haematoma formation and possible infected thrombophlebitis, and the cutting down of the skin prevents damage to the catheter tip by the tough subcutaneous connective tissue and subsequent tearing and damage to the vein. If sterile needles, stilettes and catheters and a surgically clean technique are used, the jugular vein can be catheterized repeatedly over several years with no problems of infection occurring. The procedure is humane and causes only minor disturbance to the animal at the time of catheterization.

Urine Urine samples can obviously be collected when natural voiding occurs or is provoked by digital stimulation of the perineum. Continuous sampling from female animals can be performed by the use of a temporary bladder catheter. At Babraham this approach has been used repeatedly in the sheep for experiments of 12 to 24 hours duration. Again, sterile catheters and equipment are used with clean technique. The perineal region and vulva are first cleaned with an antiseptic solution (such as dilute chlorhexidine*). The vulva is then locally anaesthetized with xylocaine jelly or anaesthetic spray. The urethral orifice is also anaesthetized under visual inspection using an observation light and vaginal speculum (modified Cusco type). A sterile Foley latex catheter (14FG) with 5 ml balloon is gently inserted and positioned in the bladder. After inflation of the balloon with 5 ml sterile water, the free end of the catheter is fixed to the skin or tied to the fleece.

* Hibitane: ICI Ltd.

Dosing and injection procedures

Experimental treatment may involve oral administration of drugs or injection of serum or hormone preparations. Oral dosing can be carried out either with a stomach tube or by simple drenching *per os*. Injections will usually be subcutaneous, intramuscular or intravenous and should be made with clean technique, sterile syringes and sharp needles. An area of skin at the site of injection should be cleansed with antiseptic solution such as 70 per cent alcohol and the animal should be firmly restrained while the injection is being made. When the injection is intended to be intramuscular or subcutaneous, a check should be made, by gentle aspiration on the syringe plunger, that the needle has not entered a blood vessel. Drugs (such as the tetracyclines) which produce pain or irritation when administered by subcutaneous or intramuscular injection, should be given intravenously.

Disease control

Routine prophylactic treatment of healthy farm animals kept under laboratory conditions is minimal provided that the animals are not transferred from their indoor environment to pasture nor subjected to frequent changes in diet. Treatment for intestinal parasites will depend on the existence of sub-clinical or clinical infestation in the stock and the advice of a qualified veterinary clinician should be sought.

Sheep and goats should be protected against the clostridial diseases by routine inoculation

TABLE IV

*Notifiable Diseases**

ANTHRAX in four-footed mammals
BRUCELLOSIS MELITENSIS in ruminating and equine animals and swine
CATTLE PLAGUE in ruminating animals and swine
EPIZOOTIC LYMPHANGITIS in equine animals
FOOT-AND-MOUTH DISEASE in ruminating animals and swine
FOWL PEST in poultry
GLANDERS or FARCY in equine animals
PARASITIC MANGE in equine animals
PLEURO-PNEUMONIA in cattle
RABIES in ruminating and equine animals, swine, dogs and cats
SHEEP POX in sheep
SHEEP SCAB in sheep
SWINE FEVER in swine
TUBERCULOSIS (certain forms only) in cattle

* From Thomas (1969).

and sheep should be vaccinated against orf which can be transmitted to man. Again, the advice of a veterinary surgeon should be obtained regarding the appropriate treatment.

Legislation under the Diseases of Animals Act 1950 (see also Thomas, 1969) lists the diseases which must be compulsorily notified to the police when their existence or suspected existence is encountered (see Table IV).

Experiments on animals are also covered by legislation under the Cruelty to Animals Act 1876 (see also Chap. 1).

REFERENCES

Balch, C. C., Johnson, V. W. & Machin, C. (1962). Housing and equipment for balance studies with cows. *J. agric. Sci.*, **59**, 355–8.

Hall, L. W. (1966). *Wright's Veterinary Anaesthesia & Analgesia*, 6th ed. London: Baillière, Tindall & Cassell.

Harrison, F. A. (1964). The anaesthesia of sheep using pentobarbitone sodium and cyclopropane. In *Small Animal Anaesthesia*, ed. Graham-Jones, O. Oxford: Pergamon Press.

Harrison, F. A. (1974). The Babraham metabolism cage for sheep. *J. Physiol.*, **242**, 20–22P.

Hill, K. J. & Perry, J. S. (1959). A method for closed-circuit anaesthesia in the pig. *Vet. Rec.*, **71**, 296–9.

Linzell, J. L. (1964). Some observations on general and regional anaesthesia in goats. In *Small Animal Anaesthesia*, ed. Graham-Jones, O. Oxford: Pergamon Press.

Miller, W. C. & Robertson, E. D. S. (1952). *Practical Animal Husbandry*, 6th ed. Edinburgh: Oliver & Boyd.

Perry, J. S. (1964). Anaesthesia of the adult sow. In *Small Animal Anaesthesia*, ed. Graham-Jones, O. Oxford: Pergamon Press.

Potter, B. J. (1958). Haemoglobinuria caused by propylene glycol in sheep. *Br. J. Pharmac. Chemother.*, **13**, 385–9.

Seldinger, S. I. (1953). Catheter replacement of the needle in percutaneous arteriography. *Acta radiol.*, **39**, 368–76.

Spector, W. S. (1956). *Handbook of Biological Data*. Philadelphia: Saunders.

Tavernor, W. D. (1964). The use of phenylcyclohexylpiperidine hydrochloride in the pig. In *Small Animal Anaesthesia*, ed. Graham-Jones, O. Oxford: Pergamon Press.

Thomas, J. L. (1969). *Diseases of Animals Law*, 6th ed. London: Police Review Publishing Co., Ltd.

UFAW (ed.) (1971). *Handbook on the Care and Management of Farm Animals*. Edinburgh: Livingstone.

31   Primates

GENERAL

R. N. T-W-FIENNES

Introduction

As laboratory animals primates are in a class by themselves, and it is important that those who use them in biomedical research should become acquainted with their uses and general characteristics. Primates are expensive animals in a number of different ways: they are expensive to breed and rear; when imported, they are expensive to buy; and importation on the scale of recent years is a drain on the natural reserves, which many observers believe to be diminishing.

Very few of the primates used either in research or vaccine production are laboratory-bred. By far the greater number are trade animals, coming from the Far East, Africa, or South America. They have usually been trapped in the wild, and nothing is known of their history with regard to age or health. If the animal is a young one, as often as not the mother will have been shot in order to acquire it; if it is an old one, it is probably the easiest of the troop to capture and may be in poor health or stressed. The mortality between capture, shipment, and receipt at the laboratory is often high, and deaths continue during the quarantine period. These animals, furthermore, constitute a potential source of danger to the human beings with whom they come in contact and to other simian species with which they associate during shipment and quarantine (see p. 423).

If they were of the standard laboratory species (rats, mice, guinea pigs and so on) such animals would not be accepted, nor would they be regarded as suitable for most types of research. At the present time most of them are required for provision of cell lines for virus culture in vaccine production, especially poliomyelitis, although this demand is diminishing. Many primary cultures have to be discarded owing to the presence of simian foamy and other viruses. Many such viruses have been incorporated unwittingly into live vaccines, particularly the SV40 adenovirus, which is known to have oncogenic properties in baby hamsters. Fortunately none of these viruses is known to have caused disease in human recipients, although there have been rare and serious episodes resulting from human contact with imported monkeys (see p. 425). These facts highlight the unsuitability of this type of animal for biomedical purposes.

Nevertheless, the great upsurge in the use of primates in both medical research and vaccine production in recent years has brought important advances, which could not have been achieved without them. This very circumstance has blinded authorities to the pressing need to find some 15 or 16 species that will breed well in captivity and also provide a nucleus of reliable research animals of known strains and backgrounds, thereby eliminating the worst dangers and providing far more reliable results. It is greatly to be hoped that in the future more attention will be paid to this, but meanwhile any researcher who contemplates the use of primates should study the position objectively. He should decide whether the use of primates is really necessary and, if it is, what species of primate is most suited to the object he has in mind. Primates are enormously variable in size, character, and requirements for nutrition and maintenance. The smallest are probably the Mouse Lemurs (*Microcebus* spp.), which weigh only a few grams; a gorilla, on the other hand, weighs upwards of 400 kg when adult. Nutritional requirements vary from those of the leaf-eating species (such as howlers (*Alouatta*), *Colobus*, and langur (*Presbytis* spp.) which have compartmented stomachs like cattle and are entirely vegetarian) to wholly omnivorous animals. Many species have rather specialized requirements with regard to vitamins and other items and are difficult to keep in full health. With regard to caging requirements, some (such as *Callicebus* (titis)) are solitary creatures and like to live alone or in pairs, whereas others prefer to live in large social groups and will remain in better health if kept in this way.

Even when primates are being bred in captivity, they will still be expensive animals to use. The females of most species produce only one offspring a year, so that a large colony of females must be kept to provide a significant number of young, which themselves mature slowly.

The higher primates, such as chimpanzees and baboons, are intelligent and sensitive animals. If handled with sympathy and understanding they can become more than research tools—even co-operative partners in experimentation. To achieve this, a little time, often time well spent, must be spent on conditioning the animal to his circumstances. A chimpanzee, for instance, will sit quietly and hold his arm out for a blood sample to be taken. In space-travel studies, the animal's co-operation in lever-pressing experiments is essential when reflex responses and so on are being investigated. An animal treated unsympathetically is liable to become aggressive and unco-operative; furthermore, unless care is taken over its comfort and needs, it is liable to become stressed and the results of the experiment may be vitiated for this reason.

Three simple rules may be proposed for the use of primates in research. First, never use a primate when another animal would do, and then select the right primate for the job. Secondly, use primates only at the penultimate stage in investigations, after studies have been made throughout the range of other laboratory animals, and as a preliminary to clinical and hospital trials in man. Thirdly, give all primates as comfortable conditions as possible during the period of an experiment; if possible bring them in to small cages in the laboratory only for the duration of the experiment, and keep stock not in use in holding-areas where spacious conditions can be provided.

General biology

Taxonomic position

Phylogenetically primates, especially the higher primates, closely resemble man, and are thus likely to provide more satisfactory models of human diseases. Although this is so, the evolutionary period which separates man from them (and especially from the lower primates) is considerable. Man's own evolution, from the time when australopithecines first appear in the fossil record, occupied the whole of the Pleistocene epoch and amounted to one million years or more. Primate evolution itself has occupied a period of some fifty million years, from the Eocene epoch until the present, and it appears that the most primitive primates (such as tree shrews and tarsiers) must have diverged from the human stock at the beginning of this period. In this sense, the earliest primates are no closer to man than they are, say, to dogs or whales, although anatomical, and possibly physiological and immunological, resemblances persist. It is, nevertheless, as well to be clear that the lower primates, even Old World species, are distant relations of man and not necessarily such a good model as might be supposed. A simple classification of living primates is given in Table I, and we can give a brief survey of the more important groups and species used in biomedical research. *Super-family Hominoidea.* This contains three families; Hominidae, with the single genus *Homo*; Pongidae, containing three genera *Pongo* (orang utan), *Pan* (chimpanzee) and *Gorilla*; and Hylobatidae, containing two genera, *Hylobates* (gibbons), and *Symphalangus*, comprising the siamangs. There is, however, some evidence to suggest that *Pan* and *Gorilla* are more closely related to man than is *Pongo*. So close indeed is this relationship that, but for reasons of prejudice, both chimpanzees and gorillas should possibly be grouped with man as Hominidae instead of with the orang utans as Pongidae. Of

TABLE I

Classification of the Primates (order Primates)

Suborder Prosimii
  Infraorder Lemuriformes
    Family Lemuridae
      Genus Lemur (lemurs) 5–8 species, Madagascar
        Hapalemur (gentle lemur) 2 species, Madagascar
        Lepilemur (sportive lemur) 1 species, Madagascar

TABLE I *(Continued)*

Family Daubentoniidae
  Genus Daubentonia (aye-aye) 1 species, N. W. Madagascar
Family Cheirogaleidae
  Genus Cheirogaleus (dwarf lemurs) 2 species, Madagascar
  Allocebus (hairy-eared dwarf lemur) 1 species, E. Madagascar
  Phaner (fork-crown dwarf lemur) 1 species, W. Madagascar
  Microcebus (mouse lemur) 2–4 species, Madagascar
Family Indriidae
  Genus Indri (indri) 1 species, N.E. Madagascar
    Propithecus (sifaka), 2 species, Madagascar
    Avahi (woolly indri), 1 species, Madagascar
Infraorder Lorisiformes
  Family Lorisidae
  Subfamily Lorisinae
    Genus Loris (slender loris) 1 species, Ceylon and S. India
      Nycticebus (slow loris) 2 species, S.E. Asia
      Perodicticus (potto and angwantibo) 2 species, Africa
  Subfamily Galaginae
    Genus Galago (bushbabies) 5–6 species, Africa
Infraorder Tarsiiformes
  Family Tarsiidae
    Genus Tarsius (tarsier) 3 species, Indonesia and Philippines
Suborder Anthropoidea
Infraorder Platyrrhini
  Family Callitrichidae
    Genus Callithrix (common marmoset) 3 species, S. America
      Cebuella (pygmy marmoset) 1 species, Upper Amazon
      Saguinus (tamarins) 10–15 species, C. and S. America
      Leontopithecus (lion marmoset) 3 species, S.E. Brazil
      Callimico (Goeldi's marmoset) 1 species, Upper Amazon
  Family Cebidae
    Genus Aotus (night monkey) 1 species, S. America
      Callicebus (titi) 3 species, S. America
      Saimiri (squirrel monkey) 2 species, C. and S. America
      Pithecia (saki) 4 species, northern S. America
      Cacajao (uakari) 2 species, northern S. America
      Cebus (capuchin monkey) 4 species, northern S. America
      Alouatta (howler monkeys) 5 species, C. and S. America
      Lagothrix (woolly monkeys) 2 species, northern S. America
      Brachyteles (woolly spider monkey) 1 species, S.E. Brazil
      Ateles (spider monkeys) (including Rhinopithecus) 4 species, C. and S. America
Infraorder Catarrhini
Superfamily Cercopithecoidea
  Family Cercopithecidae
  Subfamily Colobinae
    Genus Colobus (colobus monkeys) 4–8 species, Africa
      Presbytis (langurs) 13 species, S.E. Asia
      Pygathrix (including Rhinopithecus (douc and snub-nosed monkeys)
      4 species, Vietnam and S.W. China
      Nasalis (proboscis monkeys) 2 species, Borneo and Mentawei Islands.
  Subfamily Cercopithecinae
    Genus Cercopithecus (guenons) 12–20 species, Africa
      Erythrocebus (patas monkeys) 1 species, Africa
      Cercocebus (mangabeys) 4–7 species, Africa
      Papio (baboons) 3–4 species, Africa
      Theropithecus (gelada) 1 species, Ethiopia
      Macaca (macaques) 10–16 species
Superfamily Hominoidea
  Family Pongidae
    Genus Hylobates (gibbons) 6 species, S.E. Asia
      Pongo (orang-utan) 1 species, Borneo and Sumatra
      Pan (chimpanzee) 2 species, Africa
      Gorilla (gorilla) 2 species, Africa.
  Family Hominidae
    Genus Homo (man) 1 species, world-wide

Adapted from Groves (1972).
  Note: This author does not include tree shrews with primates, and he classifies the siamangs
    (*Symphalangus*) with the gibbons as *Hylobates*, included with the Pongidae.

the Great Apes, only the chimpanzees have been used or suggested to be used seriously as experimental animals. In this animal, however, the experimenter has a subject very closely related to man and resembling him in many physiological characteristics.

The average weight of male chimpanzees is around 50 kg and of females 40 kg. The young develop much more quickly than do human babies and at certain ages are more advanced. They come to sexual maturity at around 8 years old, and may live for as long as 40 years. Young chimpanzees are docile, malleable and affectionate, but after the age of puberty they may become aggressive and dangerous, though this does not seem to happen where an intimate relationship has developed with their master. They like best to live in troops and, although social hierarchies are established, there is little fighting or aggressiveness between males, and the females are completely promiscuous. These animals breed well in captivity, especially if kept in groups, although the females sometimes quarrel over their babies and one that has lost her baby may steal one from another. This sometimes leads to neglect, so that the baby is not properly cared for.

The chimpanzees are divided into two species, *Pan troglodytes* and *P. paniscus*, the pygmy chimpanzee. *P. troglodytes* has three sub-species, *P. troglodytes*, *P. t. verus*, *P. t. schweinfurthii*. Their diet is mainly vegetarian, supplemented by insect and animal food. In captivity, a vegetarian diet will not contain sufficient protein and should be supplemented.

Neither the orang utans nor the gorillas are likely to be useful experimental animals. Both are scarce, difficult to obtain, and expensive, and the orang utan is classed as an endangered species and export from the native habitat in the Far East is forbidden. Although neither breed well in captivity some captive births have occurred in both species.

Gibbons breed well in captivity and could possibly be useful experimental animals for some purposes. They are, however, extremely active and require much space in which to move around, and are intractable, aggressive and not easy to control. They live in monogamous groups and one young is produced annually. Strange gibbons introduced to the cage of a breeding pair are liable to be attacked and killed. Gibbons are light in weight (the males averaging some 6 kg and the females 5 kg) and their diet is mostly vegetarian, but includes some animal and insect food, particularly birds' eggs and young birds.

The diet of the orang utan is similar, but the gorilla is stated to be entirely vegetarian. Since this animal has no adequate fermentation chamber for the digestion of vegetarian foods it is difficult to see how this can be so, and in captivity, gorillas will certainly take both insect and animal foods.

Family Cercopithecidae. The Cercopithecidae comprise the Old World monkeys, which provide many of the favoured laboratory species.

For this purpose, the leaf-eating monkeys (Colobinae) may be ignored; for some reason these animals do not thrive well in captivity.

The most important groups for experimental purposes are macaques, which include the rhesus, cynomolgus and other useful laboratory species; *Papio*, the baboons; *Cercopithecus*, comprising the grivets, guenons and vervets; and *Erythrocebus*, the patas monkeys which are becoming popular. None of these groups approach the great apes in size, weight, or intelligence, and they are much less closely-related to man himself. Nevertheless, they are sufficiently close to him to provide useful models for many lines of research. Their immunological responses are similar, and human pathogens will normally grow in cell cultures derived from these animals. They are also sufficiently far removed from man for many of their natural pathogens to be both infectious and extremely dangerous for him. They vary in size from the very small swamp talapoin to the patas and baboons, both of which are savannah species which spend little time, if any, in trees. Amongst these groups will be found a range of animals suitable for most experimental work for which monkeys are required. In space research demanding response tests both baboons and some species of macaque can replace chimpanzees. Immunologically, responses may vary surprisingly between quite closely-related species of e.g. macaque, so that when these animals come to be bred for such purposes it will be necessary to develop a range rather than one species. They vary also in temperament and in the ease with which they may be handled. Moreover, some species breed readily in captivity (as for instance the rhesus, *Macaca mulatta*) whereas others are more difficult (for instance the talapoins). This is unfortunate, because the small size of the talapoins renders them particularly suitable for some purposes, but they are now difficult to obtain as well as difficult to breed and supplies are uncertain.

Unlike other groups, the baboons in Africa are in many areas regarded as vermin and shot or poisoned to control their numbers. They are, therefore, probably the one group of simians of which supplies can be assured for some time to come, and attempts are being made to encourage the local people to trap and sell them rather than to kill them. Furthermore, the baboons are probably the most co-operative and useful of the larger cercopithecids and are excellent research subjects for use where a larger monkey is required. They have been much used in surgical experiments relating to organ transplants, cross-circulation, and other such procedures.

The advantages of the different species for different purposes vary quite considerably, and some time spent on study and consultation before proceeding with an experiment will be amply repaid.

Family Cebidae. These comprise the New World monkeys, which are demarcated from Old World species by a number of anatomical and physiological features which place them clearly in a category of their own. They evidently branched away from the Old World stock at an early date in primate evolution, probably long before the hominoids diverged from the cercopithecids. Thus they are more distantly related to man and, as such, are probably a less reliable model for medical studies. Some Cebidae show features resembling those found in prosimian species and may thus be regarded as rather primitive forms of primates. The more advanced members have evidently proceeded on lines of evolution parallel with those of the Old World stock, which account for superficial resemblances. In spite of these facts, because of certain advantages which they possess, some species have become popular and extensive use is made of them in both the United States and other countries. They are mostly small in size and

relatively easy to handle. In addition, marmosets will breed readily in captivity and, unlike all Old World primate species, the females bear their young twice a year, usually as twins. In this animal, therefore, we do have at least one primate which produces young both quickly and in reasonably large numbers, so that it should be possible to establish good lines of laboratory-bred animals for experimental purposes.

Until recently, only three genera of Cebidae have been used extensively in medical research, these being the marmosets, the squirrel monkeys, and the capuchins. However, the discovery that the owl monkeys could be infected with human subtertian malaria has resulted in interest in this group which, although nocturnal, proves to be easily kept and managed.

Many species of New World monkeys are by no means easy to keep in captivity, and few of them breed readily. These remarks apply particularly to the howlers, the uakaris, and the sakis, and possibly also to *Callicebus* and *Callimico*, a very small South American monkey. *Callicebus* are being used extensively at one well-known Primate Center in the United States, although losses amongst newly-imported animals have been extremely high. Research has shown that this group has an unusually high requirement for the B vitamins 1–5, and that when these are supplied the death rate is lowered and the animals survive their quarantine period, becoming attractive, docile and easily-handled. Many species of New World primates have rather specialised dietary requirements and, once these are understood, fewer difficulties will be experienced in maintaining and breeding them. Since the early 1920s the exceptional susceptibility of many species to diseases of the skeleton has puzzled scientists. Those particularly affected are the woolly monkeys, the capuchins and the marmosets. The full story of the unusual mineral metabolism which these monkeys possess has still not been evaluated, but enough is now known to provide adequate diets to keep them healthy.

Most of these monkeys come from the dense forest areas around the Amazon river system, and they often harbour heavy parasite burdens. In their natural habitat no doubt they are in balance with these parasites, but with the stress of capture and transport the parasites are inclined to get the upper hand and a great many deaths may occur before the quarantine period has been completed. A particular feature of these monkeys is the possession of a number of different strains of herpes virus, which are antigenically different so that dangers exist from inter-specific transmission (see p. 425).

In spite of the various difficulties encountered with these monkeys, they are of particular value in certain areas of research (particularly perhaps virological and nutritional) and they offer some advantages not to be found with Old World species. It appears also that possibilities exist of developing them as true laboratory animals, bred in known lines; this applies particularly perhaps to the marmosets, squirrel monkeys, and capuchins.

Sub-order Prosimii. Apart from the tree shrews, the prosimians appear monkey-like and all come from arboreal habitats.

The tree shrews comprise very primitive genera, which share characters with both primates and insectivores. Taxonomists are divided as to whether they should be classified as primates or insectivores, or in a special class of their own. The females bear one to four young, usually two, with a gestation period of 41–50 days. Breeding proceeds throughout the year and at least two litters may be born annually. With suitable rather specialised conditions, tupaias (*Tupaia glis*) can be maintained and bred successfully in captivity. It would appear, therefore, that these small primate-like animals might be of great value in laboratory work. The doubt, however, remains as to whether when using them one really is using a primate with the special advantages which only primates can give, or whether one is employing a more primitive animal too far removed from man to be of especial value.

The other prosimians, although they are undoubtedly true primates, are also far removed from human stock. The lemuriformes from Madagascar are classed as endangered species and importation is not permitted. However, many of the true lemurs breed well in captivity and form attractive exhibits in zoos. Possibly, as with the owl monkeys, some special properties may be found in them which would make their use desirable for certain purposes. They breed only once a year, and single births are the rule,

though occasionally they may bear twins. Unlike marmosets they cannot, therefore, provide an abundant source of young animals.

The true lorises are slow-moving nocturnal animals. They are both delicate and bad-tempered, do not breed readily in captivity, and have not been used seriously as experimental animals apart from behavioural observations. The galagos or bush-babies, on the other hand, are attractive nocturnal animals and attempts have been made from time to time to use them in the laboratory, although they have little to commend them for any particular purposes. They breed quite readily in captivity, usually bearing single young once a year; however, twins and even triplets are not uncommon. The smallest species, *Galago (Galagoides) demidovii*, is a very small animal, the head and body length being only 125–160 mm. It appears to breed twice a year, but no births have been recorded in captivity. Possibly it could be a valuable laboratory animal if means could be found to induce reproduction under captive conditions.

Origins of primates

Primates radiated during the Eocene period from primitive insectivore ancestors into the great equatorial forests which girdle the earth. This habitat has been subject to much variation of extent, covering far greater areas during the Pluvial Periods of the tropics which accompanied the Ice Ages of temperate areas. During warmer times such as the present, therefore, the primate habitat has been much restricted, as a result of which some groups have become either totally or partially terrestrial. Such groups, robbed of the protection of the forest trees, have developed special characteristics seen particularly in baboons and patas monkeys. Not only have they learned to move quickly in troops over the land surfaces, but have also become endowed with dangerous canine teeth for defence. Man himself is such an animal; he originally learned to hunt for food and to protect himself by means of social organisation, and this led to the development of powers of organisation and intelligence far higher than those of any other animals. It is sometimes said that the chimpanzees originally belonged to the proto-hominid stock, but secondarily reverted to an arboreal existence. Many asiatic species of macaques, which are closely related to the baboons, also spend much of their time on the ground and even seek part of their sustenance in river waters. Nevertheless, with few exceptions, all primate species (man apart) are still confined to hot tropical areas.

The present distribution of primates is shown in Figure 31.1. Their habitat is today becoming further restricted owing to forest clearing for human habitation, and many formerly continuous groups are now becoming fragmented and their survival is precarious. These remarks apply particularly to the Great Apes, of which gorillas, chimpanzees and orang utans no longer cover a continuous habitat. Of the more popular laboratory species, the macaques are derived from the Far East, extending from India throughout Indonesia to Viet Nam, with outposts in China and Japan. The trade in these animals has been well organised and up to now has been able to supply demands. However, there is concern that wild populations are being seriously depleted and that control measures are necessary. Members of the genus *Cercopithecus* belong to Africa over which continent, south of the Sahara, they are widely distributed in a number of closely-related species, which differ mainly in colouration. The baboons also are widely distributed in Africa, where they thrive and are by no means endangered, rather constituting something of a nuisance. The numbers of *Cercopithecus*, on the other hand, could well become diminished. The popular patas is widely distributed in sub-Saharan Africa, in the savannah areas as far as the edge of the rain forests. It is common in Uganda, and extends north into the Sudan. These animals do not appear to be endangered at present, but as they breed well in captivity it is hoped that farm production of the number required will be found possible.

The important species of Cebidae (the marmosets, capuchins and spider monkeys) are widely distributed in South America and until now have been relatively abundant. The true marmosets (*Callithrix* spp.) occur in Brazil south of the River Amazon and in East Bolivia. The related tamarins (*Saguinus* spp.) are found in Central South America, the Canal Zone, North-West Columbia, and the entire Amazon Basin. Capuchins (*Cebus* spp.) occur in Central and

South America, from Honduras to South-East Brazil, Paraguay and North Argentina. The related squirrel monkeys (*Saimiri* spp.) are found in Central South America, Costa Rica and Panama, south and east of the River Orinoco and in the Amazon Basin. The other species which is of interest, the owl monkey or dourou-couli (*Aotus* sp.)*, occurs in Central and South America, from Panama to Paraguay, the Amazon Basin and North-East Brazil. It has a similar range to the titis (*Callicebus* spp.). None of these species can be regarded as immediately endangered, although the South American Governments have become nervous about the position in some localities and restrictions have been placed on the export of their primates. Thus supplies are likely to be restricted in the future, and this makes it more urgent that these animals should be bred in captivity.

Size ranges

The size-ranges within any one species, apart from sex differences, are not well-known. This is a matter which will require attention when each species comes to be laboratory-bred and lines of animals with suitable characteristics established. Of the primates suitable for laboratory uses, sizes vary from the small marmosets, averaging around 250 g for the males and slightly less for the females, to the chimpanzees, the males of which may weigh nearly 50 kg. Between these extremes occur animals of almost any required size, and size will be one of the factors to be considered when a species suitable for breeding can be selected. The African *Cercopithecus* species tend to be rather smaller than the Asian macaques, and the South American species are, on the whole, smaller still. Both patas and baboons are large animals, and thus provide useful experimental subjects for uses where the larger animals are required. There is, however, a need for a small monkey at the lower end of the Old World primate range, and the only possible candidate seems to be the talapoin (*Miopithecus*); these are quite small animals, the males weighing from 1,230 to 1,280 g and the females from 745 to 820 g. Unfortunately, as stated above, they do not breed well in captivity,

* It has recently been pointed out (Brambell, R. A., 1973, *J. med. Prim.*, **2**, 284–289) that there are two distinctive types of owl monkeys with different karyotypes. Workers and breeders should have their animals classified.

but attempts are being made in some centres to overcome this difficulty.

Behaviour

Ground-living species such as baboons and patas, and even the young of *Homo*, still have an instinct to climb, and therefore cage construction for all primates should allow sufficient height for vertical movements. The use of the hands is more advanced in primates than in any other group of animals, and their food is carefully picked over and inspected before being placed in the mouth. All primates also have the habit of grooming; they pick over each other's skins to remove blemishes or parasites, thus promoting cleanliness. At the top end of the scale, simian primates are very intelligent and can, up to a point, reason for themselves. Experimenters can

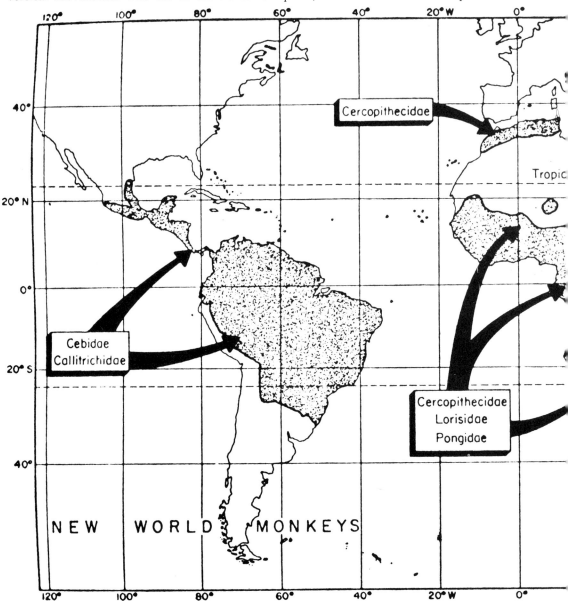

Fig. 31.1. Distribution of non-human primates by families.

use this higher degree of intelligence advantageously if they will exercise sufficient patience in order to get friendly co-operation. A vicious monkey can be dangerous, and some species, such as the patas, have razor-sharp canine teeth which can inflict severe wounds. In all the primate groups temperament varies greatly between individuals, although group characteristics are evident and may be important in the selection of the right animal for any particular experiment. Such species differences occur even among close-related groups such as macaques; rhesus, for instance, are somewhat unfriendly and nervous, whereas the stump-tailed are more phlegmatic and are easy to handle. The black apes, which are really macaques (*Cynopithecus niger* and *C. maura*), are, in spite of their rather ugly looks, most co-

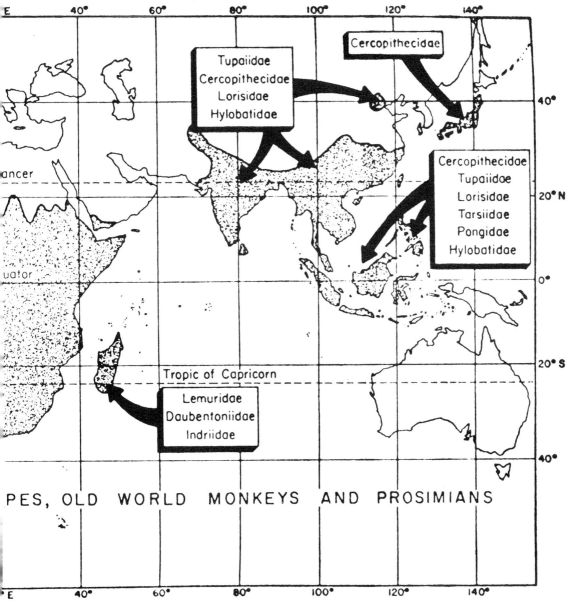

operative and friendly animals, with a ready smile for visitors whom they like. Intra-specific variations of character are perhaps most marked among the chimpanzees, each of which is an individualist in its own right. No primates can be kept in quite the same way as can rats and mice and guinea-pigs. They appreciate kindness and friendliness, and only those animal-keepers who have this approach should be selected for the charge of primates.

Breeds and strains

Hitherto no recognised breeds and strains have been bred from primates. This may be accomplished with marmosets, which are now being intensively bred in a few laboratories. It should not be impossible, with intelligent culling, to produce animals likely to be useful in experimentation; namely, placidity, docility, and cooperativeness. As a longer-term measure, there is no doubt that strains that are homozygous or heterozygous for certain properties (for example cancer susceptibility) can be produced.

Uses

In routine pharmacological investigations primates should be used only as the last step before introduction of a drug to clinical testing on man himself. In general, a primate should never be used for experiment where some other animal would do equally well, nor until it has been shown that other animals are unsatisfactory. Despite these restrictions, there is a large range of subjects for which primates have been indispensable and for which they could probably not be replaced. Although the list given under is by no means comprehensive, it will nevertheless serve to show the types of investigation for which primates have been used and in which their use has led to significant advances in technique or knowledge.

Virology. The investigation of human viruses demands cell cultures derived from animals that are sufficiently close to man to accept the viruses. Fortunately, cell cultures and cell lines derived from Old World monkeys provide a medium in which most human viruses grow readily. Thus, the primary investigations into the aetiology of such diseases as poliomyelitis and the subsequent development of vaccines have been made possible by the use of simian cell cultures, especially those derived from the kidneys. One difficulty has been the presence in these cultures of many simian viruses that are apparently inactive in the entire host but which cause cytopathogenic effects in the cultures derived from them. This has had two effects: first, it has led to alarm that some of these viruses, if included in vaccines, might have delayed pathogenic effects in man; secondly, it has led to the investigation of latent simian viruses and to the realisation that most animals, including man, are likely to carry many such latent viruses hitherto unrecognised and of unknown significance. It may be said, therefore, that a new science of latent virology has developed. It is obviously important that investigations should be actively pursued, because it is known that although the viruses are usually latent they may sometimes be activated and, when shed, have unpredictable effects, including carcinogenic ones, in heterologous hosts.

Foetal, new-born and other primates of different ages have similarly been used extensively in attempts to isolate oncogenic viruses, and two results have accrued from this work. First, Rous sarcoma viruses have proved oncogenic in new-born macaques, and in marmosets of any age. This has laid one myth: namely, that oncogenic viruses are host-specific and will not infect alternative hosts. Secondly, strong evidence now exists that human leukaemias have a viral origin and leukaemia-like diseases have been introduced into macaques by the injection of cells from humans suffering from leukaemia. Since man himself is not available as an experimental animal for this type of research, it is difficult to see how proof can be obtained without using the higher primates as experimental animals.

Parasitic diseases. Among the diseases of parasitic origin, the simians have an especial part to play in malaria investigations. No completely satisfactory hosts, other than man, have so far been found for human malignant subtertian malaria (*Plasmodium falciparum*). However, the owl monkey (*Aotus*) is receptive of the blood forms of infection, though the exo-erythrocytic forms do not develop. Drugs that are effective against both the blood and resting forms of parasite are still sought, for both prophylaxis and treatment of human malarias, and there are hopes that some primate will be found in which both forms will develop.

Immunology and immuno-suppression. The immunogenic systems of the higher primates are sufficiently close to those of man to make investigations on immune mechanisms in these subjects rewarding. In particular, the present great interest in organ transplantation has led to the extended use of primates for investigations of immuno-suppressive regimes. Furthermore, it has been found that both baboons and chimpanzees can be used for blood exchange and subsequent cross-circulation with humans of equivalent blood groupings. Such exchanges may become useful in prolonging the lives of persons awaiting liver transplants, until a suitable organ becomes available.

Surgical techniques. The larger primates have obvious uses in experimental surgery, particularly organ transplants and the use of artificial hearts and other organs. Dogs have been much used in experimental chest surgery, but the incomplete separation of the two halves of the thorax presents difficulties that are not found in, for example, baboons.

General pathology. In many areas, primates undoubtedly provide the best models available to the experimenter of some human diseases. Particular mention may be made of diseases of the cardiovascular system, the clotting mechanisms of the blood, the nervous system, and the skeleton. Much useful work is also being done on the aetiology of dental caries, for which baboons and macaques are used. They are also being used for investigations into conditions of shock and stress and the relationship of such states to other general medical conditions. In such investigations, primates have advantages over animals that are less closely related to man.

Nutrition. Primates as a whole have rather special nutritional requirements, well worthy of investigation in their own right. Moreover, these requirements seem to vary very widely between the different groups. Of particular interest is the high susceptibility, particularly of New World species, to nutritional bone disease; this fact could lead to useful investigations of sterol metabolism as a whole. They share with man a high requirement for good quality protein, and may be found of value in investigations of such diseases as kwashiorkor and other nutritional states.

Drug-testing and toxicology. It is becoming increasingly realised that tests in primates must constitute the last stage in the appraisal of drugs before clinical trials are made in human subjects. It is well established that many drugs affect different groups of animals in different ways: for instance, some drugs may prove toxic in rodents, but are safe and useful in man; conversely, some drugs safe in rodents may be toxic in man, and tests done on man's closest relatives have the best chance of revealing a drug's probable effect in the human subject.

Other uses. Other subjects for which primates are being used may be mentioned briefly: namely reproduction, ageing, irradiation, and behaviour. The important part played by the higher primates in space exploration is a case in point.

Standard biological data

The following Tables* (II–XV) of standard biological data have been adapted from Lapin *et al.* (1972), and from Kruckenberg *et al.* (1972). The figures given have been selected to some extent, in order to give the most probable range of normals in the various groups of monkeys. All figures given by Lapin *et al.* have been taken from large numbers of primates—certainly in excess of twenty, and in some instances running into hundreds. Those of Kruckenberg *et al.* are mostly taken from individual animals. It cannot be said that norms have yet been established for simian primates, and the figures should be taken only as a guide to experimenters. In most cases they are likely to represent the normal ranges but normals are difficult to establish for such highly-strung animals which readily become excited when restrained.

TABLE II
Rectal temperatures

Species	Minimum	Maximum
M. mulatta	36°C	40°C
M. irus	37°C	40°C
Papio	36°C	39°C
Aotus	37·3°C	38·5°C
Callithrix	35·4°C	39·7°C

After Lapin *et al.* (1972).

* All tables in this section are reproduced by courtesy of S. Karger AG, Basel.

TABLE III
Respiratory system

Species	Frequency/min	Minute volume cm³	Tidal volume
M. mulatta	30–31/36–49	(1)970–1050/1300–2100(2) (315 ml/kg)	17–47 ml
Papio	29 /18–22	(1)1890 (315 ml/kg)(2)	—

(1) ... Unanesthetized animals
(2) ... Anaesthetized animals
After Lapin et al., (1972).

TABLE IV
Circulation

Species	Heart rate per min. (restrained animals)	Heart rate (by ECG telemetry)	Blood pressure (direct method) mm/Hg	Blood pressure (cuff method)
M. mulatta	150–333	98–108	158/101	125/75
M. fascicularis	240	—	—	—
M. radiata	197–260	—	—	—
M. speciosa	208	—	—	—
Papio	74–200	85–90	—	135/80
Cynopithecus	169–231	—	—	—
C. aethiops	200–260	—	—	120/70
Cebus	165	—	—	—
Ateles	210	—	—	—
Saimiri	225–350	—	—	—

After Lapin et al. (1972).

TABLE V
Blood volumes

Species	Total blood volume ml/kg	Red cell volume ml/kg	Plasma volume ml/kg
M. mulatta	50–96	16–31·8	32–53
M. fuscata	70–97	—	—
P. hamadryas	62·4–65	—	—

After Lapin et al. (1972).

TABLE VI
Blood coagulation

Species	Prothrombin %	Fibrinogen mg %	Prothrombin time (sec)	Recalcification time (sec)	XIII factor %	Platelets × 10³/mm³	Index of adhesiveness	Tolerance to heparin (min)	Fibrinolytin activity (min)
M. mulatta	60–180	180–440	12±0·3	117–129	106	229–530	1·0–1·62	3′ 05″–4′ 55″	464
M. nemestrina	145–161	316–364	12·6±0·3	—	—	385–571	—	—	—
Papio	70–105	268–406	14·7±17·0	145	100	217–381	1·24	4′ 50″	385–585
Saimiri sciureus	151–163	253–283	8·6±0·2	—	—	659–741	—	—	—

After Lapin et al. (1972).

TABLE VII
Normal blood values

Species	Red cells million/mm³	Haemoglobin g/100 ml	Reticulocytes %	Sedimentation rate mm/hr	Leucocytes thousands/mm³
M. mulatta	5·5 (3·1–8·8)	11·5 (8·7–14·7)	0·1–20	1–5	10–16
Papio	5·6–6·0	11·5	—	2–5	13±1

After Lapin et al. (1972).

TABLE VIII
Leucocyte counts

Species	Lymphocytes %	Neutrophils segmented %	Neutrophils staff-like %	Eosinophils %	Basophils %	Monocytes %
M. mulatta	36–78	35–50	2–7	1–6	0–1	0–6·4
Papio	27–38	53–61	2–7	1–6	0–1	4–8

After Lapin *et al.* (1972).

Haematocrit volumes

In *M. mulatta, M. nemestrina, M. fuscata, M. speciosa* and *M. radiata* the range is 41–47 per cent, as for man (Lapin *et al.*, 1972).

Extracellular fluid volumes

These vary between winter and summer. According to Lapin *et al.* (1972) the volumes for *M. mulatta* were:

winter   184 ml/kg±25
summer   250 ml/kg±41

Serum proteins

According to Lapin *et al.* (1972) the levels of these in *M. mulatta, P. hamadryas, C. aethiops* and *S. sciureus* differ little from those in humans. Total protein is 7·0–8·0 g/100 ml; albumin is 47–59 per cent of the total and globulin 41–53 per cent of the total. There is sometimes an alpha globulin present in *M. mulatta, P. hamadryas* and *C. aethiops* which is not present in human sera.

Kruckenberg *et al.* (1972) also give serum protein levels from several species (see Table IX).

TABLE IX
Serum proteins

Species		Total proteins	Albumin	Globulin	A:G ratio
Chimpanzee	(adult)	7·4	4·8	2·6	1·86
,,	(pre-adolescent)	7·1	3·8	3·2	1·18
,,	(post-adolescent)	7·2	4·4	2·9	1·49
,,	(juvenile)	8·1	4·6	3·5	1·32
M. mulatta		6·9–8·3	4·0–5·1	3·2	1·43
M. speciosa		7·2	4·54	1·94	2·34
Papio		6·2–7·7	2·6–3·7	2·5–3·9	0·9–1·3
Marmoset		5·0–7·2	1·9–3·7	3·5–3·7	1·1

After Kruckenberg *et al.* (1972).

TABLE X
Non-protein nitrogen components of primate sera—mg %

Species	Total N	Residual N	Urea	Uric acid	Creatinine
M. mulatta	800–1020	24–27	8·9–13·8	0·3–0·9	0·9–2·3
Papio	800–1020	24–27	8·6–13·8	0·9–3·4	0·9–1·6
Saimiri	–	12–50	11–42	0·2–2·1	–
Marmoset	–	–	5–21	0·6–0·8	0·25–2·0

After Lapin *et al.* (1972).

TABLE XI

Inorganic ions of primate sera—mEq/L

Species	Sodium	Potassium	Chloride	Calcium	Magnesium	Inorganic Phosphate
M. mulatta	153	4·2	110	5·3	2·1–2·7	2·2–2·7
M. radiata	–	–	–	6·3	1·2	2·7
Papio	137–145	3·5	102–115	3·3–5·9	–	2·5–5·1
Marmoset	122–187	3·5–8·6	95–125	3·6–6·5	1·4–3·7	1·3–5·5
Capuchin	–	–	110–122	–	–	–

After Lapin *et al.* (1972).

TABLE XII

Serum bilirubin levels

Species	Total bilirubin mg/100 ml	Conjugated bilirubin mg/100 ml
Chimpanzee	0·0–1·5	0·0–0·5
M. mulatta	0·10–0·15	–
Papio	–	0·23–0·30
Saimiri	0·0–1·9	0·2
Alouatta	0·75	0·17
Marmoset	0·20	0·10

After Kruckenberg *et al.* (1972).

TABLE XIII

Serum glutamic oxaloacetic transaminase and glutamic transaminase levels

Species	SGOT	Units	SGPT	Units
Chimpanzee	20–21	Sigma–Frankel	13–16	Sigma–Frankel
,,	28–31	DADE modification	22–32	DADE modification
,,	48	Reitman–Frankel	37	Reitman–Frankel
,,	12	International	10	International
M. mulatta	31–58	Sigma–Frankel	22–45	Sigma–Frankel
,,	39–41	DADE modification	27–28	DADE modification
,,	20·4±9·9	Karmen	37·9±19·1	Karmen
Papio	10–51	DADE modification	17–33	DADE modification
C. aethiops	54–61	Sigma–Frankel	38–47	Sigma–Frankel
E. patas	20	Reitman–Frankel	15	Reitman–Frankel
Saimiri	70–138	Reitman–Frankel	117	Reitman–Frankel
Marmoset	143	DADE modification	20	DADE modification

After Kruckenberg *et al.* (1972).

TABLE XIV

Serum alkaline phosphatase levels

Species	Average SAP level	Units
M. mulatta	16	Bessey, Lowry and Brock
,,	18–24	King–Armstrong
,,	2·5–9·5	Sigma
,,	20–57	Klein–Babson–Read
,,	2·5–20·3	Bodansky
Papio	9·3–44·7	Sigma
C. aethiops	30 (10·3–111)	Bodansky
Saimiri	21·8 (6–49)	Bessey, Lowry and Brock
Marmosets	12–29	Bessey, Lowry and Brock
Cebus	5·0	Bodansky

After Kruckenberg *et al.* (1972).

TABLE XV

Cell composition of bone marrow

The influence of orientation response upon cell composition of the bone marrow (%) in *M. mulatta* (mean data obtained from 17 animals).

Bone marrow cells	Days of aspiration		
	1	2	3
Promyelocytes, myelocytes, juvenile	12·0	13·7	16·0
		p = 0·002	
Staff-like neutrophiles	16·3	14·2	16·7
Segmented neutrophiles	23·8	18·5	21·0
		p = 0·108	
Red cell series	31·5	32·5	25·9
		p = 0·062	

After Lapin *et al.* (1972).

The extraction of the animal from the cage during 3 days for bone marrow puncture is accompanied by statistically significant increase of the percentage of promyelocytes, myelocytes and juvenile cells and a decrease of red cell series.

REFERENCES

Groves, C. P. (1972). In *Pathology of Simian Primates*, ed. Fiennes, R. N. T-W-. Basel: Karger.

Kruckenberg, S. M., Cornelius, C. E. & Cook, J. E. (1972). Liver function and disease. In *Pathology of Simian Primates*, ed. Fiennes, R. N. T-W-. Basel: Karger.

Lapin, B. A., Cherkovich, G. M., Kuksova, M. I. & Annenkov, G. A. (1972). Biological normals. In *Pathology of Simian Primates*, ed. Fiennes, R. N. T-W-. Basel: Karger.

Napier, J. R. & Napier, P. H. (1967). *A Handbook of Living Primates*. London: Academic Press.

FEEDING

R. N. T-W-FIENNES

Natural food

Primates, specialised leaf-eaters apart, are all naturally omnivorous, and have the teeth and anatomical structure that go with this way of life. Their habitat (see Fig. 31.1) can be divided into three separate vegetational zones, as shown in Table XVI. The rain forest habitat can be sub-classified vertically into three strata, namely (1) the forest floor, (2) the intermediate and (3) the top forest canopies. In these habitats wild primates have their various niches and find the food on which they subsist.

The steppe habitat forms the transition between savannah and desert and only *Papio* and *Erythrocebus* are found there, with occasional incursions of grivets (*Cercopithecus aethiops*). The food resources of such habitats are meagre and natural foods attributed to *Erythrocebus* are

grasses, fruits, beans and seeds, together with insects and occasionally mushrooms, small lizards, birds' eggs and pieces of red mud (Hall, 1966a). The same author (1966b) gives the diets of *Papio anubis*, *P. cynocephalus* and *P. ursinus* as including grasses, seeds, roots, fruits, bulbs, leaves, tree bark and gum, as well as cultivated crops, mealies and fruits such as paw-paws and bananas. He adds that the physical strength and powerful jaws of these animals enable them to eat fruit such as the very hard sausage-shaped fruits of *Kigelia* which no other monkeys can obtain. They are able to dig deep into the ground in search of bulbs and tubers. Baboons also eat insects of various kinds, which may temporarily form a large part of the diet. Chacma groups inhabiting the South African coastal regions frequently eat shellfish and other marine foods; all species occasionally eat eggs of ground-nesting birds, and catch and eat the young. They also sometimes eat lizards and the eggs of crocodiles on the banks of the Victoria Nile. Infrequently it appears that they catch and eat small wild mammals or the young of larger species. It is extremely doubtful whether either patas or baboons can consume much in the way of grass, except the very young shoots, neither genus being supplied with either the teeth or the digestive apparatus required to tackle this very indigestible type of food. What is to be noted is the omnivorous nature of the diet characteristic of all except the leaf-eaters. Napier & Napier (1967) reviewing the diets of the Cercopithecines, give them as leaves, green shoots, fruit and cultivated native crops. *C. mitis* and *C. aethiops* are said to be partly insectivorous; as judged by stomach contents *C. diana* is entirely frugivorous. For *Macaca* they give fruit, roots, young leaves, insects, grubs, crops such as rice, maize, potatoes and sugar cane, and molluscs and crustaceans. These accounts, though based on information from such studies as have been published, fail to give any idea of the enormously varied types of diet available to wild primates in natural habitats which they share with no other mammalian species. Probably no review of primate feeding habits could be anything like complete unless conducted over a whole annual period so as to take account of seasonal variations. The discovery that chimpanzees and baboons will both catch and consume prey is comparatively recent, and one must suppose that many other primate species will do so too, if only birds, small reptiles and insects. The myth that primates are vegetarian has died hard, and the lessons have been learned the hard way by inflicting inadequate diets on captive animals which have subsisted only with ill-health and suffering. Primates can thrive only on high protein intakes, and it is for this reason certain that under natural conditions they must acquire high protein foods. Possibly this means animal foods on one day only in a week or fortnight, or at even longer intervals, in which case studies of stomach contents on six days in the week will reveal only vegetarian foodstuffs; the animals are then, quite erroneously, classified as vegetarian and are given vegetarian diets in captivity.

Strong doubts must also remain about the diets of animals such as gorillas, which are stated

TABLE XVI

A summary of the vegetational zones and related primate fauna of sub-Saharan Africa

	Vegetation zone	Sub-type	Vegetation and climate	Primate genera
Type I	Tropical rain forest Alternative terms: Moist forest Lowland rain forest Tropical high forest Forêt dense	 Mangrove Secondary forest Swamp forest	3 strata constituting an open and closed canopy with emergents Temp.—steady with narrow range Rainfall—high Rel. humidity—high Specialized mangroves lining estuaries and creeks to tidal limits Tropical rain forest that has been cultivated and subsequently abandoned Similar but more open and irregular in structure	*Cercopithecus* *Colobus* *Pan* *Gorilla* *Cercocebus* *Mandrillus* *Perodicticus* *Galago* *Arctocebus*
		Montane rain forest Alternative terms: Highland forest Cloud forest Bamboo forest	3,000 ft up to 8,000 ft (depending on climatic conditions). Varies from evergreen forest to woodland with tree fern and bamboo thickets. Lianes 7,000 ft–10,000 ft. Stands of bamboo from 20–35 ft. Ground cover sparse	*Pan* *Gorilla* *Cercopithecus* *Colobus* *Papio*
Type II	Savannah Alternative terms: Sour veldt (S.Af.) High grass (E.Af.)	Woodland Open savannah Forest outliers Alternative terms: Bowl forest Kurmi Copses Gallery forest Alternative terms: Riverine forest Fringing forest	Trees 20 ft–50 ft high especially *Isoberlinia* Grass 6 ft–15 ft high Trees widely spaced Grass 6 ft–15 ft high Islands of tropical rain forest. Occurs in hollows and ravines where edaphic conditions are favourable Tropical rain forest on river banks	*Cercopithecus* (esp. *C. aethiops*) *Erythrocebus* *Colobus* *Papio* *Cercocebus* *Galago*
Type III	Steppe Alternative terms: Thornland Sweet veldt (S.Af.) Short grass (E.Af.) Desert grass Orchard steppe	Wooded steppe	Open and closed woodlands or thickets. *Acacia* and *Commiphora* Short grasses	*Cercopithecus* (esp. *C. aethiops*) *Erythrocebus* *Papio*

Reproduced with permission from Napier & Napier (1967).

to be solely vegetarian although they do not possess the fermentation apparatus of vegetarian species, and their dentition is characteristically that of an omnivore. This is in strong contrast to the teeth and digestive apparatus of those primates known to be truly herbivorous—the leaf-eating monkeys, colobus, langurs and howlers. Although these animals do not ruminate, all have to a greater or lesser degree compartmented stomachs in which fermentation can take place, and capacious colons and caeca in which the process can continue. Only these species can functionally survive on purely vegetarian diets.

South American species other than howlers appear to be even more dependent on an adequate intake of high protein foods than the Old World species. The more primitive of these, such as the marmosets, tend to retain the old tritubercular dentition of the ancestral Insectivora, and rely to a great extent on insect foods. As with the Old World species, their habitats offer them an infinite variety of foods of which it must be supposed that they take full advantage under wild conditions, the diet varying from day to day and throughout the seasons. Many South American species have rather limited ranges and it would appear that they occupy specialised niches in the ecosphere; this would again suggest rather specialised feeding habits, depending upon the specialised foods present in their area, and it would account for the difficulties in finding just what foods suit each species when kept in captivity, and for the known high requirements of certain food supplements in some groups.

Special requirements

To state that many, if not most, simian groups, have some specialised feeding requirement is a self-evident truth. To state what these are is another matter. Lack of certain substances may lead to ill-health which can manifest itself in many ways which do not suggest a basic nutritional origin—failure to feed, susceptibility to infection, or just moping and failure to thrive. In the lesser event, it must be asked to what extent the failure of certain species to breed under captive conditions is due to the lack of some essential factors in the diet which operate either on the urge to mate or on the survival of the foetus to the time of parturition. Systematic research on these matters is long overdue and must be done, if the objective of producing primates for research by breeding in captivity is to be achieved.

The primatologist cannot assume that other primates have the same requirements as man. At the London Zoo, an orang utan gave birth to three babies in successive years. She was apparently in good health, but each of the first two babies required treatment for rickets at the end of the weaning period. The third baby was born at term, but died from injuries shortly after birth and the mother died two days later from septicaemia. The baby suffered from advanced foetal rickets although the skeleton of the mother was in perfect condition. Foetal rickets does occur occasionally in human infants but, according to available information on the subject, the mother's skeleton is also always subject to demineralisation processes. This serves as an instance of differences which may occur even in species so closely related as man and orang utan and which, if we place too much reliance on results derived from the human subject, may lead us to a false sense of security.

As is well known, rickets is a commonly encountered disease in the young of both New and Old World monkeys, and osteomalacia is not infrequently encountered in the adults. In many New World species, it has been such a serious scourge as to militate against their maintenance in captivity. Recent work has shown that the trouble can be avoided and, if already present, cured provided that adequate doses of vitamin D_3 are provided, vitamin D_2 being ineffective. Many years ago it was recognised that these troubles could be overcome in marmosets if the animals were subjected daily to quite small doses of ultra-violet irradiation. Indeed, so much so was this the case that, unless dosage was controlled rather strictly, multiple births led to dystokia. It is evident that the kinetics of sterol metabolism differ in many primate species, including man. It is unlikely that, in spite of the discovery of the effectiveness of vitamin D_3 in New World monkeys, the last has been heard of this. The case of the London Zoo orang utan is disturbing, and the high requirement of vitamin D_3 in New World species and the lack of toxic symptoms from it when given in high dosage, suggest that even this sterol is not fully utilised.

Vitamin D and rickets apart, what is the reason that certain species are difficult to maintain in captivity, while others present no such difficulties? Why is it that some species breed well in captivity (as do marmosets and rhesus) while others will not breed at all? It has recently been suggested that squirrel monkeys during parturition have an exceptionally high requirement of folic acid and that if this is given plentiful young are acquired. If this is true of squirrel monkeys, it may well be true of other species also. Another notoriously difficult species, *Callicebus*, was investigated by this author and it was found that they required an exceptionally high intake of the B vitamins in the B_{1-5} range. Once this was given, they thrived as well as other species. There are suggestions that other species, particularly some of the Old World, require high intakes of vitamin B_{12}.

There can be little doubt that many primate species are specialised feeders and have specialised requirements which are readily available to them in their own habitats. Once these are known, many species today that are regarded as difficult, either because they do not thrive or because they do not breed, may well become available for production in captivity and may serve as useful laboratory animals for special procedures to which they are adapted. Little is known, and research is urgently necessary.

Recommended diets, including compounded foods

The long continued argument between the merits of natural as opposed to compounded diets for captive primates often leads to somewhat heated controversy. Natural foods supplied usually consist of a variety of vegetables and fruits, together with nuts, milk products such as cottage cheese, and eggs, sometimes with added casein. The protagonists of this kind of feeding point out that, although wasteful, the diet provides the monkeys with interest and occupation in selecting and picking over the items offered, and that they can themselves select what is most palatable and thus, no doubt, most desirable.

In zoos, it is aesthetically more desirable for visitors to see animals engaged in this kind of feeding activity than for them to be seen gnawing at hard pelleted compounds which appear uninteresting. It will also be pointed out that in natural conditions monkeys will not eat the same food throughout the year. Many Directors of primate medical facilities also prefer to supply animals in the quarantine and holding sections, where space is more abundant, with the natural foods which they feel their animals will enjoy better. This, they feel, will provide them with research animals that are more contented and better adjusted and which may therefore be in better general health.

Unfortunately, it is almost impossible to select a fully omnivorous diet of natural foods for monkeys which contains all the vitamins and other elements required, and which is in balance with respect to mineral intake and so on. Possibly, in the wild, monkeys derive much of their vitamin D requirements from the effects of the sun on the skin, and the enormous variety of foods eaten supplies the rest of what is needed. For instance, with South American monkeys, a great deal of insect food is eaten whole and complete with gut content, so that there is likely to be a generous intake of nutrients in wide diversity. It thus follows that natural diets—though excellent in themselves—will almost always require to be supplemented by additional vitamins, particularly B and D groups, and adjusted for calcium and phosphorus balance by presenting additional calcium in some suitable form. A particular disadvantage of natural diets for monkeys on experiment is the great difficulty of discovering just what they eat, and so of ascertaining their actual intake of essential nutrients.

For caged monkeys on experiment it is, therefore, preferable that compounded diets with a known formula should be used. These can either be manufactured on the premises to a suitable formulation, or commercial monkey chows may be used. A chow, such as Purina 25 monkey chow, containing 25 per cent protein and especially devised for primate feeding, can give excellent results and the pellets appear to be palatable for most species. Nevertheless, such chows are devised mainly for rhesus monkeys and are not ideally suited for all species. As with natural foods, for some species the chows must be supplemented with additional protein, B vitamins, or other substances that may seem necessary. In the light of present knowledge no definite rules can be suggested, and experimenters must discover for themselves what additions are required. Any compounded foods that are stale or have come into contact with damp may be dangerous, and care should be taken when introducing new brands of pellet, even from well-known and reputable manufacturers. Pellets that are mouldy, or in which the fat content has become rancid, are dangerous, leading to intestinal disturbances or even death. This comment is made in the light of personal experience; a change was made to a new brand of pellet from a well-known manufacturer, but was subsequently found to contain a high proportion of free fatty acid indicating rancidity; the food proved toxic, deaths of some animals occurred, and experimental results were vitiated for nearly two months.

Sacks containing food, including vegetable foods such as peanuts, must be stored in such a way that the bottom of the sack cannot get wet; if this happens the food at the bottom may become mouldy and losses may result. Furthermore, where food is stored in bins, the bin must never be topped up; it must be completely emptied and cleaned before further food stores are placed in it, for otherwise the food at the bottom will become stale and rancid, or even mouldy.

Commercial chows are, of course, based on formulated diets originally developed in zoos or laboratories where primates are kept. The pioneer of formulated diets in zoos was Herbert Ratcliffe of the Philadelphia Zoo, whose methods were widely copied and developed in other institutions. Lang (1966) has described the formulated diet given to anthropoid apes at the Basel Zoo and his formulation is given in Table XVII. Wackernagel (1968) describes an adaptation of this formulation, given at Basel to omnivorous animals, including apes and monkeys.

Pioneers of cube diets for laboratory primates in Britain were D. J. Short and A. S. Parkes at the National Institute for Medical Research at Mill Hill, London. Their work produced the well-known diet 41b, described by Short (1968).

TABLE XVII
Composition of monkey cake (Lang, 1966)

Basic mixture (%)		Composition (%)	
Ground plate maize	16	Crude protein *ca*	24
Ground wheat	12	Crude fat	7·5
Ground barley	10	Crude fibre	2·5
Rolled oats	10	Calcium	0·9
Groundnut oil meal	10	Phosphorus	0·8
Soya bean oil meal	8		
Wheat germ	5		
Dried yeast	10		
Skimmed milk powder	10		
Fat	5		
Bone meal	2		
Salt	0·9		
Trace elements mixture*	0·1		
Vitamins mixture†	1		
	100		

* Trace elements added per kg of feed (mg)				† Vitamins added per kg of feed	
Fe	20	Iodine	2	A	40,000 i.u.
Cu	2	Zn	10	D_3	6,000 i.u.
Mn	50	Co	1	B_2	40 mg

Method: To prepare the cake, mix 9 parts of the basic mixture with 1 part of minced cooked meat to a stiff mash with meat broth or water. Press the mash into a shallow pan. This mash quickly hardens. The cake can easily be cut or broken into pieces for feeding. Keep in the refrigerator.

The composition of the finished cake does not differ much from that of the basic mixture:

Crude protein	*ca*	25%
Crude fat		7·5%
Crude fibre		2·25%
Calcium		0·8%
Phosphorus		0·7%

N.B. The cake is not baked. A more correct name for it would be 'dough'.

The formulation of this and of the further developed diet FP1 are given in Tables XVIII and XIX.

TABLE XVIII
Diet 41B

Per cent composition		Theoretical composition (%)	
Wheatmeal	47	Protein	13·70
Sussex ground oats	40	Fat	3·50
White fish meal	8	Carbohydrate	49·00
Dried skimmed milk	3	Fibre	1·50
Dried yeast	1		
NaCl	1		

To each ton of diet is added 2½ lb of stabilized vitamin supplement, which supplies:

Vitamin A 4,000,000 units		Pantothenic acid	0·5 g
Vitamin D_3 1,000,000 units		Nicotinic acid	2·5 g
Vitamin B_2	1·5 g	Vitamin E	1·25 g
Vitamin B_{12}	0·00325 g	Vitamin K	0·5 g
Vitamin B_1	0·5 g	Choline chloride	25·0 g

2–3% of molasses to bind the cubes
3·8 cal per g of diet

Calcium content	0·49 g per 100 g
Phosphorus	0·58 g per 100 g
Iodine	40 mg per 100 g

TABLE XIX
Diet FP1

	kg	%
Ground wheat	250·00	78·38
Soya bean meal	40·00	12·51
Alfalfa meal	6·40	2·00
Bone meal	2·40	0·75
Dried skimmed milk	5·00	1·57
Calcium carbonate	5·00	1·57
Sugar	1·00	0·32
Sodium chloride	1·60	0·51
Liver powder	0·20	0·07
Brewer's yeast	1·20	0·375
Wheat germ	5·00	1·57
Vitamin mixture	1·20	0·375
	319·00	100·000

Theoretical composition

Protein	15·75
Fat	2·03
Carbohydrate	57·12
Fibre	2·89

Composition of vitamin mixture

	g	%
Vitamin A and D oil (Achyfral oil)	30·00	2·50
Thiamine	3·00	0·25
Niacin	6·50	0·54
Folic acid	35·00	2·91
Riboflavin	3·00	0·25
Pyridoxine	4·00	0·33
Calcium pantothenate	50·00	4·19
Ascorbic acid	400·00	33·33
Wheat meal	668·50	55·70
	1,200·00	100·00

3·7 cal per g

These diets were devised mostly for Old World species, and New World monkeys may require rather special treatment. We shall examine some diets for New World monkeys below, but meanwhile it will be of interest to compare these diets with those given to rhesus and allied macaques at a commercial breeding colony, the Bionetics Institute near Washington, D.C. These are described by Valerio *et al.* (1969). Adult monkeys are fed 200–300 g of standard monkey chow (Ralston Purina or Wayne) twice daily, morning and afternoon. They are also given a specially prepared vitamin/mineral supplement mixed together in the blender to form a paste and spread between slices of white bread to make a vitamin sandwich. The vitamin/mineral mix is made as follows:

To one gallon (3,800 ml) of Pervinal Syrup is added 250 capsules of duo-CVP vitamin preparation. Each capsule contains: citrus bioflavonoid compound, 200 mg, and ascorbic acid, 200 mg. To this is added 4,800 mg folic acid and the mixture blended in a Waring blender. One gallon of the above mixture provides approximately 700 doses which represent:

5·5 ml Pervinal; 0·35 capsule duo-CVP and 6·0 mg folic acid per vitamin sandwich. Each vitamin sandwich has been calculated to represent a daily supplement per monkey of: vitamin A, 2,860 USP Units; ascorbic acid, 71·5 mg; vitamin D, 572 USP Units; vitamin E, 2·75 USP Units; citrus bioflavonoid compound, 88 mg; methionine, 33 mg; choline, 33 mg; inositol, 11 mg; thiamine HCl (B_1), 0·78 mg; riboflavin (B_2) 1·56 mg; pyridoxine HCl (B_6), 1·56 mg; vitamin B_{12} activity, 1·1 mcg; niacin, 6·05 mg; d-pantothenic acid, 5·5 mg; calcium (Ca–3·4%), 249·7 mg; phosphorus (P–3·1%), 231 mg; potassium (K–1·54%), 114·4 mg; sodium (from 3·15% sodium chloride, Na–1·24%), 92·4 mg; magnesium (Mg–0·22%), 10·6 mg; iron (Fe_5–0·053%), 3·96 mg; copper (Cu–0·0081 %), 0·605 mg; zinc (Zn–0·0081%), 0·605 mg; manganese (Mn–0·0053%), 0·396 mg; cobalt (Co–0·00030%), 0·022 mg; iodine (I–0·00015 %), 0·011 mg; folic acid, 7·15 mg.

For precise nutritional or metabolic work, it is essential that diet should be specially prepared in the laboratory since the ingredients of commercial feeds are not standardized, even though their analyses remain constant. Excellent chows, in the form of pellets, cake, doughs or pastes, can be made quite easily with reasonably cheap equipment. At the Pathology Department in this Institute, an excellent diet devised by Mr Richard Herbert and Dr Michael Barker for capuchins is made weekly in the form of hard pellets (see Appendix, p. 427). An interesting feature of this diet is the addition of chromic oxide, which passes unchanged through the gut and thus on analysis can act as an indicator of the amount of food consumed. The water used for mixing is de-ionised and intakes are thus precisely known. The capuchins to which this diet is fed are being used for precise metabolic studies, and one feed a day has been found adequate. The cages are cleaned two hours after feeding, and in this way good separation of faeces and urine is obtained with little loss of either.

Deinhardt & Deinhardt (1966) briefly described their methods of feeding marmosets, tamarins, and other New World species. These workers, who use their monkeys chiefly for investigation of infectious agents, feed a basis of high protein monkey chow with twice-weekly servings of fresh fruit and vegetables. The drinking water is sweetened by the addition of a paediatric vitamin preparation which serves to stimulate fluid intake and to supplement the vitamin intake. They state that daylight is not necessary but ultra-violet light is supplied by standard sunlamps for three hours daily.

Breeding records at Bionetics and at Deinhardt's laboratories are good, and it is evident, therefore, that formulated diets are adequate both for maintenance and reproduction in Old and New World primates. High protein intakes are essential, and all species appear also to have high requirements of certain vitamins and of balanced mineral intake. Poor reproduction results of monkeys on low protein intake are described by Eckstein (1966) in the primate colony at Birmingham, England.

Feeding twice daily for limited periods is adequate, and even once daily, at any rate for capuchins. Ad lib. feeding with natural foods is not satisfactory unless the diet is reinforced by supplements of vitamin D (D_3 for New World species), probably some of the B-group vitamins, and calcium. For some (perhaps all) species, as for *Callicebus*, commercial chows should be supplemented with added B-group vitamins and possibly other substances. Recent work on squirrel monkeys (Rosenblum & Cooper, 1968), leads to the supposition that these monkeys, and probably others, require additional folic acid during pregnancy. There is no knowledge as to why some species, such as talapoins, do not breed well in captivity and a strong possibility exists that some primates may require special dietary supplements, hitherto unknown.

REFERENCES

Deinhardt, F. & Deinhardt, J. (1966). The use of platyrrhine monkeys in medical research. In *Some Recent Developments in Comparative Medicine*, ed. Fiennes, R. N. T-W-. *Symp. zool. Soc., Lond.*, **17,** 127–59.

Eckstein, P. & Kelly, W. A. (1966). A survey of the breeding performance of rhesus monkeys in the laboratory. In *Some Recent Developments in Comparative Medicine*, ed. Fiennes, R. N. T-W-. *Symp. zool. Soc., Lond.*, **17,** 94–112.

Hall, K. R. L. (1966a). Behaviour and ecology of the wild Patas monkey, *Erythrocebus patas*, in Uganda. *J. Zool.*, **148,** 15–87.

Hall, K. R. L. (1966b). Distribution and adaptation of baboons. In *Some Recent Developments in Comparative Medicine*, ed. Fiennes, R. N. T-W-. *Symp. zool. Soc., Lond.*, **17,** 49–71.

Lang, E. M. (1966). The care and breeding of anthropoids. In *Some Recent Developments in Comparative Medicine*, ed. Fiennes, R. N. T-W-. *Symp. Zool. Soc., Lond.*, **17,** 113–25.

Napier, J. R. & Napier, P. H. (1967). *A Handbook of Living Primates*. New York: Academic Press.

Rosenblum, L. A. & Cooper, R. W. (1968). *The Squirrel Monkey*. New York: Academic Press.

Short, D. J. (1968). Experience with cubed diets for laboratory primates. In *Comparative Nutrition of Wild Animals*, ed. Crawford, M. A. *Symp. Zool. Soc., Lond.*, **21,** 13–20.

Valerio, D. A., Miller, R. L., Innes, J. R. M., Courtney, K. D., Pallota, A. J. & Guttmacher, R. M. (1969). *Macaca mulatta: Management of a Laboratory Breeding Colony*. New York: Academic Press.

Wackernagel, H. (1968). Substitution and prefabricated diets for zoo animals. In *Comparative Nutrition of Wild Animals*, ed. Crawford, M. A. *Symp. zool. Soc., Lond.*, **21,** 1–12.

HUSBANDRY

C. R. COID

Housing and caging

The basic principles applicable to the housing and caging of laboratory animals in general apply also to simians. It is, for example, essential to ensure that they have a clean environment and are kept at an ambient temperature and humidity suitable for the species. The facilities must be well maintained and staffed by appropriately-trained personnel.

Apart from these basic requirements it is also important to remember that most monkeys available today may be infected with organisms transmissible to man, simians of different species, and other laboratory animals. In view of disease risks, therefore, it is essential that buildings and cages are designed or modified to reduce these risks to an acceptable minimum. Soon after animals arrive at the laboratory a health examination should be carried out. Information on diseases and their control is given in this volume by Fiennes (pp. 421–426) and detailed recommendations on procedures for the safe handling of simians are given in Perkins & O'Donoghue (1969).

Housing

As a general guide, the accommodation for simians may be based on that of units used for holding animals in quarantine or for carrying out experiments with agents transmissible to man and other animals (Fig. 31.2). There should be changing-rooms for staff and, preferably, a complete set of clean clothing should be available daily; it may therefore be desirable to have a small self-contained laundry room within the ancillary area. The rooms for receiving new arrivals should, ideally, be well separated from the permanent stock of animals. This part of the accommodation should have a small subsidiary changing-area where items such as gowns, rubber boots, masks, gloves and caps may be obtained before entering and discarded after leaving the quarantine rooms. The quarantine

unit should be as self-contained as possible, with areas for storage of food, cage-washing and changing.

The animal rooms should be adequately ventilated and have 10 to 15 air changes per hour. Although it is not possible to be precise about the optimum amount of air-space required for each animal, species weighing about 3 to 4 kg remain healthy and in good condition when 1·3 cubic metres of air space is provided for each animal. Rooms used for holding monkeys should be kept under negative pressure in relation to other parts of the building. Where there is a significant risk of airborne infection it is necessary to contain the infected animals in units designed to remove the air away from personnel.

The method of heating rooms is not critical, but in the event of a breakdown of heat source the thermal lag associated with underfloor heating provides a useful safeguard against a sudden drop in temperature. A temperature of 21° to 24°C and a relative humidity of about 50 per cent seems to be suitable for most species of Old World monkeys. The smaller New World monkeys (marmosets (*Callithrix jacchus*) for

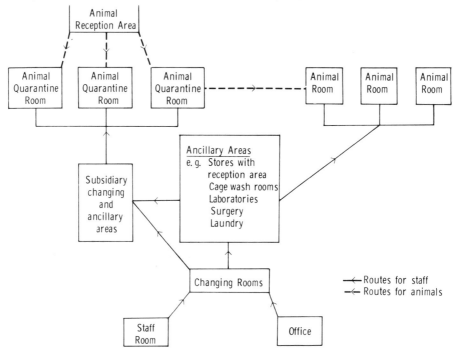

Fig. 31.2. Schematic layout of accommodation for laboratory primate facilities.

example) should be kept at a temperature of 25·6–26·8°C and a relative humidity of 60 per cent (Kingston, 1969). Recently the view has been expressed that a relative humidity of 50 per cent is satisfactory for this species (Hetherington, 1974). If economy of floor space is an important consideration, the height of rooms should be sufficient to allow for two tiers of cages. Experience shows that a height of 2·7 metres allows for this arrangement when cages not exceeding 0·9 metres in height are used (Coid, 1967). The space between opposing rows of cages is an important factor in the layout of a room and it should be sufficient to allow for comfortable working, particularly where large cages have to be handled. Too narrow a space could, for example, preclude the use of a fork-lift truck for moving and lifting heavy cages. It is not possible to give exact figures desirable for all conditions, but a distance of 2·1 metres between opposing cages is satisfactory, while it should not be less than 1·8 metres. It is also worth remembering that staff should be able to walk between cages without fear of being molested by the monkeys (Coid, 1968).

The doors of monkey-rooms should be not less than 1·3 metres wide and corridors should be not less than 1·5 metres wide (preferably 1·8 metres) to allow for easy movement of trolleys and other equipment. The windows should consist either of glass bricks or reinforced glass protected by wire mesh, and be strong enough to withstand the full force of an animal propelling itself at considerable speed against this possible escape route. The doors of the room in which monkeys are being handled should always be locked and there should be a security trap so that, should a monkey escape from a cage, it may still be contained within the building. In the United Kingdom, premises where monkeys are kept must be approved by the Ministry of Agriculture, Fisheries and Food (Rabies (Importation of Dogs, Cats and Other Mammals) Order, 1974).

Cages

In many laboratories the system of caging monkeys singly or in pairs has been found to be satisfactory.

Macaca irus and *Macaca mulatta* weighing under 4 kg may be kept in pairs in cages measuring 78 cm long × 78 cm deep × 91 cm high, with doors 41 cm high × 33 cm wide. Single animals over this weight, but under 7 kg, may also be kept in a cage of this size. Where it is necessary to increase the size of a group it is obvious that the cage should be increased in size or that a pair or more of cages should be joined together to provide the extra space (Coid, 1967). According to Davis *et al.* (1969) the larger primates (such as baboons and chimpanzees of up to 40 kg) may be held singly in cages measuring 183 cm high × 122 cm wide × 91 cm deep.

Smaller primates (such as marmosets) may be

Fig. 31.3. Cages on a free-standing rack. (Photograph: R. Bowlby.)

maintained in pairs in cages measuring 91 cm high × 45 cm wide × 45 cm deep. For breeding purposes it is desirable to provide marmosets with nest-boxes measuring 23 cm × 23 cm × 23 cm. These may be attached to the side of the cage, with access through a 13 cm square opening (Kingston, 1969).

There are many different designs of cages available and many views on which type is the most suitable. However, most workers agree that, for reasons of safety, cages for the majority of species should have a movable partition (squeeze back) which may be used to restrain animals at the front of the cage to allow easy injection of anaesthetic or sedative. This method of restraint reduces significantly the risk of injuries which may so easily occur to staff if animals are handled without sedation. The doors may be hinged, but the guillotine type of sliding door is very satisfactory (Coid, 1967; Valerio *et al.*, 1969). Such doors may be kept closed by a catch, but a padlock provides the best security.

Opinions vary about the need for a perch in the cages, but it is likely that the animals find perches more comfortable to sit on than the wire mesh floor and perches should, therefore, preferably be incorporated as part of the cage structure. Wooden perches are highly desirable for marmosets. These will have to be renewed frequently since this species enjoys gnawing its perch. This procedure probably contributes to healthy dentition.

Fig. 31.5. Mechanical handling of a cage. (Photograph: R. Bowlby.)

In both holding and experimental rooms cages may be either suspended on cantilevers over a long continuous tray or placed on racks which accommodate cages with droppings trays incorporated; the latter may be either fixed or mobile, as shown in Figures 31.3 and 31.4.

For ease of lifting the design of heavy cages (which, if made from steel, frequently weigh over 50 kg) should include an arrangement which will enable them to be handled mechanically with a fork-lift truck (Fig. 31.5).

Pens

It may, in certain circumstances, be necessary to keep simians in large rooms or pens. Compared to the method of holding them in cages this system allows more freedom of movement and space and is particularly desirable where the

Fig. 31.4. Cages suspended on cantilevers fixed to the wall. (Photograph: R. Bowlby.) Reproduced with permission from *Lab. Anim.*, **1** (1967).

larger simians are used (e.g. adult baboons and chimpanzees). Large pens may also be employed for holding groups of healthy monkeys which are free from infectious diseases. However, this method should not be used for holding groups of recently-captured animals. Such animals are likely to be in poor condition, infested with parasites and micro-organisms and should be held, preferably individually, in cages for an appropriate conditioning and quarantine period.

Certain difficulties in safe-handling arise from holding animals in areas where they have much free movement. It is necessary, therefore, to adopt techniques which reduce the risk of staff being bitten when attempts are made to catch and restrain the animals. One method is to use a capture-gun which makes it possible to inject the animals with anaesthetic from a distance of several metres. When a capture-gun is not used the animals may be induced or driven into a cage where they can be forcibly restrained by a movable squeeze-partition. Alternatively, animals may be caught in a net of suitable size by an operator who enters the pen for this purpose, but this procedure may be too dangerous to use on large monkeys or apes. Both this method of handling, and that involving the use of a capture-gun, must only be carried out by suitably trained and experienced personnel.

Food and watering utensils

Pelleted food may be presented to monkeys in a hopper fixed to the front of the cage. Other types of food, such as fruit and bread, can be given through either the mesh of the cage or some other opening.

Drinking-water can be administered by inverting a bottle with a suitable nozzle which passes through an opening in the cage. It is essential to ensure that the bottle is fixed securely for otherwise it may be detached by the monkeys. A suitable arrangement for a water-bottle holder is shown in Fig. 31.3 and a food hopper may be fixed in a similar position. Automatic watering devices may also be used, but these must be firmly fixed so that they cannot be removed by the monkeys. Regular checking is essential, to ensure that the apparatus is functioning properly.

Identification

The best method of identification is by the tattooing of the name or number of the monkey on the chest or forehead. To do this, it may be necessary to sedate the animal. For short-term experiments it may be quite suitable simply to number the cage or, where there are two animals, to mark one of them with a non-toxic dye.

Transfer of monkeys

When it is necessary to transfer simians over short distances, usually within the laboratory, a catching-box may be used. This is placed over the opening of the cage and the animal is induced to enter; the upward-sliding door of the box is then replaced. A suitable design for macaques has been described by Hartley (1964).

REFERENCES

Coid, C. R. (1967). A system of caging monkeys. *Lab. Anim.*, **1**, 25.

Coid, C. R. (1968). Building design in relation to function of a laboratory primate unit. *Lab. Anim. Symp.*, **1**, 113.

Davis, J. H., Bruce, R. McP. & Moor-Jankowski, J. (1969). Maintenance and handling of primate animals for medical research. *Ann. N.Y. Acad. Sci.*, **162**, 329.

Hartley, E. G. (1964). A container for the short distance transfer of laboratory monkeys. *Lab. Anim. Care*, **14**, 103.

Hetherington, C. M. (1974). Personal communication.

Kingston, W. R. (1969). Marmosets and tamarins. *Lab. Anim. Handb.*, **4**, 243.

Perkins, F. T. & O'Donoghue, P. (1969) (Eds). *Hazards of Handling Simians*. London: Laboratory Animals Ltd.

Valerio, D. A., Miller, R. L., Innes, J. R. M., Courtney, K. D., Pallotta, A. J. & Guttmacher, R. M. (1969). *Macaca mulatta: Management of a Laboratory Breeding Colony*, p. 18. New York: Academic Press.

BREEDING

C. J. MAHONEY

Selection of breeding stock

In selecting breeding stock for a primate colony it is easier to discard animals which fail to attain the reproductive performance required than to predict future breeding potential. This is especially so if the monkeys are newly-arrived from their native habitat when the history of the individual animal is lacking. The following account relates mainly to *Macaca mulatta* and *Macaca irus* but is probably relevant to most species of primates.

Examination for health

A preliminary inspection of unsedated monkeys can be made while they remain caged. The animals should be alert and active with bright eyes and lustrous coat; the condition of the tail is a good indication of general health. Some allowance must be made for underweight in newly-imported stock, but emaciated specimens should not be considered. A more detailed examination can be performed on monkeys sedated with an intra-muscular injection of phencyclidine HCl* at a dosage of 2 mg/kg bodyweight. It is wise to exclude animals that show overt signs of disease. In particular they should be free from enteritis, pneumonia and oral vesicular or ulcerative lesions suggestive of virus B (simian herpes complex). An estimate of the age of the monkey can be made by examination of the teeth. Sexual maturity is not reached until the monkeys have at least 28 teeth, whether these be entirely permanent or a mixture of permanent and deciduous teeth. In *Macaca mulatta* sexual maturity is attained at $4\frac{1}{2}$ years of age in the female (Hartman, 1932; Haigh & Scott, 1965) and at $4\frac{1}{2}$ to 5 years in the male (Haig & Scott, 1965). Marked attrition of the molar teeth suggests that the animal could be aged and may therefore have a limited breeding life ahead.

Examination of the genital system

External genitalia. Some idea of the previous reproductive status of females may be gained from an inspection of the mammae. Nulliparous and primigravid females tend to have small pink nipples. In multiparous females the nipples are usually elongated and pigmented. The presence of copious white milk in the mammae of non-pregnant females suggests that parturition was of recent occurrence. A watery grey secretion in

* Sernylan; Parke Davis.

the breasts indicates that parturition occurred some weeks previously.

The vulva should be examined for the presence of a partial or complete prolapse of the vagina through the vulval lips. Chronic infections of the cervix often result in such cases and fertility may be impaired.

In the sexually mature male the testes are permanently scrotal in position (Haig & Scott, 1965). Apart from inspecting the scrotum and penis for signs of disease or injury little information can be gained from a physical examination of the male. A full evaluation of semen, which can be obtained by electro-ejaculation, will enable an assessment to be made of the breeding potential (see p. 412).

Internal genitalia. The remainder of the genital tract is examined by rectal palpation.

Non-pregnant female. In the normal, non-pregnant female monkey the cervix uteri is a firm barrel-shaped structure. Except during the first few days post-partum, softening or malleability of the organ should be regarded as abnormal and suggestive of uterine or cervical infection. In such cases digital massage of the cervix towards the vagina usually results in the discharge of a purulent malodorous secretion. Recovery is prolonged and often incomplete even after antibiotic therapy.

The non-pregnant uterus in the normal-cycling monkey (*Macaca irus*) is 2·5 to 4 cm in

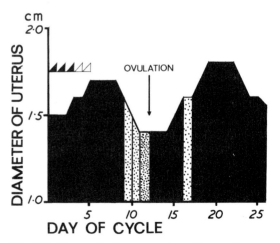

Fig. 31.6. Changes in diameter and consistency of uterus during menstrual cycle in an individual *Macaca irus*. Consistency: firm (black), resilient (stippled), soft (white). Days of menstrual bleeding indicated by triangles.

length. The diameter of the uterine body, as measured trans-abdominally, varies during different stages of the menstrual cycle, ranging from 1·3 to 2·1 cm (Fig. 31.6). Except during the peri-ovulatory period, the uterus is firm in consistency. Marked softening of the uterus, outside this period, is indicative of uterine infection which may have been introduced at the time of the previous parturition. In females with chronic amenorrhoea or repeated low-grade menstrual cycles, the uterus may become progressively atrophied until the diameter is less than 1·3 cm. Spontaneous recovery of normal reproductive activity is unlikely in such cases.

It is difficult to assess the reproductive potential of the female monkey from a single palpation of the ovaries. During the ovulatory cycle each gonad undergoes profound changes in shape, size and consistency. However, if both gonads are small, and the uterine diameter is less than 1·3 cm, it is probable that the female is chronically amenorrhoeic. Adhesions between the ovaries, fallopian tubes and adnexa, which may result in mechanical blockage to ovum transport, are usually detectable during palpation of the genital tract.

Pregnant female. In selecting pregnant female monkeys one must ascertain whether the foetus is viable at the time of examination or whether the pregnancy is liable to terminate before full term. A high incidence of foetal death in newly-imported pregnant *Macaca mulatta* was reported by Valerio *et al.* (1968). The rate of stillbirths and abortions was 59 per cent in contrast to a 12 per cent incidence of foetal loss in laboratory-bred females.

Difficulty may be experienced in ascertaining the viability of the foetus for the following reasons: in *Macaca irus* it is not easily palpable before the thirteenth week of gestation; foetal movements are usually not detected until late in pregnancy; foetal heart sounds may be inaudible, and their absence does not necessarily indicate death of the foetus.

If death has occurred some time prior to examination the foetus may be found tightly bound within the uterus, owing to loss of amniotic fluid and contraction of the uterine muscles. In such cases the spinal column of the foetus may be abnormally flexed. If degeneration of the foetal tissues has occurred digital pressure applied to the skull may disclose separation and grating of the cranial bones. Advanced cases of foetal maceration can be detected easily on palpation of the uterus.

Discarding of breeding stock

Apart from culling required as a result of infectious diseases (e.g. tuberculosis) it may be necessary to discard sexually mature monkeys which have failed to reproduce. In the male, semen may be of substandard quality (see p. 412). Valerio, Pallotta & Courtney (1969) recommended the culling of female macaques that fail to conceive after 12 to 13 consecutive matings.

Apart from the numerous reports of such conditions as ovarian cysts, endometriosis and a variety of benign and malignant conditions of the uterus, cervix and vagina (Hendrickx & Kraemer, 1970; Hendrickx & Nelson, 1971), few attempts have been made to classify the types of infertility that may be found in primate colonies. A brief description is given of the most common types of infertility existing in the colony of *Macaca irus* at the Royal College of Surgeons of England (Mahoney, 1975).

Amenorrhoea

Amenorrhoea is the term used to describe cessation of periodic menstrual bleeding. Uterine bleeding may cease either entirely, for many months or years, or may occur irregularly at 3–6 month intervals. A period of physiological amenorrhoea follows parturition, lasting up to several months (45 days on average in macaque species, according to Valerio, Pallotta & Courtney (1969)). Lactation may inhibit the return to normal cycles in species of macaques (Hartman, 1932), but this has not been proved conclusively. Hartman described the re-establishment of sexual function in post-partum females (*Macaca mulatta*) as occurring in a 'stair-case' fashion. Menstruation is the first function to be restored. The uterus and ovaries then regain their normal physical characteristics although ovulation may not occur for several cycles. If amenorrhoea persists beyond several months it should be regarded as pathological.

Chronic amenorrhoea has been noted in several post-partum *Macaca irus* but it is also quite common in females which have resided in the colony from immaturity to the age when sexual maturity would be expected. In either case, the uterus gradually atrophies and the ovaries diminish in size. It is noteworthy that in none of these cases did the vaginal smear exhibit the characteristics of an atrophic vagina such as would be found in ovariectomised females.

Anovulatory cycles

Monkeys with either regular or irregular menstrual cycles may repeatedly fail to ovulate. Anovulation can be diagnosed by inspection of the ovaries at laparotomy or by frequent palpation of the ovaries per rectum during the cycle (Hartman, 1932; Mahoney, 1970). In extreme cases of ovarian failure, little or no growth of the ovaries is palpable during the cycle and the uterus does not undergo the physical changes associated with ovulation. In less extreme cases, enlargement of the ovary and the developing follicle may be palpable but ovulation does not occur. An eventual return to ovulatory cycles is more likely in the latter instance than in females with extreme depression of ovarian function.

Cystic ovaries

Multiple small follicular cysts in the ovaries of sexually mature *Macaca irus*, both young and old, have been noted. At first these females had regular cycles, many of which were ovulatory. The first noted sign of abnormality was repeated failure to ovulate although regular menses continued. Gradually enlargement of the clitoris occurred until it became penis-like. Menstrual cycles eventually became increasingly irregular until prolonged periods of amenorrhoea set in. During this phase the uterus hypertrophied and a diagnosis of pregnancy was mistakenly made in some instances. Profuse and protracted uterine bleeding then occurred, in one case lasting 30 days.

Frigidity

Occasionally females resident in the colony since adolescence are found to be frigid, resisting all attempts by the male to mate. It has been noted that in such animals menstrual cycles are frequently anovular. Whether frigidity is a consequence of deranged pituitary-ovarian function, or is, in some way, the cause of it, has not been determined. Sedation of the male and/or female is not successful. Induction of ovulation by the use of fertility agents and impregnation by artificial insemination seems the only possible remedy for this situation.

Infection of the genital tract

Infection of the uterus and cervix may develop following parturition or abortion. This is especially so if the placenta, or fragments of it, are retained in the uterus beyond 48 hours. It is essential that all females are examined rectally within 24 hours of parturition to ascertain whether the placenta has been delivered in its entirety. Oxytoxin, if administered by intramuscular injection (10 units) soon after parturition, will induce contraction of the uterus and separation of the placenta. If this is ineffective after 2 days' treatment attempts should be made to remove the placenta through the dilated cervix or by hystero-laparotomy.

Uterine infections occasionally develop after Caesarean section. Unless antibiotic therapy is initiated promptly the prognosis is often poor and future fecundity impaired.

Post-partum and surgical trauma

Physical damage to the genital tract may follow parturition or abortion. Partial prolapse of the cervix and vagina have resulted in permanent loss of fertility in a female which aborted during the second month of pregnancy.

Post-operative adhesions involving the ovaries and fallopian tube may prevent ovulation or transport of ova down the tubes into the uterus. The uterus may become adherent to the bladder, thus endangering the animal in future pregnancies.

Data on the sexual cycle in some species of primate

The average age at menarche, the time of the first menstruation, is 3 years 7 months in *Macaca mulatta*, and sexual maturity is reached at $4\frac{1}{2}$ years of age (Hartman, 1932; Haig & Scott, 1965). The period between menarche and the attainment of sexual maturity is characterised by anovulatory cycles of irregular length (Hartman, 1932). In 103 females of the species *Macaca irus* the average age at menarche was 3 years 6 months (range 1 year 6 months to 5 years). Regular menstrual cycles were established on average at 4 years 7 months (range 3–7 years). The average age at the first conception in 6 females exposed continuously to the male was 4 years 9 months (range 3 years 11 months to 5 years) (Mahoney, 1975b).

The modal length of the menstrual cycle in *Macaca mulatta* is 28 days but there is a wide individual variation (Hartman, 1932). Ovulation occurs between days 8 and 21 of the cycle but is most frequent on days 12 and 13 (Hartman, 1939). Similar figures have been obtained for *Macaca irus*. Data on several species of primates is given in a review by Lang (1967a).

A distinct breeding season has been described for *Macaca mulatta* living in the wild (Southwick *et al.*, 1965), and for those living in semi-free colonies (Koford, 1965; Vandenburgh & Vessey, 1968). In outdoor caged colonies, a peak period of conception is described although breeding does occur throughout the year (Rowell, 1963). Hartman (1931) found that anovulatory menstrual cycles in *Macaca mulatta* maintained in outdoor cages are characteristic of the summer months in the Northern Hemisphere. The existence of true breeding seasons in macaque species maintained under the standardised environmental conditions of the laboratory has been denied by Valerio *et al.* (1968). Newly-imported females may demonstrate a residual effect of the natural breeding season, but this is soon lost as the monkeys acclimatise to the laboratory.

Table XX

Lengths of menstrual cycles in some species of non-human primates

Species	Range in days	Modal in days	Author
Macaca radiata	25–38	31	Zuckerman (1937)
Macaca nemestrina	24–66	30	Zuckerman (1937)
Papio spp. (baboons)	22–46	31–33	Zuckerman (1947)

The detection of ovulation

There is no conclusive evidence in any primate species which indicates the duration of the sperm's capacity to fertilise an ovum, nor how long ova, once shed from the follicle, remain capable of being fertilised. Evidence accrued from studies in several mammalian species does suggest, however, that fertilisation of the ovum occurs within a few hours after ovulation, provided that capacitated spermatozoa are present in the upper portion of the fallopian tube. Thus, the accurate determination of the time of ovulation is the best estimate available of conception time.

In many species of primates there is no distinct period of sexual receptivity (oestrus). Matings occur throughout the menstrual cycle in *Macaca mulatta* (Rowell, 1963) and in *Macaca irus*. The detection of ovulation in these species presents many problems.

Sex skin

Reddening and tumescence of the sexual skin of the perineum marks the periovulatory period in the baboon (*Papio sp.*) (Hendrickx, 1965), in the pigtail macaque (*M. nemestrina*) (Kuehn *et al.*, 1965; Bullock, Paris & Goy, 1972) and in the chimpanzee (*Pan troglodytes*) (Young and Orbison, 1944).

In the rhesus monkey, an intensification of colour, rather than oedema, of the sex skin occurs in various parts of the body, especially on the thighs and perineum. In the Wisconsin Regional Primate Research Center (Czaja, Eisele & Goy, 1975) and in the Oregon Center (Ediger & Mahoney, unpublished), this method is used routinely to estimate the time of conception.

Exfoliate vaginal cytology

The cyclical pattern of cornification of the desquamated vaginal epithelial cells was studied in *Macaca mulatta* by De Allende *et al.* (1945). A peak in the cornification index occurred at mid-cycle in ovulatory cycles. The method is similar to that used in laboratory rodents but the interpretation of smears requires a high degree of skill. In the baboon, Gillman (1937) recognised 5 phases in the menstrual cycle from the changing pattern of the vaginal smear. The copulative phase, lasting from day 10 to 26, is marked by flattened cornified cells and a gradual diminution in the number of leucocytes. A simpler method is the measurement of the karyopycnotic index (KPI) of the vaginal smear, stained by the Papanicolaou technique and viewed under the phase contrast microscope (Wied, 1955) (Fig. 31.7). In a study of the menstrual cycle in *Macaca irus* (Mahoney, 1970) a poor correlation was found between the mid-cycle peak of the KPI and the day of ovulation (Fig. 31.8).

Basal body temperature

A rise in the basal body temperature (BBT) occurs in the human in association with ovula-

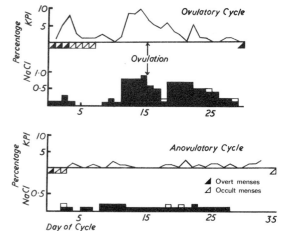

Fig. 31.8. Correlation between karyopycnotic index (KPI) of vaginal smear and apparent daily variation in chloride concentration of vaginal mucus expressed as equivalent NaCl concentration during cycle in *Macaca irus*.

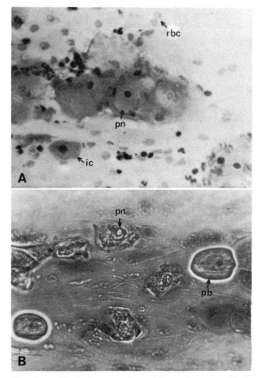

Fig. 31.7. Vaginal smear from *Macaca irus*—modified Papanicolaou stain—X400. A. Ordinary light microscopy—3rd day of cycle. B. Phase-contrast—8th day of cycle. Superficial cell with pycnotic nucleus (pn). Parabasal cell (pb); intermediate cell (in); red blood cell (rbc). (Photographs: Mahoney.)

tion (Rubenstein, 1937). Recording the BBT in the monkey by rectal thermometer is not practicable since the excitement caused by restraining the animal will invariably raise the body temperature. By the use of a telemeter implanted into the peritoneum in *Macaca mulatta*, Balin & Wan (1969) recorded a biphase shift in BBT at the time of ovulation.

Dated conception

The establishment of pregnancy is irrefutable proof that ovulation has occurred. By employing restricted periods of mating it is possible to estimate, retrospectively, the probable time of ovulation. The various systems used for obtaining dated pregnancies will be discussed later.

Estimation of the chloride ion content of cervical mucus

A simple chemical test was devised by Mc-Sweeney & Sbarra (1964) for estimating apparent daily changes in the chloride content of cervical mucus in the human. The test is carried out by applying cervical mucus to a test paper impregnated with silver nitrate and potassium chromate. Chloride ions in the mucus are precipitated as silver chloride, the intensity of the spot being approximately proportional to the amount of chloride present. The fertile period in the cycle is indicated by the most intense mucus spots. This test was used by Wilson *et al.* (1970) and Mahoney (1970) to predict imminent ovulation in *Macaca mulatta* and *Macaca irus*. Although it gives several days' warning of imminent ovulation, the test does not indicate precisely the day of ovulation in *Macaca irus* (Fig. 31.9). A series of dull spots is obtained in amenorrhoeic and some anovulatory monkeys.

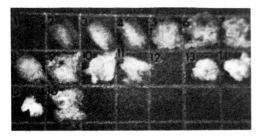

Fig. 31.9. Cervical mucus chloride test during first 16 days of cycle in *Macaca irus*. No recording made on day 12; ovulation on day 11. (Photograph: Mahoney.)

Rectal palpation of the genital tract

By using his technique of rectal palpation of the ovaries and uterus, Hartman (1932; 1933) was able to determine accurately the time of ovulation in *Macaca mulatta*. This method has proved successful in *Macaca irus* when used in conjunction with the mucus chloride test (Mahoney, 1970; 1975a). The uterus, which is firm throughout most of the cycle, becomes distinctly resilient at the time of ovulation (Fig. 31.6). Progressive enlargement of one ovary, caused by the development of a ripe follicle, is palpable. The definitive diagnosis of ovulation is made when the active ovary undergoes a sudden reduction in size following rupture of the Graafian follicle (Fig. 31.10).

Physical Changes in Ovary during Ovulatory Cycle (M. irus)

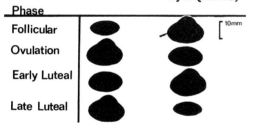

Fig. 31.10. Physical changes in ovary during ovulatory cycle (*Macaca irus*).

Surgical observation of the ovaries

Although impractical for certain types of scientific investigation, the time of ovulation can be determined retrospectively in the monkey by serial observation of the ovaries for evidence of a corpus haemorrhagicum at laparotomy (Betteridge, Marston & Kelly, 1970; Weick et al., 1973) or by laparoscopy (Jewett & Dukelow, 1971).

Hormone assays

The development of radioimmunoassays for determining serum levels of total oestrogens, as reported by Hotchkiss et al. (1971) provides a method which may prove reliable for predicting the time of ovulation. In most cases in *M. mulatta*, ovulation occurs within 48 hours after the mid-cycle peak in oestrogens, as confirmed by serial observation of the ovaries at laparotomy (Weick et al., 1973).

Polygamous and monogamous systems of breeding monkeys

Polygamous units

Under polygamous systems of breeding monkeys, several females are run continually with one male to form a harem. The extreme type of polygamous system is exemplified in the semi-free colonies of *Macaca mulatta* established on the islands of Cayo Santiago (Koford, 1965) and Parguera (Vandenburgh & Vessey, 1968) off the coast of Puerto Rico. The colony on Cayo Santiago comprises 6 groups, each ranging from 21–130 members.

On a smaller scale, harem groups can be maintained in large outdoor cages. For *Macaca mulatta* 3 to 5 females may be run with one male (Robinson, 1964; Rowell, 1963). The system described by Kriewaldt & Hendrickx (1968) permitted 15–25 female baboons to run with the male.

Several advantages are to be found in using polygamous units. In species with no marked period of oestrus and where matings occur throughout the cycle, as in *Macaca mulatta* and *Macaca irus*, maximum opportunities exist for obtaining fertile matings. Koford (1965) reported an annual incidence of conception ranging from 73–83 per cent in the semi-free colony of *Macaca mulatta*. In a comparative study of fertility between monogamous pairs of monkeys housed indoors, and polygamous units in outdoor cages, Banerjee & Woodard (1970) reported pregnancy rates of 25 per cent and 64 per cent respectively. They concluded that the freedom of movement and choice of females open to the male were responsible for the high fertility in harem systems.

Since a hierarchical structure can be established between the male and the females of his harem, especially in semi-free colonies, the results of behavioural studies carried out in polygamous units are particularly meaningful.

The cost of maintaining colonies housed in outdoor cages is less than the cost of running indoor units where systems of heating, lighting, ventilation and humidity control are expensive. Banerjee & Woodard (1970) calculated that labour costs were 25·6 per cent less per monkey in maintaining outdoor than indoor establishments.

There are however many inherent disadvantages in polygamous systems. Fighting may occur between the male and his least favoured females. The earliest signs of developing disease may not be noticed owing to the difficulty of observing animals at close range. Menstrual bleeding may be difficult to detect in females with scant or occult vaginal flows. Estimates of dates of conception can be made only if precise menstrual data are available. Female monkeys may not only have regular occult menstrual episodes but overt bleeding may be preceded by 1 to 3 days of occult haemorrhage (Mahoney, 1970). The modal time of ovulation in such cases would, therefore, be miscalculated.

Monogamous pairs

In monogamous systems the female is caged alone with a single male for prolonged or restricted periods of time. The advantage of using prolonged mating intervals is particularly apparent in species where a distinct oestrus does not occur. The main disadvantage is that pregnancies cannot be dated precisely. In species where colour and oedematous changes affect the sexual skin during the peri-ovulatory period, an estimate of the time of ovulation can be made by observing the female closely. This system calls for a high ratio of males to females, thereby increasing costs.

If restricted periods of mating are employed, precise timing of ovulation is rendered less difficult. Such systems require accurate menstrual cycle data. In the 'calendar' method, it is essential to identify day 1 of the cycle—the first day of menstrual bleeding. Matings are arranged to cover the fertile period of the cycle.

Van Wagenen (1945) demonstrated that the optimum time of the cycle for obtaining conceptions in *Macaca mulatta* was during a 48 hour period commencing on the morning of day 11. Approximately one third of females conceived after one mating. Less success was attained when matings were arranged at pre-fixed times outside this period. In an attempt to improve the rate of conception, some workers extend the period of mating from day 8 or 10 of the cycle through to day 16 or 20 (Dede & Plentl, 1966). However, estimation of the time of ovulation becomes inaccurate with such regimes. A more flexible schedule of mating is employed for *Macaca mulatta*, *Macaca irus* and *Macaca radiata* by Valerio, Pallotta & Courtney, (1969) in an attempt to improve the incidence of conception while retaining maximum accuracy in assessing the time of ovulation. Breeding periods, planned on an individual basis, are calculated from the data of previous menstrual cycles. Females with regular 26–30 day cycles are caged with the male for 72 hours beginning on the morning of day 11. Other 72 hour periods of mating are utilised for females of differing cycle lengths. Encompassing the extremes of menstrual cycles, breeding is initiated on the 6th day for females having 18–19 day cycles, and on the 14th day for those with 35–38 day cycles. Annually, 70 per cent of females become pregnant although the majority require 2–3 consecutive matings to ensure impregnation.

In the methods so far described, matings are planned by the calendar and do not take into account the events occurring in the individual female during current menstrual cycles. The mucus chloride test was used in *Macaca mulatta* in an attempt to predict ovulation (Wilson et al., 1970). The animals were mated for 48 hours during the period of brightest mucus spots and as a result a 50 per cent incidence of conception followed in contrast to only 30 per cent when the calendar method was used.

By palpation of the ovaries and uterus per rectum Hartman (1932; 1933) was able to establish the time of ovulation to within 0–24 hours of its occurrence. Mating periods of 12 hours' duration were planned in accordance with the findings on rectal examination, and successful impregnations resulted. The same technique has been employed in *Macaca irus* and *M. mulatta* in conjunction with the mucus chloride test (Mahoney, 1970; 1975a). The method is as follows:

A rectal examination of the females is performed on day 1, 2 or 3 of the cycle in order to determine (a) the comparative sizes and shapes of the gonads, and (b) the diameter and texture of the uterine body. Beginning on day 7 or 8 of the cycle, the chloride test is carried out daily on the cervical mucus. When the first bright spot is recorded the female is again palpated to determine which ovary has enlarged as a result of the developing Graafian follicle. The female is then caged with the male for mating. Each day the female's genital tract is palpated. During ovulatory cycles the uterus changes from a firm texture to one of distinct resilience on the day of ovulation and for 1 or 2 days preceding it. After ovulation the uterus again becomes firm. The definitive diagnosis of ovulation is made when the previously enlarged ovary has undergone a distinct reduction in size during the preceding 24 hours. The female monkey is then separated from the male. Pregnancy may be diagnosed by rectal palpation of the uterus at 21–28 days gestation, or earlier if the mouse bioassay is employed (Valerio, Miller, Innes, Courtney, Pallotta & Guttmacher, 1969).

With the calendar method conception dates must, of necessity, be calculated from the mating interval employed. With a 48-hour period of mating, Wilson *et al.* (1970) took the first day on which spermatozoa are found in the vaginal smear as time 0 of conception. They state that estimates of conception time have, therefore, an error of ±1 day. Valerio, Miller *et al.* (1969) employing a 72-hour breeding interval, calculated the error as plus 1 day to minus 3 days since spermatozoa may be viable for up to 48 hours. As stated earlier, it is not known for certain how long spermatozoa and ova retain the capacity to conjugate. It has been shown in several species that in the presence of capacitated sperm the ovum is fertilised within a few hours of ovulation. The method described for obtaining dated pregnancies in *Macaca irus* provides that spermatozoa are present before, during and after ovulation so that ovulation time is a close estimate of the time of conception.

Inbreeding and outbreeding

The breeding of monkeys in captivity has not reached the degree of refinement attained in the breeding of other species of laboratory animals. However, successes in obtaining second to fifth generations of rhesus monkeys, baboons and marmosets are now being reported from various parts of the world (Spiegel, 1975). Of more common occurrence, however, is the breeding of laboratory-born females to wild-trapped males.

Diagnosis of pregnancy

Physical changes affecting the genital tract during pregnancy

Pregnancy can be diagnosed in *Macaca irus* by rectal palpation of the uterus as early as 21–28 days gestation (Mahoney, 1975c). A series of physical changes affects the uterus and cervix during the course of pregnancy which enables the stage of gestation to be determined approximately at any given moment.

Increase in lateral diameter of the uterus during the first 10 weeks of pregnancy is shown in Figure 31.11. The range fluctuates between individual diameters beyond the 8th week so that determination of the stage of gestation by this criterion becomes inaccurate. Beyond the 12th week of pregnancy the uterus becomes so soft and shapeless that measuring the diameter is impossible.

During the first 20 days post-conception the uterus is firm. By day 21 a small area of softening develops in the lower segment of the uterine body. The process of softening progresses

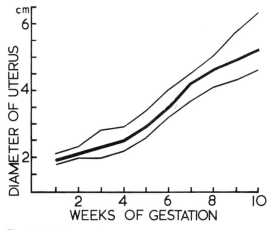

Fig. 31.11. Increase in lateral diameter of uterus (median and range) in 24 of 29 dated pregnancies in *Macaca irus*.

cranially until by days 45–50 the whole uterus is distinctly resilient (Fig. 31.12). Between days 21 and 28 the uterine fundus assumes a globular shape. This, together with the softening of the body, are diagnostically more dependable in determining early pregnancy than is the increase in diameter. The uterus is distinctly spherical by days 45–50 of gestation; its extension into the abdominal cavity increases so that by day 55 it is almost out of reach of the finger per rectum.

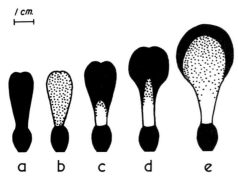

Fig. 31.12. Physical characteristics of uterus at different stages of sexual cycle in *Macaca irus*. Consistency: firm (black), resilient (stippled), soft (white). a. menstrual phase; b. ovulatory phase; c. 21 days gestation; d. 28 days gestation; e. 38–40 days gestation.

During the first 9–12 weeks of pregnancy the cervix uteri is a firm, barrel-shaped organ. It then undergoes a complex series of physical changes until, 2 weeks prior to term, it has become completely soft. During the last 7 days of pregnancy even the outline of the cervix becomes indistinguishable (Mahoney, 1969) (Fig. 31.13).

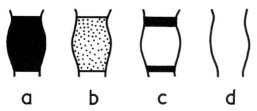

Fig. 31.13. Some of the physical changes affecting the cervix uteri during pregnancy in *Macaca irus*. Consistency: firm (black), resilient (stippled), soft (white). a. 0–12 weeks gestation; b. approximately 14 weeks gestation; c. 18–19 weeks gestation; d. approximately 2 weeks before term.

Growth of the foetus

The foetus in *Macaca irus* is first visible radiographically by day 58. Van Wagenen & Asling, (1964) studied the sequence of ossification of epiphyseal centres in ex-utero foetuses of *Macaca mulatta*. These centres are not always clearly visible in radiographs of the intra-uterine foetus and dating gestation by this means alone has proved difficult in *Macaca irus*. (See Fig. 31.14 for details of skeletal development.) The expected time of parturition in *Macaca mulatta* can be estimated by radiographical measurement of the long bones of the foetus (Hutchinson, 1966). This method has proved reasonably accurate in most of the pregnancies studied in *Macaca irus* except during the last 25 days of gestation.

Ballotment of the foetus in *Macaca irus* is possible around day 78 of gestation. The foetal mass is first felt through the wall of the uterus about day 84 but the details of the foetal skeleton are not palpable until around day 90–100. At this stage the foetal head is disproportionately large compared to the body, but by day 140 of pregnancy the size of the body has greatly increased (Mahoney, 1975c).

The placental sign

Implantation bleeding (the so-called placental sign) occurs 15 to 21 days after conception in *Macaca mulatta* and *Macaca irus*. At first it may be confused with menstruation but it is more protracted, lasting 10 to 30 days. The physical characteristics of the uterus at 21 days gestation are distinct from the non-gravid organ during the menstrual phase and enable one to distinguish menstruation from pseudomenstruation.

Biological tests

Bioassays have been employed to detect chorionic gonadotrophin (the hormone of pregnancy) in the serum and urine of several species of primates. The procedure is to inject serum or urinary extracts subcutaneously into a group of immature female mice or rats on 3 consecutive days. A control group of rodents is similarly treated with normal saline. On the 4th day both test and control animals are sacrificed. The ovaries and/or uteri are dissected out and weighed. An increase in weight of 100 per cent

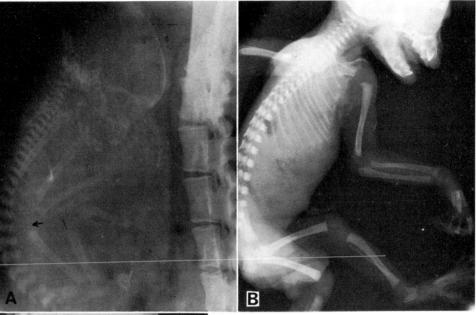

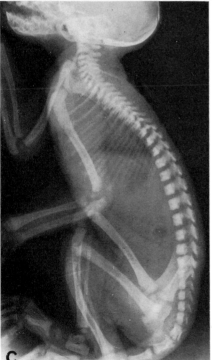

in the test group over the controls confirms pregnancy.

Several workers claim that chorionic gonadotrophin is detectable in the serum and urine of *Macaca mulatta* for only a limited period in the gestation. Tullner & Hertz (1966) demonstrated its presence from day 15 to 34 of pregnancy. The sensitivity of the bioassay was increased by augmenting urine extracts with human eunuch or menopausal gonadotrophin (Arsland *et al.*, 1967). These workers were thus able to demonstrate chorionic gonadotrophin from days 12 to 38 of pregnancy. Hobson (1970) claims to have detected chorionic gonadotrophin during the last week of gestation in *Macaca mulatta*.

Chorionic gonadotrophin was found in the serum and urine throughout the whole or major portion of pregnancy in the chimpanzee (Elder & Bruhn, 1939), the gorilla (Tullner & Gray, 1968) and the baboon (Hobson, 1970). Detection of pregnancy by biological assay was done in the squirrel monkey (Nathan *et al.*, 1966) and in the marmoset (Hampton *et al.*, 1969).

Fig. 31.14. Skeletal development of foetus of *Macaca irus*. A. In-utero foetus at 128 days gestation. The arrow indicates the faint shadow of the distal femoral epiphyseal centre. B. Ex-utero foetus at 121 days gestation. Note absence of epiphyseal centres. C. One-day-old baby. Note following epiphyseal centres; distal humeral, distal femoral, proximal tibial, carpals and tarsals. (Photographs: Mahoney.)

Immunoassays

Pregnancy in the human can be detected by using agglutination-inhibition tests. Anti-sera to human chorionic gonadotrophin (HCG) cross-react with monkey chorionic gonadotrophin (MCG). Several workers have attempted to identify MCG in the urine and blood of a few species of non-human primates but results are conflicting. Kimzey (1965) detected MCG in the urine of pregnant *Macaca mulatta* as early as day 11 and through to day 46 by using a latex particle test. Using a haemagglutination-inhibition test, Glass & Van Wagenen (1970) reported positive results in 85 per cent of pregnant monkeys. A false positive rate of 12·5 per cent was obtained in non-pregnant females. Recently, a more sensitive agglutination-inhibition test has been described (Hodgen & Ross, 1974). A high rate of false positives and false negatives was found when immunoassays were used in the marmoset (Hampton *et al.*, 1969).

An antiserum to the β-subunit of ovine luteinizing hormone (LH) has been used in a radioimmunoassay to differentiate MCG from monkey LH (Hodgen, Tullner, Vaitukaitis, Ward & Ross, 1974). This assay enabled determination of pregnancy between 8 and 12 days after conception.

Semen collection in primates

Two electrical methods have been developed for the collection of semen from species of non-human primates, one depending on direct stimulation of the penis (Mastroianni & Manson, 1963; Valerio, Ellis, Clark & Thompson, 1969; Van Pelt & Keyser, 1970), the other utilizing a rectal probe as the means of excitation (Bennett, 1967; Fussel *et al.*, 1967; Kraemer & Vera Cruz, 1969; Lang, 1967b; Weisbroth & Young, 1965).

Characteristics and evaluation of semen

The ejaculated semen of non-human species of primate is composed of two liquid fractions, one of which begins to coagulate into a dense rubber-like mass within seconds of emission. Within a few minutes the coagulum commences to liquefy at a slow rate but complete liquefication never

occurs. The liquid portion of the seminal fluid contains a high concentration of active spermatozoa.

Before selecting male monkeys for purposes of breeding a full evaluation of the semen should be made. Fertility is assessed from the following criteria (Valerio, Leverage & Munster, 1970):

(1)   Volume of ejaculate (coagulum and liquid fractions): this varies with the individual and within the same animal during consecutive ejaculations. In the squirrel monkey (*Saimiri sciureus*) Bennett (1967a) found that the volume of coagulum decreased and the volume of the liquid fraction increased in successive electro-ejaculations. Table XXI gives data on the volume of ejaculates from three species.

(2)   Concentration of spermatozoa: this is measured by using a Naubauer haemocytometer counting-chamber. The range given by different workers is very variable (Table XXII).

(3)   Sperm motility: a drop of liquid semen is placed on a glass slide and viewed under the lower power objective lens of a microscope. A system of grading 0–5 is applied. Zero indicates no motility at all, 5 indicates maximum viability of the sperm with the formation of forward progressing wave motion and eddy currents.

(4)   Percentage of live sperm: smear preparations of semen are stained with eosin-nigrosin. Dead spermatozoa take up the eosin whereas live cells remain unstained.

TABLE XXI
Volumes of semen obtained by electro-ejaculation

Species	Semen volume in cc Mean	Range
M. mulatta	0·2–4·5	1·1
M. irus	0·6–3·0	1·2
Saimiri sciureus	0·2–1·5	0·4

(Data for some species given by Roussel & Austin, 1968).

TABLE XXII
Sperm concentration (million/ml) in electro-ejaculates

Species	Conc.	Author
Macaques	200–5,000	Van Pelt & Keyser, (1968)
Macaques	93–807	Mastroianni & Manson (1963)
Macaca mulatta } *Erythrocebus patas* }	3,600	Roussel & Austin, (1968)

Artificial insemination

Attempts to impregnate female monkeys by means of artificial insemination have met with varied success. Valerio *et al.* (1969) reported 4 pregnancies in 2 species of macaques after 35 attempts to inseminate artificially. Recently, Czaja, *et al.* (1975) reported that approximately 40 per cent of single artificial inseminations carried out in 218 rhesus monkey cycles resulted in pregnancy. Semen, diluted in a buffered egg-yolk solution, was injected intracervically. In 33 spontaneous and 15 hormonally induced ovulatory cycles in *Macaca mulatta*, Dede & Plentl (1966) obtained 4 and 2 pregnancies in each group respectively. Induction of ovulation with gonadotropic hormones, followed by artificial insemination, resulted in 3 of 5 squirrel monkeys becoming pregnant (Bennett, 1967b).

Several advantages attend the use of artificial insemination in species of non-human primates. Breeding between incompatible males and females can be achieved. Precise dating of the time of conception can be facilitated. A single ejaculate can be used to impregnate several females, thus reducing the number of males required in the breeding colony. Certain types of scientific investigation can be more easily conducted—e.g., the determination of the physiological conditions for sperm transport and fertilization within the female reproductive tract.

Several technical difficulties must be overcome before artificial insemination in non-human primates is as successful a procedure as it is in some species of domestic animals. Methods of handling and storage of semen must be improved to avoid injury and loss of viability of spermatozoa. In domestic species and in the human subject, the semen is usually placed into the cervical canal. Owing to the anatomical structure of the cervix uteri in *Macaca mulatta* (Wislocki, 1933) and *Macaca irus* this route of insemination is technically difficult. Semen is usually deposited deep into the vagina with a Pasteur pipette or 1 ml syringe. The coagulated fraction of the semen is then placed into the vagina by means of a pair of forceps. Van Pelt (1970) obtained 3 pregnancies in 8 female *Macaca mulatta* by intraperitoneal inoculation of diluted semen. For maximum opportunity to obtain conception it is probably desirous that

artificial insemination be performed within a short period prior to ovulation.

Delivery of infant monkeys
Natural birth

Under laboratory conditions baby monkeys are normally born during the night. Natural delivery during daylight hours was recorded in only 1 per cent of pregnancies in *Macaca mulatta* by Valerio *et al.* (1969). By reversing the period of artificial light, Jensen & Robbitt (1967) induced 11 of 12 parturient *Macaca nemestrina*, housed in windowless rooms, to give birth during the daytime.

First and second stage labour in monkeys are rarely witnessed although Krohn (1960) described the course of labour in an individual *Macaca mulatta* and Phillipp (1931) presented a series of 14 photographs showing clearly the course of the three stages of labour in *Macaca mulatta*. More commonly the birth of the infant is discovered by laboratory personnel on arrival in the morning. The baby is found clinging to the mother's breast, having already been washed and dried. The umbilical cord is usually severed and the placenta is frequently ingested by the mother. If the baby is too weak to cling to the breast, or is being maltreated by the mother, separation should be effected immediately. It may be necessary to remove the infant in any event. The safest way of accomplishing this is to sedate the mother with phencyclidine HCl, given by intramuscular injection (2 mg/kg body-weight). Occasionally a nursing mother will severely or fatally attack the offspring on intervention by staff.

The umbilicus should be treated with tincture of iodine or an alcoholic solution of chlorhexidine if the infant is separated from its mother in the first few days of life. Localised or systemic infection may find a portal of entry via the umbilicus and may prove fatal if not treated accordingly. The infant is taken to an incubator or isolator, the temperature of which is maintained at 24–29°C.

Delivery by Caesarean section

Caesarean section may be performed in pregnant monkeys for experimental or clinical reasons. Hysterotomy is clinically warranted if the foetus is in the breach presentation 5 or 6 days prior to the expected time of parturition or if the mother becomes moribund and death would ensue if the pregnancy were allowed to continue. The technique for Caesarean section in the monkey is described by Valerio *et al.* (1969).

In dated pregnancies, the expected time of parturition can be calculated. The length of pregnancy in *Macaca mulatta* and *Macaca irus* is given as 164±4 days and 168 days respectively by Valerio, Pallotta & Courtney (1969) and Fujiwara & Imamichi (1966). The time of parturition can be assessed by study of the sequence of ossification of the epiphyseal centres in the foetus (Van Wagenen & Asling, 1964) or by a mathematical formula incorporating the length of the foetal long bones (Hutchinson, 1966). Approximately 1 week before term the cervix becomes completely softened and non-palpable.

Rearing and weaning of baby monkeys

Figures given for the birth weights of infant *Macaca mulatta* born naturally range from 350 to 575 g. The range given by Fleischman (1963) is 440–536 g and that of Robinson (1964) is 350–550 g (average 450 g). Birth weights of babies derived by Caesarean section are somewhat lower than those born by the vaginal route. For 30 Caesarean-derived babies Fleischman reported a range in birth weights of 438±50 g. The lower weights of babies delivered by Caesarean section may be explained, in part, by the difficulty of deciding at what stage hysterotomy should be performed. The growth rate of the foetus in the terminal stages of gestation is steep. Interruption of the pregnancy some time before parturition would result in the delivery of a less heavy infant.

The average birth weight for male infants (*Macaca mulatta*) is significantly greater than for females (Van Wagenen, 1956; Van Wagenen & Catchpole, 1956; Fujikura & Niemann, 1967) (Table XXIII). Valerio, Darrow & Martin (1970) compared the birth weights of infant *Macaca mulatta* conceived in the wild and those conceived in the laboratory. The latter were significantly heavier (Table XXIV).

TABLE XXIII
Birth weights in g of infant Macaca mulatta

Male	Sex Female		Author
490±60	465±70	(N= 78)	Van Wagenen, 1956
470·9±70·5	456·3±62·9	(N= 179)	Fujikura *et al.*, 1967

TABLE XXIV
Comparisons of birth weights in Macaca mulatta *between infants conceived in-nature and in-laboratory*

	Male	Sex Female	Combined
In-nature	440±69 g	402±87 g	421±78 g (N= 21)
In-laboratory	500±75 g	477±77 g	488±76 g (N=152)

Reproduced with permission from Valerio, Darrow & Martin (1970).

For the baboon, the range in birth weights given by Vice *et al.* (1966) is 539–737 g (*Papio papio*) and 539–1,077 g (*P. cynocephalus*).

Baby monkeys reared by the mother tend to grow at a slower rate than artificially fed infants. In the Taiwan macaque (*Macaca cyclopis*) Yang *et al.* (1968) demonstrated that breast-fed infants doubled their birth weight by 3 months of age and trebled it at 6 months. A greater rate of increase was observed in artificially-fed babies. Birth weights were doubled at 2 months, trebled at 5 months and infants were 4 times heavier at 1 year old than at birth. A similar trend has been noted in the Royal College of Surgeons colony of *Macaca irus* (Cohen, 1971).

During the first few days of life baby monkeys lose 5–10 per cent of their birth weight, according to Valerio, Darrow & Martin (1970), but regain this by the second week of life. Body growth continues at 5–8 g per day during the early months of life.

The daily calorific requirements of baby monkeys have been calculated by several workers. From the second to the sixth day of life the daily requirements of the baby *Macaca mulatta* rise from 119 cal/kg body-weight to 287 cal/kg. The level slowly drops to 238 cal/kg/day by the fifth week (Fleischman, 1963). Beyond this point it is difficult to assess daily values as infants then begin to ingest solid food. Infant *Macaca cyclopis* (Taiwan macaque) require a daily consumption rising to 250 cal/kg body-weight during the first 7 days (Yang *et al.*, 1968). During the following 11 weeks of post-natal development the daily requirements are maintained at 250–304 cal/kg. In a study carried out by Vice *et al.* (1966) on infant baboons a comparison of body-weight gains was made between babies fed the standard milk-substitute formula

TABLE XXV
Daily calorific intake and weight gain during first 14 days of life of infant baboons using 2 different milk-substitute preparations

Preparation	No. infants	Days 1 & 2 average cal/kg/day	Days 3–14 average cal/kg/day	Average weight gain
13 cal per 30 cc.	11	139	234	35·5 g
26·7 cal per 30 cc.	7	175	258	58·5 g

After Vice *et al.* (1966).

used for *Macaca mulatta* (the preparation containing 13 cal per 30 cc solution) and a preparation containing 26·7 cal per 30 cc. Table XXV shows the results. The gain in body-weight 14 days after birth was significantly greater on the higher calorie diet.

The daily protein requirement of infant *Macaca mulatta* has been given as 3 g/kg body-weight (Day, 1962). For *Macaca cyclopis* Yang

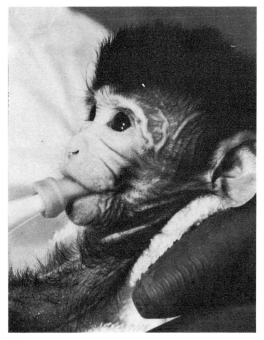

Fig. 31.15. Artificial feeding of newborn infant. Reproduced with permission from Cohen (1971).

et al. (1968) calculated an increase rising to 8 g/kg by the seventh day of life, this daily level being maintained during the following 11 weeks.

Feeding schedules

Great care and patience must be exercised in the feeding of newborn monkeys and the passage of fluid, especially milk, into the trachea must be avoided at all costs. Inhalation of milk is likely to result in pneumonia which may prove fatal.

During the first few weeks of life fluids are offered from bottles fitted with teats designed for premature human babies (Figs. 31.15 and 31.16). As the sucking ability of the infant monkey improves it may be necessary to enlarge the hole in the teat but changes should be geared to individual need. The suckling reflex may be poorly developed in the newborn monkey. If after some perseverance success is not attained, fluids may be given from a 1 ml syringe taking care that no more than one drop at a time enters the mouth. During the time of feeding frequent rests must be taken and the baby gently slapped on the back to allow ingested air to be released from the stomach via the oesophagus. If 'winding' is not performed vomiting may occur with the risk of inhalation of regurgitated material into the trachea. Once the infant has gained strength and can suck fluid with ease the transition to a larger teat may be made.

The methods described for the artificial feeding of baby monkeys vary one from another in detail only, the general principles being much the same.

Day 1. During the first 24 hours of life infants are fed a 5–10 per cent solution of glucose 3 to 4 times starting 3–4 hours after birth. Three to 5 ml of the solution are offered at each feed. In preparation for the feeding of human milk substitutes, Valerio, Darrow & Martin (1970) effect a gradual transition from glucose solution to the milk formula within the first day of life. Following the first feed of glucose solution, a mixture of half 5 per cent dextrose solution and half milk substitute is given at the second feed. Undiluted milk substitute is offered at the third feed of the day.

Day 2. Some workers change to full-cream preparations of milk substitute in the second day of life whilst others feed a 50 per cent dilution

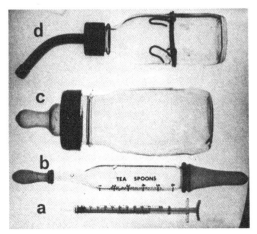

Fig. 31.16. Types of feeding bottles used in the rearing of infant *Macaca irus* from birth to weaning. A. 1 ml plastic syringe. B. premature human baby feeding bottle and teat. C. bottle and teat for older infants. D. self-feeding bottle with spout for attachment to cage wall. (Photograph: Mahoney.)

of the formula to allow a gradual introduction to the diet.

Day 3 onwards. The milk formula adopted by most workers is prepared at a dilution containing 0·7 to 1 cal per ml of solution. Five daily feeds, of 10–15 ml of the formula at each occasion, are given over the next 3 days. The frequency of feeding is then reduced to 4 times daily but the amount offered at each feed is increased to 30 ml (Valerio, Darrow & Martin, 1970). By day 7–10 three feeds of 80 ml each are given per day. Self-feeding is encouraged from the second or third day of life. This is achieved by suspending the feeding bottle low down on the side of the cage. The infant's mouth is introduced to the nipple and the head is held in position. Eventually the baby learns to suck unaided. Over the next few days the bottle is gradually moved higher up the cage. Infants are self-sufficient at 7 to 15 days of age (Kaye *et al.*, 1966; Fleischman, 1963). Once the baby can take unaided 80 ml of milk formula at each feed solid food is introduced. This may be in the form of a mash composed of biscuit and strained fruit, or as banana and boiled egg.

Weaning of the baby simian entirely from milk to solid food and water begins at 60 to 100 days of age. The dilution of the milk preparation is gradually increased over a period of several days to several weeks until the transition to plain water and solid food is made (Kaye *et al.*, 1966; Valerio, Darrow & Martin, 1970).

REFERENCES

Arslan, M., Meyer, R. K. & Wolf, R. C. (1967). Chorionic gonadotrophin in the blood and urine of pregnant rhesus monkeys (*Macaca mulatta*). *Proc. Soc. exp. Biol. Med.*, **125**, 349–52.

Balin, H. & Wan, L. S. (1969). Basal body temperature as an index of ovulation in the rhesus monkey. *Abstr. Soc. Stud. Reprod.*, **2**, 7.

Banerjee, B. N. & Woodard, G. (1970). A comparison of outdoor and indoor housing of rhesus monkeys (*Macaca mulatta*). *Lab. Anim. Care*, **20**, 80–2.

Bennett, J. P. (1967a). Semen collection in the squirrel monkey. *J. Reprod. Fert.*, **13**, 353–5.

Bennett, J. P. (1967b). Artificial insemination of the squirrel monkey. *J. Endocr.*, **37**, 473–4.

Betteridge, K. J., Kelly, W. A. & Marston, J. M. (1970). Morphology of the rhesus monkey ovary near the time of ovulation. *J. Reprod. Fert.*, **22**, 453–460.

Bullock, D. W., Paris, C. A. & Goy, R. W. (1972). Sexual behaviour, swelling of the sex skin and plasma progesterone in the pigtail macaque. *J. Reprod. Fert.*, **31**, 225–236.

Cohen, B. (1971). The monkey colony at Downe. *Ann. R. Coll. Surg.*, **48**, 46–53.

Czaja, J. A., Eisele, S. G. & Goy, R. W. (1975). Cyclical changes in the sexual skin of female rhesus: relationships to mating behaviour and successful artificial insemination. *Fed. Proc.*, **34**, 1680–1684.

Day, P. L. (1962). Nutrient requirements of the young monkey. In *Nutrient Requirements of Domestic Animals*, no. 10. Lab. Anim. Publ., 990, pp. 31–8. Washington: National Research Council.

De Allende, I. L. C., Shorr, E. & Hartman, C. G. (1945). A comparative study of the vaginal smear cycle of the rhesus monkey and human. *Contr. Embryol.*, **31**, 1–26.

Dede, J. A. & Plentl, A. A. (1966). Induced ovulation and artificial insemination in a rhesus colony. *Fert. Steril.*, **17**, 757–64.

Elder, J. H. & Bruhn, J. M. (1939). Use of Friedman test for pregnancy with chimpanzee. *Yale J. Biol. Med.*, **12**, 155–60.

Fleischman, R. W. (1963). The care of infant rhesus monkeys (*Macaca mulatta*). *Lab. Anim. Care*, **13**, 703–9.

Fujikura, T. & Niemann, W. H. (1967). Birth weight, gestational age and type of delivery in rhesus monkeys. *Am. J. Obstet. Gynec.*, **97**, 76–80.

Fujiwara, T. & Imamichi, T. (1966). Breeding of Cynomolgus monkeys as an experimental animal. *Jap. J. med. Sci. Biol.*, **19**, 125–6.

Fussell, E. N., Roussel, J. D. & Austin, C. R. (1967). Use of the rectal probe method for electrical ejaculation of apes, monkeys and a prosimian. *Lab. Anim. Care*, **17**, 528–30.

Gillman, J. (1937). The cyclical changes in the vaginal smear in the baboon and its relationship to the perineal swelling. *S. Afr. J. med. Sci.*, **2**, 44–56.

Glass, R. H. & Van Wagenen, G. (1970). Immunologic test for chorionic gonadotrophin in the serum of the pregnant monkey (*Macaca mulatta*). *Proc. Soc. exp. Biol. Med.*, **134**, 467–8.

Haig, M. V. & Scott, A. (1965). Some radiological and other factors for assessing age in the rhesus monkey using animals of known age. *Lab. Anim. Care*, **15**, 57–73.

Hampton, J. K., Levy, B. M. & Sweet, P. M. (1969). Chorionic gonadotropic secretion during pregnancy in the marmoset, *Callithrix jacchus*. *Endocrinology*, **85**, 171–4.

Hartman, C. G. (1931). The breeding season in monkeys with special reference to *Pithecus* (*Macacus*) *rhesus*. *J. Mammal.*, **12**, 129–42.

Hartman, C. G. (1932). Studies in the reproduction of the monkey *Macacus* (*Pithecus*) *rhesus*, with special reference to menstruation and pregnancy. *Contr. Embryol.*, **23**, 3–76.

Hartman, C. G. (1933). Pelvic (rectal) palpation of the female monkey with special reference to the ascertaining of ovulation time. *Am. J. Obstet. Gynec.*, **26**, 600–8.

Hartman, C. G. (1939). Studies on reproduction in the monkey and their bearing on gynecology and anthropology. *Endocrinology*, **25**, 670–82.

Hendrickx, A. G. (1965). The menstrual cycle of the baboon as assessed by the vaginal smear, vaginal biopsy and perineal swelling. In *The Baboon in Medical Research*, Vol. II. Ed. Yagtborg, H. Austin & London: University of Texas Press.

Hendrickx, A. G. & Kraemer, D. C. (1970). Primates. In *Reproduction and Breeding Techniques in Laboratory Animals*. Ed. Hafez, E. S. E.

Hendrickx, A. G. & Nelson, V. (1971). Reproductive failure. In *Comparative Reproduction of Laboratory Primates*. Ed. Hafez, E. S. E.

Hobson, B. M. (1970). Excretion of gonadotrophin by the pregnant baboon (*Papio cynocephalus*). *Folia Primat.*, **12**, 111–5.

Hodgen, G. D. & Ross, G. T. (1974). Pregnancy diagnosis by a hemagglutination inhibition test for urinary macaque chorionic gonadotropin (MCG). *J. clin. Endocr. Metab.*, **38**, 927–930.

Hodgen, G. D., Tullner, W. W., Vaitukaitis, J. L., Ward, D. N. & Ross, G. T. (1974). Specific radioimmunoassay of chorionic gonadotropins during implantation in rhesus monkeys. *J. clin. Endocr. Metab.*, **39**, 457–464.

Hotchkiss, J., Atkinson, L. W. & Knobil, E. (1971). The time course of serum estrogen and luteinizing hormone concentrations during the menstrual cycle of the rhesus monkey. *Endocrinology*, **89**, 177–183

Hutchinson, T. C. (1966). A method for determining expected parturition date of rhesus monkeys (*Macaca mulatta*). *Lab. Anim. Care*, **16**, 93–5.

Jensen, G. D. & Robbitt, R. A. (1967). Changing parturition time in monkeys (*Macaca nemestrina*) from night to day. *Lab. Anim. Care*, **17**, 379–381.

Jewett, D. A. and Dukelow, W. R. (1971). Follicular morphology in *Macaca fascicularis*. *Folia Primatol.*, **16**, 216–220.

Kaye, H., Povar, M. L. & Schrier, A. M. (1966). Rearing of *Macaca mulatta* from birth. *Lab. Anim. Care*, **16**, 476–86.

Kimzey, W. G. (1965). Concentration of gonadotrophin in the urine of the pregnant monkey, *Macaca mulatta*. *Anat. Rec.*, **151**, 372–3.

Koford, C. G. (1965). Population dynamics of rhesus on Cayo Santiago. In *Primate Behavior—Field Studies of Monkeys and Apes*, ed. de Vore, I. New York: Holt, Rinehart & Winston.

Kraemer, D. C. & Vera Cruz, N. C. (1969). Collection, gross characteristics and freezing of baboon semen. *J. Reprod. Fert.*, **20**, 345–8.

Krohn, P. L. (1960). The duration of pregnancy in rhesus monkeys (*Macaca mulatta*). *Proc. zool. Soc. Lond.*, **134**, 595–9.

Kuehn, R. E., Jensen, G. D. & Morrill, R. K (1965). Breeding *Macaca nemestrina*: a programme of birth engineering. *Folia Primat.*, **3**, 251–62.

Lang, C. M. (1967a). The oestrous cycle of non-human primates: a review of the literature. *Lab. Anim. Care*, **17**, 172–9.

Lang, C. M. (1967b). A technique for the collection of semen from squirrel monkeys (*Saimiri sciureus*), by electroejaculation. *Lab. Anim. Care*, **17**, 218–21.

Mahoney, C. J. (1969). Detection of imminent parturition in *Macaca irus*. *Lab. Anim. Handb.*, **4**, 235–8.

Mahoney, C. J. (1970). A study of the menstrual cycle in *Macaca irus* with special reference to the detection of ovulation. *J. Reprod. Fert.*, **21**, 153–63.

Mahoney, C. J. (1975a). The accuracy of bimanual rectal palpation for determining the time of ovulation and conception in the rhesus monkey (*Macaca mulatta*). In *Breeding Simians for Developmental Biology*. Eds Perkins, F. T. & O'Donoghue, P. N. *Lab. Anim. Handb.*, **6**, 127–138.

Mahoney, C. J. (1975b). Aberrant menstrual cycles in two species of macaques. In *Breeding Simians for Developmental Biology*. Eds Perkins, F. T. & O'Donoghue, P. N. *Lab. Anim. Handb.*, **6**, 243–255.

Mahoney, C. J. (1975c). Practical aspects of determining early pregnancy, stages of foetal development and imminent parturition in the monkey. (*Macaca fascicularis*). In *Breeding Simians for Developmental Biology*. Eds Perkins, F. T. & O'Donoghue, P. N. *Lab. Anim. Handb.*, **6**, 261–274.

Mastroianni, L. & Manson, W. A. (1963). Collection of monkey semen by electroejaculation. *Proc. Soc. exp. Biol. Med.*, **112**, 1025–7.

McSweeney, D. J. & Sbarra, A. J. (1964). A new cervical mucus test for hormone appraisal. *Am. J. Obstet. Gynec.*, **88**, 705–9.

Nathan, T. S., Rosenblum, L. A., Limson, G. & Nelson, J. H. (1966). Diagnosis of pregnancy in the squirrel monkey. *Anat. Rec.*, **155**, 31–6.

Phillipp, E. (1931). Physiologie und Pathologie der Geburt bei Affen. *Zentbl. Gynäk.*, **22**, 1776–82.

Robinson, J. F. (1964). The care and breeding of the rhesus monkey. *J. Anim. Techs Ass.*, **15**, 60–70.

Roussel, J. D. & Austin, C. R. (1968). Improved electro-ejaculation of primates. *J. Inst. Anim. Techns*, **19**, 22–32.

Rowell, T. E. (1963). Behaviour and female reproductive cycles of rhesus macaques. *J. Reprod. Fert.*, **6**, 193–203.

Rubenstein, B. B. (1937). The relation of cyclic changes in human vaginal smears to body temperature and basal metabolic rates. *Am. J. Physiol.*, **119**, 653–41.

Southwick, C. H., Beg, M. A. & Siddiqi, M. R. (1965). Rhesus monkeys in North India. In *Primate Behavior—Field Studies of Monkeys and Apes*, ed. de Vore, I. New York: Holt, Rinehart & Winston.

Spiegel, A. (1975). Statement on ICLA inquiry. In *Breeding Simians for Developmental Biology*. Eds. Perkins, F. T. & O'Donoghue, P. N. *Lab. Anim. Handb.*, **6**, 19–20.

Tullner, W. W. & Hertz, R. (1966). Chorionic gonadotropin levels in the rhesus monkey during early pregnancy. *Endocrinology*, **78**, 204–7.

Tullner, W. W. & Gray, C. W. (1968). Chorionic gonadotropic secretion during pregnancy in a gorilla. *Proc. Soc. exp. Biol. Med.*, **128**, 954–6.

Valerio, D. A., Courtney, K. D., Miller, R. L. & Pallotta, A. J. (1968). The establishment of a *Macaca mulatta* breeding colony. *Lab. Anim. Care*, **18**, 589–95.

Valerio, D. A., Ellis, E. B., Clark, M. L. & Thompson, G. E. (1969). Collection of semen from macaques by electroejaculation. *Lab. Anim. Care*, **19**, 250–2.

Valerio, D. A., Miller, R. L., Innes, J. R. M., Courtney, K. D., Pallotta, A. J. & Guttmacher, R. M. (1969). *Macaca mulatta: Management of a Laboratory Breeding Colony*. New York: Academic Press.

Valerio, D. A., Pallotta, A. J. & Courtney, K. D. (1969). Experiences in large-scale breeding of simians for medical experimentation. *Ann. N.Y.*

Acad. Sci., **162**, 282–96.

Valerio, D. A., Darrow, C. C. & Martin, D. P. (1970). Rearing of infant simians in modified germ-free isolators for oncogenic studies. *Lab. Anim. Care*, **20**, 713–19.

Valerio, D. A., Leverage, W. E. & Munster, J. H. (1970). Semen evaluation in macaques. *Lab. Anim. Care*, **20**, 734–40.

Vandenburgh, J. G. & Vessey, S. (1968). Seasonal breeding of free-ranging rhesus monkeys and related ecological factors. *J. Reprod. Fert.*, **15**, 71–9.

Van Pelt, L. F. (1970). Intraperitoneal insemination of *Macaca mulatta*. *Fert. Steril.*, **21**, 159–62.

Van Pelt, L. F. & Keyser, P. E. (1970). Observations on semen collection and quality in macaques. *Lab. Anim. Care*, **20**, 726–33.

Van Wagenen, G. (1945). Optimal mating time for pregnancy in the monkey. *Endocrinology*, **37**, 307–12.

Van Wagenen, G. & Catchpole, H. R. (1956). Physical growth of the rhesus monkey (*Macaca mulatta*). *Am. J. phys. Anthrop.*, **14**, 245–73.

Van Wagenen, G. & Asling, C. W. (1964). Ossification in the foetal monkey (*Macaca mulatta*): estimation of age and progress of gestation by roentography. *Am. J. Anat.*, **114**, 107–32.

Vice, T. E., Britton, H. A., Ratner, I. A. & Kalter, S. S. (1966). Care and raising of newborn baboons. *Lab. Anim. Care*, **16**, 12–22.

Weick, R. F., Dierschke, D. J., Karsch, F. J., Butler, W. R., Hotchkiss, J. & Knobil, E. (1973). Periovulatory time course of circulating gonadotropic and ovarian hormones in the rhesus monkey. *Endocrinology*, **93**, 1140–1147.

Weid, G. L. (1955). Suggested standard for karyopyknosis. Use in hormonal reading of vaginal smears. *Fert. Steril.*, **6**, 61–5.

Weisbroth, S. & Young, F. A. (1965). The collection of primate semen by electro-ejaculation. *Fert. Steril.*, **16**, 229–35.

Wilson, J. G., Fradkin, R. & Hardman, A. (1970). Breeding and pregnancy in rhesus monkeys used for teratological testing. *Teratology*, **3**, 59–71.

Wislocki, G. B. (1933). The reproductive systems. In *The Anatomy of the Rhesus Monkey*, ed. Hartman, C. G. & Staus, W. L. (1961). New York: Hafner.

Yang, C-S., Kno, C-C., Del Favero, J. E. & Alexander, E. R. (1968). Care and raising of newborn Taiwan monkeys (*Macaca cyclopis*), for virus studies. *Lab. Anim. Care*, **18**, 536–43.

Young, W. C. & Orbison, W. D. (1944). Changes in selected features of behaviours in pairs of oppositely sexed chimpanzees during the sexual cycle and after ovariectomy. *J. comp. Psych.*, **37**, 107–143.

Zuckerman, S. (1937). The menstrual cycle of primates; the part played by oestrogenic hormones in the menstrual cycle. *Proc. zool. Soc. Lond.*, **123**, 457–71.

Zuckerman, S. (1947). Duration of reproductive life in the baboon. *J. Endocr.*, **5**, 220–1.

LABORATORY PROCEDURES

W. H. Bowen

Animals to be subjected to experimental procedures should be in good health. They should have been through a rigorous quarantine procedure and screened for obvious infection with virus B (Herpes virus simiae); they should also be tuberculin tested and, if necessary, have a chest radiograph.

An accurate dietary and drug history of the animals should be ascertained. During their quarantine period monkeys are frequently given drugs which may interfere with subsequent investigations. For example, the administration of tetracycline may affect the microbial flora of the gastrointestinal tract, and the drug may be incorporated into calcifying tissues.

Anaesthesia and euthanasia

Anaesthesia is not difficult to carry out in monkeys provided the basic rules are observed. The animals should not receive food for at least 6 hours prior to the administration of a general anaesthetic. It is also essential to keep a clear air-way at all times. (For a detailed description of anaesthesia in the monkey the reader is referred to Cohen, 1964.)

Rapid induction of anaesthesia can be achieved by the intra-muscular injection of sodium methohexitone* at a dose of 20 mg/kg body-weight. An endotracheal tube (2 or 3 mm) can be readily passed intranasally or by mouth if the vocal cords are sprayed with 2 per cent lignocaine.† Anaesthesia is maintained by the administration of 1 per cent halothane‡ with nitrous oxide (2–3 litres/min) and oxygen (approximately 1 litre/min). Animals anaesthetised by this method recover rapidly even when they have been anaesthetised for as long as 2–3

* Brietal: Eli Lilly.
† Xylocaine: Astra Hewlitt.
‡ Fluothane: ICI.
§ Nembutal: Abbott.

hours. It is also more economical than previously described techniques because halothane is not used as the main inducing agent.

Euthanasia can be achieved by the administration of 60 mg/kg of sodium pentobarbitone§.

Experimental feeding procedures

Monkeys are often selective in their eating habits and they tend to scatter their food widely, thus rendering the study of the effect of a particular diet or dietary component difficult. A satisfactory way to overcome these problems is to administer either all or part of the animal's diet by stomach tube. Monkeys can be maintained in good health for prolonged periods while receiving their entire diet by stomach tube. A diet which the author has found satisfactory has the following composition.

Complan (Glaxo Ltd)	240 g
Eggs (hard boiled)	2
Water	200 ml

This mixture is homogenised and approximately 100 ml is given to a 3 kg animal once daily.

To administer the diet the animal is lightly tranquillised with phencyclidine*. A small Ryles tube is then passed down the oesophagus (Fig. 31.17). Care must be exercised to ensure that the tube does not enter the trachea or penetrate the oesophagus. The fluid diet should be administered slowly through the tube by means of a large syringe.

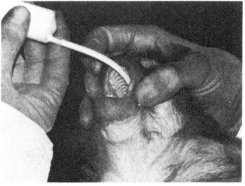

Fig. 31.17. A monkey being tube-fed. (Photograph: Bowen.)

* Sernylan: Parke Davis.

The Ryles tube can also be used to aspirate the contents of the stomach or even the duodenum.

Collection of specimens

Blood

Blood is most conveniently withdrawn from monkeys through the femoral vein. Many operators have difficulty at first, mainly because the vein is situated comparatively superficially. To withdraw blood the animal should be placed on its back with its feet tied to cleats on either side of a suitable table. It is important not to have the animal askew. The pulse in the femoral artery is next detected in the femoral triangle (Fig. 31.18). The vein is situated slightly medial to the artery. A 19 G 3·2 cm needle is introduced into the vein and blood is withdrawn slowly. Occasionally the operator may inadvertently introduce the needle into the artery; when this occurs great care must be taken to prevent the development of a haematoma by applying pressure to the area until bleeding ceases.

Cerebrospinal fluid

To withdraw cerebrospinal fluid the animal is placed on its abdomen. The lumbar and sacral area is thoroughly shaved, and a suitable disinfectant applied. The outer edges of the iliac crests are palpated. An imaginary line joining these two points will cross the posterior edge of the spine of the last lumbar vertebra. The gap, which is about 5 mm between the last lumbar vertebra and the first sacral, is then sought. A 23 G 2·5 cm needle is inserted and the fluid withdrawn slowly.

Saliva

Whole saliva can be readily collected by means of a Pasteur pipette. Profuse salivation occurs in some animals following the administration of phencyclidine. In the unanaesthetised animal saliva can be induced to flow by painting the tongue with a weak solution of acetic acid or sodium chloride.

Secretions from the individual glands can be collected by direct cannulation of the separate ducts. The animals are anaesthetised. An endotracheal tube is passed and a throat-pack inserted to ensure that saliva does not enter the trachea. The orifices of the ducts are then sought using finely tapered glass or steel rods with burnished tips. Rods of increasing diameter are used until the orifices of the ducts are sufficiently dilated to permit the introduction of fine nylon tubing (Portex No. 1). The parotid ducts are more readily cannulated than the submandibular or sublingual. Cannulation of the submandibular and sublingual ducts is facilitated if the tongue is retracted by means of tube-grasping forceps or similar instrument. The sublingual papilla is immobilised by means of fine butterfly tweezers. During the dilation procedure care must be exercised to ensure that the ducts are not penetrated. Following the introduction of the cannulae saliva is induced to flow by the subcutaneous administration of pilocarpine up to a dosage of 1 mg/kg.

Milk

Although breast pumps are available milk can be readily expressed from the breasts by means of gentle manipulation with fore-finger and

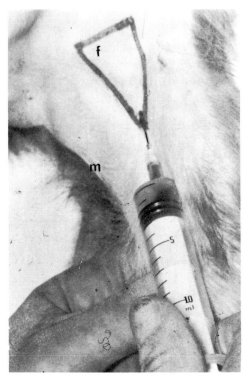

Fig. 31.18. Placing of syringe and needle for the removal of blood. f = femoral triangle; m = mesial aspect of leg. (Photograph: Bowen.)

thumb. It can be collected from some hours after parturition to 3–4 days post-partum, if the infant has been removed shortly after birth or has been delivered by Caesarean section. Up to 9–10 ml can be collected at any one time, although normally the amounts expressed are much smaller than this. The expression of milk is facilitated by the intra-muscular administration of 10 units of oxytocin to an animal weighing 3–4 kg.

Urine and faeces

Urine and faeces can be collected from most animals without having to resort to anaesthesia. A sloping tray (Fig. 31.19) is fitted beneath the animal's cage. The top of the tray can be covered with fine mesh to collect faeces while permitting urine to flow through to the tray beneath, where it can be collected in a suitable container; this container can be placed in crushed ice to keep the urine cold.

Injection procedures

Most of the routine injection procedures are easily carried out on monkeys. Intravenous and intradermal injections can be problematical on occasions because the skin of the animal may be so tough as to render the correct placing of a needle difficult. Only the finest quality needles should be used. For intradermal injections a fine gauge needle is used (23 or 26 G). The solution

Fig. 31.19. Device for the collection of faeces and urine from monkey. g = normal cage grid; st = sloping tray which has been partly removed; b = collection bottle for urine. (Photograph: Bowen.)

being injected should be deposited slowly and should raise a 'bleb' in the skin.

Solutions to be administered intravenously may be given through either the femoral vein or the recurrent tarsal. The latter is situated at the lower end of the posterior aspect of the leg. Some constriction of the leg is required to engorge the vein and, in order to identify it, it is desirable to have a good oblique light. A 23 G needle is the most suitable for gaining access to the vein.

Considerable care is required when administering substances intraperitoneally or intrathoracically. For intraperitoneal injections the skin above the abdomen should be suitably disinfected and pulled away from the underlying organs. The needle should be on a slant and the syringe aspirated before the solution is administered.

When carrying out injections into the thorax care must be taken to avoid penetration of the diaphragm. It is frequently not realised how far the diaphragm extends into the rib cage.

REFERENCE

Cohen, B. (1964). Anaesthesia in monkeys. In *Small Animal Anaesthesia*, ed. Graham-Jones, O. Oxford: Pergamon.

DISEASE CONTROL

R. N. T-W-FIENNES

Disease control has two aspects which must be constantly borne in mind by Directors of zoos or laboratories where primates are kept: first, the protection of staff from physical injury and from cross-infection; secondly, the preservation of the health of the animals themselves and their own protection from cross-infection. The importance of protecting staff needs no emphasis in view of the well-publicised tragedies that have arisen in recent years as a result of diseases transmitted to humans by captive simians. It should be remembered also that even small monkeys have vicious little teeth which can cause unpleasant injuries; some of the bigger monkeys, such as patas, have razor-sharp canine teeth which can cause very nasty wounds. Laboratory primates are high-cost animals and losses represent a drain on budgets. The loss of a whole colony can easily occur through carelessness in control measures, and this loss can represent not only the value of animals but also the loss of months of careful work. Any staff who will not conform to sensible routines of handling and management are not fitted to be in charge of primate colonies or to use primates experimentally.

The primate's life as a laboratory animal starts with his capture in places usually remote from his ultimate destination. He may be injured at the time of capture and will in any case be under stress from that time until he has undergone a conditioning period at his ultimate destination. In the less advanced countries in which his capture takes place feeding regimes are often inadequate, and he is liable to come in contact with human primates infected with a variety of diseases which can be transmitted to him, such as tuberculosis, alimentary infections (chiefly shigellosis and salmonellosis) and some comparatively inapparent diseases such as infectious hepatitis. The effect of stress is often to lower resistance to parasites which the animal tolerates in the jungle but which now cause symptoms of ill-health. These effects are enhanced if the diet is inadequate in any respect, and there is a further danger in that when in captivity or in transit the animal may come in contact with, or be caged with, primates of other species whose natural parasites are pathogenic to him.

The result of these hazards is seen in the very high mortality which overtakes many primates during the first few months of capture. It is also seen in the many instances in which these animals have passed infectious diseases both to their handlers and to other primates in established colonies with whom they come in contact too soon.

The suggestion has been made from time to time that newly-captured primates should be held in quarantine in the country of origin prior to shipment, so as to condition them to their new circumstances in natural surroundings before they are subjected to the additional strains of shipment. This idea has much to commend it, if it should prove practicable; however, it has not yet been widely tested. Indeed, the whole business of capture and shipment is somewhat uncontrolled and bad practices contribute a great deal to the death-rate. However, many of the more reliable animal-shippers are much concerned about these matters and take steps to obtain their animals from reliable sources. When laboratory primates are purchased, it is well to obtain them from a dealer with a good reputation; if this is done, far less trouble will be encountered.

The larger airlines are well accustomed to the shipment of consignments of primates and show great concern for their well-being and safety. Nevertheless, it too often happens that consignments arrive in which some of the monkeys have died and others are sick. Usually in such cases the monkeys were already sick on shipment and suffering from dysentery or some other disease of that nature. Trans-shipment at airports presents a further hazard, since the animals often require to be kept overnight, thus coming into contact with other consignments and other species of primates.

Some consignments of primates go direct to the laboratories where they are required; others go to animal-dealers or to holding centres. The more conscientious dealers hold the animals for a period during which they condition them, have them tested for tuberculosis and other diseases, and ship them only when they are satisfied that they can provide a healthy animal of the specification required. Other dealers are in great haste to unload animals as soon as possible—in fact, to get rid of them before they die. In such cases the animals often arrive dead or they die within a few days of being received. Provided an animal arrives with some flicker of life in it, dealers will usually accept no responsibility and will expect payment.

On arrival, therefore, all animals should be given a proper medical check as soon as possible, and such treatment as is necessary should be administered. This check should include X-ray examination to locate any fractures and to ascertain whether any form of disease of the skeleton is present. The X-ray will further serve to reveal the presence of buckshot or other missiles used in capturing the animal; such may not be evident at first, but may cause illness or death at a later date. X-ray of the chest will also serve to reveal the presence of respiratory disease. Haematological tests should also be done to reveal the presence of parasites and to ascertain that the animal has a normal blood picture. Faeces should be examined for evidence of parasites or helminth eggs, and cultures should be made for intestinal pathogens, such as shigella or salmonella.

In most laboratories, animals are quarantined for a period of around 3 months. During this time their general health is watched, tests are repeated at intervals, and tuberculin tests are applied at intervals of not less than one month, preferably two weeks.

Each species is best quarantined on its own, but this is not always possible. Certain general rules should, however, be observed. There appears to be little danger in placing Old and New World primates in the same quarantine room, but some species of New World primates offer dangers to each other. The full extent of these is probably not yet known; however, squirrel monkeys are known to be dangerous to both marmosets and owl monkeys, because they sometimes shed strains of herpes virus which prove pathogenic in the other species.

It sometimes happens that newly-arrived monkeys of certain species (*Callicebus* are a case in point) seem to thrive when first received, but later go into a gradual decline, refuse food and die after some weeks. Often in such cases, the autopsy and the bacterial examinations that go with it reveal little or show a variety of causes of death without any coherent pattern that can be successfully tackled. In such cases, the probability is that something is lacking in the diet and 'shotgun' methods of dietary supplementation with added vitamins and minerals should be adopted until the actual deficiency can be defined.

After a three-month period the monkeys, if fit, can be released to the holding centre or issued for experimentation. Experience has shown that during this period most natural pathogens that have been causing trouble will be shed, and the animals should be well-adjusted and conditioned to new feeding regimes. In-apparent diseases such as infectious hepatitis, transmitted by captors in the country of origin, will have run their course, and the monkey will be a relatively safe animal. These remarks seem to apply also to latent viruses, since shedding is likely to have ceased and, though still a carrier, the monkey will be relatively safe. It may still,

however, harbour latent shigella infection, and all dangers of rabies cannot be considered as eliminated until past the 12-month, possibly 14-month, period. These monkeys should still, therefore, be treated with respect and those handling and using them should take the necessary precautions.

We may now survey briefly the more important diseases from which monkeys may suffer and which may constitute a danger to other monkeys or to humans who come in contact with them.

Bowel pathogens

Shigelloses. Shigella infections occur in all groups of primates and are usually caused by *Shigella flexneri*; *S. sonnei* and *S. schmitzii* are sometimes also involved. The cause of acute human bacillary dysentery, *S. dysenteriae*, has been isolated from chimpanzees on only one occasion. The shigellosis may be acute or chronic. or apparently latent with or without bacterial shedding. Where the disease is latent relapses may occur, sometimes with death of the host and often with spread of the disease to other monkeys. A monkey that has once been infected appears never to rid itself of shigellae unless treatment is given. The shigelloses are insidious diseases which can cause a great deal of trouble in a primate colony, and every effort should be made to detect the presence of infection by faecal cultures during the quarantine period. They can be very dangerous to human beings, especially children.

Salmonelloses. Salmonella infections of monkeys are usually due to *Salmonella typhimurium*. Occasionally the normal food-poisoning organisms of man infect simian primates, but this is somewhat uncommon. Simians are rarely, if ever, carriers of *S. typhi* and *S. paratyphi*. As with the shigelloses, salmonelloses may be acute or chronic, or latent with or without bacterial shedding. An important difference is that latent infections with salmonella seem to be eliminated naturally in the course of time, and they are, therefore, far less insidious than those of the shigelloses.

Other bacterial pathogens. Enteric diseases are commonly caused by *Proteus* spp. especially *Proteus morgani*. The latter organism is the cause of summer diarrhoea in children. Since *Proteus* is usually regarded as non-pathogenic, care should be taken not to overlook its possible significance. In addition, sometimes when shigellae cannot be isolated *Proteus* carrying shigella antigens are found; these occur by antigen exchange and their presence may be taken as indicative of shigella infection. This phenomenon is found mostly at autopsy, since it appears that shigellae are quickly overgrown and eliminated after death. Another organism which has caused sporadic, but serious, trouble in recent years in many zoos is *Pasteurella pseudotuberculosis*. This ubiquitous organism causes disease in rodents, birds, snakes, tigers, primates and other animals, and care must always be taken to protect the primates' food from contamination with bird and rodent faeces.

Protozoal pathogens. The two most important protozoal pathogens of the bowel are *Entamoeba histolytica* and *Balantidium coli*. It appears also that some South American primates carry an un-named *Entamoeba*, which sometimes causes symptoms and loss. Both these organisms can cause acute symptoms and death. Both cause ulceration of the large bowel, and cysts can be found in the bowel wall.

Helminth parasites. In New World primates, thorny-headed worms (*Acanthocephala*) often cause a great deal of trouble. They are acquired by the eating of cockroaches or other insect life, which are intermediate hosts. They become lodged at the ileo-caecal junction and cause bowel stoppage. *Strongyloides* worms are serious pathogens and may be found in almost all groups of monkeys, particularly South American. These worms produce two kinds of larvae—rhabditiform and filariform. Normally the rhabditoid larvae are free-living until they metamorphose into the long filariform larvae. The latter penetrate the host skin and start the parasitic stage. However, sometimes the rhabditiform larvae metamorphose into the filariform while in the bowel, and thus renew the cycle in the same host. Thus an initial infection, without reinfection, can build up to serious proportions. The worms penetrate the bowel wall, causing abscesses and sometimes perforation. Apes and monkeys sometimes become infected with true hookworms, but the usual bowel infection of this type is caused by the nodular worm *Oesophagostomum* spp. The worms penetrate the bowel wall

and cause haemorrhagic nodules, sometimes in such enormous numbers as to cause death. Because of their protected position the worms are very difficult to treat, and the animals can easily be reinfected with larvae which may be present on the floor of their cages. The common tape worms of primates are *Moniezia* spp.; these are not highly pathogenic. The lower primates are only rarely found infected with the tape worms of man.

There are, of course, many other bowel pathogens of simian primates, but space permits mention only of the more important.

Tuberculosis

Although tuberculosis is a disappearing disease in more advanced countries, the incidence in newly-imported primates is not diminishing, and for this very reason assumes a greater importance. The most susceptible species are the Old World primates including the Great Apes, the disease usually being transmitted by infected persons in the country of capture. In Old World primates, the disease is of the pulmonary type and resembles that formerly encountered in children and known as 'galloping consumption'. Infected animals rarely survive more than 5 or 6 months, but during this time they represent a serious danger to both other monkeys and persons who have no immunity to tuberculosis. In such cases, the infecting organism is *Mycobacterium hominis*, but occasionally bovine-type tuberculosis occurs. With the latter disease, the course is chronic and often regressive; only rarely does it become acute or fulminating. Often monkeys show apparent recovery from bovine tuberculosis but succumb to the disease some years later because of reinfection, under adverse circumstances, from an indurated mesenteric or other gland. In South American primates, tuberculosis is rare. That caused by human-strain bacilli resembles the bovine type in Old World primates and often causes little ill-health. In the past, difficulties have been experienced with the tuberculin test in monkeys. Latterly, however, the intra-palpebral test has been giving very satisfactory results. All newly-imported monkeys should be tested as soon as possible after arrival and those with positive or doubtful reactions should be sacrificed. Except in very exceptional circumstances, attempts should not be made to treat infected monkeys. Treated animals will always be a potential source of danger, and it appears that the results of treatment are less satisfactory than in man. For a review of simian tuberculosis see Fiennes (1972).

Respiratory pathogens

No simian primates are naturally susceptible to the common cold, caused by rhinoviruses. Moreover, the myxoviruses, which commonly cause influenza in man, are not natural pathogens of any simian species. There is also some doubt as to whether influenza can be transmitted from human sources to simians. Viral pneumonias (as indicated by giant cells in pneumonic tissues) are occasionally encountered in monkeys, which are also susceptible to some of the para-myxoviruses, including measles. The majority of the respiratory pathogens seem to arise from the commensal nasal flora, comprising pneumococci, streptococci, staphylococci, and various other organisms such as *Bordetella*. All the groups of pneumococci appear to be involved indiscriminately in pneumonias, no one group being predominant as in man. One may suppose therefore that cross-infection between man and primate does not occur readily. Pneumococcal infections can be serious in all groups of primates and account for heavy losses. Invariably, young chimpanzees which are newly-imported succumb to respiratory troubles and even to fatal pneumonias. They respond readily to treatment with the correct antibiotic, but it is essential that this should be selected by a plate-sensitivity test. For speed in selection, a throat swab should be smeared across a sensitivity plate, and the antibiotic selected should be that which is effective against all growth on it. In this way (that is, without first identifying the individual organism) correct treatment can be introduced within 24 hours in time to save the animal's life. When trouble is encountered with respiratory pathogens, this is in many cases, perhaps the majority, secondary to other circumstances, such as nutritional inadequacy.

Blood parasites

Blood parasites, though often found in primates, do not appear to be of especial importance, though occasional deaths from malaria are encountered. The parasites more commonly encountered are malaria parasites, trypanosomes, and microfilariae. Filariae are usually regarded as non-pathogenic, but in this author's view they sometimes adopt a pathogenic role if the host is also affected with some disease of viral origin. The appearance of microfilariae in the blood may, of course, occur at all hours, or be diurnal or nocturnal.

Toxoplasmosis

Toxoplasma is sometimes found in liver or brain *post mortem*, and has caused losses through transmission in a colony, particularly of New World species.

Viral diseases

As mentioned above, many simians are hosts to latent viruses and under adverse conditions they often become virus-shedders. Herpes viruses are undoubtedly the most important group. These are all very host-specific, but cause unpredictable effects in heterologous hosts.

Of the poxviruses, simians are susceptible to small-pox, and occasional outbreaks of true monkeypox occur with a high fatality rate. Two other poxviruses, Yaba and Yaba-like viruses, cause regressive skin tumours and can be transmitted to man. Infectious hepatitis causes inapparent disease in simian primates, the species so far involved having been chimpanzees, woolly monkeys, and patas. The disease can be detected by liver function tests and liver biopsy. It is almost certainly acquired from human sources in the country of origin and, when transmitted to humans in the country of acceptance, causes a serious (but not so far fatal) disease. The well-known Marburg disease was caused by a hitherto unknown virus, which cannot be assigned to the established viral groups. No Marburg cases occurred among persons in contact with the transmitting primates; only those who performed autopsies and handled the tissue cultures acquired the infection. The last human case to occur was the wife of a recovered human patient who transmitted the virus in his semen.

In many eastern and South American countries, monkeys are regarded as one of the major hazards with regard to rabies. Although little trouble has been encountered with this disease in captive simians, workers should always be on their guard against it until after at least a 12-month period. It should be remembered that affected monkeys often do not show symptoms diagnostic of classical rabies.

General remarks

The very brief survey of simian diseases given above is intended merely to highlight the more important conditions which may occur. Simian diseases are by no means easy to diagnose and in case of trouble the assistance of some person well acquainted with them should be sought.

Further reading

Fiennes, R. N. T-W-. (1966). *Some Recent Developments in Comparative Medicine. Symp. zool. Soc., Lond.,* **17.**

Fiennes, R. N. T-W-. (1967). *Zoonoses of Primates.* London: Weidenfeld & Nicolson.

Fiennes, R. N. T-W-. (1972). *Pathology of Simian Primates,* vols I & II. Basel: Karger.

Harris, R. S. (1970). *Feeding and Nutrition of Non-human Primates.* New York: Academic Press.

Lapin, B. A. & Yakovleva, L. A. (1963). *Comparative Pathology in Monkeys.* Springfield, Ill.: Thomas.

Napier, J. R. & Napier, P. H. (1967). *A Handbook of Living Primates.* New York: Academic Press.

Pickering, D. E. (1962). *Proceedings of Conference on Research with Primates.* Beaverton, Oregon: Tektronix Foundation.

Rosenblum, L. A. & Cooper, R. W. (1963). *The Squirrel Monkey.* New York: Academic Press.

Ruch, T. C. (1959). *Diseases of Laboratory Primates.* Philadelphia: Saunders.

Valerio, D. A., Miller, R. L., Innes, J. R. M., Courtney, K. D., Pallotta, A. J. & Guttmacher, P. M. (1969). *Macaca mulatta: Management of a Laboratory Breeding Colony.* New York: Academic Press.

Vastborg, H. (1965, 1967). *The Baboon in Medical Research,* vols I & II. Austin: University of Texas Press.

Whitelock, O. V. S. (1960). Care and maintenance of the research monkey. *N.Y. Acad. Sci.*

APPENDIX
Composition and manufacture in laboratory of diet for capuchins

Mineral mixture

Sufficient minerals are weighed out to make 5 kg of mineral mixture. The minerals are then carefully mixed to give a homogeneous mixture.

Weighed amounts of the mineral mixture are taken each time a batch of food is made up.

Vitamin mix

Sufficient vitamins are taken to make 170 g vitamin mix. This amount of vitamin mix would make 50 kg of dry diet. The vitamin mix is stored by adding it to enough glucose to make 50 kg of dry diet (i.e. 5·03 kg). The whole is mixed thoroughly in the mixer and then stored in an airtight jar in the refrigerator.

Weighed amounts of the vitamin glucose mixture are taken each time a batch of food is made up.

Making up the diet

For 8 capuchin monkeys, 6·0 kg of diet are made up each week, once a week. The constituents are weighed out and then added to de-ionized water (1 litre water per 1 kg dry food mixture). The resulting mixture is then mixed in a Hobart food mixer (SE401) with chopper attachment and filler tube, until it has a solid consistency.

It is important to allow time for the cellulose powder to take up water. Although these proportions at first give a very wet mixture, after 20–30 minutes mixing the food is firm and semi-dry.

This semi-dry food is then extruded through a 16 mm filler tube attached to the Hobart food mixer. The resulting pellet is broken into suitable sized pieces which are placed on racks in a Gallenkamp moisture extraction oven (OV–440). The food is left in the oven for 12–18 hours at 40°C.

The firm pellets resulting are stored in an airtight tin in the refrigerator.

Composition of diet

Material	% dry weight	To make 1 kg dry mix	
Casein	27·00	270·0 g	
D-Glucose Monohydrate	10·06	100·6 g	
Corn Oil	5·00	50·0 g	
Corn Starch	40·00	400·0 g	
Cellulose Powder Grade B.W.40 ('Socka-Flok')	15·60	156·0 g	
Cupric Sulphate A.R.		2·0 mg	
Ammonium Ferric Citrate (Brown)		304·0 mg	
Manganous Sulphate AR		4·0 mg	
Aluminium Ammonium Sulphate AR		2·0 mg	
Potassium Iodide AR		1·0 mg	
Sodium Fluoride AR		10·0 mg	
Zinc Sulphate AR		1·0 mg	
Cobaltous Chloride AR	2·0	1·0 mg	20·0 g Mineral mix*
Calcium Carbonate AR		1·372 mg	
Calcium Citrate		6·166 g	
Calcium Orthophosphate		2·256 g	
Magnesium Oxide AR		704·0 mg	
Magnesium Sulphate AR		766·0 mg	
Potassium Chloride AR		2·494 g	
Di-Potassium Hydrogen Orthophosphate		4·376 g	
Sodium Chloride AR		1·541 g	
Meso-Inositol		110·0 mg	
Menapthone Sodium Bisulphate (Vit. K_3)		60·0 mg	
Aneurine Hydrochloride (Vit. B_1)		30·0 mg	
Riboflavin (Vit. B_2)		30·0 mg	
Nicotinic Acid (Vit. PP)		100·0 mg	
Dextro-Pantothenic Acid Calcium Salt (Vit. B_3)		100·0 mg	
Choline Chloride		1700·0 mg	
Pteroylglutamic Acid (Folic Acid)		5·0 mg	3·4 g Vitamin mix†
Pyridoxine Hydrochloride (Vit. B_6)	0·34	34·0 mg	
(+) Biotin (Vit. H)		0·400 mg	
Cyanocobalamin (Vit. B_{12})		0·40 mg	
Vitamin A Acetate 2·9 × 10^6 i.u./g.		10·0 mg	
D–L α –Tocopheryl Acetate (Vit. E)		110·0 mg	
L–Ascorbic Acid (Vit. C)		1000·0 mg	
P–Aminobenzoic Acid (Vit. H^1)		110·0 mg	
Vitamin D_3 4 × 10^7 i.u./g.		0·20 mg	
(To Label. Chromic Oxide 0·5 g/kg dry mix)			

* This Mineral Mix gives Ca^{2+} : PO^4 of 1·88

† This Vit. Mix works out for Vit. D_3 at $\simeq$ 3·0 i.u./g food

32 The Fowl

E. G. HARRY AND D. M. COOPER

The poultry industry is the second most important branch of agriculture in Great Britain and has an annual turnover in excess of £300 million. An industry of this magnitude has naturally attracted a great deal of scientific investigation in its own right, and there is now probably as much accurate information on the genetics, nutrition and pathology of the fowl as on those of any other domesticated animal. There is a mass of detailed information on these aspects of commercial poultry-farming in textbooks and scientific journals, and in government and semi-scientific leaflets and bulletins. The Ministry of Agriculture through the Agricultural Development and Advisory Service offers services freely available to the experimentalist. Food compounders, chemical manufacturers and a number of commercial poultry enterprises maintain research and advisory services, while the Agricultural Research Council have established two centres, namely the Poultry Research Centre, Edinburgh and the Houghton Poultry Research Station which are entirely devoted to poultry research.

Standard biological data

Although studies into the physiology of the fowl began much later than those in genetics, nutrition and pathology, there has been a considerable effort in this field in recent years. The reader is referred to Bell and Freeman (1971) for an indepth review of the subject.

Husbandry

Housing and caging

In recent years there has been considerable improvement in the design and construction of poultry houses and the reader is referred for details to the *UFAW Handbook on the Care and Management of Farm Animals** and to the Codes of Recommendations for the Welfare of Livestock issued by the Ministry of Agriculture,

*Published by E. & S. Livingstone, 1971.

Fisheries and Food (1971). The modern poultry-house is a well insulated building with controlled ventilation and lighting. Such buildings can readily be adapted for experimental work by using solid divisions to make separate rooms, or by the fitting of moveable wire pens (Fig. 32.1). If one ensures that the ceiling height is at least 2·5 m then tier brooders and battery cages can be installed, thus increasing the flexibility of the house. The interior lining of many prefabricated houses is of flexible asbestos, which is a vermin-proof material that will withstand power-hosing. Drains are essential, and it is of advantage to provide a trap before the drain enters the main system so that any litter or other material can be removed to prevent clogging.

Chicks. Artificial brooding of chicks has been raised to a high level of efficiency in commercial production. A variety of methods are in use and include whole-house heating, usually by oil or bottled gas, and electric or gas brooders. For experimental purposes electric tier-brooders, infra-red lamps and small gas brooders are most suitable. Tier-brooders have the advantage that they can be increased in numbers as requirements increase. There are several makes on the market and the capacity varies considerably. One of the most useful is divided into two sections and may contain from two to five tiers. Each horizontal section will accommodate 50 day-old chicks, but it should be remembered that chicks grow rapidly and if it is intended that they shall be kept to four weeks of age it is more satisfactory to place not more than 25 in each section. This avoids having to move them as they grow.

When brooded on the floor the day-old chicks are placed inside a surround of metal, plastic or cardboard so that they do not stray from the heat source—normally an infra-red lamp or gas hover-brooder. Many commercial hovers are too large for experimental groups but an infra-red lamp of 250 w capacity is ideal for groups of 20 to 25 chicks. When they are 4–5 days old the

Fig. 32.1. Movable pen partitions.
(a) Front view showing trap nests.
(b) Side view.
(Photographs: Houghton Poultry Research Station.)

surround is enlarged and at the end of the week removed entirely. Recommended space is 62·5 cm²/chick at two weeks old, 155 cm²/chick at two to four weeks old and 323 cm²/chick from four to six weeks old.

Growing and adult birds. In experimental work the accommodation of growing stock which have been removed from heat at five weeks of age presents difficulties. Tier-brooders without heat can accommodate chicks to six weeks of age (Fig. 32.2), but unless a similar system is constructed which allows additional head room the alternative method is to place the birds in pens on litter or weldmesh wire floors. If the latter are used 2·5 cm × 7·5 cm weldmesh of 10 gauge is the material of choice but one must ensure that the floors are strengthened by 6 mm rods and are well supported, otherwise leg weakness, bumblefoot and breast blisters are likely to occur. An allowance of 0·27 m²/bird is adequate for adult birds on deep litter and 0·23 m²/bird if on a wire floor or a combination of litter and wire floor. For much experimental work single-cage laying batteries are to be preferred, but it is becoming increasingly difficult to obtain such

Fig. 32.2. Tier brooder 4–6 weeks. (Photograph: Houghton Poultry Research Station.)

units from the manufacturers because the poultry industry has turned to multi-bird cage units. Perhaps the most satisfactory method is to have cages specially made, and a suitable design for both light and heavy breeds is shown in Figure 32.3. These cages can be fitted into racking made from 2·5 cm square steel which, if fitted with heavy-duty castors, may be easily moved.

Breeding stock. Few experimental units other than those specifically concerned with poultry research will wish to maintain their own breeding stock. However, it may sometimes be desirable to keep a small flock and the most satisfactory method is to maintain the birds on deep litter or in wire-floored pens. Single breeding pens holding one male to 9–10 females (heavy breeds) or 14–15 females (light breeds) can be established if full pedigrees are required. These ratios are also used in mass mating, but the number of males must be more than two for otherwise interference will take place and the level of fertility will be likely to be lower than expected.

If birds are maintained in multi-bird units natural mating can be arranged but it is essential that the height of the unit be sufficient to allow the male to stand fully erect, as for crowing, without his comb touching the top of the cage (or floor of the tier above). The wire spacing on the front of the cage must be sufficiently wide to permit the male to reach food and water with ease, and without causing damage to comb and wattles.

When birds are maintained in individual cages and fertile eggs are required they can be artificially inseminated. This technique has found an increasing use in experimental breeding and is readily learned by an efficient poultry attendant. It will be necessary to house the males in individual cages, and a cage size of 46 cm × 61 cm × 61 cm high with plastic coated floors has been found most useful. Recent reviews on factors affecting fertility and of the use of artificial insemination have been published by Lake (1969) and Cooper (1969).

Ducklings. Ducklings may be brooded as are chicks, with infra-red lamps, with hovers heated by gas or electricity, or in tier-brooders. Ducklings grow extremely rapidly, quadrupling their weight in four weeks; floor space recommendations are given in Table I.

TABLE I
Space per duckling

Age (weeks)	Space (m²)
0–2	0·045
2–3	0·069
3–4	0·09
4–5	0·116
6–10	0·27

Experimental units. There has been an increased use of specialised units for maintaining chicks for particular experimental purposes, mainly in disease studies and in the production of birds free of specified pathogens. A fairly cheap unit suitable for rearing coccidia-free chickens was developed in 1957 at Houghton Poultry Research Station (Fig. 32.4). The air of the room containing the units is filtered

Fig. 32.3. Adult cage units. (a) Male, (b) Female. (Photographs: Houghton Poultry Research Station.)

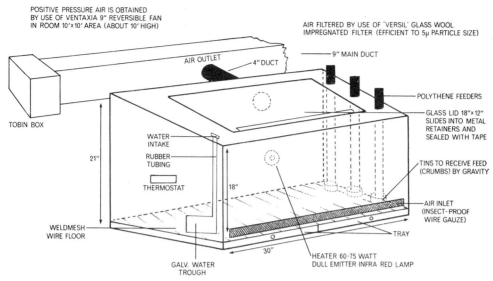

Fig. 32.4. Isolation box.

Fig. 32.5. Parasitology isolator for adult birds. (Photograph: Houghton Poultry Research Station.)

through a glass-wool filter in front of a Ventaxia fan. The ventilation of the cages is achieved by air movement created by the room air. Food and water can be given without opening the unit, but faeces must be removed from the trays at least twice a week. The unit will house about 30 chickens at day-old, 20 at one week, 15 at three weeks, 10 at four weeks or 8 at five weeks. A modification of this unit, designed to be housed in a room behind a barrier system, has been made to house adult birds (Fig. 32.5).

A modified Horsfall-Bauer system, which is more expensive to fabricate, has been success-fully used to maintain chicks used in tumour studies (Fig. 32.6). In these units 24 chicks can be kept to 3 weeks of age, 12 chicks to 5 weeks or 1 or 2 adult birds.

Isolator systems have also been used to maintain breeding birds for the supply of fertile eggs used in the manufacture of vaccines. In the United States these units are made of heavy-weight plastic in the form of a tent, and fitted over a water-filled tank into which the faeces fall. The slurry is emptied into the sewage system, but this method of disposal is not permitted in the UK. In this country studies on the use of isolator systems for the maintenance of breeding birds are being carried out. These systems are made of plastic film or fibreglass and faeces are removed by the use of a plastic film which can be drawn out of the isolator (Fig. 32.7). The filtration of incoming air and the procedures for the management of such systems are similar to those used in germ-free units (see Chap. 11).

Food and water utensils

There are a variety of trough and tubular feeders available from the suppliers of poultry equipment. Some of these are suitable for chicks from day-old. However, to ensure that the chicks start to eat as soon as they are placed in the tier brooder or on the floor, a new fibre

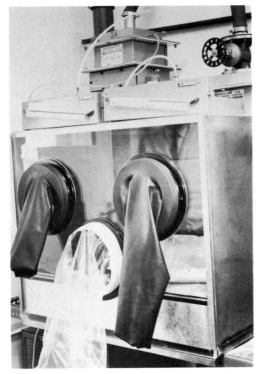

Fig. 32.6. Modified Horsfall-Bauer unit. (Photograph: Houghton Poultry Research Station.)

egg flat in which the depressions are filled with food makes a most satisfactory feeder. This can be turned over and used again on the second day, and by the third day the chicks will have learned to find the food in the trough or tubular feeder. If trough feeders are chosen, the type with a no-perch bar should be selected to prevent the birds from perching and excreting into the food. Tubular feeders must be carefully adjusted to ensure an even flow of food but to prevent wastage. They should be so adjusted that the height of the feeding-rim is in line with the height of the back of the bird; this prevents the scooping of feed into the litter.

Waterers can be of the fount type with capacities from 1 to 18 litres. Automatic drinkers are also available in both galvanized steel and plastic. Although nipple drinkers have been very popular they can cause problems of damp litter if the birds play with the valve. More recently a plastic cup drinker has been developed which is activated when the bird touches the valve, and which does not overflow.

There are two models, one for chicks and one for growing birds and adults, the difference being the ease with which the valve can be activated.

Allowances for feeding and watering space for chicks and ducklings are shown in Tables II–IV.

TABLE II
Feeding and watering space for chickens

Age (weeks)	Food containers	Water containers
Day-old to 2 wks	2·5 cm/chick	50 cm trough or 2×4.5 litre per 100 chicks
Chicks 3–6 wks	4·5 cm/chick	100 cm trough or $2 \times$ 13·5 litre per 100 chicks
Chicks 7–12 wks	7·5 cm/chick	170 cm trough or $4 \times$ 13·5 litre per 100 chicks
Adults	10–15 cm/bird	236 ml/bird/day

Tubular feeders—according to maker's instructions—generally 1 per 25 birds.
Automatic drinkers—according to maker's instructions—generally 1 per 125 birds.
Nipple and cup-type drinkers—according to maker's instructions.
Attention should be paid to distance between watering points so that the bird does not have far to go to reach the nearest drinker.

TABLE III
Lengths of feeding troughs for ducklings

Age (weeks)	Crumbs (m/100 ducklings)	Wet mash (m/100 ducklings)
0–3	4·2	4·8
4–8	9	10·7
9 onwards	10·5	13·7

Fig. 32.7. Fibreglass isolator. (Photograph: Houghton Poultry Research Station.)

TABLE IV

Water space for ducklings

Age	Trough	Quantity
0–3	105 cm	80 litres/day/100 ducklings
4–8	180 cm	280 litres/day/100 ducklings

Trough must be sufficiently deep for the duck to be able to immerse the head. It must not be more than 1·5 m from any point in the brooder until the ducklings are 3 weeks of age.

Litter

White-wood shavings provide suitable litter for chickens, but the shavings must be dry. Other materials which can be used include chopped straw, peat moss or sand.

Environmental conditions

Ventilation. The ventilation system should be designed to deliver 7·5 m³/hour/kg body-weight. It is preferable to have a system with two or more fans, so that if one fails some air will be provided while the repairs are completed. Controls can be manual or on thermostat but it is advisable to have at least some of the fans on manual control. Modern control systems provide adjustment of ventilation-rate to as little as 10 per cent of the fan output, again giving considerable flexibility to the system. Ventilation systems can be of various types and include side inlets and ridge outlets, ridge inlets and side outlets, cross ventilation with intake and extract fans on opposite sides of the house, and end intakes with side outlets. The latter provides a positive pressure system and is the most suitable if the fitting of filters on air intakes is contemplated. Whatever system is used, however, it is essential to keep the air velocity as low as possible. Birds 9–10 weeks old will feel a draught when the air velocity is 20 metres/min with an ambient temperature of 15·5°C, but will not notice an air velocity of 45 metres/min if the temperature is 21°C (Payne, 1961). If air intakes are fitted with a roughing filter, the face velocity should not exceed 60 metres/min and if a lower velocity can be achieved this will be reflected in an increased filter life. Although roughing filters (5 μ) will remove considerable quantities of dust from the atmosphere there is no guarantee that their use will prevent the ingress of pathogens (Cooper, 1970). Improvement of the filtration

system by the use of high-efficiency filters is expensive in installation and running costs, and should something more than roughing filters be necessary it may be of value to consider the housing of birds in isolator systems.

Temperature. Brooding temperature for day-old chicks should be 32°C and lowered by 3°C per week to 18°C at five weeks for optimum growth and feed efficiency. When brooding chicks on the floor the temperature of the room should not fall below 15·5°C for otherwise the birds will crowd into corners and considerable mortality from smothering may occur. Ideally they should spread out in a circle near the edge of the heat source. If tier brooders are used to house young chickens to 4 weeks of age, the temperature of the room in which they are kept should be between 18–21°C. A thermostat is fitted to each tier, the temperature set at 32°C and the heating pad lowered so that the back of the chicks just touch the pad when they are standing up. As they grow the thermostat can be lowered by 3°C per week and the heating pad raised.

The optimum environmental temperature for laying fowls has not been precisely determined but a range of 13–18°C is suitable. Chickens will lay at temperatures as low as 4·5°C but production declines if temperatures fall much below 10°C.

Optimum egg production will be obtained if the house temperature does not fall below 15·5°C. When birds are kept on deep litter the relative humidity should be kept below 70 per cent to prevent caking of the litter surface.

Lighting. Research on the lighting required for maximum growth and egg-production has been well reviewed by Morris (1967). There are two main lighting patterns: that of a constant day (normally 14 hours light : 10 hours darkness); or an increasing day (normally 6 to 8 hours light from day-old to 18 weeks of age and then an increase of 20 minutes per week until a maximum of 16 hours is reached). With controlled lighting it is essential that the buildings are lightproof. This is normally achieved by ensuring that the inlets (or outlets, depending on the ventilation system) are well baffled to exclude light. In terms of light-intensity layers will produce eggs on as little as 0·5 fc, but if birds are being used for breeding 3 fc at bird head height is essential to ensure gonadal development and semen production.

To reduce the amount of disturbance when lights are turned off it is advantageous to have a dimmer incorporated in the lighting system. This should reduce the light intensity by 50 per cent for a period of 30 minutes before all lights are extinguished. Fluorescent lights can be used in poultry units, but warm-white lamps should be chosen. These have a colour temperature of 3,000°K, very similar to that of tungsten lamps (2,850°K). A disadvantage of fluorescent lighting is that a dimmer cannot be incorporated into the system.

Identification

Chicks can be identified by the use of wing bands. Two types are available; the safety-pin which is easily inserted but more readily lost, and the clincher-type band which is permanent but requires some experience in its insertion to ensure that it is placed through the wing-web and not through the muscle. An older method of identification is that of toe punching, but this method is limited in use and the holes may grow out in time.

Adult birds which are being handled regularly may be easily identified by a numbered, coloured plastic wing tag which is either attached by a press stud or slipped over the wing. This number is recorded together with the birds' wing-band number in a book or an individual card. The plastic tag number is then used for egg or other records since it is considerably more quickly and easily read than a wing band.

Handling

Chicks. Grasp the chick loosely but firmly by placing the hand over the back of the bird and around the body. (See method illustrated for quail, Chap. 34, Fig. 34.5).

Growing and adult birds. To catch a bird in a cage, use both hands to lift it out, placing the thumbs over the wings to hold them close to the body and prevent flapping (Fig. 32.8). Birds should be held firmly but not lightly, and confidently enough to ensure there is no struggling.

Birds kept in pens should be driven into a small enclosure formed by setting up a hurdle in one corner of the pen. The following method of

catching and handling should then be adopted.

With the bird facing the handler, slide the right hand, palm upwards, under the breast and place the first finger between the legs which are then held firmly in the right hand. Place the left hand across the back to stop the wings from flapping and to steady the bird. Lift the bird over the hurdle and cradle it in the right arm, the legs still being held in the right hand.

Alternatively, the birds can be driven into a catching crate, when they are caught in the manner described above for birds in cages.

Transport

Hatching eggs should be packed in cardboard cartons suitable for lots of 15 dozen eggs. Fibre flats should be used to hold the eggs, with two flats placed on the bottom layer and two on the top layer to ensure that there is no movement.

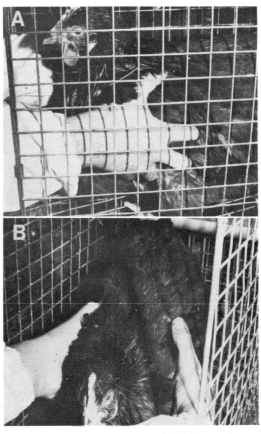

Fig. 32.8. Method of lifting adult bird out of cage. (Photograph: UFAW.)

The box should be marked 'Fragile—hatching eggs. Please keep at 13°C', and the top of the box should be indicated.

Day-old chicks should be placed in cardboard boxes designed for the purpose (25 and 50 chick size) or polypropylene chick boxes (50 chick size). Wood-wool is the normal material used to ensure that the chicks are kept comfortable. It must be remembered that with cardboard boxes sufficient air holes must be punched out and wooden strips at least 16 mm thick be placed underneath and between boxes when they are stacked to ensure adequate ventilation. The polypropylene chick box has been so designed that when it is stacked ventilation is adequate. Although more expensive these have the advantage of long life and may be easily cleaned and autoclaved.

Crates for transporting growing chickens and adult birds may be of wood, metal or plastic. The overall dimensions are approximately 75 cm long, 52 cm wide and 30 cm high. The polypropylene crate has the advantages of ease of cleaning, long life and a non-slip grid floor, thus reducing the likelihood of damage to birds in transit. The number of birds which can be held in such a crate is 8–10 adults of approximately 20 kg total live-weight.

Recommendations for the carriage of day-old chicks and turkey poults by air have been issued by the British Standards Institution, B.S.3149: Part 5, 1961.

Feeding

There is little that need be said about feeding poultry since there are many proprietary foods marketed by the leading feeding stuffs manufacturers. Foods can be obtained as mashes, crumbs or pellets and in each case special formulae are available for chicks, growers, layers and breeding stock. It is as well to point out that many compounders may incorporate additives or drugs in their standard commercial diets, and the research worker would be well advised to ask if there are any additives in the ration he is buying. Those wishing to compound their own diets are referred to Bolton (1967) and Ewing (1963).

Incubation

Two types of incubator are in common use—the natural draught incubator, usually 100–150 egg size, heated electrically and controlled by thermostatic capsule, and the forced draught incubator with a capacity of from one thousand to many thousands. The natural draught incubator is now being made by very few manufacturers, mainly for use in schools. It can have a place in research, however, in that it is a very convenient size for using as a separate hatcher. The difficulty with it is the necessity to turn the eggs by hand at least five times a day for the first 18 days of incubation. As will be appreciated it is also more affected by changes in room temperature and this may result in higher embryonic mortality.

The smallest forced draught incubator which can be purchased is the 1,000 egg size which has 6 setting trays holding 120 eggs each and a hatching tray holding 220 eggs. All forced draught incubators have a mechanical device controlled by a time clock which turns the eggs through an angle of 90° every hour or 90 minutes. Heat is supplied electrically and the air is circulated by paddles or fans. Humidity is supplied by water pans and/or drip feed or spray systems. The machines are equipped with an alarm system which comes into action when the temperature varies from the normal.

Before using any incubator it is advisable to run it empty for at least 24 hours to ensure that it is working at the required temperature and humidity and to make any necessary adjustments. The machine must stand level, and should preferably be adjusted with a spirit level. The incubator room need not be of any particular type but should be sufficiently large to allow a minimum of 30 cm between the top of the machine and the ceiling and the same clearance between machines. There must be sufficient room for removing egg trays and attending to the eggs. Lighting of the room is a matter of convenience. Ventilation should ensure a rate of 0·7 m³/hour/100 eggs. Room temperature should be 21–22°C.

Manufacturers' instructions give temperature and humidity requirements but a relative humidity of 60 per cent (32°C wet bulb) is generally considered satisfactory during the setting period. If there is doubt about the correct humidity, a tray of eggs can be weighed daily and the loss in weight determined. Humidity is correct when the weight loss is 0·4 g/egg/day, and under these conditions the air space will reach the correct volume of 7–8 ml at the end of the incubation time. The eggs are transferred to the hatching compartment (or hatching incubator) on the 18th day or the morning of the 19th day. They should be tray-candled by placing the whole tray on a testing table consisting of a box containing a series of lights (60–75 watts) (Fig. 32.9). Infertile eggs called 'clears' appear as new-laid eggs, while fertile eggs exhibit a dark spot or shadow of an embryo with or without radiating blood vessels. Eggs which show a red line or 'blood ring' contain embryos which have died in the early stages of incubation, often as a result of improper temperature in the incubator or during storage before setting.

Fig. 32.9. Candler. At the other end of the box is a second set of lights (not shown in photograph). The tray rests on top of the box. (Photograph: Houghton Poultry Research Station.)

Proper storage before setting is important, optimum temperature and humidity being 13–15°C and 85 per cent respectively. Fluctuations in storage temperature must be avoided.

Hatching takes place during the 21st day and it is advisable to remove all chicks from the incubator on the morning of the 22nd day, by which time they should have thoroughly dried off. After removal of the hatch the incubator must be thoroughly cleaned and disinfected.

Experimental techniques

Anaesthesia

The choice of anaesthetic must take into account the unique anatomy and mechanism of the respiratory apparatus of birds. Although ether

can be successfully used for minor operations, the presence of air sacs complicates the use of inhalation anaesthesia. The technique of Marley & Payne (1964) using halothane administered from a Godman or Fluote vaporizer calibrated to give up to 4 per cent (v/v) or 7 per cent (v/v) respectively of halothane in oxygen as the carrier gas has proved successful, and more recently Anderson (1967) has shown that halothane used with a direct oxygen supply can give an effective method for controlling anaesthesia in an animal in which there is a need for rapid reduction in anaesthetic concentration when an overdose has been given. Scott and Stewart (1972) found Brietal sodium administered by venocath into the wing vein to be satisfactory in maintaining birds under anaesthesia for periods up to two hours. A dose of 0·5–1·0 ml of a one-per-cent (10 mg/ml) solution induced anaesthesia and a maintenance dose of 0·5–1·0 ml was given as needed during the operation. Recovery times ranged from 20 to 40 minutes for long operations. Webster and Hollard (1973) reported that the injection of chlormethiazole (Hemineurin: Astra) and ketamine hydrochloride (Ketalar: Parke Davis) was a safe and simple method of anaesthesia for pigeons. Levinger et al. (1973) found Bay Va 1470 (2-(2,6-xylidino)-5,6-dihydro-4 H-1, 3-thiazinhydrochloride) to be a very safe and effective anaesthetic for chickens, turkeys, quail, pigeons and five other species of birds. The onset of sedation was very fast, usually within one or two minutes, and the drug has a very wide margin of safety in that even a dose ten times that required for sedation is not lethal. The use of pentobarbitone sodium* given intraperitoneally at a dose rate of 30 mg/kg bodyweight is a satisfactory anaesthetic for day-old chicks.

Euthanasia

See Chapter 14.

Obtaining blood samples

From the wing vein. The bird is placed on its side with the ventral surface towards the operator. The wing lying uppermost is turned back and the needle is inserted into the brachial vein where it crosses the elbow. The needle

* Sagatal: May & Baker Ltd.

gauge suggested is 21 gauge for collecting from chicks to 2 weeks of age; 20 gauge for chicks 3–6 weeks old; 19 gauge for chicks 6–9 weeks old and 17 gauge for adult birds. Since fowl blood clots very quickly it is necessary to transfer the blood to a tube as quickly as possible. Anticoagulant may be used. Where it is not necessary to collect sterile samples, a bleeding stylet is used to puncture the vein and the blood is collected on the hollow formed by the web of the wing, and transferred into a tube.

By heart puncture. The procedure of MacArthur (1950) is very successful. The bird is placed ventral side upwards and the legs are held in one hand while the neck is stretched forward over the edge of the table on which the bird is resting. To obtain blood from week-old chicks a 2 ml syringe is used together with a No. 20 gauge needle 3 cm long. The needle is inserted medially through the dorsal opening of the clavicles. The direction taken by the needle must be posterior and a little downward and in line with the carina of the sternum, which will take the needle directly to the heart from which sterile blood can be obtained. For older birds a No. 17 gauge 5 cm needle should be used. The volume of blood obtained by this method can be 2 ml from a three-day-old chick and 25 ml from adult birds.

Disease control

Disease status of breeding flock

As emphasised in Chapter 10, for reasons of standardisation it is advantageous to use for experimental purposes SPF stock which are known to be free from disease. At present, however, SPF breeding stock is not freely available and in many cases it will be necessary to take eggs and progeny from commercial breeding flocks. These should be as free from disease as practicable, particularly from those diseases which can be transmitted through the egg to the chick and which include the viral diseases avian encephalomyelitis, lymphoid leucosis, and possibly infectious bronchitis and Marek's disease, and the bacterial diseases fowl typhoid, pullorum disease, salmonellosis and mycoplasmosis. The breeding flock should therefore be kept under observation for signs of infection and regularly monitored for birds with latent infections.

Protection of eggs from microbial contamination after they have been laid

In addition to transovarian infection, microbial contamination of the egg contents can occur as a result of soilage of the shell if conditions are favourable for shell-penetration, but it can be reduced to a minimum by using roll-away-type nest boxes. If conventional nest boxes are used these should be lined with wood shavings, sawdust, wood fibre or synthetic materials, and not by hay or straw (Harry, 1963). The lining should be renewed at least twice a week and whenever visibly soiled. Eggs should be collected at least three times a day, the first collection being made early in the day when most of the eggs are laid. A collection should be made after dark of eggs laid at the end of the day, so that they are not included in the collection made the following morning. These eggs, which should be few in number, should be rejected. By this means the duration of contact between the egg and its unsterile environment at a stage when conditions are most favourable for shell-penetration will be minimised. Eggs must be collected gently to avoid hair cracks forming in the shell, as these facilitate the entry of bacteria and moulds. Eggs should be rejected if they have hair cracks or if they are soiled. Shell-penetration in clean eggs can be limited by disinfecting the shell surface as soon as possible after collection or at least within 12 hours of being laid (Gordon & Tucker, 1954). The shells can be disinfected either by formaldehyde gas or by means of a liquid egg dip. The former method is the more suitable for routine use as it is less likely to be carried out incorrectly, but where there is evidence that a disease problem exists it may be advantageous to use an egg dip as this can disinfect small specks of dirt which may escape formaldehyde fumigation. It is also possible to select a disinfectant which is particularly active against the disease agent concerned. Misuse of this technique, however, may lead to an increased amount of contamination and for this reason it requires an operator of a higher calibre than may be generally available.

Fumigation on the farm is carried out in a cabinet in which the eggs are exposed on non-absorbent plastic trays. Cabinets designed for this purpose are commercially available in a number of sizes (Fig. 32.10). The cabinet should

Fig. 32.10. Fumigation cabinet. (Photograph: Houghton Poultry Research Station.)

be reasonably air-tight and should be provided with an extraction fan which must be sealed during fumigation. Provision of a means of circulating the gas during fumigation is also advantageous, particularly when the eggs are stacked close together, and it is desirable to pre-heat the cabinet before fumigation so that its internal surfaces are slightly warmer than the eggs. Eggs should be fumigated for 20 minutes using 50 ml of formalin per m³ of fumigating space (Lancaster *et al.*, 1954). Formalin is added to potassium permanganate in the ratio of 3:2. Alternatively formaldehyde can be generated from solid paraformaldehyde at a rate of 10 g per m³ by means of an electric hot plate. The atmosphere in the fumigating cabinet should always be above 15°C and more than 32 per cent RH, for otherwise fumigation will be ineffective. The ambient temperature and humidity of a hatchery are usually satisfactory in this respect but some adjustment of conditions may be required on the farm. Eggs should be fumigated again at the hatchery immediately before they are set in the incubators. If they have been stored in a cool room they should be allowed to warm up to the temperature of the hatchery before fumigation.

Wet disinfection is carried out in a thermostatically controlled tank provided with some device to keep the disinfectant moving over the eggs. In addition to proprietary egg-disinfecting compounds, quaternary ammonium compounds such as benzalkonium chloride 0·1 per cent, cetyl trimethyl ammonium bromide 1 per cent, solutions containing 250–1,000 ppm of chlorine or iodine, or sodium pentachlorophenate 0·5 per cent can be used as general disinfectants (Gordon *et al.*, 1956). If contamination with sporing bacteria becomes a problem a 3 per cent solution of hypochlorite with 4 per cent sodium hydroxide is suitable. For removing mould contamination, a 0·2 per cent solution of phenyl-mercury dinaphthylmethane disulphonate concentrate* has been found effective. The dip solutions should be used at a temperature of 27°C with a contact period of 15 minutes. It is possible to use a higher temperature for a shorter period of time but this presents practical difficulties because of the more critical temperature regulation required. The disinfectant solution should be renewed before it loses its activity and at least once a day. Advice should be sought from the manufacturer on this aspect and recommendations must be carefully followed.

In addition to the beneficial effect on hatchability it is advantageous from a hygiene aspect to store eggs at a constant temperature below 15°C as described on page 436.

Disease control in the hatchery

This is directed towards preventing disease agents from entering the hatchery from outside and spreading within. The effectiveness of the hygiene measures used is greatly influenced by the layout of the hatchery which, if badly planned, will make disease control almost impossible. (For suggested layout see MAFF Bulletin No. 148).

A receiving room should be provided for the eggs so that egg boxes and persons who have been in contact with farms are excluded from

*Zeetagen: Jeyes, Animal Health Division, Thetford, Norfolk.

the incubator rooms. Flies and vermin must also be excluded from the hatchery.

Transmission of disease within the hatchery results mainly from the dispersal of fluff from diseased chicks when they hatch. In order to prevent the widespread dispersal of fluff the use of incubators with separate setting and hatching sections is imperative. These can be accommodated in separate rooms which will further reduce the possibility of one hatch contaminating subsequent hatches, particularly if the air from the hatchers is exhausted to the outside of the hatchery building well away from air intakes. When the chicks are taken from the trays some dispersal of fluff is inevitable so the ventilation of the hatchery should be such that air does not flow from the hatching room to the setting room and movement of personnel from room to room should be restricted. The dispersal of fluff is also reduced by increasing the humidity of the hatching room when the chicks are being taken off. This can be accomplished by dispersing an aerosol of water. Alternatively, dust-adhesive films can be deposited on surfaces by dispersing an aerosol containing a 40 per cent solution of triethylene glycol. If a nontoxic volatile phenolic disinfectant is included, such a mixture can disinfect the smaller fluff particles in atmospheric suspension as well as having a residual disinfectant activity when it eventually settles on surfaces (Harry, 1955).

The setter does not require frequent disinfection unless it is soiled by broken eggs but the hatcher requires thorough cleaning and disinfection after each hatch. After removal of the chicks from the trays the debris should be gently transferred to plastic bags for incineration. Unhatched eggs must not be broken or their contents dispersed, as such eggs may be infected. After use the hatcher and its trays should be cleaned and disinfected as indicated in the MAFF Poultry Health Scheme Regulations (MAFF, 1966).

Transmission of disease can also occur in the hatchery through contact with the hands of workers. This can be restricted by the washing of hands or application of a barrier cream before and after handling separate batches of eggs and chicks.

As an extra precaution eggs can be fumigated on transfer to the hatcher with 15–20 ml formalin per m^3 of incubator space. Fumigation in the setter, however, may cause embryonic mortality (Harry & Blinstead, 1961). Chicks can also be fumigated when the majority have hatched and humidity in the incubator is at a peak level, using not more than 15 ml of formalin per m^3.

The hatchery should be kept clean at all times and at regular intervals it should be subjected to an overnight period of fumigation with 10 ml of formalin per m^3.

The air in the hatchery should be monitored at monthly intervals to determine the degree to which it is contaminated with bacteria and moulds. This practice provides an indication of the general hygiene standard of the hatchery whereby any deterioration can be detected and appropriate corrective measures applied.

The control of disease in conventional poultry houses

The removal of disease agents from poultry accommodation after use. Efforts spent in obtaining chicks as free from disease as practicable will be wasted if sufficient care is not taken to ensure that accommodation is disinfected before restocking.

The three stages involved in preparing accommodation are (1) dry cleaning (2) wet cleaning (3) chemical disinfection. Chemical disinfection without prior dry and wet cleaning is of little value.

Dry cleaning is carried out by using brushes, scrapers and vacuum cleaners. Bulky material such as litter should be placed in plastic bags and sealed in each of the units cleaned so as to minimise contamination of the surrounding environment. Dusty litter should be dampened before it is disturbed and fans should be switched off to minimise the amount of contaminated material expelled through the ventilators. Staff engaged on cleaning work should not enter units containing birds before washing or changing their clothes. Particular care must be taken when chickens of different ages are kept in close proximity.

Wet cleaning is usually necessary after dry cleaning to reduce microbial contamination a stage further. Details of the methods suitable are given in the MAFF leaflet No. 514 (MAFF, 1973).

Cleaning is the most essential part of the procedures aimed at reducing the microbial contamination of buildings and it is therefore important that buildings, their fittings, cages and equipment are designed to facilitate this process.

The floors, walls and ceilings of poultry buildings should be surfaced with materials that are waterproof and not readily corroded. Dust traps should be avoided as far as possible in the construction of the buildings but those which cannot be avoided (such as ventilation shafts and fans) should be given special attention during cleaning.

Where there has been disease or an infestation with ectoparasites, further treatment of the building with chemical disinfectants or disinfestants is required.

Disinfectants vary widely in their activity against the different groups of microbial disease agents, namely viruses, bacteria and moulds, and to a lesser extent in their activity against individual species within these groups. It is desirable therefore, that the chemical compound chosen should be particularly effective against the disease agent to be eliminated.

Coccidial oocysts are resistant to most disinfectants. Certain phenolic and chlorinated phenolic compounds have some anticoccidial activity, but this is relatively low compared with that of ammonia (usually used in a 1 per cent solution), to which coccidial oocysts are particularly susceptible.

The external parasites of poultry can be destroyed by a number of chemical compounds in both the adult or growing stages. The eggs of these parasites, however, are more resistant and unless a type of disinfestant is used which will persist until the parasites hatch, treatment has to be repeated at intervals.

The suitability of a disinfectant is dependent also on characteristics which influence its practical application such as the effect of organic matter, its speed of action, toxicity, corrosiveness, stability and inflammability. Information must be sought on these aspects where they are relevant.

It is essential that the chemicals used should come into contact with the disease agent in a concentration which is lethal within the duration of the contact period.

The most effective method of dispersal is probably the use of a high pressure spray lance, by which a fan-shaped swathe of droplets can be projected for a distance of several feet. By this means the complete inner surface of the house can be cleaned and exposed to a wet film of disinfectant in a relatively short time.

Gaseous disinfection or fumigation is possible with certain disinfectants. Formaldehyde, at present, is the most convenient gas for poultry-house disinfection, although it lacks the penetrability of the non-polymerising gases such as ethylene oxide and methyl bromide. The methods of applying formaldehyde are described in the MAFF Leaflet No. 514.

Disinfestants can be applied as sprays or dusts where localised application is required at points of heavy infestation. Lindane, which is heat stable, can be dispersed in empty houses by means of smoke generators. Dichlorvos can be vaporised continuously for the control of insect pests, and strings impregnated with fenchlorphos can be suspended in houses for the same purpose. Pyrethrin, diazinon, malathion and carbaryl are usually applied as sprays. In the presence of livestock, disinfestants should be used only as recommended by the manufacturer.

Further reduction in the microbial contamination of buildings occurs when they are left depopulated, particularly if they are wet initially and are allowed to dry by ventilation. The reduction in contamination achieved in this way is progressive, the greatest reduction being obtained in the first week. (Harry & Hemsley, 1964.)

The protection of experimental birds from external infection. In addition to reducing the level of contamination in the environment to non-infectious levels before the birds are housed, it is necessary to take precautions against the entry of disease on vectors. This involves eradication and exclusion of pests, while in the case of human carriers protection can be obtained by the use of special clothing and footwear for attendants and observation windows for visitors. Where infected and healthy birds are kept in different rooms the attendant should attend to the healthy birds first. Some disease agents, the more important of which are salmonellae, can be introduced as food contaminants. Food should always be protected from rodents and in particular cases the use of pasteurised foods may be justified. Pelleted food is less likely to be contaminated than meal.

Another route of infection is by the air, particularly if other animals are kept nearby. Some degree of control can be obtained by spacing units and houses well apart and providing filters on the inlets and in some cases on the outlets also. To be effective these measures can be costly and are only justified if the other more likely routes of infection are eliminated.

Factors influencing the susceptibility of poultry to infection. With conventional housing and management it is not always possible to prevent the occurrence of disease. For this reason measures must be taken to keep the birds as resistant as possible to infection to avoid its spread. In young birds resistance to specific diseases is derived mainly from maternal antibodies which are later supplanted by antibodies acquired by vaccination or by natural contact with disease agents.

Vaccines are available for fowl pox, fowl typhoid, fowl cholera, erysipelas, Newcastle disease, infectious bronchitis, epidemic tremor, infectious laryngotracheitis and Marek's disease. These should be used whenever experimental conditions permit and if the risk of contracting the disease is considered likely. Resistance to infection is also influenced by psychological and physiological stresses. The following treatments are known to lower resistance to infection by causing general debilitation or localised injury to certain tissues, and they should be avoided: deviations from optimal temperature and humidity, deficiency of food and water or oxygen and the presence of a toxic concentration of carbon dioxide or ammonia. Litter dust can also aggravate respiratory diseases as a result of its irritant effect on the respiratory tract. Subjection to stress should be avoided not only in the case of housed birds but also in the transport of chicks from the hatchery. Malnutrition, with a deficiency of certain vitamins, apart from causing general devitalisation and specific deficiency diseases, can result in a lowered resistance to infectious diseases.

The control of disease transmission within groups of birds. The routes whereby infection is transmitted include the food, water and the air. In the case of poultry kept in pens, infection can also be transmitted by the litter and by direct contact between birds.

Food and water may be contaminated by the respiratory and intestinal excretions of sick birds. Faecal contamination of the food and water is particularly likely in young chicks but if properly designed feeders and drinkers are provided it should be less likely where older birds are concerned. Some degree of control may be achieved by adding nontoxic disinfectants to the water. Prevention of transmission of infection through the air is particularly important in the case of respiratory diseases such as Newcastle disease (Report of the Committee of Fowl Pest Policy, 1962). The survival of airborne bacteria and viruses is greatly affected by the temperature and humidity (Akers *et al.*, 1966), loss of viability being more rapid in most cases at intermediate humidities at elevated temperatures. Further reduction may be achieved by means of disinfectant aerosols but the effectiveness of these in practice has not yet been convincingly demonstrated (Harry, 1956; Williams, 1960). The litter is a particularly likely means of transmitting intestinal infections. It should be sufficiently deep and loose to allow dilution of faecal matter to occur and should be renewed at sites where contamination is heavy, i.e. around drinkers and feeders. Dust from excessively dry litter may be associated with transmission of respiratory diseases. Excessively moist litter, however, will encourage mould growth which may lead to aspergillosis and will also aid the survival of coccidial oocysts and nematode parasites and promote microbial decomposition, thus increasing the liberation of ammonia and other noxious gases.

Transmission of infection by direct contact would be expected to occur with all infectious disease agents. Opportunity for direct contact depends not only on the number of birds in the house but on how evenly they are distributed. Distribution will depend largely on the extent to which adequate warmth and sufficient feeders and drinkers are provided. Further limitation of direct contact can be achieved by subdividing the house by removable partitions.

Birds of different breeds and ages show considerable differences in susceptibility and carrier rates to various disease agents. For this reason they should be kept in separate accommodation.

Diseases

The diseases of poultry are adequately covered in standard textbooks and the reader is referred to Hofstad *et al.* (1972). This work deals with the subject in great detail but is intended primarily for the specialist in poultry pathology. Adequate information is also given by the following: British Veterinary Association (1965), Barger *et al.* (1958) and Fritzsche & Gerriets (1962).

Advice can always be obtained from the MAFF's various Veterinary Investigation Centres. There are also a few independent veterinary laboratories which specialise in poultry diagnostic work.

Although many poultry diseases produce typical clinical symptoms or pathological lesions in affected birds these are not invariably consistent or specific. In most cases diagnosis necessitates the examination of the sick birds or carcases by a laboratory that specialises in diagnostic work. Ideally every bird which dies or shows some abnormality in an experimental unit should be examined in this way so that a decision can be made as to whether the remaining birds are still suitable as test animals or whether they should be protected by vaccination or preventive medication.

The common diseases of poultry can be classified as follows:

Bacterial diseases. Pullorum disease, fowl typhoid and salmonellosis (*Salmonella* spp.).

Tuberculosis (*Mycobacterium tuberculosis* var. *avium*).

Fowl cholera, pseudotuberculosis, anatipestifer septicaemia (*Pasteurella* spp.).

Erysipelas (*Erysipelothrix rhusiopathiae*).

Coli septicaemia (*Escherichia coli*—certain serotypes only).

Staphylococcosis, yolk sac infection, oomphalitis (*Staphylococcus aureus*).

Infectious coryza (*Haemophilus gallinarum*).

Mycoplasmosis, infectious synovitis, infectious sinusitis (*Mycoplasma* spp.).

Viral diseases. Infectious laryngotracheitis, infectious bronchitis, fowl pest* (Newcastle disease and fowl plague), fowl pox, psittacosis (ornithosis), virus hepatitis, avian encephalomyelitis, lymphoid leucosis, myeloid leucosis,

* Notifiable disease (ref. Diseases of Animals Act 1950).

erythroleucosis, Marek's disease (fowl paralysis), gumboro disease, inclusion-body hepatitis.

In addition to the diseases listed, complexes occur which involve two or more diseases. An example is a chronic respiratory disease in which infectious bronchitis is complicated by a secondary mycoplasma or *E. coli* infection. Conditions referred to as syndromes, the aetiology of which is not fully understood, also occur; these include haemorrhagic syndrome and turkey syndrome 65.

Fungal diseases. Aspergillosis caused by *Aspergillus fumigatus* and moniliasis caused by *Candida albicans*.

Parasitic diseases. Protozoan infections; coccidiosis, histomoniasis (blackhead) and hexamitiasis.

Helminth infestations caused by *Ascaridia*, *Heterakis*, and *Capillaria* spp.

External parasites. Infestations involving lice, fleas and various species of mite.

Nutritional diseases. Diseases caused by lack of certain vitamins and mineral salts are not uncommon in birds on restricted feeding regimes or on diets incorrectly formulated or stored for too long or at elevated temperatures. With the exception of occasional cases of crazy chick disease (avian encephalomalacia) and the fatty liver and kidney syndrome, nutritional diseases seldom occur in poultry fed normally on proprietary feeds stored in accordance with the manufacturers' instructions.

REFERENCES

Akers, T. G., Bond, S. & Goldberg, L. J. (1966). Effect of temperature and relative humidity in virus survival. *Appl. Microbiol.*, **14**, 361–4.

Anderson, J. C. (1967). A simple method of anaesthesia in the fowl. *Vet. Rec.*, **81**, 130–1.

Barger, E. H., Card, L. E. & Pomeroy, B. S. (1958). *Diseases and Parasites of Poultry*. London: Kimpton.

Bell, D. J. & Freeman, B. M. (1971). *Physiology and Biochemistry of the Domestic Fowl*. Vols. I–III. London: Academic Press.

Bolton, W. (1967). *Poultry Nutrition*. MAFF Bull. No. 174. London: HMSO.

British Veterinary Association (1965). *Diseases of Poultry*. London: BVA.

Cooper, D. M. (1969). The use of artificial insemination in poultry breeding, the evaluation of semen and semen dilution and storage. In *The Fertility and*

Hatchability of the Hen's Egg, ed. Carter, T. C. & Freeman, B. M. Edinburgh: Oliver & Boyd.

Cooper, D. M. (1969). The use of artificial insemina-pathogen-free stock by management-environment control. *Vet. Rec.*, **86**, 388–96.

Ewing, W. R. (1963). *Poultry Nutrition*.

Fritzsche, K. O. & Gerriets, E. (1962). *Geflügelkrankheiten*. Berlin: Parey.

Gordon, R. F. & Tucker, J. F. (1954). Behaviour of *Pseudomonas* sp. and the natural occurrence of the organism in the fowl and its environment. *Proc. 10th Wld's Poult. Congr.*, pp. 348–50.

Gordon, R. F., Harry, E. G. & Tucker, J. F. (1956). The use of germicidal dips in the control of bacterial contamination of the shells of hatching eggs. *Vet. Rec.*, **68**, 33–8.

Harry, E. G. (1955). The application of aerosols to atmospheric and surface disinfection in the poultry industry. 1. The disinfection of surfaces and dust deposit using the Microsol Aerosol Generator. *Vet. Rec.*, **67**, 1109–16.

Harry, E. G. (1956). The application of aerosols to atmospheric and surface disinfection in the poultry industry. 2. Atmospheric disinfection and its value as a means of controlling cross infection. *Vet. Rec.*, **68**, 334–9.

Harry, E. G. (1963). The relationship between egg spoilage and the environment of the egg when laid. *Br. Poult. Sci.*, **4**, 91–100.

Harry, E. G. & Binstead, J. A. (1961). Studies on disinfection of eggs and incubators. 5. The toxicity of formaldehyde to the developing embryo. *Br. vet. J.*, **117**, 532–9.

Harry, E. G. & Hemsley, L. A. (1964). Factors influencing the survival of coliforms in the dust of deep litter houses. *Vet. Rec.*, **76**, 863–7.

Hofstad, M. S., Calnek, B. W., Hembolt, C. F., Reid, W. M. & Yoder, H. W., Jr. (1972). *Diseases of Poultry*. Ames: Iowa State University Press.

Lake, P. E. (1969). Factors affecting fertility. In *The Fertility and Hatchability of the Hen's Egg*, ed. Carter, T. C. & Freeman, B. M. Edinburgh: Oliver & Boyd.

Lancaster, J. E., Gordon, R. F. & Harry, E. G. (1954). Studies on disinfection of eggs and incubators. 3. The use of formaldehyde at room temperature for the fumigation of eggs prior to incubation. *Br. vet. J.*, **110**, 238–46.

Levinger, I. M., Kedem, J. & Abram, M. (1973). A new anaesthetic-sedative for birds. *Brit. vet. J.*, **129**, 296–300.

MacArthur, F. X. (1950). Simplified heart puncture in poultry diagnosis. *J. Am. vet. med. Ass.*, **116**, 38.

Marley, E. & Payne, J. P. (1964). Halothane anaesthesia in the fowl. In *Small Animal Anaesthesia*, ed. Graham-Jones, O. Oxford: Pergamon Press.

Ministry of Agriculture, Fisheries and Food (1965). *The Disinfection and Disinfestation of Poultry Houses*. Advisory Leaflet 514. London: HMSO.

Ministry of Agriculture, Fisheries and Food (1966). *Poultry Health Scheme Regulations*. London: HMSO.

Ministry of Agriculture, Fisheries and Food (1968). *Incubation and Hatchery Practice*. Bulletin No. 148. London: HMSO.

Ministry of Agriculture, Fisheries and Food (1969). *Codes of Recommendations for the Welfare of Livestock. 3. Domestic Fowls*. London: HMSO.

Morris, T. R. (1967). Light requirements of the fowl. In *Environmental Control in Poultry Production*, ed. Carter, T. C. Edinburgh & London: Oliver & Boyd.

Payne, C. G. (1961). Studies on the climate of broiler houses. I. Air movement. *Br. vet. J.*, **117**, 36–43.

Plant, Sir A. (1962). (Chairman) *Report of the Committee on Fowl Pest Policy*. Cmnd. 1664. London: HMSO.

Scott, H. A. & Stewart, J. M. (1972). A new anaesthetic for birds. *Brit. Poultry Sci.*, **13**, 105–106.

UFAW (1971). *Handbook on the Care and Management of Farm Animals*. Edinburgh: Livingstone.

Webster, D. M. & Holland, V. D. (1973). A safe and simple anaesthetic for birds. *Physiol. Behavior*, **10**, 831.

Williams, R. E. O. (1960). Intra-mural spread of bacteria and viruses in human populations. *A. Rev. Microbiol.*, **14**, 43–46.

33   The Chick Embryo

L. N. PAYNE

Introduction

Since the days of Aristotle the chick embryo has been the subject of extensive and detailed scientific investigation, and there now exists a great body of information on the embryology of the chick, for which the reader is referred to the works of Needham (1931), Patten (1950), Romanoff & Romanoff (1949), Hamilton (1952a) and Romanoff (1960). This article is concerned with a limited aspect of the study of the chick embryo, namely its use as a research tool and experimental animal in other scientific disciplines. An excellent account of the application of chick-embryo techniques to virus research, covering the literature up to 1945, is given in a report by Beveridge & Burnet (1946). Reviews of other fields in which the chick embryo has been used may be found in a monograph by Miner (1952). Another American publication (National Academy of Sciences, 1971) contains much that is of interest to research workers using chick embryos. In this present article a general account is given of the use of the chick embryo in biological research, with details of the more commonly used techniques. No attempt has been made to survey the numerous reports of the application of established techniques to new problems within the fields broadly mentioned. Unfortunately there are very few recent articles which review the application of embryo techniques to particular fields of study, and the reader must perforce find these individual reports with the help of abstracting journals such as *Biological Abstracts*.

The special characteristics possessed by the embryonated hen's egg which are responsible for its versatility as a biological tool are not hard to seek. Each fertile egg is a self-contained unit possessing, with the exception of oxygen, all that is needed for the development of the fully formed chick. The embryo is provided with a relatively constant immediate environment in the form of the extra-embryonic fluids and membranes, enclosed within a hard, protective shell. Although environment outside the shell has great influence on the development of the embryo it is one which may be optimally maintained by the operator during incubation. Facilities for technical procedures are provided by the structure of the embryonated egg, which allows ready access to a number of different sites for the injection of substances under investigation. The absence of an effective immunological defence mechanism in the embryo means that there is no active immunological interference with the growth of foreign micro-organisms and tissues. As a result, the embryo is susceptible to infection by a wide range of organisms, and has the added advantage of responding to many of these in a characteristic way. In certain circumstances, however, there may be some immunological limitation due to the presence of antibodies transferred from the hen to the egg in the yolk (p. 458). From the practical viewpoint another great advantage of the embryo is the ready supply and relative cheapness both of fertile eggs and of the equipment required for their incubation.

Uses of the chick embryo

Apart from experimental embryologists, Rous & Murphy (1911) were the first to use the embryo as a research tool, when they grew normal and neoplastic tissues and virus on the extra-embryonic membranes. It is in virology that the chick embryo has found its principal popularity, notably as a result of the exploitation of the techniques involved by Goodpasture in America and Burnet in Australia (Beveridge & Burnet, 1946). The importance of the embryo in virus work is a consequence of its susceptibility to many different viruses and of its convenience as an experimental animal. Its main uses are in the production of live and dead virus vaccines, in the isolation of virus from pathological material in diagnosis, for the propagation of strains of virus and the titration of viruses and antisera, and in fundamental studies on virus growth and action.

The prominence of the virological uses of the embryo is reflected in the accounts of the various techniques described in this article, which are drawn mainly from virological sources. Usually little or no modification of the techniques is required for their use in other fields of study. The embryo is also used in studies on rickettsia, bacteria, fungi (Beveridge & Burnet, 1946; Cox, 1952; Buddingh, 1952), protozoa (Pipkin & Jenson, 1958; Long, 1966; Ryley, 1968, Goedbloed & Kinyanjui, 1970) and even trematode larvae (Fried, 1962).

Probably the next most frequent use of the embryo is in toxicology, for it provides a convenient and sensitive indicator of toxic and teratogenic properties of drugs and other chemicals (Ridgway & Karnofsky, 1952; Landauer, 1954; Karnofsky, 1955; Murphy et al., 1957). It has the possible advantage over mammalian embryos that the substance being studied may be easily applied direct to the embryo, without the intervention of a placental membrane or reaction from the maternal host. For recent applications of chick embryo techniques in toxicology the papers of Walker (1967) and Khera & Lyon (1968) should be consulted.

In addition to virology and toxicology, embryos are used in such fields as cancer research, endocrinology, nutrition, pharmacology, and tissue transplantation. In this last discipline the embryo, and in particular the chorioallantoic membrane, has been used on numerous occasions for the grafting of embryonic (Rawles, 1952) and neoplastic tissues (Karnofsky et al., 1952; Dagg et al., 1956), and to a lesser extent, adult tissues (see for example, Converse et al., 1958).

The inoculation technique appropriate to any particular study depends on the nature of the inoculum, the purpose of the investigation, and the experience of previous workers. The types of substance which are introduced into the egg may be divided into three groups: (1) living organisms (viruses, bacteria, fungi, protozoa); (2) chemicals (drugs, hormones, toxins, vitamins etc.) and (3) living tissues (adult, embryonic or neoplastic). Table I shows the routes of inoculation which are mainly used for these groups.

In many investigations the effect of the inoculum on the embryonic host, or vice versa, is the subject of study. At other times, the embryo may be a vehicle for studying the interaction between two inocula. For example, a foreign tissue grafted on to the chorio-allantoic membrane may be used to cultivate a virus which does not grow in chick tissues (Goodpasture & Anderson, 1944; Blank et al., 1948), or the effect of a drug on an implanted tumour (Dagg et al., 1955) or on an infection (Robbins et al., 1950; Lyles et al., 1963) may be studied.

Supply and management of fertile eggs

Choice of eggs

While it may be desirable sometimes to produce fertile eggs from a laboratory's own breeding

TABLE I

Inoculation routes

	Organisms	Chemicals	Tissues
Chorio-allantoic membrane	*	*	*
Allantoic cavity	*	*	†
Amniotic cavity	*	†	†
Yolk sac	*	*	*
Albumen	†	*	†
Air sac	†	*	†
Embryo body	*	*	*
Blood-stream	*	*	*
Brain	*	†	†
Coelom	†	†	*
Eye	*	†	*
Flank	†	†	*
Diffusion through shell	†	*	†

* Commonly used.   † Rarely or not used.

flock, it is usually more convenient to obtain them from a commercial hatchery. As far as possible the source of eggs should be kept constant as different strains of eggs may vary in response to an agent. Eggs of genetic uniformity are not generally available and response variations may be noted from egg to egg. The influence of genetic factors on response of embryos to viruses is reviewed by Payne (1968). For some work it may be desirable to screen different strains of eggs for suitability. Very few inbred lines of fowl are available. Four lines of Leghorns (RPL lines 15I, 6 and 7, and HPRS-BrL) free from lymphoid leukosis, Newcastle disease, mycoplasmosis, and salmonellosis are kept in isolation at the Leukosis Production Unit of Houghton Poultry Research Station, Huntingdon. For most purposes, however, inbred or special lines are not necessary. The vaccination and disease history of the flock providing the eggs must be considered in some work; for example, the presence of a specific antibody may interfere with the growth of virus in the yolk sac. The yolk of eggs from hens given therapeutic levels of antibiotics may contain sufficient antibiotic to interfere with the growth of microorganisms; it is possible also that dietary prophylactic levels of antibiotic may be high enough to be significant in this respect.

It is usually necessary to transilluminate the egg to determine inoculation sites and for this reason eggs with white or only slightly tinted shells are recommended. With practice, however, more deeply tinted eggs can be used. Eggs should be clean and of standard size, and weigh about 53 to 57 g.

Incubation

Before incubation fertile eggs should be held at a steady temperature of between 10 and 15°C and at a relative humidity of 75 to 85 per cent. Embryonic development starts at about 27°C. Eggs should preferably not be stored for more than 10 days, as fertility may be decreased after this time.

The principles of incubation are given on pp. 435 to 436 and need not be repeated here. However, brief mention should be made of the types of incubator suitable for laboratory use. For the incubation of eggs up to the time of inoculation a commercial incubator is usually

employed. Such incubators may be obtained in a capacity range of from 50 to 1,000 or more eggs. The smaller-capacity machines are usually of the natural-draught type. Various heating methods are available, but electricity is normally used in the laboratory. The eggs are turned by hand, three times a day. The larger-capacity incubators are usually electrically operated, with forced air circulation by means of fans, and automatic turning. The choice of machine is mainly determined by the number of eggs which have to be set. Details of the models available can be obtained from the manufacturers.

In microbiological work it is preferable that the eggs be incubated in a second incubator after inoculation. With certain techniques this is essential as the eggs have to be incubated in one position without turning. In some incubators special trays may have to be made to maintain the eggs in the correct position; the standard commercial compressed-paper or plastic egg trays can often be used for this purpose. Ordinary bacteriological incubators fitted with water trays are frequently used for incubation of eggs after inoculation. This method is satisfactory when the eggs are to be incubated for no more than a few days after injection, but for longer periods the use of machines with controlled humidity and ventilation is desirable. Particularly useful for all stages of incubation when small numbers of eggs are used are the Curfew tabletop observation incubators (Curfew Appliances Ltd., Ottershaw, Chertsey, Surrey), which are inexpensive and are manufactured with egg capacities of 50, 100 and 130.

Candling (transillumination) of the eggs is necessary during incubation to remove clear (infertile) eggs and those with dead embryos, to mark inoculation sites and to detect dead embryos after inoculation. A candling box may be made by mounting a 40-W electric light bulb in a small wooden box with an oval aperture on one side. The edge of the aperture should be rubber-lined so that the egg fits closely during candling. Alternatively, commercial candlers are available which are suitable for experimental work. Candling and marking of the eggs must be done in a darkened room.

Normally, commercial hatching eggs should have a fertility of about 90 per cent and a hatchability of about 80 per cent of the fertile eggs. Two peaks of natural mortality occur during normal incubation. About one quarter of the total mortality occurs at 3 to 5 days of incubation, and one half at 18–20 days. These peaks are associated with periods of increased sensitivity and are accentuated by abnormal conditions which may occur during incubation, particularly faulty incubation techniques affecting temperature, humidity or turning. A third sensitive period at 13 to 14 days may be associated with increased mortality, particularly of embryos from hens with nutritional deficiency, especially of riboflavin. The subject of sensitive periods during incubation has been reviewed by Hamilton (1952b).

In addition to mortality, gross developmental abnormalities of embryos may be encountered. Faults in both incubation and nutrition may be responsible, and a number of genetic causes are also known (Landauer, 1967; National Academy of Sciences, 1971). Some knowledge of such factors is useful when incubation or other faults of management have to be traced. Also, an awareness of the mortality patterns and abnormalities so caused is important when the eggs are used in tests which themselves bring about similar changes, e.g. in the screening of drugs for toxic or teratogenic effects.

Development of the extra-embryonic membranes

Most of the techniques described in this article make use of the extra-embryonic membranes or the cavities they enclose and a simple account of the development of these structures is therefore given. Detailed accounts may be found in embryological textbooks (Patten, 1950; Hamilton, 1952a; Romanoff, 1960). The structures of interest in this respect are the yolk sac, chorion, amnion and allantois.

The main structures of a fertile, newly laid egg are the albumen (56 per cent), yolk (32 per cent), and shell and shell membranes (12 per cent) (Fig. 33.1). The albumen is an important source of water to the developing embryo and the yolk the main source of nutritive material. The chalazae, and the chalaziferous layer surrounding the vitelline membrane which encloses the yolk, are composed of dense albumen and help to maintain the yolk in the correct position within the albumen. Before fertilization occurs a small whitish disc, the blastodisc, lies on the surface of the yolk. This is the protoplasm of the ovum and is in direct continuity with the yolk. The yolk and blastodisc are surrounded by the vitelline membrane. Only the yolk, blastodisc,

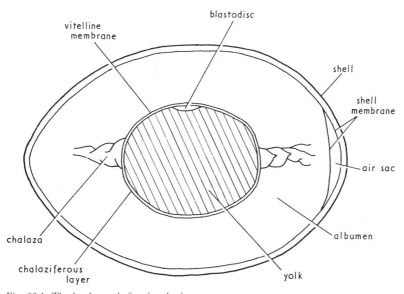

Fig. 33.1. The hen's egg before incubation.

and vitelline membrane originate in the ovary of the hen, the other structures being laid down as the ovum passes down the oviduct of the bird.

The first stage of the development of the embryo after fertilization is a division of the blastodisc to form a multicellular disc, the blastoderm, which expands over the yolk. The embryo proper arises from the central part of the blastoderm; the peripheral parts are extra-embryonic and give rise to the amniotic, chorionic and yolk-sac membranes. As the blastoderm expands the cells differentiate to form three layers of cells, the ectoderm (outer-most), the mesoderm, and the endoderm (inner-most). A horizontal cleavage of the mesoderm occurs, to form an outer (or somatic) layer, and an inner (or splanchnic) layer. The ectoderm and the somatic mesoderm together form the somatopleure, and the endoderm and splanchnic mesoderm form the splanchnopleure. The cavity between the somatopleure and splanchnopleure is the coelom, of which the extra-embryonic body cavity is a part. The well-vascularized splanchnopleure spreads over the surface of the yolk to form the yolk sac (Fig. 33.2); the en-velopment is complete in the fifth day. The yolk sac is connected to the gut of the developing embryo by the narrow yolk stalk, and the yolk thus comes to lie in a bag-like appendage to the gut. A transfer of water from the albumen to the yolk increases the volume of the yolk up to the seventh day, after which the amount decreases until hatching. In terms of weight the yolk sac increases until the sixteenth day. Just before hatching the yolk sac is drawn into the body of the embryo, and the remaining yolk is absorbed during early post-embryonic life.

The amnion and the chorion also arise from extra-embryonic blastoderm. The amnion is formed by the inner parts of folds of the somato-pleure which unite, forming a fluid-filled sac in which the embryo lies (Fig. 33.2). Formation of the amniotic cavity is complete by about the end of the third day of incubation. The volume of amniotic fluid increases to a maximum of 3 to 4 ml at about the thirteenth day, disappearing during the last few days of incubation. The outer parts of the folds of the somatopleure form the chorion. The inner, mesodermal, layer of the chorion and the outer, mesodermal, layer of the amnion are separated by the extra-embryonic body cavity. During the fourth day a hollow bud formed from the posterior part of the embryo gut begins to project outside the body of the embryo into the extra-embryonic coelom. This bud is the allantois and, like the yolk sac, is formed from splanchnopleure (Fig. 33.2). The allantois grows out to form a large fluid-filled sac which almost obliterates the extra-embryonic coelom. During the fifth day the outer, mesodermal,

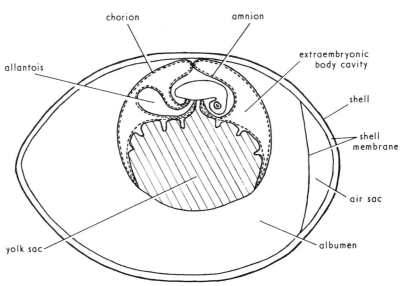

Fig. 33.2. The embryonic membranes on the fourth day of incubation.

THE CHICK EMBRYO 449

layer of the allantois begins to fuse with the inner, mesodermal, layer of the chorion to form a composite structure, the chorio-allantoic membrane (Fig. 33.3). By the twelfth day the allantois has grown to enclose completely the contents of the egg in a double-walled, fluid-filled envelope. The outer wall, fused with the chorion, lines the inner surface of the shell; it is well vascularized and forms the respiratory organ of the embryo. The inner wall of the allantois fuses with the amnion. The allantoic cavity receives the excretions of the embryonic kidneys. The volume of allantoic fluid reaches a maximum of about 6 ml at thirteen days, after which water is absorbed and practically no fluid remains at the twentieth day. At hatching most of the allantois is left in the shell. The albumen becomes enclosed in folds of the chorio-allantoic membrane which forms the albumen sac.

Any description of the development of the embryo itself would be out of place here; however, certain features of embryonic development are useful in determining the age at which embryonic mortality occurs. A detailed series of normal stages has been published by Hamburger & Hamilton (1951), and a simple set of coloured photographs of embryos of various ages is also available (Ministry of Agriculture, Fisheries and Food, 1972). Davis & Garrison (1968) have correlated the mean weights of chick embryos with the stages of Hamburger & Hamilton.

Inoculation techniques

Chorio-allantoic inoculation

Technique for virus inoculation. For virological work eggs are usually used at 11 or 12 days of incubation. The egg is candled, the position of the air sac noted, and a spot marked on the shell 15 to 20 mm below the edge of the air sac over an area where the chorio-allantoic membrane

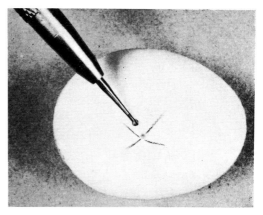

Fig. 33.4. Dental burr used for preparing egg for chorio-allantoic inoculation. (Photograph: Payne.)

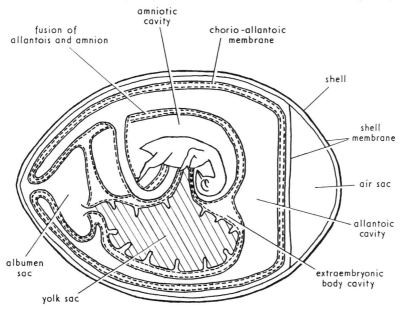

Fig. 33.3. The embryonic membranes on the twelfth day of incubation.

(CAM) is well developed, taking care to avoid large blood-vessels. The shell around the marked area is lightly swabbed with 70-per-cent alcohol, and a hole 1·5 mm across cut in the shell with a No. 6 round dental burr (Fig. 33.4) driven by a dental motor. (The burr is obtainable from C. Ash, Son & Co., Amalco House, 26–40 Broadwick Street, London W.1., and the dental motor from C. J. Plucknett & Co. Ltd., Charlton Village, London S.E.7.) The shell membrane must not be damaged during this operation. Any shell dust should be wiped away from the hole with cotton-wool moistened with alcohol. With a sharp dissecting needle a small hole is made into the air sac at the blunt pole of the egg. The egg is placed on its side with the hole over the CAM uppermost, and a small drop of sterile saline is placed on the hole. A blunt-pointed needle is then gently pushed at an oblique angle through the saline drop and a small tear pro-

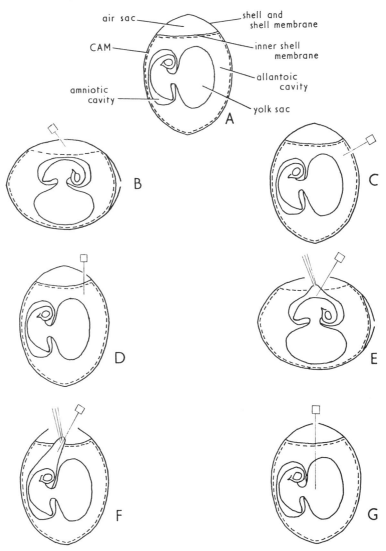

Fig. 33.5. The more commonly used inoculation sites. A, principal parts of the embryonated egg; B, chorio-allantoic inoculation; C, allantoic inoculation (method 1); D, allantoic inoculation (method 2); E, amniotic inoculation (method 1); F, amniotic inoculation (method 2); G, yolk-sac inoculation.

duced in the shell membrane. It is essential that the under-lying CAM be not damaged at this stage. By means of a rubber teat, or a suction tube from the mouth, suction is applied to the hole into the air sac. As the natural air sac is obliterated, air is sucked through the tear in the shell membrane to form an artificial air-space over the dropped CAM. The saline on the shell membrane passes between the shell membrane and the CAM forming a fluid wedge which assists the dropping of the CAM. The egg should be candled as suction is applied to check that the CAM falls away from the shell.

A 0·1-ml volume of inoculum is delivered on to the dropped CAM with a 1-ml tuberculin syringe and a 6-mm 26-gauge needle (Fig. 33.5B). The hole over the CAM is sealed with collodion (e.g. Collodion flexible, methylated, from Hopkin & Williams Ltd., Chadwell Heath, Essex) which is allowed to dry, and the egg is then stood with the hole into the natural air sac uppermost and this hole is sealed. The eggs are then incubated in this upright position without turning for about three to eight days, depending on the virus which has been used. The advantage of this technique over the older methods (Beveridge & Burnet, 1946) is its speed and the improved distribution of the virus over the CAM which occurs when the egg is stood upright.

To harvest the CAM, the embryo is killed by chilling to about 4°C, and the egg cut with scissors from pole to pole round one side, and split into two halves. The embryo and yolk sac are tipped out, and the CAM, in one piece, is stripped from the shell with forceps. The membrane is rinsed in physiological saline or formal saline, spread out in a petri dish and examined for lesions against a dark background. A pair of artists' paint brushes may be used to spread out the membrane in the dish.

Technique for tissue implantation. For tissue transplantation the CAM technique is modified to allow a relatively large fragment of tissue to be placed on the membrane.

Instead of a small hole over the CAM, a triangular window, with sides about 1 cm long, is cut in the shell (Fig. 33.6) with a carborundum disc (obtainable from C. Ash, Son & Co., Amalco House, 26–40 Broadwick Street, London, W.1.). The triangle of shell is lifted off with a fine needle, a drop of saline placed on the

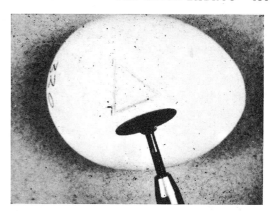

Fig. 33.6. Carborundum disc used for removing a triangle of shell. (Photograph: Payne.)

exposed shell membrane, and the shell membrane pricked through. The CAM is dropped as before, and the shell membrane is then removed from the window to provide a clear view of the dropped CAM. The tissue is placed in position with forceps. After implantation the triangular window is sealed with transparent adhesive tape, such as Sellotape. Beveridge & Burnet (1946) describe alternative methods of sealing. The egg is then reincubated, window uppermost, without turning. In transplantation work 8-day-old eggs are often used to provide a post-implantation period of 10 or 11 days. Modifications of this technique are described by Ballantyne (1959) and Zwilling (1959).

Techniques for localising transplanted tumours on the CAM by use of compartments in plastic grids or discs are described by Hellman & Tucker (1965a & b). The cultivation of leucocytes in diffusion chambers placed on the CAM is described by Bell (1960) and Owen & Harrison (1967).

Other modifications. Nadeje *et al.* (1955) have described a simple instrument with which the shell and shell membrane can be punctured in one operation. Egg punches are available commercially (e.g. Virtis egg punch—Virtis Co. Inc., Gardiner, New York) which will remove a small area of shell without damaging the underlying membranes. A combination egg candler and chorio-allantoic-membrane dropping apparatus is described by Carey & Hahon (1960).

A technique for CAM inoculation in which the CAM is not dropped is described by Litwin (1957). The inoculum is injected into the air sac and allowed to diffuse on to the CAM through a hole in the inner shell membrane made with a blunt needle. This technique is not recommended for accurate work. Gorham (1957) has described a similar technique in which a combined inoculation of the CAM, allantoic cavity, and yolk sac is made.

A method for inoculating several dilutions of virus on to a single CAM through windows in the inner shell membrane of the air sac is described by Andrzejewski & Bogunowicz (1966).

D'Arcy & Howard (1967) described the use of impregnated filter paper discs placed on the CAM of 8-day-old embryos to measure the anti-inflammatory activity of various chemicals. *Non-specific lesions on the CAM.* The chorioallantoic membrane readily develops lesions in response to physical or chemical trauma, and these non-specific lesions may be confused with specific lesions in attempts at primary virus isolation or may interfere with the development of specific lesions. Non-specific focus formation has been studied and the lesions are described by Wyler & Van Tongeren (1957), Voss & Henneberg (1957) and Camain *et al.* (1960). The ultrastructure of the CAM and its reaction to inoculation trauma is described by Ganote *et al.* (1964). Metaplastic changes in the CAM associated with tissue grafts are described by Moscona (1960). It has been suggested that the process of dropping the CAM is sufficient to damage the membrane and produce non-specific lesions. Using a modified CAM technique, Hahon *et al.* (1957) reduced the occurrence of such lesions from 70 per cent to 4 per cent and obtained higher estimates of virus titre on undamaged membranes.

Allantoic inoculation

Eggs of 9 to 11 days' incubation are used for virological purposes. The egg is candled and a point is marked on the shell, about 10 mm from the edge of the air sac, over an area where the CAM is well developed, avoiding large blood-vessels. The area is swabbed with 70-per-cent alcohol and a small hole is drilled in the shell with a No. 6 round dental burr. The inoculation of 0·1 to 0·2 ml of fluid is made through this hole into the allantoic cavity with a 1 or 2-ml syringe and a 12·5-mm 22-gauge needle and the hole is sealed with collodion (Fig. 33.5c).

Back flow of allantoic fluid through the hole at the time of injection may be prevented, if necessary, by making a second hole into the air sac at the blunt pole, thereby allowing the extra volume to displace air from the air sac. The egg is incubated without turning for 2 to 3 days. Bleeding into the allantoic fluid at the time of collection can be prevented by first chilling the embryos at 4°C. The shell and shell membrane over the air sac are removed with scissors or a trephine. The inner shell membrane and underlying CAM are cut away with scissors, the embryo pushed to one side, and the allantoic fluid aspirated with a Pasteur pipette.

In a modification of this technique (Fig. 33.5D), the inoculation hole is drilled into the air sac 3 mm from its margin, avoiding large blood-vessels. The inoculation is made by passing a 25-mm 22-gauge needle for half its length through the air sac and into the allantoic cavity in a direction parallel to the long axis of the egg.

The successful large scale inoculation of tylosin into the allantoic cavity of embryos is reported by Tudor & Woodward (1968).

A method for taking repeated samples of allantoic fluid from individual embryos by means of a cannula is given by DuBose (1965) and Jordan *et al.* (1967).

Amniotic inoculation

Embryos of 10 to 14 days' incubation are usually used. The method has been used particularly in influenza virus work, for the route allows access of the virus to the embryonic respiratory tract. The egg is candled, the air sac is delineated, and a rectangular area 2 by 1·5 cm is marked on the long axis of the egg overlying an area where the embryo appears densest. Within the rectangle a triangle is cut with the dental disc as in the CAM transplantation technique, and the CAM is dropped as already described for this method. The triangle is then enlarged by removing both shell and shell membrane within the rectangular area. The dropped CAM is held with forceps and a 5 mm cut is made in it; the amnion is grasped and drawn out through this cut. While the amnion is held, 0·05 to 0·25 ml of the inoculum is injected into the amniotic cavity with a tuberculin syringe and a 25-mm 30-gauge needle (obtainable from Shrimpton &

Fletcher Ltd., Premier Works, Redditch, Worcestershire), or a fine capillary pipette (Fig. 33.5E). When about 0·05 ml of air is injected at the same time the bubble may be observed within the amniotic cavity if the technique has been properly performed. The shell window is sealed with adhesive tape and the egg is incubated, window uppermost, without turning for 3 to 5 days.

Infected amniotic fluid may be collected by cutting away the shell to the level of the dropped CAM, which is ruptured and the allantoic fluid spilled out. With the amnion held with forceps amniotic fluid may then be removed with a Pasteur pipette.

In an alternative method of inoculation the amnion is approached via the natural air sac (Fig. 33.5F). A circle of shell and shell membrane, 3 cm across, is removed from the blunt end of the egg, and the inner layer of shell membrane is cleared with a drop of liquid paraffin. A pair of fine forceps is pushed through the cleared membrane into the allantoic cavity and the amnion is grasped and lifted upwards. Inoculation is made as before. The inoculated egg is incubated blunt end uppermost. The amniotic fluid is collected as before after the shell has been cut away to the level of the CAM and the allantoic fluid drained off.

Yolk sac inoculation

Eggs are used for virus studies after 5 or 6 days' incubation. In toxicology the yolk sac may be injected before incubation starts, or between 4 and 6 days, or as late as 12 days. A small hole is drilled into the air sac at the blunt pole, and the inoculation of 0·1 to 0·2 ml is made after a 32-mm 23-gauge needle has been passed down the long axis of the egg to a point just beyond the centre (Fig. 33.5G). The hole is sealed with collodion and the egg is incubated in an upright position without turning. To harvest the yolk sac the egg is placed upright and the shell over the air sac removed to reveal the inner shell membrane, which is in turn cut away together with the underlying CAM. The embryo is removed after the yolk-sac stalk has been cut through, and the yolk sac is lifted out with forceps by the yolk stalk. The yolk may be washed away from the membrane with saline.

A method for making repeated injections into the embryonated egg, by means of a glass needle left in situ in the yolk sac, is given by Collier (1951). A technique for the complete removal of the yolk at 3 days, and replacement by various media, is described by Klein et al. (1958) and Grau et al. (1962). A method for the partial interchange of yolk between eggs at 5 days of incubation is given by Ryle (1957).

The yolk sac may also be used for growing tumour tissue (Taylor et al., 1948).

Intravenous inoculation

This method is used when virus suspensions or other fluids have to be introduced directly into the embryonic blood-stream. Eggs from about 8 days of incubation onwards may be used. The egg is candled and a triangle with sides 1 cm long is marked on the shell around a large chorio-allantoic vein. The triangle of shell is removed and the shell membrane is cleared with a drop of sterile liquid paraffin. The vein is illuminated, either from below the CAM by transillumination through the egg, or from above. A tuberculin syringe with a 25-mm 30-gauge needle is used to make the injection. The needle should be held at as acute an angle as possible to the vein, and the injection is preferably made in the direction of the blood flow. (Blood flows in the direction of the air sac; veins become larger as the air sac is approached, arteries smaller). The technique is made considerably more easy if the syringe can be held in a carrier with mechanical means of forward and lateral movement. A mechanical stage from a microscope, held in retort clamps, can be adapted for this purpose (Fig. 33.7). A similar syringe-carrier is described by Goldwasser & Shelesnyak (1953). The egg should also be firmly held while the injection is made. A volume of 0·05 to 0·1 ml is usually injected. The window should be sealed with adhesive tape or melted candlewax after injection, particularly when the egg is to be incubated for more than 3 or 4 days, otherwise excessive moisture may be lost through the window. Eggs injected by this route may be turned. If eggs are to hatch, any tape applied to the window should be removed a day or so before hatching is due, or pipping may be interfered with. If candlewax is used to seal the window it may be left in place during hatching.

Fig. 33.7. A simple syringe-carrier for intravenous inoculation. (Photograph: Payne.)

Barnes & Julian (1958) have described the use of a combined egg-holder and candler. Blood may be withdrawn from the embryo by the inoculation technique, except that it is preferable to insert the needle against the blood flow. Two further methods, in which the embryo is killed, are described by Beveridge & Burnet (1946). A method for the continuous infusion of fluids into the chorio-allantoic circulation, used for up to four days in embryos 6 to 17 days old, is described by Drachman & Coulombre (1962).

Grossgebauer & Langmaack (1967) recommended that the needle should be left in the vein during the 24 to 48 hr period of incubation after injection to avoid bleeding caused by extraction of the needle.

Intracerebral inoculation

Embryos 8 to 13 days old are usually used. The inoculation technique is similar to that for amniotic injection except that the needle is passed through the amnion and into the skull of the embryo. It may be helpful to immobilize the embryos as the injection is made by holding an eyelid with forceps. The route has been used in virological studies.

Injection of chick embryos

A method of injecting a virus suspension beneath the vitelline membrane over the 45-hour blastoderm is given by Blattner & Williamson (1951) and Williamson et al. (1953). Yoshino (1967) has reviewed studies on the growth of several viruses in the blastoderm of one-day eggs. A modification described by Groenendijk-Huijbers (1961), in which a false air sac is formed over the blastoderm, may be found useful when preparing eggs for injection. A toxicity test in which 50-hour embryos are exposed to the test substance is described by Goerttler (1962). Older embryos may be injected with the approach used for amniotic inoculation, passing the needle into the body of the embryo.

Injection into the albumen

This route has been used for the injection of toxic chemicals by Cravens & Snell (1949) and Blackwood (1960). During the early stages of incubation the inoculation may be made from any convenient point on the shell, either directly or via the air sac. As embryonic development proceeds and the volume of albumen decreases, the injection is made at the pointed end of the egg (Beveridge & Burnet, 1946). A technique for partial interchange of albumen between eggs at 2 to 3 days is described by Ryle (1957).

Injection into the extra-embryonic body cavity

A technique is described by Beveridge & Burnet (1946). As yet it has found no application.

Inoculation into air sac

This route has been used to diffuse antibiotics (Bloom & Gordon, 1955), radioactive chemicals (Marrian *et al.*, 1956; Ruggiero & Skauen, 1962), vital dyes (Freericks, 1954) and toxic chemicals (Ehmann, 1963) through the inner shell membrane of the air sac into the embryo.

De-embryonated egg technique

Bernkopf (1949) described a technique in which the embryo was removed after 14 to 15 days' incubation, leaving the shell lined by chorio-allantoic membrane. The cavity of the egg was then filled with Tyrode's solution. This preparation was used to study the growth of influenza virus for up to 48 hours. Fragments of CAM may be incubated in artificial media while attached to the shell, and used for virus infectivity studies (Fazekas de St. Groth & White, 1958; Beard, 1969).

Cultivation of rickettsia in dead chick embryos

Rabinowitz *et al.* (1948) found that after killing chick embryos by chilling to 4°C for 24 hours living cells were demonstrable after subsequent incubation for 16 days at 37°C. *Rickettsia prowazeki* multiplied on inoculation into the yolk sac of the dead embryo.

Intracoelomic implantation

Techniques for introducing tissue fragments into the coelom of the chick embryo are described by Dossel (1954) and Hamburger (1960).

Intra-ocular implantation

See May & Thillard (1957). Intra-ocular inoculation has also been used to a limited extent in virus studies.

Flank implantation

See Hamburger (1960).

Parabiosis

Hasek & Hraba (1955) and Lazzarini (1960) have described techniques for bringing embryos into parabiotic vascular union, either in pairs or in multiple egg groups, for use in studies on acquired tolerance to foreign antigens.

Cross-transfusion of blood between embryos

See Terasaki & Cannon (1957).

Diffusion through the shell

A variety of substances, including sex hormones, may be introduced into embryos by dipping the egg before incubation in an aqueous or oily suspension of the test substance (Seltzer, 1956). This technique has been used to influence the sexual development of birds, and it has been found that the introduction of testosterone and other steroids by dipping or injection will depress or inhibit the development of the bursa of Fabricius and antibody production (Glick & Sadler, 1961; Aspinall *et al.*, 1961). An egg-dipping technique for introduction of erythromycin and tylosin, for control of *Mycoplasma gallisepticum*, is described by Hall *et al.* (1963).

Other chick-embryo techniques

Use in tissue culture

Tissues and cells from chick embryos have been extensively used in tissue culture. The techniques are described by Parker (1961) and Paul (1961).

Growing chick embryos in vitro

Boone (1963) has described a method for growing chick embryos up to eight days old in glass beakers covered with transparent wrapping materials. Quisenberry (1961) and Quisenberry and Dillon (1962) have briefly reported the growth of embryos to six days old in clear plastic shells. See also Elliott & Bennett (1971), Corner & Richter (1973) and Dunn (1974).

Irradiation of chick embryo

Recent discussions and further references may be found in the articles of Reyss-Brion (1956) and Goff (1959) on X-irradiation, and of Stearner *et al.* (1960) and Tyler & Stearner (1960) on γ-irradiation.

Graphic recording of chick embryo movements

See Voino-Iasenetskii & Moskalenko (1961).

Bioelectrical activity of chick embryo

Terzin *et al.* (1963) describe specific characteristics for electro-ovograms of embryos infected with mumps virus and embryos cooled at room temperature.

Electrocardiography

See Lazzarini & Bellville (1956), Hama (1959), Mohsen *et al.* (1962), Deboo & Jenkins (1965).

Electrical and mechanical activity of amnion

See Prosser & Rafferty (1956), and Cuthbert (1962).

Electric polarity of CAM

See Scott (1963).

Electrical potential on shell of embryonated egg

See Ookawa *et al.* (1968).

Determination of embryonic orientation in the egg

Use of radiographs is described by Robertson (1961).

Determination of volume of allantoic fluid

Talman *et al.* (1957) describe a dilution technique using radio-iodinated human serum albumen.

Use of chorio-allantoic membrane for feeding insects and ticks

Techniques for feeding blood meals to mosquitos and other flies by letting them feed on the CAM through the inner shell membrane of the air sac are described by Haas & Ewing (1945), Akins (1950), Ferris & Hanson (1952) and Wallis (1962). A similar technique was used by Burgdorfer & Pickens (1954) to feed argasid ticks. Wallis (1962) also describes a procedure in which the contents of the infertile egg are removed and replaced by a test substance, which the mosquito feeds on through the inner shell membrane of the air sac.

Castration of chick embryos

Groenendijk-Huijbers (1957) has described a method of destroying the gonadal primordium in the 4-day-old chick embryo by electro-coagulation. A castration technique by irradiation is also cited.

Determination of blood volume in chick embryos

See Yoshpe-Purer *et al.* (1953) and Rychter *et al.* (1955).

Microscopical observation of living CAM

A device for the continuous microscopical observation of the CAM *in ovo* is described by Meinecke (1959). See also Narayanan (1970).

Genital and extra-genital contamination of eggs by micro-organisms

The belief held by early workers that the embryonated egg provides a germ-free environment cannot be fully upheld. There is ample evidence that at times viruses and bacteria are present in untreated, newly laid eggs, and may interfere with their experimental use. The occurrence of contamination is of great importance in vaccine-production, where undesirable organisms in the embryonated egg or in embryo cells in tissue culture may interfere with the production of the vaccine or become disseminated during the use of the vaccine. A second danger from egg contaminants occurs in investigational work in which material is being examined for the presence of an infective agent. Here the possibility exists of accidentally picking up a contaminating organism during egg passage and subsequently attaching a false significance to its importance. The subject has been reviewed by Cottral (1952), Luginbuhl (1968), Payne (1968), Starke *et al.* (1968) and Biggs (1970), and the practical aspects of the problem are dealt with fully by the National Academy of Sciences (1971).

Genital infection of the egg may occur in the ovary or in the oviduct. Disease-producing agents which have been shown to infect eggs by these routes are shown in Table II. Of these diseases, Newcastle disease, pullorum disease, fowl typhoid and tuberculosis are infections against which control measures are operated by the poultry industry and government agencies. Provided eggs are obtained from a reputable producer contamination by the agents causing these diseases is unlikely. The remaining diseases are ones in which the complete eradication of the agent from an infected flock is difficult. Listed in order of incidence amongst flocks in this country at the present time, they are lymphoid leukosis, infectious bronchitis, chronic respiratory disease, infectious encephalomyelitis and infectious synovitis. Synovitis is not common and the agent of this disease is an unlikely contaminant. Encephalomyelitis is not uncommon, but egg transmission does not occur in immune birds. Infectious bronchitis virus and mycoplasma are of wide occurrence amongst flocks so that these agents are possible contaminants. The agent of lymphoid leukosis is ubiquitous and contamination by this agent is likely to occur occasionally. If necessary, flocks free from any of these infections may be produced by selection in serological and virus-screening programmes (Hlozanek & Vesely, 1969). Technically the most difficult problem is obtaining flocks free from exogenous lymphoid leukosis virus. Fertile eggs from flocks free from infection by subgroup A and B leukosis viruses are available commercially from several sources in the UK and further information can be obtained from the author. It has recently been discovered that probably *all* embryos and chickens carry in a repressed form a non-pathogenic endogenous leukosis-type virus belonging to subgroup E (Weiss *et al.*, 1971). The response of chick embryos to experimental infection by several of these agents is described by Casorso & Jungherr (1959).

Bacteria and viruses apparently not associated with specific diseases may also infect eggs in the reproductive system of the hen. Harry (1963a) isolated bacteria from the developing ova of seven out of nine laying birds; however, the low infection rate (2·5 per cent) of the yolks of newly laid eggs found in another study (Harry, 1957) suggested that yolk possesses antibacterial properties which may control this source of infection. The subject of bacterial contamination of eggs has been reviewed in detail by Board (1966, 1968).

An agent of low pathogenicity, the chicken-embryo-lethal-orphan (CELO) virus, has recently been shown to be widespread amongst chickens, in which it stimulates antibody production. (Yates & Fry, 1957; Chomiak *et al.*, 1961). Evidence indicates that it can exist as a latent infection in chick embryos or tissue culture, but that it may become unmasked on passage causing embryonic deaths or cytopathic effects in tissue cultures. It is considered to be the same as quail bronchitis virus (DuBose & Grumbles, 1959). Another orphan virus, named Gallus adeno-like (GAL) virus, was found as a contaminating virus in RPL 12 lymphoma filtrates, and was found to produce changes in the parenchymatous organs of chick embryos and cytopathic effects in chick-embryo tissue cultures (Burmester *et al.*, 1960). Serological evidence suggests that this virus may also be common under natural conditions (Kohn, 1962). It is probable that in addition to CELO virus and GAL virus, other viruses exist which may affect embryos and tissue cultures in a similar fashion (Taylor & Calnek, 1962; Kawamura *et al.*, 1964).

Extra-genital infection of the egg may occur from contamination of the shell with faeces or floor litter. Pathogenic organisms such as salmonellae may gain entry from infected faeces; or non-pathogenic bacteria, such as *Pseudomonas* and *Proteus*, or fungi may be involved, in which case the system of management of the laying birds may be an important factor (Harry, 1963b). Washing or sanding of eggs to remove dirt from the shell increases

TABLE II
Diseases of the fowl in which egg transmission of the agent occurs

Disease	Agent	Disease	Agent
Avian infectious encephalomyelitis	Virus	Chronic respiratory disease	Mycoplasma
		Infectious synovitis	Mycoplasma
Infectious bronchitis	Virus	Fowl typhoid	*Salmonella gallinarum*
Lymphoid leukosis	Virus	Pullorum disease	*Salmonella pullorum*
Newcastle disease	Virus	Tuberculosis	*Mycobacterium tuberculosis, avian type*

bacterial infection of eggs, as does pre-incubation storage at high temperatures and humidities. However, the dipping of eggs in germicidal detergent solution at a warmer temperature than that of the egg should remove chance contaminants from the shell. The eggs should remain for about 15 minutes in the dipping fluid which must be at a temperature of about 31°C (Gordon *et al.*, 1956). The natural antibacterial action of the albumen, due to lysozyme, plays an important role in the resistance of the egg to contamination from the shell (Romanoff & Romanoff, 1949). See also Board & Fuller (1974).

Immunity and the chick embryo

Active immunity

The long-held concept that the chick embryo is incapable of an active immunological response to foreign antigens requires modification in the light of recent discoveries. Numerous observations have been made on the inability of the chick embryo to produce humoral antibodies in response to antigens or to reject grafted tissues (Ebert & DeLanney, 1959). Not only is the usual primary response lacking, but a state of specific immunological unresponsiveness may be induced which is manifested as an inability to respond to the antigen in a normal fashion when the animal is immunologically mature. This phenomenon has been termed 'actively acquired tolerance' (Billingham *et al.*, 1956). Studies by Howard & Michie (1962), Kalmutz (1962) and Solomon (1963) suggest however that there is no qualitative difference between immunologically competent cells in the embryo and adult, but that the response to an antigenic stimulus depends upon the relative amounts of antigen and responsive cells. In proportion to the numbers of cells available, large doses of antigen lead to tolerance whereas small doses stimulate immunity. By use of a sensitive technique for detecting antibody Solomon (1966) found that antibody synthesis was readily induced at 15 days of incubation and could occur as early as 12 days. Antibody production was relatively poor between the 15th day of embryonic life and 15 days after hatching, and this was apparently due to the interfering action of passive antibody

The lack of an immune response found in earlier studies when chick embryos were challenged antigenically may perhaps be explained in terms of tolerance induction due to the relatively large amounts of antigen administered. The ease with which the tolerant state can be produced in the chick embryo is an important factor in the susceptibility of the embryo to infective agents and tissue transplants. The embryological development of the immune mechanism in chicks has been reviewed by Miller & Davies (1964).

Passive immunity

The transfer of antibodies from the hen to the chick through the yolk has been reported by many investigators (Buxton, 1952; Ebert & DeLanney, 1959). Diseases of the fowl for which antibody transfer has been shown include Newcastle disease and fowl plague (Brandly *et al.*, 1946), infectious bronchitis (Jungherr & Terrell, 1948), Rous sarcoma (Andrewes, 1939), avian encephalomyelitis (Sumner *et al.*, 1957; Calnek *et al.*, 1960), infectious laryngotracheitis (Benton *et al.*, 1960), lymphoid leukosis (Burmester, 1955; Rubin *et al.*, 1962), pullorum disease and fowl typhoid (Buxton, 1952), and Marek's disease (Chubb & Churchill, 1969).

The transfer of antibody from the hen to the yolk has been studied by Patterson *et al.* (1962a) by using injected homologous γ-globulin and actively formed antibody against bovine serum albumen and influenza virus. After a single injection of serum albumen into the hen, antibody appeared in the yolks of eggs laid 4 days after it appeared in the serum, and peak yolk levels occurred 5 to 6 days after peak serum levels; similar results were obtained with antibody against influenza virus. In experiments in which agglutinins against *Salmonella pullorum* were studied, Brierley & Hemmings (1956) found that yolk titres were about one-eighth of the serum titres of the hens.

Most evidence suggests that antibody begins to pass from the yolk into the serum of the embryo from about 10 days of incubation. Jungherr & Terrell (1948) found neutralizing antibody against infectious bronchitis virus in the serum of 10-day-old embryos. Buxton (1952) found non-agglutinating antibodies for *Salmonella pullorum* on the eleventh day, but agglutinating antibodies were not detectable until the fifteenth to the seventeenth day. Serum antibody levels reached a maximum at hatching time. Neutralizing antibodies against Newcastle disease virus appeared in the serum between 12 and 15 days of incubation (Brandly *et al.* 1946). Schechtman & Knight (1955) injected bovine γ-globulin into the yolk sac of embryos of various ages and studied the appearance of the protein in the embryo or serum. The globulin was first found in the serum at 8 days of incubation.

Patterson *et al.* (1962b) found that the half-life of injected γ-globulin was 72 hours in chicks 1 to 7 days old, compared with 35 hours in adult hens.

Serum protein synthesis by embryonic and neonatal chicks has been studied by Asofsky *et al.* (1962), who used immunoelectrophoresis and autoradiography to distinguish between passively acquired protein and protein formed by the chick. No formation of γ-globulin or β_2-macroglobulin could be shown.

The presence of specific antibody in the egg may become of significance when certain routes of infection or embryos of particular ages are employed to study the growth of an infectious agent. For instance, the work of Brandly *et al.* (1946) suggests that the presence in the yolk of antibody to Newcastle disease virus may interfere with the growth of the virus in the yolk sac at any age, or in the allantoic cavity after 16 days of incubation. See also Payne (1968).

Interferon production

The mechanism of cellular resistance to viral infection has been the subject of much study since the discovery by Isaacs & Lindenmann (1957) of a substance, interferon, which interferes with the propagation of virus in cells. Interferon is produced by fragments of chorioallantoic membrane (Isaacs & Lindenmann, 1957) in the allantoic fluid (Wagner, 1960) and in chick embryo fibroblast cultures (Ho, 1962) exposed to virus (Wagner, 1963). Lampson *et al.* (1963) have given a detailed analysis of the biological and biophysical properties of a purified chick-embryo interferon. Recent reviews on interferon were published by Baron & Levy (1966) and Lockart (1967).

Baron & Isaacs (1961) pointed out that the interferon mechanism begins to function in the chick embryo at 8 days of incubation, and that there is a rapid increase in sensitivity to the antiviral action of interferon and in ability to recover from viral infections between the seventh and tenth days of incubation.

Recognition that the embryo is a potent producer of interferon thus gives another dimension to the study of the response of the embryo to virus infection, and it is certain that much information will be forthcoming which will be of significance in relation to the use of the embryo for growing viruses.

REFERENCES

Akins, H. (1950). A method of infecting *Aedes aegypti* with *Plasmodium gallinaceum* from chick embryos. *J. natn. Malar. Soc.*, **9**, 248.

Andrewes, C. H. (1939). The occurrence of neutralizing antibodies for Rous Sarcoma virus in the sera of young 'normal' chicks. *J. Path. Bact.*, **48**, 225–7.

Andrzejewski, J. & Bogunowicz, A. (1966). Eine modifizierte beimpfungsmethode der chorioallantoismembran für die anzüchtung von viren. *Zentbl. Bakt. ParasitKde* **201**, 432–4.

Asofsky, R., Trnka, Z. & Thorbecke, G. J. (1962). Serum protein synthesis by embryonic and neonatal chicks. *Proc. Soc. exp. Biol. Med.*, **111**, 497–9.

Aspinall, R. L., Meyer, R. K. & Appaswami Rao, M. (1961). Effect of various steroids on the development of the bursa Fabricii in chick embryos. *Endocrinology*, **68**, 944–9.

Ballantyne, D. L. (1959). Modifications of technique used for tissue-grafting to chorio-allantoic membrane of the chick embryo. *Transplantn Bull.*, **7**, 110–3.

Barnes, C. M. & Julian, L. M. (1958). A technique for the intravenous injection of embryonating eggs. *Am. J. vet. Res.*, **19**, 759–60.

Baron, S. & Isaacs, A. (1961). Mechanism of recovery from viral infection in the chick embryo. *Nature, Lond.*, **191**, 97–8.

Baron, S. & Levy, H. B. (1966). Interferon. *A. Rev. Microbiol.*, **20**, 291.

Beard, C. W. (1969). The egg-bit technique for measuring Newcastle disease virus and its neutralizing antibodies. *Avian Dis.*, **13**, 309–20.

Bell, E. (1960). Culture of human leucocytes on the chorioallantoic membrane of the chick. *Nature, Lond.*, **186**, 403–4.

Benton, W. J., Cover, M. S. & Krauss, W. C. (1960). Studies on parental immunity to infectious laryngotracheitis of chickens. *Avian Dis.*, **4**, 491–9.

Bernkopf, H. (1949). Cultivation of influenza virus in the chorio-allantoic membrane of deembryonated eggs. *Proc. Soc. exp. Biol. Med.*, **72**, 680–2.

Beveridge, W. I. B. & Burnet, F. M. (1946). The cultivation of viruses and rickettsiae in the chick embryo. *Spec. Rep. Ser. med. Res. Coun.*, No. 256.

Biggs, P. M. (1970). Production of pathogen-free avian cell substrate for production of vaccines. *Lab. Pract.*, **19**, 45–9.

Billingham, R. E., Brent, L. & Medawar, P. B. (1956). Quantitative studies on tissue transplantation immunity. III. Actively acquired tolerance. *Phil. Trans. R. Soc. Ser. B*, **239**, 357–414.

Blackwood, U. B. (1960). Selective inhibitory and teratogenic effects of 2,5-alkylbenzimidazole homologues on chick embryonic development. *Proc. Soc. exp. Biol. Med.*, **104**, 373–8.

Blank, H., Coriell, L. L. & Scott, T. F. McN. (1948). Human skin grafted upon the chorioallantois of the chick embryo for virus cultivation. *Proc. Soc. exp. Biol. Med.*, **69**, 341–5.

Blattner, R. J. & Williamson, A. P. (1951). Development abnormalities in the chick embryo following infection with Newcastle disease virus. *Proc. Soc. exp. Biol. Med.*, **77**, 619–21.

Bloom, H. H. & Gordon, F. B. (1955). Introduction of antiviral drugs into eggs by the air sac route. *J. Bact.*, **70**, 260–4.

Board, R. G. (1966). Review article: The course of microbial infection of the hen's egg. *J. appl. Bact.*, **29**, 319–41.

Board, R. G. (1968). Microbiology of the egg: a review. In *Egg Quality: A Study of the Hen's Egg*, ed. Carter, T. C. Edinburgh: Oliver & Boyd.

Board, R. G. & Fuller, R. (1974). Non-specific anti-microbial defences of the avian egg, embryo and neonate. *Biol. Rev.*, **49**, 15–49.

Boone, M. A. (1963). A method of growing chick embryos *in vitro*. *Poult. Sci.*, **42**, 916–21.

Brandly, C. A., Moses, H. E. & Jungherr, E. L. (1946). Transmission of anti-viral activity via the egg and the role of congenital passive immunity to Newcastle disease in chickens. *Am. J. vet. Res.*, **7**, 333–42.

Brierley, J. & Hemmings, W. A. (1956). The selective transport of antibodies from the yolk sac to the circulation of the chick. *J. Embryol. exp. Morph.*, **4**, 34–41.

Buddingh, G. J. (1952). Bacterial and mycotic infections of the chick embryo. *Ann. N.Y. Acad. Sci.*, **55**, 282–7.

Burgdorfer, W. & Pickens, E. G. (1954). Technique employing embryonated chicken eggs for the infection of argasid ticks with *Coxiella burnetii*, *Bacterium tularense*, *Leptospira icterohaemorrhagiae* and western equine encephalitis virus. *J. infect. Dis.*, **94**, 84–9.

Burmester, B. R. (1955). Immunity to visceral lymphomatosis in chicks following injection of virus into dams. *Proc. Soc. exp. Biol. Med.*, **88**, 153–5.

Burmester, B. R., Sharpless, G. R. & Fontes, A. K. (1960). Virus isolated from avian lymphomas unrelated to lymphomatosis virus. *J. natn. Cancer Inst.*, **24**, 1443–7.

Buxton, A. (1952). On the transference of bacterial antibodies from the hen to the chick. *J. gen. Microbiol.*, **7**, 268–86.

Calnek, B. W., Taylor, P. J. & Sevoian, M. (1960). Studies on avian encephalomyelitis. IV. Epizootiology. *Avian Dis.*, **4**, 325–47.

Camain, R., Bres, P. & Plagnol, H. (1960). Aspects histo-pathologiques des membranes chorio-allantoïdiennes soumises à des agressions expérimentales non virales. *Annls Inst. Pasteur, Paris*, **98**, 846–60.

Carey, L. E. & Hahon, N. (1960). Combination egg candler and chorio-allantoic membrane dropping apparatus. *J. Bact.*, **79**, 310–1.

Casorso, D. R. & Jungherr, E. L. (1959). The response of the developing chick embryo to certain avian pathogens. *Am. J. vet. Res.*, **20**, 547–57.

Chomiak, T. W., Luginbuhl, R. E. & Helmboldt, C. F. (1961). Tissue culture propagation and pathology of CELO virus. *Avian Dis.*, **5**, 313–20.

Chubb, R. C. & Churchill, A. E. (1969). Effect of maternal antibody on Marek's disease. *Vet. Rec.*, **85**, 303–305.

Collier, L. H. (1951). A method for administering chemotherapeutic agents to chick embryos. *Nature, Lond.*, **168**, 832.

Converse, J. M., Ballantyne, D. L., Rogers, B. O. & Raisbeck, A. P. (1958). A study of viable and non-viable skin grafts transplanted to the chorio-allantoic membrane of the chick embryo. *Transplantn Bull.*, **5**, 108–20.

Corner, M. A. & Richter, A. P. J. (1973). Extended survival of the chick embryo *in vitro*. *Experientia*, **29**, 467.

Cottral, G. E. (1952). Endogenous viruses in the egg. *Ann. N.Y. Acad. Sci.*, **55**, 221–34.

Cox, H. R. (1952). Growth of viruses and rickettsiae in the developing chick embryo. *Ann. N.Y. Acad. Sci.*, **55**, 236–47.

Cravens, W. W. & Snell, E. E. (1949). Effects of desoxypyridoxine and vitamin B_6 on development of the chick embryo. *Proc. Soc. exp. Biol. Med.*, **71**, 73–6.

Cuthbert, A. W. (1962). Electrical and mechanical activity of the chick amnion. *Nature, Lond.*, **193**, 488–9.

Dagg, C. P., Karnofsky, D. A. & Roddy, J. (1956). Growth of transplantable human tumors in the chick embryo and hatched chick. *Cancer Res.*, **16**, 589–94.

Dagg, C. P., Karnofsky, D. A., Stock, C. C., Lacon,

C. R. & Roddy, J. (1955). Effects of certain triazenes on chick embryos and on tumors explanted to the chorio-allantois. *Proc. Soc. exp. Biol. Med.*, **90**, 489–94.

D'Arcy, P. F. & Howard, E. M. (1967). A new anti-inflammatory test, utilizing the chorio-allantoic membrane of the chick embryo. *Br. J. Pharmac. Chemother.*, **29**, 378–87.

Davis, J. E. & Garrison, N. E. (1968). Mean weights of chick embryos correlated with the stages of Hamburger and Hamilton. *J. Morphol.*, **124**, 79–82.

Deboo, G. J. & Jenkins, R. S. (1965). A technique for recording a noise-free electrocardiogram from a chick embryo still in its shell. *Med. Electron Biol. Eng.*, **3**, 443–5.

Dossel, W. E. (1954). A new method of intracoelomic grafting. *Science, N.Y.*, **120**, 262–3.

Drachman, D. B. & Coulombre, A. J. (1962). Method for continuous infusion of fluids into the chorio-allantoic circulation of the chick embryo. *Science, N.Y.*, **138**, 144–5.

Dubose, R. T. (1965). A method for periodic harvests from individual avian embryos. *Avian Dis.*, **9**, 598–603.

Dubose, R. T. & Grumbles, L. C. (1959). The relationship between quail bronchitis virus and chicken embryo lethal orphan virus. *Avian Dis.*, **3**, 321–44.

Dunn, B. E. (1974). Technique for shell-less culture of the 72-hour avian embryo. *Poult. Sci.*, **50**, 974–975.

Ebert, J. D. & Delanney, L. E. (1959). Ontogenesis of the immune response. *In* Symposium on normal and abnormal differentiation and development. *Natn. Cancer Inst. Monogr.*, **2**, 73–111.

Ehmann, B. (1963). Teratogenic effects of thalidomide. *Lancet*, **1**, 772.

Elliott, J. H. & Bennett, J. (1971). Growth of chick embryos in polyethylene bags. *Poult. Sci.*, **50**, 974–975.

Fazekas, de St. Groth, S. & White, D. O. (1958). The dose-response relationship between influenza viruses and the surviving allantois. *J. Hyg., Camb.*, **56**, 523–34.

Ferris, D. H. & Hanson, R. P. (1952). A technique employing embryonating egg to test virus-transmitting ability of certain insect vectors. *Cornell Vet.*, **42**, 389–94.

Freericks, R. (1954). Uber die Vitalfärbbarkeit von Hühnerembryonen. *Biol. Zbl.*, **73**, 155–69.

Fried, B. (1962). Growth of *Philophthalmus* sp. (Trematoda) on the chorio-allantois of the chick. *J. Parasit.*, **48**, 545–50.

Ganote, C. E., Beaver, D. L. & Moses, H. L. (1964). Ultrastructure of the chick chorio-allantoic membrane and its reaction to inoculation trauma. *Lab.*

Invest., **13**, 1575–9.

Glick, B. & Sadler, C. R. (1961). The elimination of the bursa of Fabricius and reduction of antibody production in birds from eggs dipped in hormone solutions. *Poult. Sci.*, **40**, 185–9.

Goedbloed, E. & Kinyanjui, H. (1970). Development of African pathogenic trypanosomes in chicken embryos. *Exptl. Parasit.*, **27**, 464–478.

Goerttler, K. (1962). Der 'teratologische Grund-versuch' am bebrüketen Hühnchenkeim seine Möglichkeiten und Grenzen. *Klin. Wschr.*, **40**, 809.

Goff, R. A. (1959). The acute lethal response of the chick embryo to x-radiation. *J. exp. Zool.*, **141**, 477–97.

Goldwasser, R. & Shelesnyak, M. C. (1953). A syringe carrier and egg clamp for intravenous inoculation of chick embryo. *Science, N.Y.*, **118**, 47–8.

Goodpasture, E. W. & Anderson, K. (1944). Infection of human skin, grafted on the chorio-allantois of chick embryos, with the virus of herpes zoster. *Am. J. Path.*, **20**, 447.

Gordon, R. F., Harry, E. G. & Tucker, J. F. (1956). The use of germicidal dips in the control of bacterial contamination of the shells of hatching eggs. *Vet. Rec.*, **68**, 33–8.

Gorham, J. R. (1957). A simple technique for the inoculation of the chorio-allantoic membrane of chicken embryos. *Am. J. vet. Res.*, **18**, 691–2.

Grau, C. R., Fritz, H. I., Walker, N. E. & Klein, N. W. (1962). Nutrition studies with chick embryos deprived of yolk. *J. exp. Zool.*, **150**, 185–95.

Groenendijk-Huijbers, M. M. (1957). A method for castration of chick embryos. *Acta morph. neerl.-scand.*, **1**, 241–5.

Groenendijk-Huijbers, M. M. (1961). Technical improvements to reduce bleeding of operated chick embryos. *Acta morph. neerl.-scand.*, **4**, 61–2.

Grossgebauer, K. & Langmaack, H. (1967). Eine Verbesserung der intravenösen Injektionstechnik beim embryonierten Hühnerei. *Z. med. Mikrobiol. Immunol.*, **153**, 225–7.

Haas, V. H. & Ewing, F. M. (1945). Inoculation of chick embryos with sporozoites of *Plasmodium gallinaceum* by inducing mosquitoes to feed through the shell membrane. *Rep. U.S. Dep. Hlth Educ. Welf.*, **60**, 185–8.

Hahon, N., Louie, R. & Ratner, M. (1957). Method of chorio-allantoic membrane inoculation which decreases nonspecific lesions. *Proc. Soc. exp. Biol. Med.*, **94**, 697–700.

Hall, C. F., Flowers, A. I. & Grumbles, L. C. (1963). Dipping of hatching eggs for control of *Mycoplasma gallisepticum*. *Avian Dis.*, **7**, 178–83.

Hama, K. (1959). The ECG of chick embryo and effects of various drugs upon it. *J. physiol. Soc. Japan*, **21**, 753–7. (Jap. with Engl. Summary).

Hamburger, V. (1960). *A Manual of Experimental Embryology*. Chicago University Press.

Hamburger, V. & Hamilton, H. L. (1951). A series of normal stages in the development of the chick embryo. *J. Morph.*, **88**, 49–92.

Hamilton, H. L. (1952a). *Lillie's Development of the Chick*, 3rd ed. New York: Holt.

Hamilton, H. L. (1952b). Sensitive periods during development. *Ann. N.Y. Acad. Sci.*, **55**, 177–87.

Harry, E. G. (1957). The effect on embryonic and chick mortality of yolk contamination with bacteria from the hen. *Vet. Rec.*, **69**, 1433–9.

Harry, E. G. (1963a). Some observations on the bacterial content of the ovary and oviduct of the fowl. *Br. Poult. Sci.*, **4**, 63–70.

Harry, E. G. (1963b). The relationship between egg spoilage and the environment of the egg when laid. *Br. Poult. Sci.*, **4**, 91–100.

Hasek, M. & Hraba, T. (1955). Immunological effects of experimental embryonal parabiosis. *Nature, Lond.*, **175**, 764–5.

Hellman, K. & Tucker, D. F. (1965a). Tumor growth in grids on chorio-allantoic membranes of chick embryos. *Cancer Chemotherap. Rep.*, **47**, 73–9.

Hellman, K. & Tucker, D. F. (1965b). Localization of tumours and drugs on the chorio-allantoic membrane of the chick embryo. *Nature, Lond.*, **207**, 1408–9.

Hložánek, I. & Veselý, P. (1969). Project for the preparation and use of defined chicken cells for biological experiments including production of vaccines. *Inst. exp. Biol. Genet. Monogr.*, **1**, Prague, 1–84.

Ho, M. (1962). Kinetic considerations of the inhibitory action of an interferon produced in chick cultures infected with Sindbis virus. *Virology*, **17**, 262–75.

Howard, J. G. & Michie, D. (1962). Introduction of transplantation immunity in the newborn mouse. *Transplantn Bull.*, **29**, 91–6.

Isaacs, A. & Lindenmann, J. (1957). Virus interference. 1. Interferon. *Proc. R. Soc.*, Ser. B., **147**, 258–67.

Jordan, C. E., Monk, B. M. & Dubose, R. T. (1967). A modified method for cannulation of the avian allantoic sac. *Avian Dis.*, **11**, 255–7.

Jungherr, E. L. & Terrell, N. L. (1948). Naturally acquired passive immunity to infectious bronchitis in chicks. *Am. J. vet. Res.*, **9**, 201–5.

Kalmutz, S. E. (1962). Antibody production in the opossum embryo. *Nature, Lond.*, **193**, 851–3.

Karnofsky, D. A. (1955). Use of developing chick embryo in pharmacologic research. *Stanford med. Bull.*, **13**, 247–59.

Karnofsky, D. A., Ridgway, L. P. & Patterson, P. A. (1952). Tumor transplantation to the chick embryo.

Ann. N.Y. Acad. Sci., **55**, 313–29.

Kawamura, H., Shimizu, F. & Tsubahara, H. (1964). Avian adenovirus: its properties and serological classification. *Natn. Inst. Anim. Hlth Q., Tokyo*, **4**, 183–93.

Khera, K. S. & Lyon, D. A. (1968). Chick and duck embryos in the evaluation of pesticide toxicity. *Toxicol. appl. Pharmac.*, **13**, 1–15.

Klein, N. W., Grau, C. R. & Green, N. J. (1958). Yolk-sac perfusion of chick embryos; effects of various media on survival. *Proc. Soc. exp. Biol. Med.*, **97**, 425–8.

Kohn, A. (1962). Gallus adeno-like virus in chickens —studies on infection, excretion and immunity. *Am. J. vet. Res.*, **23**, 562–7.

Lampson, G. P., Tytell, A. A., Nemes, M. M. & Hilleman, M. R. (1963). Purification and characterization of chick embryo interferon. *Proc. Soc. exp. Biol. Med.*, **112**, 468–78.

Landauer, W. (1954). On the chemical production of developmental abnormalities and phenocopies in chicken embryos. *J. cell. comp. Physiol.*, **43**, Suppl. 1, 261–305.

Landauer, W. (1967). The hatchability of chicken eggs as influenced by environment and heredity. *Storrs Agric. Exp. Stn Monograph 1.* (Revised edition.) Storrs: University of Connecticut.

Lazzarini, A. A. (1960). Immunological effects of multiple experimental embryonal parabiosis in birds. *Ann. N.Y. Acad. Sci.*, **87**, 133–9.

Lazzarini, A. A. & Bellville, J. W. (1956). Method for study of electrocardiogram of early chick embryo within the shell. *Proc. Soc. exp. Biol. Med.*, **93**, 27–30.

Litwin, J. (1957). A simple method for cultivation of viruses and rickettsiae in the chorio-allantoic ectoderm of the chick embryo by inoculation via the air sac. *J. Infect. Dis.*, **191**, 100–8.

Lockart, R. Z. (1967). Recent progress in research on interferons. *Progr. Med. Virol.*, **9**, 451–75.

Long, P. L. (1966). The growth of some species of *Eimeria* in avian embryos. *Parasitology*, **56**, 575–81.

Luginbuhl, R. E. (1968). Viral flora of chick and duck tissue sources. *Natn. Cancer Inst. Monogr.*, **29**, 109–18.

Lyles, S. T., Nabors, R. E. & Potter, P. P. (1963). *In vivo* penicillin resistance tests, using the chick embryo. *Tex. Rep. Biol. Med.*, **21**, 102–3.

Marrian, D. H., Hughes, A. F. W. & Werba, S. M. (1956). Nucleic acid metabolism of the developing chick embryo. *Biochim. biophys. Acta*, **19**, 318–23.

May, R. M. & Thillard, M. J. (1957). Nouvelle technique de transplantation chez l'embryon de poulet: la greffe intraoculaire. Application à l'épiphyse. *C.r. hebd. Séanc. Acad. Sci., Paris*, **244**, 1553–5.

Meinecke, G. (1959). Vorrichtung für mikroskopische Lebendbeobachtungen an bebrüteten Hühnereiern.

Z. wiss. Mikrosk., **64,** 227–35.

Miller, J. F. A. P. & Davies, A. J. S. (1964). Embryological development of the immune mechanism. *A. Rev. Med.*, **15,** 23–36.

Miner, R. W. (1952). The chick embryo in biological research. *Ann. N.Y. Acad. Sci.*, **55,** 37–344.

Ministry of Agriculture, Fisheries & Food (1972). *Avian Embryo Development, Tech. Bull., No. 23.* London: HMSO.

Mohsen, T., Carricaburu, P. & Filliol, M. H. (1962). Evolution de l'eléctrocardiogramme au cours du développement de l'embryon de poulet. *C.r. Séanc. Biol. Soc.*, **156,** 331–4.

Moscona, A. (1960). Metaplastic changes in the chorio-allantoic membrane. *Transplantn Bull.*, **26,** 120–4.

Murphy, M. L., Dagg, C. P. & Karnofsky, D. A. (1957). Comparison of teratogenic chemicals in the rat and chick embryos. An exhibit with additions for publication. *Pediatrics*, **19,** 701–14.

Nadeje, T., Tamm, I. & Overman, J. R. (1955). A new technique for dropping the chorio-allantoic membrane in embryonated chicken eggs. *J. Lab. clin. Med.*, **46,** 648–50.

Narayanan, C. H. (1970). Apparatus and current techniques in the preparation of avian embryos for microsurgery and for observing embryonic behavior. *Bio Science*, **20,** 869–870.

National Academy of Sciences (1971). *Methods for the Examination of Poultry Biologics and for Identifying and Quantifying Avian Pathogens.* Washington: NRC.

Needham, J. (1931). *Chemical Embryology.* London: Cambridge University Press.

Ookawa, T., Bures, J. & Mysliveckova, H. (1968). Electrical potential on the surface of fertilized hen's egg as an indicator of embryonic development. *Poult. Sci.*, **47,** 1862–70.

Owen, J. J. T. & Harrison, G. A. (1967). Studies on human leucocytes in diffusion chambers on the chick chorio-allantois. *Transplantation*, **5,** 643–51.

Parker, R. C. (1961). *Methods of Tissue Culture*, 3rd ed. London: Pitman.

Patten, B. M. (1950). *Early Embryology of the Chick*, 4th ed. London: Lewis.

Patterson, R., Younger, J. S., Weigle, W. O. & Dixon, F. J. (1962a). Antibody production and transfer to egg yolk in chickens. *J. Immun.*, **89,** 272–8.

Patterson, R., Younger, J. S., Weigle, W. O. & Dixon, F. J. (1962b). The metabolism of serum proteins in the hen and chick and secretion of serum proteins by the ovary of the hen. *J. gen. Physiol.*, **45,** 501–13.

Paul, J. (1961). *Cell and Tissue Culture.* 2nd ed. Edinburgh: Livingstone.

Payne, L. N. (1968). Eggs in Virology. In *Egg Quality: A Study of the Hen's Egg*, ed. Carter, T. C.

Edinburgh: Oliver & Boyd.

Pipkin, A. C. & Jensen, D. V. (1958). Avian embryos and tissue culture in the study of the parasitic protozoa. I. Malarial parasites. *Expl. Parasit.*, **7,** 491–530.

Prosser, C. L. & Rafferty, N. S. (1956). Electrical activity in chick amnion. *Am. J. Physiol.*, **187,** 546–8.

Quisenberry, J. H. (1961). New approaches to the study of the early embryology of the chick. *Poult. Sci.*, **40,** 1446.

Quisenberry, J. H. & Dillon, E. J. (1962). Growing embryos in plastic shells. *Poult. Sci.*, **41,** 1675.

Rabinowitz, E., Aschner, M. & Grossowicz, N. (1948). Cultivation of *Rickettsia prowazeki* in dead chick embryos. *Proc. Soc. exp. Biol. Med.*, **67,** 469–70.

Rawles, M. E. (1952). Transplantation of normal embryonic tissues. *Ann. N.Y. Acad. Sci.*, **55,** 302–12.

Reyss-Brion, M. (1956). La sensibilité différentielle de certaines ébauches de l'embryon de poulet aux rayons X, à différents stades due développement. *Archs Anat. microsc. Morph. exp.*, **45,** 342–53.

Ridgway, L. P. & Karnofsky, D. A. (1952). The effects of metals on the chick embryo: toxicity and production of abnormalities in development. *Ann. N.Y. Acad. Sci.*, **55,** 203–15.

Robbins, M. L., Bourke, A. R. & Smith, P. K. (1950). The effect of certain chemicals on *Rickettsia typhi* infections in chick embryos. *J. Immun.*, **64,** 431–46.

Robertson, I. S. (1961). Studies of chick embryo orientation using X-rays. I. A preliminary investigation of presumed normal embryos. *Br. Poult. Sci.*, **2,** 39–47.

Romanoff, A. L. (1960). *The Avian Embryo.* New York: Macmillan.

Romanoff, A. L. & Romanoff, A. J. (1949). *The Avian Egg.* New York: Wiley.

Rous, P. & Murphy, J. B. (1911). Tumour implantations in the developing embryo; experiments with a transmissible sarcoma of the fowl. *J. Am. med. Ass.*, **56,** 741–2.

Rubin, H., Fanshier, L., Cornelius, A. & Hughes, W. F. (1962). Tolerance and immunity in chickens after congenital and contact infection with an avian leukosis virus. *Virology*, **17,** 143–56.

Ruggiero, J. S. & Skauen, D. M. (1962). Chick embryo technique for the evaluation of absorption of radio-active sodium iodide from ointment bases. *J. Pharm. Sci.*, **51,** 233–5.

Rychter, Z., Kopecky, M. & Lemez, L. (1955). A micromethod for determination of the circulating blood volume in chick embryos. *Nature, Lond.*, **175,** 1126–7.

Ryle, M. (1957). Studies on the influence of the embryonic environment on the post hatching development of chickens. *J. exp. Biol.*, **34,** 529–42.

Ryley, J. F. (1968). Chick embryo infections for the evaluation of anti coccidial drugs. *Parasitology*, **58**, 215–20.

Schechtman, A. M. & Knight, P. F. (1955). Transfer of proteins from the yolk to the chick embryo. *Nature, Lond.*, **176**, 786–7.

Scott, B. I. (1963). Electric polarity in the chorio-allantoic membrane of the chick embryo. *Proc. Soc. exp. Biol. Med.*, **113**, 337–9.

Seltzer, W. (1956). The method of controlling the sex of avian embryo, improving embryo hatchability, and improving viability of the hatched chick. *US Pat. Syst. Leafl.* 2,734,482.

Solomon, J. B. (1963). Actively acquired transplantation immunity in the chick embryo. *Nature, Lond.*, **198**, 1171–3.

Solomon, J. B. (1966). Induction of antibody formation to goat erythrocytes in the developing chick embryo and effects of maternal antibody. *Immunology*, **11**, 89–96.

Starke, G., Glathe, H. & Hlinak, P. (1968). Zur Problematik der Freundviren bei Virusimpfstoffen, die auf Hühnerembryobasis produziert werden. *Pharmazie*, **23**, 669–78.

Stearner, S. P., Sanderson, M. H., Christian, E. J. & Tyler, S. A. (1960). Modes of radiation death in chick embryo. 1. Acute syndromes. *Radiat. Res.*, **12**, 286–300.

Sumner, F. W., Luginbuhl, R. E. & Jungherr, E. L. (1957). Studies on avian encephalomyelitis. II. Flock survey for embryo susceptibility to the virus. *Am. J. vet. Res.*, **18**, 720–3.

Talman, E. L., Hutchens, T. T. & Aldrich, R. A. (1957). A radioisotope technic for determining allantoic fluid volume of embryonated eggs. *Proc. Soc. exp. Biol. Med.*, **96**, 130–3.

Taylor, A., Carmichael, N. & Norris, T. (1948). A further report on yolk sac cultivation of tumor tissue. *Cancer Res.*, **8**, 264–9.

Taylor, P. J. & Calnek, B. W. (1962). Isolation and classification of avian enteric cytopathogenic agents. *Avian Dis.*, **6**, 51–8.

Terasaki, P. I. & Cannon, J. A. (1957). A technic for cross-transfusion of blood in embryonic chicks and its effect on hatchability. *Proc. Soc. exp. Biol. Med.*, **94**, 103–7.

Terzin, A. L., Zec, N. R. & Bokonjic, N. J. (1963). Recording of bioelectrical activities of intact chick embryo *in ovo*. *Br. J. exp. Path.*, **44**, 88–100.

Tudor, D. C. & Woodward, H. (1968). Research note: a mass method for chicken embryo inoculation with tylosin. *Avian Dis.*, **12**, 379–82.

Tyler, S. A. & Stearner, S. P. (1960). Modes of radiation death in the chick embryo. II. A model of lethal mechanisms. *Radiat. Res.*, **12**, 301–16.

Voino-Iasenetskii, A. V. & Moskalenko, Iu. E. (1961). Graphic recording of chick embryo movements. *Fiziol. Zh. SSSR*, **47**, 106–8.

Voss, H. & Henneberg, G. (1957). Kritische Hinweise für die Beurteilung der Spezifität histologischer Reaktionen der Chorio-Allantois-Membran des Hühnchens im Rahmen der Virusdiagnostik. II. Mitteilung. Die Reaktionen der Chorio-Allantois-Membran auf unspecifische Reize. *Virchows Arch. path. Anat. Physiol.*, **329**, 765–93.

Wagner, R. R. (1960). Viral interference. *Bact. Rev.*, **24**, 151–66.

Wagner, R. R. (1963). Cellular resistance to viral infection, with particular reference to endogenous interferon. *Bact. Rev.*, **27**, 72–86.

Walker, N. E. (1967). Distribution of chemicals injected into fertile eggs and the effect upon apparent toxicity. *Toxicol. appl. Pharmac.*, **10**, 290–9.

Wallis, R. C. (1962). Chicken egg chorio-allantoic membrane for mosquito feeding. *Mosquito News*, **22**, 305–6.

Weiss, R. A., Friis, R. R., Katz, E. & Vogt, P. K. (1971). Induction of avian tumor viruses in normal cells by physical and chemical carcinogens. *Virology*, **46**, 920–938.

Williamson, A. P., Blattner, R. J. & Robertson, G. G. (1953). Factors influencing the production of developmental defects in the chick embryo following infection with Newcastle disease virus. *J. Immun.*, **71**, 207–13.

Wyler, R. & Van Tongeren, H. A. E. (1957). Reactions of the chorio-allantoic membrane of the developing chick embryo to inoculation with various sterile solutions, dispersion media, suspensions and some antibiotics. *J. Path. Bact.*, **74**, 275–9.

Yates, V. J. & Fry, D. E. (1957). Observations on a chicken-embryo-lethal-orphan (CELO) virus. *Am. J. vet. Res.*, **18**, 687–90.

Yoshino, K. (1967). One-day egg culture of animal viruses with special reference to the production of anti-rabies vaccine. *Jap. J. med. Sci. Biol.*, **20**, 111–25.

Yoshpe-Purer, Y., Fendrich, J. & Davies, A. M. (1953). Estimation of the blood volumes of embryonated hen eggs at different ages. *Am. J. Physiol.*, **175**, 178–80.

Zwilling, E. (1959). A modified chorio-allantoic grafting procedure. *Transplantn Bull.*, **6**, 115–6.

34 The Japanese Quail

D. M. COOPER

Introduction

In Japan, the Japanese quail (*Coturnix coturnix japonica*) was domesticated as early as the 12th century and kept mainly for song, but by the beginning of this century was also being bred for meat and eggs. Breeding stock was selected for factors such as plumage colour, body size and egg production. During the Second World War many of the lines were lost and the song lines completely disappeared.

The species was introduced unsuccessfully into the United States as potential game birds on three occasions—the late 1870's, the 1920's and again in 1955 (Wetherbee, 1961; Stanford, 1957; Labinsky, 1961). They were introduced into Great Britain in the late 1940's and commercial breeders have selected lines for egg production and/or body size.

The use of the Japanese quail as a laboratory animal was suggested by Padgett & Ivey (1959), Wilson *et al.* (1959) and others. Since that time they have been used in such fields of research as avian diseases, behaviour, embryology, endocrinology, genetics, physiology, reproductive physiology, pharmacological and toxicological research, tumour studies and germ-free studies. A list of laboratories together with fields of research in which quail are being used has appeared in the Quail Quarterly (1964, 1966).

In the United States some of the oldest established quail lines for research are at the Universities of Alabama and California, and in the UK lines have been established for some time at the Institute of Animal Genetics, Edinburgh, Houghton Poultry Research Station, the Department of Experimental Psychology, University of Cambridge, and the National Institute for Research in Dairying, Shinfield, Reading.

Advantages and limitations

The advantages of the Japanese quail as a laboratory animal include an incubation period of 16/17 days, early sexual maturity with egg-production commencing at approximately 42 days, a high egg-yield, and a natural hardiness once the chicks have passed the brooding stage. They may be kept with ease in laboratory animal accommodation or in rooms, and are inexpensive to maintain. The main disadvantage appears to be the wide variation in fertility and hatchability, and there is also some variation in incubation time. This is rather surprising considering that some strains were started with relatively few birds and in many instances a closed-flock policy has been maintained.

Standard biological data

To utilise quail for research purposes base lines of biological, biochemical and nutritional data must be established from which an evaluation can be made of experimental deviations from the normal. For this reason it is well to request information from the supplier as to the genetic and other background of the birds being used. A very useful list of physiological values has recently been compiled by Woodard *et al.* (1973). Because the Japanese quail is a relatively novel animal in biological research data have not always been published, but experience has shown that suppliers are very willing to provide all they have available.

The Japanese quail was suggested very early on as a valuable tool in the study of poultry genetics since, as the first egg is laid at 42 days, it would be possible to produce four or more generations per year. Results have not been equal to the original expectations since the most serious problem of inbreeding programmes has been the loss of reproductive fitness in sib-matings and the work of Sittman & Abplanalp (1965) has shown that the Japanese quail is two to four times as sensitive to inbreeding as are chickens. Thus any inbreeding experiments would have to be conducted with larger populations than are necessary for comparable studies with chickens if significant inbreeding effects are to be avoided (Abplanalp, 1967). A control population has been set up for the study of stability of genetic characteristics (Marks, 1967a).

Several deleterious genes have been identified. Hill *et al.* (1963) have identified an autosomal recessive gene which results in general micromelia in the embryo. Early embryonic chondrodystrophy has been shown by Collins *et al.* (1968) to be due to a single autosomal recessive gene which is lethal at 8–10 days' incubation. Albinism, in which the adult plumage is light buff and the eggs laid show normal pigmentation of shell and yolk, and which is due to a sex-linked recessive gene, has been described by Lauber (1964). Hatchability is similar to that of normal quail but many of the albinos die at 3–4 days of age.

Normal embryology of the Japanese quail has been described by Padgett & Ivey (1960) and comparisons made with the normal development of the chick, pheasant and bobwhite quail. An incubation procedure developed by Abbott & Craig (1960) involved the eggs being set in a standard forced-draught incubator and transferred to a still-air machine on the morning of the 16th day of incubation. Using this procedure with the University of California strain of quail the total time required for incubation was $393 \cdot 6 \pm 1 \cdot 2$ hours.

Various aspects of haematology have been studied by Atwal *et al.* (1964), Atwal & McFarland (1966) and Shellenberger *et al.* (1965). Blood pressure has been shown to be similar for both males and females (Ringer, 1968) in contrast to the chicken and turkey, in which the male has a blood pressure some 15–20 per cent higher than the female (Sturkie *et al.*, 1953; Ringer & Rood, 1959).

Although it has been suggested that Japanese quail are sensitive to temperature until well grown, the work of Freeman (1967) has shown that though the critical environmental temperature for birds one week of age is 35°C, this is reduced to 31°C at two weeks and to 23°C at three weeks of age. By the time the birds are five weeks old the critical environmental temperature of individuals has been reduced to below 19°C. For groups these figures would be reduced by approximately 2°C (personal communication).

The mean body-weight of the day-old quail is 6–7 g. There is a marked increase in body-weight by the time the chicks are 7 days old (as much as 3–6 times) and weights are similar for both males and females until the birds approach sexual maturity. At this time there is an increasing body-weight for females due primarily to heavier gonads, liver and intestines (Wilson *et al.*, 1961).

The Japanese quail is a prolific egg-layer and some birds lay as many as 300 eggs in the first year of production (Wilson *et al.*, 1961). In a flock which has been closed for 34 generations the average egg-production over a 19-week production period was 70 per cent (Cooper, unpublished). Average egg weight is between 9 and 10 g.

In a study of egg yolk weights in relation to the position of the egg sequence Woodard & Wilson (1963) found that, unlike the chicken, succeeding follicles in egg sequences are not apparently ovulated at progressively earlier stages of development. Shell weight of intrasequence eggs was significantly greater than that of initial eggs, and the first egg in a sequence was smaller than succeeding eggs.

There is a considerable variation in shell colour, ranging from white, flesh-tinted or light brown, but generally speckled blue and/or brown and these shell pigments were found to be ooporphyrin and biliverdin (Poole, 1965). This pigment is laid down about 3·5 hours before oviposition (Woodard & Mather, 1964). It presents a problem if the eggs are to be used for virus inoculation, but the pigment can be easily removed prior to incubation with a steel-wool soap-pad (Rauscher *et al.*, 1962). Size, shape and colour-pattern of eggs have been suggested as a means of identifying hens when the mating system involves several females mated to a single male (Jones *ei al.*, 1964) for, as yet, an efficient trap-nest system for quail has not been devised.

Husbandry

Housing and caging

Japanese quail may be kept in several ways: small laboratory units such as mouse and rat cages can be adapted providing they are of the opaque type so that the birds do not see their neighbours (Spivey Fox, 1963); small colony

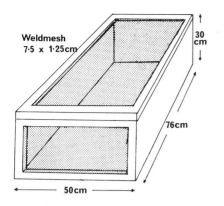

Fig. 34.1. Small unit for 4 male and 10 female adult quail.

units can be constructed of wire mesh or combination of chipboard and wire mesh; (Fig. 34.1); wire mesh battery cages can be constructed (Fig. 34.2); chick battery brooders can be adapted to their requirements, or they may be kept in rooms on deep litter in a similar manner to that used for chickens. Quail brooders, colony and individual cages are obtainable from a commercial manufacturer (GQF Manufacturing Co., PO Box 152, Savannah, Georgia, USA). Other cages which may be constructed have been described by Woodard *et al.* (1973).

Perhaps the simplest method is that of the room with deep litter system, when some 150 quail chicks may be retained in a metal, plastic or cardboard surround 105 cm diameter and provided with a 250 W infra-red lamp. This system has been found to be most successful in terms of lower mortality with the average mortality to three weeks of age being about 10 per cent. One of the most frequent causes of death during this period is from drowning, despite precautions taken with the provision of pebbles in the water founts. The room system has the advantage over the cage system in that the birds are more docile, scalping does not occur, and in breeding pens bullying of the males by the females is absent.

When quail are kept in colony cages a difficulty is the provision of adequate heat without overheating certain positions in the cage. Mortality is higher, and the birds become more excited and fly upward hitting their heads against the lid of the cage. Bullying of the males by females is commonplace and may occur to

such an extent that the male becomes incapable of mating.

Debeaking has been suggested as a method of reducing the antagonistic behaviour exhibited when mating groups are established. This consists of removing approximately one-half of the premaxilla and Marks (1967b) and Mahn & Blackwell (1968) found that it reduced the degree of antagonism without affecting fertility and hatchability.

Overcrowding is to be avoided for it not only results in an increased number of hair cracks and dirty eggs but in poorly feathered birds. Ernst & Coleman (1966) found that densities of 43 to 215 birds per m² did not affect growth rate significantly but when birds were housed at densities exceeding 86 per m² the incidence of hair cracks and dirty eggs increased. The highest levels of fertility and hatchability of fertile eggs were obtained when the birds were housed at 46/m², the lowest density used in these experiments.

Food and water utensils

Varied types of utensils can be used for feeders, small glass, china or plastic pots and shallow

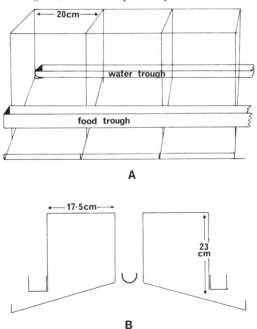

Fig. 34.2. Battery cages for individual quail
(a) side-view
(b) cross-section.

metal trays all being suitable, and wastage can be reduced by fitting wire mesh over the food. One of the cheapest and most efficient feeders for large numbers of day-old quail is a new fibre egg tray (a Keyes tray). When it is remembered that the 7-day body-weight is 3–6 times that of day-old quail the necessity for providing adequate feeding space becomes obvious. From 7 days onward quail chicks can cope with a trough-type chick feeder covered with wire mesh to prevent wastage. The amount of trough *space* rather than volume is the essential factor in ensuring a satisfactory growth rate and low mortality from starve-outs.

In small units, plastic water-drinkers similar to those used for cage birds (Fig. 34.3) are satisfactory. In battery cages, a small V-shaped trough can be fitted at the back of the cage. In floor pens, a 0·5-litre chick fount is suitable provided that sterilized pebbles are placed in the fount to prevent drowning. Nipple drinkers can be used to supply water but the siting of the drinker must be such that it is directly over

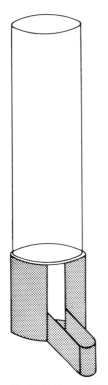

Fig. 34.3. Cage bird drinker.

the head of the bird and not set to the side or rear of the cage.

Litter

White-wood shavings are suitable for litter but care is needed to ensure that the material is free from pieces of wire, nails and staples since quail have an inclination to seek out these pieces with resulting perforation of the gizzard.

Environmental conditions

A room temperature of 27–29°C is required for quail chicks together with additional heat from an infra-red lamp or from a small gas brooder. The temperature should be decreased by 2°C per week but, as with all animals, observation as to the comfort of the bird under the particular environment provided must be considered and it will be necessary to regulate the heat according to environmental conditions. The temperature should be reduced until an ambient temperature of 21–22°C is reached by the time the birds are 5–6 weeks old. This temperature also appears to be optimum for maximum egg-production. There is no information available on optimum humidity but quail seem to do well under a wide range of humidities.

A reasonable ventilation rate should be allowed for since quail have a characteristic odour, and unless there are sufficient air changes this can be objectionable.

Lighting

Abplanalp (1961) has shown that the amount of light influences egg production as it does in chickens, and that this can be changed by reversing the lighting pattern. A 14-hour day will induce Japanese quail to lay at a high rate at 7–8 weeks of age, and earlier sexual maturity can be obtained by increasing the day length. Similarly sexual maturity can be delayed by subjecting the birds to a shorter day-length (Wilson *et al.*, 1962). Recently there have been interesting papers by Woodard *et al.* (1968, 1969) on the effect of wavelength on growth and sexual maturity. Growth to five weeks of age was found to be lower in females reared under green (5,500 Å) or blue (4,500 Å) light with an intensity of 10 lux or more than when reared under red (6,500 Å) or white light. Sexual maturity could be hastened when birds were brooded under red light with high or low intensity, and in males brooded under red light testes development was two and three times greater than in males brooded under green and blue light respectively. Egg-production in females reared under red light occurred some two weeks earlier than in those reared under blue and green light, and they also maintained a higher rate of lay. In addition, fertility in hens kept under blue light was significantly lower than in hens kept under red, green or white light.

Wilson & Huang (1962) found that hens subjected to a 14-hour day in which the time period was 05.00 to 19.00 hours laid some 80 per cent of the eggs in the afternoon, mostly after 16.00 hours. With our strain of Japanese quail, using a 14-hour day with a time period of 06.00 to 20.00 hours and an intensity of 20–30 lux, some 60 per cent of the eggs were laid before 12.30 hours, with the majority of these being laid by 09.00 hours. The birds had been maintained on this light pattern from one day of age.

Identification

The male is smaller than the female (100–130 g as compared with 120–160 g). Visually the breast feathering of the male is an even 'strawberry brown' colour while the breast feathers of the female are greyish with black speckling. Sexing by feathering can be carried out at two weeks of age and at three to four weeks old the birds can be easily sexed even by inexperienced personnel. For those interested in sexing quail at an earlier age, a method involving cloacal examination has been described by Homma *et al.* (1966).

Newly-hatched quail chicks can be identified by leg-banding, using a coloured and/or a numbered canary leg-band. It is, however, essential to remove these leg-bands when the chicks are 7–10 days old, before they become

Fig. 34.4. Clincher tag.

too tight; after that age a permanent clincher-type wing band (Fig. 34.4) can be used.

Transport

Quail eggs are notoriously poor travellers and unless they can be carried as hand baggage it is preferable to arrange for a supply of four-week-old birds. By this time they are able to adjust to the ambient temperature but care should be taken that they are not transported during extremes of temperature, nor when they are likely to arrive at weekends. Cardboard chick-boxes are suitable containers.

Handling

Quail chicks are very small and require only to be picked up using thumb and forefingers. Growing and adult quail kept in pens may be driven into a catching crate or contained by a hurdle in the corner of the pen. Care must be taken to ensure that large numbers are not driven together so as to cause suffocating. When quail are kept in cages or in small pens, it is possible to pick them up by placing the hand over the back and grasping the body firmly but not tightly (Fig. 34.5).

Feeding

Only a limited amount of information on the nutritional requirements of Japanese quail has been published. The majority of workers feed their birds on turkey starter-crumbs with a crude protein content of 25–28 per cent. Howes (1965) reared quail on a corn-soya diet containing 23 per cent protein and 2,090 kcal productive energy/kg. At nine weeks of age there were no significant differences between groups fed iso-caloric diets with 18 to 36 per cent protein. Increasing the energy level to 2,530 kcal/kg resulted in a greater rate of mortality and this was attributed to the inability of quail to metabolise high concentrations of dietary fat. Methionine supplementation significantly increased egg-production but there was no response to added glycine or lysine. When chicks were fed protein levels of 16 and 20 per cent body-weights were consistently lower than in chicks fed 24 per cent protein. Egg production in birds fed 20 per cent protein was delayed only slightly but almost inhibited in birds fed 16 per

Fig. 34.5. Method of holding quail. (Photograph: Houghton Poultry Research Station.)

cent protein. Supplemental glycine, lysine and methionine in the 16 per cent ration enhanced egg-production. Weber & Reid (1967) studied the protein requirement of quail to five weeks of age using a soya-bean meal diet with protein levels from 13 to 35 per cent at approximately 5 per cent increments. Their results indicate that the protein requirement of a soya-bean meal diet supplemented with methionine is 24 per cent. The calorie (productive)/protein ratio of 36–38 was found to be adequate, in that this c/p ratio resulted in most efficient protein retention, food conversion and body-weight gain. Recent work by Svacha et al. (1969) suggests that when a 26-per-cent protein diet is fed, the lysine requirement for chicks to 3 weeks of age is 1·37 per cent, methionine plus cystine 0·69 per cent and glycine 1·74 per cent. Requirements for birds over 3 weeks of age is 1·20, 0·67 and 1·17 per cent respectively.

Shellenberger & Lee (1966) found that birds fed high and low-energy diets containing 3,300 USP units vitamin A/kg reproduced normally. There would appear to be a sex difference in their vitamin D requirement (Chang & McGinnis, 1967), since mature male quail fed a vitamin-D-deficient diet for a year were still in good condition whilst the majority of females died.

Spivey Fox & Harrison (1964) used Japanese quail in a study of zinc deficiency and found that at four weeks of age the chicks were severely affected by zinc deficiency and showed slow growth, abnormal feathering, incoordinated gait, laboured respiration and had a low ash content of the tibia. Controls were fed zinc at 90 mg/kg. Thus it is essential to ensure that sufficient zinc is present in the ration if birds are to be fed and watered from plastic, glass or stainless steel containers.

Calcium and phosphorous requirements have been studied by Nelson et al. (1964) using a basal diet containing 25 per cent protein. Egg-production decreased after 8 weeks in quail fed 1·0, 1·5 and 2·0 per cent calcium but reached a peak at 90 per cent production and maintained a high level for 13–21 weeks when the concentration was increased to 2·5 and 3·0 per cent. There was little difference in fertility and embryonic mortality, but groups fed 2·5 or 3·9 per cent calcium and 0·8 per cent phosphorus tended to perform better. Spivey Fox et al. (1966), feeding purified diets, have shown that quail require 15 mg/kg diet of calcium pantothenate to overcome mortality and slow growth and 30 mg/kg to prevent abnormal feathering. If, however, chicks were fed 40 mg/kg calcium pantothenate for one week they then required only 10 mg/kg for normal development to four weeks of age.

Gough *et al.* (1968) compared two purified diets with a commercial game-bird starter and found that the body-weights of birds fed on purified diets were significantly lower, their normal activity reduced, feather-development delayed, feed consumption markedly reduced and mortality rate doubled. Furthermore the eggs laid were smaller and egg-production and hatchability significantly lower.

Calvert (1965) reported that diets deficient in linoleic acid reduced egg-production, mean egg-weight and percentage fertility. These differences are similar to those occurring in chickens suffering from linoleic acid deficiency.

A summary of the nutrient requirements of Japanese quail, together with the composition of some commercial-type and purified diets, has been published by Woodard *et al.* (1973).

Breeding

The normal mating ratio is three females to one male but for certain studies birds may be paired. Woodard & Abplanalp (1967) compared sex ratios of from 1:1 to 1:6 with a total number of birds in colony units ranging from 20 to 24. Highest fertility was obtained with a ratio of 1:1 or 1:2 and the duration of fertility after removal of the male from the unit was found to be 9–10 days. Fertility was reduced in both males and females aged six months or more and hatchability of fertile eggs was reduced in old females but not in young females which were mated to ageing males. It was noted that young females mated more frequently than older females and young males completed twice as many matings as old males.

Artificial insemination using a modification of the massage technique for collection of semen was found to be applicable to Japanese quail (Wentworth & Mellen, 1963) but insemination was carried out by depositing semen by means of a hypodermic needle into the uterus (shell gland). This procedure resulted in 75 per cent of hens being fertilized, fertile eggs being produced for an average of 4·6 days. The procedure

did not affect egg-production and caused no mortality. Marks & Lepore (1965) reported a simple intravaginal technique for insemination, and by using this technique were able to fertilize 83 per cent of the females. There was no subsequent depression of egg-production nor was there any mortality.

Incubation Eggs for incubation should be selected and handled with care as the shells are thin and easily broken. They should be clean and of uniform size and should have been stored at 13–15°C for not more than seven days.

Quail eggs can be incubated in most commercial incubators; those with wire-mesh setting trays need no modification, but incubators with metal trays require wire inserts made from 6·25-mm mesh into which the eggs can be placed. Non-absorbent cotton-wool is used to hold the eggs firmly in the setting trays. Hatching can be carried out either in the hatching compartment or in small observation incubators.

Hatching time varies between strains; it generally occurs after 16–17 days, but some inbred strains may take up to 18 days to hatch. There are two peaks of embryonic mortality—at three days and again just before hatching. Eggs can be candled when they are transferred from setter to hatcher (on the fourteenth day), but better hatching results were obtained in this Institute by transferring all the eggs at this time, omitting the candling, and breaking out those eggs which did not hatch at the completion of the incubation period. It is presumed that this was due to a consequent reduction in the cooling which occurs when eggs are being transferred.

The age of the parent stock markedly affects hatchability, maximum results being obtained when the birds are 8–20 weeks of age.

Laboratory procedures

Blood samples are often required for various purposes and the most efficient method is that of cardiac puncture, the technique being similar to that used for the chicken. For bleeding quail chicks of 2–3 weeks old a No. 12 hypodermic needle is recommended; on average 1·5 ml can be obtained without deleterious effects. For adults a No. 4 serum needle is most useful; a volume of 3 ml or more can be obtained without causing death. A volume of 5–7 ml can be taken from adult birds if they are to be killed immediately the blood is withdrawn (Freeman, personal communication).

Euthanasia

Quail may be killed by dislocation of the vertebrae or by carbon dioxide. If the latter is used the birds are placed in a wire cage which is put into a plastic bag filled with carbon dioxide. The bag is then topped up with gas by introducing it via a rubber tube from a CO_2 cylinder. When the bag has been filled the tube is withdrawn and the neck tied. Birds must be exposed to the gas for at least five minutes to ensure death. (See also Chapter 14).

Diseases

Japanese quail have been shown to be susceptible to many of the diseases affecting poultry. Edgar *et al.* (1964) found them to be susceptible to fowl pox, Newcastle disease (B–1 and Grumbles-Boney), infectious bronchitis, several strains of *Salmonella*, *Pasteurella multocida*, one pathogenic strain of *Escherichia coli*, *Aspergillus fumigatus*, *Trichomonas* spp., *Histomonas meleagridis*. They were also susceptible to *Eimeria dispersa* but not to eight of the nine known species of chicken coccidia nor the three most pathogenic species of turkey coccidia. Infections of *Capillaria obsignata* and *Heterakis gallinae* occur occasionally but *Ascaridia galli* is unable to mature in the quail. The first report of aspergillosis in a colony of Japanese quail is by Olson (1969). Larson & Hansen (1966) and Lund & Ellis (1967) infected Japanese quail with *Heterakis gallinarum* and *Histomonas meleagridis* but neither parasite thrived. Hill & Raymond (1962) reported a natural infection of avian encephalomyelitis in Japanese quail hens held in a room with chickens which had been inoculated with IAE virus. Rauscher *et al.* (1962) found that a number of viruses could be grown in the quail chick embryo. These included influenza A, B, C and D, mumps, Newcastle disease, infectious bronchitis, laryngotracheitis, fowl pox, vaccinia, vesicular stomatitis and Rous sarcoma (Bryan strain). Quail bronchitis virus would appear to be similar if not identical to chicken embryo lethal orphan (CELO) virus since DuBose & Grumbles (1959) were unable to differentiate

between the two. Dutta & Pomeroy (1967), using electron microscopy, found two types of particles present, the large particles (70–75 mμ in diameter) being structurally similar to CELO virus. Ultrastructure of the small particles (20–25 mμ in diameter) was not clear.

When Rous sarcoma virus (Bryan strain) was injected into 24-hour-old chicks they developed tumours within 6–8 days (Reyniers & Sacksteder, 1960). Shipman & Levine (1966) produced tumours in quail by inoculating Rous sarcoma virus (Schmidt-Ruppin strain) into the wing web. There appeared to be a selective response to various strains of Rous sarcoma viruses as although they could initiate tumours with Bryan high-titre, Carr-Zilborn, Harris, Prague, Schmidt-Ruppin and '29' strains, only the Bryan high-titre and Harris strains could be serially passaged with cell-free tumour extracts.

From the information which is available on susceptibility to avian disease it is apparent that procedures for management and hygiene suitable for chickens can equally be applied to Japanese quail.

REFERENCES

Abbott, U. K. & Craig, R. M. (1960). Observations on hatching time in three avian species. *Poult. Sci.*, **39**, 827–30.

Abplanalp, H. (1961). Response of Japanese quail to restricted lighting. *Nature, Lond.*, **189**, 942–3.

Abplanalp, H. (1967). Genetic studies with Japanese quail. *Zuchter/Genet. Breed. Res.*, **37**, 99–104.

Atwal, O. S., McFarland, L. Z. & Wilson, W. O. (1964). Hematology of Coturnix from birth to maturity. *Poult. Sci.*, **43**, 1392–401.

Atwal, O. S. & McFarland, L. Z. (1966). A morphologic and cytochemical study of erythrocytes and leucocytes of *Coturnix coturnix japonica*. *Am. J. vet. Res.*, **27**, 1059–65.

Calvert, C. C. (1965). The performance of female Japanese quail on linoleic acid deficient diets. *Poult. Sci.*, **44**, 1358 abs.

Chang, S. I. & McGinnis, J. (1967). Vitamin D deficiency in adult quail and chickens and effects of estrogen and testosterone treatments. *Proc. Soc. exp. Biol. Med.*, **124**, 1131–5.

Collins, W. M., Abplanalp, H. & Yoshida, S. (1968). Early embryonic chrondrodystrophy in Japanese quail. *J. Hered.*, **59**, 248–50.

DuBose, R. T. & Grumbles, L. C. (1959). The relationship between quail bronchitis virus and chicken embryo lethal orphan virus. *Avian Dis.*, **3**, 321–44.

Dutta, S. K. & Pomeroy, B. S. (1967). Electron microscopic studies of quail bronchitis virus. *Am. J. vet. Res.*, **28**, 296–9.

Edgar, S. A., Waggoner, R. & Flanagan, C. (1964). Susceptibility of Coturnix quail to certain disease producing agents common to poultry. *Poult. Sci.*, **43**, 1315 abs.

Ernst, R. A. & Coleman, T. H. (1966). The influence of floor space on growth, egg production, fertility and hatchability of the *Coturnix coturnix japonica*. *Poult. Sci.*, **45**, 437–40.

Freeman, B. M. (1967). Oxygen consumption by the Japanese quail (*Coturnix coturnix japonica*). *Br. Poult. Sci.*, **8**, 147–52.

Freeman, B. M. (1969). Personal communication.

Gough, B. J. Shellenberger, T. E. & Esuriex, L. A. (1968). Responses of Japanese quail fed purified diets. *Quail Quart.*, **5**, 2–6.

Hill, R. W. & Raymond, R. G. (1962). Apparent natural infection of Coturnix quail hens with the virus of avian encephalomyelitis—case report. *Avian Dis.*, **6**, 226–7.

Hill, W. G., Lloyd, G. L. & Abplanalp, H. (1963). Micromelia in Japanese quail. An embryonic mutant. *J. Hered.*, **54**, 188–90.

Homma, K., Siopes, T. D., Wilson, W. O. & McFarland, L. Z. (1966). Identification of sex of day-old quail (*Coturnix coturnix japonica*) by cloacal examination. *Poult. Sci.*, **45**, 469–72.

Howes, J. R. (1965). Effects of season on ovulation rate and egg composition of Coturnix quail. Paper presented at the 61st Ann. Conven. Ass. Southern Agricultural Workers. Poultry Section, Atlanta, Ga. Feb. 1964. Cited in *Quail Quart.*, **2**, 24–5.

Jones, J. M., Maloney, M. A. & Gilbreath, J. C. (1964). Size, shape and color pattern as criteria for identifying Coturnix eggs. *Poult. Sci.*, **43**, 1292–4.

Labinsky, R. F. (1961). Reports of attempts to establish Japanese quail in Illinois. *J. Wildl. Mgmt*, **25**, 290–5. Quoted in Shellenberger, T. E. (1968). *Lab. Anim. Care*, **18**, 244–50.

Lauber, J. K. (1964). Sex-linked albinism in the Japanese quail. *Science, N.Y.*, **146**, 948–50.

Larson, I. W. & Hansen, M. F. (1966). Susceptibility of Japanese quail (*Coturnix coturnix japonica*) to *Heterakis gallinarum* and *Histomonas meleagridis*. *Poult. Sci.*, **45**, 1430–2.

Lund, E. E. & Ellis, D. J. (1967). The Japanese quail, *Coturnix coturnix japonica*, as a host for Heterakis and Histomonas. *Lab. Anim. Care*, **17**, 110–13.

Mahn, S. & Blackwell, R. L. (1968). Debeaking in Japanese quail. *Wld's Poult. Sci. J.*, **24**, 58.

Marks, H. L. & Lepore, P. D. (1965). A procedure for artificial insemination of Japanese quail. *Poult.*

Sci., **44**, 1001–3.

Marks, H. L. (1967). A Japanese quail control population. *Quail Quart.*, **4**, 2–4.

Marks, H. L. (1967). The effect of debeaking on fertility of Japanese quail. *Quail Quart.*, **4**, 43–5.

Nelson, F. E., Lauber, J. K. & Mirosh, L. (1964). Calcium and phosphorous requirement for the breeding Coturnix quail. *Poult. Sci.*, **43**, 1346 abs.

Olson, L. D. (1969). Aspergillosis in Japanese quail. *Avian Dis.*, **13**, 225–7.

Padgett, C. S. & Ivey, W. D. (1959). Coturnix quail as a laboratory research animal. *Science, N.Y.*, **129**, 267–8.

Padgett, C. S. & Ivey, W. D. (1960). The normal embryology of the Coturnix quail. *Anat. Rec.*, **137**, 1–11.

Poole, H. K. (1965). Spectrophotometric identification of egg shell pigments and timing of superficial pigment deposition in the Japanese quail. *Proc. Soc. exp. Biol. Med.*, **119**, 547–551.

Quail Quart. (1964). **1**, 2–6.

Quail Quart. (1966). **3**, 2–10.

Rauscher, F. J., Reyniers, J. A. & Sacksteder, M. R. (1962). Japanese quail egg embryo as a host for viruses. *J. Bact.*, **84**, 1134–9.

Reyniers, J. A. & Sacksteder, M. R. (1960). Raising Japanese quail under germ-free and conventional conditions and their use in cancer research. *J. Nat. Cancer Inst.*, **24**, 1405–21.

Ringer, R. K. & Rood, K. (1959). Hemodynamic changes associated with aging in the broad breasted bronze turkey. *Poult. Sci.*, **38**, 395–7.

Ringer, R. K. (1968). Blood pressure of Japanese and Bobwhite quail. *Poult. Sci.*, **47**, 1602–4.

Shellenberger, T. E., Adams, R. F., Virgin, H. & Newell, G. W. (1965). Erythrocyte and leukocyte evaluations of Coturnix quail. *Poult. Sci.*, **44**, 1334–5.

Shellenberger, T. E. & Lee, J. M. (1966). Effect of vitamin A on growth, egg production and reproduction of Japanese quail. *Poult. Sci.*, **45**, 708–13.

Shipman, C. & Levine, A. S. (1966). Selective response of the Japanese quail to various strains of Rous sarcoma virus. *J. Bact.*, **92**, 161–3.

Sittman, K. & Abplanalp, H. (1965). Inbreeding depression in Japanese quail. *Genetics, Princeton*, **52**, 475–6.

Spivey Fox, M. R. (1963). Care of the Japanese quail (*Coturnix coturnix japonica*) in the laboratory. Personal communication.

Spivey Fox, M. R. & Harrison, B. H. (1964). Use of Japanese quail for the study of zinc deficiency. *Proc. Soc. exp. Biol. Med.*, **116**, 256–9.

Spivey Fox, M. R., Hudson, G. A. & Hintz, M. E. (1966). Pantothenic acid requirements of young Japanese quail. *Wld's Poult. Sci. J.*, **22**, 346 abs.

Stanford, J. A. (1957). A progress report of *Coturnix*

quail investigations in Missouri. *Proc. 22nd Am. Wildl. Conf.*, pp. 316–59. Quoted in Shellenberger, T. E. (1968). *Lab. Anim. Care*, **18,** 244–50.

Sturkie, P. D., Weiss, H. S. & Ringer, R. K. (1953). The effects of age on blood pressure in the fowl. *Am. J. Physiol.*, **174,** 405–7.

Svacha, A. J., Weber, C. W. & Reid, B. L. (1969). Lysine, methionine and glycine requirements of growing Coturnix quail. *Poult. Sci.*, **48,** 1881 abs.

Weber, C. W. & Reid, B. L. (1967). Protein requirements of Coturnix quail to five weeks of age. *Poult. Sci.*, **46,** 1190–4.

Wentworth, B. C. & Mellen, W. J. (1963). Egg production and fertility following various methods of insemination in Japanese quail (*Coturnix coturnix japonica*). *J. Reprod. Fert.*, **6,** 215–20.

Wetherbee, D. K. (1961). Investigations in the life history of the common *Coturnix*. *Am. Midl. Nat.*, **65,** 168–86. Quoted in Shellenberger, T. E. (1968). *Lab. Anim. Care*, **18,** 244–50.

Wilson, W. O., Abbott, U. K. & Abplanalp, H. (1959). Developmental and physiological studies with a new pilot animal for poultry—Coturnix quail. *Poult. Sci.*, **38,** 1260–1.

Wilson, W. O., Abbott, U. K. & Abplanalp, H. (1961). Evaluation of Coturnix (Japanese quail) as pilot animal for poultry. *Poult. Sci.*, **40,** 651–7.

Wilson, W. O., Abplanalp, H. & Arrington, L. (1962). Sexual development of Coturnix as affected by changes in photoperiods. *Poult. Sci.*, **41,** 17–22.

Wilson, W. O. & Huang, R. H. (1962). A comparison of the time of ovipositing for Coturnix and chicken. *Poult. Sci.*, **41,** 1843–5.

Woodard, A. E. & Wilson, W. O. (1963). Egg and yolk weight of Coturnix quail (*Coturnix coturnix japonica*) in relation to position in egg sequences. *Poult. Sci.*, **42,** 544–5.

Woodard, A. E. & Mather, F. B. (1964). The time of ovulation, movement of the ovum through the oviduct, pigmentation and shell deposition in Japanese quail. *Poult. Sci.*, **43,** 1427–32.

Woodard, A. E. & Abplanalp, H. (1967). The effects of mating ratio and age in fertility and hatchability in Japanese quail. *Poult. Sci.*, **46,** 383–8.

Woodard, A. E., Abplanalp, H., Wilson, W. O. & Vohra, P. (1973). *Japanese Quail Husbandry in the Laboratory (Coturnix coturnix japonica)*. Dept. of Avian Sciences, University of California, Davis, CA95616. 22 pp.

Woodard, A. E., Moore, J. A. & Wilson, W. O. (1968). Effect of wave length of light on growth and reproduction in Japanese quail (*Coturnix coturnix japonica*). *Poult. Sci.*, **47,** 1733–4 abs.

Woodard, A. E., Moore, J. A. & Wilson, W. O. (1969). Effect of wave length of light on growth and reproduction in Japanese quail (*Coturnix coturnix japonica*). *Poult. Sci.*, **48,** 118–23.

35 The Canary

I. F. KEYMER

General biology

Taxonomic position

The canary (*Serinus canaria*) is placed in the class Aves, order Passeriformes (passerine or perching birds), family Fringillidae.

Origin

It is believed that the canary has been kept in captivity since the sixteenth century and is the domesticated form of the serin or wild canary, which occurs in the Canary Islands, Madeira and the Azores and was until recently regarded as the subspecies (*Serinus canarius canarius*). It is now generally accepted, however, that this bird should be a species in its own right, i.e. *S. canaria*.

Size ranges

The domesticated canary is generally larger than the wild bird and varies in size according to the breed (see below).

Behaviour

Generally speaking, if birds are correctly housed and fed, normal behaviour patterns will be maintained. Abnormal behaviour can be caused by overcrowding or isolation, especially of healthy mature birds, or by disease. Those interested in pursuing the study of avian ethology in more detail should read the paper by Shoemaker (1939) and the relevant sections in *A New Dictionary of Birds* (ed. Landsborough Thomson, 1964) as well as the book by Hinde (1966); see also page 479 for breeding behaviour.

Breed and types

Most varieties are bred for characteristics such as colour (of which there are several), shape and plumage variations, but Roller canaries are kept purely for their song. Popular and hardy breeds are the Border, Yorkshire and Norwich. The Border does not exceed 14 cm in length but the Yorkshire and Norwich are larger and are preferable for most research projects, weighing approximately 24 g, whilst the smaller breeds average 14 g (Blackmore, personal communication). The weight range of the species is about 12 to 29 g.

Many canary fanciers cross their birds with other finches especially the goldfinch (*Carduelis carduelis*), greenfinch (*C. chloris*), bullfinch (*Pyrrhula pyrrhula*) and linnet (*Acanthis cannabina*) and these hybrids, which are usually sterile (Gray, 1958) are called mules.

Uses

Although canaries are not widely kept in laboratories they have been used for many years in the study of malaria and occasionally other protozoal infections. They are also used in virological studies of pox and for toxicity tests on therapeutic substances.

Standard biological data

Knowledge of the physiology of small passerines is very incomplete and the data regarding physiological normals quoted by various workers often vary considerably. A good general review of the subject is provided by Farner (1969).

Body temperature

The cloacal temperature is stated by Gelineo (1955) to be 41–42°C.

Cardio-vascular system

The number of erythrocytes in millions per ml of blood has been given as 4·5 to 5·0 (Ben-Harel, 1923), 4·06 (Yamamoto, 1959). 3·5 (Worden, 1956), and 4·516 (Young, 1937). Worden (1956) states that 100 ml of blood contains 10·4 g of haemoglobin whilst Young (1937) gives a figure of 9·5 g. The different types of blood cells are described and illustrated by Hewitt (1940). Yamamoto (1959) stated that the total white cell count is 10·9 thousands per mm³ (lymphocytes 53·1 per cent; monocytes 8·0; heterophils 21·6; eosinophils 2·4 and basophils 14·9) and thrombocytes 37·4 thousands per mm³. He also gave the haemoglobin (Sahli°) figure of 80·5. Woodbury & Hamilton (1937) found the average heart-rate of four canaries to be 690 beats per minute and state that the average systolic blood-pressure is 220 mm of mercury and the diastolic 154 mm. Arnall (1969) stated that the heart-rate varies from 560 to 1,000 beats per minute.

Respiratory system

The respiratory rate is given as 96 to 120 per minute by Groebbels (1932) and 96 to 144 by Arnall (1969).

Reproductive system

Clutch size may vary from one to six eggs, but normally three are laid (Hinde, 1959); Romanoff & Romanoff (1949) stated that a hen will lay a maximum of 60 eggs per annum if they are continually removed. The incubation period is approximately 14 days. The eggs are pale bluish-green in ground colour and have pale reddish or brown spots which are confined chiefly to the larger end where they sometimes form a fairly distinct band. Witherby *et al.* (1940) state that the average measurements of 100 serin (*S. canaria*) eggs were $16·17 \times 11·86$ mm.

The effect of oestrogen and progesterone on nest-building activities has been studied by Warren & Hinde (1959).

Special senses

According to Brand & Kellogg (1939) the frequency range of sounds audible to the canary is 1,100 to 10,000 cycles per second.

Longevity

The average length of life is approximately 5 to 6 years, although there are reports of birds living for as long as 22 years (Flower, 1938). Any age exceeding 14 years, however, is exceptional (Hasholt, 1969).

Weight

See p. 475.

Husbandry

Housing and caging

Breeding cages. Cages are available with two compartments separated by either a wire or a solid moveable partition. Suitable dimensions for a double breeding-cage with wire front and unpainted metal or wooden sides are: length 63 to 100 cm, depth 30 cm, and height 45 cm although smaller cages are frequently used. Sliding doors are preferable to those which open outwards as birds are less likely to escape through these.

A sliding sand tray on the floor will facilitate cleaning without disturbing the birds, and if wooden cages are used all crevices and cracks in the wood and corners should be sealed. In order to avoid accumulation of dust, excreta and seed husks, cages must not be placed on the top of each other, but on tubular shelving.

Flight cages. Breeding-cages can be converted into flight cages by removing the partitions or, if more space is required, an aviary can be used.

Seed hoppers and water receptacles. These are made of plastic, earthenware or glass and should be in one piece to facilitate cleaning. It is customary to hang them on the outside of the cage, the birds gaining access through holes in the wire front. Nevertheless, provided that the receptacles are not placed under perches and are cleaned regularly, there is no reason why they should not be placed in the cage as is customary for budgerigars. Plastic bird baths, as they are rather large, are preferably hung on the outside, and attachments with sliding partitions in the wire front are provided on some cages. Water for bathing should be removed by midday, to prevent birds from roosting at night whilst still wet.

Egg-trays are sometimes used as food-hoppers when youngsters are being reared.

Metal clips for cuttlefish bone should be attached to the wire front of the cage. It is inadvisable to leave cuttlefish bone on the cage floor, where it can become contaminated.

Perches. It is important that perches are not slippery or insecure, for if they are coition may be unsuccessful. Corran & Edgar (1957) recommend a grooved perch fixed to the wire on the front of the cage by means of a slot and to the back of the cage by a small nail inserted in the other end which can be attached to the cage in dart fashion or by means of a hook. The sectional dimensions they recommend are 16×19 mm for large canaries such as Norwich or Yorkshire and 13×19 mm for smaller breeds, although it is advisable to vary the diameters to provide proper exercise for the leg muscles and toes. To avoid contamination by excreta, perches must never be situated near food and water receptacles.

Nesting receptacles. These are either cup-shaped earthenware vessels or shallow pans made of wood or plastic, approximately 125 mm square and 50 mm deep, with a flat base. They are hung on hooks at the back of the cage.

Environmental conditions

The bird room should be situated in a quiet position and be light, well-ventilated, draught-proof and, if necessary, mosquito-proof. Ideally the temperature should be thermostatically controlled between 10 and 16°C.

Identification

Ringing or banding the legs is the only satis-factory method of marking small birds. Coloured plastic split rings are obtainable together with an expanding-tool for use when placing the ring on the leg (Cornwallis & Smith, 1960). The bird should be held as described below with the thumb and index or middle finger grasping the leg at and above the hock (tibiotarsus-tarsometa-tarsal) joint. The ring is placed in the expanding-tool and opened sufficiently widely to slip easily over the shank or tarsometatarsus; it is then transferred to the leg by placing the leg in the groove of the tool and sliding the tool away. When the ring is in position the split closes.

Handling

When attempting to catch a bird it is essential to act quietly and swiftly, and in an aviary a net is indispensable. If the bird is in a cage it is advisable to remove the perches as these hinder movement and may cause injury. Care must be taken to grasp the whole bird and not one of the limbs, and it must be held sufficiently firmly to prevent struggling, the wings being held gently against the sides of the body. The neck should be situated between the first and second fingers and the back against the palm of the hand with the thumb and the third finger encasing the body (see Fig. 35.1). If it is necessary to hand the bird to another person, the neck is suspended between the first and second fingers and speedily transferred to those of the receiver. A small bird

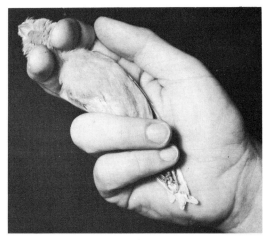

Fig. 35.1. Method of holding a canary or other small passerine bird.

such as a canary should never be restrained for longer than necessary and if for any reason it has to remain out of its cage for more than a few minutes it should be placed in a black cloth bag with a purse string aperture, from which it can easily be recaptured.

Transport

Within the laboratory. Birds may be carried in a small box or cage, or in an ordinary flight cage covered with a dark cloth. They can also be placed in a black cloth bag for short periods, but care is needed to place it in a safe position where the bird cannot be injured.

Rail. Ventilated cardboard boxes 11 cm high by 22 cm long are available. These are fitted with a wooden raised central strut a little under 12 mm wide and deep, extending across the floor from either side and with a curved upper surface; this enables the bird to obtain a secure foothold.

Road. Birds are seldom sent by public road transport, but they can be easily carried in a car. They should be left in their cage or placed in a box as recommended for rail transport. In warm weather they must not be placed in direct sun-light and a free current of air must be main-tained by leaving windows partially open.

Sea and air. For sea and air transport small sturdy cages are available which are made of wood and fitted with a canvas blind which can be pulled down over the wire front. The British Standards Institution (1961) has published a small booklet of recommendations for the carriage of small and medium-size seed-eating birds by air, and this together with the IATA Live Animal Regulations (1975) should be consulted for further details.

For long journeys, food must always be provided together with water in special non-splash receptacles with curved lips.

Feeding

Virtually no scientific information is available although the composition of some of the seeds normally fed to canaries has been studied (compare with budgerigar, p. 488). By far the best account of the nutrition of small birds is that by Tollefson (1969).

Natural foods

Canary breeders usually feed a basic diet of two to three parts by weight of canary seed (*Phalaris canariensis*) and one part of red rape (*Brassica* spp.), the former providing the bulk of the carbohydrate and the latter providing mainly fat and protein. A small proportion of canary seed is often replaced by millet (*Panicum* spp.) but its value as a substitute is debatable because, with the exception of the yellow variety, canaries often refuse to eat it. Tollefson (1969) reports the results of two free-choice feeding studies. In one experiment it was found that the percentages of seeds selected and the quantities consumed per day over a period of four months were similar for both adult non-breeding canaries and youngsters just removed from the nesting cages. Adults consumed about 3·4 g of seeds per day selected in the following proportions: 47 per cent canary seed, 27·7 rape seed, 5·6 red proso millet, 5·4 white proso millet and 14·3 per cent yellow proso millet (Bice, 1955). Different food was offered in the other free-choice feeding study, including niger (*Guizotia abyssinica*), oat groats and a larger variety of oil seeds. Niger was the most popular (26·1 per cent), followed by canary seed (16 per cent), and the birds selected a diet containing over 50 per cent of oil seeds compared with 28 in the other experiments.

In addition to canary seed, niger, rape and millet, other seeds are often fed by fanciers for special purposes, being credited (probably erroneously in some cases) with various properties. Hemp (*Cannabis sativa*) because of its high fat content, is fed mainly during the winter months and to hens which are rearing youngsters. Poppy or mawseed (*Papaver somniferum*) and teasel (*Dipsacus* spp.) are credited with tonic properties, whilst linseed (*Linum usitatissimum*) is fed to moulting birds. Niger is also given to hens when breeding as it is said to prevent egg-binding.

Fresh green-food is usually provided once or twice a week. Lettuce, watercress and Brussels sprouts are popular, but wild plants such as chickweed (*Stellaria media*), dandelion (*Taraxacum officinale*) and groundsel (*Senecio vulgaris*) are also relished. Care must be taken, however, not to feed frosted plants and those contaminated with fungicides, insecticides or herbicides. Plants from road verges should be avoided as they are seldom clean.

Insoluble grit is provided in the form of sand, and soluble grit as cuttlefish bone, oystershell or sterilized ground eggshell.

Special requirements

Fanciers feed special foods for various reasons. Many of the diets are of an empirical nature and Tollefson (1969) discussed this subject.

A fairly typical procedure used by canary-breeders for rearing purposes is to provide a small quantity of soft food to the hen at 4-hourly intervals during daylight, commencing on the thirteenth day of incubation and continuing until the youngsters are 10 days old. A popular soft food is egg-yolk made crumbly-moist by mixing it with powdered plain biscuits and wholemeal bread previously dried in the oven and then ground up. During this period hard seed is usually withdrawn, but when the young are able to feed themselves they are provided with cracked hemp and canary seed, and soaked teasel. The adult diet of hard seed and green-food is gradually introduced when the youngsters are 6 weeks old.

Proprietary rearing foods are available and save time and labour, but it is advisable to avoid those which contain dried egg products because they may harbour pathogenic *Escherichia coli* and *Salmonella* bacteria (Graham-Jones & Fiennes, 1959).

In order to enhance the orange coloration of

the birds the feeding during the moulting period of a carotenoid pigment (e.g. canthaxanthin) or proprietary colour foods containing paprika is a common practice with breeders of Norwich, Yorkshire and Lizard canaries.

Method of presenting food

(See also p. 476, seed hoppers and water receptacles.)

Insoluble grit is often scattered on the cage floor but should preferably be placed in a shallow vessel, and green-food should be hung in bundles by string from the perches or wire front.

Watering

Clean water must be provided daily, both for drinking and bathing, and the vessels regularly washed. Tollefson (1969) stated that canaries will die within about 48 hours if deprived of water. Bice (1955) found that adults drank 5·4 ml per day.

Breeding

Methods

In this country breeding usually starts in late March or early April, when the environmental temperature is in the neighbourhood of 10°C. Canaries may be mated when they are one year old and used for breeding for two or three years. Some will remain virile for much longer, but six years is usually regarded as the maximum age for breeding.

Before the breeding season the hens are housed together in flight cages and the cocks kept in single cages. During February suitable pairs are selected and placed in breeding cages, but separated from one another by a solid partition. Alternatively the cock may be placed in a cage nearby, but where it cannot be seen by the hen. By March both sexes should be in breeding condition, and the next procedure is to substitute a wire partition for the solid partition. If the cock is in a separate cage this is then placed next to the hen's cage. When the birds feed each other through the wires they are ready for pairing, and are allowed to meet. The nest box and a small quantity of nesting materials such as cow hair, small feathers, moss, felt,

meadow hay, grass or cotton-wool (all of which can be purchased) are provided. A further supply of nesting material is made available when the pair have settled down. The hen by this stage should show nest-building behaviour. At first she simply carries material in her beak and executes a limited number of stereotyped movements before starting actual nest construction. The cock plays little or no part in nest building. A little harmless fighting may occur when the pair first meet and the hen will dominate the cock, either pecking or threatening to peck him. The cock will then either move away or retaliate, chasing the hen around the cage, until she permits copulation, usually on the perch.

Egg-laying usually begins one week after pairing, although three weeks may elapse from the time the birds are put together. Eggs are laid at 24-hour intervals, and it is customary to replace them with china dummies. The eggs should be stored at room temperature (13°C) until the afternoon on which the third one is laid, when they are replaced and the dummies removed, thus ensuring that all hatch on the same day. Removal of the eggs in this way is not, however, entirely necessary and the practice is not observed by all breeders. The cock is usually allowed to remain with the hen, but if (as occasionally occurs) he interferes with her and disturbs incubation he must be separated by a wire partition or removed to another cage, which must be placed so that he is able to feed her through the wires. The cock is re-introduced when the young are 6 to 9 days old in order that he may assist with rearing. Some hens sit so closely that the young are not fed adequately, and in this case the cock must be removed so that the hen will have to leave the nest to be fed and to feed the nestlings, which is done by regurgitation of the crop contents.

It is important not to disturb the hen during incubation or rearing, especially in her first breeding season; otherwise she may desert the eggs or young. The cage floor, however, must be inspected regularly in case youngsters are accidentally dragged from the nest. This sometimes happens if the hen's claws are overgrown and become attached to the chicks. As in some other passerines, the hen removes egg shell remains and faecal sacs, although after about a week the nestlings will excrete over the edge of the nest.

Youngsters normally leave the nest at three weeks of age and are soon able to feed themselves, although at first they prefer soft food. They become independent about 36 days after hatching, when they should be transferred to flight cages. If the parents feather-peck the youngsters they must be separated from them by a wire partition.

If a second clutch is required the hen must be provided with a new nesting receptacle and nesting material when the youngsters of the first clutch are 16 days old. When rearing is finished the pairs are separated. It is inadvisable to continue breeding after July, as the parents are likely to start moulting before they have finished rearing the young.

Sometimes colony breeding in aviaries is practised and social behaviour under these conditions has been studied by Tsuneki (1960a; 1960b, 1961a, 1961b). Unless plenty of space is available, however, breeding in colonies results in restlessness which is attributed by Tsuneki (1961b) to disturbance and confusion of mates, leading to mortality of eggs and young. The best results are likely to be obtained with one cock per three hens, but plenty of nesting receptacles, or small bushes as nesting sites, should be provided.

Hinde (1965) has studied the breeding behaviour of the canary in great detail, especially the effect and interaction of internal and external factors on reproduction, and he gives numerous references to earlier work.

Genetic considerations

Dunker (1924) has shown that there is a dominant white which is lethal in the homozygous state, the birds apparently dying at an early stage of development. The colours cinnamon, fawn and dilute (recognised by canary fanciers) follow a sex-linked manner of inheritance and further information on this and other aspects of genetics is provided by Gill (1951, 1955).

Laboratory procedures

Anaesthesia and euthanasia

(This section is also relevant to the budgerigar and other small birds.)

Restraint. For parenteral administration of anaesthetic no special method of restraint is necessary, and after the injection the bird is placed in a small box until narcotized. When employing inhalation anaesthesia handling or the use of a face mask is not recommended because of the inevitable struggling and danger of shock. This can be avoided by using the apparatus described and illustrated by Graham-Jones (1962), which in its simplest form consists of a transparent plastic salad box with opaque lid. The box is used upside down and has a gas entry-hole drilled in one end; the gas can escape through the loose-fitting lid. The bird is placed in the box and observed as the gas of choice is introduced through the inlet tube. A more elaborate chamber has a heating element in the base and double gas entry holes to facilitate the mixing of anaesthetics or administration of oxygen if required.

General anaesthesia. This subject has been dealt with in detail by Arnall (1961) who recommended ether or cyclopropane for inhalation and pentobarbitone sodium for parenteral administration. Several other agents have also been used with varying degrees of success. Warren (1961) recommended divinyl ether as a satisfactory short-acting inhalation anaesthetic for the budgerigar. Grono (1961) obtained good results with this species using halothane* and Graham-Jones (1966) considered this to be the best anaesthetic. Judging from the results obtained by Ryder-Davies (1973) using metomidate† intramuscularly, this substance would probably prove to be a satisfactory narcotic for canaries.

Ether is a short-acting anaesthetic and deep narcosis to medium anaesthesia can be obtained within from ten seconds to two minutes, depending upon the size of the bird and other factors such as age. In budgerigars, however, the narcosis may not become sufficiently deep for surgery and in such cases pentobarbitone sodium or another suitable agent should be given parenterally to produce narcosis and followed by ether to maintain anaesthesia. This is a useful method when periods of anaesthesia exceeding 30 minutes are required.

By using 25 to 45 per cent cyclopropane in

* Fluothane: ICI.

† R7315 or methoxynol: Parke Davis.

oxygen graded planes of anaesthesia from deep narcosis to deep anaesthesia can be induced in half to five minutes and can be safely maintained for at least an hour without weakening the mixture (Arnall, 1961).

One of the safest parenteral anaesthetics is probably pentobarbitone sodium (Donovan, 1958), although response is variable, especially in budgerigars. Arnall (1961) considered subcutaneous injection to be the best route of administration. The concentration of the solution should be not less than 60 mg/10 ml for otherwise the volume would be too large for safety. The recommended dose is 4 mg/100 g body-weight, which produces full anaesthesia in ten minutes. By this route, however, rapid detoxication occurs and, to maintain anaesthesia, the administration of ether or halothane may be necessary (Graham-Jones, 1960).

Gandal (1969) was convinced that Equi-Thesin* is the drug of choice for most avian surgery. A single dose given intramuscularly on the basis of 0·2 to 0·25 ml per 100 g body-weight produces general anaesthesia lasting from 30 to 75 minutes. The lower dosage-rate should be used for weak or debilitated birds and the higher for those in good condition. If necessary, in order to prolong anaesthesia, a supplementary dose equivalent to 25 per cent of the original amount of Equi-Thesin may be given 30 minutes after the initial injection, although for this purpose Gandal prefers methoxyflurane† given in the same manner as other volatile anaesthetics such as ether.

Local anaesthesia. There are virtually no indications for local anaesthesia in canaries and other small birds. Most local anaesthetics are toxic and any containing adrenaline are contra-indicated. General anaesthetics are readily available and much more reliable, greatly diminishing the risk of shock and injury which may so easily occur under local anaesthesia.

Post-operative care. During recovery from general anaesthesia there is always the danger that a bird may injure itself by struggling. This can be minimized by placing it in a loosely-

fitting roll of paper secured with tape (Donovan, 1958). The bird is put in its cage and gradually works its way out of the tube as it recovers. Warmth is essential and is conveniently provided by using a dull-emitter infra-red lamp directed into a corner of the cage and positioned as required. Depending upon the condition of the bird, the temperature should be between 24 and 32°C but should be reduced as improvement takes place or if the heat causes distress. Administration of oxygen or normal saline containing 5 per cent dextrose may sometimes be indicated; 0·5 to 1 ml of the latter, depending upon the size of the bird, may be injected intraperitoneally. *Euthanasia.* Injection of an overdose of pentobarbitone sodium is both swift and efficient.

Experimental feeding procedures

In order to administer measured quantities of fluids for experimental purposes the method described by Bishop (1942) is suitable for healthy birds. A fine catheter tube is warmed in water and attached to the end of a graduated syringe. By holding the beak open with forceps the tube is passed down the oesophagus. Alternatively a similar but more refined method devised by Blackmore & Lucas (1965) for budgerigars can be used.

Collection of specimens

Collection of blood. Small quantities of blood may be obtained from a wing vein by merely pricking it with a Hagedorn needle, the bird being held on its back with the wing extended and pressed downwards. As blood escapes from the brachial vein it accumulates over the ventral surface of the anterior patagium (membrane of skin) in front of the humero-ulna joint where it can be collected with a pipette or small tube. Frequently, however, a subcutaneous haematoma forms and it may be necessary to use another method.

To obtain blood from the jugular vein the bird is held on its back and the neck is extended by placing the index and middle fingers on either side of the head. The skin overlying the right jugular (which is usually larger than the left) is swabbed with alcohol. Slight pressure with the thumb at the base of the neck will distend the vein into which the needle is inserted by holding the syringe in the right hand. The

* Equi-Thesin, Jensen-Salsbery Laboratories, Kansas City, Mo. USA. Each 500 ml contains 328 g of chloral hydrate, 75 g of pentobarbital, and 164 g of magnesium sulphate in aqueous solution of propylene glycol 35 per cent; with 9·5 per cent alcohol.

† Metobane, Pitman-Moore Co., New York, USA.

syringe should first be moistened with a few drops of 0·001 per cent heparin in 0·85 per cent saline (Hewitt, 1940) to prevent coagulation of the blood. For a small canary, it is suggested that a No. 27 gauge needle be used, while No. 26 is recommended for a large canary and No. 24 for a budgerigar. If no more than 0·1 ml is taken from a canary and 0·5 ml from a budgerigar it should be safe to bleed the bird at weekly intervals or even more frequently.

For heart-puncture the anaesthetized bird is placed on its back on a table so that the head can hang over the edge (McIntosh, personal communication). A No. 24 or No. 25 gauge needle with syringe attached is then inserted horizontally in the mid-line of the supra-sternal passage, tilted slightly towards the sternum, and then pushed deeper until it is felt to penetrate the heart. Recovery after heart-puncture cannot be expected.

Leonard (1969) preferred the simplest method of all, namely clipping a toenail. Certainly in a small bird such as a canary this is undoubtedly the safest way, but only a few drops of blood can be obtained. It is also necessary to work quickly, because the blood clots rapidly. A fingernail clipper should be used to clip the nail about halfway between the nail bed and the end of the vessel visible in the nail. The nail should not be clipped sideways (because this compresses the vessel) but in an antero-posterior direction, which tends to dilate it. Care must be taken to hold the limb relatively gently and not to press the toe between the thumb and forefinger before blood has been taken as this will restrict the flow of blood. Finally the tip of the amputated nail should be cauterised with a silver nitrate stick. Subsequent blood samples should be taken from different nails.

Collection of excreta. No special methods have been devised, but relatively clean samples of faeces and urates can be obtained by placing strong paper or cardboard under the perches.

Dosing and injection procedures

Oral administration of drugs. Seed impregnated with different antibiotics is available (Graham-Jones, 1961) for the treatment of some diseases, and is a convenient and reasonably accurate method of administration provided that there is no inappetence. Sometimes, however, it may be necessary to mix drugs in the drinking water or food but this method is obviously not ideal because it is impossible to control the dosage accurately. If a bird is sick it may stop drinking, or drink erratically or excessively. Unfortunately there are occasions when for certain reasons it is not possible to use a more satisfactory route of administration and then it is necessary to choose a drug which is completely soluble, long-acting, efficiently absorbed by the intestinal tract, relatively tasteless, preferably colourless, harmless if taken in large quantities and effective if taken in small doses. Such drugs are uncommon. Obviously direct oral administration is the method of choice, at least when treating alimentary disorders and when using drugs which are readily absorbed from the intestinal tract. The method is less suitable for the canary, however, than for the more robust budgerigar, especially if the bird is weak, because handling in such cases may do more harm than good. The bird should be restrained as described above (p. 477), making sure that it is held in such a way that the head is above the horizontal plane to prevent any inhalation of fluid. Measured quantities can be administered as described on p. 481.

Subcutaneous injection. One good site is the dorsal surface of the neck, whilst Stone (1969) recommended the ventral surface of the wing. A hypodermic syringe with a 2·5-cm No. 27 or 26 gauge needle is suitable for a canary and No. 24 or 25 for a budgerigar. A maximum of 1 to 1·5 ml of fluid can be injected, depending upon the size of the bird. If too much fluid is inoculated it will seep out through the puncture wound because the avian skin is lacking in elasticity and closely applied to the underlying subcutaneous tissues.

Intramuscular injection. Intramuscular injection into the pectoral muscles overlying the sternum is a simple and usually satisfactory route, although Arnall (1961) warned of the danger of puncturing one of the large venous sinuses in this region. Needles of the same gauge as recommended for subcutaneous injection are suitable, but there is less danger of puncturing a venous sinus if a 6·25-mm needle is used.

Intravenous injection. Intravenous injection requires skill and practice, and has been described in the canary by Hewitt (1940), who used the tarsometatarsal vein of the leg between the foot and the hock joint; one of the small toe veins may also be used. Blackmore (personal communication) states that in the budgerigar the jugular vein can be used. (See also p. 481.)

Intraperitoneal injection. Intraperitoneal injection is safe only if the needle is inserted a few millimetres deep in the mid-line posterior part of the abdomen, thus avoiding penetration of the abdominal air-sacs, liver or kidneys. A 6·25-mm needle of No. 24 to 27 gauge is suitable, depending upon the size of the bird.

Disease control

Before any disease can be diagnosed it is essential to be conversant with the appearance and behaviour of healthy birds.

General characteristics of healthy birds

A healthy bird appears alert, inquisitive, bright-eyed and sleek in plumage and holds its wings close to the body. During sleep one leg is usually drawn up beneath the abdomen and the bill is tucked under the top of the wing, although Eckstein (1940) described other postures.

All recently-acquired birds should be isolated and kept under close observation for at least a fortnight and the cage should be inspected daily for signs of excessive moulting and diarrhoea. Normal droppings, composed of faeces and white urate excretion, are slightly moist but firm.

Newly-hatched young are blind and naked, except for a small amount of down growing from the feather tracts. The abdomen is distended but gets progressively smaller as the yolk sac is absorbed and the nestling grows. The eyes open and the feathers begin to grow about seven days after hatching. Up to this time the excreta is enclosed in a gelatinous sac.

Feathering should be completed within a month and the juvenile plumage is moulted between six and eight weeks of age. So-called unflighted birds have pale wing feathers; they are less than a year old and have not passed through the adult moult which, in Great Britain, normally occurs during July and August and lasts for about six weeks. The breast feathers are shed first, followed by those on the back and wings. The head and neck feathers are moulted last. The plumage of old birds usually lacks the well-groomed appearance of the young, and the feathering may be less dense. The shanks or tarsometatarsals, instead of being smooth and flesh-coloured, are dull and covered with elongated scales.

Except in the breeding season, adult canaries may be difficult to sex. At five weeks of age the male's throat may swell slightly and feeble attempts at singing are made, but full song does not develop until after the first annual moult. The adult cock tends to be more thickset, active and bolder than the hen. The male's song is strong and musical, but the hen utters only a quiet cheep. During the breeding season the cock's vent and lower portion of the abdomen become rather prominent and protrude ventrally.

Macroscopically, the internal organs are similar to those of the domestic fowl, but the caeca are rudimentary and the spleen is elongated and sausage-shaped.

Clinical signs of common ailments

In small birds such as the canary clinical signs are seldom specific and can be regarded only as an indication of the type of disease present. If a bird is excessively drowsy and sleeps with both feet on the perch, or rests for long periods on the floor of the cage, it is unwell and should be isolated. As its condition worsens it will close its eyes, ruffle its feathers and often shiver. The appetite may be lost, although frequently a sick bird will continue to eat and drink, often excessively. Excessive thirst may be associated with nephritis or enteritis. A craving for grit frequently indicates the latter, especially if the droppings are fluid and greenish in colour, whilst with nephritis or nephrosis there is an increase in urate excretion.

Respiratory disorders are usually manifested by rapid breathing and gaping, accompanied by rhythmic squeaking sounds as the bird gasps for breath. In severe cases a pumping action of the tail may be seen, which keeps in unison with the respiratory movements. These clinical signs may occur in advanced aspergillosis and in acute pox, when they are associated with inflammation of the eyelids. Canary pox, however, is not commonly encountered.

Straining and eversion of the cloaca are associated with egg binding, constipation and diarrhoea, especially when the feathers around the vent become matted with excreta and cause irritation.

If more than one bird becomes ill those affected should be killed and *post-mortem* and bacteriological examinations carried out to determine if an infectious agent is present.

Immediate aid

Warmth is essential for sick birds and may be provided either as recommended on p. 481 or by using a commercially-produced hospital cage which, however, has no advantage over the judicious use of an infra-red lamp.

Fractured limbs require speedy attention and various methods of immobilization have been described; for further information Gandal (1969) and Altman (1969) should be consulted.

Nutritional deficiencies

In the opinion of Taylor (1969) canaries are less likely to be affected than budgerigars. Only very little has been published on the subject (see Keymer, 1959a).

Common diseases

The more common pathological conditions encountered have been described by Keymer (1961, 1973). In Great Britain pseudotuberculosis (*Yersinia pseudotuberculosis*; Syn. *Pasteurella pseudotuberculosis*) is relatively common and salmonellosis, usually caused by *Salmonella typhimurium*, is occasionally seen. In both infections the internal organs are usually congested and small yellowish necrotic lesions occur in the liver and spleen. There is no satisfactory treatment for pseudotuberculosis although salmonellosis may respond to nitrofurans in the drinking-water (Burkhart *et al.*, 1962; Keymer, 1959b). Under laboratory conditions destruction of all infected and in-contact birds is recommended, because recovered birds may remain carriers.

The red mite *Dermanyssus gallinae* and various species of lice are commonly found on canaries, but can be eliminated by either dusting the feathers with 0·5 per cent pyrethrum powder (Buxton & Busvine, 1957) or lightly spraying with an aerosol containing pyrethrum synergized with piperonyl butoxide (Keymer, 1959b). Red mites breed in cracks, nesting-pans, etc. and therefore these places should be inspected

regularly and sprayed with an aerosol. If mites are discovered well away from the birds an insecticide such as gamma benzene hexachloride may be used. Lice merely cause irritation and restlessness, but red mites often produce anaemia and death. Mites are also vectors of certain protozoal infections such as lankesterellosis (atoxoplasmosis). These and other parasitic diseases are discussed by Keymer (1969) in a textbook dealing comprehensively with all types of diseases of small birds commonly kept in captivity (Petrak, 1969).

REFERENCES

Altman, I. E. (1969). Disorders of the skeletal system. In *Diseases of Cage and Aviary Birds*, ed. Petrak, Margaret L. Philadelphia: Lea & Febiger.

Arnall, L. (1961). Anaesthesia and surgery in cage and aviary birds. Part I. *Vet. Rec.*, **73**, 139–42.

Arnall, L. (1969). Diseases of the respiratory system. In *Diseases of Cage and Aviary Birds*, ed. Petrak, Margaret L. Philadelphia: Lea & Febiger.

Ben-Harel, S. (1923). Studies on bird malaria in relation to the mechanism of relapse. *Am. J. Hyg.*, **3**, 652–85.

Bice, C. W. (1955). Millets for cage birds. *All-Pets Magazine*, **26**, (3), 72–84; **26** (4), 109–26.

Bishop, A. (1942). Chemotherapy and avian malaria. *Parasitology*, **34**, 1–54.

Blackmore, D. K. & Lucas, J. F. (1965). A simple method for the accurate oral administration of drugs to budgerigars. *J. small Anim. Pract.*, **6**, 27–9.

Brand, A. R. & Kellogg, P. P. (1939). The range of hearing of canaries. *Science, N.Y.*, **90**, 354.

British Standards Institution (1961). *Carriage of live animals. 3149: Recommendations for the carriage of live animals by air. Part 2: Small and medium-sized seed eating birds.*

Burkhart, D. M., Wolfgang, R. W. & Harwood, P. D. (1962). Salmonellosis in parakeets and canaries treated with nitrofurans in the drinking water. *Avian Dis.*, **6**, 275–83.

Buxton, P. A. & Busvine, J. R. (1957). Pests of the animal house. In *The UFAW Handbook on the Care and Management of Laboratory Animals*, 2nd ed. London: Universities Federation for Animal Welfare.

Cornwallis, R. K. & Smith, A. E. (1960). *The Bird in the Hand*. Field Guide No. 6. Oxford: British Trust for Ornithology.

Corran, J. W. & Edgar, S. H. (1957). The canary. In *The UFAW Handbook on the Care and Management of*

Laboratory Animals, 2nd ed. London: Universities Federation for Animal Welfare.

Donovan, C. A. (1958). Restraint and anaesthesia of caged birds. *Vet. Med.*, **53**, 541–3.

Dunker, H. (1924). *Z. indukt. Abstamm.—u. Vererb-Lehre*, **32**, 363–76. Quoted by Hutt, F. B. (1949). In *Genetics of the Fowl*, p. 171. New York: McGraw-Hill.

Eckstein, G. (1940). The sleep of canaries. *Science, N.Y.*, **92**, 577–8.

Farner, D. S. (1969). Some physiological attributes of small birds. In *Diseases of Cage and Aviary Birds*, ed. Petrak, Margaret L. Philadelphia: Lea & Febiger.

Flower, S. S. (1938). Further notes on the duration of life in animals—IV. Birds. *Proc. zool. Soc. Lond.*, **108**, ser. A, part II, 195–235.

Gandal, C. P. (1969). Surgical techniques and anaesthesia. In *Diseases of Cage and Aviary Birds*, ed. Petrak, Margaret L. Philadelphia: Lea & Febiger.

Gelineo, S. (1955). Temperature d'adaptation et production de chaleur chez les oiseaux de petite taille. *Archs Sci. physiol.*, **9**, 225–43.

Gill, A. K. (1951). *Cinnamon Inheritance in Canaries*. London: Iliffe.

Gill, A. K. (1955). *New-coloured Canaries*. London: Iliffe.

Graham-Jones, O. (1960). Halothane anaesthesia in small animal practice. *Vet. Rec.*, **72**, 673–4.

Graham-Jones, O. (1961). Administering medicine to small birds. *Vet. Rec.*, **73**, 331–2.

Graham-Jones, O. (1962). Anaesthesia of small non-domesticated animals: some simple apparatus. *Vet. Rec.*, **74**, 987–8.

Graham-Jones, O. (1966). The clinical approach to tumours in cage birds—III. Restraint and anaesthesia of small cage birds. *J. small Anim. Pract.*, **7**, 231–9.

Graham-Jones, O. & Fiennes, R. N. (1959). Dried egg in birds' food. *Vet. Rec.*, **71**, 245.

Gray, A. P. (1958). Bird hybrids: a check-list with bibliography. In *Commonwealth Bureau of Animal Breeding and Genetics. Techn. Comm. No. 13*. Farnham Royal.

Groebbels, F. (1932). *Der Vogel*. Vol. I: *Atmungswelt und Nahrungswelt*. Borntraeger: Berlin. Quoted by Sturkie, P. D. 1954. In: *Avian Physiology*, p. 95. Ithaca, N.Y.: Comstock.

Grono, L. R. (1961). Anaesthesia in budgerigars. *Aust. vet. J.*, **37**, 463–4.

Hasholt, J. (1969). Senility. In *Diseases of Cage and Aviary Birds*, ed. Petrak, Margaret L. Philadelphia: Lea & Febiger.

Hewitt, R. (1940). Bird malaria. *Am. J. Hyg.*, Monographic Ser. No. 15.

Hinde, R. A. (1959). Seasonal variations in clutch size and hatching success of domesticated canaries. *Bird Study*, **6**, 15–19.

Hinde, R. A. (1965). Interaction of internal and external factors in integration of canary reproduction. In *Sex and Behaviour*, ed. Beach, F. A. New York: Wiley.

Hinde, R. A. (1966). *Animal Behaviour. A Synthesis of Ethology and Comparative Psychology*. 2nd ed. New York: McGraw-Hill.

International Air Transport Association (1969). *Manual for the Carriage of Live Animals by Air*. Montreal: IATA.

Keymer, I. F. (1959a). The diagnosis and treatment of some diseases of seed eating passerine birds. *Mod. vet. Pract.*, **40** (7), 30–4; (8), 34–7.

Keymer, I. F. (1959b). Specific diseases of the canary and other seed eating passerine birds. *Mod. vet. Pract.*, **40** (17), 32–5; (18), 45–8; (24), 16.

Keymer, I. F. (1961). Post-mortem examination of pet birds. *Mod. vet. Pract.*, **42** (23), 35–8; (24), 47–51.

Keymer, I. F. (1969). Parasitic diseases. In *Diseases of Cage and Aviary Birds*, ed. Petrak, Margaret L. Philadelphia: Lea & Febiger.

Keymer, I. F. (1973). Diseases of passerine birds. *Vet. Review, May & Baker*, **23** (3), 47–52.

Leonard, J. L. (1969). Clinical laboratory examinations. In *Diseases of Cage and Aviary Birds*, ed. Petrak, Margaret L. Philadelphia: Lea & Febiger.

Petrak, M. L. (1969). *Diseases of Cage and Aviary Birds*. Philadelphia: Lea & Febiger.

Romanoff, A. L. & Romanoff, A. J. (1949). *The Avian Egg*. London: Chapman & Hall.

Ryder-Davies, P. (1973). The use of metomidate, an intramuscular narcotic for birds. *Vet. Rec.*, **92**, 507–509.

Shoemaker, H. (1939). Social hierarchy in flocks of canaries. *Auk*, **56**, 381–406.

Stone, R. M. (1969). Clinical examination and methods of treatment. In *Diseases of Cage and Aviary Birds*, ed. Petrak, Margaret L. Philadelphia: Lea & Febiger.

Taylor, T. G. (1969). Nutritional deficiencies. In *Diseases of Cage and Aviary Birds*, ed. Petrak, Margaret L. Philadelphia: Lea & Febiger.

Thomson, Sir A. Landsborough (1964). *A New Dictionary of Birds*. London: Nelson.

Tollefson, C. I. (1969). Nutrition. In *Diseases of Cage and Aviary Birds*, ed. Petrak, Margaret L. Philadelphia: Lea & Febiger.

Tsuneki, K. (1960a). Social organisation in flocks of canaries. *Jap. J. Ecol.*, **10**, 177–89.

Tsuneki, K. (1960b). Pair formation in flocks of the cage canaries. *Jap. J. Ecol.*, **10**, 238–44.

Tsuneki, K. (1961a). Territory in flocks of the caged canaries, with special reference to its causation and defence. *Jap. J. Ecol.*, **11**, 66–75.

Tsuneki, K. (1961b). Breeding in a dense population of the caged canaries. *Jap. J. Ecol.*, **11**, 142–6.

Warren, A. G. (1961). Some recent ideas on animal

anaesthesia. *Vet. Rev.*, **12,** 111–20.

Warren, R. P. & Hinde, R. A. (1959). The effect of oestrogen and progesterone on the nest-building of domesticated canaries. *Anim. Behav.*, **7,** 209–13.

Witherby, H. F., Jourdain, F. C. R., Ticehurst, N. F. and Tucker, B. W. (1940). *The Handbook of British Birds.* **1,** p. 82. London: Witherby.

Woodbury, R. A. & Hamilton, W. F. (1937). Blood-pressure studies in small animals. *Am. J. Physiol.*,

119, 663–74.

Worden, A. N. (1956). *Functional Anatomy of Birds.* London: Poultry World.

Yamamoto, T. (1959). Study on the blood cells of birds. *Acta med., Fukuoka* (Igaku Kenkyû), **28,** 1057–9.

Young, M. D. (1937). Erythrocyte counts and haemoglobin concentration in normal female canaries. *J. Parasit.*, **23,** 424–6.

36 The Budgerigar

I. F. KEYMER

General biology

Taxonomic position

The budgerigar (*Melopsittacus undulatus*) is placed in the class Aves, order Psittaciformes (parrot-like or psittacine birds), family Psittacidae.

Origin

The wild budgerigar is a native of Australia where it is found mainly in arid areas. It was first kept in captivity over 100 years ago, but has been bred extensively under domestic conditions only since the 1930's.

Size ranges

Although there are many different colour varieties these are similar in size.

Behaviour

(See also Canary). Brockway and others have studied reproductive behaviour (see p. 490), and the ability of this species to learn to mimic the human voice is well known (Thorpe, 1964). Thorpe also discusses experiments on counting ability of budgerigars and other species.

Breeds and types

The wild budgerigar is green in colour, but in domesticated birds there are four main colour types, viz.: normal colours in the basic shades of blue, green, yellow, white and grey with dark eyes; lutinos which are yellow and devoid of markings and have red eyes; true albinos; and fallows, which also have red eyes, but in which the black marking of the normal colour varieties are replaced by dark brown. Detailed information regarding the various colours is given by Watmough (1960).

Uses

The use of budgerigars in research is mainly confined to investigations of their nutritional requirements for the preparation and manufacture of pet-bird foods and to the study of neoplasms, which are common in this species, an incidence of 24·2 per cent being recorded by Beach (1962); 14·8 by Blackmore (1967); 8·8 by Keymer (1960); 4·4 by Kronberger (1962) and 15·81 by Ratcliffe (1933).

Standard biological data

In recent years a great deal has been learned about the physiology of this species, especially concerning haematology and the anatomy has been well described by Evans (1969).

Body temperature

Böni (1942), using a thermocouple, found the cloacal temperature to vary from 40 to 42°C. Beach (1961), using a mouse-type rectal thermometer, obtained the same results with two adult hens and stated that the average varied between 41·2 and 41·3°C. Arnall (1969) gave the temperature as 42·4°C. The subject of temperature regulation is discussed by Aulie (1971) and Lang (1972).

Cardio-vascular system

Normal haemoglobin values have been found by Leonard & Gallagher to vary between 15·5 and 16·7 g of haemoglobin per 100 ml in adults about 1 year old, whilst Gallagher's results indicate that the values are lower in birds under a month old (Leonard, 1969). Gallagher also obtained lower haematocrit values in youngsters (Leonard, 1969). There appear, however, to be no significant differences in haematocrit values related to sex, colour, or weight in birds over 2 months of age; mean values ranging from 51 to 53·7 per cent were quoted by Leonard & Gallagher in birds from 2 to 12 months old. Similarly red cell counts are lower in birds under a month old: 3·15 million erythrocytes, compared with a mean ranging from 4·49 to 5·66 per mm³ in old birds (Leonard, 1969). Tables showing various other haematological values have been compiled by Leonard (1969), based on his work and that of Gallagher. For budgerigars over a month old the mean values can be summarised as follows: normal reticulocyte count 6·7 per cent; normal thrombocyte count 29,450 to 33,800 per mm³; MCV, 90·0 to 120·1; MCH, 26·5 to 38·2; MCHC, 29·8 to 31·6.

The mean total white cell count varied from 3,803 to 6,530 cells per mm³. Heterophil counts ranged from 2,000 to 3,090 per mm³ with mean percentage values varying from 52·4 to 64·4; corresponding figures for other leucocytes are: lymphocytes 721 to 2,320 (21·9 to 33·3 per cent); monocytes 124 to 323 (3·0 to 5·5 per cent) and basophils 478 to 550 (3·2 to 13·3 per cent).

The mean blood glucose level in 10 budgerigars after fasting for 18 to 24 hours was found by Leonard (1969) to be 153 mg/100 ml, whilst Schlumberger (1956) found the average level of 30 non-fasted birds to be 210 mg/100 ml.

The normal blood uric acid level is quoted by Schlumberger & Henschke (1956) as being within the range of 5·3 to 8·0 mg/100 ml.

The heart rate varies from 250 to 600 beats per minute (Arnall, 1969).

Respiratory system

Arnall (1969) stated that the respiratory rate varies from 80 to 100 per minute. Respiratory exchange and evaporative water loss in the flying budgerigar have been studied by Tucker (1968).

Reproductive system

The normal clutch size is five or six, eggs being laid at intervals of two days (British Veterinary Association, 1964). The incubation period is approximately 18 days and the white eggs are stated by Blackmore (personal communication) to measure approximately 2·0 × 1·6 cm and to weigh from 2·0 to 2·8 g, the average being about 2·2 g.

Longevity

The average length of life is probably in the region of 3 to 4 years or longer, depending upon the methods of management. Budgerigars more than 10 or 11 years old are seldom seen (Hasholt, 1969). There are records, however, of birds living to 21 years (Flower, 1938).

Weights

Adults vary in weight within the approximate range of 35 to 60 g or more (British Veterinary Association, 1964), the average probably being from 45 to 50 g. A healthy chick at hatching should weigh nearly 2 g, at seven days about 12 to 15 g, fourteen days 30 g and twenty-one days 40 to 45 g. These figures, however, which are based on those of Blackmore (personal communication), British Veterinary Association (1964) and Taylor (1958), are only approximate, especially after two weeks of age when a good deal of variation occurs.

Husbandry

Housing and caging

Breeding cages. For a single pair, a canary double breeding cage with the partition removed is suitable.

Flight cages. Approximately two dozen non-breeding birds can be housed in a wire netting flight 2·5 metres long, 0·9 m wide and 1·5 m high, or they may be kept singly in each compartment of the double breeding cages.

Breeding aviaries. Size depends upon the number of pairs to be housed and cannot be conveniently calculated on the basis of cubic volume per bird. An aviary, however, of 6·7 m³ (e.g. 1·8 m long, 0·6 m wide and 1·8 m high) is suitable for housing four pairs. Information on types of aviaries is given by Armstrong (1961), on cages by Dix (1961) and on construction by Luke & Silver (1955).

Seed hoppers and water receptacles.   (See also Canary). It is unnecessary to provide bathing facilities for budgerigars as these birds do not bathe in the same manner as canaries. They do, however, appreciate an occasional spraying.

Perches. Perches of the dimensions recommended for large breeds of canary are suitable and in aviaries small branches may also be provided.

Nest boxes. There are several types, but the British Veterinary Association (1964) recommends wooden boxes with a concave interior base of about 14 cm diameter. The box should be 15 to 17 cm square and 23 cm high. Entrance is made via a hole 4 cm in diameter, situated about 18 cm from the base, with a small outside perch below it. A hinged roof is provided for observation, and a sliding base to facilitate cleaning.

Environmental conditions

Birds are usually housed in a special birdroom, although they may be kept outside in aviaries all the year round if adequate weatherproof shelters are provided. Aviary flights should be constructed of 1·25 cm mesh wire-netting, preferably fixed to the inside of the wooden framework to prevent damage from the birds' beaks.

Identification

(See canary).

Handling

Budgerigars can be handled safely for longer periods than is possible with canaries, and the same technique is used. Hens, however, often bite and it is therefore advisable to hold the head by exerting pressure with the thumb and index finger on each side of the mandibular articulations (Fig. 36.1).

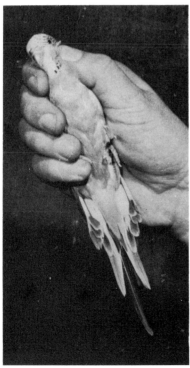

Fig. 36.1. Method of holding a budgerigar or other small psittacine.

Transport

(See canary).

Wooden boxes are preferable to cardboard ones, since budgerigars may bite through the latter.

Feeding

The nutritional requirements of the budgerigar have received more attention than those of the canary, largely as a result of the work of Taylor (1958, 1959, 1960, 1965, 1969a) and his co-workers in their investigations into the cause of French moult. Budgerigar nutrition has also been reviewed by Bice (1956a), Worden (1960) and Tollefson (1969).

It is important to realise that, unlike the canary, the budgerigar dehusks seeds with its tongue and beak before swallowing them whole. The husk is not eaten and if feeding receptacles are not emptied and the contents renewed regularly the husks may accumulate and bury the whole seeds in the bottom of the hopper.

Natural foods

In the wild, grass seeds of various species appear to be the staple diet. Tollefson (1969) illustrated and described the seeds normally fed to budgerigars, canaries and other small birds in captivity. The composition of most of these seeds, including those fed to budgerigars, has been investigated by several workers, and Tollefson (1969) provided a table showing the composition of some common seeds. Massey et al. (1960) have estimated the amino-acid content and compared it with that of the carcase, liver and whole egg of the normal bird. They stated that the usual diet of canary seed, small yellow millet and large white millet is seriously deficient in lysine and arginine, while Bishop & Taylor (1963) have found this mixture to be deficient in vitamins of the B group, especially riboflavin and biotin. Blackmore (1963) was also of the opinion that the usual seed mixture is deficient in iodine and commonly produces thyroid dysplasia. Partington & Sellwood (1961) reported that, although millets are a poor source of manganese, there is no evidence of a deficiency in the mixture that is usually fed.

Waterhouse et al. (1961) found that when non-breeding budgerigars were offered a free choice of seeds they selected a mixture composed of approximately 40 parts canary seed and 60 parts millet. The daily intake of whole seed (i.e. including husk) was 7 g per bird. Waterhouse *et al.* also found that, although whole canary seed was preferred to that which had been dehusked, both whole and dehusked millet were eaten. Tollefson (1969) also demonstrated a preference for whole seed and found that adult non-breeding birds consumed 6 g of seed per day. The birds were given a free choice of various seeds over a period of four months and selected the following percentages: 31·7 canary seed; 21·7 red proso millet (*Panicum miliaceum*, red variety); 20·7 yellow proso millet; 19·3 white proso millet and 6·6 chopped oat (*Avena sativa*). Although millets and canary seed are obviously preferred, adult non-breeding budgerigars will eat other seeds if these are available. Bice (1956b) conducted a feeding test using 12 different seeds and wheat germ and found that the birds consumed 80 per cent canary seed and millets and 20 per cent of other types of seed.

Various plants are appreciated at all times, expecially turves of grass and green twigs of deciduous trees. In addition to the green food reommended for canaries, rye-grass (*Lolium perenne*), plantains (*Plantago* spp.) and shepherd's purse (*Capsella bursa-pastoris*) are suitable. When these are not available salad vegetables, sliced cabbage stalks, apple and sprouted grains may be fed.

Insoluble grit is provided in the form of sand and soluble grit as cuttlefish bone.

Special requirements

Budgerigars, especially non-breeding birds, have conservative tastes and often have to be trained gradually to accept unusual foods. Waterhouse *et al.* (1961) have carried out experiments on the palatability of various seeds and supplements such as biscuit and rusk crumbs and crushed broiler pellets and powders, and found that all the unfamiliar substances are rejected by non-breeding birds. Although non-breeding adults will often remain in a reasonably satisfactory condition for long periods on the conventional millet and canary seed mixtures, there is no doubt that they do benefit if certain supplements are eaten. Tollefson (1969) stated that the basic

seed mixtures are inadequate for youngsters and breeding birds and result in low egg-production, poor hatchability, high chick mortality and poor growth-rate of surviving youngsters. Bice (1956b) used a mash supplement for breeding birds which contained dried egg-yolk, dried bakery products, alfalfa leaf meal, milk, iodised salt, parsley flakes, wheat germ, yeasts and vitamins A, D_3 and B_{12}. He also added a little canary seed and millet to encourage the birds to take the mash. The basic seed mixture was fed at the same time and Bice (1956b) found that the amount of mash eaten was approximately 14 per cent of the total food consumption. Tollefson (1969) also used this mash supplement and reported extremely good breeding results.

For breeders Taylor (1960) recommended the provision of cod-liver oil fortified with pure vitamin E at a level of 0·1 per cent, mixed with the seed. The oil is not only a rich source of vitamins A and D, thus ensuring proper utilization of calcium for egg-shell formation, but is also a source of iodine. Blackmore (1963) states that when only the usual seed mixture is fed a level of 4 per cent iodine is necessary for the normal development of the thyroids. Only a few days' supply should be mixed at once, however, to prevent the oil from going rancid. Taylor (1958) advocated the use of a special protein-rich soft food for rearing youngsters and provides recipes; alternatively a commercial supplement can be used. Milk may also be substituted for drinking water.

Methods of presenting food

(See also Canary).

When young budgerigars leave the nest they may have difficulty in finding food in the ordinary receptacles. It may therefore be necessary to place a shallow food tray on the cage floor. This can be removed when the youngsters have learned to feed themselves, but care must be taken to see that all receptacles are placed in positions acceptable to the birds. Sometimes birds of any age may ignore food placed in certain positions.

Watering

Clean water should be provided daily, unless it is replaced by milk for breeding birds. The rate of water-consumption varies considerably under different conditions (Valle, 1971). Bice (1956b) found that adult non-breeding birds drank 3 ml of water per day. Cade & Dybas (1962) calculated that the mean daily water-consumption of a budgerigar weighing 30 g was 5 per cent of its body-weight and could be less than 1 per cent. They also showed that budgerigars could survive without water for up to 38 days at a temperature of 30°C and a relative humidity of 30 per cent. Tollefson (1969) stated that two birds have been reported in good health after 130 days without drinking and Greenwald *et al.* (1967) have studied the effects of dehydration conditions.

Breeding

Methods

Budgerigars are extremely prolific and easy to breed. Both sexes mature at approximately 3 to 4 months of age, but cocks should not be used for breeding until they are 10 months old and hens until they are 11 months. The upper age-limits are six and four years respectively.

In outdoor aviaries breeding may be commenced in March. Selected pairs, however, should be caged together in the bird-room for at least a week before being released with others in an aviary; otherwise, owing to the polygamous nature of the cocks, multiple matings will occur. Pairs which show signs of incompatability must be separated and introduced to other mates. Suitably-matched pairs will show affection by rubbing their beaks together and kissing, and the hen will allow the cock to feed her with regurgitated seed.

It is usually necessary to provide a choice of at least two nest boxes per pair. No nesting material is needed as the eggs are laid on the bare concave base of the nest box. Breeding hens become very pugnacious and zealously guard their boxes against the intrusions of others, although occasionally two hens will share one box. New hens should not be introduced into an established breeding colony, for serious fighting may occur. Single pairs may also be housed in breeding cages fitted with one nest box.

The hen commences incubation after laying the first egg, and the young hatch at intervals. Although many pairs will satisfactorily rear five or six youngsters the best results are obtained by restricting them to four. Breeders of exhibition birds often allow only two broods per season; this is done by removing nest boxes when the second brood are out of the nest. According to Blackmore (1966) laboratory birds can safely be allowed to raise 6 clutches of 4 chicks per year.

During incubation the cock feeds the hen, but when the youngsters hatch both parents feed the nestlings with regurgitated, partially-digested seed. The hen also produces a proventricular secretion during this period which has been erroneously termed 'crop-milk' by analogy with that produced by pigeons. This secretion has been shown by Taylor (1958) to be rich in protein.

Youngsters leave the nest between 4 and 6 weeks of age and are fed by the cock for a few days until they learn to fend for themselves. They should not be removed from the breeding colony earlier than seven days after leaving the nest. By then most hens will be incubating their second clutch, and youngsters must be removed before these hatch.

Vaugien (1951), Ficken *et al.* (1960) and Brockway (1961, 1962, 1964a, 1964b) should be consulted for detailed studies on reproductive behaviour.

Genetic considerations

The inheritance of colour has been closely studied and is described by Armour (1956), Taylor & Warner (1961) and Watmough (1960).

Laboratory procedures

Anaesthesia and euthanasia

(See Canary, p. 480).

Experimental feeding procedures

(See Canary, p. 481).

Collection of specimens

(See Canary, p. 481).

Dosing and injection procedures

(See Canary, p. 482).

Disease control

(See also Canary, p. 483).

General characteristics of healthy birds

The signs of health are similar to those of the canary. Regurgitation of the crop and pro-ventricular contents in budgerigars in breeding condition must not, however, be confused with vomiting.

Newly-hatched young have a distended abdomen and are blind and almost naked. The eyes open at about six days of age, when the feather growth becomes visible. At two weeks the nestling is able to lift its head and move about. Wing and tail feathers have developed by three weeks, when the quills of other feathers are well formed. The juvenile plumage is complete at about a month, when the youngsters leave the nest box. At this stage the forehead and back of the head have a slightly dark, barred appearance. The overall colouring is also paler than in adults. The first moult begins between 10 and 12 weeks of age, when the barring is gradually lost. In this country adults usually moult in October although, unlike canaries, there is often no complete annual moult but only a partial one, two or three times a year.

There is no satisfactory method of ageing an adult, although often the upper mandible and nails become friable and increase in length and obesity develops as the bird gets older. Old budgerigars, however, unlike canaries, do not normally develop scaly proliferation of the skin of the feet and legs.

The sex of normally-coloured adult birds is easily recognised because the cere of the cock is blue and that of the hen brown, although in both sexes the ceres of young birds when they leave the nest are pinkish lilac. The ceres of the adult cocks in the red-eyed varieties are pinkish, although sometimes a bluish tinge may show. In addition to these features the cock's head is larger and more domed than that of the hen.

Internally, the budgerigar differs from the canary by having a thick, fleshy, cylindrical tongue, an almost round spleen and no caeca or gall-bladder.

Clinical signs of common ailments

(See Canary).

Immediate aid

(See Canary).

Nutritional deficiencies

Budgerigars, especially breeding birds, laying hens and fast-growing chicks are more likely than canaries to be affected by nutritional deficiencies because the conventional seed mixture of canary seed and millet is deficient in a number of nutrients (Taylor, 1969b). The nutritional requirements of breeding birds has been discussed above (p. 485).

Iodine deficiency is commonly encountered and severely-affected birds often emit high-pitched squeaking noises associated with the respiration (Beach, 1962; British Veterinary Association, 1964; Schlumberger, 1955). Black-more (1963) recently found 85 per cent of pet budgerigars to be affected by thyroid disease and in a survey of *post-mortem* examinations diagnosed thyroid dysplasia as the cause of death in 23·8 per cent.

Taylor (1969b) stated that various unexplained leg deformities, similar to perosis of the young of domestic fowls and turkeys, sometimes also affect young budgerigars. It is possible, therefore, that a manganese deficiency may be implicated.

Common diseases

(See also Canary). These are similar to those affecting the canary, although pox and pseudo-tuberculosis are rare. In fact, the budgerigar, as stressed by Blackmore (1966) appears to have a low susceptibility to most infectious diseases with the exception of psittacosis (ornithosis). This disease was uncommon in Great Britain (Beach, 1962; Keymer, 1959; MacCallum *et al.*, 1961) until about 1962, since when it has been increasingly encountered (Blackmore, 1968; Keymer, 1972) in aviaries of breeding birds. Sometimes mainly the hens are affected, presumably as the result of stress caused by egg-laying and rearing young. The disease is infectious to human beings and whenever a number of birds develop general malaise, greenish diarrhoea and ocular or nasal discharge, special hygienic precautions should be taken to prevent spread of the infection. The causal virus may be demonstrated by mouse inoculation (Meyer, 1952).

Vomition of greenish mucus associated with necrosis of the oesophageal and crop epithelium is not uncommon (Beach, 1962; Keymer, 1958) but the cause is unknown.

In fledglings, French moult characterized by moulting especially of the wing and tail feathers is a problem in many breeding establishments (Feyerabend, 1957). Taylor (1958, 1959, 1960 and 1969a) has investigated the disease, but the cause remains unknown.

As previously stated, budgerigars are particularly susceptible to neoplasia, an especially good account of which is provided by Petrak & Gilmore (1969).

Although red mites (*Dermanyssus gallinae*) are pathogenic to canaries, they seldom trouble budgerigars and Blackmore (personal communication) has demonstrated experimentally that these mites are apparently harmless even to youngsters in at least some strains of budgerigars in England.

For further information on diseases Petrak (1969) should be consulted.

REFERENCES

Armour, M. D. S. (1956). *Exhibition Budgerigars*, 2nd ed. London: Iliffe.

Armstrong, E. M. (1961). Aviary planning. Supplement to *Cage and Aviary Birds*, p. 119, (3090) i–vi. London: Farmer & Stockbreeder.

Arnall, L. (1969). Diseases of the respiratory system. In *Diseases of Cage and Aviary Birds*, ed. Petrak, Margaret L. Philadelphia: Lea & Febiger.

Aulie, A. (1971). Body temperatures in pigeons and budgerigars during sustained flight. *Comp. Biochem. Physiol.*, **39**, 173–176.

Beach, J. E. (1961). Temperatures of budgerigars. *Vet. Rec.*, **73**, 513.

Beach. J. E. (1962). Diseases of budgerigars and other cage birds. *Vet. Rec.*, **74**, 10–17; 63–8; 134–40.

Bice, C. W. (1956a). Nutrition of the budgerigar. In *Proceedings of the Second World Budgerigar Congress*. Birmingham: The Budgerigar Society.

Bice, C. W. (1956b). Observations on budgie feeding. *Budgerigar Bulletin*, No. 113, pp. 19–27.

Bishop, C. & Taylor, T. G. (1963). Studies on the vitamin content of bird seeds. *Vet. Rec.*, **75**, 688–91.

Blackmore, D. K. (1963). The incidence and aetiology of thyroid dysplasia in budgerigars (*Melopsittacus undulatus*). *Vet. Rec.*, **75**, 1068–72.

Blackmore, D. K. (1966). The budgerigar as a laboratory animal. *J. Inst. Anim. Techns*, **17**, 151–4.

Blackmore, D. K. (1967). *The Pattern of Disease in Budgerigars. A Study in Comparative Pathology*. Ph.D. Thesis, London University.

Blackmore, D. K. (1968). Some observations on ornithosis, part II. The disease in cage and aviary birds. In *Some Diseases of Animals Communicable to Man in Britain*. Proc. of Symposium organised by the BVA and BSAVA, pp. 151–60. June 1966, ed. Graham-Jones, O. Oxford: Pergamon Press.

Böni, A. (1942). Über die Entwicklung der Temperaturregulation bei verschiedenen Nesthockern (Wellensittich, Neuntöter und Wendehals). *Archs suisses Orn.*, **2**, 1–56.

British Veterinary Association (1964). *Handbook on the Treatment of Exotic Pets. Part One: Cage Birds*. London: British Veterinary Association.

Brockway, B. F. (1961). The effects of visual and vocal stimuli upon the reproductive ethology and physiology of paired budgerigars (*Melopsittacus undulatus*). *Am. Zool.*, **1**, 64.

Brockway, B. F. (1962). The effects of nest-entrance positions and male vocalisations on reproduction in budgerigars. In *The Living Bird First Annual of the Cornell Laboratory of Ornithology 1962*, pp. 93–101. Published by The Laboratory of Ornithology at Cornell University, Ithaca, New York.

Brockway, B. F. (1964a). Social influences on reproductive physiology and ethology of budgerigars (*Melopsittacus undulatus*). *Anim. Behav.*, **12**, 493–501.

Brockway, B. F. (1964b). Ethological studies of the budgerigar: reproductive behaviour. *Behaviour*, **23**, 294–324.

Cade, T. J. & Dybas, J. A., Jr. (1962). Water economy of the budgerygah. *Auk*, **79**, 345–64.

Dix, D. A. (1961). Cages and Appliances. Supplement to *Cage and Aviary Birds*, p. 119. (3094), i–vii.

Evans, H. E. (1969). Anatomy of the budgerigar. In *Diseases of Cage and Aviary Birds*, ed. Petrak, Margaret L. Philadelphia: Lea & Febiger.

Feyerabend, C. (1957). *Your Budgie's Health Book*. Wisconsin: All-Pets Books.

Ficken, R. W., Tienhoven, A. van, Ficken, M. S. & Sibley, F. C. (1960). Effect of visual and vocal stimuli on breeding in the Budgerigar (*Melopsittacus undulatus*). *Anim. Behav.*, **8**, 104–6.

Flower, S. S. (1938). Further notes on the duration of life in animals—IV. Birds. *Proc. zool. Soc. Lond.*, **108**, ser. A., part II, 195–235.

Greenwald, L., Stone, W. B. & Cade, T. J. (1967). Physiological adjustments of the budgerigar (*Melopsittacus undulatus*) to dehydrating conditions. *Comp. Biochem. Physiol.*, **22**, 91–100.

Hasholt, J. (1969). Senility. In *Diseases of Cage and Aviary Birds*, ed. Petrak, Margaret L. Philadelphia: Lea & Febiger.

Keymer, I. F. (1958). The diagnosis and treatment of common psittacine diseases. *Mod. vet. Pract.*, **39**, (21) 22–30.

Keymer, I. F. (1959). Ornithosis (psittacosis). *Vet. Rec.*, **71**, 354.

Keymer, I. F. (1960). Cage and aviary bird surgery. *Mod. vet. Pract.*, **41** (11), 28–31; (12) 32–6.

Keymer, I. F. (1972). The unsuitability of non-domesticated animals as pets. *Vet. Rec.*, **91**, 373–381.

Kronberger, H. (1962). Geschwulste bei Zootieren. *Nord. VetMed.*, **14**, suppl. 1, 297–304.

Lang, P. L. (1972). Temperature regulation in the parakeet (*Melopsittacus undulatus*). *Comp. Biochem. Physiol.*, **32**, 483.

Leonard, J. L. (1969). Clinical laboratory examinations. In *Diseases of Cage and Aviary Birds*, ed. Petrak, Margaret L. Philadelphia: Lea & Febiger.

Luke, L. P. & Silver, A. (1955). *Aviaries, Birdrooms and Cages*, 6th ed. London: Iliffe.

MacCallum, F. O., McDonald, J. R. & Macrea, A. D. (1961). Psittacine birds as a source of virus infection for man in England. *Mon. Bull. Minist. Hlth.*, **20**, 114–6.

Massey, D. M., Sellwood, E. H. B. & Waterhouse, C. E. (1960). The amino-acid composition of budgerigar diet, tissues and carcase. *Vet. Rec.*, **72**, 283–6.

Meyer, K. F. (1952). Ornithosis and psittacosis. In *Diseases of Poultry*, 3rd ed., ed. Biester, H. E. & Schwarte, L. H. Iowa: State College Press.

Partington, H. & Sellwood, E. H. B. (1961). The manganese content of the diet of the budgerigar. *J. small Anim. Pract.*, **1**, 281–5.

Petrak, M. L. (1969). *Diseases of Cage and Aviary Birds*. Philadelphia: Lea & Febiger.

Petrak, M. L. & Gilmore, C. E. (1969). Neoplasms In *Diseases of Cage and Aviary Birds*, ed. Petrak, Margaret L. Philadelphia: Lea & Febiger.

Ratcliffe, H. L. (1933). Incidence and nature of tumours in captive wild mammals and birds. *Am. J. Cancer*, **17**, 116–35.

Schlumberger, H. G. (1955). Spontaneous goiter and cancer of the thyroid in animals. *Ohio J. Sci.*, **55**, 23–43.

Schlumberger, H. G. (1956). Neoplasia in the parakeet, II, Transplantation of the pituitary tumor. *Cancer Res.*, **16**, 149–53.

Schlumberger, H. G. & Henschke, U. K. (1956). Effect of total body X-irradiation on the parakeet. *Proc. Soc. exp. Biol. Med.*, **92**, 261–6.

Taylor, T. G. (1958). *Feeding Exhibition Budgerigars*, 2nd ed. London: Iliffe.

Taylor, T. G. (1959). Narrowing down causes of French moult. *Cage Birds*, **116** (3010) 339–40.

Taylor, T. G. (1960). Important findings in French moult probe. Report from team engaged on NCA research at Reading. *Cage Birds*, **118** (3066), 535–6.

Taylor, T. G. (1965). Nutrient requirements of budgerigars. *Mod. vet. Pract.*, **46** (9), 60–66.

Taylor, T. G. (1969a). French moult. In *Diseases of Cage and Aviary Birds*, ed. Petrak, Margaret L. Philadelphia: Lea & Febiger.

Taylor, T. G. (1969b). Nutritional deficiencies. In *Diseases of Cage and Aviary Birds*, ed. Petrak, Margaret L. Philadelphia: Lea & Febiger.

Taylor, T. G. & Warner, C. (1961). *Genetics for Budgerigar Breeders*, London: Iliffe.

Thorpe, W. H. (1964). Mimicry, vocal. In *A New Dictionary of Birds*, ed. Thomson, Sir A. Landsborough. London: Nelson.

Tollefson, C. I. (1969). Nutrition. In *Diseases of Cage and Aviary Birds*, ed. Petrak, Margaret L. Philadelphia: Lea & Febiger.

Tucker, V. A. (1968). Respiratory exchange and evaporative water loss in the flying budgerigar. *J. exp. Biol.*, **48**, 67–87.

Valle, R. C. (1971). Water economy in relation to food consumption in budgerigars and canaries. *Diss. Abs.*, **32B** (1), 635–636.

Vaugien, L. (1951). Ponte induite chez la Perruche ondulée maintenue à l'obscurité et dans l'ambience des volières. *C.r. hebd. Séanc. Acad. Sci., Paris*, **232**, 1706–8.

Waterhouse, C. E., Hutcheson, L. M. & Booker, K. M. (1961). Food consumption and palatability studies in budgerigars. *J. small Anim. Pract.*, **2**, 175–88.

Watmough, W. (1960). *The Cult of the Budgerigar*, 5th ed. London: Iliffe.

Worden, A. N. (1960). Nutrition of the budgerigar. Annotation of paper delivered at 3rd World Budgerigar Congress, 1959. *Vet. Rec.*, **72**, 205.

37   Reptiles*

D. J. BALL AND A. d'A. BELLAIRS

General biology

Members of the class Reptilia can be distinguished fairly easily from most other vertebrates by the fact that their bodies are covered by dry horny scales; in tortoises and their relatives these are modified to form the horny plates of the shell. They breathe air by lungs at all stages in life and do not pass through a gill-breathing tadpole stage like many amphibians. They are said to be cold-blooded (poikilothermic), in that their body temperature usually varies with that of their immediate surroundings. The eggs are fertilized internally and, except in viviparous forms, are laid on land, even though the adults (for instance, turtles) may be aquatic. They are divided into four orders: the Chelonia (tortoises, etc.), Crocodilia, Rhynchocephalia (containing only the lizard-like tuatara, *Sphenodon*, of New Zealand) and Squamata (lizards and snakes).

Further general information on reptiles is given in the books by Oliver (1955), Pope (1956, 1962), Schmidt & Inger (1957), Hellmich (1962), Bellairs & Carrington (1966), Bellairs (1969, 1970) and M. Smith (1973). Information on keeping reptiles may be found in the publications by Burghardt (1963), Vogel (1964), H. M. Smith (1965), Pawley (1966), Knight (1970) and Jocher (1973a), and in the volumes of the *International Zoo Yearbook*.

The life of reptiles is greatly influenced by environmental conditions of which temperature is probably the most important. They require fairly high temperatures to keep them active, so that species which live in temperate countries must hibernate during the winter. At the same time, they are rapidly killed by overheating and an increase in body temperatures to much over 45°C is lethal even to most desert species; certain reptiles may die at body temperatures even below 40°C (see Cloudsley-Thompson, 1971).

Many reptiles have specific preferences for particular ranges of temperature and seek out environmental conditions suitable for keeping their bodies within the optimum thermal limits.

They bask in the sun or lie on warm surfaces in order to raise their temperature to the level requisite for activity, and when in danger of overheating they shelter under vegetation, below ground or in water. In this way they are able to practise a measure of temperature-control, relying primarily on external sources of heat (such as solar radiation) for temperature maintenance.

In captivity reptiles may often be denied facilities such as basking sites or shade necessary for keeping their bodies at the optimum degree of warmth. Other conditions which probably play important roles in their lives, such as humidity, light and variety of diet, may also be unsuitable. Cyclical activities such as mating, skin-shedding and hibernation are likely to be upset in captivity. Furthermore, many reptiles in the wild show territorial behaviour during the breeding season, and this may be inhibited by lack of space and overcrowding.

It is hardly surprising that the behaviour of reptiles in captivity is often a travesty of that in nature. They are liable to become sluggish and refuse to feed, eventually dying of starvation. They seldom breed, so that stock can be replenished only by purchase or collection of fresh specimens. These are some of the reasons why reptiles are used in experimental work less often than other vertebrates.

Nevertheless, many species can be maintained in the laboratory, although the ordinary animal house, designed primarily for small mammals, may not be suitable for their accommodation.

Uses of reptiles

Most live reptiles which enter laboratories in England are probably killed fairly soon as material for students to dissect, but others may be used for various types of research. For

* Including some matter from the chapter by the late J. W. Lester in the 2nd edition of this *Handbook*.

example, they are suitable for certain types of experimental work on the nervous system or, in the case of lizards, for studies on tail regeneration. In such work the small size of many species is an asset, since it facilitates the preparation of serial microscopic sections through quite large regions of the body.

Because of their remote relationship to man and domestic animals, reptiles are little used in medical and veterinary research except where the study of snake venoms is concerned. The majority of biologists who work on reptiles have some primary interest in this group of animals as such, and are willing to take considerable trouble over their care and maintenance.

For many types of experimental research it is necessary to keep reptiles in the laboratory for a matter of only a few weeks or months. With many species this is not a very difficult problem so long as they can be induced to feed. No great attempt to reproduce a semblance of natural conditions is necessary and indeed it is probably better not to make one. Neither is it advisable to allow the animals to hibernate; they can be kept active and feeding at a more or less constant temperature throughout the year. If the animals are to be kept for longer periods, however, either for scientific purposes or as pets, greater attention must be given to their care and accommodation.

The different kinds of reptiles

The species most often imported to Great Britain are probably the Mediterranean tortoise (*Testudo graeca*), the wall and green lizards (*Lacerta muralis* and *L. viridis*) and the grass snake (*Natrix natrix*). The common lizard (*L. vivipara*), the slow-worm (*Anguis fragilis*), the grass snake and the adder (*Vipera berus*) are inhabitants of Britain and can be obtained, during the summer at least, from local collectors. All these species may be very difficult to obtain during the winter. A wide range of other species including crocodilians and pythons, can often be obtained from specialized animal dealers. Useful notes on the more readily available types are given by Vogel (1964).

Order Crocodilia

This contains about 23 species of crocodiles, alligators, caimans and gharials. The different types vary only slightly in habits and appearance as, for example, in the relative length and width of the snout. They reproduce by laying eggs, in either craters dug in the sand or nests built of vegetable debris and some species, at least, show a degree of maternal care. For identification the keys by Wermuth & Mertens (1961) and Brazaitis (1974) should be consulted.

Most crocodilians offered for sale are young, under 1·0 m long. The forms most often sold are certain caimans such as the 'spectacled' *Caiman crocodilus* from southern America. Although they may be labelled 'baby alligators', the true American *Alligator mississippiensis* is protected by law in the US, and is seldom imported. It is, however, perhaps the most suitable crocodilian for laboratory work and for keeping as a pet, being comparatively unaggressive and resistant to low temperatures. For directions on maintenance in the laboratory, see Coulson & Hernandez (1964).

The sex of crocodilians may not be obvious from outside inspection, although the males of most species grow bigger than the females. Sex can be determined, however, by gently opening the cloaca with blunt forceps or a finger. The comparatively rigid penis of the male can then be felt (Brazaitis, 1968).

Order Chelonia or Testudinata (tortoises etc.)

This order contains about 240 living species and nine or more families, depending on the system of classification used. They are found throughout the warmer parts of the world. Broadly speaking, chelonians can be divided into three groups on the basis of habits, although this arrangement does not conform with strict zoological classification. Those which are terrestrial are usually called tortoises and are mainly herbivorous. Amphibious freshwater chelonians are often called turtles or terrapins; they are more or less carnivorous. Finally, there are the almost entirely aquatic marine turtles with paddle-like limbs, such as the green turtle (*Chelonia mydas*) and the hawksbill (*Eretmochelys imbricata*), the source of commercial tortoiseshell. Adult green turtles are almost entirely herbivorous but the other species feed on both animal and vegetable material. Sea turtles must be kept in large salt-water aquaria. The names 'tortoise', 'terrapin' and 'turtle' are not used consistently and American writers may apply the name 'turtle' to any type of chelonian.

The structure of chelonians is greatly modified in connection with the shell. They have no teeth, but a horny beak instead, They reproduce by egg-laying, the eggs of tortoises being hard-shelled while those of many terrapins and turtles have a leathery or parchment-like shell. The eggs are usually laid in holes dug by the female in earth or sand, the nest of some species being moistened with urine.

It is often possible to distinguish the sex of a chelonian by external features. In males the tail is generally long and thick, while in the female it is relatively small. The plastron or under-part of the shell of the male tends to be concave in shape, fitting over the carapace or upper shell of the female during coitus. The male has a single penis.

Many species show some form of courtship behaviour. For example, a male tortoise will follow the female about during the breeding season, butting her shell and biting at her feet. Some terrapins of the genera *Chrysemys* and *Pseudemys* have an elaborate courtship display, the male swimming backwards in front of the female and stroking her face with his elongated front claws.

Chelonians are traditionally long-lived and even some of the smaller species have been known to survive for periods of over 50 years.

The common Mediterranean tortoise, *Testudo graeca*, (family Testudinidae) is a native of southern Europe and North Africa. Large individuals have a carapace nearly 0·3 metre long. This animal is sometimes called the 'spur-thighed tortoise' because it possesses a prominent spur on the back side of each thigh; the related Hermann's tortoise (*T. hermanni*), with which it is easily confused, lacks these spurs, but has a horny tubercle on the tip of the tail instead.

The terrapins most commonly available are the European pond-terrapin (*Emys orbicularis*) from central and southern Europe and North Africa, and the North American red-eared or elegant terrapin (*Pseudemys scripta elegans*); babies of the latter are often sold but need special care. Both species belong to the family Emydidae. The European form has a blackish carapace, sometimes 20 cm long; the shell, head and limbs have small yellow streaks or dots. The red-eared terrapin is more handsomely marked, having a prominent red or orange stripe on the side of the head behind the eye and yellow markings on the shell and extremities. This terrapin seems very suitable for laboratory work, and directions for keeping it are given by Boycott & Robins (1961). There are other related forms and correct identification may be important.

Larger species which might be kept in the laboratory are the snapper (*Chelydra serpentina*; family Chelydridae), a fierce predatory terrapin from North America, and some of the soft-shelled turtles (family Trionychidae); these are adapted for life at the bottom and have flat shells and long necks which can be darted out at great speed to capture prey.

The books by Carr (1952), Wermuth & Mertens (1961), Pritchard (1967) and Ernst & Barbour (1972) are helpful in the identification of chelonians.

Order Squamata (lizards and snakes)

Lizards are classified in the suborder Sauria or Lacertilia, snakes in the suborder Serpentes or Ophidia. Since many lizards (for example, the slow-worm) lack external limbs, the distinction may be difficult. It may be stated as a rough guide that a limbless reptile with a single row of large scales down the belly in front of the vent and a transparent spectacle-like eye-covering is a snake, whereas one with small belly scales and movable eyelids is a lizard.

The lizards and snakes are by far the largest groups of modern reptiles, each containing nearly 3,000 species and about 17 and 10 families respectively. The males possess paired organs of copulation known as hemipenes, instead of a single median penis as in chelonians and crocodilians. In all snakes and many lizards the tongue is protrusible and its tip is forked to a greater or lesser extent. It acts in conjunction with a pair of important sense organs, the organs of Jacobson, which lie above the roof of the mouth. The sense which they serve is akin to smell, scent particles being carried to them by the tongue. This mechanism explains the constant tongue movements made by snakes and certain lizards when they are exploring their surroundings or investigating their food.

Most lizards and snakes lay eggs with parchment-like shells. These are usually deposited in holes or under vegetation and are hatched by the warmth of their surroundings; in a few species, (for example some pythons) the eggs are brooded by the mother. A substantial number of both snakes and lizards produce their young alive, retaining the eggs within their bodies until development of the embryos is complete.

Mating behaviour shows much variation; certain lizards, notably those in the agamid and iguanid groups, go through elaborate forms of courtship display in which the erection of crests and other appendages, bobbing movements of the head and forequarters, and colour change may play a conspicuous part. Sex distinction is often possible on the basis of size or colour, especially in lizards, but may be difficult in snakes. In males the base of the tail may appear somewhat swollen owing to the presence of the hemipenes, while in snakes the tail of the male is often relatively longer than that in the female. A further way of determining the sex of a living snake is to insert a blunt probe backwards into the cloaca. In the male the probe passes back into the tail through the cavity of one of the hemipenes for about 1 cm or more, depending on the size of the individual. In the female the probe cannot be inserted for any distance since its passage is checked by the posterior wall of the cloaca (see Fitch, 1960).

Most lizards shed their skins periodically in large flakes but in snakes the skin is usually shed as a single piece or slough. The frequency of moulting varies with the age and health of the animal, but may be several times a year. The skin of a snake goes dull some time before it is shed and the transparent spectacle over the eye becomes bluish and opaque, regaining its transparency a few days before the act of moulting. Snakes often stop feeding during the pre-moulting period and normally tame animals may become irritable. Captive snakes which have difficulty in shedding should be bathed and gently helped by hand, the old spectacle being detached if necessary.

Many lizards have a special mechanism for breaking off the tail, which will regenerate.

Many lizards and snakes have lived for over 5 years in captivity and some species have been known to survive for over 20 years.

It is possible to mention here only the more important types of lizards and snakes, in particular those most likely to be used for laboratory work.

Lizards

Family Gekkonidae (geckos) (Fig. 37.5): includes genera *Hemidactylus*, *Tarentola*, *Gekko*. A large family of (mostly) small lizards, common in all warm countries, some species entering buildings. Many are adapted for climbing and have specialised digital pads which enable them to cling to smooth surfaces, but some live on the ground. Many are partly nocturnal. Mostly oviparous, laying hard-shelled eggs. Many species are probably suitable for laboratory work.

Family Agamidae (agamids): e.g. *Agama*, *Uromastyx* (spiny-tailed lizards), *Chlamydosaurus* (frilled lizard), *Draco* (flying lizards). Old World, including Australasia. Many are desert-living, some arboreal, a few amphibious. Sexual differences are often marked, the males tending to be bigger, more brightly coloured and having better developed skin appendages such as crests and throat-fans. Oviparous. The more active species are not easy to keep. The anatomy and behaviour of *Agama* are described by Harris (1963, 1964).

Family Iguanidae (iguanids). Includes some large forms such as *Iguana* (1·5 m) and many smaller ones such as *Phrynosoma* (horned lizards) and *Sceloporus* (spiny lizards, swifts). Nearly all found in New World, paralleling the agamids in adaptive radiation. The arboreal *Anolis* lizards have a well developed throat-fan and striking powers of colour change. They are often called 'chameleons' in the USA and have been used a good deal in laboratory studies. Mostly oviparous.

Family Chamaeleonidae (chameleons). Specialised arboreal lizards with remarkable eyes, powers of colour change, prehensile limbs and tail, and long extensible tongues with which they capture insect prey. Old World, mostly oviparous. If given proper attention and sufficient food, chameleons do better in captivity than is often believed and may even breed. They show territorial behaviour. Some of the dwarf species such as *Microsaura* (= *Lophosaura*) *pumila* from southern Africa, a viviparous type, might be fairly easy to keep in the laboratory (Bustard, 1959).

Family Lacertidae. Small or medium-sized terrestrial lizards from the Old World, including the common lizard (*Lacerta vivipara*), the sand lizard (*L. agilis*, also found in England), the wall lizard (*L. muralis*), the green lizard (*L. viridis*). The last is probably the easiest to keep in the laboratory and its size (24–45 cm) makes it suitable for many types of experimental work. *L. dugesii*, the Madeira wall lizard, if obtainable, may also be a useful laboratory species. Most lacertids reproduce by egg-laying, *L. vivipara* being an exception. The eggs of this lizard will develop in culture after being removed from the mother (see R. Bellairs, 1971).

Family Scincidae (skinks): e.g. *Scincus, Chalcides, Tiliqua.* A large, world-wide group; many species secretive or burrowing, some live in sand. Body often rather long with shiny scales; legs short, sometimes tiny or absent. A few, such as the blue-tongue (*Tiliqua scincoides*) from Australia are quite big (40 cm). Many skinks do well in captivity. Many are viviparous, some have well developed placentation.

Family Anguidae. Includes limbless forms such as *Anguis fragilis* (slow-worm), found in Britain, and others such as *Gerrhonotus* (alligator lizards) from USA in which limbs are present. *Anguis* is easy to keep, very long-lived (record 54 years), and eats slugs and earthworms. It is viviparous and its eggs can be explanted and cultured (see R. Bellairs, 1971).

Family Varanidae (monitors) (Fig. 37.1): one genus, *Varanus.* Old World, including Australasia. Mostly large, with predatory habits, feeding on carrion, eggs and small animals. *V. komodoensis*, the Komodo dragon is the largest living lizard, reaching about 3 m. The tongue is long, forked and highly protrusible, much as in snakes. Oviparous. Monitors have lived for a long time in captivity. Their size makes them suitable for some kinds of laboratory work and their skeletons are useful for class demonstration.

Family Helodermatidae. This contains the only two poisonous lizards known, the Gila monster (*Heloderma suspectum*) from parts of the southern USA, and the Mexican beaded lizard, *H. horridum.* Anyone contemplating work on these lizards should consult the fine monograph by Bogert & del Campo (1956).

Snakes

Family Boidae (boas and pythons). Primitive snakes, many species retaining vestiges of hind limbs. Non-poisonous, killing prey by constriction in coils. Can often be tamed. Pythons found in Old World, including Australasia, boas mainly in New World. The species range from smallish burrowers such as *Eryx* to the huge anaconda (*Eunectes*) and reticulated python (10 m). Pythons lay eggs which in some species are brooded in the coils of the female; most boas produce living young. Pope (1962) gives much information on the large species.

Family Colubridae. Some genera are listed in Table I, p. 511. This group (at least as recognised in traditional classification) contains the great majority of snakes, including most of the familiar harmless types. It is often divided into two groups, the harmless aglyphs, which have simple teeth, and the more or less poisonous opisthoglyphs (back-fanged snakes) in which one or more of the posterior maxillary teeth are grooved to transmit venom. Only a few opisthoglyphs such as the boomslang (*Dispholidus*) are dangerous to man but a bite from any of the large species should be treated seriously. Oviparous and viviparous, according to species.

Family Elapidae: e.g. *Naja* (cobra), *Bungarus* (krait), *Dendroaspis* (mamba). Found in most warm countries; includes all venomous snakes in Australia. Front maxillary teeth modified to form fangs and canalised for venom conduction. Some species very dangerous. Mostly oviparous.

Family Hydrophiidae (sea snakes). Highly poisonous, found mainly in warm oriental seas. Most species are entirely aquatic and give birth to living young in the water. Notes on keeping them are given by Klemmer (1967).

Family Viperidae: e.g. *Vipera* (e.g. adder), *Bitis* (e.g. puff adder), *Crotalus* (rattlesnakes), *Agkistrodon* (moccasin, Figs 37.7–37.9: copperhead). Divided into two subfamilies, the Viperinae (Old World vipers) and the Crotalinae or pit-vipers, including the rattlesnakes. The pit-vipers are found mainly in the New World and Asia; they are so-called because they possess a pit on either side of the head between the eye and nostril which functions as a heat receptor, probably detecting warm-blooded prey.

In all Viperidae the maxillary bones are hinged to the skull so that the canalised fangs can be erected; the fangs may be very long and when not in use they are folded back along the roof of the mouth. Mostly viviparous.

General. It should be emphasised that in general there is no easy way of distinguishing harmless snakes from poisonous ones without examining the teeth, and all unidentified snakes should be treated with caution. Since the poison fangs of snakes, like reptilian teeth in general, are renewed throughout life, fang-extraction is only a temporary method of rendering a snake harmless. A more permanent method is to block the venom duct by coagulation with an electric cautery; this technique has been used successfully on rattlesnakes and allegedly has little effect on their general health (Jaros, 1940). Generally speaking, however, the use of live poisonous snakes in the laboratory is inadvisable unless their presence is really necessary, as in venom research. Special precautions for keeping them are described by Gans & Taub (1964).

Handling

Most reptiles will bite if they are unaccustomed to handling, but where the smaller non-poisonous species are concerned such injuries seldom have untoward effects. Large lizards such as monitors, the bigger terrapins and snakes such as pythons can all produce severe wounds which require treatment; the recurved teeth of snakes may break off in the wound and must be removed.

The secret of handling reptiles is to grasp them firmly, confidently and rapidly; a hesitant approach often provokes a bite. Potentially vicious reptiles such as monitors, small crocodiles and the bigger non-poisonous snakes should be held firmly behind the head; a snake-stick may be used to pin the neck. It is important that the creature's body should be supported comfortably by the other hand or by an assistant, to prevent the body and tail from thrashing (see Figs. 37.1–37.3). The jaws of crocodilians can be tied or muzzled; vicious turtles can be held by the back of the carapace (Fig. 37.4). Lizards with fragile tails should not be grasped by these appendages; climbing forms such as geckos may be pinned with a duster against the side of the cage (Fig. 37.5). Leather gloves give useful protection from the bites and scratches of larger reptiles.

Poisonous snakes, even when apparently tame, should be handled as little as possible and can be lifted by means of a snake rod (Fig. 37.6A). This is particularly useful for sluggish heavy snakes

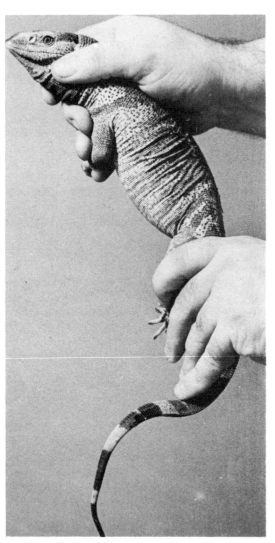

Fig. 37.1. Correct method of holding large lizard with body supported. (African monitor: *Varanus albigularis = V. exanthematicus.*) (Photograph: UFAW.)

Fig. 37.2. Correct method of holding snake liable to bite. The species is the harmless four-lined snake, *Elaphe quatuorlineata*. (Photograph: UFAW.)

such as puff adders and rattlesnakes and can be slipped under the middle of the snake's body so that it balances when lifted. If it is essential to handle these reptiles the head can be firmly pinned with a stick and gripped with the free hand (Figs. 37.6B, 37.7, 37.8). Large viperid snakes have extremely long fangs and if loosely held may twist round and use one fang by biting sideways. These snakes are best lifted or re-strained by means of the special 'stick' shown in Figure 37.6c. Further information on handling snakes is given by Ball (1968).

It may be convenient to immobilise a snake for purposes of examination or operation by inducing it to enter a transparent tube (Fig. 37.9). Such a tube can be made by rolling a sheet of cellulose acetate and securing it with tape or elastic. Holes can then be cut in the walls of the tube to allow access to the snake's body. If one corner of a sack is cut away and the edges of the opening are fastened round the mouth of the tube, the snake can be dropped into the sack and will probably enter the tube of its own accord.

Identification of individuals

It is often necessary to identify individuals among a number of animals kept in a single cage, and this cannot always be done on the basis of such features as size and colour. Marking with spots of paint or quick-drying enamel is useful as a short-term means of recognition, but such spots may be rubbed off, or shed when the animal moults.

Chelonians can easily be marked by filing a notch in the edge of the carapace; this will remain visible for many years. Lizards and crocodilians can be identified by clipping off the outer phalanges from one or more of the toes with sharp scissors. Careful notes should be kept of the position of the extremities clipped (e.g. right hind foot, 1st digit). This is a more or less permanent method often used in field studies. If any regeneration does occur it is slow and often incomplete. Snakes can be marked by clipping the ventral scales. Other methods of identification, such as tattooing, have been recommended.

Snake-bite precautions

The treatment of snake-bite, especially first aid treatment, presents many controversial problems and ideas about it have changed considerably during recent years. Current views on the matter have been expressed by Chapman and by Reid (in Bücherl *et al.*, 1968), and in an American handbook (see reference to US Navy Bureau). It is very difficult to make hard and fast recommendations for bites by all species of snake, but the following measures are suggested.

1. Reassure the victim. Bites by poisonous snakes sometimes have no serious conse-quences, even if untreated.

Fig. 37.3. Incorrect method of holding snake (same specimen shown in Fig. 37.2). Here the body is not held by the other hand; the snake is difficult to control and may pull its head away. (Photograph: UFAW.)

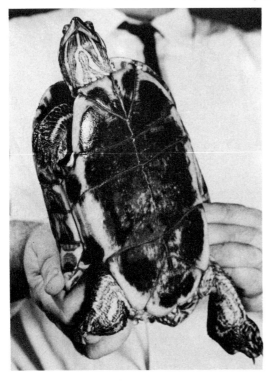

Fig. 37.4. Correct method of holding terrapin (*Pseudemys scripta*) liable to bite. The hands are out of reach of the beak and claws. This method is more important in the case of fiercer species such as the snapper (*Chelydra*). (Photograph: UFAW.)

2. Wipe excess venom from the wound.
3. Make the victim rest, and if possible immobilise the bitten part by a sling or splint. Remove rings, bracelets, etc. from the region of the injury so that they are not involved in any swelling.
4. Admit the victim to hospital as soon as possible for observation and treatment with antiserum if necessary. Generally speaking, routine administration of antiserum immediately after the bite is not advised. The administration of antiserum should be a matter for decision by physicians at the hospital concerned (see below).
5. Incision and suction of the wound are not advised after bites by most kinds of snakes and the value of tourniquet seems to be doubtful. If this last measure is followed, it should be carried out as soon as possible. The US Manual recommends in some cases

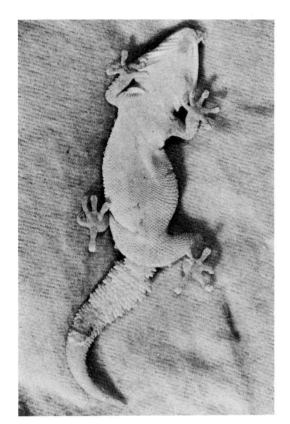

the application of the tourniquet 2–4 inches above the bite and tight enough to block the superficial venous and lymphatic return but not the arterial flow. It should be released for 90 seconds every 10 minutes and removed either at the time when anti-serum is given, or after a maximum of 4 hours.

Prevention of snake-bite is much better than cure. It should be emphasised that long con-tinued dealing with poisonous snakes may engender a dangerous familiarity and that it is the accident which can occur after years of impunity that should particularly be guarded against. Stringent precautions should always be taken (see Gans & Taub, 1964), and certain routine safety measures, such as ascertaining the positions of all the snakes in a cage before it is opened, should be enforced. Cages should be thoroughly secure and no unauthorised people should be allowed access to them. The presence of poisonous (as opposed to non-poisonous) snakes should be clearly indicated by labels.

Fig. 37.5. Method of pinning gecko to glass with soft duster. The delicate skin and fragile tail will not be injured. (Photograph: UFAW.)

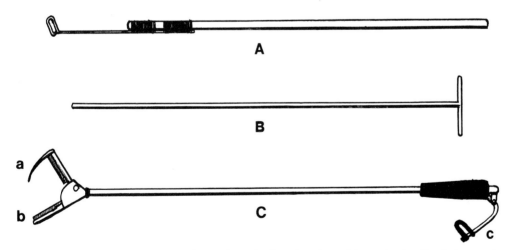

Fig. 37.6. A. Snake rod. This consists of a wooden pole about 1 m. long with metal rod ending in loop, and bound to pole with wire.

B. Snake stick with T-end for pinning down snakes. (See also Fig. 37.7.)

C. Special type of snake stick used for lifting or restraining dangerous snakes. The jaw (a) is hinged and operated by the handle (c); (b) is fixed. Both jaws are padded with rubber. Manufactured by the CEE-VEE Co., Bexhill, Sussex, from a design by J. Ashe (late of the Nairobi Snake Park).

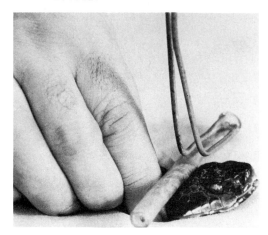

Fig. 37.7. Snake stick applied to neck of cottonmouth moccasin (*Agkistrodon pisicivorus*) held on plastic foam sheet to prevent slipping. (Photograph: UFAW.)

It is very important that a special relationship with a nearby hospital should be established by any institution where poisonous snakes are kept. Routine procedure for the transport, admission and treatment of snake-bite victims should be worked out in mutual consultation, before any casualties occur. For example the reactions to serum and the blood groups of staff handling snakes should be investigated and records of such data should be kept readily available. Supplies of serum in adequate quantities for the appropriate species should be stored either in the hospital or where the snakes are, so that they can be sent along to the hospital with the victim. The identity of the snake inflicting a bite should of course be noted.

Transport

For journeys of up to a few days, small reptiles can be placed in boxes or tins with ventilation holes, packed fairly tightly with moss or hay. It is important to avoid overcrowding as reptiles may injure each other with their claws. Snakes will also travel conveniently in bags which are permeable to air; several bags, each containing one or two snakes, may be packed close together in a single wooden box. Larger reptiles such as monitor lizards, crocodilians and tortoises should if possible be packed individually in separate containers or in one partitioned container. On journeys lasting more than a few days they will need water and should be hosed down

at frequent intervals, the water being introduced through a hole at the top of the crate and allowed to drain away through holes in the bottom. Except perhaps in the case of small lizards, feeding is unnecessary.

Tropical reptiles in transit should not be exposed to temperatures of much under 20°C for more than a few hours; journeys by air in unheated freight compartments may be fatal, and permission should be sought from airline companies for the reptiles to enjoy passenger conditions of temperature and pressure. Reptiles in transit may also be in danger from overheating, and their containers should never be left in strong sun.

Special precautions should be taken over the transport of poisonous snakes, and all containers should be thoroughly checked for holes through which young snakes could pass. Many poisonous snakes (e.g. big vipers and rattlesnakes) are viviparous, and young may be born during a journey. It should be noted that the very long fangs of some viperine snakes can easily penetrate the cloth of bags or ventilation

Fig. 37.8. Correct method of holding moccasin. The angle of the mouth is firmly gripped between the thumb and index finger. (Photograph: UFAW.)

Fig. 37.9. Moccasin immobilized in acrylic tube. (Photograph: UFAW.)

holes in box containers; the latter should therefore be covered with wire gauze. Accurate and conspicious labelling of boxes containing venomous reptiles is, of course, essential. Suitable conditions for transit should also be indicated.

Breeding

Under favourable conditions reptiles sometimes mate and reproduce in captivity, while animals inseminated before capture will lay eggs or produce their young. Most do not breed frequently or regularly enough, however, to provide a reliable source of stock for laboratory work. In certain chelonians, lizards and snakes sperm can be stored alive for long periods within the reproductive tract of the female and fertile eggs can be laid after months, or exceptionally years of isolation. Parthenogenetic reproduction has also been shown to occur in a few species of lizards and may possibly be more widespread among reptiles than is at present realised (see Bellairs, 1970, for account of recent work).

The hard-shelled, relatively impermeable eggs of tortoises and the Nile crocodile can be incubated in dry sand, while those of geckos should also be kept dry and left where they are laid, usually adhering to objects such as bark or stones. The eggs of alligators and of those crocodilians which make nests of vegetation probably require rather more moisture (see Chaffee, 1969).

Striking the right degree of humidity is important in incubating the parchment-shelled eggs laid by most other reptiles. If they are kept too dry they become dessiccated and collapse, while if they become too moist they are prone to fungus infection. The eggs can be shallowly imbedded in clean, slightly damp sand, sawdust or sphagnum moss and placed in a clean container. Alternatively they may be placed between layers of moist paper towels. Zweifel (1961) and Hoessle (1969) who review the various methods, recommend keeping eggs in damp sand or sterile moss in polythene bags, slightly inflated and sealed, for instance, with an elastic band. Temperatures of 25° to 30°C seem suitable for the eggs of most subtropical and tropical species. Many reptile eggs have an incubation period of two to three months, the duration being more prolonged at lower temperatures. Once the eggs have been placed in their artificial nest they should be disturbed as little as possible. It is usually advisable to remove them from the parent's cage, except in the case of species such as pythons which brood their eggs. Brooding females must be isolated from other individuals which might damage the eggs.

Accommodation

An attempt to simulate the animals' normal surroundings may be necessary if some aspect of its natural behaviour, such as courtship, is to be studied or in attempts at breeding. Elaborate cages, however, require much time and trouble to maintain and if not kept really clean may harbour parasites; for most laboratory purposes cages or enclosures should be simple.

Outdoor enclosures

If there is space in the laboratory grounds, British and some temperate-zone reptiles can be kept out of doors, at least in the summer. The additional space available and the possibility of the animals obtaining some natural food such as insects are advantages, and breeding is more likely to occur. On the other hand, snakes and lizards kept in this way tend to become very wild, darting for cover when approached, and in winter problems of hibernation arise. If not required during the winter, some species can be left to hibernate in deep well-drained holes filled with loose earth and covered over. Reptiles from warmer temperate zones should probably be brought indoors and either allowed to hibernate in a cold room or shed under frost-proof conditions or in boxes filled with straw, or kept active artificially at raised temperatures.

The continental reptiles most suitable for keeping out of doors are the Mediterranean tortoise (*Testudo graeca*) and the green and wall lizards (*Lacerta viridis* and *L. muralis*).

A wire-netting fence about 30 cm high makes a suitable pen for tortoises, and can be erected on a lawn. It should contain a dish of water deep enough to allow the tortoise to immerse its head when drinking and sunk to the level of the ground. There must also be some kind of waterproof shelter into which the tortoise can retreat from wet, cold, or excessive heat. This can be made from a wooden box roofed with linoleum.

Outdoor cages for lizards can be made from four sheets of glass or any really smooth material, the ends of each sheet being let into grooves in wooden uprights. Triangular pieces of glass should be placed over the tops of each corner and fixed on the inside with Sellotape. The enclosure may be placed on a smooth surface such as a flat roof or strip of concrete, or it may be dug a couple of inches into the ground. Good drainage is essential to prevent the floor of the pen from becoming waterlogged. The height of the glass should be at least twice the length of the lizard (including tail) and no vegetation which might help the animal to climb out should be allowed to grow near the sides. Tins with perforated bottoms and sunk in the ground are useful repositories for meal-worms and other living food. A shallow dish of water, sunk to ground level, should always be provided. There should also be some kind of retreat such as a dry, straw-filled wooden box which is easily moved and opened.

Other types of enclosure can be made by bolting together sheets of galvanized metal to form a circle, or from Netlon gauze (a finely perforated material obtainable from Spicer's Ltd., 19 New Bridge Street, London, E.C.4) attached to a wooden frame. The gauze forms a roof as well as sides so that the cage must be furnished with a door. Roofed enclosures have the advantage of excluding predators such as birds, cats and rats, the latter being liable to

Fig. 37.10. Fibre-glass tank about 1.2×0.3 metres used for housing active, dangerous snakes. The top is made of wood and perspex. (Photograph: UFAW.)

attack reptiles during hibernation. They are also more suitable for snakes which may be able to climb out of roofless cages unless these have smooth walls somewhat higher than their own length. Pits and enclosures with raised opaque walls (for example, made of tiles) are unsuitable unless they are large enough to receive sun during the greater part of the day.

The essential requirements for all outdoor enclosures are good drainage, a sufficiency of suitable basking sites and shelter from cold, wet, and powerful sun. Generally speaking, snakes require less sun than lizards (except the slow-worm and other burrowers) and tortoises, but more water.

Indoor cages

Aquatic. Terrapins and crocodilians should be kept in covered aquarium tanks partly filled with water; a depth of 5 to 7·5 cm is adequate for small specimens. About half of the space inside the tank should be converted into land. Bricks placed on each side of the tank to support a board just above water-level are very suitable since the animals often like to lie under cover in the water. It is not necessary to put sand or shingle in the tank, and, indeed, such material renders cleaning difficult. Terrapins are messy feeders, generally eating under water. Cleaning their tanks is easier if plugs are fitted, but it may save trouble if they are transferred to a separate tank when food is given.

Terrestrial. Small lizards and snakes can be kept in containers of almost any kind provided these admit plenty of light and have some ventilation. Perspex may be used instead of glass and a small hole closed by a cork may be bored in the top to allow food to be dropped in without risk of lizards escaping. The glass-fronted, metal vivaria sold in shops are satisfactory and fibre-glass tanks and plastic propagators made for plants (e.g. propagators made by Stewart Plastics Ltd, Purley Way, Croydon) can be highly recommended (Figs. 37.10, 37.11, 37.12).

The vivarium floor should be kept as clean and dry as possible; slight tilting may help to drain off any water which escapes from the dish. The placing of sheets of paper on the floor facilitates cleaning: removal of excreta, pieces of shed skin, dead insects, etc.

The amount of 'furniture' in the cage should

Fig. 37.11. Tank shown in Fig. 10 turned on its side and containing large poisonous snakes. Supplementary heating is provided by the lamp at the left. The card in the perspex window shows directions for the first-aid treatment of snake-bite. (Photograph: UFAW.)

be minimal. Climbing reptiles such as chameleons and many snakes require twigs or a branch; geckos may be given a few pieces of bark or pot. Snake cages should also contain a rough heavy object such as a brick against which they can rub off their sloughs. Heavy objects in a vivarium should rest securely on a level floor and not on a substrate of sand or shingle, for if they shift slightly they may crush animals hiding beneath them.

Materials such as sand, leaf mould and moss are best used only in the cages of burrowing species such as certain skinks and slow-worms. The latter live well in covered glass bowls if provided with leaf mould, a small dish of water and a piece of bark under which to hide. More ambitious uses of living vegetation in zoos and the lighting problem involved in its maintenance are discussed by Logan (1969).

All cages should contain a dish of water and in the case of snakes and certain lizards such as monitors and iguanas this should be large enough to allow the reptiles to immerse themselves without the water overflowing. Snakes often like to bathe, especially before shedding their skins, and many of the species commonly sold as pets such as grass snakes and pythons are particularly fond of water.

Reptiles kept as pets, or in the hope that they will breed, should be allowed as much space as possible; this is particularly important with

lizards such as agamas which show pronounced territorial behaviour, and may also be a factor with lacertids such as green and wall lizards. A pair of green lizards might be accommodated in a cage 60 cm long and 30 cm wide, but for breeding purposes an even bigger one might be desirable. Such reptiles will, of course, survive and feed under more confined conditions, but even for laboratory purposes the crowding of large numbers of individuals together for any length of time should be avoided. Baby reptiles should not be kept with adults and should be transferred to smaller containers from which there is no risk of escape. Useful directions for keeping snakes under simple conditions are given by Littleford & Keller (1946); further information is given by Smith (1965), and Ball (1969).

Light, temperature and humidity

All vivaria (except those containing burrowers) should if possible receive some direct sunlight for part of the day, but cages should not be placed where there is danger of overheating, as in a glass window facing south. If sunlight is not available electric-light bulbs can be used to simulate it and also provide the readiest means of increasing the temperature: the reptiles should not be able to make contact with them or they may burn themselves. Infra-red lamps are also useful for providing basking sites for heat-loving reptiles, especially in large cages. Ultra-violet light has been recommended for desert lizards (see Kauffeld, 1969) and has been found beneficial to the matamata terrapin (*Chelus*) by Pawley (1969). It is probably unnecessary for many reptiles, however, and may even be

Fig. 37.12. Polystyrene cages made from plant propagators are useful for keeping small reptiles. As used by Dr. I. W. Whimster. (Photograph: UFAW.)

harmful to certain snakes. Laszlo (1969) favours the use of fluorescent tubes which emit light resembling natural daylight.

British species and some continental forms such as the green and wall lizards can be kept during the summer at room temperature in the United Kingdom, though supplementary heating by means of an electric bulb may be desirable. Sub-tropical and tropical reptiles should always receive artificial heating. While it is not easy to lay down a general rule it is probably suitable to keep cages for lizards between 25° and 38°C and those for snakes between 25° and 30°C. European and North American terrapins should be kept between 23° and 30°C, crocodilians between 25° and 33°C. Alligators may prefer slightly lower temperature ranges than crocodiles or caimans. The temperatures should fall a few degrees at night. Many lizards, especially desert species, often expose themselves to higher temperatures for limited periods. This can be made possible by the provision of suitable basking sites such as rocks placed beneath a lamp.

Heated cages should allow for a gradient of temperature, with some parts warmer than others. This permits reptiles to practise their own temperature control by basking or retiring under rocks or into water according to choice. For this reason, lamps are more suitable than sub-floor heating which tends to raise the temperature of the whole cage uniformly. Sub-floor or radiator heating is useful, however, as background heating for large cages. Suitable equipment can be obtained from manufacturers of greenhouse equipment such as Humex Ltd, 4 High Rd, Byfleet, Surrey. A thermostatic heater suitable for a tropical fish tank can be placed in the water of vivaria containing amphibious species.

Since the natural activity rhythms of many reptiles are influenced by light as well as temperature, it seems desirable that cages should be darkened at night, or during the day in cages designed for the exhibition of nocturnal species. The possibility of simulating conditions of seasonal change deserves further study (see Peaker, 1969).

Little is known about the humidity requirements of captive reptiles and these doubtless vary according to species and habitat. It is probably better to err on the side of dryness as a general rule, so long as bathing facilities are available.

Hibernation

Captive reptiles from temperate regions such as northern Europe and northern America may be allowed to hibernate during the winter. Lizards and snakes in outdoor vivaria must be provided with suitable retreats out of reach of frost and damp. One method is to place a waterproof container such as a large glass jar on the ground, tilting it with a piece of wood so that any water drains out of its open end. It can then be filled with dry moss or other insulating material and covered with sacking, wood and earth to a depth of a foot or more to exclude frost. Ingress to the hibernating chamber can be provided through drainpipes or other tubular passages opening at ground level. A similar and more permanent retreat below ground level can be provided so long as proper drainage can be ensured. Alternatively, the animals may be kept in an unheated building in the manner suggested below for chelonians.

Although tortoises may sometimes survive an English winter by burying themselves, and terrapins by burrowing into the mud at the bottom of a natural pond, it may be better to bring these reptiles indoors say towards the end of September, when they stop feeding. They should be placed in a cool but frost-proof building such as a brick outhouse or garage, in boxes at least 0·3 metres deep, packed with moss, leaves, etc. To prevent them from wandering they may be confined individually in cloth bags within a large box. They should be left alone during mild winter spells since temperatures which are high enough to promote activity may not be adequate to induce them to feed. Around the end of March, if the weather is good, the chelonians can be released in their outdoor enclosures during the daytime but should be brought indoors at night. Captive reptiles tend to have a high mortality after they have emerged from hibernation and before they have begun to take much food; a sudden fall in temperature may render them inactive so that they do not seek shelter and fall victims to night frost. Chelonians and other reptiles can normally be left outside for the whole time after the end of May.

Some reptile keepers prefer to keep their animals active and feeding throughout the winter in warm vivaria. This is probably the most suitable procedure for temperate-climate reptiles kept for many purposes in the laboratory, and also for very young individuals, and specimens in bad health. While it may be less risky in the short term than allowing the animals to hibernate it is uncertain whether it is conducive to longevity; it is also possible that breeding is inhibited since the animals are prevented from undergoing their normal activity rhythms. Hibernation may be more important for maintaining health in some species than in others (see Peaker, 1969).

Nutrition

Some reptiles, and especially snakes, can go without food for long periods, and although they should normally be fed at regular intervals a fast of a fortnight or so (as during a holiday period when the laboratory is closed) will probably do them no harm. At such times the temperature of their cages may be lowered slightly to reduce activity.

Tortoises will eat lettuce, cabbage, dandelions, clover, grass, tomatoes and other fruit, and are said to be fond of yellow flowers. Some will take raw fish or meat; in fact they seem to be more carnivorous than has been generally realised. Cereals such as Fairex, white bread, and tinned dog meat should be offered and provide a useful source or extra minerals and vitamins. Individuals often show preference for one particular food, and it may be necessary to experiment with a newly-acquired animal in order to determine its choice. They should be fed daily.

Terrapins can be given chopped meat including heart and liver, whole fish such as sprats or chopped fish, snails, shrimps, worms and insects, or freshly killed small frogs and tadpoles if these are available. Many species also like vegetable food such as lettuce which can be dropped into the tank. Terrapins should be fed at least every other day. Small crocodilians will accept much the same food as terrapins and also small mice but will not eat plants. Reluctant feeders may be induced to eat by one or more of the following expedients:

(a) slightly raising the temperature of their tank;
(b) waving strips of food close to their mouths or;
(c) placing small pieces of food between their jaws.

If kept on an inadequate diet, chelonians and crocodiles, especially when young, are prone to develop deficiency diseases resembling rickets; these may lead to softening of the limb bones and, in the case of chelonians, of the shell. Such animals should be given fish-liver oil and bone meal smeared on the food or wrapped inside a small piece of meat. Oil can sometimes be introduced into the back of the mouth with a dropper. The provision of cuttlefish 'bone' and chips of vertebrate bone has been recommended, but it is uncertain how far these are utilised. If an adequate diet and access to sunlight are available it is probably unnecessary to give vitamin supplements.

The majority of lizards feed on any small creatures which they can overcome, and should be given as much variety of food as can be obtained. Mealworms (larvae of *Tenebrio* beetles) and gentles (fly larvae) are most valuable as a stand-by, and can be placed in the lizards' cages in a petri dish to stop them crawling about; they may be inadequate, however, as the only food for long periods, and some lizards refuse to touch them after a time. Earthworms, slugs, flies, butterflies, moths and their larvae, cockroaches, grass-hoppers, spiders and centipedes should be offered when possible; grasshoppers and spiders are particularly relished by many species. Many insects can easily be caught during the summer by shaking bushes over a tray or cloth, by drawing a net through vegetation or by the use of a moth trap. Some insects can be raised in the laboratory (Jocher, 1973b and Chap. 42 *et seq.* here).

Some of the larger lizards such as eyed lizards (*Lacerta lepida*) and monitors (*Varanus*) will eat young mice and smaller lizards. For the larger skinks, (e.g. *Tiliqua*) and for heloderms, chopped meat and raw egg are recommended, and dog meat should be tried. Slow-worms will feed entirely on earthworms and slugs, especially the white *Agriolimax*. Not many lizards eat ants, but the small spiny agamid *Moloch horridus* of Australia feeds exclusively on certain species of these insects. Baby lizards, including *Lacerta vivipara*, may prove difficult to feed, but can be offered *Drosophila*, aphids (e.g. greenfly) and tiny spiders.

A few lizards, such as the agamid *Uromastyx* and some of the big iguanas, are more or less herbivorous and should be given lettuce, fruit and other plant food as well as insects. Even some predominantly insectivorous species, such as green and wall lizards, will often eat pieces of tomato, grape and other fruit. Iguanas like meat.

All snakes feed on animals. Most eat vertebrates; some young snakes and adults of a few species feed mainly on invertebrates. Constricting and poisonous snakes usually kill their prey before eating; other species in the wild state swallow it alive. They are always swallowed whole, a procedure which is made possible by the looseness of the snake's jaws and the distensibility of the skin of its head and neck. The swallowing process may be very long-drawn-out if the prey is large and the glottis may be protruded from the mouth to prevent breathing from being obstructed.

Many colubrid snakes, including those of the genus *Natrix*, which are particularly suitable as laboratory animals, feed almost exclusively on amphibians, especially frogs, newts, and on fish. Individuals of some species, such as the grass snake (*N. natrix*) also take toads. Garter snakes (*Thamnophis*) eat earthworms, and in captivity can be given strips of raw meat and fish. Contrary to popular belief, young grass snakes may not often eat invertebrates; they feed readily on the tadpoles of frogs, toads and newts. Although in the wild snakes seldom, if ever, eat animals which have been previously killed, they will generally take dead food in captivity. Live food such as rats or mice have been known to attack and seriously damage reptiles.

Captive snakes will usually feed by day or night, but it may be advisable to offer food in the evening to newly-caught nocturnal species. The food should be removed in the morning if still uneaten. After feeding, the temperature should be maintained at a steady level and the snake left undisturbed since excitement or sudden cooling may cause it to disgorge its food. If possible snakes should not be handled for 48 hours after they have eaten, and it is inadvisable to feed them immediately before a journey.

Food need not always be offered freshly killed; at the London Zoo wild rabbits which have been thawed out after many weeks in cold storage have been accepted by pythons and puff adders. A reluctant feeder may be tempted if dead food is gently moved by means of a stick or wire to give the appearance of life. Warming the carcases of birds and mammals which have been dead for some time may also act as an inducement to feed.

Two or more snakes may attack the same food, and in their efforts to eat it one of them may damage, or even swallow, its companion. Poisonous snakes often strike at one another at feeding time, and the larger vipers may inflict serious wounds with their long fangs. Although snakes have some immunity to the venom of their own species, they may quickly succumb to a bite from another species.

Healthy snakes may do without food for several months without harm. A fast may sometimes be terminated by raising the temperature, by giving the snake a warm bath, or by repeatedly presenting different types of food in various ways. Changing the snake to a new cage will sometimes start it feeding.

If it is necessary to preserve a snake which persistently refuses to eat, force-feeding may be employed. A snake will sometimes eat if its jaws are gently opened and food is pushed into its mouth. Great care should be taken not to damage the jaws. If it still refuses to swallow, more radical measures must be adopted, the food being pushed with a blunt instrument for some distance down the snake's throat. Dipping the food in egg will help to lubricate it and assist its passage down the gullet. In force-feeding it is important to use food which is considerably smaller than that usually eaten by the snake. It is very difficult to force-feed small snakes. If other food is not available the snake may be forcibly fed on chopped meat. The administration through a pipette of egg beaten up in milk has been recommended; such a preparation, however, is liable to adhere to the snake's head and obstruct the nostrils.

A list of the main food preferences of some common genera of snakes is given in Table I. Further directions for feeding are given by Ball (1973) and Savage (1973).

TABLE I

Food preferences of snakes

Family	Genus	common names	Food
BOIDAE		Boas and pythons	Birds and mammals
COLUBRIDAE	*Natrix*	E.g. grass snake	Amphibians and fish
	Thamnophis	Garter snakes	Amphibians, fish, earthworms, chopped meat
	Heterodon	Hog-nosed snakes	Toads
	Coluber	E.g. dark-green snake	Lizards, rodents
	Elaphe	E.g. Aesculapian and four-lined snakes,	Lizards, rodents
	Ptyas	Rat snakes	Rodents
	Pituophis	Bull and pine snakes	Rodents, birds
	Coronella	Smooth snakes	Lizards, young rodents
	Lampropeltis	King snakes	Reptiles, rodents
	Dasypeltis	Egg-eating snake	Eggs only
	All colubrine tree snakes		Lizards, frogs, small birds, rodents
ELAPIDAE	*Naja*	Cobras	Frogs, birds, rodents
	Dendroaspis	Mambas	Rodents, birds
	Bungarus	Kraits	Frogs, rodents
VIPERIDAE		Vipers, rattlesnakes, etc.	Small birds, mammals; fish and frogs for aquatic species

Water

Reptiles should always have water available for drinking and bathing. Some species such as the heloderms, which live in dry regions, will spend much time lying in their bath. Chameleons seldom drink from a dish but will take water from foliage that has been sprayed.

Laboratory procedures

Anaesthesia

Ether can be used as a general anaesthetic for reptiles such as lizards and snakes, the animal being placed in a closed glass container and quickly removed as soon as it becomes relaxed and motionless. Swabs of cotton wool soaked in ether can be applied to the nose if there is any tendency for the animal to come round during the operation. It is, however, easy to kill small reptiles in this way and more refined techniques involving the use of pentobarbitone sodium (Betz, 1962; Hunt, 1964) and 2-bromo-2-chloro-1,1,1-trifluorethane (fluothane) (Gans & Elliott, 1968; Jackson, 1970) are recommended. Kennedy & Brockman (1965) have used deep freezing for 25–60 mins to produce anaesthesia in the alligator for purposes of heart surgery.

Euthanasia

Most reptiles can be killed easily with ether or chloroform, or by decapitation, though the latter may cause the body to go into violent spasms which persist for some time. Tortoises, however are very resistant to inhaled anaesthetics. The following methods for the humane killing of chelonians are recommended by G. M. Hughes & C. Gans (personal communications):

1. Injection through a long needle of an overdose of pentobarbitone sodium (say 50 mg/kg of weight of tortoise) via a hind limb pocket.
2. If a chelonian is covered with crushed ice it will rapidly be immobilised and may then be killed by placing it in a deep freeze (or the freezing compartment of a refrigerator). The cooling with crushed ice is reversible and does not appear to cause unusual trauma; it can be used as a standard method of anaesthesia and the animals recover from it quite quickly. (Gans, personal communication.)

Experimental techniques

Various techniques for operating on reptiles, collecting their blood and other body fluids (Coulson & Hernandez, 1964; Dessauer, 1970) have been developed in recent years. Methods of venom extraction have been described by many workers (see Klauber, 1956, Belluomini, in Bücherl, 1968; Gans & Elliott, 1968: the last article contains many references to recent work on venoms).

Bragdon (1953) describes various surgical techniques which can be applied to snakes.

Embryos of viviparous species at successive stages of development can be obtained from a single mother by Caesarian section (see Zehr, 1962). In experiments on the brain of reptiles parts of the skull can be removed with a dental drill; the wound can subsequently be closed with gelatin sponge and sealed with 5 per cent celloidin (Boycott & Guillery, 1962) or with a solution of perspex in acetone (Gamble, 1956). Techniques used in temperature studies are given by Bradshaw & Main (1968) and by Stebbins & Barwick (1968), and in studies on sense organs, psychology and behaviour by Boycott & Guillery (1962). Various aspects of reptilian biochemistry are described in a book edited by Florkin & Scheer (1974).

Injuries and diseases

Only some of the more important pathological conditions likely to occur among laboratory reptiles will be mentioned here. Reptile diseases in general are described by Mertens (1965) and by Reichenbach-Klinke & Elkan (1965), while Wallach (1969) and Fry (1973) deal with treatment.

Treatment of wounds and external infections

Wounds should be washed and treated with antiseptic ointments or powders; if extensive they should be stitched. Antibiotics such as streptomycin and penicillin should be used in the treatment of severe wounds; Streptopen (Glaxo) and Streptocillin (Willows & Francis) have been found useful. Abscesses and cysts should be incised and packed with such preparations.

Snakes are very susceptible to infections of the mouth and jaws which may result from injuries such as those caused by striking the nose against the cage. If taken in time such conditions can usually be cured by gently removing the dead tissue and pus and powdering the mouth with Cicatrin (neomycin-bacitracin-amino-acid powder) made by Calmic. At the same time a solution of chloramphenicol should be given orally by pouring it into the animal's throat with a dropper. This should be done each day until the animal recovers, the dose being 50 mg/kg body-weight, given in the proportion of 2 mg/ml water (M. Buchanan-Jones, personal communication). As a preventive measure against mouth infections the use of a solution of chloramphenicol powder* (750 mg/4·5 litre water) is recommended. The solution is given daily in the reptile's water vessel for 7 consecutive days and the treatment repeated every six weeks.

Diseases due to unfavourable environmental conditions

Persistent refusal to feed and eventual death through starvation may be caused by keeping reptiles at temperatures which are too low, or denying them access to sufficient sunlight. Diseases resembling rickets may be arrested by allowing greater exposure to sunlight and by giving cod-liver-oil preparations and supplementary calcium (p. 509).

Blisters often develop beneath the superficial layers of the scales of snakes kept under too damp conditions; they clear up when the conditions are remedied.

Ophthalmia

Reptiles, especially tortoises emerging from hibernation, often develop swollen and inflamed eyes which are often due to vitamin A deficiency. Hime (1972) recommends treatment by injection of 5,000 i.u. of vitamin A in oily solution into a hind limb; this should be repeated after 14 days if there is no improvement. A full diet should also be given.

External parasites

Wild lizards, snakes and tortoises are often infested with ticks. These can be killed by painting with paraffin or methylated spirits, and then picked off with forceps, care being taken not to leave the head in the host's body.

Squamata, especially snakes in poor health, are frequently attacked by mites such as *Ophionyssus* spp. which may carry pathogenic microorganisms capable of producing fatal disease. Mite infection is transmitted from reptile to reptile and, since the mites have good powers of locomotion, can travel from cage to cage. It is important to recognize early a mite-infested animal so that it can be segregated and the condition prevented from spreading through the collection. In snakes the first sign of infesta-

* Chloromycetin: Parke Davis.

tion is usually the appearance of silvery-white deposits of mites' faeces on the scales. A few fully-grown females may be seen as shiny red or black beads up to about 1 mm in diameter. Immature mites are like specks of fast-moving dust and may be detected on the hands after handling an infested snake.

Mites can be eliminated by repeatedly washing the entire animal in warm water and wiping it down with cotton-wool. It is best to do this in a special bath outside the cage; all mites must be removed from the bath before another reptile is immersed. An infested cage should be thoroughly cleaned with boiling water, or with a strong disinfectant such as lysol followed by soapy water. No reptiles should be placed in it for at least a week. Mite-infestation is favoured by dirty cages containing pieces of sloughed skin etc. In clearing out infested cages or handling infested reptiles care should be taken to prevent the mites from spreading around the neighbourhood.

Dri-die powder (obtainable from the Grace Chemical Coy, Baltimore, USA) will often rid lizards and snakes of mites if the reptile is allowed to crawl around in a bag containing about 60 mm of powder at the bottom. The cage will still need disinfestation.

Perhaps the most effective anti-mite measure is the hanging of Vapona insecticide strips (made by Shellstar Ltd., 70 Brompton Road, London, S.W.3) in the cage; mites will usually disappear in about three days and no other treatment of the cage is necessary. This method was suggested by Mr Ernest Wagner of the Woodlands Park Zoo, Seattle (personal communication) and has been used with good results in the collection of the Zoological Society of London. Reptiles should not be exposed to preparations containing DDT.

Internal parasites

Reptiles are subject to many types of internal infection—by bacteria, by amoebae which may cause a fatal enteritis, by coccidia (a form of protozoan parasite), and by linguatulids or 'tongue-worms' which are parasites related to the Arachnida. Infestations by round worms and/or tape worms are also common. Treatment for some of these conditions is shown in Table II, for which we are indebted to Mr J. M. Hime, MRCVS, of The Zoological Society of London.

These doses have been calculated for mammals, but may also be applied to reptiles. However, owing to the small size of most reptiles it is usually necessary to break the tablets into pieces, each representing the approximately correct proportion. The broken tablets, or powder obtained from them, can be incorporated into the reptile's food or dusted over it.

REFERENCES

Ball, D. J. (1968). Handling reptiles. *J. Inst. Anim. Techns*, **19**, 143–66.
Ball, D. J. (1969). Housing reptiles. *J. Inst. Anim. Techns*, **20**, 137–54.
Ball, D. J. (1973). Feeding reptiles. *J. Inst. anim. Techns*, **24,** 83–90.
Bellairs, A. d'A. (1969). *The Life of Reptiles*. 2 vols London: Weidenfeld & Nicolson.
Bellairs, A. d'A. (1970). *Reptiles*, Third Edition. London: Hutchinson.
Bellairs, A. d'A. & Carrington, R. (1966). *The World of Reptiles*. London: Chatto & Windus.
Bellairs, R. (1971). *Developmental Processes in Higher Vertebrates*. London: Logos Press.
Belluomini, H. E. (1968). Extraction and quantities of venom obtained from some Brazilian snakes. In *Venomous Animals and Their Venoms*. 1. *Venomous Vertebrates*, ed. Bücherl, W., Buckley, E. & Deulofeu, V. London & New York: Academic Press.

TABLE II
Treatment of parasitic infestations

Parasite	Name of Preparation	Dose
Amoebae	'Furamide' (diloxanide furoate) (Boots)	0·5 g tablet/kg body-weight
Round worms	'Thibenzole' (thiabendazole) (Merck, Sharpe & Dohme)	0·5 g/4·5 kg body-weight. Tablets weigh 2 g each and should be broken
Tape worms	'Dicestal' (dichlorophen) (May & Baker)	0·5 g tablet/2·7–4·5 kg body-weight
Coccidia	'Bifuran' (nitrofurazone) (Smith, Kline & French)	1 tablet/4·5 kg body-weight

Betz, T. W. (1962). Surgical anaesthesia in reptiles, with special reference to the water snake, *Natrix rhombifera. Copeia*, 284–7.

Bogert, C. M. & Del Campo, R. M. (1956). The Gila monster and its allies. The relationships, habits and behaviour of the lizards of the family Helodermatidae. *Bull. Am. Mus. Nat. Hist.*, **109**, 1–238.

Boycott, B. B. & Guillery, R. W. (1962). Olfactory and visual learning in the red-eared terrapin, *Pseudemys scripta elegans* (Wied.). *J. exp. Biol.*, **39**, 567–77.

Boycott, B. B. & Robins, M. W. (1961). The care of young red-eared terrapins (*Pseudemys scripta elegans*) in the laboratory. *Br. J. Herpet.*, **2**, 206–10.

Bradshaw, S. D. & Main, A. R. (1968). Behavioural attitudes and regulation of temperature in *Amphibolurus* lizards. *J. Zool., Lond.*, **154**, 193–221.

Bragdon, D. E. (1953). A contribution to the surgical anatomy of the water snake, *Natrix sipedon sipedon*; the location of the visceral endocrine organs with reference to ventral scutellation. *Anat. Rec.*, **117**, 145–62.

Brazaitis, P. J. (1968). The determination of sex in living crocodilians. *Br. J. Herpet.*, **4**, 54–8.

Brazaitis, P. J. (1974). The identification of living crocodilians. *Zoologica*, N. Y., **58**, 59–83, 101.

Bücherl, W., Buckley, E. & Deulofeu, V. (1968–71). *Venomous Animals and Their Venoms*. 3 vols. London & New York: Academic Press.

Burghardt, G. E. (1963). *Iguanas . . . as pets*. Jersey City. T. F. H. Publications.

Bustard, R. (1959). Chamaeleons in captivity. *Br. J. Herpet.*, **2**, 163–5.

Carr, A. (1952). *Handbook of Turtles: the Turtles of the United States, Canada and Baja California*. Ithaca, New York: Comstock; London: Constable.

Chaffee, P. S. (1969). Artificial incubation of alligator eggs at Fresno Zoo. *Int. Zoo Yb.*, **9**, 34.

Chapman, D. S. (1968). The symptomatology, pathology, and treatment of the bites of venomous snakes of Central and Southern Africa. In *Venomous Animals and Their Venoms. 1. Venomous Vertebrates*, ed. Bücherl, W., Buckley, E. & Deulofeu, V. London & New York: Academic Press.

Cloudsley-Thompson, J. L. (1971). *The Temperature and Water Relations of Reptiles*. Watford: Merrow.

Coulson, R. A. & Hernandez, T. (1964). *Biochemistry of the Alligator*. Louisiana State University Press.

Dessauer, H. C. (1970). Blood chemistry of reptiles: physiological and evolutionary aspects. In *Biology of the Reptilia*, ed. Gans C. & Parsons, T. S. Morphology C, vol. 3, pp. 1–72. London: Academic Press.

Ernst, C. H. & Barbour, R. W. (1972). *Turtles of the United States*. University Press of Kentucky.

Fitch, H. S. (1960). Criteria for determining sex and breeding maturity in snakes. *Herpetologica*, **16**, 49–51.

Florkin, M. & Scheer, B. T. (1974) (Eds.) *Chemical Zoology*, **9** (Amphibia and Reptilia). New York & London: Academic Press.

Fry, F. L. (1973). *Husbandry, Medicine and Surgery in Captive Reptiles*. Bonner Springs, Kansas: V. M. Publications.

Gamble, H. J. (1956). An experimental study of the secondary olfactory connexions in *Testudo graeca*. *J. Anat.*, **90**, 15–29.

Gans, C. & Elliott, W. B. (1968). Snake venoms: production, injection, action. *Adv. oral. Biol.*, **3**, 45–81.

Gans, C. & Taub, A. M. (1964). Precautions for keeping poisonous snakes in captivity. *Curator*, **7**, 196–205.

Harris, V. A. (1963). *The Anatomy of the Rainbow Lizard Agama agama* (*L*). London: Hutchinson.

Harris, V. A. (1964). *The Life of the Rainbow Lizard*. London: Hutchinson.

Hellmich, W. (1962). *Reptiles and Amphibians of Europe*. London: Blandford Press.

Hime, J. M. (1972). Eye diseases in terrapins. *Vet. Rec.*, **91**, 493.

Hoessle, C. (1969). Simple incubators for reptile eggs at St. Louis Zoo. *Int. Zoo Yb.*, **9**, 13–14.

Hunt, T. J. (1964). Anaesthesia of the tortoise. In *Small Animal Anaesthesia*, ed. Graham-Jones, O. London: Pergamon Press.

Jackson, O. F. (1970). Snake anaesthesia. *Brit. J. Herpet.*, **4**, 172–5.

Jaros, D. B. (1940). Occlusion of the venom duct of Crotalidae by electrocoagulation: an innovation in operative technique. *Zoologica*, *N.Y.*, **15**, 49–51.

Jocher, W. (1973a). *Turtles for Home and Garden*. Jersey City: TFH Publications.

Jocher, W. (1973b). Live Foods for the Aquarium and Terrarium. Jersey City: TFH Publications.

Kauffeld, C. (1969). The effect of altitude, ultraviolet light and humidity on captive reptiles. *Int. Zoo Yb.*, **9**, 8–9.

Kennedy, J. P. & Brockman, H. L. (1965). Open heart surgery in *Alligator. Herpetologica*, **21**, 6–15.

Klauber, L. M. (1972). *Rattlesnakes. Their Habits, Life Histories, and Influence on Mankind*. 2 vols. 2nd edn. Berkeley & Los Angeles: University of California Press.

Klemmer, K. (1967). Observations on the sea-snake *Laticauda laticauda* in captivity. *Int. Zoo Yb.*, **7**, 229–31.

Knight, M. (1970). *Tortoises and How to Keep Them*. (Revised and edited by D. Ball). Leicester: Brockhampton Press.

Laszlo, J. (1969). Observations on two new artificial lights for reptile displays. *Int. Zoo Yb.*, **9**, 12–13.

Littleford, R. A. & Keller, W. F. (1946). Observations on captive pilot black snakes and common water snakes. *Copeia*, 160–7.

Logan, T. (1969). Experiments with Gro-Lux light and its effect on reptiles. *Int. Zoo Yb.*, **9,** 9–11.

Mertens, R. (1965). Beiträge zum Thema: Krankheiten der Reptilien. *Zool. Gart., Lpz.,* **31,** 133–43.

Oliver, J. A. (1955). *The Natural History of North American Amphibians and Reptiles.* Princeton: Van Nostrand.

Pawley, R. (1966). *Geckos . . . as pets.* New Jersey: TFH Publications.

Pawley, R. (1969). Observations on the reactions of the Mata mata turtle *Chelys fimbriata* to ultra-violet radiation. *Int. Zoo Yb.,* **9,** 31–2.

Peaker, M. (1969). Some aspects of the thermal requirements of reptiles in captivity. *Int. Zoo Yb.,* **9,** 3–8.

Pope, C. H. (1956). *The Reptile World.* London: Routledge & Kegan Paul.

Pope, C. H. (1962). *The Giant Snakes.* London: Routledge & Kegan Paul.

Pritchard, P. C. H. (1967). *Living Turtles of the World.* Jersey City, New Jersey: TFH Publications.

Reichenbach-Klinke, H. H. & Elkan, E. (1965). *The Principal Diseases of Lower Vertebrates.* London & New York: Academic Press.

Reid, H. A. (1968). Symptomatology, pathology and treatment of land snake bite in India and Southeast Asia. In *Venomous Animals and Their Venoms.* 1. *Venomous Vertebrates,* ed. Bücherl, W., Buckley, E., Deulofeu, V. London & New York: Academic Press.

Savage, S. B. (1973). The housing and force-feeding of emerald tree boas (*Corallus carinus*). *Int. Zoo. Yb.,* **13,** 156–157.

Schmidt, K. P. & Inger, R. F. (1957). *Living Reptiles of the World.* London: Hamish Hamilton.

Smith, H. M. (1965). *Snakes as Pets.* Jersey City: TFH Publications.

Smith, M. (1973). *The British Amphibians and Reptiles* 5th. edn. London: Collins.

Stebbins, R. C. & Barwick, R. E. (1968). Radiotelemetric study of thermoregulation in a lace monitor. *Copeia*, 541–7.

US Dept. of the Navy Bureau of Medicine and Surgery (1968). *Poisonous Snakes of the World.* Washington.

Vogel, Z. (1964). *Reptiles and Amphibians. Their Care and Behaviour.* London: Studio Vista.

Wallach, J. D. (1969). Medical care of reptiles. *J. Amer. vet. med. Ass.,* **155,** 1017–1034.

Wermuth, H. & Mertens, R. (1961). *Schildkröten. Krokodile. Brückenechsen.* Jena: Fischer.

Zweifel, R. G. (1961). Another method of incubating reptile eggs. *Copeia*, 112–3.

The *Zoological Record,* published annually by the Zoological Society of London, is an invaluable classified bibliography and contains many references to the care of reptiles in captivity.

M. C. Downes (1973) has compiled a bibliography of the literature on crocodile husbandry (*Proc. 2nd Working Meeting of Crocodile Specialists*) International Union for Conservation of Nature and Natural Resources: Morges, Switzerland.

38 Anura (Frogs and Toads)

J. F. D. FRAZER

General biology

Frogs and toads are known as the Anura or tailless amphibians, to distinguish them from the tailed newts and the caecilians. They form one of the three groups of amphibia and originated from more primitive species, the earliest amphibians having developed in the Devonian era from lobe-finned fishes.

Frogs and toads vary greatly in size, from species 1 cm or less in length to giant ones 25 cm long. Their skin is moist and water can pass through it in either direction, so that they must therefore be protected from both water-loss and the uptake of toxic substances through the skin. Some species of Salientia avoid the risk of desiccation by spending their whole life in water (as in *Xenopus* and other genera), while others remain either within a frog's jump of the pond edge or in damp habitats, and similar conditions must be provided for these in the laboratory. Others (e.g. toads and tree-frogs) have adapted themselves to drier regions, although even these species will become desiccated if kept too dry, and certain Australian species of water-holding frogs maintain their own damp environment internally during the dry season.

The special problem of toxins and their absorption through the skin may be considered from various aspects. It must be remembered that the skin secretions of some anurans are toxic to other species. This applies not only to rare tropical species: common frogs (*Rana temporaria*) kept for as little as half an hour in a sack with toads (*Bufo bufo*) will be found dead at the end of this period. Such species must therefore be kept apart, and only those known to be mutually compatible allowed to mingle.

There is great variation in the sexual distinctions among different species, and the only sure way of telling that the right sex has been supplied is to know the peculiarities of the species being used. British male frogs and toads have black swellings on the thumb and first two fingers and have thicker forearms than the female. These nuptial pads may occur elsewhere in other species and are most apparent during the breeding season. At other times they are rudimentary and can only be seen by the practised eye. The tree-frog (*Hyla arborea*) has no pads, and the male is distinguished by the discoloured underside of the throat caused by the deflated vocal sacs. The blackish throat distinguishes the male of the common African toad *Bufo regularis*. In *Xenopus* species the distinction between the sexes lies in the small labial folds found by the cloaca of the female.

In the wild, anurans feed on a wide variety of invertebrates, such as insects ranging from ants to moths and locusts, worms and spiders; the larger species will eat vertebrates as large as adult mice and lizards.

Uses

There are a number of laboratory uses for frogs and toads. They are in demand for the teaching of biology, and may be required in simple routine anatomical dissections, for class experiments in the physiology of nerve and muscle and of cardiac function, or in an extension of this type of work for pharmacological purposes. Females of *Xenopus laevis* and males of a large number of species were formerly used for tests of human pregnancy: they can also be used for the assay of gonadotrophins. Changes in the skin colour of tree-frogs have been shown (Sulman, 1952) to follow the use of ACTH under standard conditions: the response to standard doses has been followed and the sensitivity found to be very great (Frazer, unpublished). The responses of individual melanophores in frog skin have been studied *in vitro* and researches have been carried out on the permeability of frog skin to various ions.

In toads there is an impressive volume of South American research on the processes of spermiation and ovulation, as well as on the secretions of the oviduct. Houssay used the giant species for much of his early work on the pituitary and pancreas, Taylor & Ewer (1956) have investigated the sloughing of the skin in *Bufo regularis*, while laboratory workers have been investigating the processes of reproduction in an ever-growing number of species.

If accurate research is to be carried out, it is essential that the species used, their size and the conditions under which they have been housed and fed are recorded. It has been shown (Jeffree, 1953; Frazer, 1956) that the spermiation response of toads to gonadotrophin shows minute variations with different seasons even when the animals are kept under standard conditions, while differences in the temperature at which they are housed may cause major differences in response. For work on changes in skin colour the animals must have been acclimatised over a period of time to the colour of their background.

Standard biological data

Little has been recorded. Although frogs and toads appear at a breeding site for only a few seasons, if kept in captivity under proper conditions they can live for 20 years. Sexual maturity occurs at 2–3 years of age, but may occur earlier in well-fed specimens. Spawning of the common British species occurs during the spring months, but in other parts of the world the time of spawning will vary from one species to another. Local variation in any species occurs with variation in the ecological conditions. The length of the egg and larval stages of a species vary with temperature, the time of metamorphosis depending upon the nutritional state.

Husbandry

Housing and caging

If this is to be adequate, due regard must be taken of certain of the animals' characteristics. Owing to the permeability of the skin they must be protected from water-loss and from the uptake of toxic substances. Frogs are very sensitive to the products of decomposition, and a dead frog in the water-container will soon result in other deaths. It is therefore axiomatic that the cage be examined daily and any dead or obviously diseased animals removed: while

this is normal practice in the laboratory, it cannot so easily be carried out when frogs and toads are in transit over long distances. Frogs may also be sensitive to normal substances, such as salt and certain metals, and should not be allowed to come in contact with them.

The type of container required depends on the species, the number of individuals to be housed and the period for which they are to be kept. Over a short period a relatively small container will suffice for a small number of frogs or toads, but a period of a week or more requires more specialised conditions, which will vary depending on whether the animals are being kept at room temperature, under artificial hibernation or at tropical temperatures.

Frogs. At room temperature frogs are usually kept in large sinks from which slowly running water removes any excreta. Galvanized metal or stainless steel tanks or glazed porcelain sinks may equally well be used. Bricks rising above the water surface give footing, while damp moss can supply shelter. Alternatively, they can be kept in aquaria where the water is changed every few days. Whatever the container, it must have a close-fitting lid of either wire mesh, expanded metal or perforated zinc. The holes should be of such a size that flies cannot escape through the mesh, while maximum strength is retained. A hole of 2·5 to 4 cm diameter should be cut in the lid and kept plugged with a cork: this can then be used for introducing bluebottles, flies, moths or other live food.

When shy species such as the edible frog are kept, the sink must contain some deep water as well as land, and must be in a room remote from disturbance. Without these conditions the frogs will not feed.

Recently Nace (1968) has given details of his technique for housing adults, suitably grouped, in plastic containers. These have a lower compartment through which water flows, and an upper one floored with rubber matting and containing pieces of broken flowerpot to give the necessary hiding places. Racks are designed to hold these cages in batteries. Each cage is roofed with stainless steel wire cloth. Larger colonies are kept in troughs some 4 m long, which can be divided by partitions. Small stools inside allow the frogs to come out of the water which flows shallowly round them.

Toads. Under this heading are included for present purposes not only members of the genus *Bufo* but also certain other species with terrestrial habits and dry skins, that do not therefore require such damp habitats as frogs. Similar types of container may be used for them, but when large numbers are kept it is convenient to provide special containers on the plan of giant vivaria. These can be made of galvanized metal, about 120 cm long by 75 cm deep, the front being about 15 cm high and the back 27 cm; the side walls then vary from 15 cm at the front to 27 cm at the back. In one of these sides is a hole of about 4 cm diameter, which is normally kept closed by a cork or by a metal plate pivoting on a pin, the position of equilibrium being the closed one. This hole is for the introduction of food or individual amphibia on occasions when it is not desired to lift the lid. The vivarium is roofed by two slabs of 0·5 cm plate glass which fit on to a ledge 0·5 cm wide running round the inside of the vivarium, 1 cm from the top. This ledge must be made with care, so that when the lid is on there are no gaps through which small invertebrates can escape. Ventilation holes 1–1·5 mm in diameter should be drilled at 2·5 cm intervals along the sides, roughly 2·5 cm below the ledge. Once it is made, the vivarium should be stove-enamelled inside.

The two slabs of plate glass which roof the vivarium must be cut to fit exactly. After cutting, the edges are ground smooth and two holes drilled in each slab, roughly 15 cm from the front and back edges, and midway between the two sides. In these holes are bolted handles of a suitable size for lifting the lid. It may be desirable to replace one of the two lids by a similar one of perforated zinc or expanded metal and in this case the usual precautions must be taken to prevent the escape of food insects. Such a lid has an advantage over a completely solid one in allowing water vapour to pass through it, thus preventing the humidity from becoming too high.

Vivarium-type cages may be kept in racks or on shelves, or it may be more convenient to use a low trolley with several shelves. This can be made in skeleton form, for lightness and ease of handling.

For different species of toads the floor of the cage may be covered with earth or sand, but this may make cleaning difficult. Cover can best be provided by placing moss of sufficient depth for the toads to hide in all over the floor of the cage. Shelter may also be given in the shape of broken flowerpots, laid on their sides. Water should be supplied in all cages, either in a large enamel dish or in a big and stable earthenware

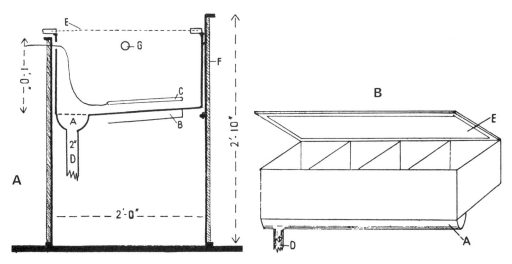

Fig. 38.1. Galvanized-steel tank for *Xenopus*. (a) Section seen from right in (b); (b) Tank with asbestos (F) and support removed. A, Gulley with perforated tray to collect dirt; B, Pocket for heating element of 100 to 200 W; C, Phial for distant-control thermometer; D, Drain-pipe, 5 cm in diameter, with full-gate valve; E, Lid fitted with windolite; F, Asbestos millboard, insulating the tank; G, Water inlet.

dog-bowl. The water level must be such that toads will not drown before they can find their way out.

Tree-frogs. These are more akin to toads than frogs, but they require somewhat damper conditions, although they can survive with only ground moisture. It is most convenient to keep them either in an aquarium with 5 to 10 cm of water, roofed with perforated zinc, or else in the cylindrical type of caterpillar cage. This has a round metal or plastic base, topped with a cylinder of transparent plastic and bearing a perforated lid above this. Metal cages should receive an extra coat of paint as soon as they are obtained. After this has dried, 2–3 cm of water are placed in the cage. Either type of cage can be furnished with a spray of foliage if desired, and the lid must have a hole cut and plugged as usual with a cork, for the introduction of flying insects as food. The tree-frogs will remain either on the foliage or on the vertical walls of the cage.

Platannas. *Xenopus laevis* is the usual species kept, but others may be kept in a similar manner. It will survive at normal room temperature, so that heating needs to be provided only in very cold weather. Any kind of aquarium tank or even a glass jar may be used. The platannas are almost wholly aquatic, so about 25 cm of water should be provided. The tank should be kept clear of vegetation as *Xenopus* causes a lot of disturbance as it moves about. In the wild it lives in muddy ponds, so muddy water is no disadvantage. If a heater is provided it should be outside the tank. Elkan (1957) has described a galvanized steel tank approximately 250×120 cm (Fig. 38.1), with a sloping floor to facilitate the draining out of any uneaten food elements and containing a heating element and thermostat.

Tadpoles of *Xenopus laevis* can easily be kept in an ordinary laboratory sink if no aquarium tank is available.

Animals required for pregnancy tests. Not so many amphibians as were formerly used are now employed in pregnancy tests. The experimental cage will vary with the species used. For *Xenopus* females a large jar will take an individual in 8–10 cm of water, with a grating on corks below that, so that any eggs produced can drop through this and will not be eaten. A pair of male toads or treefrogs can be placed in a similar jar or a

400 ml beaker covered with a tile and containing approx. 1 ml of water. If several treefrogs are required, a larger beaker should be used.

Food utensils

Since usually only live foods are taken, it is not easy to find a container which will hold these and also allow the amphibians to reach the food. It is sometimes convenient to put maggots in a petri dish, which will localise them temporarily.

Bedding

None is normally required, but dry moss can provide cover when a large number of individuals are kept in a single cage.

Environmental conditions

Most laboratory species can be kept at normal room temperatures, special conditions being required only for a few from warm climates. It is sometimes convenient to maintain frogs and toads under artificial hibernation. Before starting, they must be in good condition and well fed. Those obtained in spring (which will have only just come out of hibernation before being collected from their breeding ponds) are not suitable for immediate cooling and they must first be fed well under normal conditions, until they are in good condition and then left unfed for a few days. Frogs can thereafter be kept in the normal laboratory refrigerator, although deaths may occur if the shallow water in their containers is not changed regularly. Allison (1957) has described a successful technique of artificial hibernation for toads, using trays 8–10 cm deep, covered with fine mesh wire netting and containing slightly damp moss instead of water. She has pointed out that while frogs will tolerate a constant temperature of 3·3°C, toads should not be kept at temperatures below 4·4–6·7°C.

Tropical species requiring warm conditions are usually best kept in a thermostatically-controlled room. Where only a few cages are involved, or where small numbers have to be maintained at different temperatures, it is often most convenient to heat cages individually. Over a narrow range of temperature, thermostatic control using an electric heater is normally best: for aquatic species, it can conveniently be of the normal aquarium type. Where a wider tempera-ture range can be tolerated it is sometimes possible to use a carefully-adjusted spirit or paraffin lamp to heat the base of the cage, if this is made of a suitable material.

Identification

Individual ground colour and/or pattern of markings may sometimes be used. Clipping of toes or tattooing of the web between them has been used in the wild, but in captivity the easiest method is to sew small coloured beads with cotton thread to the skin over one shoulder. The beads must be chosen so that their colour does not wash out in water; up to four or five of differing colours may be used at once, to give a wide variety of combinations. Never under any circumstances use a ring of metal, thread or wool; these will either be too loose and come off, or too tight, causing damage to the limb on which they have been placed.

Handling

Frogs and toads may be clasped by the hind legs or (with less assurance) by the body. If the skin is moist or slippery (and invariably for *Xenopus*) a rough cloth or duster should be used. Handling for pregnancy tests is described on page 523.

Transport

Within the laboratory, amphibians may be transported in large earthenware or glass jars. For the more active species such as treefrogs, a lid is essential. The more lethargic ones, such as toads, may be carried in buckets.

Closed boxes or cans are best for transport by road, rail or air. No food is required during such journeys, but if a long, slow trip is en-visaged, arrangements should be made for pouring in a little water occasionally. Tree-frogs have travelled well by air over long distances in small tins with a few holes in the lid, containing some damp cloth, and with a few pieces of twig wedged across to afford footholds. If holes are punched through the tin lid, these must have the sharp protuberances on the outside.

Feeding

Natural foods

A variety of foods, both living and dead, may be given. Purely aquatic species such as *Xenopus* will take various aquatic animals, including invertebrates, tadpoles and small fish. Worms can be a useful standby. Other species will take a variety of suitable live foods.

Bluebottles and houseflies. Bluebottles and to a lesser extent houseflies are very useful, and some species of anurans have survived for years on these alone. They may be reared from maggots. The most convenient way to do this is to leave the maggots in fresh sawdust to pupate. A container such as a conical flask can be used, but moisture will condense in it, so that not only is wing development prevented, but the flies themselves stick to the walls of the container and cannot easily be fed to the amphibians. When flies are reared in bulk, it is better to place maggots and sawdust in a large biscuit-tin from which most of the centre of the lid has been removed: perforated zinc of a suitable mesh is soldered over the resulting hole. An opening on one side, near the top of the tin, should be fitted with a piece of glass tubing which is plugged with cottonwool: alternatively the tube can be of Perspex and fitted with a Perspex slide to close it (Fig. 38.2).

The edges of the lid must be sealed to the tins with adhesive tape to prevent the egress of newly emerged flies. When flies are seen to have emerged they can be fed to the amphibians by inserting the tube through the hole in the side or roof of the cage, then removing the plug or slide which blocks it. A piece of wood or card-board is placed over the lid of the flybox, and the flies then move towards the light and enter the cage. It is inadvisable to leave the cage and flybox permanently together, since toads have been known to enter the flybox and die there.

In the case of toads, it is advisable to alternate flies with maggots, and an excess of either must be avoided. *B. bufo* in particular will pass un-digested maggots if too many are given, while an excess of bluebottles in the cage will cause panic, sometimes resulting in death. An excess of flies may be fed, however, provided that the moss is deep enough for the toads to take cover in it. The Green Toad (*B. viridis*) is not affected

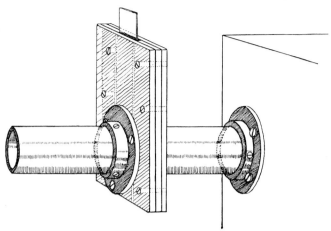

Fig. 38.2. Perspex tube for admitting emergent flies from emergence cage into amphibians' vivarium.

by bluebottles in this way, and will merely continue eating until all the food is gone.

Mixed insects. A varied diet is of greater value than one based on a single species. Many moths and other insects may be taken in a light-trap, and form a very valuable source of food. Alternatively, sweeping a net through herbage or beating foliage over a sheet will yield a good haul.

Locusts. These can be very useful for any species large enough to take them (e.g. *Bufo marinus*, *B. ictericus* and *B. arenarum*).

Other invertebrates. Spiders and woodlice can form a useful part of a mixed diet, and can be collected like insects or by searching under stones, beneath bark, under logs and the like. Worms can also be important. It is advisable that any secretive and wingless species be given in a special feeding cage kept free of vegetation or soil in which they might hide, since amphibians will not search for concealed prey. Since worms drown in water, care must be taken to remove the corpses of any given to frogs or *Xenopus*. White worms (Enchytridae) can be given to small species.

Special requirements

A few species will take only one form of food—for example, ants. These idiosyncrasies will be known to the experimenter and do not require special mention here. Meat or liver can be used to feed *Xenopus*, but it must be remembered that if young are being reared on liver, horse liver is inadequate and rat liver must be provided.

Terrestrial species will also take meat or liver, if it is cut to a suitable size and moved about, e.g. by a fine wire. A more satisfactory way of giving these is by forced feeding, as for compounded foods.

Compounded foods

Artificial diets have been recommended, these being made up to a firm consistency as for rats. Portions are force-fed twice a week to each frog individually. The frog is held in one hand, the thumb being on one side while the fingers on the other side are arranged so that the middle and ring ones are on either side of the frog's forelimb, clasping this firmly. With the free hand the edge of a spatula is gently inserted between the frog's jaws. Once the mouth is open, the other fore-finger is used to keep it in this position, by placing it in the angle of the jaw. The free hand is then used to insert the food, the frog's mouth is held shut and the throat is massaged gently until the food is swallowed. Thus, by expending some of the technician's time it is possible to ensure that all the laboratory stock of frogs are kept in a reasonably-fed condition.

Breeding

Little has been recorded about breeding frogs and toads in captivity. It is most easily carried out with freshly-caught animals, which in many cases will have come straight from the breeding ponds. If kept under conditions suitable for the species concerned mating and oviposition will occur. The commoner laboratory species lay their eggs in water, where they hatch into free-swimming tadpoles. The requirements of these differ from those of the adults and in some ways are easier to satisfy and in others more difficult. Water is normally required for them to live in; this must be sufficiently aerated and, while containing adequate food, be kept free from decaying food particles and faecal matter. These requirements are best provided in aquaria; if necessary a small pump can be used to aerate the water. Conditions inside the aquarium will depend on the species concerned: some survive well in clear water, while other tadpoles prefer weed for shelter. Some live harmoniously together, while others do so only when food is plentiful and in times of scarcity may turn to cannibalism, a mode of feeding which in turn may be regularly adopted by others. Species like *Rana ridibunda*, which feed voraciously on other tadpoles, must therefore be kept isolated in small individual containers.

The feeding of tadpoles is relatively easy. During their early stages they will thrive on vegetation or on the diatoms which occur on this (Savage, 1962). This diet may be natural in the form of waterweed or algae, or alternatively a substitute such as a piece of lettuce leaf may be left to float on the surface. Under these conditions, little cleaning of the tank is needed. However, in the later stages of tadpole life when the limbs appear and the tail regresses a protein diet is required. This can be provided by giving small pieces of meat, but small amounts of high-protein artificial foods, which can be sprinkled on the surface are equally good.

Nace (1968) keeps newly-hatched larvae in shallow enamel pans of water, changed every three days. Once the tadpoles are swimming freely they are placed in inverted, bottomless 4·5 litre bottles, through which is a constant and steady flow of water. These are placed in racks so as to form batteries of 130 bottles.

When the tadpoles have almost lost their tails, access to land must be provided. If the tank has no land surface, the water depth can be reduced to 5 cm and a couple of bricks placed on the bottom. Once the froglets or toadlets have metamorphosed completely the phase of major

difficulty is reached. Unless there are strong reasons for rearing them further, it is advisable to release the young now. Lees (1962) has described a method of raising the young in special outdoor frog pits covered with nylon net, using *Drosophila* and aphids as food. He points out that the provision of an excess of such food seems to be the necessary stimulus for regular feeding in captivity. Where frogs have been reared under vivarium conditions it is recommended that they be fed on flies from an early date.

Nace places his newly-metamorphosed amphibians in smaller versions of his adult containers: these are tilted, providing a pool of shallow water at one end. At this stage he finds mosquitoes are an excellent food supply. Larger flies and even immature crickets may be taken by the bigger amphibian species.

Baby platannas should be fed on *Daphnia* and then on bloodworms and meat. A diet of horse liver alone has proved unsatisfactory (Bruce & Parkes, 1950) and causes young *Xenopus* to develop severe rickets. In order to avoid this, rat or rabbit liver must be given at regular intervals. A better diet consists of small invertebrates and other natural foods.

Laboratory procedures

Selection of animals for experiment

Healthy frogs and toads should be well-nourished. Although bodily bulk may vary with fluid content, good nourishment shows in plumpness of the thighs: wasting here is an indication of previous starvation. The animals should be lively, and should not have ulceration or abnormal reddening of the skin, since these may be signs of the early stages of red-leg, a disease which is particularly fatal. Signs of lymphatic blockage or other infestation by parasites may be manifested by superficial oedema or by prominence of the lateral line organs in *Xenopus* (Elkan, 1957).

Anaesthesia and euthanasia

Anaesthesia is seldom used except for short demonstrations of the capillary circulation. For this, the frog is immersed in a solution of 1 per cent urethane until anaesthetised. This solution can be used for recovery experiments, the frog being placed for the recovery phase under a dripping tap, when the urethane becomes dialysed through the frog's skin. The same technique can sometimes be used with success for recovery of toads which have been injected with pregnancy urine which has proved to be contaminated with sedatives or other toxins.

Euthanasia can be carried out by holding the amphibian by the hind legs and tapping the back of the head sharply on the edge of a sink. This can then be followed by decapitation and/ or pithing, as required. (See also Chap. 14.)

Experimental feeding procedures

Although not frequently required for experimental purposes, the technique described above (p. 521) can be used.

Collection of specimens

The usual specimen required is cloacal urine from males being used for tests of human pregnancy. To sample this, the toad or tree-frog is held by the body with the hind legs flexed and a fine glass pipette (1 mm or less in external diameter) is inserted gently into the cloaca for a short distance. Urine will often track up this at once, but if it does not, gentle manipulation of the pipette backwards and forwards will soon produce it. The need for gentleness in this technique must be stressed.

Dosing and injection procedures

While some drugs such as urethane may be given by absorption through the skin, fluids are generally injected into the dorsal or ventral lymph sac, some experts advising that the hypodermic needle be introduced through the thigh muscles in order to prevent leakage. If this technique is not employed, a long fine needle (No. 12 or 17), already attached to the syringe should be inserted about the middle of the back, the point being passed as far as possible towards the tip of the urostyle. The injection is made slowly and the point of the needle withdrawn very slowly in order to prevent reflux of fluid.

Disease control

Prevention of disease is a matter of quarantining newly-arrived stock for a week or two and inspecting them daily for red-leg. During this period it may pay to keep the amphibians in 0·15 per cent saline solution, to leave a piece of copper wire in the water or to sprinkle sulpha-diazine powder on the surface. Any infectious disease should become apparent during the quarantine period.

Nace (1968) keeps newly-received frogs for ten minutes in calcium hypochlorite of such a strength as to yield 6 p.p.m. of chlorine.

The prevention of rickets in young growing animals (as in *Xenopus laevis*—see above) depends upon a proper diet. Apart from red-leg and similar infectious diseases, this is the only one common enough to merit consideration.

The question of curing disease in frogs and toads calls for setting the cost of the cure against the value of the animals. With common species it is frequently better to kill those affected and isolate their cage mates in quarantine, but there are occasions when every effort should be made for a cure.

In red-leg, it is essential to control the infection, but the sensitivity of the frog to drugs makes this difficult. Kaplan & Licht (1955) have recommended exposure to 1–2 per cent copper sulphate for at least an hour: others believe in the injection of streptomycin and Miles (1950) recommended a total of 1,000 units given subcutaneously over 9 days. He found that this was accompanied by the recovery of frogs treated in the early stages of infection, while it delayed the death of one in which the disease had reached a more advanced stage.

Smith (1950) has cured *Bufo marinus* by giving 5 mg/100 g body-weight of chloramphenicol by stomach tube, and following this up with 3 mg/ 100 g twice daily for five days. Hunsaker & Potter (1960) used the same treatment for a mixture of fifteen species of toads: this was successful at first, but a chloramphenicol-resistant strain of the infective organism appeared. The disease was then successfully treated by giving 10 mg/g of tetracycline* intraperitoneally, followed by twice-daily doses of 5 mg/g for a week.

Kaplan has suggested that 0·05 per cent copper sulphate applied for a matter of minutes could be used to treat *Saprolegnia* infection.

* Achromycin V: Cyanamid.

REFERENCES

Allison, Rhoda M. (1957). The care and maintenance of male toads and frogs used in the bioassay of chorionic gonadotrophin. In *UFAW Handbook on the Care and Management of Laboratory Animals*, 2nd ed. London: UFAW.

Bruce, H. M. & Parkes, A. S. (1950). Rickets and osteoporosis in *Xenopus laevis*. *J. Endocr.*, **7,** 64–81.

Elkan, E. (1957). *Xenopus laevis* Daudin. In *UFAW Handbook on the Care and Management of Laboratory Animals*, 2nd ed. London: UFAW.

Frazer, J. F. D. (1956). The sperm-shedding response of male toads and tree-frogs after the injection of two types of gonadotrophin. *Br. J. Pharmac. Chemother.*, **11,** 249–56.

Hunsaker, D. H. & Potter, F. E., Jr. (1960). 'Red-leg' in a natural population of amphibians. *Herpetologica*, **16,** 285–6.

Jeffree, G. M. (1953). The comparative sensitivity of the male toad pregnancy test and the Friedman test. *J. clin. Path.*, **6,** 150–4.

Kaplan, H. M. & Licht, L. (1955). The valuation of chemicals used in the control and treatment of disease in fish and frogs caused by *Pseudomonas hydrophila*. *Am. J. vet. Res.*, **16,** 342–4.

Lees, E. (1962). Rearing of frogs for parasitological research. *Br. J. Herpet.*, **3,** 25–7.

Miles, E. M. (1950). Red-leg in tree-frogs caused by *Bacterium alkaligenes*. *J. gen. Microbiol.*, **4,** 434–6.

Nace, A. W. (1968). The amphibian facility of the University of Michigan. *Bioscience*, **18,** 767–75.

Savage, R. M. (1961). *The Ecology and Life History of the Common Frog (Rana temporaria temporaria)*. London: Putnam.

Smith, S. W. (1950). Chloromycetin in the treatment of 'Red Leg'. *Science*, **112,** 274–5.

Sulman, F. G. (1952). Chromatophorotropic effect of adrenocorticotrophic hormones. *Nature, Lond.*, **169,** 588.

Taylor, S. & Ewer, D. F. (1956). Moulting in the Anura: the normal moulting cycle of *Bufo regularis* Reuss. *Proc. zool. Soc. Lond.*, **127,** 461–78.

39 Urodeles

Elze C. Boterenbrood and Romee Verhoeff-de Fremery

Introduction

In maintaining urodeles in the laboratory we may distinguish between (1) collecting the animals in nature and keeping them only temporarily, and (2) maintaining a permanent stock. In the first case the animals are usually collected during the breeding season, when they concentrate in ponds and streams. The husbandry then serves only to keep them in good condition for a certain period and, if necessary, to allow them to spawn. The permanent maintenance of a colony puts high demands on accommodation and care, since spawning will occur only when the animals are in healthy condition.

There are only a few species of urodeles that are known at present to adapt themselves successfully to continuous life in the laboratory. All these are species which can adapt to a permanently aquatic mode of life, viz. *Ambystoma mexicanum* (Mexican axolotl), *Pleurodeles waltlii* (Spanish salamander), and *Triturus pyrrhogaster* (Japanese newt).

THE MEXICAN AXOLOTL (*Ambystoma mexicanum*)

General biology

Ambystoma mexicanum (Shaw) belongs to the family Ambystomatidae, characterized by the rather frequent occurrence of partial or complete neoteny: a condition in which sexual maturity is attained although several essentially larval features, such as external gills, tail-fin, etc., are retained. Some American authors use the name *Siredon mexicanum*.

Most European laboratory stocks are progeny originating from a Paris import from Mexico in 1864, descendants of which were shipped to laboratories all over the world. This stock is characterised by consistent neoteny, but in nature the axolotl seems to metamorphose frequently (Smith, 1969).

In nature the axolotl is found only in Mexico, in the Xochimilco and Chalco Lakes. Gadow (1903) has described the ecological conditions. The lakes are situated at a height of 2,250 m in a mountainous and fertile environment, and get their main water supply from deep, clear springs. The lakes are 1·5–3 m deep and contain many small, floating peat islands. In summer the water is covered with a layer of water plants. The aquatic fauna is very rich, with many small fishes, insect larvae, worms, etc. In nature axolotls spawn in February; the larvae grow very rapidly and reach adult size in June.

The adult axolotl has a dark brownish-black skin with numerous black dots and several diffuse dark grey-brown spots, occurring particularly on the limbs and the tail. Old or not entirely healthy animals often show an increase in size of the grey-brown spots, and generally become more greyish in colour. Healthy axolotls have well developed external gills and a long tail with a dorsal and a ventral fin, and may attain an ultimate length of 30 cm. The males are more slender than the females, and generally have a longer tail; the margin of the cloaca is markedly swollen, particularly during the reproductive season. The females are considerably plumper in appearance and have a flat cloacal region. A description of axolotl anatomy and histology was given by Brunst (1955b). Instructions for dissection were given by Heydecke (1964).

Metamorphosis can be induced artificially and may take place particularly in young axolotls of about 15 cm in length (1) when fed or injected with thyroid gland, iodine preparations or pituitary extracts; or (2) when gradually adapted to a terrestrial mode of life (see Koch, 1926, Brunst, 1955a). In some laboratory strains spontaneous metamorphosis may occur rarely. Metamorphosed young axolotls can continue aquatic life without difficulty, but older animals often drown if no terrestrial environment is provided. After metamorphosis the axolotl has a smooth, dark-grey skin with a large number of distinct small yellow spots. Regular sloughing takes place, in contrast to the neotenic form. The gills disappear and the gill-slits close. The tail assumes a roundish, stout appearance similar to that seen in land salamanders. Richter (1968) succeeded in obtaining offspring from experimentally metamorphosed axolotls.

An interesting characteristic of axolotls is their high regenerative capacity. Regeneration is observed after amputation of limbs, tail, and gills. In adult animals regeneration takes place at a slower rate than in young animals.

Several laboratories maintain a colony of white axolotls. Whiteness (not albinoism) is caused by homozygous presence of the recessive gene *d*, which leads to a restricted distribution of black and yellow pigment cells and a decreased pigment formation. A review on available axolotl mutants is given by Malacinski & Brothers (1974).

As in all urodeles insemination is internal and takes place by means of spermatophores. The spermatophore is deposited by the male during courtship; it consists of a cone of transparent jelly secreted by the cloacal glands and bearing a white mass of spermatozoa at its apex. The base of the cone adheres to the substrate and the female swims over it and takes up the spermatozoa into her cloaca. The sperm is stored in the spermathecal tubules until the eggs leave the oviduct and pass through the cloaca. Before oviposition several spermatozoa penetrate the jelly capsule of each egg. Much information on reproduction, courtship patterns, breeding behaviour and oviposition in *Ambystoma* species is given by Noble (1954), Lofts (1974) and Salthe & Mecham (1974).

The eggs of the axolotl are relatively large: 1·9–2·3 mm (depending, among other things, on the age of the female), and are surrounded by a soft jelly capsule. They are widely used for embryological research. The normal table of Harrison for *A. punctatum* is used for staging (Harrison, 1969). The larvae are often used for regeneration experiments.

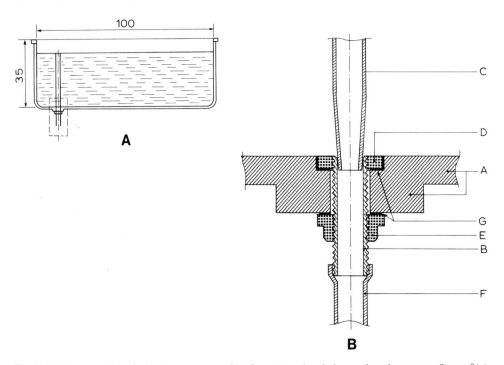

Fig. 39.1. (a) cross-section of asbestos cement container for groups of axolotls, etc. (b) enlargement of part of (a). A, bottom of container with local reinforcement. B, metal or polythene drain pipe screwed into outlet opening. C, polythene pipe for overflow outlet, fitting into B. D, and E, metal or polythene rings. F, rubber tube leading to drainage system. G, cementing compound.

Husbandry

Accommodation

Axolotls can be adequately kept in asbestos cement containers from the moment they have reached a length of 5 cm. This material is strong, durable and relatively cheap and does not give off toxic substances. Before use containers must be rinsed continuously for 3–4 weeks, in order to remove excess calcium, and subsequently the inner surface should be coated completely with a waterproof, chemically inert varnish. Aquaria constructed of glass with a metal framework and bottom are also useful, provided the metal parts are well varnished.

Each container should be fitted at the bottom with an outlet for complete drainage. In this opening a tube can be inserted to provide an overflow outlet (Fig. 39.1).

In containers with a water surface area of 50×100 cm 10 adult axolotls can be housed. Such containers can be divided into compartments by means of partitions made of perforated plexiglass, held in place by a rubber strip. The partitions should have some larger openings at the level of the water surface for removal of the surface film (see next section). The height of the water level has an influence on the condition of the gills: a level of at least 25 cm keeps the gills in good condition. When the walls and the partitions are at least 5 cm higher than the water level the axolotls will not jump out.

No sand, gravel, or aquatic plants are needed; this considerably facilitates the care of large numbers of animals. However, the presence of some algal growth on the bottom and side walls is commendable. This considerably promotes the clearness of the water and reduces the nitrate concentration. A too abundant growth of algae must be prevented, however, since axolotls suffer much from the nitrates released by putrefying algae. The entire layer of algae must be removed now and then, especially at times of decreasing day-length or temperature.

Water

In many laboratories the quality of the water is not suitable to keep the sensitive skin of axolotls in a healthy condition. The harmful influence of chlorine can be eliminated by allowing the water to stand exposed to air for some days, or by leading the water over a charcoal filter bed. Care should be taken that the water is free from copper ions, so copper taps and piping must be excluded.

Axolotls, like fishes, should be kept in relatively shallow water with a large surface area (Chaps. 40 & 41). The gas exchange can be accelerated by the use of a mechanical water-aerator, which serves a threefold purpose: (1) it causes turbulence of the water, so that water from deeper layers comes to the surface regularly, (2) the small air bubbles offer a large surface area to the water, and (3) it removes the greater part of the surface film, a thin bacterial layer which is often formed on standing water and seriously interferes with gas exchange. Also a temporary slight raising of the water level will cause the film to disappear through the overflow outlet.

Axolotls can be kept in running tap water or in well-aerated standing water. The latter condition seems to offer better protection against skin infections, and fresh tap water can be supplied two to three times a week in a quantity sufficient to replenish the amount removed during siphoning off detritus, and during removal of the surface film.

The temperature of the water should preferably be kept at 14–18°C. Axolotls do not thrive at temperatures higher than 22°C but tolerate low temperatures very well.

Lighting

Axolotls can be kept in ordinary daylight, or in artificial light which is switched on during ten to twelve hours a day. Correct illumination is obtained by using fluorescent lamps of white/pink colour (Philips TL 33) providing 40 W per 2 m² surface area, and placed 40–50 cm above the water level. Half of the container should be covered with a lid in order to provide a part sheltered from light.

Handling

Adult axolotls should never be lifted with a net. This startles them so much that they vigorously beat their tails and may damage these considerably. A suitable method for holding an axolotl is shown in Figure 39.2. After one hand has been moved gently towards the animal's head, the trunk is firmly grasped from above between the

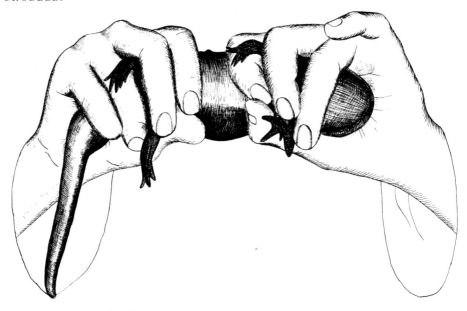

Fig. 39.2. Method for holding an axolotl.

thumb and the forefinger and middle finger, just behind the fore limbs. At the same time the two other fingers of the hand enclose the head, so that the axolotl cannot slip forward. One has to take care that the gills are not squeezed. Simultaneously the other hand grasps the caudal trunk and anterior tail region from above. In this position the animal can be lifted out of the water easily; it should not be released before being placed entirely under water.

Young animals (10–20 cm) can be held with one hand, the head and the anterior trunk being grasped from above in the same way as described for adults. Very young larvae have to be transported with a glass pipette with a wide mouth and a rubber nipple. Older ones are removed with a net made from a square of iron wire with a piece of nylon fabric stretched tautly across it.

Transport

Preparations. Every journey involving a long distance, particularly to foreign countries, must be planned carefully. The transport company involved should be consulted in advance with respect to (1) the most suitable route and time-schedule; (2) the availability of an air-conditioned luggage cabin, so that excessive temperatures are avoided; and (3) packing conditions.

Usually the shipments have to be paid in advance and cannot be insured. For transport abroad, moreover, one has to find out in advance the requirements regarding import and export permits or health certificates. Shipment over the week-end should be avoided.

At the receiving end arrangements have to be made to collect the consignment as soon as possible; hence the package should, in addition to the full address, be provided with the telephone number of the consignee, and the transport company should be requested to give notice to the consignee immediately upon arrival of the shipment. The consignee has to be informed in advance about the details of the transport and the planned hour of arrival, and may be advised to contact the receiving office in advance in order to arrange for collection.

Packing. Axolotls can be transported in firm double plastic sacs (one sac inside the other) filled with water to a level of 2–3 cm above the animals, and with at least twice this volume of oxygen or air. The sacs are tied up with string and placed in a strong cardboard box. During transport axolotls can be packed rather closely: six adults in about 3 litres of water. To minimize fouling of the water they should not be fed on the day before transport, nor during the journey. No waterplants should be added.

Feeding

Axolotls thrive on beef heart from which fat, tough fibres, etc. have been carefully removed. Addition of vitamins and minerals appears to be necessary (see Moore, 1964). Fat-soluble or water-soluble vitamin preparations may be used. Per kg beef-heart c. 2,000 i.u. each of vitamin A and D, 10 mg of vitamin B complex, and 3 g of powdered $CaCO_3$ may be added. Addition of a multi-vitamin and mineral preparation also gives good results (e.g. 2 per cent Carnicon, Trouw & Co, Amsterdam). The required amount of meat has to be determined empirically and it should be minced (or, when frozen, cut into small pieces). In order to reduce fouling of the water the meat is briefly flooded and stirred with water, mixed thoroughly with the preparation, and put into the container. If the animals are kept in groups they will seek the food spontaneously, but if kept separately they will not seek the food and must be fed individually by hand. The meat should in this case be cut into strips of $30 \times 4 \times 4$ mm and mixed with the preparations. It is important to offer strips of the proper size. Too large pieces will not be readily swallowed, and food will be refused even though the animal's appetite is not satisfied. If a strip is moved gently in front of the snout with a pair of forceps the animal will usually snap at it vigorously. It will often also snap at the forceps, and therefore forceps with blunt tips should be used. Adult axolotls should be fed two or three times a week. During spring and summer they usually eat three to four strips on each feeding day but in the autumn only one to two. Individual strips must be offered with suitable pauses, to prevent regurgitation. From the above it will be seen that the keeping of axolotls in groups considerably cuts down the time needed for feeding.

Instead of beef heart or beef, liver (not pig liver) can be given (Witschi, 1938; Humphrey, 1962). The liver is prepared in the same way as beef heart, except that it is not necessary to add vitamins. Young tadpoles and earthworms are also excellent as food.

If minced meat is given it is advisable to remove faecal matter before feeding, since the droppings, which are enclosed in a thin membrane, may be broken by the active swimming movements of the animals. A few hours after every feeding, droppings and remains of food should be siphoned out.

Breeding

Breeding season

Sexual maturity is indicated in the male by a considerable swelling of the margin of the cloaca, caused by an increase in size of the cloacal glands as they become functional. In the female a conspicuous roundish body shape indicates the presence of eggs in the ovary. External sex characters being to appear in the males at seven months, whereas a female cannot be recognized with certainty before nine to ten months. Sexual maturity can be reached in one year, but both number and size of the eggs of the first batch are relatively small.

The breeding season extends from December until June, but young animals reaching maturity in late spring often lay eggs in June, July and November as well and sometimes also in the intervening period. Newrock & Brothers (1973) were able to extend the breeding season by injecting females with follicle stimulating hormone. As males may be remated after 1–2 months and females after 2–3 months they can be used several times during each season. In order to keep axolotls in good breeding condition it is important to give them an opportunity to lay eggs at least once a year. The best breeding age is 1–5 years.

Induction of breeding

Breeding is most conveniently induced by a sudden drop in temperature. The difference in temperature required depends on the sexual condition of the animals, the time of the year, etc., but usually a difference of 8–10°C is sufficient.

All through the year males and females are kept in separate containers. The animals to be used for breeding are first kept for about one week at 22°C. Two days before the eggs are required a male and a female are put together in a clean, large container with about 100 litres of fresh water at a temperature of about 12°C. Either cold running tap water or crushed ice can be used to obtain the low temperature. The bottom of the container should be covered with clean slates or plastic plates for spermatophore deposition, and for oviposition several ordinary plastic electricity tubes (diameter 13 mm) should be fitted between the side walls at about half the height of the water level. The container must be completely sheltered from light in order to avoid disturbance of courtship and oviposition.

The low temperature should be maintained at least until the time of spermatophore deposition, which usually occurs 16–24 hours after the temperature shock. When spermatophore deposition is delayed, it may yet be induced by raising the water temperature again.

The axolotls should not be fed from the day on which they are put into cold water until the end of oviposition. For every planned spawning at least two couples should be used.

Spawning and collecting of eggs

Some time before oviposition begins the female swims around in a remarkable floating manner. She starts to deposit her eggs about one day after spermatophore deposition. Eggs should not be collected until about one hundred have been laid, since disturbance during the early phase of oviposition may stop the spawning process. The plastic tubes can be taken out of the container and the eggs in their jelly capsules can be removed easily with a pair of forceps with curved tips. They are put into dishes containing water of about the same temperature as that in the container used for oviposition. The water should not be deeper than 2·5 cm and the eggs should be placed in a single layer and not packed closely, in order to avoid lack of oxygen. The tubes can be put back into the container after having been scrubbed clean.

The duration of oviposition partly depends on the temperature. At 12–15°C it takes about two days. A low temperature not only aids in slowing down the development of the eggs, which may be essential for experimental work on early stages, but also makes freshly-laid eggs available over a longer period. A temperature lower than 12°C, though not impeding spermatophore deposition, seems to block oviposition.

The number of eggs varies with the season and with the age of the female, and may range between 200 and 2,000 per spawning. The usual number is 500–800.

Rearing of eggs and larvae

Failures in the rearing of eggs and larvae are usually caused by fouling of eggs and culture water. Hygienic conditions are therefore essential. The material must be inspected daily, and bad eggs removed immediately. In order to prevent evaporation of the water and to keep out dust the dishes should be almost completely covered, although ready gas-exchange is essential. The surface of the jelly capsules is a favourable site for growth of bacteria, fungi, etc.; therefore the outer layer of jelly, which swells considerably by water uptake, must be removed. The simplest way to do this is to place the egg on the palm of the hand and to separate the outer, softer jelly from the deeper, firmer layer with a blunt knife. It may be useful afterwards to dip the eggs into 70 per cent ethanol for about 10 seconds for partial sterilization of the jelly surface.

Eggs can be reared at room temperature, the optimum temperature being 14–18°C, although they can stand temperatures from 8–25°C. Sudden temperature changes must be avoided. Both eggs and larvae should be protected from direct sunlight. After about two weeks at 18°C the larvae slip out of their capsules. Empty capsules should be removed at once and the

TABLE I

Space requirements for growing axolotls (temperature 18–20°C)

Approximate length	Approximate age	Average bottom surface area per animal	Depths of water required
1·3 cm	2 weeks	10 cm²	3 cm
3–4 cm	5–6 weeks	40 cm²	5 cm
5 cm	7 weeks	100 cm²	10 cm
10 cm	3 months	250 cm²	15 cm
15 cm	6 months	500 cm²	25 cm
18 cm	7 months	600 cm²	25 cm
25–30 cm	18 months and older, adult	500 cm²	25 cm

newly-hatched larvae transferred to dishes containing moderately aerated water, without any plants or sand.

During culturing overcrowding should be avoided. The more space the larvae have the more quickly they grow. Moreover, in crowded conditions the larvae often attack one another. The bottom surface areas and water depths generally required for successive stages are given in Table I. These data hold for animals kept in groups and fed abundantly. The larvae should be distributed over several dishes in order to keep within bounds possible infections, which usually spread throughout a dish in a short time. Now and then the larvae should be sorted into groups of similar size.

The skin of urodele larvae is very delicate, which necessitates some precautions with respect to the water in which they are reared. Common tap water often has too low a salt concentration or contains chlorine; it leads to bad condition of the larval skin and thus increases the chances of infection. The use of clear pond water, warmed up to room temperature, is commendable. Water from aquaria which are in stable biological equilibrium can also be used provided that it has not been inhabited previously by anurans. Pond water apparently provides young larvae with a very good protection against infections (Breder, 1931). Larvae from about 5 cm upwards in length can be safely reared in tap water.

It is necessary to feed growing larvae abundantly and the best results are obtained when food is constantly present. It must, therefore, be supplied regularly, but no more should be provided than the animals can eat in a day, in order to avoid fouling of the water. The amounts have to be determined empirically.

Two to three days after hatching the larvae start to feed on small moving aquatic animals, such as larvae of *Artemia salina* or very small *Daphnia*. Eggs of *Artemia*, obtainable from aquarium supply houses, hatch in 2 per cent NaCl solution in one day at 28°C or in two days at 22°C, and can be supplied after being rinsed in tap water. Later the larvae thrive on a diet of *Daphnia* only (Fig. 39.3). Apart from *Daphnia* other small aquatic animals can be supplied, such as crustaceans and mosquito larvae, but large quantities of *Cyclops* are dangerous, since

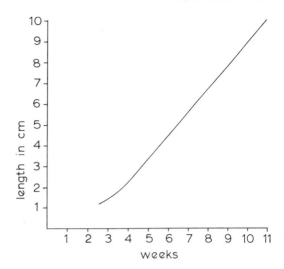

Fig. 39.3. Growth of axolotl larvae on a diet of *Daphnia* at about 18–20°C.

this crustacean attacks the larvae and may ruin a healthy culture in one hour. The diet for larvae of 3 cm and longer can be supplemented with *Enchytraeus* and *Tubifex* (preferably the former), but a diet consisting exclusively of these two genera does not give satisfactory results. At first only small worms should be fed. For larvae of 5 cm and longer the diet can be supplemented with finely-minced meat. Larvae of 10 cm in length can be gradually switched over to the diet for adult animals. They must be fed every day until they have reached a length of about 20 cm, for otherwise they will attack each other. Animals of nearly 20 cm are especially aggressive, and for that reason they need a relatively large amount of space (see Table I). The animals can be kept separate during this period, but they will have to be fed by hand and may unlearn the ability to seek the food themselves when they are again kept in groups.

Hutchinson & Hewitt (1935) compared various diets in a few *Ambystoma* species and found that a mixture of *Daphnia* and *Enchytraeus* gives the most rapid, and meat the slowest, growth. Witschi (1938) fed larvae from the beginning with small pieces of liver, offered individually on a dissecting needle. This method offers the advantage of control over the amount of food taken, while the water does not foul as a result of decaying food remains. The method is suitable in some experimental conditions, but for rearing larger groups it is too time-consuming.

Gordon (1950), Hutchinson (1950), Geyer (1957) and Needham (1959) gave methods for culturing *Daphnia, Artemia* and *Enchytraeus* indoors.

At the end of every day detritus has to be removed. In dishes with smaller larvae this can be done with a glass pipette with a rubber bulb, while in those containing larger larvae it can be siphoned out with a rubber tube. The water removed must be replenished. Whenever the water becomes cloudy or foul-smelling the larvae must be transferred immediately to a clean dish with fresh water.

Laboratory procedures

Anaesthesia and euthanasia

The most satisfactory narcotic is MS 222 (tricaine methane sulphonate*) (Rothlin, 1932), which has only a slight effect on heart-rate (Copenhaver, 1938). For larvae up to 15 cm in length the recommended concentration varies from 0·03–0·07 per cent, depending on the size of the animal. The animals should be immobile and numb in about 10 minutes. Recovery takes place in 15–30 minutes in tap water. Larvae can be kept under narcosis during more than two days in tapwater containing 0·007 per cent MS 222, without cumulative effects (Schotté & Butler, 1941). For adult axolotls a solution of 0·1 per cent can be used.

Chloretone is used in concentrations of 0·005–0·05 per cent for larvae and of 0·1 per cent for adult axolotls.

For euthanasia the animal should be placed in a tin containing cotton-wool soaked in chloroform; there must, however, be no direct contact between the chloroform and the animal's skin. It is also possible to give an overdose of MS 222 (0·2 per cent); the animal must be left in this situation for at least three hours to be sure that the treatment is fatal. Sometimes it is preferable to kill the animal by decapitation with extremely sharp scissors.

Artificial ovulation and fertilization

Oviposition can be stimulated artificially by intramuscular injection of 180–220 i.u. of

* Sandoz Chemical Works, Basel, Switzerland.

follicle-stimulating hormone (Humphrey, 1962). However, the eggs obtained will be unfertilized. Methods for artificial fertilization of eggs, involving the sacrifice of both the female and the male, were given by Brunst (1955a) and by Humphrey (1962).

Disease control

When an animal in a group shows symptoms of a disease it should be isolated at once. If an attempt is made to save it, precautions must be taken to ensure that in spite of isolation the infection does not spread. Separate siphons, etc., should be used for infected animals, and the person who attends them should wash his hands afterwards. It is very difficult to maintain complete isolation between separate containers in the same room, and it is therefore preferable to kill sick animals.

Dead animals must be removed immediately. If dissection does not yield a clear indication that death was not the result of infection the container must be cleaned and sterilized thoroughly, e.g. with a strong solution of chloramine (paratoluol sodium sulphonchloramide). Other animals that were in the same container must be watched closely afterwards. If more dead animals appear in the same container within a short time it is advisable to kill all the remaining ones and to sterilize the container and associated apparatus thoroughly.

Axolotl larvae are very susceptible to fungal infections, which show as small tufts of whitish threads protruding from small openings in the skin. These appear initially at the base of the tail and on the gills and head, but spread very rapidly all over the body and cause death within two days. This infection may occur when the larvae are cultured in dusty and dirty places. Treatment consists of placing the larvae in cleaner surroundings and in a fresh solution of 0·001 per cent chloramine. Its chief action is due to oxygen liberated *in statu nascendi*; the liberated chlorine is immediately bound as NaCl, so that the animals can stay in the solution and can be fed normally. One day after treatment the mould-threads will have disappeared completely, and in the formerly infected spots only a skin-thickening or a haemorrhage will remain. No abnormalities in growth or development have been observed. In severe cases a new dose of chloramine may be added after three to seven days. However, one has to be careful with chloramine treatment as the skin may be weakened, making the animals more susceptible to subsequent infections.

Mercurochrome is used to cure fungal as well as non-fungal skin diseases. The animals are put in a concentration of 0·0002–0·0004 per cent for three days, then in pond water for a week, and finally in a fresh mercurochrome solution of the same concentration for another three days. To treat mould infection Detwiler & McKennon (1929) kept larvae older than stage 37 (Harrison) continuously in a mercurochrome solution of 0·00013 per cent in spring water for a period of two months without noticeable effects on growth.

Wounds and local fungal infections (*Saprolegnia*) can be treated by local painting with a 2 per cent mercurochrome solution, followed after a few minutes by rinsing with tap water.

Very little is known about diseases apparently caused by bacterial or virus infections, which may bring considerable damage to usually healthy stocks of axolotls. See Reichenbach-Klinke & Elkan (1965). Large whitish spots on the skin (appearing first in the border region of the trunk and tail and later on the head) are symptoms of infection of internal organs, e.g. the liver. A continuous stay of several months in 0·6 per cent NaCl was found to remove the above symptoms without producing any ill-effect. This treatment, employed every year during September and October, appeared to be a valuable prophylactic.

Brunst (1955b) has reviewed various tumours and other malformations observed in axolotls.

TRITURUS SPECIES

The eggs of various *Triturus* species are widely used for experimental embryological research.

The main difficulty in keeping continuously-breeding stocks of European and American *Triturus* species in the laboratory lies in the poor quality of eggs produced by animals which have lived in captivity for a lengthy period, and in the troubles encountered in the management of the animals after metamorphosis. Therefore these species are usually collected during the breeding season, when they come in large numbers to ponds, ditches, etc. for spawning. Little is known about their terrestrial period of life.

Only the Japanese newt, which in nature does not leave the water except for a brief hibernation period, can be kept continuously in the laboratory.

Data on localities and breeding season of American newts were given by Bishop (1943). For the species most frequently kept in the laboratory such as *T. torosus* (= *Taricha torosus*, California newt) and *T. viridescens* (= *Diemictylus viridescens*, Eastern newt), data about care and management were given by Fankhauser (1963).

General biology

The European newts were described by Gadow (1901), Boulenger (1910) and Klingelhöffer (1956). *T. taeniatus* (= *T. vulgaris*, common newt) is common throughout most of Europe, has a length of about 7 cm, and an egg diameter of 1·25 mm. *T. alpestris* (alpine newt) occurs mainly in mountainous regions up to 2,000 m in central Europe; it has a length of 7–10 cm, and an egg diameter of 1·5 mm. *T. cristatus* (crested newt) occurs in northern and middle Europe; its length is 12–16 cm, and the diameter of the egg 1·5–2 mm. *T. helveticus* (= *T. palmatus*, palmate newt) occurs in western central Europe up to 1,000 m and is common in France; its length is 6–8 cm; the diameter of the egg is 1·5 mm.

The Japanese newt, *T. pyrrhogaster* (= *Cynops pyrrhogaster*) was described by Sawada (1963). Adult animals are dark brown or black with a red-orange belly beautifully patterned with black spots. They have markedly protruding parotids. The length is 12–17 cm; the diameter of the egg is 2 mm.

In *Triturus* species neoteny occurs occasionally. The limbs, tail and gills have a high regenerative capacity.

Accommodation
Aquatic period

Males and females are kept together throughout the year. About 30 animals can be kept in large containers, as described for the axolotl, containing standing or slowly-running tap water to a depth of 20–30 cm. Couples kept in individual containers need at least 6 litres of water.

Several rough stones should be put on the bottom of the container, as rubbing against them facilitates moulting. The animals like to sit on pieces of floating cork and these should be provided. It is also necessary to put some plants into the water; *Elodea* spp. are most suitable. The leaves of the plants should not be in contact with the side walls, in order to prevent the newts from using them as a support for climbing out. One has to avoid large populations of Hydra, which may frighten the newts. It is advisable to aerate the water with a mechanical aerator. A smooth metal strip about 5 cm broad should be fixed horizontally to the rim, projecting over the container, to prevent escape of the animals without preventing ready access of light and fresh air.

The containers should preferably be kept in a light place, although small ones should be sheltered from light coming through the side walls, since this disturbs the animals. The temperature should be 14–18°C for European newts, and 16–20°C for the Japanese newt.

It is possible to keep adult European newts continuously in the laboratory if after the breeding season they are adapted to a permanently aquatic mode of life and kept in a container as described above, provided with rather more plants. The water should be kept cool, and for feeding (which should go on as usual) the animals should be driven to the bottom of the container. Under these conditions they may spawn the next spring, but the egg-material obtained will generally be of poor quality.

Hibernation period

In order to obtain fertilized eggs it is advisable to give Japanese newts a hibernation period of 6–8 weeks in winter by keeping the animals at a temperature of ± 3°C without food. For this purpose the vegetable compartment of a refrigerator is a convenient place. Only some very damp moss or wet pieces of polyester sponge need to be added. The compartment must be inspected every week, care being taken that the animals are disturbed as little as possible. The moss or sponge must be moistened again and dead animals removed. Only healthy and well-fed animals can bear hibernation. They should be left unfed for two days beforehand to prevent intestinal disorders from undigested food. The

temperature should be lowered gradually to prevent a shock.

For general hibernation conditions see Klingelhöffer (1955, 1956).

Transport

During the breeding season it is preferable to transport the animals either in a tin or in a plastic sac filled with water, as described for the axolotl (see p. 528). At other times some damp moss or wet polyester sponge is enough to keep the atmosphere humid, and the animals can be transported in a wooden or plastic box provided with adequate ventilation holes and covered by fine-mesh wire gauze, positioned so that the animals cannot damage themselves by rubbing against it. The use of wood-wool, although sometimes recommended, is wrong since the animals are easily damaged by the sharp edges.

Feeding

Newts can be fed three times a week with meat prepared as for axolotls. As with axolotls, animals kept in groups seek the food spontaneously but those kept individually must be fed by hand with very thin strips. Earth-worms can also be given; small worms can be given whole but larger ones should be cut into pieces. Some varieties of earth-worms are not readily accepted, and these should not be given. At the end of every feeding day remains of food, decaying leaves, etc. should be removed. The better newts are fed the longer they will lay and the more eggs will be produced.

Breeding

Sexual activity, breeding season

In European newts sexual activity is indicated by considerably intensified colouration of the skin. The males develop a dorsal crest. In *T. helveticus* a cutaneous fold is formed on each side of the body, the tail is prolonged into a thin black filament, and the toes of the hind legs become webbed. In *T. pyrrhogaster* sexual activity is indicated in the male by a greyish-violet colour of the skin in the region of the parotid glands, the flanks, and the tail; the digits become more pronounced, the tail more pointed, and the margin of the cloaca swells considerably. In the female there are no external sex characters except for a swelling of the trunk region.

In *Triturus* the male deposits spermatophores at the end of a courtship dance. The eggs are laid singly, fixed inside folded leaves or on strips of nylon lace curtain and are surrounded by a rather stiff and tough jelly capsule.

The breeding season extends for about two months, some time during the period of March to July, depending on local temperature conditions. In *T. pyrrhogaster* egg-production can be regulated in the laboratory by shifting the hibernation period. For more information on breeding in *Triturus* spp. refer to Lofts (1974) and Salthe & Mecham (1974).

Spawning and induction of breeding

Animals collected during the breeding season lay eggs in the laboratory without any particular stimulation. When collected early in the breeding season *T. alpestris* will lay eggs in the laboratory for a period of one to two months, provided the temperature is kept at about 15°C. The average daily number of eggs laid is about ten. *T. taeniatus* and *T. helveticus* will lay eggs for 3–4 weeks, the average number of eggs being ten and three to four respectively. *T. cristatus* lays only a few eggs a day for a short period.

From a healthy stock of *T. pyrrhogaster*, in which males and females are kept together and have previously been subjected to hibernation, one can usually obtain some fertilized eggs; this shows that males can produce spermatophores in captivity. However, one can rely neither on the time at which the eggs will be laid, nor on their number. For induced oviposition see below.

Rearing of eggs and larvae

In general the same methods as described for the axolotl can be used. The best temperature is 15–25°C. The development of *T. alpestris* becomes abnormal below 10°C. Data on the rate of early development were given by Knight (1938) for *T. alpestris*, by Glaesner (1925) and Glücksohn (1931) for *T. taeniatus*, by Gallien & Bidaud (1959) for *T. helveticus*, and by Anderson (1943) for *T. pyrrhogaster*. The larvae can remain continuously in water, provided that some water-plants are placed in the container during and after metamorphosis, so that they can easily reach the water surface. It is difficult to keep the animals after metamorphosis. During metamorphosis they do not feed, and it is often necessary afterwards to feed them by hand for several weeks before they get re-accustomed to feeding themselves. *Daphnia*, small worms, etc. can be fed, and gradually the diet can be switched over to finely minced meat. When food is given the animals should be driven to the bottom.

Laboratory procedures

Anaesthesia and euthanasia

The following solutions are used for anaesthesia: MS 222 in a concentration of 0·02–0·03 per cent for larvae and of 0·03–0·1 per cent for adult animals; chloretone in a concentration of 0·03–0·05 per cent for larvae and of 0·1–0·2 per cent for adult animals. For euthanasia see page 532 (axolotls).

Artificial ovulation and fertilization

In *T. pyrrhogaster* egg-production can be accelerated and controlled by implanting into the female small grafts of fresh pituitary glands from cows, female frogs, newts, etc. The best results are obtained by implantation, under anaesthesia, of a fragment of about 3 mm³ in a pocket made underneath the skin of the lateral trunk region or the lower jaw. If too large a piece is inserted the female may lay a number of abnormal eggs. The animals begin to spawn after a few days and continue for a period of one to three weeks. When implantation fails to lead to egg-production, or when egg-production stops, another piece may be implanted. In this way one can work with one female for three to four weeks. After a rest period of two months (with normal feeding) new implantations can be made. The optimum temperature for oviposition is 18–20°C. Eggs are laid most conveniently on *Vallisneria* leaves.

Pituitary stimulation has not been successful in males and so the method described above yields fertilized eggs only when the males deposit spermatophores spontaneously, which usually occurs when they have been subjected to hibernation (see above). Animals may be kept temporarily in an inactive condition at 10°C; it is then necessary to feed them. In order to get eggs in autumn the animals should be stored in hibernation until about one month before the eggs are required (data from K. Hara, personal communication, and Streett, 1940). According to Hamburger (1960), in *Triturus* species, collected in the wild, fertile eggs can be obtained outside the breeding season by injecting the female with frog pituitaries, relying for fertilization on the presence in the spermatheca of spermatozoa, which remain functional over a long period. According to Stefanova (personal communication), spawning can be induced in European newts by injecting the female on two successive days with 100 i.u. of gonadotropic hormone and giving the male 50 i.u. on the second day. The injections should be intraperitoneal, using a short needle inserted anteriorly into the belly region. Fertile eggs will then be laid for about a week.

Artificial fertilization (for instance, as required for hybridization) involves the sacrifice of both the female and the male. The method was described by Hamburger (1960) and Rugh (1962).

Disease control

The diseases known to occur in *Triturus* species have been described by Klingelhöffer (1955) and Reichenbach-Klinke (1965). The most dangerous and most frequently occurring disease is designated in German literature as 'Molchpest' ('Newt pest'). It is characterized by various symptoms, such as sloughing, blisters on the snout, mouldy spots, and a smell of parsley. The disease is epidemic in character and is always fatal. So far the cause is unknown, but it occurs frequently when the animals have been temporarily in unfavourable environmental conditions. Its propagation can be arrested by promptly killing the affected animals and removing the remaining ones to thoroughly sterilized containers, after bathing them in a 0·0002 per cent solution of mercurochrome for two hours, or after treatment with chloramine as described for axolotl-larvae.

OTHER URODELES

There are a number of other urodeles which can be kept in the laboratory either continuously

(e.g. *Pleurodeles waltlii*) or temporarily. Owing to lack of space it is not possible to treat their care and management in detail; references are as follows:

Pleurodeles waltlii michah (the Spanish newt). Boterenbrood (1967).

Ambystoma tigrinum (tiger salamander). Hutchinson (1950), Cohen (1968).

Necturus maculosus (mud puppy). Kaplan & Glaczenski (1965).

Salamandra salamandra (fire salamander). Lester (1957).

Data on other Salamandra species can be found in Klingelhöffer (1956).

REFERENCES

Anderson, P. L. (1943). The normal development of *Triturus pyrrhogaster. Anat. Rec.*, **86**, 59–73.

Bishop, S. C. (1943). *Handbook of Salamanders.* Ithaca, N.Y.: Comstock.

Boterenbrood, E. C. (1967). Newts and salamanders. In *The UFAW Handbook on the Care and Management of Laboratory Animals*, 3rd ed. London: UFAW.

Boulenger, G. A. (1910). *Les Batraciens et principalement ceux d'Europe.* Paris: Doin et Fils.

Breder, C. M. (1931). On the organic equilibria in aquaria. (Abs.) *Copeia*, 1931 (2), 66.

Brunst, V. V. (1955a). The axolotl (*Siredon mexicanum*). I. As material for scientific research. *Lab. Invest.*, **4**, 45–64.

Brunst, V. V. (1955b). The axolotl (*Siredon mexicanum*). II. Morphology and pathology. *Lab. Invest.*, **4**, 429–49.

Cohen, N. (1968). A method for mass rearing of *Ambystoma tigrinum* during and after metamorphosis in a laboratory environment. *Herpetologica*, **24**, 86–7.

Copenhaver, W. M. (1939). Initiation of beat and intrinsic contraction rates in the different parts of the ambystoma heart. *J. exp. Zool.*, **80**, 193–224.

Detwiler, S. R. & McKennon, G. E. (1929). Mercurochrome (di-brom oxy mercuri fluorescin) as a fungicidal agent in the growth of amphibian embryos. *Anat. Rec.*, **41**, 205–11.

Fankhauser, G. (1963). Amphibia. In *Animals for Research. Principles of Breeding and Management.* London & New York: Academic Press.

Gadow, H. (1901). *The Cambridge Natural History. 8, Amphibia and Reptiles.* London: Macmillan.

Gadow, H. (1903). The Mexican Axolotl. *Nature, Lond.*, **67**, 330–2.

Gallien, L. & Bidaud, O. (1959). Table chronologique

du développement chez *Triturus helveticus* Razoumowsky. *Bull. Soc. zool. Fr.*, **84**, 22–32.

Geyer, H. (1957). *Praktische Futterkunde*. Stuttgart: Kernen.

Glaesner, L. (1925). *Normentafel zur Entwicklungsgeschichte des gemeinen Wassermolches (Molge vulgaris)*. Jena: Fischer.

Glücksohn, S. (1931). Äussere Entwicklung der Extremitäten und Stadieneinteilung der Larvenperiode von *Triton taeniatus* Leyd. und *Triton cristatus* Laur. *Wilhelm Roux Arch. EntwMech. Org.*, **125**, 341–405.

Gordon, M. (1950). Fishes as laboratory animals. In *The Care and Breeding of Laboratory Animals*, ed. Farris, E. J. London: Chapman & Hall.

Hamburger, V. (1960). *A Manual of Experimental Embryology*. Chicago: University Press.

Harrison, R. G. (1969). Harrison stages and description of the spotted salamander, Amblystoma punctatum (Linn.). In *Organization and Development of the Embryo*, ed. Wilens, S. New Haven & London: Yale University Press.

Heydecke, R. (1964). Präparationsanleitung für den Axolotl (*Siredon mexicanum*, Shaw). *Wiss. Z. Humboldt Univ. Berl., Math.-Naturwiss. Reihe*, **13**, 599–604.

Humphrey, R. R. (1962). Mexican axolotls, dark and mutant white strains: care of experimental animals. *Bull. Philad. Herpet. Soc.*, April-September, 21–5.

Hutchinson, R. C. (1950). Amphibia. In *The Care and Breeding of Laboratory Animals*, ed. Farris, E. J. London: Chapman & Hall.

Hutchinson, R. C. & Hewitt, D. (1935). A study of larval growth in *Ambystoma punctatum* and *Ambystoma tigrinum*. *J. exp. Zool.*, **71**, 465–82.

Kaplan, H. M. & Glaczenski, S. S. (1965). Salamanders as laboratory animals: *Necturus*. *Lab. Anim. Care*, **15**, 151–5.

Klingelhöffer, W. (1955). *Terrarienkunde I. Allgemeines und Technik*. Stuttgart: Kernen.

Klingelhöffer, W. (1956). *Terrarienkunde II. Lurche*. Stuttgart: Kernen.

Knight, F. C. E. (1938). Die Entwicklung von Triton alpestris bei verschiedenen Temperaturen, mit Normentafel. *Wilhelm Roux Arch. EntwMech. Org.*, **137**, 461–73.

Koch, M. (1926). Zur Umwandlung des Mexikanischen Axolotls mittels Schilddrüsen-futterung. *Bl. Aquar.-u. Terrarienk.*, **37**, 245–52.

Lester, J. W. (1957). Amphibia I. In *The UFAW Handbook on the Care and Management of Laboratory Animals*, 2nd ed. London: UFAW.

Lofts, B. (1974). Reproduction. In *Physiology of the Amphibia* Vol. II. Ed. Lofts, B. London & New York: Academic Press.

Malacinski, J. & Brothers, A. J. (1974). Mutant genes in the Mexican axolotl. *Science*, **184**, 1142–1147.

Moore, J. A. (1964). *Physiology of the Amphibia*. London & New York: Academic Press.

Newrock, K. M. & Brothers, A. J. (1973). The artificial induction of out of season breeding in the Mexican axolotl, *Ambystoma mexicanum*. *Amer. Soc. Zool., Div. Dev. Biol.* Newsletter October 1973.

Noble, G. K. (1954). *The Biology of the Amphibia*. New York: Dover Publications.

Reichenbach-Klinke, H. & Elkan, E. (1965). *The Principal Diseases of Lower Vertebrates*. London & New York: Academic Press.

Richter, W. (1968). Nachzucht bei vollständig metamorphosierten Axolotln (*Ambystoma mexicanum*). *Salamandra*, **4**, 10–15.

Rothlin, E. (1932). MS 222 (Lösliches Anaesthesin), ein Narkotikum für Kaltblüter. *Schweiz. med. Wschr.*, **45**, 1042–3.

Rugh, R. (1962). *Experimental embryology*. Minneapolis: Burgess.

Salthe, S. N. & Mecham, J. S. (1974). Reproduction and courtship patterns. In *Physiology of the Amphibia* Vol. II. Ed. Lofts, B. London & New York: Academic Press.

Sawada, S. (1963). Studies on the local races of the Japanese newt, *Triturus pyrrhogaster* Boie. I. Morphological characters. *J. Sci. Hiroshima Univ.*, **21**, 135–65.

Schotté, O. E. & Butler, E. G. (1941). Morphological effects of denervation and amputation of limbs in urodele larvae. *J. exp. Zool.*, **87**, 279–322.

Smith, H. M. (1969). The Mexican Axolotl: some misconceptions and problems. *Bioscience*, **19**, 593–7.

Streett, J. C. (1940). Experiments on the organization of the unsegmented egg of *Triturus pyrrhogaster*. *J. exp. Zool.*, **85**, 383–408.

Witschi, E. (1938). Aufzucht und Haltung der gebräuchlichen Laboratoriumtiere. Amphibien und Reptilien. *Handb. biol. ArbMeth.*, **9**, 611–51.

40   Freshwater Fish

H. G. VEVERS

Introduction

Freshwater fish may, for laboratory purposes, be divided into those from temperate waters and those from the tropics. This is a purely arbitrary division and there are, of course, many intermediates. It is, however, a convenient division, since the two groups demand somewhat different conditions in the laboratory.

Temperate freshwater fishes and invertebrates can mostly be kept in laboratories in circulation systems of the type described under Marine Aquaria (Chap. 41). Many species of tropical freshwater fishes, on the other hand, can be kept more simply and without water circulation, and a short guide to this is given at the end of this chapter.

TEMPERATE FRESHWATER FISH

Temperate freshwater fish are mainly kept in laboratories for physiological research, as subjects for the study of pollution, and sometimes for testing the effects of various drugs. The most suitable species for these purposes are brown trout (*Salmo trutta*), rainbow trout (*Salmo gairdneri = S. irideus*), goldfish (*Carassius auratus*), carp (*Cyprinus carpio*), orfe (*Leuciscus idus*), sticklebacks (*Gasterosteus* spp.) and minnows (*Phoxinus phoxinus*). These species do well in captivity and can withstand a considerable amount of handling, whereas others are very sensitive.

Accommodation

Tanks

Temperate freshwater fish required for laboratory work are normally kept in tanks indoors, although if very large numbers are required outside ponds will be necessary.

The type of tank chosen depends on the numbers of fish to be kept, the amount of floor space available and the volume of water to be used. It is important to remember that the larger the surface area of water in relation to the volume the better, always providing that there is adequate depth of water to allow the fish to move about normally. The tanks should not normally be smaller than $1\cdot5 \times 1\cdot0$ m in area and $0\cdot6$ m high, and larger tanks should have the same proportions.

The tanks should be installed in a cool place well away from any source of heat. Tanks can be made of glass, glass and metal, metal, wood, or waterproofed concrete. Wood tends to warp and rot. Metal tanks should be thoroughly painted with a good bitumastic paint or with epoxy-resin in order to protect them from corrosion. All-glass tanks need no preservative, but they can only be obtained in relatively small sizes suitable for housing a few species of small fish. In general the most practicable tanks are those with a strong angle-iron frame, the four sides being of plate glass, and the bottom of slate or wired glass. Here again the metal portions must be protected against rust. Asbestos or fibreglass tanks can be used when there is no need to observe the fish from the side.

Water circulation

In general it is best to arrange for a continual flow of water through the tanks and there are several ways of achieving this.

(1)   Bore a hole in the base of the tank near to one corner, and fit to it a stand pipe, the length being cut to the desired height of the water. When the tank is in use, the water will escape through the upper end of the stand pipe. When, however, the tank needs to be emptied, the pipe can be removed and the water will then drain away. The top of the stand pipe must be provided with a wire mesh or gauze cage to prevent the escape of fish. The inlet piping supplying the tank should reach to the bottom,

preferably in the corner farthest from the outlet stand pipe.

(2) The water is pumped up from a reservoir into a gravity tank which is situated above the fish tanks. From there it is fed by gravity to the aquarium tanks. The outgoing water should leave from an overflow situated at the top of the tank, whence it passes through piping either directly back to the reservoir or to a filter. After passing through the filter, it then returns to the reservoir. Immediately the level of the water in the top gravity tank falls below a certain height, a float resting on the surface of the water drops and operates a switch to start an electric pump which raises more water from the reservoir to the gravity tank. This method is, in effect, the same as that recommended for marine aquaria (p. 543).

The water supply will most easily be obtained from the laboratory tap. In order to get as much benefit from aeration as possible it should be delivered above and not below the water surface and as fine a jet as possible should be used. Alternatively a filter pump can be fitted to the water supply which permits air to be drawn down with the current of water (Pentelow, 1957). The mixture of air and water in this case should enter the tank at the bottom.

Free chlorine in the public water supply may sometimes cause trouble in freshwater aquaria. This is quite easily overcome by passing the water through a column of activated carbon before it enters the aquarium tanks. Pentelow (1957) described a simple apparatus consisting of two 1-m lengths of earthenware pipe joined end to end and filled with about 30 ml of activated carbon. The bottom of the lower pipe is sealed with an expanding pipe-stopper to which an outlet pipe has been attached. The outlet pipe leads to a rubber or plastic hose and the treated water is carried up to the level of the top of the activated carbon before being fed to the aquarium tank. Thus the container of carbon is always filled with water.

Tanks should always be covered, as fish, particularly trout, often jump out of the water. The cover can be of nylon netting supported on a wooden frame; if the area to be covered is large the frame should be so constructed that it can be raised vertically from the tank by means of a pulley, thus allowing ample room for working.

Lighting

It is a good general rule that aquarium tanks should not be exposed to the direct full light of the sun, for this will encourage an excessive growth of algae which will form a green film on the glass. Moreover the sunlight will probably overheat the water during the middle of the day. Freshwater tanks are not quite so sensitive to temperature change as those with sea water, nevertheless temperate species do better when the water is kept fairly cool, and all freshwater tanks do best when the light is not too intense. Freshwater plants are not normally kept in laboratory freshwater aquaria, but if they are used the light should be just sufficient to keep them in healthy condition without encouraging algal growth.

Artificial lighting should not be so bright as to cause the fish discomfort. An 80-W fluorescent strip light kept on for 8 to 10 hours a day is sufficient for a tank of 350 litres capacity.

Feeding

In nature trout and sticklebacks are almost entirely carnivorous, whereas minnows are omnivorous. Ponds and watercress beds are a rich source of invertebrate food, such as the larvae of many aquatic insects together with snails, freshwater shrimps (*Gammarus*) and other small crustaceans. Extensive use should be made of these, as it is always best to supplement an artificial food with as much natural food as possible. Most laboratories, however, will have to rely to a great extent on artificial foods and these are readily available. In addition most fish will eat minced horseflesh or horse heart cut to size according to the species, but pike and perch are best fed on whole fish.

Trout can be fed on minced horseflies, sandhoppers, shrimps, fly larvae; minnows on minced liver, horse heart and *Daphnia*; goldfish do well on many proprietary foods, together with biscuit meal, chopped earthworms and an occasional supplement of very finely minced beef.

Breeding

Trout do not usually breed naturally in indoor tanks, but fry can be produced artificially by expressing the sperm and eggs from ripe male and female fish. The eggs are collected in a dry bowl and the milt (sperm) is then added; the mixture is stirred and the eggs are then covered with about 7·5 cm of water.

After about an hour the eggs are washed and transferred to perforated earthenware or painted zinc trays placed in shallow troughs in a gentle but continuous flow of water. Water will thus be circulating continuously round the eggs, which must not be disturbed except to remove those which die and turn opaque in colour. Up to the time when the eyes of the embryos can be seen, any form of mechanical shock will be fatal. After this the eggs may be washed or transferred.

The young fish will hatch from 30 to 50 days after fertilization. They will take no food for about a month, during which period they live on the yolk sac. After this they require very careful attention or many losses will occur. It is usually not difficult to hatch 95 per cent of all the eggs laid down, but to rear 10 per cent of the resulting alevins to the fingerling stage is no mean achievement. For this reason it is infinitely less trouble and much cheaper to obtain young fish from trout farms, rather than to produce them in the laboratory.

Sticklebacks will breed in the laboratory in small tanks provided with sufficient vegetation in which to build nests. These fish are, however, very pugnacious, and to prevent fighting only one pair should be kept in each tank.

Goldfish can be bred in large tanks or preferably in ponds. Here again there should be submerged plants or bundles of willow twigs to which the fish can attach the eggs. Large numbers of fry will be produced, but the parent fish must be fenced off from the young as soon as they have hatched.

Diseases

Saprolegnia (*white-fungus disease*). This is possibly one of the most common diseases in goldfish and other cold-water fish and is responsible for many deaths. The trouble usually starts on the tail and other fins, eventually spreading over the body until it reaches the gills, when it is invariably fatal since it quickly suffocates the fish. The disease can be recognized by the presence of a white scum which destroys the fins and prevents the skin from functioning normally. The fungus spores are present in practically all natural waters, but a healthy fish will not be troubled at all by this complaint. The disease only becomes apparent when the fish becomes weak or bruised and damaged. Anything likely to result in the weakening of stock, such as overcrowding, overfeeding or sudden changes in temperature, should be avoided. The disease was investigated by Patterson (1903) who found that the killing agent was a bacillus (*B. salmonis pestis*) which entered the fish body through a wound, and that the fungus (*Saprolegnia*) was living saprophytically on a necrotic tissue resulting from the bacterial attack.

If the fins are extremely frayed it may be necessary to trim these with the aid of a very sharp knife and a wooden cutting board. Do not use scissors. Treat the fresh cut edges with a dilute solution of potassium permanganate (230 mg/l of water). Sometimes the fungus can be removed by means of a piece of soft cloth dipped in salt water. Occasionally this treatment will result in raw patches being left, and Innes (1947) recommends that the affected spot be first dried with a soft piece of cotton material and then dabbed with Turlington's Balsam. The gills of the fish must be kept moist during this treatment.

The more usual treatment for fungus is to place the fish in a salt bath, using either sea water or a 3 per cent solution of sodium chloride; the fish can be left in this for 15 to 20 minutes, and the treatment should be repeated daily for three or four days. The sick fish should be kept in shallow water and the light must not be too intense. The temperature can be very gradually brought up to about 20°C every day.

Swimbladder disease. This is only loosely termed a disease, as the causes are not understood. The principal symptom is a loss of equilibrium, and this is very commonly found in goldfish, particularly in those belonging to the highly selected breeds. No certain cure is known, but the condition can sometimes be relieved by placing the fish in a shallow container with weak salt water.

White spot. This is a common disease of temperate and tropical freshwater fishes. It is seen as a coating of tiny white spots on the fins, skin and gills. The causative agent is the ciliate *Ichthyophthirius multifiliis*. In most cases the disease can be cured by placing the affected fish in a solution of methylene blue. Some writers specify definite quantities, but others recommend that the colour of the treatment water should be inky blue-black. Scaleless fish should not be treated by this method. Other authors have used mercurochrome at the rate of 4 drops to 4·5 litres of water, but care should be taken as some species may be more susceptible than others to this extremely poisonous substance.

Furunculosis. This is a deadly disease sometimes found in populations of trout; it is caused by *Bacterium salmonicida*. It occurs mainly in old fish and can be cured by the addition to the food of 6 g sulphamerazine per day for each 45 kg of diseased fish (Sniesko, Gutsell & Priddle, 1948).

More detailed information on fish diseases is given by van Duijn (1956) and by Reichenbach-Klinke & Elkan (1965).

TROPICAL FRESHWATER FISH

The smaller tropical fishes, e.g. cyprinodonts, anabantids and many cyprinids, can be kept quite satisfactorily without water circulation. In some cases a small amount of aeration should be given, but this is quite unnecessary for the anabantids which come to the surface to breathe air. The temperature of the water should be kept at 22 to 24°C.

Larger tropical freshwater fish, e.g. electric eels and many cichlids, can be kept in a circulating system of exactly the same type as those detailed above for temperate freshwater fish, but of course at a higher temperature.

There are numerous books on tropical aquaria which will supply all the information required by the laboratory worker. The following may be specially recommended: Sterba (1967), Hervey & Hems (1952), Axelrod & Schultz (1955), McInerny & Gerard (1958).

REFERENCES

Axelrod, H. R. & Schultz, L. P. (1955). *Handbook of Tropical Aquarium Fishes*. New York: McGraw-Hill.

Gutsell, J. S. (1946). Sulfa drugs and the treatment of furunculosis in trout. *Science*, **104,** 85–6.

Hervey, G. F. & Hems, J. (1952). *Freshwater Tropical Aquarium Fishes*. London: Batchworth.

McInerny, D. & Gerard, G. (1958). *All about Tropical Fish*. London: Harrap.

Patterson, J. H. (1903). *The Cause of Salmon Disease*. Fishery Bd for Scotland: Salmon Fisheries.

Pentelow, F. T. K. (1957). Freshwater Fish. In *UFAW Handbook on the Care and Management of Laboratory Animals*, 2nd ed. London: UFAW.

Reichenbach-Klinke, H. & Elkan, E. (1965). *The Principal Diseases of Lower Vertebrates*. London & New York: Academic Press.

Sniesko, S. E., Gutsell, J. S. & Priddle, S. B. (1948). Various sulphonamide treatments of furunculosis in brook trout. *Trans. Am. Fish. Soc.*, **78,** 181–8.

Sterba, G. (1967). *Aquarium Care*. London: Studio Vista.

Van Duijn, C. (1956). *Diseases of Fishes*. London: Water Life.

41 Marine Aquaria

H. G. Vevers

Introduction

Marine aquaria are being used more and more in laboratories, particularly for research on physiological problems. As a general rule they are more difficult to maintain than freshwater aquaria. The most important points to be watched are the condition and temperature of the sea water and the types of materials used in the construction of the tanks, pumps, and ancillary apparatus.

Sea water

Natural sea water can be used, provided that it is not collected from the shore in areas that are highly populated and industrialized. Coastal waters near to busy ports are nearly always polluted by sewage and industrial wastes.

For a small number of tanks the sea water can be collected, from shores distant from human habitations, in plastic buckets or double plastic bags packed in cardboard containers. For larger quantities most laboratories rely on supplies of offshore water from marine biological stations. Newly acquired natural sea water should be stored in the dark for 2 or 3 weeks. This allows plankton organisms to die and drop to the bottom of the reservoir and gives time for silt to settle.

Synthetic sea water can also be used successfully and is often indispensable when the laboratory is a long distance from the sea. Several formulae have been published, but one of the most successful is that shown in Table I, which has been used in the University of Illinois for maintaining marine invertebrates (Knowles, 1953).

TABLE I

Synthetic sea water

Salt	g/litre	Salt	g/litre
Sodium chloride	27·2	Potassium sulphate	0·9
Magnesium chloride	3·8	Calcium carbonate	0·1
Magnesium sulphate	1·6	Magnesium bromide	0·1
Calcium sulphate	1·3		

Whenever possible a small amount of natural sea water (preferably 10 per cent) should be added to this kind of mixture to provide a supply of trace substances.

A good commercial synthetic salt product is marketed by Westchester Supply Company, White Plains, New York under the name of 'Neptune Salts' (Simkatis, 1958).

As an alternative some laboratory workers and aquarists use a concentrate which is really the mixture of salts remaining after evaporation of sea water. A suitable sea-water concentrate is marketed by the San Francisco Aquarium Society, San Francisco, California under the name of 'Eden Brine'. This has been successfully used for the notoriously difficult coral-reef fishes (Simkatis, 1958).

In general the density of aquarium sea water should be kept at approximately 1025, and this should be checked once a week with a suitably graduated hydrometer. In most aquaria the tendency is for the density to rise, owing to evaporation. This must be corrected by the gradual addition of natural fresh, or preferably distilled, water.

The pH of the water should be kept at approximately 7·8 to 8·2. In nearly every case the pH of aquarium tanks decreases owing to the metabolic processes of the fish and invertebrates. Some aquaria add lime to correct the pH, but this eventually leads to an increase in the relative proportion of calcium. On the whole sodium hydroxide, added gradually and well-diluted, is probably the most satisfactory additive for the correction of low pH values.

Tanks and piping

These should be carefully chosen to avoid introducing substances which are either unstable in sea water or injurious to the animals. All forms of copper (including brass, gun-metal and bronze) and zinc must be completely excluded from the aquarium, whether from the tanks or other fittings. It is also advisable to avoid the use

542

of aluminium or lead. Iron can be used but it quickly rusts. There is, however, no need to resort to any of these metals as all aquarium components can be constructed of either glass, polythene, stainless steel, asbestos, porcelain, fibreglass, polyvinyl chloride, or waterproof cement.

Slate should not be used as it easily cracks and is very expensive. Moulded glass tanks are satisfactory for certain small animals, but they are fragile and do not lend themselves to the installation of water-circulation equipment.

Asbestos tanks, now marketed for domestic purposes, and ordinary procelain sinks are perfectly satisfactory when there is no need to view the animals from the side through glass. The most suitable tank for most laboratory purposes is, however, one constructed with a solid angle-iron frame supporting sheets of plate glass. The metal frame should be painted with three coats of good epoxy resin to prevent corrosion. The requisite thickness of the glass depends upon the depth of water and on the area of glass involved. Table II gives the recommended thicknesses of glass.

In assembling the tank the bottom plate should be fitted first, then the two long sheets (back and front) and finally the end sheets. The glass should be bedded against the metal frame with a mastic compound, preferably one that does not set hard. In Britain the product Glasticon marketed by Industrial Engineering Ltd., Vogue House, Hanover Square, W.1., has been satisfactorily used for many years by laboratories and in large public aquaria. This compound is made in several grades, and the sizes of the tanks should be specified when ordering. Thus large aquarium tanks approximately 2 m long and 1 m deep require a thick grade, which can just be kneaded by hand. If a thinner grade were used in such a tank the pressure of water would squeeze it out, and the glass would then be in contact with the frame and liable to crack. The thinner grades can be used for smaller tanks, e.g. with area about 70 × 50 cm and 30 cm deep.

The size of the tank depends, of course, on the number and size of the animals to be kept. In general the length:breadth:depth proportions should be approximately 15:10:6, thus giving a container in which the surface area of the water is relatively large compared with the volume. Tanks holding less than 70 litres should be avoided.

Until recently most marine aquaria had to use glass-lined or porcelain-lined piping for conveying the circulating water. This was costly and difficult to obtain. Nowadays the problem is much simpler. Plastic tubing is cheap and readily available in a variety of sizes. Hard polyvinyl chloride and polythene have proved very successful, provided the whole circulation is flushed with fresh water for about a week. Jointing of polythene is best done by welding with a torch delivering hot nitrogen, a technique now available in most large cities. Taps and valves should also be of polythene, but these should be inspected to ensure that they contain no metal parts which might come into contact with the circulating water.

Circulation of the water

Marine aquarium circulations may be classified into three types: open, semi-closed and closed.

Open circulations are used in marine laboratories situated on the coast in areas where there is no pollution. The sea water is pumped up directly into a gravity tank installed above the aquarium tanks. The water flows down from the

TABLE II

Thickness of aquarium glass required (mm)

Height of tank (cm)	Length of tank (cm)													
	20	30	40	50	60	70	80	90	100	110	120	130	140	150
30	2·5	2·8	3·3	3·8	4·1	4·2	4·4	4·6	4·9	—	—	—	—	—
40	—	3·4	4·3	5·1	5·6	6·0	6·3	6·5	6·7	6·9	7·0	7·1	—	—
50	—	4·4	5·1	5·8	6·5	7·2	7·7	8·2	8·4	8·7	8·9	9·1	9·2	—
60	—	—	6·0	6·5	7·5	8·5	9·3	9·7	10·2	10·7	11·1	11·4	11·6	11·7
70	—	—	6·6	7·3	8·2	9·0	10·0	10·9	11·6	12·2	12·7	13·1	13·4	13·6
80	—	—	7·4	8·2	8·8	9·3	11·0	12·2	13·1	13·7	14·3	14·9	15·6	16·1

(After Loderstedt, from Sterba, 1967)

gravity tank, through the aquarium tanks, and then out to sea again.

Semi-closed circulations are used in laboratories situated on the coast, but in areas where the tidal water is polluted for most of the time, e.g. Plymouth or Naples, the water is pumped up only at the time of a high spring tide. It is stored in underground reservoirs, whence it is pumped into the aquarium tanks, with or without an intervening filter. In this type of circulation each batch of water is used for periods of 3 to 6 months and is then discarded and replaced. This avoids the use of water which has become increasingly polluted by the aquarium animals.

Most laboratories, and indeed all those which lie inland, must of necessity use a closed circulation. In this system a supply of sea water is brought from the coast in glass carboys or in clean, suitably lined tanker lorries, and stored in a reservoir. Alternatively the reservoir can be stocked with water made from a concentrate or artificially from salts.

The stored sea water must now be conveyed through a carefully designed system to the aquarium tanks and then back to the reservoir; the system may or may not include a filter at some point in the circulation. The design of the system can take many forms and indeed among

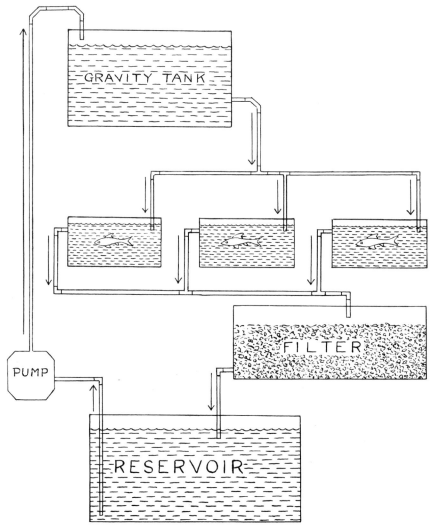

Fig. 41.1. Elevation of closed-circulation system for sea-water or fresh-water aquaria.

public aquaria there are scarcely two which use exactly the same method. The following system (Fig. 41.1) has been used for 40 years in the Aquarium of the Zoological Society of London and, with variants, elsewhere. In recent years it has also been used, on a reduced scale, in several biological laboratories in Britain. The water is pumped from the reservoir to a gravity tank situated above the aquarium. From there it falls through piping into the aquarium tanks, preferably to the bottom of each. The water moves through each tank and escapes through an outlet at the top, whence it flows by gravity to the filters, which will be described below. After having passed through the filters it flows, again by gravity, back to the reservoir.

As a variant, to suit space requirements, there is no reason why the filter should not be built between the gravity tank and the aquarium tanks.

In such a closed circulation the water is used over and over again, for periods of several years, although it is usually necessary to add about 10 or 20 per cent of new water per annum to make good losses caused by siphoning, tank-cleaning and other maintenance operations.

It has, however, been shown that extensive changes take place in the composition of the water in the course of time. After the London Zoo Aquarium had been running for some 30 years, Oliver (1957) found that, in spite of the small annual supplements, the sea water in circulation had radically changed. Calcium and phosphorus had greatly increased, but the most striking feature was the enormous build-up of nitrate. This apparently had no effect on the teleost and selachian fishes, but it was becoming increasingly difficult to keep certain invertebrates, particularly echinoderms and octopus.

This problem has been solved by taking certain tanks off the main circulation and supplying them with unused sea water from a separate reservoir. Every week about 80 per cent of the water in these special tanks is siphoned off into the main circulation and replaced with a fresh supply of unused sea water. As a result the invertebrates, in particular octopus, have thrived as never before (Vevers, 1961).

This brief account of an experiment is given here to emphasize the kind of changes which will also take place in laboratory circulations, and it is suggested that a regular check should be made on the nitrate content of the water. When this exceeds 10 or 20 parts per million a percentage of the water should be replaced if the tanks are being used for marine invertebrates.

Hale (1957) has described an extremely interesting method of producing a fast circulation of sea water in small marine tanks. This gave a rate of water flow of up to 60 cm per minute and proved particularly successful in maintaining marine invertebrates over long periods.

In temperate climates the main hazard is the overheating of the circulating water during summer. This may necessitate the installation of a cooling plant at some point in the circulation. If this is done the cooling coil must be made of a non-toxic and non-corrosible material. It is often difficult to explain to refrigeration engineers that this is an essential condition. Stainless steel is satisfactory but expensive. An ordinary iron coil can be used, provided that it is coated inside and outside with epoxy resin.

Filtration

A clear distinction must be made between sieving and biological filtration. In sieving a proportion of the relatively large particles in the water can be removed by trapping them with glass wool or nylon nappe or, on a small scale, with filter paper. Nylon is preferable to glass wool, which sometimes fragments so that particles of glass enter the circulating water and become lodged in the gills of fishes. Sieving is of only minor importance in aquarium work.

A true aquarium filter works on the same principle as a sewage farm. The water is passed through a filter bed of grit supported on pebbles. The grit particles become coated with bacteria, heterotrich ciliates, harpacticid copepods and other organisms which work on the organic debris and wastes. Tests have shown that, in an efficient filter, water entering with an oxygen content of 7 mg/litre will leave the filter with a content of about 4 mg/litre; the exact figures depend, of course, on the temperature. The oxygen lost in the passage of the water through the filter has been used by the organisms principally, it is believed, in oxidizing ammonia (which is very toxic to aquatic animals) to nitrite and nitrate.

The surface area of the filter should be large in relation to its volume. The bottom layer may be of pebbles approximately 1 or 2 cm in diameter, supporting the main filtering medium, which should be grit with a particle-size of 2 mm. If finer grades (sand) are used the filter becomes too close-packed and ceases to function efficiently.

The water can enter at the bottom and ascend through the filter, but it is more usual to distribute it over the surface and allow it to descend through the filter bed. It can then be collected in a perforated hard polythene pipe lying 2 cm above the bottom, whence it is conveyed by gravity to the reservoir.

Filtration is not used in all aquaria. Some aquarists rely, for instance, on settling-tanks to remove suspended matter. This may, indeed, be sufficient for semi-closed circulations where the water is changed periodically. In a fully-closed system, however, there are obvious advantages in being able to oxidize ammonia and possibly other toxic substances and at the same time remove particulate waste.

Temperature

The temperature of the aquarium water must naturally depend upon the temperature of the environment from which the animals come. In theory this could mean that a large number of temperatures might have to be catered for, but in practice a temperate circulation in the range 10 to 14°C and a tropical circulation at 22 to 25°C are usually sufficient.

When new animals are introduced into an aquarium tank, care should be taken to ensure that the temperature of the latter corresponds, within 1°C, with the temperature of the water in which the animals have travelled.

Light

Aquaria can be lit by natural or artificial light. With natural light the tanks tend to become coated with algal growth and this is not desirable unless the occupants include browsing herbivores such as limpets or certain sea-urchins. Tungsten lamps also tend to encourage algae and produce a lot of heat.

Fluorescent tubes are probably the most satisfactory form of lighting. They produce no perceptible heat, have a high lumen output and are available in a wide range of tones. For most purposes tubes which cover the whole visible spectrum are most suitable, preferably with an increase in the amount of red light.

Tropical coral fishes need bright light and so do sea-anemones, such as *Anemonia*, which have symbiotic algae.

So far as possible light fittings should be protected from corrosion by two to three coats of good epoxy resin. For small tanks the fittings can be hung over a glass or transparent plastic sheet which is allowed to rest, with an air gap, on the rim of the tank. Lighting from the side is not recommended, except as a temporary measure when observations have to be made.

Aeration

With an efficient circulating system there should be little need for aeration in the tanks. It is, however, useful to have an air supply available in case of emergency. If the laboratory has a piped supply of compressed air this can be used satisfactorily. If there is no air-supply the tanks can be aerated with a small air pump. In either case care should be taken that the delivered air is free from oil.

Only a proportion of the air will actually be dissolved in the water before the bubbles have reached the surface, but this does not matter because one of the main functions of aeration is to move the water in such a way that new layers are continually reaching the surface, where a natural exchange of gases can take place; this involves the loss of carbon dioxide and the acquisition of oxygen.

Feeding

The type of food required depends entirely on the species kept, and the variation in feeding habits is enormous. The following notes are intended as a guide.

Most adult marine fishes will take mussels, ragworms, earthworms or raw squid cut up into suitable sizes. Mammalian flesh can be used in emergency, either minced or diced, but should never be given as the sole food. Some fishes thrive best on moving live food and these can be fed on sandhoppers (Amphipoda), shrimps, prawns or newly hatched tropical freshwater fish, such as guppies (*Lebistes*). Fish with powerful jaws and teeth such as wrasse (*Labridae*) and trigger-fish (*Balistes*) will attack and eat crustaceans.

Young fishes are more difficult to feed and they require to be fed more frequently. The standard diet consists of a culture of brine-shrimp (*Artemia*) nauplii. Brine-shrimp eggs can be obtained from the San Francisco Aquarium Society. Take $4\frac{1}{2}$ litres of a $3\frac{1}{2}$ per cent solution of sodium chloride, add $\frac{1}{2}$ a teaspoonful of the eggs, and aerate. If the temperature is kept at 21°C the eggs will hatch in about 24 hours.

The newly hatched nauplii must be separated from the empty egg-cases and may then be fed directly to very young fish. Or they may be grown to larger sizes by feeding on a culture of microscopic algae or on a very dilute suspension of yeast. Brine-shrimps are eagerly taken by sea-horses (*Hippocampus*).

Crabs, lobsters and most other crustaceans will feed on fish or squid cut into suitable sizes. Starfishes of the *Asterias* type feed on bivalve molluscs of which the most readily available will probably be mussels (*Mytilus*). Most sea-anemones will take tiny pieces of chopped fish, squid or mussel, but there are some, e.g. *Metridium*, which must be fed on a suspension of finely-ground mussel.

Octopus feed primarily on crustaceans, particularly crabs. In the aquarium, small to medium-sized octopus feed well on crabs, e.g. *Carcinus*, with a carapace-breadth up to about 6 cm. Larger octopus will take a lobster, and all sizes will capture and feed on small fish. The legs and carapace of the crustaceans will be rejected and will fall down in front of the octopus's favourite hiding-place, whence they must be regularly removed to avoid fouling.

Filter-feeding invertebrates, such as lamellibranchs or tunicates, are a more difficult problem. Ideally they should be fed on a culture of microscopic algae. If this is unobtainable attempts should be made to feed them on very dilute suspensions of yeast or fresh ground mollusc flesh. When this method is used great care should be taken to ensure that the quantity of suspension introduced is not sufficient to cloud and foul the water.

Treatment of marine invertebrates before fixation or preservation

Molluscs, Actinia, hydroids, bryozoans, Amphioxus. Leave overnight in 7 per cent solution of magnesium chloride.

Most aneomones (not Actinia), marine worms, holothurians, ascidians. Leave overnight in sea water after sprinkling menthol crystals on the surface.

Starfish, crinoids. Place in fresh water for 10 minutes.

REFERENCES

Hale, L. J. (1957). The problem of keeping marine animals. *Proc. R. phys. Soc. Edinb.*, **26**, 19–24.

Knowles, F. G. W. (1953). *Freshwater and Salt-Water Aquaria.* London: Harrap.

Oliver, J. H. (1957). The chemical composition of the sea water in the aquarium. *Proc. zool. Soc. Lond.*, **129**, 137–45.

Simkatis, H. (1958). *Salt-water Fishes for the Home Aquarium.* Philadelphia & New York: Lippincott.

Sterba, G. (1967). *Aquarium Care.* London: Studio Vista.

Vevers, H. G. (1961). Observations on the laying and hatching of octopus eggs in the Society's Aquarium. *Proc. zool. Soc. Lond.*, **137**, 311–15.

42 Fruitflies

MARSHALL R. WHEELER

The ease with which *Drosophila* flies can be raised in the laboratory has made them favourite animals for scientific investigations. This is especially true of the cosmopolitan species, usually associated with man's garbage, since these species are readily cultured on rather simple media. Examples are *Drosophila melanogaster, D. simulans, D. hydei, D. immigrans,* and *D. funebris,* all of which have been used extensively in studies of genetics, cytology, physiology, etc. With extra effort and minor modifications of the basic food media, another hundred species, more or less, can be kept in culture quite readily. At the other extreme, at least two hundred species of this genus can be maintained in culture simultaneously with about three basic food media plus a few modifications.

Culture media

Most workers use the cornmeal-agar medium (Table I), especially for species of the subgenus *Sophopora* (e.g. *D. melanogaster, D. simulans, D. obscura*).

To prepare, mix the agar with about three-fourths of the water, boil, and continue heating until the agar has gone into solution completely. Add ethanol to the dry yeast to wet it. Add the syrup and cornmeal to the yeast mixture, then gradually stir in the remaining water. Add this mixture to the boiling agar-water solution. Boil gently for 10 to 15 minutes. Allow to cool partially and then add the propionic acid (cool preferably to 43°C or less; the effectiveness of the acid as a mould inhibitor is greatly reduced if it is added when the medium is too hot). Pour into culture containers and allow to cool completely. Cotton plugs should not be added at this time since water of condensation must have a chance to evaporate. It is wise to drape a layer of thin sterile cloth over the open vials to avoid contamination. When the food is cool, spray it with a live yeast suspension, plug with cotton (non-absorbent cotton is required) and store for a day at cool room temperature before use.

If Tegosept* (Nipagin) or Moldex* is used instead of propionic acid as the mould inhibitor, it should be added to the medium in alcoholic solution during cooking. These substances are only slightly water-soluble; hence the usual practice is to prepare a 10 per cent solution in 95 per cent ethanol, and then add 7 to 10 ml of this solution per litre of medium.

A banana-agar medium (Table II) is widely used for members of the subgenus *Drosophila* (*D. immigrans, D. funebris, D. virilis,* etc.). To prepare, mix the agar with the water and bring to a boil. Continue heating until the agar has gone into solution completely. Add mashed bananas

* *Editors' note:* These are derivatives of *p*-hydroxy benzoic acid, Tegosept being the methyl ester and Moldex the sodium salt.

TABLE I
Cornmeal-agar medium

Water	1,000 ml	Dark corn syrup ('Karo')	100 ml
Agar	15 g	Propionic acid	5 ml
Yeast (dried brewer's)	30 g	Ethanol (95 per cent)	5 ml
Cornmeal (maize meal)	100 g		

TABLE II
Banana-agar medium

Water	1,000 ml	Sucrose	25 g
Agar	20 g	Propionic acid	4 ml
Banana (ripe, crushed)	300 g	Ethanol (95 per cent)	4 ml
Yeast (dried brewer's)	50 g		

(about three medium-sized bananas weigh 300 g). Stir and cook for a few minutes. Then add the yeast and sugar, wetting the yeast first with the ethanol. Stir and cook until the mixture is homogeneous and the sugar is completely dissolved. Allow to cool partially, then add propionic acid and stir well. Pour into culture containers and allow to cool gradually. Spray with yeast suspension and plug with cotton as described above.

The bananas should be very thoroughly mashed, preferably in a blender, before being added to the mixture. However, bananas can also be kept frozen for long periods, and when they are allowed to thaw at room temperature their consistency is such that very little mashing is needed.

If Tegosept (Nipagin) or Moldex is used as the mould inhibitor, the addition should be made while the medium is being cooked (see remarks above, under cornmeal-agar medium). It should also be noted that the amount of agar required tends to change with the relative humidity, so that during the most humid months as much as 4 g more per litre are required than during very dry periods. The particle-size of the agar must also be considered. Finely powdered agar such as is routinely used in bacteriology is more easily put into solution than is agar of larger particle-size.

Sang & Burnet (1963) have stressed the important fact that agar from different sources does not support *Drosophila* equally well—in fact, some agar preparations seemed to be toxic. This should be kept well in mind when preparing media with newly purchased agar supplies.

High-protein diet

A medium especially rich in proteins, vitamins, and minerals was described by Wheeler & Clayton (1965). Basically, one or more commercial, dry-type breakfast cereals containing such supplements are used, while greater precision of the sugar source is gained by the use of commercial canned baby foods (e.g. Gerber's Strained Banana Baby Food). A successful formula for such a medium is given in Table III. To prepare, blend the dry cereals in a Waring Blender, allowing 3 to 5 minutes for thorough mixing. Mix the agar with the water and bring to the boil. Continue heating until the agar is in solution. Wet the yeast with ethanol, add this to the blended ingredients, and add the jar of baby food. Stir this mixture into the water and allow to cook for at least 8 minutes. Allow the mixture to cool, and plug with cotton as with other types of medium.

A larger number of species can be maintained on the Wheeler-Clayton medium than on any other so far devised. Several substitutions in the formula are possible. At the time of writing there are many kinds of enriched breakfast cereals for children; without doubt most of these would serve as well as the ones listed. Other types of canned fruits for babies may also be used; e.g. the junior size jar of Gerber's banana-pineapple baby food is as useful as the one listed. Perhaps other brands, in other areas, will work equally well.

Availability of stocks

All aspects of *Drosophila* culture are treated in the annual editions of *Drosophila Information Service* (DIS*), including lists of stocks in the various laboratories of the world. Two formal centres for the maintenance and distribution of genetic mutants of *Drosophila melanogaster* are established in the United States and will supply specific stocks upon request:
Drosophila Stock Center, Division of Biology, California Institute of Technology, Pasadena, California, USA, 91109.
Mid-America Drosophila Stock Center, Department of Biology, Bowling Green State University, Bowling Green, Ohio, USA, 43402.

Similar collections of stocks are available at a

* See footnote in the references to this chapter.

TABLE III
High-protein diet

Water	1,000 ml	Kellogg's 'Special K' cereal	10 g
Agar (fine grain)	13.5 g	Kellogg's 'Concentrate' cereal	5 g
Yeast (dried brewer's)	50 g	Gerber's 'High Protein' cereal	15 g
Banana Baby Food	1 jar (4¾ oz)	Kretschmer's 'Wheat Germ' cereal	15 g
Ethanol	5 ml	Propionic acid	5 ml

wide variety of laboratories throughout the world; examples are as follows:

Institute of Genetics, University of Stockholm, Stockholm, Sweden.

National Institute of Genetics, Shizuoka-ken, Misima, Japan.

Department of Genetics, Cambridge University, Cambridge, England.

Department of Genetics, University of Melbourne, Melbourne, Australia.

Istituto di Genetica, Universita di Milano, Milano, Italy.

Available stocks of the different species of *Drosophila* are also listed in DIS, but relatively few laboratories maintain more than a few such stocks. The two largest collections are the following:

Genetics Foundation, Zoology Department, University of Texas, Austin, Texas, USA, 78712.

Department of Biology, University of Chicago, Chicago, Illinois, USA, 60637.

Collection of wild flies

With just a little experience, a variety of *Drosophila* species can be collected locally by using either naturally-occurring baits such as fallen fruits, fungi, bleeding sap flows on trees, etc., or artificial baits placed strategically in the area. The most frequently-used bait is crushed, well-fermented banana, either placed in containers (the size is not critical, but in general larger bait can attract more specimens) or smeared on to tree trunks and the like. A novel baiting method has been reported by Lakovaara *et al.* (1969) who soaked rye or barley malt in hot water for some hours, then added yeast to it when it was cooled. Flies will accumulate at baits most strongly just after dawn and again near twilight; in most climates they show little activity during the rest of the day. Banana baits seem to attract more species than any others that have been tried, but other fruits, crushed and fermented, often attract a good variety of species, with a few being predominant. As a bait ages, its attractiveness for certain species tends to change so that a species which was rare on the first day or two may become common near the week's end. Baits left in a given locality for a number of days attract *Drosophila* species much better than do those moved around too often.

After about ten days of use, baits should be carefully destroyed; otherwise a new generation of flies may begin emerging from the bait itself, and upset the collection data.

Most workers remove the flies from the baits with an insect net, although modified vacuum cleaners can sometimes serve (McMahon & Taylor, 1967). Nets may have a large bag, in which case an entomologists' aspirator is used to remove individuals from the net. Other nets have been devised to simplify the collection and removal of flies from baits; two commercial nets of this type are the *Turtox Wheeler Insect Net*, available from General Biological Supply House, Chicago, Illinois, and the *Drosophila Insect Net* made by Ward's Natural Science Establishment, Rochester, New York. In practice, however, individual collectors usually modify such nets to suit their own needs. An automatic trapping device has been described by Hooper (1967).

Handling wild-caught flies

Newly-collected material should be sorted as soon as possible, in part to discard unwanted insects, debris, etc., and in part to segregate the wanted specimens as early as practical. Females should be isolated in individual culture vials, rather than being put together in masses, since such 'iso-♀' lines are far more valuable in research than are heterogeneous mass stocks. Etherization and sorting in the field is possible but difficult. Working in a shelter, or in a nearby hotel room, is more practical but one must be extremely cautious to ensure that insecticides are not in use in such rooms.

Transporting collections back to the laboratory may be an easy matter if short distances and moderate temperatures prevail. Harsher situations require more careful treatment; for example, flies may have to be carried in polystyrene or polyfoam plastic boxes or insulated picnic ice-chests, to protect them from excessive heat.

For longer journeys, especially when great extremes of temperature are to be encountered, Spieth (1966) devised a novel sugar-vial method of feeding the flies en route without using the customary culture media. Eight-dram vials were lined with dampened 6·5 × 6·5 cm pieces of chromatography paper (or blotting-paper) which was later impregnated with the special medium and autoclaved. This simple medium contains only one litre of water, 15 g of fine agar, and 50 ml of dark syrup (Karo). The agar is added to the water, then heated to dissolve the agar completely. The syrup is added, the mixture is heated gently for another three minutes, and then poured into the paper-lined vials to about 8 mm depth per vial (the exact amount is determined by the paper's absorbency). After being stoppered with non-absorbent cotton the vials are autoclaved (15 min at 9 kg pressure), after which the autoclave is slow-exhausted. As a result of autoclaving, all of the liquid medium should be absorbed into the paper lining and only a very thin film should remain on the bottom of the vial. Such vials, being sterile, do not need refrigeration, and hence are especially useful for field use.

When the flies are placed in the vials they feed readily upon the surface of the impregnated paper and can cling to it easily. Condensation does not form readily, and few flies become stuck in transit. For long distances, it is best to add to the insulated box or container a frozen can or two of Scotch Ice, Magic Cold, or similar chilling agent, wrapping the vials in paper to avoid excessive chilling. Flies can be kept safely for about one week in vials of the type described.

Establishment and maintenance of stocks

Newly-collected flies should be examined carefully for parasitic mites before being added to the stock shelves. Heavily infested individuals should be discarded at once; a single mite can often be removed by hand with fine forceps. Careful record-keeping of collections and newly-acquired material is also important. One should always make a permanent record of collection data—place (in detail), date, baits (or other methods used), collectors, climate information, as well as any other observations that might later prove useful. Most laboratories use a code system for listing such data. At the University of Texas each collection is assigned a number, each species is given a decimal number, and iso-♀ lines which are established are given an

additional number. For example, 2199·1–5 indicates a collection made August 13, 1951, at Morgan, Utah (= 2,199 entry), the species *flavomontana* was listed as ·1, and of the iso-♀ lines which were established, the one referred to was number 5. This identification number is firmly attached to the stock in the laboratory. At the University of Hawaii, where many collectors may be involved in the same trip, the system uses a letter-number combination for the place-date information, a letter for the collector (each has been assigned a personal letter in advance) and a number for the species or iso-♀. Thus the code L31P19 indicates a collection at Waikamoi, Maui, Hawaii, on Feb. 10, 1968, P shows that the collector was Dr. M. Kambysellis, and 19 indicates the species *discreta*.

Newly-acquired stocks should be tested on several types of food media to determine which seems to work best. Thereafter a routine schedule of stock-changing can be followed e.g. every 7 days, or every 10 days. When changing cultures on to a fresh medium, it is important not to try to change more than one at a time—otherwise identification tags may get accidentally interchanged, or vials of one might get placed with those of another stock. Such errors are not tolerable.

Stock-changers need to be familiar with the appearance of mite infections, nematode infections, and some of the worst mold infections, Details on the methods of combatting these infections were described in the third edition of this *Handbook* and some additional methods are described below. It is equally important that stock-changers be quite familiar with the other characteristics of the stocks for which they are responsible; thus, should an accidental contamination of one stock occur with specimens of another, the stock-changer should note this at once and report it to the supervisor for correction. Such contaminations are not at all rare, largely due to the fact that cotton plugs are not always tightly fitted, or may become partly dislodged, allowing the contaminants to gain entry. It is also helpful to leave a few opened vials of medium placed around the laboratory to attract those free, escaped specimens which might be bothersome as contaminants; after a few days such vials should be plugged and discarded, and fresh ones put out in their place.

It is also important that stock-handlers become familiar with the behavioural differences of the various species. Some cultures need to be handled gently, for otherwise the adults may fall on to the food in a kind of apoplectic fit, and thus get stuck to the medium. With some species, adults survive readily in vials with many well-developed larvae, while in others, the adults become mired in the food mass if left as long as this. Species differ widely in the number of days of ageing required before mating and egg laying begin. There are great differences in the number of eggs laid per day, or per female, which helps to determine the proper number of adults to be kept in a single container.

There are striking differences between species in the manner of pupation. There appear to be four principal ways in which this occurs:

1. Mature larvae do not leave the food source but pupate in it, usually on the surface. Since younger larvae are still churning the food mass, this often results in many puparia getting buried within the food. In the wild, perhaps, the food mass may dry up at about the time most larvae pupate.

2. Mature larvae of many species, at least in standard culture vials or bottles, crawl out of the food and pupate on the sides of the container, at distances ranging from a centimetre or two to ten or more. In some cases, they tend to pupate side-by-side while others pupate randomly. This pupation habit has also been observed in some sap-flow inhabiting species; when mature these larvae crawl away from the flux and pupate in crevices of the nearby bark.

3. In some species the larvae leave the food and penetrate some well-formed but soft material to pupate. In the laboratory, this habit is seen when they enter folded pieces of soft paper placed in the containers, seeming to prefer this situation to any other. Various types of soft tissues are frequently added to culture vials, in part to aid in absorbing excess moisture, but also to facilitate pupation.

4. The fourth pupation type is shown by species whose larvae leave the food when mature and crawl and skip or jump considerable distances (up to several metres) before pupation. In the usual laboratory situation, such larvae often end up within the cotton plugs, and a great many of them die here. While working with endemic Hawaiian species of *Drosophila*, Wheeler & Clayton (1965) found that allowing such larvae to dig their way into moistened sand (preferably washed beach sand) solved the problem completely. Mature larvae would burrow into the sand, in some instances as deeply as six inches, and pupate there. When fully developed, the emerging imago would expand the ptilinum—an expandable sac-like structure on the top of the head—and use this to force its way upward through the sand. On reaching the surface, the fly would walk a short distance, then stop and finish the process of adult maturation—wing expansion, etc. It seems likely that this type of soil or sand pupation is very widespread in *Drosophila* and related genera. Some species of *Mycodrosophila*, for example, living as larvae in fungi, can later be recovered as pupae in soil near the fungus. It is interesting to note that mature larvae of species of 'pupation type one', described above, when placed on sand, will burrow into it readily, then turn back up toward the surface, and finally pupate with only the anterior tip of the puparium protruding above the sand surface.

The principal difficulties in maintaining healthy cultures continue to be undesirable growths of bacteria, molds, and, occasionally, yeasts. The discussion of their control in the third edition of this *Handbook* is still quite pertinent. A few innovations have, however, been made more recently. Hanks *et al.* (1968) and Felix (1969) described newer, more efficient methods for the control of bacterial infections of the food medium. Mold control has been improved by Cooper (1968), and Gonzales & Abrahamson (1968) described a method of controlling undesirable yeast growth on the food.

Mention was made earlier of the necessity of avoiding insecticides around *Drosophila* operations. This has become a serious problem since so many vapour types are now marketed. A new aspect of this problem was described by Strömnaes (1968) whose *Drosophila* stocks were failing because of insecticide contaminations of the cornmeal used in preparation of the culture medium.

Other useful references

The most recent survey of all aspects of *Drosophila* culture was that of Wheeler (1967). No single publication can, however, cover all aspects adequately, and since techniques change year after year, any such publication is a bit out-of-date by the time it appears. Reading the Technical Notes section of *Drosophila Information Service* is the most satisfactory means of keeping aware of the changes. On the other hand, many basic techniques have not changed much; these time-tested procedures are well covered in the following references: Demerec & Kaufmann (1962); Spencer (1950); and Strickberger (1962).

REFERENCES

The *Drosophila Information Service*, usually referred to as *DIS*, provides an annual compendium of stock lists, research and technical notes, and an inventory of laboratories engaged in various types of *Drosophila* research. Information concerning subscriptions to this service can be obtained from the Department of Biology, University of Oregon, Eugene, Oregon, USA, 97403.

Cooper, K. W. (1968). Freeing *Drosophila* of mold. *Drosoph. Inf. Serv.*, **43,** 194.

Demerec, M. & Kaufmann, B. P. (1962). *Drosophila Guide*, 7th ed. Washington: Carnegie Institute of Washington.

Felix, R. (1969). Control of bacterial contamination in *Drosophila* food medium. *Drosoph. Inf. Serv.*, **44,** 131.

Gonzalez, F. W. & Abrahamson (1968). Acti-dione, a yeast inhibitor facilitating egg counts. *Drosoph. Inf. Serv.*, **43,** 200.

Hanks, G. D., King, A. L. & Arp, A. (1968). Control of a gram negative bacterium in *Drosophila* cultures. *Drosoph. Inf. Serv.*, **43,** 180.

Hooper, G. B. (1967). An automatic trapping device for *Drosophila*. *Drosoph. Inf. Serv.*, **42,** 115.

Lakovaara, S., Hackman, W. & Vepsäläinen, K. (1969). A malt bait in trapping Drosophilids. *Drosoph. Inf. Serv.*, **44,** 123.

McMahon, J. & Taylor, D. (1967). Another method for removing *Drosophila* from traps. *Drosoph. Inf. Serv.*, **42,** 118.

Sang, J. H. & Burnet, B. (1963). The importance of agar. *Drosoph. Inf. Serv.*, **38,** 102.

Spencer, W. P. (1950). Collection and laboratory culture. In *Biology of Drosophila*, ed. Demerec, M. New York: Wiley; London: Chapman & Hall.

Spieth, H. T. (1966). A method for transporting adult *Drosophila*. *Drosoph. Inf. Serv.*, **41,** 196.

Strickberger, M. W. (1962). *Experiments in Genetics with Drosophila*. New York: Wiley.

Strömnaes, O. (1968). Insecticides. *Drosoph. Inf. Serv.*, **43,** 188.

Wheeler, M. R. (1967). The fruitfly (*Drosophila*). In *The UFAW Handbook on the Care and Management of Laboratory Animals*, 3rd ed. Edinburgh: Livingstone.

Wheeler, M. R. & Clayton, F. (1965). A new *Drosophila* culture technique. *Drosoph. Inf. Serv.*, **40,** 98.

43 The Housefly

G. J. ASHBY

With a geographical range extending from the sub-polar regions to the tropics the housefly, *Musca domestica*, is probably the most widely distributed insect in the world. It is also perhaps the most important insect available to the laboratory worker for the routine testing of insecticides, insect repellants and attractants. There is no doubt that it has come to be regarded as indispensable for the standardization of contact sprays (West, 1951). Houseflies are also employed extensively in zoos and laboratories as a live food for certain carnivorous insects, reptiles and birds.

The culture of *M. domestica* in the laboratory is simple. The flies breed rapidly and are extremely prolific. Disease is no problem and healthy stocks can be built up and maintained over long periods without difficulty. Several media for rearing flies have been used, ranging from pasteurized horse manure and yeast to quite complicated formulae. It is, however, important to emphasise the point made by Parkin & Green (1957) that once a medium and methods of rearing and handling have been found satisfactory, they should be adhered to, or excessive variations in the physiological behaviour of the adults towards insecticides may complicate tests.

Obtaining stocks

Stocks to start the cultures may be obtained either by acquiring pupae or adults from a laboratory already engaged in their production, and this is the safer method, or by setting traps, at least during the warmer months of the year, to catch specimens around farms, poultry houses and stables.

Descriptions and keys for identifying the wild-caught flies are given by Hewitt (1914), Patton (1930), Herms (1950), and West (1951).

Housing the adults

If flies are to be produced on a large scale and for insecticide-testing, the culture room should be sufficiently spacious to accommodate the whole of the equipment and have a controlled temperature of $27 \cdot 5°C \pm 1°C$. The relative humidity should be maintained at between 50 and 55 per cent, but should not vary more than 3 per cent. The ceiling and the walls of the room should be free from cracks and crevices and preferably sealed with several coats of hard gloss paint or enamel to facilitate regular cleaning with detergent and warm water. White or cream paint has the advantage of showing up dirt or unwelcome insects or parasites which may find their way into the room.

The floor should be of concrete, preferably coated with either asphalt or bitumastic, thus avoiding any holes or cracks in which spilt food could lodge. Sunlight or even daylight is not really necessary, in fact windows are to be avoided as they give rise to unnecessary heat-loss and become damp with condensation during cold weather. Fluorescent lighting is to be preferred because of the low output of heat and should be used in conjunction with an automatic time switch giving nine hours light each day. It is important to avoid any tendency for the atmosphere in the room to vary in temperature and humidity at different levels, and this can be overcome by the use of a small fan to keep the air continually circulating. For the same reason racking should be used rather than solid shelving.

Mating and oviposition

The simplest and cheapest type of cage for mating and oviposition is made of wood, and for convenience is not more than 28 cm³ in volume. Wire or terylene gauze should be used for the sides and top, and the bottom should be of plywood. One or two sleeved holes should be fitted into the gauze to allow the flies to be put in and food to be changed. Cages of this size will accommodate about 600 flies. They are cheap and easy to construct and are extremely satis-

Fig. 43.1. Wooden-framed oviposition cage. Note projecting bolts and wing-nuts. (Photograph: Zoological Society of London.)

factory except that they are difficult to keep clean for long periods without deterioration, and so have to be periodically replaced.

Basden (1947) described a more robust type of wooden cage which can be dismantled for cleaning (Fig. 43.1). The measurements are approximately $45 \times 45 \times 45$ cm and it consists of two side frames with bolts protruding from the front, back and top. The other three corresponding frames are fixed on to these bolts with the aid of wing nuts. The plywood floor fixes into retaining grooves at the base of the frames. Brass wire gauze (6 meshes per cm) or perforated zinc is used to fill in the side, top and rear frames. The front consists of an upper portion made of clear glass, while the lower half consists of plywood into which two holes, approximately 13 cm in diameter, are cut. These holes are each fitted with a metal flange 13 cm in diameter and 2·5 cm in height on to which muslin sleeves 30 cm long are fixed by means of elastic bands.

Although these are excellent cages, they are rather heavy to handle and the tendency of the wood to warp sometimes makes them difficult to assemble. Parkin & Green (1957) therefore used Basden's original principles and designed an improved cage of 18-gauge aluminium. The sides, front and back are cut and bent from a single sheet, the overlapping 1·25 cm of the two free edges being riveted; the jointed corners of the cage are then filled with solder to give a smooth internal finish. The solid metal bottom and the top frame are cut 1·25 cm oversize on all edges. These edges are then bent at right angles and riveted to the upright sides of the cage. A sliding glass door forms the upper half of the cage front, while the lower half of the front is similar to Basden's, being finished off with two 13-cm flanged holes to which muslin or terylene sleeves are attached. The same material is used to fill in the top, back and side frames where it is held in place by angled strips and springs (Fig. 43.1).

Care and feeding of adults

The number of flies needed to stock the oviposition cages is not critical. A minimum of 50 cm³ of space per individual is a good guide. The easiest way to stock the cage with a given

Fig. 43.2. Metal-framed oviposition cage showing method of securing muslin panels. (Photograph: Zoological Society of London.)

number of flies is to insert a shallow open-topped container into which clean pupae have been counted or measured. Immediately the flies start emerging, fresh food and water must be given and kept constantly available.

The water can be given in the form of a fountain. Peterson (1959) describes various types but the one most commonly used is an upturned beaker of water resting on a shallow pad of cotton-wool which is pressed into a petri dish. The flies can rest upon the damp cotton-wool between the edge of the petri dish and the beaker, and will not drown. Alternatively a jar of water can be used, through the lid of which a thick absorbent cotton wick protrudes.

Sugar should be provided as a thin layer in a shallow dish. Cotton-wool pads soaked in a mixture of equal parts of milk and water should be placed inside the cage for oviposition. These should be changed daily as they quickly become sour in the high temperature. Souring can be delayed by adding 1 part of 40 per cent formaldehyde to 40 parts of the diluted milk (McGovran & Gersdorff, 1945).

The first eggs will be laid on the cotton-wool between 4 and 5 days after the flies emerge and egg-laying should continue for a period of up to 10 days.

Treatment and transfer of eggs

The eggs laid on the milk pads must now be transferred to the rearing medium. To estimate the number of eggs available, they may be detached by gently shaking the cotton-wool pad under water. The supernatant liquid is decanted and replaced with clean warm water. The eggs in suspension are transferred to a calibrated tube and allowed to settle, when the number present can be closely calculated. A volume of 0·1 ml contains approximately 700 eggs (Parkin & Green, 1957). Care should be taken not to stir or shake the suspended eggs unnecessarily as this may result in high mortality.

If there is no need to estimate the number of eggs, the soaked pads with their attached eggs are removed daily from the oviposition cages and transferred to jars of the rearing medium. The jars are then closed with muslin tops secured with elastic bands; the depth of medium in the jars should be about 10 cm. Glass pickle jars are

used and they allow the progress of the larvae to be followed carefully.

Rearing-media for the larvae

Various media have been used since 1927, when Glaser discovered that continuous rearing of houseflies throughout the year was possible with a mixture of horse manure and bakers' yeast. Richardson (1932) published details of a synthetic medium consisting of wheat bran, alfalfa meal, bakers' yeast, malt extract and water which is more pleasant to handle and less liable to infestation by mites. The official method of the National Association of Insecticide and Disinfectant Manufacturers for rearing houseflies for Peet Grady tests (Soap & Sanitary Chemicals, 1951) is substantially as follows: to a mixture of 1,200 g of soft wheat bran and 600 g of alfalfa meal add a suspension containing 30 g of bakers' yeast and 50 ml of malt extract in 2,700 ml of water and mix thoroughly to give a loose texture. The prepared medium is placed loosely in cylindrical battery jars 15 cm in diameter and 23 cm high. Each jar is seeded with about 2,000 eggs and covered with a single thickness of cheesecloth. In Great Britain grass meal has successfully been used as a substitute for alfalfa (lucerne) meal.

A formula recommended by many laboratories consists of 3 parts by weight of dried yeast powder, 4 parts dried malt extract, 60 parts grass meal, 120 parts toppings, mixed with 390 parts by weight of water.

The medium which has proved successful at the London Zoo for some considerable time, and is still in use, consists of a mixture (in parts by volume) of 2 parts fish meal, 2 parts ground rabbit pellets, 2 parts dried yeast, 1 part sugar or molasses, 6 parts bran. This is mixed with enough water to make the whole damp but not wet.

The jars should all be clearly marked with the date when the eggs were laid, the number of the oviposition cage, the strain of flies and the date when the pupae and flies are expected to be available.

Pupation, and care of pupae

About 8 days after the eggs are laid, the larvae migrate to the upper layer of medium and pupate. If the correct quantity of water was used in preparing the medium, the top centimetre or so in each jar should by then be dry. One or two days later, the upper portion of the medium containing the pupae is removed. Either the flies can then be allowed to emerge directly from the medium into an emergence cage, or the medium can be allowed to dry thoroughly before being broken up so that the pupae can be separated by either sieving, rolling on a sloping board, or winnowing with the aid of an electric fan.

Alternatively the contents of the rearing jars can be emptied gently into a basin of warm water (25 to 28°C) nine days after the eggs were laid. If the whole is allowed to soak for a few minutes, most of the crusty portions of the medium can then be easily broken up by hand. This frees the pupae, most of which will float at the surface. They can then be easily collected with the aid of a small scoop made from wire gauze or perforated zinc. They should be washed and spread on absorbent paper to dry off. When floated off, the pupae will mostly be reddish-brown but will darken as they dry. At this stage the pupae are in a very sensitive condition and must be handled with great care to avoid damage and the production of abnormal adults.

Transferring the adults

It is possible to avoid much of the handling of adult flies when stocking oviposition and holding cages, by counting pupae and placing these in the cages to emerge where they are required. Parkin & Green (1957) describe a useful holding cage for approximately 100 to 150 flies. They use circular-knitted mutton cloth (cotton stockinette) which can be bought in rolls. A piece of this is drawn over a cylindrical 23 × 15 cm wire frame and then knotted at the top and bottom to retain the flies. The necessary food, i.e. milk diluted with 50 per cent water, is given in a glass tube plugged with cotton-wool soaked in the mixture.

When the flies have to be examined, counted or otherwise handled and this cannot be easily accomplished, their lively movements can be reduced by chilling. It is possible to use certain anaesthetics but this requires careful judgement to avoid losses, and recovery time is usually lengthy. Anaesthetization can, however, be easily achieved and controlled with the use of carbon dioxide. Recovery in fresh air is fairly quick.

Although all precautions are taken to avoid escapes into the fly room, it is inevitable that the odd one or two find their way out. These can be trapped in the balloon type of trap made of wire mesh by using a small piece of meat or fish as bait, or by providing a saucer of milk in which has been mixed 2 per cent by volume of 40 per cent formalin, or by a syrup containing 1 per cent of malathion applied to glass plates or cotton gauze.

Parasites

It is important to prevent parasites or predators from entering the breeding room, and so it is best to avoid using wild-caught pupae to start a culture as there is always a possibility that these will come already infected with parasitic Hymenoptera. Wild-caught flies also often carry mites which are sometimes very difficult to eradicate from the breeding room. Newly caught flies should therefore be quarantined and the eggs they produce should be used only after careful washing and examination under the microscope.

Certain species of fungi are parasitic on Diptera, but the only form encountered commonly is *Empusa muscae*, which occurs both north and south of the equator, and frequently assumes epidemic proportions among wild flies during the autumn.

Most of the hymenopteran parasites that attack *Musca* and its relatives belong to the Cynipidae or the Chalcidoidea. The former mostly attack the larval stages, whereas the Chalcids are pupal parasites. In Great Britain much trouble has been caused by the small Chalcid *Mormoniella vitripennis* in laboratories producing *Musca*, and also on commercial bait farms breeding *Calliphora* and *Lucilia*. Fortunately it cannot locate pupae covered by 1 mm of sand (West, 1951), and attacks have been overcome at the London Zoo by this means together with the use of a substitute breeding-room for a period of six weeks; meanwhile the original breeding-room must be thoroughly cleaned and washed out with hot water and detergent.

REFERENCES

Basden, E. B. (1947). Breeding the housefly (*Musca domestica* L.) in the laboratory. *Bull. ent. Res.*, **37**, 381–7.

Brown, A. W. A. (1938). Nitrogen metabolism of an insect. *Biochem. J.*, **32**, 895–903.

Glaser, R. W. (1927). Note on the continuous breeding of *Musca domestica*. *Bull. ent. Res.*, **20**, 432–3.

Glaser, R. W. (1938). A method for the sterile culture of houseflies. *J. Parasit.*, **24**, 177–9.

Haub, J. G. & Miller, D. F. (1932). Food requirements of blowfly cultures used in the treatment of osteomyelitis. *J. exp. Zool.*, **64**, 51–6.

Herms, W. B. (1950). *Medical Entomology*, 4th ed. New York: Macmillan.

Hewitt, C. G. (1914). *The Housefly, Musca domestica Linn*. Cambridge: University Press.

McGovran, E. R. & Gersdorff, W. A. (1945). The effect of food on resistance to insecticides containing DDT or pyrethrum. *Soap sanit. Chem.*, **21**, (12), 165, 169.

Parkin, E. A. & Green, A. A. (1957). Houseflies and blowflies. In *UFAW Handbook on the Care and Management of Laboratory Animals*, 2nd ed. London: UFAW.

Patton, W. S. (1930). *Insects, Ticks, Mites and Venomous Animals of Medical Importance*, Part II. London: Grubb.

Peterson, A. (1959). *Entomological Techniques*. Michigan: Edwards.

Soap & Sanitary Chemicals (1951). Peet-Grady method. *Soap Sanit. Chem.*, 1951; *Blue Book*, 237–9. New York: MacNair Dorland.

West, L. S. (1952). *The Housefly*. Ithaca, N.Y.: Comstock.

White, G. F. (1937). Rearing maggots for surgical use. In *Culture methods for invertebrate animals*, ed. Galtsoff, P. S. Ithaca, N.Y.: Comstock.

44 Blowflies

G. J. ASHBY

Blowflies of the genera *Calliphora, Lucilia,* and *Phormia* are extensively bred for use in toxicological and physiological investigations, although perhaps not quite so commonly as the housefly. Possibly the biggest demand for blowfly larvae is as bait for anglers or as live food for birds, fish, and reptiles. Like the housefly's, their short life cycle and the fact that they can be bred all the year round makes them ideal for laboratory work.

Obtaining stock

Stock can be obtained either from a laboratory engaged in culturing blowflies, or from a fish-bait farm, or from nature. Adults can be caught with a balloon-type trap exposed near a slaughterhouse or boneyard, and eggs can be obtained by exposing pieces of liver, beef or fish head for oviposition by wild flies. Whichever method is used, a number of different species will probably be obtained and identification will be necessary. The keys and descriptions published by Seguy (1923 *et seq.*), Richards (1926), Hall (1948) and Roback (1951) will be found useful.

Housing the adults

Accommodation similar to that used for the housefly is required, but it is important that all inlets and outlets to the room should be screened to avoid contamination by gravid females of *Calliphora erythrocephala* from outside. In order to reduce the smell of the larval food all air should be discharged outside the building, or a large adequately ventilated cabinet should be used for rearing. The room should be maintained at a temperature of 25 to 27°C with a relative humidity of 50 to 60 per cent.

Mating and oviposition

It is possible to breed certain species of blowfly under conditions which avoid the smell and other undesirable features associated with putrifying food. Elaborate methods have been developed by White (1937) and by Haub & Miller (1932) for the aseptic culture of *Phormia regina* larvae. Hill, Bell & Chadwick (1947) describe a method based upon the work of Brown (1938) but with extensive modifications for the breeding of *Phormia regina.* The medium consists of powdered casein, brewers' yeast powder and powdered agar in the proportions of 30:3:1, to which is added lanolin and a solution containing phosphate.

The adult females are allowed to deposit their eggs on ground horse meat, from which they are removed with forceps or needles; when handling eggs it is advisable to use moistened instruments. At 25°C and a relative humidity of 70 per cent the eggs will hatch in less than 24 hours. The larvae will cease feeding at about the eighth day and pass into a prepupal or resting stage, after which they will soon pupate. The pupal stage lasts five to six days and the flies emerge on the fifteenth or sixteenth day after the eggs were laid. It is, however, only fair to say that, although this method is quite successful, the flies breed more prolifically on a meat or fish diet.

Breeding-cages are similar to those used for houseflies (Figs. 43.1 & 43.2). The number of eggs required will determine the number of adults used to stock a cage, but it is suggested that these should be not less than 200 nor more than about 800. A pan of sugar must always be available in addition to a water fountain of the type used for houseflies, or a water bottle provided with an extruding cotton wick. A piece of moist ox liver or ground lean beef weighing about 40 g is placed in the cage daily. The surface of the liver can be scored with a knife to provide moist furrows in which the eggs will be laid. In *Lucilia* and *Calliphora* the first eggs are laid four to five days after the flies emerge from the pupae, in *Phormia* after seven days.

Treatment and transfer of eggs

The eggs are attached to each other in batches and will not usually separate in water, but if necessary they can be freed by immersion for not longer than 10 minutes in a 1 per cent solution of sodium hydroxide. A suspension of eggs in water can be measured in a culture tube. A volume of 0·1 ml contains approximately 650 eggs of *Lucilia sericata* or 350 eggs of *Calliphora erythrocephala* (Parkin & Green, 1957).

The eggs are usually left on the liver or meat to hatch, the pieces being cut up into portions, each carrying about 600 to 700 eggs; these are kept in closed dishes until the eggs hatch. Each piece is then added to a culture jar containing more food. The eggs should be covered in order to maintain the humidity essential for hatching. The time taken for the eggs to hatch varies according to species. At 24°C *C. erythrocephala* takes 3 to 8 hours and *L. sericata* 10 to 14 hours, whereas *Phormia terrae-novae* takes as long as 19 to 25 hours. *Calliphora* sometimes deposits first-stage larvae.

Rearing-media for the larvae

Various high-protein foods can be used for larval cultures. Frings (1948) used fermenting dog biscuits, and Dorman, Hale & Hoskins (1938) used fish heads to breed *Callitroga*, *Lucilia*, *Sarcophaga* and *Phormia*. Minnich (1937) used fish heads lying in soup plates. Three such plates were stacked one above the other on a wooden rack which rested on a layer of soil inside a rearing-can; the latter was fitted at the top with a manually operated damper to regulate the ventilation and humidity. Pupation took place in the soil at the bottom of the can. Ox liver or ground horse meat is an excellent food for the larvae. The resulting smell can be kept at a relatively low level by feeding fresh or chilled food, and not giving more food than the larvae will consume. Several small cultures are more easily managed than a large one. Glass pickle jars can be used for culturing, the small pieces of liver or meat carrying the eggs being placed on a larger piece (about 300 g in weight) which rests in the middle of the base of the culture jar. A small quantity of sawdust, dry peat, or vermiculite should then be sprinkled around the food to absorb any exuding liquid. More food can be added if necessary.

Pupation and care of pupae

In most species the larvae migrate from the food when fully grown. Just before this stage they should be provided with sufficient material in which to pupate. Here again a further quantity of sawdust, peat or vermiculite should be added to a depth of about 3 cm. *Calliphora* and *Lucilia* larvae feed for four to five days, *Phormia* for six to seven days. Pupae can be separated by the methods described for the housefly.

Transferring the adults

Here, as with houseflies, it is possible to avoid undue handling of adult blowflies, by counting out pupae and placing these in the selected cages to emerge where they are required. If it is really necessary to handle the adults, and they are very lively, they may be quietened down by chilling or by the use of carbon dioxide.

Parasites

See Housefly, Chapter 43.

REFERENCES

Dorman, S. C., Hale, W. C. & Hoskins, W. N. (1938). The laboratory rearing of flesh flies and the relations between temperature, diet and egg production. *J. econ. Ent.*, **31**, 44–51.

Frings, H. (1948). Rearing houseflies and blowflies on dog biscuit. *Science*, **107**, 629–30.

Hall, D. G. (1948). *The Blowflies of North America.* Thomas Say Foundation (United States).

Hill, D. L., Bell, V. A. & Chadwick, L. F. (1947). Rearing of the blowfly *Phormia regina* Meigen on sterile synthetic diet. *Ann. ent. Soc. Am.*, **40**, 213–16.

Minnich, D. E. (1937). The culture of blowflies. In *Culture methods for invertebrate animals*, ed. Galtsoff, P. Ithaca, N.Y.: Comstock.

Richards, O. W. (1926). Notes on British species of *Lucilia* (Diptera). With a supplementary note by J. E. Collin. *Trans. R. ent. Soc. Lond.*, **74**, 255–60.

Roback, S. S. (1951). A classification of the Muscoid Calyptrate Diptera. *Ann. ent. Soc. Am.*, **44**, 327–61.

Seguy, E. (1923 *et seq.*). *Faune de France.* Paris: Lechevalier.

45　The Stablefly

L. C. STONES

The stablefly or biting housefly (*Stomoxys calcitrans* L.) superficially resembles the housefly though its blood-sucking habit on both man and animals provides a practical demonstration of a fundamental difference. The adult fly of both sexes, after suitable starvation, can be relied upon to feed and it is in this connection that the fly is usually maintained in the laboratory. Although the stablefly has been used for general insecticide tests, in which it is usually found to be more susceptible than *Musca*, its chief use has been in the testing of systemic insecticides in animals since it provides a ready bio-assay method for detecting and estimating levels of insecticide in blood. It has also been widely used as a test for insect repellents on livestock and in the experimental transmission of various animal pathogens. In its own right, the fly is of considerable importance as a major source of fly worry; it is the intermediate host of several nematode parasites and has been implicated as a mechanical vector in the transmission of many pathogens, notably trypanosomiasis in horses.

Although this is a prolific species, thriving under a wide variety of climatic conditions all over the world, it has nevertheless acquired a reputation for being difficult to breed in the laboratory. This is perhaps supported by the numerous accounts of breeding techniques which have been published. Earlier workers described methods for which success was claimed yet others subsequently found them to be unsatisfactory. Some accounts show fairly fundamental differences in the methods employed while others represent quite small modifications designed to overcome some particular difficulty. In fact these flies are not unduly difficult to rear and several different methods may be satisfactory, but experience, a knowledge of basic requirements and close attention to detail are all important. Once rearing is well established, the technique should be rigidly followed if variations in vigour and development rates are to be avoided. It is often comparatively easy simply to

maintain a species but sometimes more difficult to rear it in large numbers to a set timetable.

As might be expected, in the earliest attempts to rear the stablefly, natural breeding conditions were closely followed (Glaser, 1924) fresh horse manure being used as the larval medium, whilst attempts were made to feed the adult flies on living animals. This method was not satisfactory and later workers turned to artificial media for larval cultures and the use of blood, presented in dishes or tubes, for adult feeding.

Two rather dissimilar methods, those of Doty (1937) and Eagleson (1943), form the basis of modern breeding techniques, though neither in its original form has proved entirely satisfactory. The method described here incorporates features taken from each of the above methods.

Stock

The simplest and most convenient means of establishing a culture is to obtain several hundred adults or pupae from another laboratory which rears stableflies. Young pupae have the advantage that they can be kept for several days under temperate conditions and can be sent by post. In the writer's opinion these flies may become adapted to laboratory culture, probably from the point of view of adult feeding, and it is therefore easier to start with such a strain rather than with wild flies. Certainly it is sometimes difficult to establish a colony from a few individuals brought in from the field.

Stableflies are normally plentiful around livestock yards and buildings during warm weather from May to October, particularly during the last three months. They bask on sunny walls and woodwork and may be roughly selected from other flies by their habit of sitting head uppermost and by the presence of a dark proboscis projecting horizontally in front of the head. *Stomoxys calcitrans* is the only European representative of the genus and so identification presents no problem in this area. If the flies are

561

Fig. 45.1. Cage rack, showing the two types of fly-cage and overhead fluorescent lights. (Photograph: Cooper Technical Bureau.)

to be used for insecticide testing, it is important that consideration should be given to their resistance status. At present, there seems to be some doubt as to how far, and where, resistance has developed in this species (McDuffie, 1960) and it would therefore be prudent to start with a known primitive strain or, at least, to collect wild specimens from an area where they are unlikely to have encountered insecticides.

When adults are not available, pupae can usually be found in old bedding in cattle yards and calf boxes. They favour sites near walls and corners and are usually found clustered in quite large numbers though it may take a little patience to locate them. Quite a high proportion of wild pupae are often found parasitized by small chalcids and ichneumons.

Accommodation

The insectary building used by the author is reasonably well heat-insulated by double windows and double fly-proof doors, and it has not been found necessary to exclude daylight. The insectary has a cage rack along one wall, the general arrangement of which is shown in Figure 45.1. Lighting is provided by fluorescent tubes fixed to the wall above the cage rack at a distance of 80 cm from the top of the large fly cages. The tubes, which are placed end to end,

are 120 cm long, 40 W, and of the daylight type. They are conveniently wired in circuit with a time switch which turns them on and off automatically at 9 a.m. and 5 p.m., respectively, on every day of the week. In addition a large fly-proof cupboard, divided into sections and well ventilated, is used for holding the larval cultures.

The temperature of the insectary is maintained continuously at $26 \pm 1°C$. The relative humidity is controlled at 50 ± 3 per cent. This humidity is low by comparison with that in most insectaries, but it has been found that higher humidities are of no apparent advantage in breeding *Stomoxys* and merely serve to encourage mould growth and render the working conditions less pleasant. Forced ventilation is maintained at a low rate by an extraction fan, fresh air being drawn in through a small port close to one of the heating radiators. This continuous ventilation is not, in itself, essential but is a valuable safeguard against the chance build-up of any low concentration of volatilized insecticide. It is felt that the importance of this danger cannot be overestimated where *Stomoxys* breeding is concerned. During our early attempts at breeding this fly, it was found to be quite impossible to rear them in another poorly ventilated insectary which had been used continuously for many years for breeding the blowfly, *Lucilia sericata*, and in which precautions had been taken to prevent any chance contamination with the newer synthetic insecticides. It was eventually established that this failure was due to the presence of a very low concentration of volatilized benzene hexachloride which, though not proving obviously toxic to the flies, was exerting a sufficient effect to inhibit the adults from feeding so that they succumbed from a combination of starvation and intoxication in 2 or 3 days. In spite of this, no trouble had ever been experienced with the blowflies, whilst *Musca domestica* exposed there in open cages showed no more than a 10-per-cent mortality after 5 days' exposure. For the successful rearing of *Stomoxys*, the mortality of newly-emerged caged *Musca* used to test the conditions should probably be less than 4 per cent after 11 days' exposure.

Equipment

Cages for adult flies

The type of cage used for holding and feeding the adult flies is shown in Figure 45.2. These cages are constructed of plywood and are made with all battens and other structural details externally. This leaves the inside of the cage as a simple cube without any projections and facilitates cleaning, an important point when dealing with blood-sucking flies which make a cage very dirty. The front of the cage consists of a large sheet of glass held in position by clips on three sides. The glass may therefore be completely removed for cleaning. In order to minimize the reflection of light internally, the inside of the cage is painted black. A matt blackboard paint has been found most suitable for this purpose, the surface being finished off with several coats of a shellac-alcohol varnish. This varnish is quick-drying and non-toxic to flies within a few hours of application, and it can be used comparatively frequently to renovate the cages, whilst a number of coats will also provide a good waterproof finish for the outside of the cages.

The internal dimensions of the cage are $45 \times 45 \times 45$ cm. Into one side of the cage a door is fitted, in an opening 12·5 cm square. On the framework of this opening fittings are arranged to hold the end of a removable sleeve made of black cloth. If a permanent sleeve is attached to these cages in the ordinary way, it rapidly becomes very foul. It has, therefore, been found much more satisfactory simply to have a door opening, and to attach a sleeve over this only when it is necessary to carry out some operation inside the cage.

Fig. 45.2. Large fly-cage with glass front removed and showing position of side door and feeding window. (Photograph: Cooper Technical Bureau.)

On the top of the cage, in the right-hand back corner, is cut a window, Figure 45.3, measuring 7.5×10 cm. A number of small brass clips allow a panel of perforated zinc to be slid into position, closing the window. Further clips hold a removable rack for blood tubes. The perforated zinc, which is easier to clean than wire gauze, is of the normal household type having approximately 13 holes to the cm², each hole being about 1.5 mm in diameter.

The glass blood-tubes used for feeding the flies are 9 cm long and 6 mm in internal diameter. The ends of the tubes are flame-polished, care being taken not to reduce the internal diameter. A small rubber bung or pipette bulb is used to close the top end of the blood-tube and so maintain a column of blood. Eight of these tubes are provided for each cage and, when full, together hold about 15 ml of blood.

When in use, a piece of dark material is hung over the front of the cage to exclude light, so that the only light available to the flies is that entering through the zinc panel from the fluorescent lights above. The floor of the cage is lightly sprinkled with sawdust. This helps to prevent the surface from becoming sticky with drops of blood and excreta.

This method of keeping the flies in total darkness save only for a brightly lit panel to which they are attracted and where they find a blood meal is basically that of Eagleson (1943). He, in fact, kept them in a normal gauze cage which was blacked-out with an external hood and this is also recommended by McGregor & Dreiss (1955). In our experience these flies do not readily locate and feed on an artificial blood meal and, in fact, can be remarkably stupid in this respect even when a living animal is introduced into their cage. We therefore favour the use of light to attract them to the vicinity of the blood and find that under these conditions few flies fail to engorge. Other workers (Doty, 1937; Campau et al., 1953; Champlain et al., 1954) have, however, kept stableflies in undarkened gauze cages and fed them successfully from cotton-wool soaked in blood and placed in dishes on the floor of the cage. Parr (1959), however, did not find this method satisfactory and fed his flies from blood-soaked absorbent-cotton plugs suspended in the cages.

Oviposition cages

These are of fundamentally the same construction as those described above but are smaller, measuring only $25 \times 25 \times 25$ cm internally. The flies appear to oviposit more readily when confined in a small cage. The floor of the cage is also modified to allow oviposition to take place through it. Instead of being solid, as in the larger cages, it is made of perforated zinc of the same type as that described above. A small external

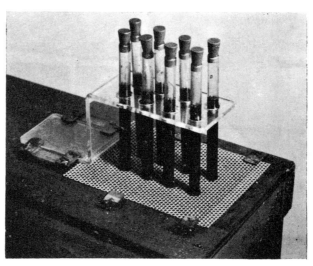

Fig. 45.3. Feeding-window with perforated zinc panel, tube rack and blood tubes. (Photograph: Cooper Technical Bureau.)

wooden beading secures the zinc in position and acts as a distance piece to hold the zinc 6·25 mm away from the surface on which the cage is resting. The top window with its fitments is identical with that already described except that it is placed in the *left*-hand back corner. This facilitates the transfer of flies from the large cages to the small. No side door is required in these small cages since no operation need be carried out in the cage once the flies are introduced.

Egg-counting

In setting up larval cultures, each culture should be implanted with a standard number of eggs, and a rapid method of egg-counting is therefore necessary. A convenient counter for this purpose can be made from two small sheets of black plastic material (Perspex) 3 mm in thickness. One sheet is drilled with 10 holes, each precisely 6 mm in diameter. The sheets and holes are then polished. By placing the drilled sheet on the top of the other sheet, 10 small pits are formed which, when filled level to the top with clean dry eggs, are found to contain approximately 1,000 each. This counter, which has an accuracy of ±10 per cent, is sufficiently accurate for setting up standardized cultures and for making a reasonable assessment of the daily egg-production.

Containers for larval cultures

The initial larval cultures are set up in 150-ml beakers with a depth of 8 cm and a diameter of 6·5 cm.

The larval cultures are later transferred to larger shallow dishes in which the feeding of the larvae is completed and pupation takes place. These dishes are made of glass with an internal diameter of 16·5 cm and a depth of 6 cm. (Phoenix soufflé dishes have been found excellent for the purpose.) For the small beakers and these larger dishes fine cotton covers, secured by rubber bands, are used to limit excessive drying of the cultures, and also to prevent any contamination with eggs from stray flies.

Containers for pupae

After separation from the culture medium the pupae are placed in cylindrical containers made of eight-mesh-to-the-inch copper gauze (known as 'hardware cloth' in USA) 20 cm high and

10 cm in diameter. In other solid-sided containers the pupae tend to sweat, resulting in mould growth and much damage to the newly emerging flies. With the gauze containers, the pupae remain dry and healthy and the mesh retains the pupae without interfering with the escape of newly emerged flies.

Feeding

Larval culture medium

The medium used for larval culture is a slight modification of the standard Peet-Grady (1951) formula, Table I, used for breeding *Musca domestica*.

TABLE I
Modified Peet-Grady medium

Coarse bran	1,200 g	Dried yeast	45 g
Dried grass meal or alfalfa	900 g	Extract of malt	75 ml
Crushed whole oats	600 g	Water	4,500 ml

The medium is prepared by mixing the bran, grass meal and oats together in the dry state and then adding the yeast and malt dissolved in the total quantity of water. The medium is mixed thoroughly until it is homogeneous, and then is packed down lightly in a closed container which is placed in a cold store until shortly before use. This quantity gives approximately one week's supply, and is not normally stored longer. Where there is doubt about the purity of the water supply, e.g. excessive chlorination, distilled water is used. Preliminary sterilization of the ingredients has not been found necessary but would be of advantage should extraneous moulds or insect contaminants become a problem. It is, of course, important to ensure that the ingredients have not been treated with an insecticide at any stage of production or storage.

Blood

The blood used for feeding the adult flies is collected fresh each week from an abbatoir. Beef blood is normally used although Pospisil (1961) states that horse and pig blood are also suitable whilst sheep, guinea-pig and mouse blood are apparently not. Coagulation is prevented by citration to give a concentration of 0·40 per cent of sodium citrate in the blood.

The level of citration given should not be exceeded. A concentration of 1·0 per cent of sodium citrate inhibited feeding to such an extent that subsequent egg-laying was inadequate, whilst a concentration of 1·5 per cent almost completely prevented feeding.

Breeding

Emergence and feeding of adults

In breeding *Stomoxys calcitrans* it has not been found possible to obtain sufficiently even development in the larval cultures to ensure that all the larvae pupate at the same time. In practice pupation takes place quite steadily over a period of from 4 to 7 days. With such pupation, it follows that the emergence of adult flies will also be somewhat extended and it normally takes about 8 days from the emergence of the first fly in a culture to the last. In order to fill a cage with flies from such pupae, all the pupae available at that time should be placed in a cage together. A fairly steady emergence of flies takes place, and the cage is filled to a good population in 24 hours. The pupae may then be moved on to fill another cage.

During the first two days of their life the flies do not engorge fully, and they therefore need only two blood-tubes in the morning and two in the afternoon. From two days onwards their appetite increases, and the cage is therefore supplied with a full battery of eight tubes in the morning and again in the afternoon. Feeding is most rapid in the morning and it is usually necessary to replenish the tubes at least once. The afternoon feed is left in position overnight even though the lights are not kept on. The highest rate of feeding occurs in flies 3 to 5 days old, when they engorge so fully that 1 ml of blood is sufficient to satisfy only 50 to 60 flies. The average weight of the unfed female fly is approximately 13 mg and thus, with a meal of about 17 mg of blood, she consumes at one engorgement roughly 1·33 times her own body-weight. After the fifth day feeding is much reduced. In preparing flies for breeding, it is important that they should feed well, since flies which have fed poorly are invariably poor egg-layers.

Oviposition

In a well-fed cage of flies the first eggs are produced at 6½ days after emergence from the pupa. They are not produced in large numbers at this time and are not usually seen. Between 7 and 9 days many of the flies are ready to lay and a small number of eggs are often seen round the blood-tubes. On the tenth day, when about 30 to 50 per cent of the flies are gravid, they are transferred from the large cage to two smaller oviposition cages. By delaying transfer until this proportion of flies are gravid, a good early production of eggs from the oviposition cages is ensured.

The transfer of flies from one cage to another is achieved by inverting one of the oviposition cages over the large cage so that the feeding panels of the two cages coincide with each other. The perforated zinc panels may now be slid aside. If the lower cage is blacked out by hanging a curtain over the glass front and a light is placed above the upper cage, the flies, attracted by the light, will pass into the oviposition cage. When half the flies in the large cage are judged to have left, the curtain is removed from this cage, the zinc panels are replaced and the filled oviposition cage is taken away. The operation is now repeated with the second oviposition cage and this is filled with the remaining flies. It is usually necessary to agitate the flies in the large cage so that not too many stragglers remain in it.

Each oviposition cage is placed on a piece of dry black cloth and stood on the cage rack under the fluorescent lights. Feeding is carried out in exactly the same manner as for the larger cages, except that with these cages four blood-tubes, supplied twice a day, are sufficient. The flies start to lay immediately, inserting their ovipositors through the perforations in the zinc at the base of the cage and depositing their eggs singly on the black cloth.

This method of collecting eggs has, for several reasons, proved more satisfactory than the practice of placing fermented medium in the cage in the hope that the flies would lay on this. They do, in fact, lay a proportion of their eggs on the medium but a far greater proportion are distributed more or less at random elsewhere over the floor of the cage and are lost. Further, when the eggs are laid on or in the medium, it is impossible to make an accurate estimate of their number, and the larval cultures cannot therefore be accurately seeded with eggs. The chief merit of the method, however, lies in the elimination from the cultures of parasitic mites. It has been found that under field conditions a high proportion of wild *Stomoxys* are parasitized by a mite which attaches itself to the ventral surface of the fly. As many as five such mites are frequently found on one fly. When the ovigerous female visits fermented medium to lay her eggs, some of the mites leave their host and remain in the medium to breed. If this is allowed to occur the insectary is likely to become heavily contaminated with mites, many of which attach themselves to the newly emerged flies of the next generation. By allowing the flies to oviposit on to dry cloth as described, mites are prevented from being carried over to the larval cultures and they soon disappear entirely from the insectary.

Other methods of collecting eggs have been described. Thus Campau *et al.* (1953) obtained them from the blood dishes used for feeding and cleaned them by washing in water and sedimenting; Champlain *et al.* (1954) collected them on sponge soaked in water, subsequently squeezing them out and sedimenting in water; McGregor & Dreiss (1955) collected them on a 2·5-cm ball of moist cotton-wool round which black cloth was wrapped and to which had been added a few drops of 5-per-cent ammonia solution.

After oviposition on dry cloth, the eggs are collected morning and evening and are transferred in the dry state to the counter already described, where they are divided into batches of 1,000 ready for seeding on to the larval culture medium. The viability of the eggs is not reduced, under the insectary conditions, provided they do not remain on the dry cloth for longer than 24 hours. Within this limit the hatching rate averages 60 to 70 per cent, regardless of their exposure time on the cloth.

Larval cultures

The culture of *Stomoxys* in the larval state has, in our experience, proved much more difficult than that of *Musca domestica*. The successful management of the cultures depends largely on observation of the following conditions:

(1) Excessive heating of the medium due to fermentation must be prevented, by using only a small bulk of medium initially and increasing it as the larvae develop. Also, during the later stages, the cultures are contained in shallow flat dishes which facilitate heat-loss. Some workers have advocated the initial fermentation of the medium before seeding the eggs, so avoiding heating at a later stage. This method we have found less satisfactory for not only is the time and type of fermentation very variable, but the larvae do not develop so rapidly or so evenly in previously fermented medium.

(2) The growth of extraneous moulds must be kept at a low level by limiting the amount of medium initially, so that there is never so much medium present that the larvae cannot work steadily through the entire mass.

(3) The moisture content of the medium must be carefully maintained by periodic watering. If it is too wet and sodden, the larvae will migrate out at the surface. On the other hand, in medium which is too dry some larval mortality occurs owing to desiccation and the remaining larvae will tend to cluster deep in the medium and develop very slowly.

(4) The pH of the medium is of importance, particularly in young cultures. The optimum pH for *Stomoxys* larvae is probably on the alkaline side of neutrality but they appear to be able to tolerate more acid conditions, probably at a level of about 5, for a short period. The pH of freshly prepared medium is about 5·8, and this changes with fermentation to 5 to 4·8 after 24 hours and then rapidly shifts to the alkaline side, ultimately reaching a consistent level of about 8·7. If perfectly fresh medium is sown with eggs, the larvae emerging after 12 to 24 hours encounter excessively acid conditions, and this sometimes results in a mass migration of the young larvae from the medium and a very high mortality. On the other hand, if the sowing of eggs is delayed until the medium has been kept at the insectary temperature for 24 hours then, though the eggs are probably implanted on to acid medium of about 5, the pH has become more alkaline by the time the larvae hatch. This practice

of using 24-hour-old medium for setting up cultures initially should not be confused with the general use of fully fermented medium advocated by other workers, e.g. Eagleson (1943).

(5) Larvae do not thrive in densely packed medium and this is likely to occur where large, deep containers are used. Under such conditions the larvae only burrow to about 2·5 cm below the surface, presumably owing to lack of aeration, and this can be only partially corrected by frequent stirring. Under our conditions, where a shallow depth of medium is used, the larvae tunnel right through it continuously and so keep it adequately light and open. Other workers have overcome this difficulty by mixing 5 parts of wood shavings with 1 part of medium (McGregor & Dreiss, 1955) or 2 volumes of wet vermiculite with 1 volume of medium (Goodhue & Cantrel, 1958). We ourselves have used both methods and found them satisfactory. The wood shavings used are from a mechanical wood-planing machine and the whole mixture is made sufficiently moist so that only a single drop can be squeezed from a handful. The original workers recommend the use of standard Peet-Grady medium from which both the malt and yeast were omitted, whilst Goodhue & Cantrel, in their vermiculite medium, retained these two ingredients. Pospisil enriched Peet-Grady medium with powdered casein.

The initial larval cultures are set up in 150-ml beakers, each half-filled with freshly prepared or refrigerator-stored medium which has been maintained at insectary temperature for 24 hours. The medium should not be tightly packed. On to the top of each beaker 1,000 eggs are brushed and these are then moistened with about a dozen drops of water. The beakers are covered with cotton covers and stored. The cultures are examined each day, a little water being added if the medium is too dry or if mould is growing to excess. The maintenance of the correct dampness is largely a matter of experience since cultures vary considerably but, in general, the larvae will migrate to the surface if too damp, and cluster together deep in the medium if too dry. No larvae are seen for the first two days, but after this they should be seen moving throughout the medium. Steady growth takes place and they are ready for transference to the larger containers by the fifth day.

The whole contents of a small beaker are turned out into one of the large dishes and the medium is broken up to lie in a shallow heap. Fresh medium is now added so that it covers the old medium and lies about 2·5 cm thick all over the bottom of the dish. The surface is evenly sprinkled with about 30 ml of water. The dish is covered and stored. Larval growth is rapid at this period and the cultures should be kept well watered. The medium should be moist throughout without being sodden, and the sound of the larvae working in it can be heard distinctly. On the fourth day after setting up these larger cultures, fresh medium is again added, the dish being filled to the top, and the medium already present being well mixed with it by gentle stirring. About 30 ml of water are again added evenly over the surface. From this time onwards, only very little watering is required. The cultures should not be interfered with unduly since the larvae at this time are congregating near the surface and preparing to pupate.

Pupation

In a good culture the first pupae are normally seen on the ninth or tenth day; that is, about the time when the last addition of fresh medium is made. From then on, pupation continues steadily and is virtually complete by the fourteenth day. The first fly also usually emerges about this time, and it is therefore necessary to pick out the pupae. Since the larvae conveniently congregate near the surface before pupation there is little difficulty in collecting clean pupae relatively free from contamination with medium. The pupae from all cultures of the same age are gently transferred to a single pupal container made of wire gauze and this is then placed, together with all other available pupae, in an adult fly cage where emergence takes place.

REFERENCES

Camfau, E. J., Baker, G. J. & Morrison, F. D. (1953). Rearing stable-flies for laboratory tests. *J. econ. Ent.*, **46**, 524.

Champlain, R. A., Frisk, F. W. & Dowdy, A. C. (1954). Some improvements in rearing stable-flies. *J. econ. Ent.*, **47,** 940–1.

Doty, A. E. (1937). Convenient method of rearing the stable-fly. *J. econ. Ent.*, **30,** 367–9.

Eagleson, C. (1943). Laboratory procedures in the studies of the chemical control of insects. Publ. *Am. Ass. Advmt. Sci.*, No. 20, pp. 60–73.

Glaser, R. W. (1924). Rearing flies for experimental purposes, with biological notes. *J. econ. Ent.*, **17,** 486–97.

Goodhue, L. D. & Cantrel, K. E. (1958). The use of vermiculite in medium for stable-fly larvae. *J. econ. Ent.*, **51,** 250.

McDuffie, W. C. (1960). Current status of insecticide resistance in livestock pests. *Misc. Publ. ent. Soc. Am.*, **2,** 49–54.

McGregor, W. S. & Dreiss, J. M. (1955). Rearing stable-flies in the laboratory. *J. econ. Ent.*, **48,** 327–8.

Parr, H. C. M. (1959). A method of rearing large numbers of *Stomoxys calcitrans*. *Bull. ent. Res.*, **50,** 165–9.

Pospisil, J. (1961). A simple method of breeding the biting house-fly *Stomoxys calcitrans* L. *Folia Zool.*, **10,** 222.

Soap & Sanitary Chemicals (1951). Peet-Grady method. *Soap Sanit. Chem.*, 24th ed.; *Blue Book*, pp. 237–40.

46   Tsetse Flies

T. A. M. NASH AND A. M. JORDAN

Introduction

The genus *Glossina* occupies some $4\frac{1}{4}$ million square miles of tropical Africa south of the Sahara, and comprises 22 species; all probably carry animal trypanosomiasis and five are known to transmit human trypanosomiasis or sleeping sickness. The recent successful colonisation of a number of tsetse species enables research to be undertaken anywhere in the world on the vector, on the life cycle of the trypanosome in the invertebrate host, and on the diseases when naturally transmitted by the fly.

The tsetse is larviparous and consequently the rate of reproduction is slow. At 25°C the female gives birth to a single larva when about 20 days old, and thereafter reproduces at about 10 day intervals; with a pupal period of about 30 days, the life cycle takes some 50 days. By maintenance at a higher temperature the cycle can be accelerated but reduced longevity leads to poorer productivity, and also to a reduction in the fat reserves available to the young fly on hatching (Bursell, 1960).

The tsetse feeds solely on blood. Until recently laboratory colonies have been successful only when maintained on living animals, but it is now possible, at least on an experimental scale, to achieve equally good fly performance by feeding them on fresh defibrinated blood presented beneath an artificial membrane. No artificial food medium has been devised as yet.

The methods described below are those devised in England by the authors (Nash *et al.*, 1968) using living animals to feed the flies, and are selected because their techniques permit of the mass-rearing of tsetse, the release of almost the maximum reproductive potential of the female, and the production of pupae having a mean weight equal to that found in the field. The techniques described below are proven for *G. austeni* and *G. morsitans* and are probably applicable to many other species of tsetse. There are two reasons for devising methods capable of expansion for the large scale rearing of *Glossina*.

Firstly, they permit of regular supplies being made to other research workers, and secondly they would be essential if the sterile male release technique is found to be applicable to the control of tsetse flies (Dame & Schmidt, 1970). Self-maintaining colonies of tsetse flies have also been established at Lisbon (Azevedo & Pinhão, 1964; Azevedo *et al.*, 1968), in Paris (Itard & Maillot, 1970), Vienna (Mews *et al.*, 1972) and Antwerp (Van der Vloedt, 1971).

Obtaining stock

Stock can be obtained direct from Africa in the form of wild pupae. These should be packed in nylon shavings, between two layers of slightly damp plastic foam, within a container made of 4 cm-thick expanded polystyrene. A box with an internal volume of 500 ml will safely take 400 pupae, is light (92 g) and has considerable insulating and shock-absorbing properties. The mean emergence rate from over 11,000 field-collected pupae of *G. austeni* despatched by air from Zanzibar was 60 per cent (Kernaghan & Nash, 1964). To prevent the introduction of parasites, the contents of the boxes should be examined under a polythene cage within a laboratory remote from the fly room, and the pupae put into transparent jars securely covered with Terylene voile, a material which will prevent the escape of even the smallest parasite.

Provided that the tsetse species required can be obtained from a laboratory colony in Europe, this should be the method of choice. Currently, the following laboratories have colonies of the species specified:
Tsetse Research Laboratory, University of Bristol, Langford, near Bristol, UK—*G. morsitans* and *G. austeni*.
Institut d'Elevage et de Médecine Vétérinaire des Pays Tropicaux, 10, rue Pierre-Curie, 94 Maisons-Alfort, Paris—*G. tachinoides*, *G. morsitans*, *G. austeni*, *G. fuscipes* and *G. palpalis*.
Zoology Department, University of Antwerp,

Middelheim, Laan, 1, Antwerp—*G. palpalis.*
Insect and Pest Control Section International Atomic Energy Agency, Kartner Ring 11–13, PO Box 590, A-1011 Vienna, Austria—*G. morsitans.*

Cages, containing twice the usual number of adult flies and enclosed in a polystyrene container, have been successfully sent through the post for journeys of short duration; the flies should be fed before despatch. However, far more pupae can be successfully sent in the same volume for journeys lasting up to 7–10 days.

Accommodation

In 1962 a specially-designed laboratory was built so that an attempt could be made to establish a large self-maintaining colony of tsetse flies which would meet demands for material by many workers in the United Kingdom and elsewhere. A centrally heated animal house connects through a heat trap with the fly-handling room which is $6 \times 5 \times 2 \cdot 3$ metres high (Fig. 46.1). A tropical climate is maintained in this room during working hours only. There are four, small double-glazed windows (for psychological reasons only) and above these 4 shuttered windows which are opened to outer air when the room is blown out and the floor washed in the evenings. Fluorescent lighting is installed except over the area used for fly feeding. Off this room are two windowless controlled-climate rooms, each $4 \cdot 3 \times 2 \cdot 5 \times 2 \cdot 3$ metres high, in which the insects are housed except when taken out for checking and feeding. There are racks on both sides of each room with wire meshed shelves. A diurnal light rhythm is provided by tungsten bulbs over the central passageway whose illumination is regulated by rheostats to give values on the shelves which range from 5–27 lux.

In the three tropical climate rooms the temperature is maintained at about 25°C by heaters in the air-conditioner ducts. The relative humidity is kept at 70–80 per cent in two of the rooms, but at 60–70 per cent in the room reserved for *G. morsitans.* The humidifiers are supplied with piped de-ionised water. An emergency generator automatically supplies an alternative source of heating in case of breakdown. Alarm bells, connected to staff houses, ring in the event of mains failure or overheating of the controlled climate rooms.

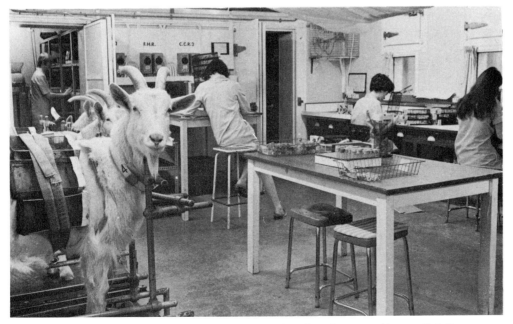

Fig. 46.1. Fly handling room. View from heat trap showing, from L. to R., goats with controlled climate rooms (doors open) behind, fly feeding table, fly checking bench, and mating table under mosquito net in foreground. (Photograph: UFAW.)

Equipment

Fly maintenance cages

The frame of the cage is made of 3 mm diameter stainless steel wire. One end is closed with a piece of 6 mm thick opaque perspex with recessed edges; the centre is drilled with a 2·5 cm diameter cork-hole for the insertion and removal of flies (Fig. 46.2A). A small hole is drilled in each of the four corners of the perspex so that the end can be pushed in and tied to the corners of the frame with strong monofilament nylon.

Two sizes of cage are made. The large size, holding 25 flies, is used for goat-fed tsetse; the external dimensions are 25·5 × 12·5 × 5 cm; the long narrow faces are strengthened by one or two struts on each side (Fig. 46.2B). The small size, holding 10 or possibly more flies, is used for rabbit-fed tsetse; the dimensions are 15 × 9 × 5 cm; no lateral struts are needed (Fig. 46.2A).

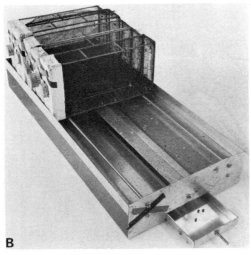

Fig. 46.2. Fly maintenance cage. A. Assembled and un-assembled (small size). B. In rack (large size). (Photograph: UFAW.)

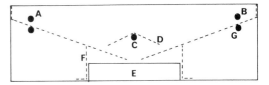

Fig. 46.3. Front end of rack for fly maintenance cages. (See text for explanation.)

The cages are covered in black Terylene netting.* A rectangle of netting is sewn on to the open end, which is then pushed into a tight-fitting sleeve and sewn to that end; the excess of sleeve is secured over the perspex with zinc oxide plaster. The netting should be 2·3–3 mm mesh to permit the larvae to fall through into a container. Suitable netting is often difficult to obtain; when found, a large stock should be laid in as the manufacturers frequently change their lines.

Such shallow cages, with their open construction, greatly assist the fly in locating the source of food.

Rack for breeding females

For mass-rearing purposes a rack has been designed by M. A. Trewern of this laboratory which holds 10 of the large-size cages and hence up to 250 goat-fed fertilized female tsetse (Figs. 46.2B and 46.3). The rack is 61 cm long, 27 cm wide and 8 cm high. The cages, each resting on a narrow face and at right angles to the length of the rack, are supported on two rods (A and B). The floor of the rack consists of two metal sheets, one on each side, which slope down at an angle of 15° to the horizontal to produce a median slot, 14 mm wide, through which the larvae fall to pupate in a 18 mm-deep tray, (E) which runs the length of the rack. A third rod (C) supports the slot protector (D) which prevents liquid tsetse faeces from falling through the slot into the tray. The three rods, provided with a thread and nut at each end, also serve to keep the structure fixed in position by tension. Two lengths of angle (F) bolted to the bottom of the rack along each side act as runners for the tray and prevent any shift in the slope of the two halves of the floor, which are also supported on

* Black because it is easy to see through, and Terylene because it is non-absorbent and non-brittle.

two bolts (G) at each end of the rack. The only materials required are 16 gauge (1·6 mm) soft aluminium sheeting and 6·35 mm stainless steel rod. To serve the same purpose, Itard & Gruvel (1969) have devised a rack for holding 10 of the small-sized cages, but their design requires greater metal-working ability.

Emergence cages

Unless flies are allowed to emerge in a large cage a proportion will fail to develop their wings, having been disturbed on hatching by those that have already emerged (Nash *et al.*, 1968). The cage must also be designed to facilitate the daily collection of young flies.

Newly emerged flies are collected after being immobilised at a temperature of 2–4°C and the emergence cages used are designed to fit a working top inserted into a specially adapted deep freeze. The cages are composed of an aluminium base, which holds the pupae, and a stainless-steel wire-framed top covered with black Terylene netting into which the newly emerged insects can

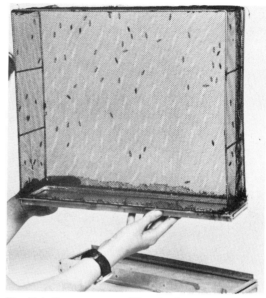

Fig. 46.4. Emergence cage. The wire-framed netting top, containing newly emerged tsetse flies, has been removed from the aluminium base and is ready for chilling. (Photograph: M. C. H. Parsons.)

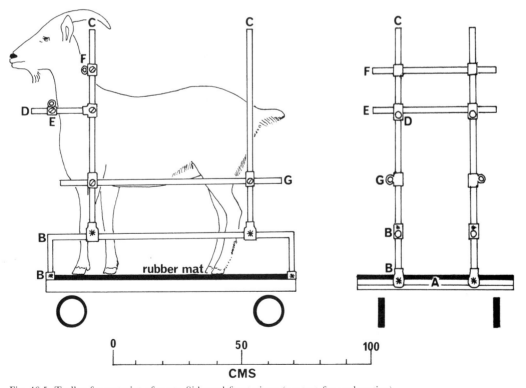

Fig. 46.5. Trolley for restraint of goats. Side and front views (see text for explanation).

fly. The sides of the base are made of 1-cm-thick aluminium bar and the bottom of aluminium sheet. The internal measurements of the base are $40 \times 8.2 \times 2.5$ cm deep. A layer of previously washed and baked sand, on which up to 2,500 pupae can be distributed, is placed in the bottom of the base. The inside of the sides and one end of the base are grooved to take an aluminium slide, which is inserted through a slit in the other end and which can isolate the base from the top of the emergence cage. The wire-framed top is built on a basal frame of aluminium bar which fits into the aluminium bar of the base. The basal frame of the top is also grooved inter-

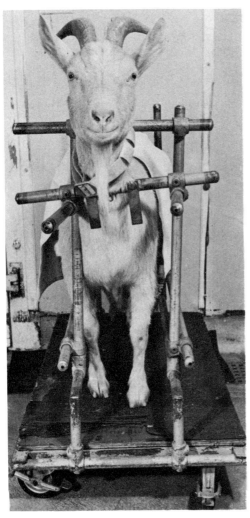

Fig. 46.6. Trolley for restraint of goats. (Photograph: UFAW.)

nally and has a slit at one end, through which a second aluminium slide can be inserted. The wire-framed top measures $40 \times 35 \times 10$ cm. Before the collection of newly emerged flies, both slides are inserted and the rubber bands used to hold together the two parts of the cage are removed; the whole of the top part of the cage can then be put into the adapted deep-freeze and chilled to $2-4°C$.

Trolley for restraint of goats

An animal is led from its pen, steps up 20 cm on to the floor of a trolley, is secured and wheeled into the fly handling room. The trolley has already been described (Nash & Kernaghan, 1964), but subsequent experience has shown that the goat is so co-operative that the degree of restraint and of adjustability of bars can be considerably reduced. The trolley described below (see also Figs. 46.1, 46.5 and 46.6) is equally suitable for 4-month old kids and adult goats weighing up to 72 kg; a large model has been successfully used for calves, but this host was abandoned as too dirty and odiferous for use in a hot and humid room.

The chassis, 107 cm long × 61 cm wide, is constructed of lengths of 3.8 cm steel angle welded together; the chassis is mounted on 13 cm diameter rubber-tyred castors supplied with brakes (Fig. 46.6). The floor of the trolley is made of 1.3 cm thick marine plywood and is 99 cm long × 61 cm wide; it is surfaced with corrugated rubber matting which is bonded to the wood with an impact adhesive. Animals panic if their footing is insecure.

All rails and rods referred to below are made of 2.5 cm diameter galvanized steel pipe and socket construction, secured by Kee Klamps at places where adjustability is required. Bars (A) extend across the front and back ends of the chassis. The two bars (B) are welded to bars (A) at a distance of 15 cm from each side of the trolley; bars (B) are 25 cm apart. The four verticals (C) are welded to bars (B) at a distance of 15 cm from the front and back of the trolley. The two bars (G) can be slid up or down the verticals (C) to suit the height of the animal and are then secured by Kee Klamps; adjustment is only necessary for animals with considerable disparity in height, the purpose being to leave the flanks readily available for the strapping on of cages of tsetse. The single cross-bar (E), which spans the two bars (D) should be about 10 cm in front of uprights (C); cross-bar (F) is a yoke which is slid down the verticals (C) until it touches the neck when it is secured. This is the only restraint required; the cross-bar also prevents a goat from inadvertently bucking its head and horning an assistant's face when the latter stoops to strap on cages. The success of minimal restraint is probably due to the affection shown by the staff for the goats, which have never been frightened. Several goats prefer to stand with their hind legs on the ground, their forelegs on the trolley and their necks in the yoke; this does not interfere with the fly feeding.

Straps for securing cages to the flanks of goats

The basic strap is 180 cm long × 4 cm wide and is made of 3 mm thick elastic belting. A piece of 'male' Velcro, 50 cm long × 3·2 cm wide, is machine-sewn to the *top* surface of one end of the strap, and a similar piece of 'female' Velcro to the *under* surface of the opposite end. Such a

standard strap can be used for securing cages to a kid or to a large goat (Nash & Kernaghan, 1964).

Box for the restraint of rabbits

Tsetse are also fed on the ears of lop-eared rabbits, using those with ears measuring 14–25 cm in length (Nash et al., 1966). A wooden box is used, the internal dimensions of which are 16 cm high, 14 cm wide and 30 cm long or 38 cm for very large rabbits, (Figs. 46.7 and 46.8). The front consists of a board in which a 6·5 cm diameter circle is cut out; the bottom of the circle is 6·5 cm above floor level. The front board is then cut laterally across the centre of the circle. The lower section (A) is a fixture. Having placed the rabbit in the box with its neck resting in the semicircle and its head sticking out of the box, the top section (B), which acts as a neck yoke, is dropped through grooves until it re-unites with the bottom. The rabbit cannot lift the yoke because this is secured on each side by a large nail (C) which fits loosely into a hole drilled through E and B. The rear end of the box is closed with sheet rubber (D), because if a rabbit

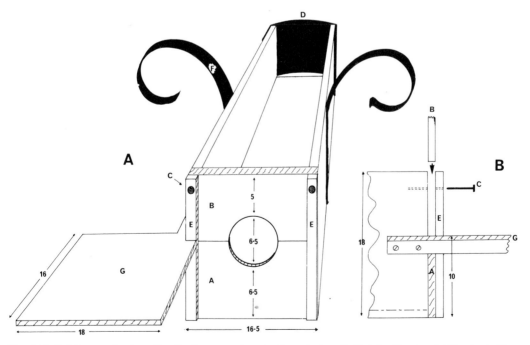

Fig. 46.7. Rabbit restraint box. Details of construction. A. General view. B. Detail of front end (side view). (For explanation see text.) (Measurements in centimetres.)

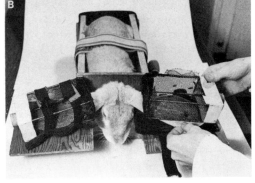

Fig. 46.8. Rabbit restraint box. A. Empty. B. With rabbit and fly cages in position. (Photographs: UFAW.)

kicks a solid end it may break its back. The rabbit is further restrained by a strap (F), similar to that described in the last section, but 120 cm long with 35 cm of Velcro at each end; the middle of the strap is nailed to the underside of the middle of the box; the ends of the straps are passed over the rabbit's back and are secured, one to the other, by pressing the Velcro ends together. From the front end, two ear-support platforms (G) extend at right angles to the box, on a level with the junction of A and B; they are 18 cm long and 16 cm wide.

Straps and pad for application of cages to the ears of lop-eared rabbits

The convex pad should be about 15 cm long × 10 cm wide and 5·5 cm thick in the centre; the core is made of cotton wool enclosed in plastic foam and covered with corduroy. The 28 cm long straps are made of 2·5 cm wide elastic; the last 8 cm of the ends are faced with 2·2 cm wide Velcro, the principle being similar to that for the goat straps described above. Two straps, 4 cm apart, are sewn on to the bottom of the pad at right angles to its long axis. The pad is placed on the ear-support platform and the rabbit's ear is then placed on top of the pad, with a cage of flies on top of the ear; the straps, which arise from below the pad, are then stretched over the cage, and their ends secured. The convexity of the pad takes up any slack in the cage face, and assures a close contact between the tsetse and the food surface.

General precautions

All insecticides should be banned from the laboratories and animal houses, and their use by adjacent departments discouraged. Stable- and house-flies can be controlled by an apparatus which electrocutes insects attracted to ultra-violet light (e.g. Insect-O-Cutor) and by adhesive fly papers. The former is also excellent for killing escaped tsetse, but should be switched on after working hours because in a low room the light causes headaches.

To guard against contamination with sub-lethal amounts of insecticide, which reduce fly longevity and fecundity, all newly-purchased goats, materials and articles which will come in contact with tsetse are first washed with hot water and soft soap (Sapo Mollis B.P.); in the case of rabbits, only the ears are washed. Subsequently goats are clipped weekly and washed at 2-monthly intervals; the ears of the rabbits are washed monthly. No chemicals or detergents are permitted in the breeding rooms, which should not be redecorated until the paints have been tested for toxicity.

Dead flies are removed from the maintenance cages on alternate days and are dropped into jars of 70 per cent alcohol; the cages are stripped, boiled and re-covered periodically. The racks and trays used for breeding females are washed weekly in a soft soap solution. To obviate any risk from entomophagous fungi, the relative humidity is not allowed to rise above 80 per cent.

To avoid the possibility of inbreeding depression developing in colonies of tsetse, males only, hatched from African-collected pupae, should be added to stock every few years (Jordan et al., 1970).

Rearing techniques

Newly emerged flies

Each day, Sundays excepted, the young flies are collected from the emergence cages. The Terylene-covered, wire-framed top of the emergence cage, sealed at its base by an aluminium slide (page 573) is placed on the working surface of a specially converted deep freeze, at a temperature of 2–4°C. Within two minutes the flies are immobilised and can be tipped out of the cage on to the working surface. Using a soft paint brush, the flies are sexed, swept into small tubes and then gently tipped into cages. The cages of immobilised flies are left at room temperature for a few minutes before further handling takes place. All newly emerged flies are offered food daily, Sundays excepted, but at least an hour is allowed to elapse between recovery of the flies from immobilisation and feeding.

Mating

Mating techniques have to vary for different species of tsetse. For *G. morsitans*, a cage of fed 3-day-old females is connected to a cage of males by a 7-cm plastic tube from one cork-hole to the other. The males should be at least 7 days old and should not have been mated more than twice before. The male cage is then covered by a slightly larger box, with the cork-hole facing the light and the males fly through the tube into the cage of females. Up to 25 females can be mated in this way; up to 5 additional males should be used. The sexes are left together for 24 hours and are then separated and recaged after immobilisation at 2–4°C.

Although almost all female *G. morsitans* are inseminated by this method, only poor insemination results if *G. austeni* is mated in this way. This species can be reluctant to mate in captivity and it is necessary to mate the females individually in tubes in order to achieve a good insemination rate. A fed 3-day-old female is placed in a tube to which is added a male that is at least 7 days old and has not been used for mating more than twice. If pairing does not take place at once another male is substituted; the few females which still refuse to pair are discarded. The sexes are separated after 24 hours by placing the tube containing two flies end to end with an empty tube. As soon as one fly has flown into the empty tube the mouth of each tube is closed with two fingers and corks are substituted using the remaining fingers of each hand. The two sexes are then appropriately caged. These operations can conveniently be carried out under a mosquito net (Fig. 46.1), when the occasional fly that escapes can be readily recaptured.

Maintenance of fertilised females

At the end of the 24-hour mating period, 25 fertilised females are placed in each of the large type of cage used for goat-fed flies. For routine breeding purposes, ten of these cages are kept on each breeding rack (Fig. 46.2B); the pupal trays are emptied daily. For smaller scale or experimental purposes, 5 large cages may be kept in a standard $30·5 \times 25·5 \times 5$ cm enamel tray. The cages are stacked, side by side, across the 25·5 cm width of the tray, with a narrow face downwards; the closed ends rest on the edge of the tray and the open ends on the floor of the opposite side, which enables the larvae to drop through on to the tray floor. If the pupal yield from each cage is required cardboard partitions may be erected. The pupae are collected daily with brush and teaspoon. The rack or tray method is equally applicable to the small type of cage used for rabbit-fed flies. For both methods the same narrow face of a cage should always face downwards, which ensures that only one non-feeding face gets soiled by excretion. As the population on a rack or tray decreases, the contents of the cages are amalgamated.

Pupal maintenance

Daily, the trays under the breeding racks (Fig. 46.2B) are emptied on to sand. Having allowed an hour for any larvae to pupate, the sand is poured through lightly stretched netting, the pupae are scooped up with a teaspoon, and are then spread on the sand-covered floor of the tray in the bottom of an emergence cage (Fig. 46.6). Such cages are kept on the bottom shelves of the controlled-climate rooms, where the temperature is lowest at about 23·5°C. The duration of the pupal period is 32–37 days. A lamp above the emergence cages provides an illumination of 7 lux on the soil surface for 12 hours daily and encourages the young flies to climb up the sides instead of disturbing the just-hatched flies on the floor, as this would result in some wastage due to failure to develop the wings. An effective emergence-rate of about 96 per cent should be obtained.

Fly feeding

When cages are transferred to the fly handling room for removal of dead tsetse and for fly feeding, they are stacked on grids over large shallow trays into which larvae fall; this saves some 3 per cent of production from being lost on the floor.

The tsetse feed by probing through the netting of one of the large faces of the cage. Flies are offered food for 15 minutes daily, Sundays excepted, because after parturition the female needs to engorge on several successive days; on average, a fly feeds every other day (Nash et al., 1967).

Goats. Goats are excellent hosts for feeding large numbers of tsetse; up to 600 flies can be applied to one adult beast in one hour. Further, if only castrated males and females are used, the goat is non-odorous and the dry faecal pellets are easily swept up. Goats are amenable, intelligent and easy to breed.

The hosts' flanks should not be shaved but clipped weekly using a thin blade designed for extra-close clipping; regular clipping is essential if the flies are to feed well, and it also prevents any build-up of insecticide from the hay and will remove hardened tsetse faeces.

The feeding procedure is carried out by two laboratory assistants, who stand facing each other with the goat between them. There are two straps under the goat's belly. Each assistant presses 2 or 3 cages to the flank and stretching the strap passes her end across to the other assistant, who presses Velcro to Velcro. The second strap is similarly secured (Fig. 46.1). At the end of 15 minutes a bell rings, and, again acting in unison, the two assistants press the cages against the flank with one hand whilst tearing the Velcro ends apart with the other. During feeding, a linen huckaback towel (112 × 46 cm) is slung across the goat's back to cover all six cages; the towel counters phototropism, and so encourages the flies to settle on the skin. To get an even distribution of the feeding flies, the closed ends of the cages should face the source of light; otherwise, concentrated fly-feeding at one end of the cage may harden the host's skin and produce a hidebound condition. The number of flies applied to any one goat in a day is limited to 700; about half will feed. Adult goats remain placid whilst the flies are feeding, but kids may become restive owing to the restraint imposed on their freedom.

By allowing a 3- or 4-day interval, during which the bite lesions can heal, mechanical damage to the hide can be avoided. Teams of goats are brought in on specified days of the week; some individuals have been in regular use for 7–8 years.

Sensitization reactions to fly-bites are reduced to negligible proportions by adopting the following procedure:—Ten flies are allowed to feed on each kid within 24 hours of birth, and then five flies twice weekly for several months; the challenge is then slowly raised (Nash, 1970). Prior to this innovation, a few goats became sensitized, as indicated by a rash of persistent papules, general erythema and alopecia; such animals were culled, not because they were unattractive to tsetse, but on humane grounds and because they became irritable and exceedingly restive when used.

It is advisable to test the herd periodically for host suitability, as an individual's configuration and its hide thickness and texture may change with age. Previously unfed 2-day old male flies are used as the test organism. Ten such flies are put in each of two large cages and after 15 minutes' application to the flank the cages are removed and two observers reach agreement on the number fed. The host suitability index is based on 5 test days; if less than 80 per cent of flies have fed, the goat is culled. The average figure for the Langford herd is over 90 per cent.

Rabbits. The ears of lop-eared rabbits provide a freely available source of blood for the tsetse, and such animals are the host of choice for the research worker undertaking a small project (Jordan et al., 1967). Tsetse can be satisfactorily maintained in an incubator with a tray of distilled water on the top shelf (Jordan et al., 1968). Provided the laboratory is heated in winter the flies can be safely taken out for 15 minutes for feeding on a boxed rabbit placed on the bench (Fig. 46.8).

Rabbits with ears 15 to 20 cm in length are best; those with much longer ears are inbred and tend to be more delicate. Those with ears that are cold to the touch do not make good hosts, presumably because the blood supply is poor. The ears are not shaved or clipped.

A rabbit is used for fly-feeding at 3- or 4-day intervals. The two cages, each holding 10 or 15 flies, are changed every 15 minutes, so that 80–120 flies can be offered food in an hour, the challenge being kept to 160 flies per feeding day. This regimen is not deleterious to the rabbit's ears and does not cause exsanguination.

The longevity and fecundity of tsetse fed on rabbits is even better than that from flies fed on goats. This may be because rabbit blood is better, but possibly the caged females, which become apterous with age, find it easier to drop on to the ears than to cross from one side of the cage to the other in order to feed on the flanks of goats.

Given a large rabbitry, with breeding facilities and extremely hygienic maintenance, the lop-eared rabbit could be a very suitable host for the mass-rearing of tsetse flies. It would, however, be wise to feed a proportion of the fly stock on goats in case an epidemic should sweep through the rabbitry.

Whether this breed of rabbit would survive under tropical conditions, without air conditioning, is unknown.

Productivity

As the large-scale production of pupae from self-maintaining colonies of tsetse flies is a recent development, it may be of interest to indicate current productivity when using the methods described above. In the case of *G. austeni* the mean number of pupae per initial number of mated females is 14 when the females are fed on rabbit ears, and 10 when fed on the flanks of goats. The mean age at death of the females when fed on the two hosts is about 175 and 146 days respectively. Comparable productivity figures for *G. morsitans* are 10·5 pupae per female when fed on rabbits and 7·5 when fed on goats.

In 1973 a mean stock of 7,300 mated female tsetse was kept at the Tsetse Research Laboratory, Langford. During the year, 201,000 pupae were produced, of which 113,400 were surplus to the requirement of maintaining numbers in the colonies and were used for research work at the laboratory or sent to workers elsewhere. Some 91 per cent of the flies were fed on goats and the remainder on rabbits.

Sources of equipment

Maintenance cages, stainless steel wire construction—North Kent Engineering Co. Ltd., Home Gardens, Dartford, Kent.

Velcro—Selectus Ltd., Biddulph, Stoke-on-Trent, ST8 7RH. $\frac{7}{8}$ inch wide (rabbits), $1\frac{1}{4}$ inch wide (goats), sold as twin yards.

Black Terylene netting (this is not always readily available in a suitable mesh)—John C. Small & Tidmas Ltd., Perry Road, Sherwood, Nottingham (100 yards minimum —nos. 6480 and 6190 are both suitable for *G. morsitans* and *G. austeni*).

Insect-O-Cutor—Henry Simon Ltd., P.O. Box 31, Stockport. Model IOC 50 for large animal house; model WM2 for fly handling room.

Clippers for goats—Wolseley Engineering Ltd., Electric Avenue, Witton, Birmingham, B6 7JA. A354 Pedigree clipper; A2 fine plate.

REFERENCES

Azevedo, J. F. de & Pinhão, R. da C. (1964). The maintenance of a laboratory colony of *Glossina morsitans* since 1959. *Bull. Wld Hlth Org.*, **31,** 835–41.

Azevedo, J. F. de, Pinhão, R. da C., Santos, A. M. T. dos & Ferreira, A. E. (1968). Studies carried out with the *Glossina morsitans* colony of Lisbon. I.— Some aspects of the evolution of the *Glossina morsitans* colony of Lisbon. *An. Esc. nac. Sáude públ. e Med. trop.*, **2,** 19–42.

Bursell, E. (1960). The effect of temperature on the consumption of fat during pupal development in *Glossina*. *Bull. ent. Res.*, **51,** 583–98.

Dame, D. A. & Schmidt, C. H. (1970). The sterile-male technique against tsetse flies, *Glossina* spp. *Bull. ent. Soc. Amer.*, **16,** 24–30.

Itard, J. & Gruvel, J. (1969). Description d'un appareil destiné au stockage des femelles de Glossines et à la récolte des pupes. *Rev. El. Med. vet. Pays trop.*, **22,** 289–92.

Itard, J. & Maillot, L. (1970). Les élevages de Glossines à Maisons-Alfort (France). *Proc. 1st Symp. Tsetse Fly Breeding Under Laboratory Conditions and its Practical Application*, Lisbon, pp. 125–36.

Jordan, A. M., Nash, T. A. M. & Boyle, J. A. (1967). The rearing of *Glossina austeni* Newst. with lop-eared rabbits as hosts. I.—Efficacy of the method. *Ann. trop. Med. Parasit.,* **61,** 182–8.

Jordan, A. M., Nash, T. A. M. & Boyle, J. A. (1968). The rearing of *Glossina austeni* Newst. with lop-eared rabbits as hosts. II.—Rearing in an incubator. *Ann. trop. Med. Parasit.,* **62,** 331–5.

Jordan, A. M., Nash, T. A. M. & Trewern, M. A. (1970). The performance of crosses between wild and laboratory-bred *Glossina morsitans orientalis* Vanderplank. *Bull. ent. Res.,* **60,** 333–7.

Kernaghan, R. J. & Nash, T. A. M. (1964). A technique for the dispatch of pupae of *Glossina* and other insects by air from the tropics. *Ann. trop. Med. Parasit.,* **58,** 355–8.

Mews, A. R., Offori, E., Baumgartner, E. & Luger, D. (1972). Techniques used at the IAEA laboratory for rearing the tsetse fly, *Glossina morsitans* Westwood. *Proc. 13th Meeting Int. Sci. Comm. Tryp. Res.,* Lagos, pp. 243–54.

Nash, T. A. M. (1970). Possible induction in goats of immunological tolerance against the saliva of un-infected tsetse flies (*Glossina* spp.). *Trop. Anim. Hlth Prod.,* **2,** 126–30.

Nash, T. A. M., Jordan, A. M. & Boyle, J. A. (1966). A promising method for rearing *Glossina austeni* (Newst.) on a small scale, based on the use of rabbits' ears for feeding. *Trans. R. Soc. trop. Med. Hyg.,* **60,** 183–8.

Nash, T. A. M., Jordan, A. M. & Boyle, J. A. (1967). A method of maintaining *Glossina austeni* Newst. singly, and a study of the feeding habits of the female in relation to larviposition and pupal weight. *Bull. ent. Res.,* **57,** 327–36.

Nash, T. A. M., Jordan, A. M. & Boyle, J. A. (1968). The large-scale rearing of *Glossina austeni* Newst. in the laboratory. IV.—The final technique. *Ann. trop. Med. Parasit.,* **62,** 336–41.

Nash, T. A. M. & Kernaghan, R. J. (1964). The feeding of haematophagous insects on goats and sheep: techniques for host restraint and cage application. *Ann. trop. Med. Parasit.,* **58,** 168–70.

Van der Vloedt, A. M. V. (1971). Ecology and population dynamics of a *Glossina palpalis palpalis* colony kept under laboratory conditions. *Trans. R. Soc. trop. Med. Hyg.* **65,** 214–20.

47   The American Cockroach

P. J. ROCHFORD

The American cockroach (*Periplaneta americana* L.) is a large insect used extensively in the teaching of biology, research and the testing of insecticides, to which, unlike *Blatella germanica*, it has not as yet shown any resistance. It also has advantages over *Blatta orientalis* in that it is not likely to establish itself as a pest in Britain, and its smell is less offensive.

General biology

In warm conditions the American cockroach is an active animal and can run fast even up a vertical surface. The claws are used if the material is rough, and the plantulae or adhesive pads at the tips of the tarsal segments if the material is smooth, e.g. glass. On an inverted smooth surface it uses an adhesive lobe, the arolium, which lies between the claws. The young nymphs are unable to climb the vertical sides of a clean glass vessel. Willis *et al.* (1958) state that this ability is acquired after the sixth or seventh moult. Although winged, cockroaches do not fly when kept in captivity under regulated conditions. Under normal conditions only the males have a capacity for sustained flight.

Cannibalism is prevalent among dense populations of young larvae, especially during the first three weeks, but isolation retards development—though Wharton *et al.* (1967) found that the adults so produced are heavier. They considered cannibalism to be a sanitary and scavenging measure rather than a form of predation. Single larvae grow best in the presence of adults (Wharton *et al.*, 1968), but if larvae are not isolated they tend to crowd together regardless of the space provided. Adults are apt to bite each other, and in a large colony there are often some animals with one or more legs missing. There is more fighting between males than between females or between males and females. Bell & Gary (1973) found that aggression is increased by the sex pheromone,

disturbance by light and disturbance of the environment.

Although in warm climates the American cockroach is found in sewers, latrines and cesspools (Cornwell, 1968), it regularly cleans itself with the mouthparts and legs. There are no grounds for assuming that it should not be kept, like other laboratory animals, in hygienic conditions.

There is a well marked diurnal locomotory rhythm which is normally fixed by regular periods of exposure to light. (Harker, 1956 and 1960).

Reproduction and life history

The adult males may be distinguished by their anal styles and the females by the boat-shaped end to the abdomen. In newly-emerged females the oocytes are relatively small and do not reach full size for a week or more. The males are attracted by the odour of a cuticular grease on both males and females. Sturckow & Bodenstein (1966), who studied the location of the pheromone, maintain that it should not be considered a sex attractant in the strict sense as it is not confined to one sex. Courtship display is shown by the male by the wings being raised, fluttered or spread out motionless; the abdomen may be either arched or extended. When contact is made with a receptive female, the male turns and copulates in the opposed position—i.e. the heads face in opposite directions; this lasts for an hour or more. The spermatozoa are transferred in a spermatophore and received by the spermatheca of the female where they may remain viable for a considerable time. If males are absent the percentage of viable eggs decreases with age.

Eggs are normally laid in two rows of eight in an egg-case, 13 to 15 per cent of the dry weight of which is calcium oxalate (Stay *et al.*, 1960). It becomes dark and hard a few days after being deposited and loses weight mainly in the form of water (Kinsella & Smyth, 1966). The female carries the egg-case for a few hours or days, depending on conditions, before burying it or attaching it by a glutinous substance to an object, leaving free the dorsal edge or keel which is modified into a series of respiratory chambers that admit air to the developing embryos. By rolling the cases between the thumb and first finger those which contain eggs can be rapidly distinguished from the empty ones which may show the remains of the embryonic membranes. Internal pressure exerted by the fully-formed nymphs causes the case to split along the keel.

The nymphs are similar in appearance to the adults, but lack wings. On emergence, and after each moult, the body is white but it acquires the normal brown colour during the next few hours. Authors differ as to the number of moults: in 13 insects reared in isolation Willis *et al.* (1958) found between 9 and 13 moults. Styles are present in nymphs of both sexes but the females can be distinguished by a median notch at the end of the ninth sternum, which is either absent or present only to a small degree in the male.

Development is greatly influenced by temperature. Gould (1941) obtained an average of 520 days for full development at about 25·5°C and 195 when the mean was 28·3°C. Similarly the mean incubation period varied from 58·5 days at 24·6°C to 48 days at 26·0°C and 34·5 days at 29·2°C. Lenoir-Rousseaux & Lender (1970) describe the stages in the development of the embryo. Development is also influenced by population density (Wharton *et al.*, 1967). Some further biological data are given in Table I.

Accommodation

Humidity

This is one of the most important factors in the keeping of cockroaches and the following points should be noted:

(1) The egg-cases can develop and hatch in very low humidities provided that they

have not been damaged in any way, but they are particularly susceptible for the first three days after formation; after this a high percentage of eggs may be hatched from cases with the keels removed if the relative humidity is 90 per cent (Roth & Willis, 1955). One of the principal functions of the case is to prevent loss of water.

(2) The first instar is very susceptible to drying and should be kept at 70 per cent R.H. for the first week (Haskins, 1962).

(3) Older nymphs and adults can live quite healthily in much lower humidities, provided they have access to water. In humidities of about 5 per cent R.H. and at temperatures of about 32°C adults may lose between 20 and 30 per cent of their body-weight in twenty-four hours (Gunn, 1935).

(4) Humidities above 60 per cent may lead to outbreaks of mites (Haskins, 1962) or psocids, but with careful sterilization of containers and food, and exclusion of crawling animals by means of an oil barrier, self-contained colonies can be maintained successfully in a R.H. of 70 per cent. (G. A. Brett, personal communication).

(5) At a R.H. of 75 per cent, moulds may infect both food and egg-cases.

Humidity can be controlled in several ways:

(a) Warm air from a heating element may be blown over wood-wool kept moist with water. A humidistat controls a heater and fan for this purpose and a thermostat controls a second heater which regulates the temperature of the breeding room.

(b) A humidistat may be connected to an aerosol humidifier.

(c) Air may be circulated over saturated salt solutions. A salt is chosen which, in a saturated solution, gives the vapour pres-

TABLE I

Comparison of records of biology of the American cockroach

Source	Adulthood in days Female Av. Max.	Male Av. Max.	Preoviposition period (days)	Days between egg-cases	Egg-cases per female	Days of incubation
Gould & Deay	438 588	— —	13·4	5·9	58·8	55·3
Griffiths & Tauber	225 706	200 362	20·0	6·8	21·1	53·0
Rau	310 330	82 175	—	7·4	11·9	43·0

(Abridged from Griffiths & Tauber, 1942)

sure corresponding to the humidity required at the correct temperature.

Temperature

For breeding at a constant temperature, a minimum of about 28°C is required (Haskins, 1962) but the animals can be maintained at a lower temperature (24 to 25°C) without difficulty. They become almost inactive at 10°C, and near freezing-point will appear to be dead but may be brought round if not subjected to this temperature for more than 2 or 3 hours.

Prevention of escape

The following methods may be used:
(1) A cage with either a well-fitting lid, or sides of polythene with a barrier of P.T.F.E. fluon dispersion. This barrier is not so effective on glass.
(2) An electric fence around the top of the chamber. This can be made from two strips of copper foil firmly tacked to a wooden frame and differing in electric potential by about 35 v. Each strip should be not less than 1 cm wide (preferably 2 cm if space permits) and they should be placed 2 or 3 mm apart. If the fence is facing downwards the animals will touch it with their antennae before they walk on it and so be spared a shock.
(3) Petroleum jelly on glass. This should be applied to form a barrier 8 cm wide on a vertical or (better still) a downward-facing surface. It is suitable for small containers but is not entirely reliable and will collect dirt.

Feeding

Cockroaches can live on very poor diets. Growth is retarded when the American cockroach is reared aseptically (Gier, 1947), and Gallagher (1963) found that a sample freed from symbionts died within 33 days when not supplied with vitamin B. The animals can secrete cellulase, but the chewing of cellulose products by the females as egg-laying begins is probably related to the need to conceal the egg-cases. They can live on dry foods but it is important in this case to maintain the water supply especially when the conditions are dry. Willis & Lewis (1957) found that at a R.H. of 36 per cent cockroaches could live as long with no food or water as they could with dry food alone, whereas they lived much longer with water alone. A watering fountain consisting of a full beaker inverted on a Petri dish with cotton wool to prevent the young nymphs from drowning is suitable for most purposes. The water should be changed when it becomes fouled by saliva. A test-tube plugged with cotton wool and lying on its side is useful where space is limited. Noland (1956) describes a watering device used in rearing-tubes.

Dry foods have the advantage that they can be sterilized by heat but the disadvantage that they may be strewn all over the cage. Many forms and mixtures have been used such as dog biscuits and rat cubes. The medium used at the Pest Infestation Laboratory, Slough, contains rolled oats, wheatfeed, fishmeal and dried yeast powder in the ratio 9:9:1:1 (Haskins, 1962). After moulting, cockroaches consume their own exuviae, but not the legs.

Breeding

The Tolworth system

The following method has been used at the Infestation Control Laboratory of the Ministry of Agriculture, Fisheries and Food at Tolworth, Surbiton, Surrey. The colonies are kept in larvae cages (Fig. 47.1), 40 cm high and 20 cm in diameter with plastic sides, metal bases, and perforated metal lids. Most of the space in each cage is occupied by 10 platforms of hardboard measuring 15×10 cm separated by 1·8 cm wooden blocks held together by two pieces of wire which pass through them; these also provide two handles at the top. Water and food are placed on the top platform. The colonies are subcultured every three months. A hole in the plastic, normally stoppered with a rubber bung, provides an inlet for carbon dioxide should anaesthetization be necessary. The cages rest on raised stands on galvanized iron trays which contain a layer of Technical White Oil. This provides a reliable barrier to cross-infestation from other insects and mites. The room is kept at a R.H. of 70 per cent and a temperature of 27·7°C.

The method is particularly suited to keeping a large number of isolated colonies of insects in one

Fig. 47.1. A larva cage used in the Tolworth system (Crown copyright).

room, and is used for *Blatta orientalis*, *Blatella germanica* and *Periplaneta australasiae*.

The Slough system (Haskins, 1962)

After the first week the adults are kept at a R.H. of 30 per cent but this does not appear to retard development provided that there is ready access to water. The egg-cases are buried in the dried food mixture which is contained in a crystallizing dish, and less than 2 per cent are damaged. They are collected weekly by passing the whole of the food medium through a sieve; they are then washed, dried and incubated in another room at a R.H. of 70 per cent.

Breeding observation-chamber
(Rochford, 1957) (Figs. 47.2 and 47.3)
A false floor of 0·6 cm wire mesh induces the females to lay the majority of their egg-cases in holes in a piece of hardwood. The main part of the colony rests on pieces of plywood hanging from a frame which can be moved to one side of the chamber to permit the removal of the plywood, and the cleaning of the chamber. The eggs are placed in a separate smaller container which can be kept at about R.H. 70 per cent by using a small tube of saturated sodium chloride solution. Water is supplied in a test-tube plugged with cotton wool and resting on its side. After a week the nymphs can be removed and put into a nymph chamber with small platforms as in the larvae cages (Fig. 47.1) or they may be released into the main chamber with the adults and older nymphs. This may help growth and development but if they are released too early they will crowd into the holes in the hardwood causing the females to drop their egg-cases through the wire mesh. As the false floor is made on two frames, it can be taken up one half at a time for the removal of droppings and debris, such as legs,

Fig. 47.2. The breeding observation chamber, with nymph cage on right, main colony on boards in centre and egg-capsule box on left (Photograph: Rochford).

from the trays underneath while the colony on the plywood is moved to the other end of the chamber; but this produces a certain amount of disturbance. If potatoes and apples are fed, little or no food falls through the wire, and the debris is kept to a minimum.

There is a temperature gradient of about 3°C between the top and bottom of the plywood in the shade. A convenient range of temperatures is from 25°C at the bottom to 28°C at the top of the plywood and 30°C directly under the lights. At these temperatures the colony can be handled without turning off the lights to allow the chamber to cool. However, individual animals are liable to fly for a short distance if disturbed too much.

Other methods

Cummings & Menn (1959) describe a cage in which the wire mesh becomes the floor and the egg-cases are collected in a tray below. A watering and feeding device is built into one side so that the lid does not need to be opened regularly. A still more compact chamber has

been illustrated by Koshy & Mallik (1968) and is suitable where space is limited, but Wharton *et al.* (1967) have drawn attention to the dangers of overcrowding.

Transport

Exposure to the cold is, perhaps, the difficulty which is most likely to be encountered, but in normal summer temperatures this should present no problem, even for a long journey.

Egg capsules should be packed with cotton wool and the container sealed. Nymphs and adults should be provided with something on which to rest, such as crumpled paper, and put into a tin with a perforated lid. If the animals are accompanied on a long journey, it may be more convenient to put them into a large glass bottle. Provided the paper does not reach the neck, the top can be removed without much difficulty for feeding. If a warm radiator is available, the bottle may be placed close to it and animals will find their own optimum between the hot and the cold sides.

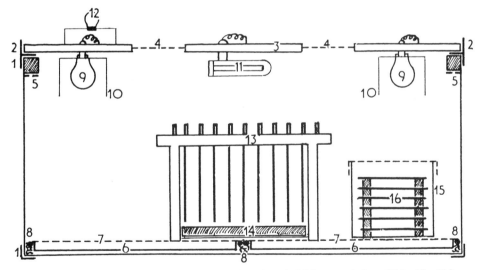

Fig. 47.3. Breeding observation chamber. 1. Aquarium frame of 2·5 cm angle iron, 92·5 · 38 · 40·5 cm; 2. 2·5 cm flange to frame to ensure lid is escape-proof; 3. Lid of five-plywood with strong diagonal batten to prevent warping; 4. Ventilation areas, 15 × 15 cm of perforated zinc; 5. Electric fence; 6. Galvanised-iron trays 2·5 cm deep; 7. 6·4 mm wire mesh on frames made to fit chamber, and resting on the trays; 8. Wire wool to fill in the gaps around the trays; 9. 60-W lamps in parallel; 10. Metal shades (tins), 15 cm in diameter; 11. Thermostat, standard aquarium model; 12. Pilot lamp for electric fence and lamps; 13. Metal stand for hanging boards, 15 × 23 cm; 14. 19 mm oak board, 23 × 23 cm with 6·4 mm holes at 2·5 cm intervals and 1·3 cm clearance above; 15. All-glass nymph chamber with perforated zinc lid; 16. Platforms of hardboard separated by small wooden blocks 9 mm thick. (Redrawn from School Science Review, **38,** 450)

Handling, control and killing

It is difficult to pick up individual cockroaches by hand without damage, but a long pair of fine forceps can be used satisfactorily if the insect is held in the middle of the abdomen. Larger numbers can be subdued by allowing the chamber or container to become chilled. Staszak & Mutchmor (1973) have described the effects of and recovery from chill-coma at different temperatures. Carbon dioxide can be used if it will not affect subsequent experimental work but its use has a growth-retarding effect on the German cockroach (Brooks, 1957).

Methods of capturing wild specimens have been described by Whitelaw & Smith (1964) and Wagner et al. (1964).

Cockroaches can be killed in chloroform vapour.

Parasites

Attention has already been drawn to mites. Though most of these are harmless in small numbers, they may over-run a colony. The species *Pimeliaphilus cunliffei* (see Field *et al.*, 1966) is, however, a true parasite. Fisk (1951) eliminated mites (possibly of this species) in a colony by using a 5 per cent dust of p-chlorophenyl, p-chlorobenzene sulfonate.

Nematodes may be present in large numbers without apparently affecting either health or fertility. Egg-cases which have been sent from abroad should be examined for signs of the eggs of parasitic wasps, especially Evaniidae. Griffiths & Tauber (1942b) dipped the egg-cases in 70 per cent alcohol for 10 seconds to prevent fungal growth.

Sources of equipment

P.T.F.E. fluon dispersion—Imperial Chemical Industries Ltd.

Larvae cages—Watkins & Doncaster, 110 Park View Road, Welling, Kent.

REFERENCES

Bell, W. J. & Gary, R. S. (1973). Aggressiveness in the cockroach *Periplaneta americana*. *Behav. Biol.*, **9**, 581–593.

Brooks, M. A. (1957). Growth retarding effect of carbon dioxide anaesthesia on the German cockroach. *J. Insect Physiol.*, **1**, 76–84.

Cornwell, P. B. (1968). *The Cockroach*. London: Hutchinson.

Cummings, E. C. & Menn, J. J. (1959). An American cockroach rearing cage. *J. econ. Ent.*, **52**, 1227–8.

Field, G., Savage, L. B. & Duplessis, R. J. (1966). Note on the Cockroach mite *Pimeliaphilus cunliffei* infesting German and American cockroaches. *J. econ. Ent.*, **59**, 1532.

Fisk, F. W. (1951). Use of a specific mite control in roach and mouse cultures. *J. econ. Ent.*, **44**, 1016.

Gallagher, M. R. (1963). Vitamin synthesis by the symbionts in the fat body of *Periplaneta americana*. *Diss. Abstr.*, **23**, 3449–50.

Gier, H. T. (1947). Growth rate in the cockroach *Periplaneta americana* (Linn.). *Ann. ent. Soc. Am.*, **40**, 303–17.

Gould, G. E. (1941). The effect of temperature upon the development of cockroaches. *Proc. Indiana Acad. Sci.*, **50**, 242–8.

Gould, G. E. & Deay, H. O. (1938). Biology of the American cockroach *Periplaneta americana* (L.). *Ann. ent. Soc. Am.*, **31**, 489–98.

Griffiths, J. T. & Tauber, O. E. (1942a). Fecundity, longevity, and parthenogenesis of the American cockroach, *Periplaneta americana* L. *Physiol. Zool.*, **15**, 196–209.

Griffiths, J. T. & Tauber, O. E. (1942b). The nymphal development for the cockroach (*Periplaneta americana* L.). *Jl N.Y. ent. Soc.*, **50**, 263–72.

Gunn, D. L. (1935). The temperature and humidity relations of the cockroach. III. A comparison of temperature preference, rates of desiccation and respiration of *Periplaneta americana*, *Blatta orientalis* and *Blatella germanica*. *J. exp. Biol.*, **12**, 185–90.

Harker, J. E. (1956). Factors controlling the diurnal rhythm of activity of *Periplaneta americana* L. *J. exp. Biol.*, **33**, 224–34.

Harker, J. E. (1960). The effect of perturbations in the environmental cycle of the diurnal rhythm of activity of *Periplaneta americana* L. *J. exp. Biol.*, **37**, 154–63.

Haskins, K. P. F. (1962). A new system for rearing the American cockroach (*Periplaneta americana*). *Entomologist*, **95**, 27–9.

Kinsella, J. E. & Smyth, T. Jr. (1966). Lipid metabolism of *Periplaneta americana* L. during embryogenesis. *Comp. Biochem. Physiol.*, **17**, 237–44.

Koshy, T. & Mallik, D. (1968). An improved method for rearing and maintaining large colonies of the American cockroach in the laboratory. *J. econ. Ent.*, **61**, 1748–50.

Lenoir-Rousseaux, J. J. & Lender, T. (1970). Table de development embryonaire de *Periplaneta americana*. *Bull. Soc. zool. Fr.*, **95**, 737–751.

Noland, J. L. (1956). An improved method for rearing cockroaches. *J. econ. Ent.*, **49**, 411–2.

Ray, P. (1940). The life history of the American cockroach (*Periplaneta americana* L.). *Ent. News*, **51**, 121–4, 151–5, 186–9, 223–7, 273–8.

Rochford, P. J. (1957). A breeding observation chamber for the American cockroach (*Periplaneta americana*). *Sch. Sci. Rev.*, **38**, 450.

Roth, L. M. & Willis, E. R. (1955). Water relations of cockroach oöthecae. *J. econ. Ent.*, **48**, 33–6.

Staszak, D. J. & Mutchmor, J. A. (1973). Influence of temperature on induction of chill-coma and movement of the American cockroach, *Periplaneta americana*. *Comp. Biochem. Physiol.*, **45**, 895–908.

Stay, B., King, A. & Roth, L. M. (1960). Calcium oxalate in the oöthecae of cockroaches. *Ann. ent. Soc. Am.*, **53**, 79–86.

Sturckow, B. & Bodenstein, W. G. (1966). Location of the sex pheromone in the American cockroach, *Periplaneta americana* (L). *Experientia*, **22**, 851–3.

Wagner, R. E., Ebling, W. & Clark, W. R. (1964). An electric barrier for confining cockroaches in large rearing or field collecting cans. *J. econ. Ent.*, **57**, 1007–9.

Wharton, D. R. A., Lola, J. E. & Wharton, M. L. (1967). Population density, survival and development of the American cockroach. *J. Insect. Physiol.*, **13**, 699–716.

Wharton, R. D. A., Lola, J. E. & Wharton, M. L. (1968). Growth factors and population density in the American cockroach *Periplaneta americana*. *J. Insect Physiol.*, **14**, 637–53.

Whitelaw, J. T. Jr. & Smith, L. W. Jr. (1964). Equipment for trapping and rearing the American cockroach, *Periplaneta americana*. *J. econ. Ent.*, **57**, 164–5.

Willis, E. R. & Lewis, N. (1957). The longevity of starved cockroaches. *J. econ. Ent.*, **50**, 438–40.

Willis, E. R., Riser, G. R. & Roth, L. M. (1958). Observations on reproduction and development in cockroaches. *Ann. ent. Soc. Am.*, **51**, 53–69.

FURTHER READING

Roth, L. M. & Willis, E. R. (1957). The medical and veterinary importance of cockroaches. *Smithson. misc. Collns*, **134** (10), 1–147.

Roth, L. M. & Willis, E. R. (1960). The biotic associations of cockroaches. *Smithson. misc. Collns*, **141**, 1–470.

48   Locusts

G. J. ASHBY

One result of the increased use of modern insecticides and more efficient methods of warehousing has been greatly to decrease the numbers of cockroaches available to zoological laboratories, where in the past considerable numbers have been used for dissection and as a live food for laboratory animals. Since Albrecht (1953) described the detailed anatomy of the Migratory Locust and Hunter-Jones (1956) gave instructions for breeding and rearing it, the cockroach has been largely superseded. The locust, being easy to breed and rear has in many ways proved to be more suitable for teaching purposes, and it is also an ideal form of live food for many laboratory and zoo mammals, birds and reptiles.

Species

The various species of locusts belong to the family Acrididae.

The African migratory locust

(*Locusta migratoria migratorioides*)
This species is by far the easiest to rear in the laboratory. The young locusts, or hoppers, are about 8·5 mm long when newly hatched, and by the end of the first day they are usually all actively feeding. Apart from the skin which the newly hatched insect leaves behind as it arrives at the surface of the sand, there are normally five moults occurring at intervals of approximately, 5, 4, 4, 5, and 8 days respectively (Hunter-Jones, 1961), thus giving an approximate total nymphal life of nearly one month. The wings are fully formed after the last moult, but the new adults are very soft and the insects will not be able to take flight until the cuticle has hardened. The adults mature in about four weeks after fledging. When young they are pale grey marked with darker grey, changing to bright yellow in the males and dark brown in the females at sexual maturity. The average female produces six egg pods in her lifetime (about one every 5 or 6 days). There are between 30 and 100 eggs per pod. The incubation period of the eggs is about 16 days at 28°C, 11 days at 32°C, or 9 days at a higher temperature, but this is not recommended.

The desert locust

(*Schistocerca gregaria*)
Although somewhat larger in size than *Locusta* this species is not so useful for teaching purposes as it has a longer life cycle and is prone to suffer from a bacterial disease. The freshly hatched hoppers of *Schistocerca* are approximately 9·5 mm long and they usually moult five times, taking 4 to 5 weeks to become adults at the temperatures recommended. Their colour is controlled to a large extent by the culturing temperature; at 28°C the average fifth-stage hoppers are almost wholly black, whereas at 40°C they are yellow with faint dark markings. Under the temperature conditions advocated for *Locusta*, i.e. 28°C during the night and 34°C during the day, the amount of black and yellow pigmentation is about the same. The young adults are pink but as they age the pink changes to creamy brown which at maturity yellows in the female and turns to bright yellow in the male. Given the recommended conditions and a population-density of approximately 40 pairs per 60 cm³ the females should lay their first eggs 4 to 6 weeks after fledging, and should produce approximately one pod per week for about 7 weeks (Hunter-Jones, 1961). The number of eggs per pod varies between 40 to 70. The incubation period is 17 days at 28°C, 12 days at 32°C and about 10 days at 34°C.

The red locust

(*Nomadacris septemfasciata*)
This species occurs in plague form over a great part of Africa south of the Equator. Normally it has only one generation a year, but three generations can be produced in the laboratory. It requires similar conditions to *Schistocerca* and the adults are of about the same size. The newly-

hatched hoppers are about 9·5 mm long and they undergo six or seven moults as opposed to the five of *Schistocerca* and *Locusta*. Some adults may have a very long maturation period, while others emerging at a different period of the year may only require about 6 weeks. The females lay from four to six pods at an average rate of one a week. The number of eggs per pod is usually about 100. At 32°C the incubation period is approximately 22 days.

Egyptian grasshopper

(*Anacridium aegyptium*)

The Egyptian Grasshopper, sometimes known as the Tree Locust, is common in the countries around the Mediterranean. Unlike the other species it does not, however, form large swarms. It is often imported accidentally into England with cargoes of vegetables. The newly hatched hoppers are pale green in colour and about 8 mm long. They moult five or six times and become adults in from 7 to 8 weeks. They are similar to the previous species mentioned in that some require a lengthy maturation period. The first egg pod is laid from 1 to 6 months after fledging. Up to ten egg pods can be laid, each containing on an average 50 eggs. The pods are laid singly at intervals of 1 or 2 weeks. At 32°C the incubation period is approximately 22 days.

Identifying the species

If the ventral surface of the thorax is pubescent and there is no peg-like projection between the front pair of legs the insect is the African Migratory Locust. If a peg-like projection is present between the front legs, and the hind wings show a blackish or smoky band when extended this is the Egyptian Grasshopper.

If a peg-like projection is present but there is no smoky band on the hind wing, examine the shape of the projection. It is curved and pointed in the Red Locust, but straight and blunt in the Desert Locust.

Phase and colour variation

Locusts are known to exhibit variations in both coloration and behaviour under different environmental conditions. This phase-variation has been shown to be associated with the appearance of the well-known locust plagues in many countries. Locusts of the plague phase, i.e. phase *gregaria*, are generally very brightly coloured and marked with a deep black pattern. They are also extremely active, excitable, gregarious and continually on the move. In the other phase, *solitaria*, the locusts are fairly inconspicuous in appearance, inactive, and either green overall or brown with no black markings. *Locusta* shows the greatest variation, and an initial colony of 1,000 hoppers in a cage measuring 30 to 60 cm^3 will give rise to typical *gregaria* coloration. The hoppers will be glossy black and orange and the young adults will be creamy-white with dark grey or black markings. Males and females will be of approximately the same size.

If, on the other hand, the young hoppers are kept strictly isolated throughout their lives they will show *solitaria* coloration.

Husbandry

Accommodation

Breeding-cages. Equipment need not be expensive or elaborate, but sound well-fitting cages are necessary as the young hoppers will easily escape through any hole much more than 2 mm in diameter. It is possible to breed locusts in a cage of almost any size, ranging from about 30 cm^3 upwards, the basic necessities being adequate light and heat and not less than 10 cm of coarse moist sand at the bottom of the cage for oviposition.

Many years of experience have shown that the most suitable type of cage for large-scale breeding is undoubtedly one based upon the principles described by Hunter-Jones (1961). Where, however, unheated rooms are to be used, two additional electric lamps can be fitted below the false floor to give the required heat.

The breeding-cage (Fig. 48.1) is in the form of a box measuring $46 \times 40 \times 24$ cm. It is made with a metal angle frame supporting perforated zinc sides and back. Access for working is provided by a hole 11 cm in diameter, cut in the flat metal front and covered with a circular flap. Observation is made through two glass windows, one on the top of the cage which is removable and the other positioned in the front wall immediately above the covered floor.

For convenience of working, the sand for oviposition is contained in tubes made of either metal or plastic, closed at one end, and measur-

Fig. 48.1. Breeding-cages. (Photograph: Zoological Society of London.)

ing not less than 30 mm in diameter and 100 mm in height. In order to accommodate these oviposition tubes in the bottom of the breeding-cage, it is necessary to provide a false floor made of either 16-gauge perforated zinc or even hardboard. The distance of the false floor from the main floor is governed by the length of the oviposition tubes. Normally two such tubes should be adequate for a breeding-cage of the size recommended. The holes should be cut in the false floor at the front of the cage, and should be 3·8 cm in diameter, just sufficiently large to accommodate the open tops of the tubes, so that when filled with sand and inserted in position they are flush with the upper surface of the false floor. A 25 W electric lamp is fixed at the rear of the top of the cage for illumination.

Rearing-cages

The rearing cage (Fig. 48.2) is basically a large metal box 60 cm high by 60 cm wide by 50 cm deep, with a completely detachable lid and having one whole side made of stout perforated zinc.

The front of the cage is provided with a large fixed window on one side; in the other is a circular hole 11 cm in diameter for introducing food, catching up hoppers etc. This is covered by a sliding Perspex flap. Illumination is provided by an electric lamp fitted in the top of the cage. If additional heating is required, extra lamps can easily be fitted. It is very important to allow the growing insects to perch if they are to moult properly, and so wire netting or dry branched twigs should be placed inside the cage. Lack of adequate perching material leads to crippled insects and unnecessary deaths.

Temperature and humidity

For breeding and rearing most species the temperature should range from approximately 28°C at night to 34°C during the day, the length of day being at least seven hours. The temperature of the cage should be read with the bulb of the thermometer about 10 cm above the false floor and about 10 cm from the sides of the cage (Hunter-Jones, 1961). For a cage equipped with electric-lamp heating and standing in a room at normal temperature this would probably require two 25 W bulbs below the false floor and one 60 W bulb in the top of the cage.

Little humidity-control is required. An average of R.H. 45 has given extremely good results, the grass fed during the day providing all the

Fig. 48.2. Rearing-cage. Top shelf, jars containing egg pods; middle shelf, breeding-cages with adult locusts; lower shelf, rearing-cages with hoppers. (Photograph: Zoological Society of London.)

humidity necessary. A very high humidity should be avoided as it tends to encourage the growth of moulds and other pathogens. When the humidity is unduly high the faeces will be wet and sloppy and condensation will appear inside the cage. This condition should be immediately corrected. Excess humidity can be caused by the presence of too much grass in the cage.

Hygiene

It is of the utmost importance that scrupulous cleanliness be observed at all times. All traces of uneaten food, dead locusts, and faeces should be removed daily. At monthly intervals, or when the cages become empty, they should be thoroughly scrubbed out with hot water and an efficient detergent disinfectant. This should be followed by a thorough rinsing in clean cold water.

Breeding

Egg-laying

In their natural habitat locusts deposit their egg pods in the soil. In the four common species these pods are rod-shaped and measure approximately $100 \times 8 \cdot 5$ mm. The lower part of the pod contains 30 to 100 eggs, cemented together with a frothy substance which solidifies shortly after being laid. This substance extends to the surface of the soil and forms the upper half of the pod.

In the laboratory, locusts will lay eggs in containers of well-packed coarse moist sand. Sharp builders' sand is suitable, providing it is washed to remove dust and then sterilized by heat. The sand should be mixed with 15 per cent of water and packed firmly into the oviposition tubes, which are inserted in position in the base of the breeding-cage. The number of pairs of adults required per cage is not critical but 30 to 40 pairs are recommended. The sand tubes should be removed every one or two days and replaced by freshly packed ones, whether egg pods have been deposited or not. It is easy to determine the presence of any pods by gently tapping the top of the tube and pouring off the top surface of the sand. The frothy plugs at the tops of the pods will then be exposed to view. The tubes are then covered with a well-fitting metal cap and are put into a 1 kg screw-topped jar, which is dated and put into an incubator set at 32°C. An incubator is not essential and any warm place will do. The number of pods laid per tube governs to some extent the number of hatchings expected per pod. If more than one or two pods are laid in a tube then the number of hatchings per pod will be reduced. Six or more pods will probably result in no hatchings at all. If too many pods are laid in each tube then the number of locusts should be reduced or the number of tubes increased. At from 28 to 32°C the eggs take from 11 to 25 days to hatch, depending on the species. After the eggs have been placed in the incubator they should not be disturbed until just before their expected hatching. The uncovered tubes can then be put into a new empty rearing cage, so that when the youngsters hatch they will be free to move about and feed; or they can be left in the incubator to hatch in the jars and be dealt with subsequently.

Feeding

For breeding and rearing, most locust species only need a daily supply of fresh grass and dry wheat bran. The grass should be cut into longish lengths and made up into bunches which are held in water-filled pots inside the cage. The actual daily requirement has to be learned by experience. If insufficient is given the locusts tend to cannibalize, and if there is too much the cage will need to be cleaned more often. Hunter-Jones (1961) recommends that a shallow tray filled with water should be given to adults, but this is not necessary if the grass is fed in water-filled pots. The bran is fed in shallow trays which

must not be allowed to become dirty or mouldy. Normally, fresh grass should be given every day, but it can be omitted on one day a week, provided that bran is always obtainable.

It is sometimes difficult to obtain green grass at certain times of the year. If supplies fail an artificial diet may be used, consisting of a dry mixture of bran, dried milk, dried grass and dried yeast in the ratio by weight of $1:1:1:0.1$ (Hunter-Jones, 1961). Dried chopped grass can be obtained commercially. With this diet the locusts must have access to water. As an alternative, some establishments use wheat grown hydroponically, which the locusts will accept as a substitute.

Diseases

From time to time, locusts cultured in the laboratory are troubled with a bacterial disease. Whereas *Locusta* is rarely affected, and for this reason is the most useful species to work with, *Schistocerca* and *Nomadacris* are much more susceptible. The symptoms are loss of appetite and an increase in mortality, particularly among hoppers in the fifth instar and young adults which have recently moulted. The body becomes bright pink and its contents almost fluid. The insects have an offensive odour and tend to break up on being touched. It has been suggested that superficial injuries may assist the onset of this disease, and for this reason Hunter-Jones (1961) recommends that locusts should not be handled when moulting and that there should be plenty of sticks and moulting perches in the cages in order to prevent unnecessary injuries at this time.

The disease is unlikely to occur if the cultures are kept at the correct temperature and humidity, are properly fed, and are kept scrupulously clean. It is possible that the bacteria are always present and only become really troublesome when the locusts are in poor health owing to some other cause. Any infected cages should be cleared and sterilized.

Occasionally the eggs of the parasitic nematode *Mermis* are picked up on the grass used as food. After being eaten by the locust, the young nematode hatches and starts to burrow through the body of the insect until it reaches the body cavity. There it may grow to a length of several centimetres. The remedy is to change the source of the grass used as food.

REFERENCES

Albrecht, F. O. (1953). *The Anatomy of the Migratory Locust*. London: Athlone Press.

Hunter-Jones, P. (1956). *Instructions for Rearing and Breeding of Locusts in the Laboratory*. London: Anti-Locust Research Centre.

Hunter-Jones, P. (1961). *Rearing and Breeding Locusts in the Laboratory*. London: Anti-Locust Research Centre.

McFerran, F. (1957). The breeding of *Locusta migratoria*. In *UFAW Handbook on the Care and Management of Laboratory Animals*, 2nd ed. London: UFAW.

Moriarty, F. (1969). The laboratory breeding and embryonic development of *Chorthippus brunneus* Thunberg. (Orthoptera: Acrididae). *Proc. R. ent. Soc. Lond.* (A), **44,** 25–34.

Stevenson, J. P. (1954). An epizootic among laboratory stocks of the desert locust, *Schistocerca gregaria*. Forsk. *Nature, Lond.,* **174,** 222.

49 The House Cricket

G. J. Ashby

The house cricket (*Acheta domesticus*) is a useful laboratory insect which is very simple to culture. The literature on its biology is not extensive, but reference may be made to the work of Cappe de Baillon (1920) on growth rates and development; Kemper (1937) on feeding habits, oviposition and general development; Bolduirev (1946) on the influence of temperature; and Busvine (1955) on the methods he used in maintaining a colony over a period of several years, together with data on the minimum periods of incubation and development in *Acheta* (synonym *Gryllulus*).

Besides being bred and reared in a number of laboratories *A. domesticus* is also cultured on a large scale in zoological gardens, e.g. London and Antwerp, where it forms part of the diet of certain birds and other animals. To prevent unwanted populations starting up in heated houses, the insects are killed before being fed to the animals.

Obtaining stocks

Stocks can be obtained either from laboratories dealing with their culture, or by trapping during the cooler months of the year in the vicinity of refuse dumps. Peterson (1959) has described a simple trap that is easy to construct. It consists of a glass or plastic jar to which is fitted a conical entrance of wire gauze. The diameter of the mouth of the cone is the same as that of the jar. The other end of the gauze cone is approximately 6 mm in diameter, and this is fitted with a flanged collar. The trap is baited with small pieces of lettuce or fruit and placed among debris where crickets are known to abound.

Housing

Crickets can be kept and bred in uncovered containers such as large glass battery jars or ordinary aquaria with clean glass walls. The height of the container should be not less than 23 cm. The glass walls must be kept clean to prevent the young from escaping. The bottom or floor should be covered with 1 cm of clean sawdust or sand, and folded cardboard or crumpled paper towels should be placed in the jar to provide shelter.

Temperature

Crickets will do well at temperatures of 25 to 35°C; they take about 5 to 6 weeks to reach the adult stage at the latter figure.

Food

The main diet consists of a mixture of biscuit meal, dried milk, Bemax, and dried yeast in the ratio of 50:5:40:5. This is fed in shallow trays placed inside the cage. Fresh lettuce, cabbage or carrot should be provided once a week.

Water

There must be a constant supply of water, lack of which leads to cannibalism. It can be provided by means of a cotton wool pad in a Petri dish with an inverted tube filled with water standing on the pad.

Eggs

The eggs are laid on the damp cotton wool round the edge of the tube. The pads are periodically removed and placed in small containers for the eggs to hatch, so that the age of the eggs can readily be ascertained. In the Locust Breeding Laboratories at the London Zoo which are kept at an average temperature of 32°C the eggs hatch in eleven days.

REFERENCES

Bolduirev, V. F. (1946). Orthopterous insects injurious under greenhouse conditions and their control. (In Russian.) *Dokl. sel'sk Acad. Timiryazeva*, **3**, 88–91. Abst. in *Rev. appl. Ent.*, (A) **35**, 346, 1947.

Busvine, J. R. (1955). Simple methods of rearing the Cricket, *Gryllulus domesticus*, with some observations on speed of development at different temperatures. *Proc. R. ent. Soc. Lond.,* (A) **30,** 15–18.

Cappe de Baillon, P. (1920). Contributions anatomique et physiologiques à l'étude de la reproduction chez les locustriens et les grilloniens. *Cellule,* **32,** 1–193.

Kemper, H. (1937). Beobachtungen über die Biologie der Hausgrille *Gryllus domesticus. Z. hyg. Zool.,* **29,** 69–86.

Peterson, A. (1959). *Entomological techniques.* Columbus, Ohio: Ohio State University.

50 Beetles

J. H. COLE

The Coleoptera or beetles are widely used as laboratory insects for a variety of purposes, and at least 90 species are routinely bred in British laboratories. The majority of these are pests of stored foods, but a few are important pests of textiles and timber and one, *Phaedon cochleariae* F., is a pest of cruciferous crops. Some of the more important laboratory uses of beetles are to provide food for other animals, for nutrition, life history and genetical studies, and for the evaluation of insecticides and other pest control measures.

It is impossible to deal in any detail here with the biology and culture of a large number of species but many of the requirements and techniques are common to a number of the pests of stored-products listed in Table I. The following widely-used species have been selected for individual treatment:

Mealworms, *Tenebrio molitor* L., and *T. obscurus* F. (Tenebrionidae).

Flour beetles, *Tribolium confusum* J. du V. and *T. castaneum* Herbst (Tenebrionidae).

Grain Weevils, *Sitophilus* spp. (Circulionidae).

Black carpet beetle, *Attagenus megatoma* F. (= *piceus* Oliv.) (Dermestidae).

Mustard beetle, *Phaedon cochleariae* F. (Chrysomelidae).

The following notes are based on experience, but techniques vary among laboratories and a fair amount of latitude in temperature, diet and handling methods is possible with many of the species concerned.

General requirements

Accommodation

Basic requirements are a well-insulated, preferably windowless, room with thermostatically controlled heating, racks or shelves for culture jars, a water supply with sink, and a work bench. Small-scale breeding may be carried out in an insulated cabinet or cupboard, but if large-scale breeding is undertaken a series of small rooms is preferable to one large room as it allows segregation of cultures in case of disease or chance chemical contamination and a range of temperatures if required. The rooms should be easily cleanable with all cracks and crevices eliminated and all surfaces washable.

Temperature control

A complete air-conditioning unit may be installed provided it is properly designed, but excellent results can be obtained in small insectaries with the simplest equipment. We have had trouble-free temperature control for many years using simple, thermostatically-controlled, asbestos heating pads, but it is essential to provide a safety device to guard against overheating. An electrical failure will cause no great harm as, provided the insulation is adequate, the temperature will fall slowly; but should a thermostat stick in the closed position the temperature will rise rapidly, killing many cultures, the critical temperature being 40–50°C, depending on the duration. To guard against this occurrence two thermostats should be fitted in series, one set to operate at a desired room temperature and the other 5°C higher. While the thermostat set at the lower temperature is working correctly the reserve instrument will be permanently closed, but if the former fails the temperature will rise only 5°C before the other thermostat operates and no damage will be done.

The temperature range within which breeding can be maintained is generally fairly wide for any given species. Within limits the higher the temperature the shorter the life cycle, and for general purposes the shortest life cycle consistent with healthy development is indicated. The majority of stored food and textile pests will breed at about 24°C or 25°C but some will require either higher or lower temperatures for optimum development (see Table I). The temperature variation should be reduced to ±0·5°C when rearing insects for critical experimental work, and a small mixing fan may be

TABLE I

Pests of stored products bred in British laboratories

Species	Common name	Food see Table II	Laboratory*		Temperature °C
ANOBIIDAE					
Lasioderma serricorne (F.)	cigarette beetle	5	S	H	25
Stegobium paniceum (L.)	biscuit beetle	5	S	H	25
BOSTRYCHIDAE					
Rhyzopertha dominica (F.)	lesser grain borer	1	S		30
BRUCHIDAE					
Acanthoscelides obtectus (Say)	American seed beetle	17	S		30
Callosobruchus chinensis (L.)	cowpea weevil	19	S		25
Callosobruchus maculatus (F.)		19	S		30
Caryedon serratus (Oliv.)	groundnut beetle	16 W	S		30
Zabrotes subfasciatus (Bok.)		18	S		30
COLYDIIDAE					
Murmidius ovalis Beck.		10	S		25
CLERIDAE					
Necrobia rufipes (Deg.)	copra beetle	15	S		30
CUCUJIDAE					
Cryptolestes ferrugineus (Steph.)	Rust-red grain beetle	7	S		30
Cryptolestes pusillus (Schon.)	Flat grain beetle	7	S	H	25
DERMESTIDAE					
Anthrenocerus australis Hope	Australian carpet beetle	14	S	H	25
Anthrenus flavipes Le Conte	furniture carpet beetle	14	S	H	30
Anthrenus verbasci (L.)	Varied carpet beetle	14	S	H	27
Attagenus pellio (L.)	fur beetle	14	S		20
Dermestes haemorrhoidalis (Kuster & Praze)		13 W	S	H	25
Dermestes lardarius L.	bacon beetle	13 W	S		25
Dermestes maculatus Deg.	leather beetle	13 W	S	H	25
Dermestes peruvianus Castelnau		13 W	S	H	25
Trogoderma granarium Everts	khapra beetle	2	S	H	30
MYCETOPHAGIDAE					
Typhaea stercorea (L.)	hairy grain beetle	3	S		25
NITIDULIDAE					
Carpophilus hemipterus (L.)	dried fruit beetle	11	S		25
OSTOMATIDAE					
Lophocateres pusillus (Klug.)	Siamese grain beetle	8	S		30
Tenebroides mauretanicus (L.)	cadelle	9	S		30
PTINIDAE					
Gibbium psylloides (Czenp.)	humped spider beetle	12 W	S		20
Mezium affine Boield.		12 W	S		20
Niptus hololeucus (Fald.)	golden spider beetle	12 W	S	H	25
Pseudeurostus hilleri (Reitt.)		12 W	S		20
Ptinus pusillus Sturm		12 W	S		20
Ptinus tectus Boield.	Australian spider beetle	12 W	S	H	25
Stethomezium squamosum Hint.	African spider beetle	12 W	S		20
Tipnus unicolor (P. & M.)		12 W	S		20
Trigonogenius globulus Sol.	globular spider beetle	12 W	S	H	25
SILVANIDAE					
Ahasuerus advena (Waltl.)	foreign grain beetle	7	S		25
Cathartus quadricollis (Guer.)	square-necked grain beetle	7	S		25
Oryzaephilus mercator (Fauv.)	merchant grain beetle	7	S		25
Oryzaephilus surinamensis (L.)	saw-toothed beetle	7	S	H	25
TENEBRIONIDAE					
Alphitobius diaperinus (Panz.)	lesser mealworm	6	S	H	25
Alphitophagus bifasciatus (Say)	two-banded fungus beetle	4	S		25
Gnathocerus cornutus (F.)	broad-horned flour beetle	12	S		25
Latheticus oryzae Waterh.	long-headed flour beetle	5	S		30
Palorus ratzeburgii (Wissm.)	small-eyed flour beetle	5	S		25
Palorus subdepressus (Woll.)	depressed flour beetle	6	S		25
Tribolium destructor Uytt.	dark flour beetle	12	S		25
LANGURIIDAE					
Pharaxonotha kirschi (Reitt.)		5 W			25

*S = Pest Infestation Control Laboratory, Slough
H = Huntingdon Research Centre, Huntingdon.

TABLE II

Key to media in Table I

Number	Food	Weight ratio
1	wheat	
2	wheat and wheatfeed	7:3
3	wheat + wheatfeed + glycerol on a damp cotton wool pad	7:3:1
4	wheat + wheatfeed on a damp pad	7:3
5	wheatfeed + yeast	10:1
6	wheatfeed + yeast on a damp pad	10;1
7	wheatfeed + rolled oats + yeast	5:5:1
8	No. 7 + groundnuts	10:1
9	No. 8 + cork	
10	wheatfeed + rolled oats on a damp pad	2:1
11	rolled oats + yeast + dates	6:1:6
12	wheatfeed + fishmeal + yeast	8:4:1
13	fishmeal + yeast	16:1
14	fishmeal + yeast + flannel	
15	No. 13 + bacon ends + cheese	
16	ground nuts	
17	haricot beans	
18	butter beans	
19	cowpeas + dried green peas	1:1
W	drinking water added	

required to prevent temperature gradients from occurring in the insectary.

Humidity control

It is generally not necessary to control the relative humidity within narrow limits as a microclimate is created within each breeding jar which is surprisingly little affected by the room humidity, unless this is extreme. We have found 50–65 per cent R.H. to be satisfactory and easy to maintain; a humidity of above 75 per cent R.H. should be avoided as it would encourage rapid growth of mould and mites.

Ventilation

Only minimal ventilation is normally required, but an exhaust fan fitted at one end of the room with inlet holes sited at the opposite end will be found useful.

Lighting

Fixed lighting regimes are not required, except for species reared on living plants (see Mustard Beetle, p. 601).

Equipment and general techniques

The only essential apparatus is a thermo-hygrograph and an oven. For most species wide-mouthed glass jars of about 2 litre capacity are used. The jars are covered with cotton cloth held by elastic bands. Where indicated, water is provided in 5 cm diameter glass bulbs drawn out into a 2·4 cm neck with a 6 mm diameter opening. A cotton wool wick is provided and the bulb is sunk in the food with the neck above the surface. Care should be taken that the wet wick does not touch the food. The outside of the neck is roughened so that species unable to climb smooth vertical surfaces can reach the water. Other equipment needed are oil trays (see Hygiene below) and various dishes, tubes, trays, a set of graded sieves, forceps, etc. for general purposes.

New cultures are usually started with young adult beetles at regular intervals of days or weeks depending on the numbers required, so that a series of jars with insects in different stages of development are always available. If larvae of known age are required for experimental purposes, the adult beetles are removed from the jar by sieving a few days after starting the culture. In cultures of mixed stages, larvae of different sizes may be at least partially separated from the food by passing through a set of graded sieves. Ten meshes to the inch decreasing in steps of ten down to sixty meshes to the inch is a suitable range. The larvae and adults of many beetle species can be separated from the food by spreading the mixture thinly on a sheet of rough paper, leaving for a few seconds, then tilting the sheet so that the food slides off leaving the insects clinging to the paper. Individuals may be handled with forceps but it is usually quicker and safer to manipulate them singly or in numbers with a No. 4 or 5 water colour paint brush by sweeping them into specimen tubes or jars.

Hygiene

Strict hygiene is of the greatest importance. It is best to start with established laboratory strains which are known to be disease-free. It is unwise to attempt to start cultures from wild sources, as parasites and mites are likely to be introduced.

Mites and other insects will frequently be present in cereals and milled products from provender mills and it is essential to sterilize all foodstuffs before starting new culture jars. An hour at 60°C is sufficient if the food is spread thinly on trays, but 3–4 hours will be required to penetrate the centre of food in a half-filled breeding jar.

All culture jars of sterilized food should be clearly separated from each other in trays of oil. Liquid paraffin or a cheap grade of engine oil about 5 mm deep is suitable. This is partly to prevent escape of the insects cultured and partly to prevent the entry of mites. In spite of precautions culture jars do occasionally become infested with scavenger mites (mainly tyroglyphids), and unless the insects are particularly required it is best to destroy the culture. If desired, beetles and many larvae may be freed of mites and food particles by placing them in a sieve of suitable mesh and washing under running water at ambient temperatures. Parasitic mites will not be removed in this way.

Mealworms—*Tenebrio* spp

Tenebrio molitor L. and *T. obscurus* F., the yellow and dark mealworms respectively, are very similar and may be taken together, but more detailed study has been devoted to the former. Both species are virtually cosmopolitan scavengers commonly found in spillage and general debris in provender mills and warehouses where hygiene is poor.

T. molitor in particular is bred in large quantities for sale in pet shops as a food for insectivorous mammals, birds, reptiles, amphibians and fish and is widely used in laboratories for the same purpose. Both species have been reared in large numbers in Germany and Czechoslovakia as a winter food for fowl (Stampfel, 1944) and *T. molitor* has been used by several workers in insect nutrition studies.

Breeding

Fraenkel *et al.* (1950) list seven component vitamins of the B complex required by *T. molitor* in addition to cholesterol and 80–85 per cent carbohydrate. The rate of growth depends very much on the water-content of the food, growth being fastest on food in equilibrium with 70 per cent R.H. and very slow at 30 per cent R.H. They conclude '. . . (it) grows well on a powdery medium with a water content of only 10 per cent or even less and . . . can be conveniently bred all the year round in the laboratory and eggs and larvae can be obtained with the greatest ease at the required time and in the required number'. We are in agreement with this and have found no

difficulty in breeding the species but McFerran (1957) states that it 'does sometimes prove to be impossible (to breed them) on a small scale and impracticable in large quantities', and we have had requests for assistance from persons who have found trouble in rearing them.

Cotton & St. George (1929) have published a detailed study of the life history of the mealworms and directions for rearing them are given by McFerran (1957), Le Ray & Ford (1937) and Mann (1937). These writers advocate bulk cultures in boxes measuring about $60 \times 40 \times 30$ cm except Mann (1937), who preferred shallow trays. When larvae of know age are required for test purposes, breeding in smaller units is essential. We combine both techniques. Larvae for experimental purposes are reared in 2 litre glass jars about two thirds filled with 95 per cent wheatfeed (toppings) and 5 per cent debittered dried yeast. Each jar is started with 30–40 adults and the life cycle is completed in about 6 months at 21°C and 4 months at 26·7°C. According to Howe & Burges (1953) it is doubtful if *T. molitor* can complete development at temperatures continuously over 30°C. Only one generation is completed in each jar; the contents are then tipped into a bulk culture box $60 \times 50 \times 50$ cm. This has a glass top and a strip of wire gauze let into the top of one side to provide ventilation.

A number of workers (Stampfel, 1944; McFerran, 1957; I.W.T.O. Technical Committee, 1964) have found it necessary to add root vegetables and lettuce, and stress the importance of keeping the culture moist to prevent cannibalism. Fraenkel *et al.* (1950) have shown, and our experience confirms, that this is not necessary. Other references to the food requirements and breeding of *Tenebrio* are Arendsen-Hein (1923), Lafon & Teissier (1939), Martin & Hare (1942), Fraenkel & Leclercq (1956).

Flour beetles—*Tribolium* spp

Tribolium confusum J. du V., the confused flour beetle, and *T. castaneum* Herbst, the rust-red flour beetle, are important cosmopolitan pests of flour and a wide range of dry food products, and although unable to attack whole grain they frequently occur with the grain beetles *Sitophilus* and *Rhyzopertha* and attack grain which has already been damaged. Flour beetles are extensively used in nutrition, genetic and population studies and for the evaluation of insecticides. A number of workers have made intensive studies of the bionomics of *Tribolium*, among them Good (1933 and 1936), Park & Frank (1948), Cotton (1963 and many other publications) and Howe (1956 and 1960).

Breeding

The specific nutritional requirements of *Tribolium* have been investigated among others by Schneider (1943), Loschiavo (1952), Frobrich (1953), Fraenkel & Printy (1954), Chirigos (1957), Charbonneau & Lemonde (1960), Chirigos *et al.* (1960) and Naylor (1964). However, a simple medium of 95 per cent wholemeal flour and 5 per cent debittered dried yeast provides all requirements. Cultures are started with 100 adults in a 2 litre jar about one third full of flour and the life cycle is completed in about 8 weeks at 21°C and 42 days at 27°C. Park (1937) gives a programme for maintaining stock cultures and a diagram of an automatic flour sifter. Other information on breeding and handling is given by Harein & Soderstrom (1966).

Grain weevils—*Sitophilus* spp

The three cosmopolitan species (*Sitophilus granarius* L., *S. oryzae* L. and *S. zeamais* Motsch., respectively known as the grain, rice and maize weevils, are major pests of harvested grains. They are primarily bred in the laboratory for research into control measures. *S. zeamais* was separated from *S. oryzae* as a distinct species only in 1961 and was previously treated as a larger race of the latter.

The bionomics of these weevils have been studied among others by Dendy & Elkington (1920), Cotton (1920), Back & Cotton (1924 and 1926) and Reddy (1950).

Breeding

The females bite small holes in cereal grains and lay a single egg in each grain. The whole larval and pupal development takes place within the grain until the emerging adult bites its way out. Breeding these species in large numbers is a simple process; all three will develop satisfactorily on almost any cereal but *S. granarius* prefers wheat, and the other two rice or maize. One hundred adults are placed in a 2 litre jar one third filled with grain; yeast and water are not required. The development period from egg-laying to adult emergence at 24°C is about 5–6 weeks for *S. oryzae* and *S. zeamais* and 6–7 weeks for *S. granarius*. Further information will be found in Harein & Soderstrom (1966).

The black carpet-beetle— *Attagenus megatoma*

All the beetles which are known to be able to digest keratin, with the possible exception of one or two species of Ptinidae, belong to the family Dermestidae, of which the two genera *Attagenus* and *Anthrenus* provide several major pests of woollen textiles and furs throughout the world. *Attagenus megatoma* F. (= *piceus* Oliv.) is the most widespread and destructive carpet beetle in the USA (Back & Cotton, 1938) and in Japan, but is less important in Europe. The larvae also attack seeds, grains and all types of dry animal products; in Japan it is an important pest in silkworm production, where it feeds on the pupae and raw silk of the cocoons. Detailed studies of the life history of *A. megatoma* have been made by Back & Cotton (1938) and Griswold (1941), while Hinton (1945) gives a comprehensive review of the literature.

Breeding

The International Wool Textile Organization (I.W.T.O. Technical Committee, 1964) and the Chemical Specialties Manufacturers Association (Soap & Chemical Specialties, 1969) have published detailed specifications for breeding this species which, although a textile pest, can be bred in large numbers more readily and cheaply on a non-keratinous diet. The food we have used with complete success for many years is a mixture of equal parts by volume of bran, rolled oats and fishmeal with dried yeast to make up 5 per cent of the total. Two hundred adult beetles are placed in a 2 litre jar one third filled with this mixture and at 24°C and 27°C the life cycle is completed in about 8–10 months and 5–6 months respectively.

The mustard beetle—*Phaedon cochleariae*

This is the only beetle of agricultural importance which is bred in British laboratories on anything other than an experimental scale. Both larvae and adults feed on the aerial parts of cruciferous plants and are a pest of mustard, turnips, swedes and brassicas.

Breeding

Breeding may be carried out in a glasshouse or plant growth room at 18°C with not less than 16 hours of daylight per day, made up by suitable artificial lighting when necessary. Cages for adults must be large enough to hold several potted plants and may consist of a metal or wooden frame with a solid base and covered with fine-mesh nylon net. A 50 cm cube is a convenient size. We have found radish plants convenient to use for adults and larvae but kale, turnip or chinese cabbage may be used. When eggs are required old plants are removed from the adults' cages and replaced by two fresh plants per 100 adults. Eggs are laid on the plants, which must be removed from the cages before the beetles have attacked them too severely. Two days should be enough to obtain 200–300 eggs. The eggs desiccate easily during the 7–8 days before hatching and this may be prevented by covering the plants loosely with a plastic bag or putting them on moist sand or peat in small cages. A few days after the eggs hatch the plants on which the young larvae are feeding must be broken off and placed on a layer of peat in containers such as plastic bread-bins covered with nylon net. Fresh detached cabbage leaves must be provided daily. Alternatively larvae can be reared to pupation on growing plants in the same type of cage as that used for adults, with peat provided for pupation.

The larval stage lasts for about 2 weeks, after which the pupae may be collected from the peat and transferred to adult cages for emergence. The adults live for 2 or 3 months.

Other beetles

Other species of beetles currently being bred in British laboratories are shown in Table I, which has been compiled largely from a list kindly supplied by the Pest Infestation Control Laboratory, Slough (Agricultural Research Council). Representatives of all genera are included, but as space does not permit a complete list, a number of less well-known species have been left out. The culture media code is given in Table II.

The Department of the Environment's Forest Products Research Laboratory, Princes Risborough, breeds three wood-boring beetles; the powderpost beetle *Lyctus brunneus* Steph., the common furniture beetle *Anobium punctatum* Deg., and the house longhorn beetle *Hylotrupes bajulus* L., but only the first of these is cultured on a large scale (Harris & Taylor, 1960). In experimental *Anobium* cultures the life cycle has been reduced from up to three years to just over a year, but the particular combination of conditions which induces mature larvae to pupate is not yet known (J. M. Baker, private communication). Techniques for culturing *Hylotrupes* have been published by Saraiva (1957) and Technau & Behrenz (1958).

Except *Phaedon*, the mustard beetle, all the foregoing are to some degree pests associated with stored foods, textiles or timber. However, a number of other species of importance to agriculture and forestry are bred in laboratories in Europe and America. Some of these (including the cotton boll weevil, *Anthonomus*, and the corn root worm *Diabrotica*) are described in *Insect Colonisation and Mass Production* (edited by Smith, 1966), and notes on many species of general interest from 18 families will be found in *Culture Methods for Invertebrate Animals*, edited by a committee of the American Association for the Advancement of Science (1937). Hurpin (1959) and Newton (1958) describe respectively the breeding of the cockchafer *Melolontha melolontha* L. and the bean weevil *Sitona hispidula*, both of which are found in Britain. The maintenance of dytiscid water beetles has been described by Hodgson (1953) and other water beetles are mentioned in *Culture Methods for Invertebrate Animals* (A.A.A.S., 1937).

Parasites

Insects

A number of Diptera and Hymenoptera parasitise beetles, but few, if any, are likely to be troublesome in laboratory cultures—although

Back (1940) notes that *Laelius voracis* Muesbeck (Hymenoptera Bethylidae) sometimes completely destroys cultures of the carpet beetle *Anthrenus flavipes*.

Mites

Among the various mites which will infest culture media if precautions are relaxed are a few species of Parasitiformes, such as *Melichares* (*Blattisocius*) *tarsalis* (Berlese), which will attack beetle eggs and young larvae. Two species of true parasitic mites are important. *Acarophenax tribolii* Newstead & Duval, a trombidiform mite, is found attached to species of *Tribolium* where the cuticle is soft, frequently under the wings. *Pyemotes ventricosus* Newport, another trombidiform mite, will rapidly destroy cultures of powder-post beetles (*Lyctus* spp.); preventive measures are described by Harris & Taylor (1960).

Protozoa

The most troublesome parasites of laboratory bred beetles are protozoans and there is now a considerable literature on the coccidians and gregarines found in various species of which *Tribolium* seems to be particularly susceptible. This, however, may be so because *Tribolium* species are the most widely-bred and studied beetles.

Harein & Soderstrom (1966) state that the coccidian *Adelina tribolii* frequently invades flour-beetle cultures, causing adults and larvae to become lethargic and unresponsive to stimuli, and the microsporidians *Nosema whitei* and *N. buckleyi* have been recorded from *Tribolium castaneum* by Weiser (1953) and Dissanaike (1955) respectively; but most trouble is caused by various gregarines.

Dr. E. V. Canning of the Imperial College of Science and Technology (private communication) notes that *Tenebrio molitor* harbours four species of Gregarina which '. . . are considered relatively harmless, but it seems unlikely that in large numbers they would cause no damage'. Dr. Canning has found a *Mattesia* sp. in *Trogoderma granarium* and Finlayson (1950) reports that *Mattesia dispora* caused heavy mortality in *Cryptolestes* cultures.

Various authors have found *Farinocystis tribolii* and *Triboliocystis garnhami* in *Tribolium* cultures including Dissanaike (1955), Marshall

Laird (1959) and Tyler (1962). Infestation of *Tribolium castaneum* cultures by *Triboliocystis* has caused serious trouble in the Pest Infestation Laboratory, Slough (Kane & Mahon, 1961) and at the Cooper Technical Bureau (Holborn, 1957). The latter writes (private communication) 'without proof I suspect that infestation is associated with nutritional deficiency, overcrowding and excessive handling of insects. Sickly cultures when allowed freedom under semi-field conditions in our tropical storage hut appear to regain their vigour and breed freely'. Techniques for ridding *Tribolium* cultures of *T. garnhami* are described by Park (1948), Sokoloff (1962) and Stanley (1961 and 1964).

REFERENCES

American Association for the Advancement of Science (1937). *Culture Methods for Invertebrate Animals.* Ithaca, N.Y.: Comstock.

Arendsen-Hein, S. A. (1923). Larvenarten von der Gattung *Tenebrio* und ihre Kultur. *Ent. Mitt.*, **12,** 121.

Back, E. A. (1940). A new parasite of *Anthrenus vorax* Waterhouse. *Proc. ent. Soc. Wash.*, **42,** 110–13.

Back, E. A. & Cotton, R. T. (1924). Relative resistance of the rice weevil *Sitophilus oryzae* L. and the granary weevil *Sitophilus granarius* L. to high and low temperatures. *J. Agric. Res.*, **28,** 1043–4.

Back, E. A. & Cotton, R. T. (1926). The granary weevil. *U.S. Dept. Agric. Bull. No. 1393.*

Back, E. A. & Cotton, R. T. (1938). The black carpet beetle *Attagenus piceus* (Oliv.). *J. econ. Ent.*, **31,** 280–6.

Charbonneau, R. & Lemonde, A. (1960). Unidentified growth factors in brewers yeast, necessity of these factors for *Tribolium confusum* Duval. *Can. J. Zool.*, **38,** 87–90.

Chirigos, M. A. (1957). Nutritional studies with the insect *Tribolium confusum* Duval. Doctoral thesis, Rutgers State University, New Jersey.

Chirigos, M. A., Meiss, A. N., Pisano, J. J. & Taylor, M. W. (1960). Growth response of the confused flour beetle, *Tribolium confusum* Duval, to six selected protein sources. *J. Nutr.*, **72,** 121–30.

Cotton, R. T. (1920). The rice weevil (*Calandra*) *Sitophilus oryzae. J. agric. Res.*, **20,** 409–22.

Cotton, R. T. (1963). *Pests of Stored Grain and Grain Products.* Minneapolis: Burgess Publ. Co.

Cotton, R. T. & St. George, R. A. (1929). The mealworms. *U.S. Dept. Agric. Tech. Bull. No. 95.*

Dendy, A. & Elkington, H. D. (1920). Report on the vitality and rate of multiplication of certain grain insects under various conditions of temperature and moisture. *R.S. Grain Comm. Report No. 7. Roy. Soc. Lond.*

Dissanaike, A. S. (1955). A new schizogregarine, *Triboliocystis garnhami* n.g., n. sp. and a new microsporidian *Nosema buckleyi* n. sp. from the fat body of the flour beetle, *Tribolium castaneum. J. Protozool.*, **2**, 150–6.

Finlayson, L. H. (1950). Mortality of *Laemophloeus* (Coleoptera, Cucujidae) infested with *Mattesia dispora* Naville (Protozoa, Schizogregarinaria). *Parasitology*, **40**, 261–6.

Fraenkel, G., Blewett, M. & Coles, M. (1950). The nutrition of the mealworm, *Tenebrio molitor* L. (Tenebrionidae, Coleoptera). *Physiol. Zoöl.*, **23**, 92–108.

Fraenkel, G. & Leclercq, J. (1956). Nouvelles recherches sur les besoins nutritifs de la larve du *Tenebrio molitor* L. (Insecta Coléoptère) *Archs int. Physiol.*, **64**, 601–22.

Fraenkel, G. & Printy, G. E. (1954). The amino acid requirement of the confused flour beetle, *Tribolium confusum* Duval. *Biol. Bull.*, **106**, 149–57.

Fröbrich, G. (1953). Die Caseindosierung in synthetischen Diäten für die Aufzucht von *Tribolium confusum* Duval (Tenebrionidae, Coleoptera). *Naturwissenschaften*, **40**, 556.i

Good, N. E. (1933). Biology of the flour beetle, *Tribolium confusum* Duval and *T. ferrugineus* Fab. *J. agric. Res.*, **46**, 1327–34.

Good, N. E. (1936). The flour beetles of the genus Tribolium. *U.S. Dept. Agric. Mech. Bull. No. 498.*

Griswold, G. H. (1941). Studies on the biology of four common carpet beetles. Part 1. The black carpet beetle (*Attagenus piceus* Oliv.), the varied carpet beetle (*Anthrenus verbasci* L.) and the furniture carpet beetle (*Anthrenus vorax* Waterhouse). *Mem. Cornell Univ. agric. Exp. Stn*, **240**, 3–57, 70–5.

Harein, P. K. & Soderstrom, E. L. (1966). Coleoptera infesting stored products. In *Insect Colonization and Mass Production.* New York & London: Academic Press.

Harris, E. C. & Taylor, J. M. (1960). Powder post beetles for test purposes. *Timb. Technol.*, **68**, 193–5.

Hinton, H. E. (1945). *A Monograph of the Beetles Associated with Stored Products.* London: British Museum (Nat. Hist.).

Hodgson, E. S. (1953). Collection and laboratory maintenance of Dytiscidae (Coleoptera). *Ent. News*, **64**, 36–7.

Holborn, J. M. (1957). The susceptibility to insecticides of laboratory cultures of an insect species. *J. Sci. Fd Agric.*, **8**, 182–8.

Howe, R. W. (1956). The effect of temperature and humidity on the rate of development and mortality of *Tribolium confusum* Duval (Coleoptera, Tene-

brionidae). *Ann. appl. Biol.*, **48**, 363–76.

Howe, R. W. (1960). The effect of temperature and humidity on the rate of development and mortality of *Tribolium confusum* Duval (Coleoptera, Tenebrionidae). *Ann. appl. Biol.*, **48**, 363–76.

Howe, R. W. & Burges, H. D. (1953). A note on the resistance of *Tenebrio molitor* L. (Coleoptera, Tenebrionidae) to tropical temperatures. *Entomologist's mon. Mag.*, **89**, 4–6.

Hurpin, B. (1959). Recherche sur l'alimentation des vers blancs ou larves de *Melolontha melolontha* L. (Coleoptera Scarabaeidae). *Annls Epiphyt.*, **11**, 35–80.

I.W.T.O. Technical Committee (1964). *Method of Test and Assessment for Proofness of Wool Fabrics Against the Black Carpet Beetle, Attagenus piceus (Oliv.).* London: International Wool Secretariat.

Kane, T. & Mahon, P. A. (1961). Sporozoan parasite of *Tribolium* spp. *Pest Infestation Research*, 1961. London: Agric. Res. Council.

Lafon, M. & Teissier, G. (1939). Sur les besoins nutritifs de la larve de *Tenebrio molitor* L. *C.r. Séanc. Soc. Biol.*, **131**, 75–7.

Le Ray, W. & Ford, N. (1937). Mealworms *Blapstinus moestus* and *Tenebrio molitor*. In *Culture Methods for Invertebrate Animals.* A.A.A.S. Ithaca, N.Y.: Comstock.

Loschiavo, S. R. (1952). A study of some food preferences of *Tribolium confusum* Duval. *Cereal Chem.*, **29**, 91–107.

McFerran, F. (1957). Use of Insects for Feeding. In *UFAW Handbook on the Care and Management of Laboratory Animals*, 2nd ed. Edinburgh: Livingstone.

Mann, W. M. (1937). *Tenebrio* culture. In *Culture Methods for Invertebrate Animals.* A.A.A.S. Ithaca, N.Y.: Comstock.

Marshall Laird (1959). Gregarines from laboratory colonies of flour beetles *Tribolium castaneum* Herbst. and *Tribolium confusum* Duval at Montreal. *Can. J. Zool.*, **37**, 378–81.

Martin, H. E. & Hare, L. (1942). The nutritional requirements of *Tenebrio molitor* L. *Biol. Bull.*, **83**, 428–37.

Naylor, A. F. (1964). Possible value of casein, gluten, egg albumin or fibrin as whole protein in the diet of two strains of the flour beetle *Tribolium confusum* (Tenebrionidae). *Can. J. Zool.*, **42**, 1–9.

Newton, R. C. (1958). Breeding *Sitona hispidula* larvae for research use. *J. econ. Ent.*, **51**, 917–8.

Park, T. (1937). The culture of *Tribolium confusum*. In *Culture Methods for Invertebrate Animals.* A.A.A.S. Ithaca, N.Y.: Comstock.

Park, T. (1948). Experimental studies of interspecies competition 1. Competition between populations of the flour beetles *Tribolium confusum* Duval. and *T. castaneum* Herbst. *Ecol. Monogr.*, **18**, 265–308.

Park, T. & Frank, M. B. (1948). The fecundity and development of the flour beetles *Tribolium confusum* and *Tribolium castaneum* at three constant temperatures. *Etiology Durham, N.C.*, **29**, 268–374.

Reddy, D. B. (1950). Ecological studies of the rice weevil. *J. econ. Ent.*, **43**, 203–6.

Saraiva, A. C. (1957). Antibioticos, vitaminos e outros factores no desenvoluimento do *Hylotrupes bajulus* (L). *Agronomia lusit.*, **19**, 161–218.

Schneider, B. A. (1943). The nutritional requirements of *T. confusum* II. The effect of vitamin B complex on metamorphosis, growth and adult vitality. *Am. J. Hyg.*, **37**, 170–92.

Smith, C. N. (1966). *Insect Colonisation and Mass Production*. London & New York: Academic Press.

Soap & Chemical Specialties. (1969). Textile Resistance Test, *Soap Chem. Spec.*, **45**, (No. 4A Blue Book), pp. 197–200.

Sokoloff, A. (1962). A simple technique for ridding

Tribolium cultures of parasites. *Tribolium Info. Bull.*, **5**, 48.

Stampfel, J. (1944). Pokusy o výživě putemnikú. (Experiments on the feeding of Tenebrionids). *Acta Soc. ent. Bohem.*, **41**, 4–12. (See *Rev. appl. Ent.* (A), **34**, 328.)

Stanley, J. (1961). Two techniques for use in the control of *Triboliocystis garnhami*. *Can. J. Zool.*, **39**, 121–2.

Stanley, J. (1964). Washing the eggs of *Tribolium* for gregarine control. *Can. J. Zool.*, **42**, 920.

Technau, G. & Behrenz, W. (1958). Erfahrung in der Zucht von *Hylotrupes bajulus* L. *Holz Roh-u Werkstoff*, **16**, 90–3.

Tyler, P. S. (1962). On an infection of *Tribolium* spp. by the sporozoan, *Triboliocystis garnhami* Dissanaike. *J. Insect. Path.*, **4**, 270–2.

Weiser, J. (1953). Schizogregariny a hmyzu skodiciho zasobam monky. *Věst. čsl. Spol. zool.*, **17**, 199–202.

51 Ticks

S. F. Barnett

Introduction

This contribution is intended to supplement the following reviews on the maintenance of tick colonies: Kohls (1959); Gregson (1966); Feldman-Muhsam (1967); and Hadani *et al*. (1969).

There are three main reasons for maintaining ticks in the laboratory:

1. to test their susceptibility to acaricides;
2. to study their biology and physiology;
3. to study them as disease vectors.

The first of these requires a continuous supply of large numbers of certain instars of a tick species of a standard age and of a known provenance, while the other two commonly require specialised knowledge of individual tick species and will not be described in detail here. Species commonly reared in the laboratory are either of importance in the husbandry of livestock or are hazardous to human health. All ticks require a meal of blood before egg-laying or moulting to the next instar, but the degree of engorgement and the success of the subsequent stages may be influenced by the species and the immune status of the host on which they feed. Ticks have host preferences which are sometimes very restricted, and that of the immature stages may be quite different from that of adults.

There are two families of ticks with marked differences in biology; the Argasidae or soft ticks, and the Ixodidae or hard ticks.

Argasidae copulate off the host, usually shortly after a blood meal. Eggs are laid in small clutches on the ground, and usually the female feeds again before laying another clutch. Copulation is not always essential for the production of subsequent clutches, but males should be present throughout. Not all larval stages feed, but those which do remain attached for 4 to 11 days. The nymphal stages, of which there are two to four, feed and detach within the hour, often within minutes, and the adults feed similarly. There are two genera, *Ornithodorus*, which are usually parasitic on mammals, and *Argas*, which are mainly bird parasites.

The Ixodidae have only three instars, larval, nymphal and adult, and each instar spends several days on a host before engorgement is complete. Some species may copulate before feeding, but commonly the males and females attach to the host and feed slightly for a few days; the male then seeks an attached female and mates. It will do this several times, and thus males can remain on the host for several weeks. Unmated females do not engorge properly, but mated females engorge in 7 to 14 days and then drop off to lay eggs on the ground. The eggs are laid in a large cluster over a period of 2 to 4 weeks, after which the female dies. The larval stages take 3 to 5 days to engorge, and their subsequent behaviour depends on whether they belong to one-host, two-host or three-host species. In the one-host tick the larva remains attached and moults on the animal to a nymph which re-attaches close by, engorges and moults to an adult, which again feeds on the same animal. After mating and engorging, the female drops off to lay eggs. All instars thus occur on the same host, the whole feeding cycle taking about three weeks for *Boophilus* species and *Dermacentor nitens*, but in the winter tick *D. albipictus* it may take several months unless given warmth and artificial daylight. In the two-host species, the larvae and nymphs feed on the same host, the engorged nymphs dropping off to moult on the ground and giving rise to adults which then seek a fresh host. *Rhipicephalus evertsi* behaves consistently as a two-host tick, but in other species, such as *Hyalomma dromadarii*, individuals from the same egg mass may feed as three-, two-, or even one-host ticks. Many hard ticks are three-host ticks, each stage normally engorging in a few days and dropping off to moult. Each stage must therefore be harvested and allowed to moult and harden before the next instar is ready to feed on a new host.

Collection and transport of strains

Ticks that have been maintained in laboratories can usually be obtained for starting new colonies and these have the advantage of known genetic history and behaviour. Strains can also be started from wild-caught ticks, usually engorged females or unfed adults. These can be transported in small tubes or polythene packets, provided that slight moisture is present. Care must be taken with exotic ticks, which may be infected with disease organisms or be resistant to insecticides, and clearance from the appropriate authorities must be obtained before importation. There may also be a danger to personnel in handling.

Ticks are found on both wild and domestic animals, and for many species this is the only available source. It is difficult to detach them without injuring the mouthparts or other tissues, but engorged stages can usually be wriggled free by gentle pulling and twisting. Dead small mammals or birds sometimes have ticks attached, and if the whole animal is stored in a cotton bag for several days the ticks will often release their hold. If only a few are present, the skin to which they are attached can be cut out and placed in a humid tube until they detach themselves. Ticks which have not completed feeding will usually re-attach to a new host. Live animals bearing ticks can be confined in small cages over water until the ticks complete feeding and drop down engorged. Free-living stages of ticks can be obtained by dragging a white woollen blanket or cotton towel over the vegetation, or by flagging a smaller square of the same material over grass and shrubs. Before attempting to breed a particular species its taxonomy should be carefully checked by a competent systematist. If experiments are to be made on the biology or physiology, the exact source of the strain should be recorded, as significant biological differences can occur within a species. Detailed records of the laboratory-established strains are essential.

Equipment and handling

A room which is used specifically for tick work, and which cannot become contaminated with insecticide is essential. It should have a sink with hot and cold water and the tables and benches must be easy to clean; a white surface helps the detection of escaped ticks. Incubators which can be set at various temperatures and a hot-air sterilizing cabinet are essential.

Adult ticks are best picked up with strip steel forceps which exert little pressure. They can be kept individually or in small groups in wool- or gauze-stoppered vials. These are emptied on to the host, or into the confining chamber fixed to it, by flicking the base of the inverted tube. Small numbers of engorged females can be handled individually, but for mass-rearing of species such as *Boophilus* the engorged females are allowed to drop through the slatted floor of the animal box and either collected in a sieve (Harrison & Mundy, 1968) or swept up with a soft brush. Nymphs and larvae are too small and active to be handled in any numbers by forceps, but can be sucked into small tubes or pooters which have a gauze screen at the suction end. The other end is then sealed with wax. Adult ticks can be counted by picking them up in this way from a surface surrounded by water, but larvae are elusive and if large numbers are required it is more satisfactory to count or weigh out the egg mass into tubes and hatch out the required number separately. Handling of eggs and larvae causes mortality, and should be done as gently and as seldom as possible. When only approximate numbers of larvae are needed they can be picked up with a camel hair brush. Movement can be limited by floating the larvae and wiping them off from the brush on to filter-paper strips in a container, or directly on to the host for feeding.

Active ticks are immobilized by cold in a household refrigerator at 0 to 4°C, but they quickly recover on exposure to air and must be maintained on an ice-cold surface for counting and collection. Anaesthetics can be used, but will result in some mortality. Air bubbled through ether soon anaesthetises the ticks but a safer, though slower, method is the use of carbon dioxide gas. In either case they should be removed as soon as they are immobile.

Transfer of immature ticks from tube to host is more difficult than with adults since it is not easy to flick them out of the tube. When putting them into an enclosed space, such as an ear bag, the stoppered tube should be inserted and the bag closed; the stopper can then be wriggled loose through the bag wall. If a bag is not used, immature ticks can be wiped on to the surface of the animal with a camel hair brush, or the tube can be taped on to the hair and the ticks allowed to move out.

Feeding

Living hosts

It is convenient to use a readily available host species which can be easily maintained and handled, and the rabbit and guinea-pig are probably the most useful general hosts. Rodents and chickens are acceptable hosts for many larvae.

It is often possible to rear a small number of ticks on an accessible but not particularly suitable host, but if large numbers are required a specific host is essential. For instance, *Boophilus* species will feed on rats and dogs but very few reach the engorged adult state on these animals and it is necessary to use cattle for rearing in quantity. When the wrong host is used ticks may feed but fail to moult or lay eggs, or the whole colony may die off after a few generations. Periodic feeding on a different host may be required to maintain a colony's viability (Nutting, 1967). Periodic introduction of wild-caught strains of ticks is sometimes required to re-vitalise or replace old colonies.

Argasid nymphs and adults, which feed rapidly, can be fed by manually holding or lightly affixing the container to an immobilized host. Larval and some nymphal argasids and all stages of ixodid ticks take several days to engorge and some kind of restraining device is necessary to prevent the host from removing them. Ixodids are usually fed on the ears and in large animals the scrotum can also be used; funnel shaped tubes of cloth being used to confine the ticks. The narrow end fits round the base of the ear, and it can be glued on with an adhesive, or retained by binding adhesive plaster round it; the open end of the funnel is closed with an elastic band. Bags with zip fasteners down the side can also be used. In guinea-pigs, the sides and belly are used, a panel of cloth being stuck with adhesive over a shaved area and a zip or sewn slit provided in the cloth for inspection and removal of ticks. The ear bags must be protected against scratching and other

sites against licking. Cattle, sheep and goats are usually head-haltered to prevent them licking, while rodents, etc. can be fitted with neck and loin collars made from polyvinyl floor covering. If the claws of rodents are covered with adhesive tape, less damage occurs. It is difficult to confine ticks on birds, and only those feeding on the head escape preening. Wire floors and dim red light must be used, especially when the ticks are dropping. For short-feeding argasid ticks the whole bird can be enveloped in a cloth bag tied around the neck. Kaiser (1966) describes a method of fitting capsules to pigeons; these have a flange at the base for sticking on to the host and a screw cap for access to the ticks and can be used on a wide variety of hosts for feeding small numbers of ticks. Capsules can be made from the tops of metal or polythene containers and stuck on with adhesive, or fitted with a soft pad at the base and held down with adhesive plaster passed around the body. The types used have been described by Gregson (1966).

When ticks are not confined by a bag or capsule, or if disease-carrying ticks are being used, the animals must be carefully confined. With large animals this is done by having a moat filled with oil or water around the stall. Smaller animals are kept in cages with mesh bottoms standing over trays of water from which the ticks can be removed daily. The edges of the trays are smeared with grease. Sometimes the cage is enclosed in a cloth bag. In all cases excreta and floor sweepings must be burnt.

Engorged argasid ticks which bury themselves in litter or hide in crevices provided in the cage can be left to breed and moult in the feeding chamber, a suitable host being put in from time to time to allow them to feed. (Avivi, 1967; Micks, 1951). This cannot be done with ixodid ticks which have more specialised environmental needs and which also crawl out of the container.

Host-reaction or immunity is a problem frequently encountered in hosts which have previously been exposed to ticks. The yield of *Boophilus* ticks from calves decreases considerably after the first infestation (Bennett & Wharton, 1968), but continued infestation with large numbers of larvae still yields sufficient engorged adults for about three months. After this the animals should be discarded. To ensure the successful feeding of a tick which is scarce or is being used for disease transmission studies it is advisable to use a host which has never had any previous infestation.

Artificial feeding

Argasidae can be induced to feed through membranes on suitable host blood maintained at host body temperature (Tawfik & Guirgis, 1969; Tarshis, 1958). Partial feeding of Ixodidae has also been achieved in this way (Chabaud, 1950; Burgdorfer, 1957). Both hard and soft ticks have been induced to feed through the air sac membrane of embryonated chicken eggs (Pierce & Pierce, 1956; Burgdorfer & Pickens, 1964).

These methods are not suitable for the mass-rearing of ticks, although Avivi (1967) proposes to do so with *O. tholozani*, but the method has great potential for the study of the transmissibility of disease agents and their development within the tick (Joyner & Purnell, 1968). It is also valuable for the study of physiological processes in the tick (Gregson, 1957; Galun & Kindler, 1965).

Post-feeding care

Engorged ticks are usually confined in small glass or polythene tubes with perforated lids to allow air and moisture to permeate. In tubes containing eggs the perforation must be small enough to retain the larvae when they hatch, and bolting silk is usually used. Filter-paper strips are inserted into the tubes to absorb excretions and secretions. Ticks which have fed in the confinement of bags often have blood or faecal granules adhering to them, which gives rise to mould and bacterial growth: such ticks should be washed in water and drained on filter-paper before being put into tubes. For the same reason the tubes should be changed after moulting, and the female ticks removed as soon as egg-laying is finished. Moulds grow readily in containers kept at high humidity and on dead ticks. It is doubtful if moulds actually kill ticks which require high humidity, but water of condensation can do so. Both moulds and water can be controlled by fairly frequent changing of the filter-paper strips, and moulds by washing the tubes and ticks with 1:10,000 merthiolate.

Humidity can be imprecisely but adequately maintained in three ways:

1. The tubes are stood in a deep container with two inches of wet sand at the bottom. To retain moisture and prevent mould growth, salt is added to the sand which is wetted every week. The method is adequate for ticks with relatively low humidity requirements of 60 to 70 per cent R.H.

2. A porous substance, usually a mixture of plaster of paris and charcoal, is poured into the base of the tick tube, and when set it is thoroughly moistened. It is difficult to add water to this base when ticks are in the tube, and it is better if bottomless cylinders are used. The tubes can then be stood in moist sand or vermiculite to maintain the moisture in the plaster at the base of the tube.

3. The tubes can be closed with cotton wool plugs which are moistened every few days with distilled water.

Precise humidity is maintained by placing the tick tubes over saturated solutions of various salts in closed containers (Winston & Bates, 1960). The salt solutions do not gain or lose water as do solutions of potassium hydroxide and sulphuric acid (Solomon, 1951) and hence do not change the vapour pressure. For humidities above 85 per cent glycerol solutions can be employed (Johnson, 1940).

Temperature conditions for moulting and egg-laying vary with the species. Most ixodid ticks are held between $22°$ and $25°C$ and argasid ticks between $25°$ and $30°C$. Temperatures of $25°C$ may overcome diapause in some of the temperate climate ticks, but it is better to subject ticks which are likely to diapause to low temperatures of 5 to $10°C$ for two months before incubating them at $22°C$. Manipulation of photoperiod is difficult, and it is best to maintain a standard 18-hour-light/6-hour-dark regime. In non-diapausing species it is usual to hold post-fed ticks in subdued light or total darkness. This is particularly necessary for argasids, which should also be provided with cavities or folds of paper in which to shelter. They can be left undisturbed since the low humidities of 50 to 75 per cent at which they are held is not conducive to moisture or mould troubles.

REFERENCES

Avivi, A. (1967). The maintenance of colonies of argasid ticks. In *The Ecology, Biology and Control of Ticks and Mites of Public Health Importance*. Geneva: W.H.O.

Bennett, G. F. & Wharton, R. A. (1968). Variability of host resistance to cattle tick. *Proc. ecol. Soc. Aust.*, **3**, 150–4.

Burgdorfer, W. (1957). Artificial feeding of ixodid ticks for studies on the transmission of disease agents. *J. infect. Dis.*, **100**, 212–4.

Burgdorfer, W. & Pickens, E. G. (1954). A technique employing embryonated eggs for the infection of argasid ticks etc. *J. infect. Dis.*, **94**, 84–9.

Chabaud, A. G. (1950). Sur la nutrition artificielle des tiques. *Annls Parasit. hum. comp.*, **25**, 42–7.

Feldman-Muhsam, B. (1967). Maintenance of colonies of ixodid ticks. In *The Ecology, Biology and Control of Ticks and Mites of Public Health Importance*. Geneva: W.H.O.

Galun, R. & Kindler, S. H. (1965). Glutathione as an inducer of feeding in ticks. *Science, N.Y.*, **147**, 166–7.

Gregson, J. D. (1957). Experiments on the oral secretion of the Rocky Mountain wood tick, *D. andersoni. Can. Ent.*, **89**, 1–5.

Gregson, J. D. (1966). Ticks. In *Insect Colonisation and Mass Production*, ed. Smith, C. N. London & New York: Academic Press.

Hadani, A., Cwilich, R., Rechav, Y. & Dinur, Y. (1969). Some methods for the breeding of ticks in the laboratory. *Refuah Vet.*, **26**, 87–100.

Harrison, I. R. & Mundy, A. E. (1968). The housing of calves for experiments with the cattle tick *Boophilus microplus. J. Inst. Anim. Techns*, **19**, 132–5.

Johnson, C. G. (1940). The maintenance of high atmospheric humidities for entomological work with glycerol-water mixtures. *Ann. appl. Biol.*, **27**, 295–9.

Joyner, L. P. & Purnell, R. E. (1968). The feeding behaviour on rabbits and in-vitro of the ixodid tick *R. appendiculatus. Parasitology*, **58**, 715–23.

Kaiser, M. N. (1966). The life cycle of *A. arboreus* and a standardized rearing method for argasid ticks. *Ann. ent. Soc. Am.*, **59**, 496–502.

Kohls, G. M. (1959). Tick rearing methods with special reference to *D. andersoni*. In *Culture Methods for Invertebrate Animals*. Ithaca, N.Y.: Comstock.

Micks, D. W. (1951). The laboratory rearing of the common fowl tick, *Argas persicus. J. Parasit.*, **37**, 102–5.

Nutting, W. B. (1967). Host specificity in parasitic acarines. *Acarologia*, **10**, 165–80.

Pierce, A. E. & Pierce, M. H. (1956). A note on the cultivation of *B. microplus* on the embryonated hen egg. *Aust. vet. J.*, **32**, 144–6.

Solomon, M. E. (1951). Control of humidity with potassium hydroxide, sulphuric acid and other solutions. *Bull. ent. Res.*, **42**, 543–54.

Tarshis, I. B. (1958). A preliminary study on feeding *O. savigny* on human blood through animal derived membranes. *Ann. ent. Soc.*, **51**, 294–9.

Tawfik, M. S. & Guirgis, S. S. (1969). Experimental feeding of *Argas arboreus* through membranes. *J. med. Ent.*, **6**, 191–5.

Winston, P. W. & Bates, D. H. (1960). Saturated solutions for the control of humidity in biological research. *Ecology*, **41**, 232–7.

52 Land and Freshwater Molluscs

C. A. Wright

Estimates of the number of molluscan species vary between 80,000 and 100,000. The phylum is divided into six classes but the members of only one, the Gastropoda (snails and slugs), are of much interest as laboratory animals. Some terrestrial snails are useful subjects for genetic and physiological studies and some slugs are of interest because of their economic importance as crop-pests, but the majority of species commonly maintained in laboratories are the freshwater intermediate hosts of trematodes of medical and veterinary importance, such as *Schistosoma* spp. and *Fasciola* spp. Even within one genus of snails the various species have different requirements in laboratory culture and it is not therefore possible to give a single comprehensive method of maintenance. This brief account mentions a number of general considerations, based largely on experience in the British Museum (Natural History). Details for particular groups are given later, together with appropriate references.

Aquatic and amphibious snails

Accommodation

The most generally used aquaria are moulded all-glass tanks the horizontal dimensions of which should be greater than the depth in order to provide a high surface-area/depth ratio. These tanks are easy to keep clean but they are fragile and expensive. Angle-iron and glass aquaria are cheaper but they are not so easily cleaned and there is a risk of contamination by metal, paint or sealing compounds. Moulded polythene tanks are useful but it is essential that they should be thoroughly weathered before being brought into use. For intensive rearing of many species shallow polypropylene instrument trays ($5 \times 24 \times 33$ cm) are excellent, as are glass crystalizing dishes and polystyrene refrigerator boxes. For some of the smaller species square polystyrene petri-dishes ($10 \times 10 \times 2$ cm) are extremely useful for mass rearing of young.

Polypropylene and polystyrene containers do not require prior weathering. All aquaria should be covered to prevent contamination by dust and water-loss by evaporation.

Water

Where supplies of natural, unpolluted water are available from wells, springs or streams these should be used provided that they do not contain any exceptional concentrations of unusual salts or gases. Rainwater can be used in rural areas where atmospheric pollution is low, but its lack of dissolved solids may be a disadvantage. Where the only available source of water is from a normal urban domestic supply it is necessary to take certain precautions. Although some snails can tolerate water that has merely been allowed to stand for a few days to permit the escape of chlorine this is not generally adequate. It is preferable to maintain a series of reservoir tanks (about 100 litre capacity) stocked with fish such as guppies (*Lesbistes reticulatus*). Tap-water is settled in these tanks for about three weeks after which it is suitable for use with most aquatic snails. Water is drawn off into aspirators through sintered glass discs to reduce the chances of pests being carried over into the culture tanks and, provided the level in the reservoirs is not reduced below half their full volume, they can be refilled directly from the mains supply and the water is fit to use after a further three days. The mains supply should be tested for traces of copper and other heavy metal ions, which are toxic to snails, and taps should always be run for some time before filling reservoir tanks in order to avoid contamination acquired by water lying overnight in copper or lead piping.

Aeration

Most aquatic snails depend partly on dissolved and partly on atmospheric oxygen. There is no need to provide aeration in shallow aquaria (5 cm or less) but if it is required in larger tanks either porous bubblers supplied from an air compressor or aquatic plants may be used. Bubblers provide gentle circulation of the water, which is useful, but excessive aeration may cause too much agitation for those species which prefer still conditions. The use of aquatic plants may introduce other animals which are detrimental to the snail cultures. Standen (1949) recommends the use of balanced aquaria with plants and a food chain including annelids (*Tubifex*) and crustaceans (*Daphnia*) to keep the water clean. The most commonly used plants are *Vallisneria spiralis*, *Ludwigia palustris*, *Myriophyllum* spp. and *Elodea canadensis*. The last species does not need to be rooted and although the others will survive without a rooting substratum they thrive better if provided with a layer of fine sand.

Light

Light is essential if plants are used for aeration of aquaria, but most snails do not require direct light and many species can live and breed in total darkness if provided with adequate food and oxygen. Daylight is beneficial but usually inadequate, and either tungsten or fluorescent lighting may be used to supplement it. Tungsten lighting has the advantage that there is an appreciable output of heat from the lamps and this can help to introduce a useful diurnal temperature fluctuation if needed; but where steady temperatures are required fluorescent lighting is to be preferred. The warmer types of tube are the best, because the bluish daylight type tends to encourage the growth of brown slimes which are detrimental to most snails.

Temperature

Maintenance temperatures of aquaria must depend entirely on the snails to be cultured. For most tropical freshwater species temperatures between 22° and 28°C are near to the optimum but tolerance ranges are much wider. Diurnal rhythms of water temperatures are always of a lesser amplitude than air temperatures, and although in natural conditions snails definitely move in response to these daily changes it is the larger seasonal rhythms that may be important. Some snails do not breed well if maintained at a uniform temperature throughout the year, and tropical species often benefit by brief periods out

of the constant-temperature room while temperate species such as *Lymnaea peregra* can be stimulated to lay eggs by exposing them to near-freezing temperatures in a refrigerator for a few days and then returning them to their normal environment. Optimum temperatures for the development of trematode parasites within snails are usually higher than the temperatures required for breeding and in order to incubate such infections it is useful to have a separate room or insulated cupboard that can be kept a few degrees warmer than the breeding room.

Crowding

Overcrowding of laboratory snail colonies is one of the commonest causes of failure. The optimum population density must be determined for each species, and it varies with the size of the individuals. Adult *Lymnaea stagnalis* require a minimum water volume of 2 litres per snail, but young of the same species can be kept at higher densities. The effects of overcrowding include stunted growth, reduced fecundity and increased resistance to trematode infection. The effects are attributed to excessive accumulation of hormone-like excretory products which are probably important in low concentrations for the growth of young snails, for there are minimum densities of juveniles below which their growth is retarded (Wright, 1960). The effects can be reduced by frequent, regular changes of water and adequate oxygenation or by use of a circulating water system such as that described by van der Steen, van den Hoven & Jager (1969). Care must be taken in the choice of activated charcoal for filters in circulating systems because the effects of some forms appear to be deleterious. The best method of overcoming crowding problems is to use very shallow water which can be changed often without straining possibly limited resources and which, because of the large surface area/depth ratio, is well oxygenated.

Nutrition

The natural food of most aquatic snails consists of algae, mud and decaying vegetable or animal matter. A certain amount of algal growth usually forms in aquaria but the quantity is rarely enough to support a healthy colony. The most usual dietary supplement is lettuce, either raw, cooked or dried, but watercress, dried maple leaves and pieces of carrot have been recommended. Standen (1951) described a food consisting of dried powdered lettuce, dried milk and Bemax (wheat germ) suspended in a gel of calcium alginate; this food is excellent for *Biomphalaria glabrata* but is less suitable for other species. For bottom-feeding prosobranchs (*Bithynia*, *Melanoides* etc.) small quantities of powdered rat-cake, dried fish-food and wheat-germ are excellent but they should not be provided in excess because of the ease with which they decay and foul the water. *Oncomelania* spp. can survive for long periods on a diet of filter-paper but this is inadequate to support egg-production (Wagner & Wong Chi, 1967). Taylor & Mozeley (1948) supplemented the algal food of the amphibious *Lymnaea truncatula* with a mixture of finely powdered chalk and oatmeal. Claugher (1960) mentions that an important article of diet for newly-hatched *Bulinus* is the faecal material of the parents. Recent experience in this laboratory has shown that improved reproductive performance and better survival of infected snails are obtained on a diet of autoclaved organically-rich mud on which a growth of *Oscillatoria* alga has formed; this is dried and fed in small pieces, and it appears to be acceptable to nearly all snails.

Pests

A variety of pests occur in laboratory aquaria and although some snails may be more affected than others nearly all are at least a nuisance. Most leeches will attack snails of any size and turbellarians (particularly *Macrostomum* spp.) are troublesome to young *Bulinus*. Ostracods (*Cypridopsis*), although basically detritus feeders, may cluster on the anal lobe of snails and irritate them so that they cease to feed. The commensal annelid *Chaetogaster* helps to keep aquaria clean, but the worms which live under the shell of the snails have a predilection for trematode miracidia and it is almost impossible to establish an infection in a grossly infested colony. Turbellaria can be removed by narcotizing the snails with a 2 per cent urethane solution, cleaning the shells with a fine brush and transferring the snails through several changes of clean water. Dr. Mandahl-Barth (private communication) uses

gammexane (gamma-isomeric hexachlorcyclo-hexan) to kill ostracods. At a concentration of 0·1 ppm the ostracods are killed but neither *Daphnia* nor snails are affected. Gammexane is made up in 96 per cent alcohol at a strength of 1 mg/1 ml. *Chaetogaster* are difficult to eliminate and the only method that is of much use is to transfer egg-masses, after thorough rinsing in distilled water, to fresh aquaria. The only satisfactory method of pest-control is rigorous aquarium-room hygiene with isolation of infested colonies, avoidance of cross-contamination with instruments, and the use of filtered water supplies.

Transport

Recently-collected snails must be kept cool and the most satisfactory method is to place each sample in a polythene bag, without water, in a pre-cooled wide-mouthed vacuum jar. Labels should be attached to the outside of the bag because most snails eat paper. For postal transport freshwater snails survive best when packed in a container between layers of damp (*not* wet) cotton wool, filter-paper or moss. The individuals should not be in contact with one another and the packing should be just firm enough to prevent movement without crushing the shells. The lid of the container should be sealed with adhesive tape; no air-holes are necessary. Snails will survive for at least several days under these conditions and material has been received in London from all over the world after up to a week in transit. Specimens with mature trematode infections do not survive well, and if infected snails are to be shipped they should be sent soon after exposure to the parasites.

Special methods

Biomphalaria glabrata (= *Australorbis glabratus*). This is the most easily maintained of all tropical freshwater snails. Standen (1951) uses ecologically-balanced aquaria and the calcium alginate gel diet mentioned above with sprigs of *Ludwigia palustris* as egg-traps. Olivier & Haskins (1960) used pieces of polythene sheeting for surface egg-traps. Rowan (1958) described a mass-culture technique in which the snails are kept in nylon-mesh hammocks suspended in 45-litre tanks; 500–800 snails per tank can be brought to maturity in 5 weeks on a diet of lettuce supplemented by equal parts of finely-ground dry dog chow and fine dry silt with 1 per cent dry calcium carbonate. For rapid rearing of young snails for molluscicide screening Hopf & Muller (1962) use shallow dishes in which autoclaved river mud is inoculated with *Oscillatoria*, covered by about 5 cm of hard water and left for a week before the introduction of snail egg-masses. Chernin (1957) and Chernin & Schork (1959) described methods for obtaining bacteriologically sterile cultures.

Bulinus and Biomphalaria spp.   Claugher (1960) described the basic methods for maintenance of *Bulinus* in pure culture. The methods in general are appropriate also for most species of *Biomphalaria* and the tropical *Lymnaea* spp. Current modifications of technique involve the use of shallow water, dried mud with algae as food for intensive work, and scalded, dried lettuce supplemented weekly with a little wheat-germ for maintenance colonies. The lettuce is prepared by brief immersion in boiling water until it is soft and it is then spread out on glass or perspex sheets in the constant-temperature (22°C) room to dry; in this form it is acceptable to most species, it does not decay rapidly in the water and it can be stored indefinitely after preparation. A general maintenance temperature of about 22°C is suitable for most species but some members of the *Bulinus truncatus* complex and the polyploid species of *Bulinus* breed better if they are kept for several weeks at lower temperatures (about 18°C) and then returned to the warm room. Steady incubation temperatures of about 30°C give the best results with snails infected with *Schistosoma haematobium*. Frank (1962) uses concrete aquaria 30 cm deep and 90 cm diameter for outdoor mass culture of *Bulinus* and *Biomphalaria* in South Africa. A substratum of organically rich mud is provided and lettuce and lucerne are used as food; lucerne alone or in conjunction with alginate-based food causes high mortality.

Lymnaea truncatula. Taylor & Mozeley (1948) bred this amphibious species on heat-sterilized mud built up into slopes in earthenware pans 30–45 cm in diameter. A pool of water is provided at the foot of the slope and the pans are covered with glass sheets and left in a greenhouse for 2–3 weeks to acquire an abundant growth of algae. This process can be speeded by the use of ultra-violet light. When snails are introduced a mixture of equal parts of fine oatmeal and powdered chalk is sprinkled on the mud each week, to supplement the algal food.

Oncomelania spp. Vogel (1948) used angle-iron aquaria with a turf slope supported on bricks, sand and loam with a pool of water at the bottom and moist loam in the corners to provide egg-laying sites. The snails were fed on lettuce and dried tropical-fish food. Wagner & Wong (1956) used unglazed earthenware dishes (12–15 cm diameter) with slopes of soil, sand and gravel in the ratio 2:1:1. An appropriate water level can be maintained by standing the dishes in shallow containers filled with water. Van der Schalie & Davis (1968) have carried out a comparative study of methods for culturing *Oncomelania* spp. and they conclude that the best results are obtained with five pairs of snails in medium-sized clay pots, with a fine-textured soil of high calcium content supporting a dense flora of diatoms. The soil serves as food, and this is supplemented with filter paper.

Terrestrial snails and slugs

General principles for setting up cultures of terrestrial snails are given by Krull & Archer in Galtsoff *et al.* (1937). Temperature and humidity are the two most critical factors and the substratum is important for egg-laying and as a medium to help to control humidity. In most cases a base of coarse gravel covered with soil and leaf-litter is suitable, with sufficient water in the gravel to keep the soil and lower layer of leaves moist. For transport of live land snails J. F. Peake, of the British Museum (Natural History), uses linen bags which can be closed with draw-strings and which have a label for collection data sewn on the outside. Provided that the bags are kept dry the contained snails will aestivate and survive for several weeks.

Special methods

Helix aspersa. Stelfox (1915) bred this species through several generations in boxes with a sand substratum and a diet of lettuce leaves and slices of carrot or turnip.

Cepaea nemoralis. Cain & Sheppard (1957) bred *C. nemoralis* in 20 cm flowerpots two-thirds filled with light, crumbly soil, damped with rainwater and covered with glass. Lettuce, carrot, cabbage and natural chalk were provided for food. Young snails were removed to plastic boxes filled with damp earth and fed on lettuce, bran and powdered chalk.

Testacella spp. The carnivorous slugs *T. scutulum* and *T. haliotidea* have been successfully bred in the laboratory in corked glass tubes (10 × 3·7 cm) two-thirds filled with soil and kept in a cool, dark place (Barnes & Stokes, 1951; Stokes, 1958). Earthworms were provided for food.

Milax spp. Stephenson (1962) reared slugs in the laboratory on a peat-and-soil substrate and a diet of carrot and raw potato. *M. budapestensis* requires an open-textured soil medium but can burrow through heavy soil by ingesting it. Reproduction is inhibited by a constant temperature of 20°C and stimulated by fluctuating temperatures (Stephenson, 1966).

Narcotization and anaesthesia

The use of pre-fixation narcotics and anaesthetics for operative procedures has been reviewed by Runham *et al.* (1965) who emphasize the variation in susceptibility to these substances shown by different molluscan species. Many of the best techniques involve the use of materials not always easily available such as pentobarbitone sodium (Nembutal); amylocaine hydrochloride (Stovaine); MS 222 (Sandoz: meta-amino-benzoic acid ethyl ester methansulphonate); or urethane, which has undesirable handling properties. Most land and freshwater species can be narcotized in cooled, boiled water from which air is excluded, but the most useful method for freshwater snails is to sprinkle menthol crystals on the water surface. Marine molluscs can usually be narcotized by a 7 per cent solution of magnesium chloride in seawater, and another (though less reliable) method is to increase the carbon-dioxide content by adding soda-water. One per cent propylene phenoxytol can be used as a narcotic for most molluscs but is particularly useful for marine bivalves.

REFERENCES

Barnes, H. F. & Stokes, B. M. (1951). Marking and breeding *Testacella* slugs. *Ann. appl. Biol.*, **38**, 540–5.

Cain, A. J. & Sheppard, P. M. (1957). Some breeding experiments with *Cepaea nemoralis*. *J. Genet.*, **55,**195–9.

Chernin, E. (1957). A method of securing bacteriologically sterile snails (*Australorbis glabratus*). *Proc. Soc. exp. Biol. Med.*, **96**, 204–10.

Chernin, E. & Schork, A. R. (1959). Growth in axenic culture of the snail *Australorbis glabratus*. *Am. J. Hyg.*, **69**, 146–60.

Claugher, D. (1960). The transport and laboratory culture of the snail intermediate hosts of *Schistosoma haematobium*. *Ann. trop. Med. Parasit.*, **54**, 333–7.

Frank, G. H. (1962). Maintenance of *Biomphalaria* sp. and *Bulinus* (*Physopsis*) sp. in outdoor aquaria in the Eastern Transvaal Lowveldt. *Wld Hlth Org.*, Mol/Inf/**7**, 1–5.

Galtsoff, P. S., Lutz, F. E., Welch, P. S. & Needham, P. S. (1937). *Culture Methods for Invertebrate Animals.* Ithaca, N.Y.: Comstock.

Hopf, H. S. & Muller, R. L. (1962). Laboratory breeding and testing of *Australorbis glabratus* for molluscicidal screening. *Bull. Wld Hlth Org.*, **27**, 783–9.

Olivier, L. & Haskins, W. T. (1960). The effects of low concentrations of sodium pentachlorophenate on the fecundity and egg viability of *Australorbis glabratus*. *Am. J. trop. Med. Hyg.*, **9**, 199–205.

Rowan, W. B. (1958). Mass cultivation of *Australorbis glabratus*, intermediate host of *Schistosoma mansoni* in Puerto Rico. *J. Parasit.*, **44**, 247.

Runham, N. W., Isarankura, K. & Smith, B. J. (1965). Methods for narcotizing and anaesthetising gastropods. *Malacologia*, **2**, 231–8.

Standen, O. D. (1949). Experimental schistosomiasis. 1. The culture of the snail vectors *Planorbis boissyi* and *Bulinus truncatus*. *Ann. trop. Med. Parasit.*, **43**, 13–22.

Standen, O. D. (1951). Some observations upon maintenance of *Australorbis glabratus* in the laboratory. *Ann. trop. Med. Parasit.*, **45**, 80–3.

Stelfox, A. W. (1915). A cross between typical *Helix aspersa* and var. *exalbida*: its results and lessons. *J. Conch., Lond.*, **14**, 293–5.

Stephenson, J. W. (1962). A culture method for slugs. *Proc. malac. Soc. Lond.*, **35**, 43–5.

Stephenson, J. W. (1966). Notes on the rearing and behaviour in soil of *Milax budapestensis* (Hazay). *J. Conch., Lond.*, **26**, 141–5.

Stokes, B. M. (1958). The worm-eating slugs *Testacella scutulum* Sowerby and *T. haliotidea* Drap. in captivity. *Proc. malac. Soc. Lond.*, **33**, 11–20.

Taylor, E. L. & Mozeley, A. (1948). A culture method for *Lymnaea truncatula*. *Nature, Lond.*, **161**, 894.

Van der Schalie, H. & Davis, G. M. (1968). Culturing *Oncomelaria* snails (Prosobranchia: Hydrobiidae) for studies of oriental schistosomiasis. *Malacologia*, **6**, 321–67.

Van der Steen, W. J., Van den Hoven, N. P. & Jager, J. C. (1969). A method for breeding and studying freshwater snails under continuous water change, with some remarks on growth and reproduction in *Lymnaea stagnalis* (L). *Neth. J. Zool.*, **19**, 131–9.

Vogel, H. (1948). Uber eine Dauerzucht von *Oncomelania hupensis* und Infektionsversuche mit *Bilharzia japonica*. *Z. Parasit.*, **14**, 70–91.

Wagner, E. D. & Wong, L. W. (1956). Some factors influencing egg-laying in *Oncomelania nosophora* and *Oncomelania quadrasi*, intermediate hosts of *Schistosoma japonicum*. *Am. J. trop. Med. Hyg.*, **5**, 544–52.

Wagner, E. D. & Wong Chi, L. (1957). Egg-laying inhibition in *Oncomelania nosophora* maintained on filter-paper. *Am. J. trop. Med. Hyg.*, **6**, 946–8.

Wright, C. A. (1960). The crowding phenomenon in laboratory colonies of freshwater snails. *Ann. trop. Med. Parasit.*, **54**, 224–32.

53 Earthworms

G. J. ASHBY

Earthworms (family Lumbricidae) are used in educational and research establishments and are also bred on a commercial basis to provide farmers with stock for certain kinds of land improvements involving, for example, the production of humus. Some zoological societies have found the demand for worms so great that it has been necessary to breed them on a large scale to provide food for certain species of birds and mammals. They are also used for fish bait, fish food, and, when poisoned, as a bait to kill moles (Guild, 1957).

There are approximately 220 species of earthworms in the family *Lumbricidae*, of which about nineteen are common over much of Europe and have been spread by man to many parts of the world. These species have become the dominant members of the local earthworm fauna and have locally often replaced the indigenous species (Gerard, 1964). In Britain there are 25 species, but the most widely distributed belong to six genera: *Eiseniella* (4 species), *Allolobophora* (3 species), *Dendrobaena* (3 species), *Bimastus* (1 species), *Octolasium* (2 species), *Lumbricus* (4 species).

These species can all be found in a wide variety of soils and environmental conditions. They differ from each other in habits, size and colour, and although there are certain endemic species which require specialized conditions of temperature, humidity and even soil acidity in order to thrive, many of the species in the genera mentioned can nevertheless be bred and successfully maintained in the laboratory.

Accommodation

Containers

Several types of container have been used with success for the culturing of worms, from jam jars, specimen tubes, oil drums, and biscuit tins to wooden boxes; the latter are undoubtedly the most practical and efficient. When considering the type of container, it is important to bear in mind the necessity of providing sufficient aeration. Soggy and water-logged areas must not be allowed to develop in the medium. This means, in fact, that if glass jars or tubes without adequate drainage are used the medium must be carefully turned over at intervals.

The size of box used is not important, but for convenience in handling and maximum production a good practical area is 46×38 cm with a depth of 20 cm. The boxes should be made of wooden boards 1·5 cm thick, jointed and screwed together. Angle irons can be used to give additional support. The wood can be preserved with either Cuprinol or aluminium paint. Drainage holes 3 mm in diameter at the rate of one per square decimetre should be drilled in the base of each box.

For the efficient use of the available space the boxes may be stacked in tiers, from five to eight boxes high. The lowest box in each tier should be supported on a wooden base about 14 cm above the ground to allow correct drainage. The individual boxes in the tiers can be separated from each other by 5×5-cm blocks of wood running lengthwise. These blocks or separators make it easier to water the cultures. The boxes need not be covered provided they are not overcrowded but, to avoid every possibility of escape, wire-gauze or perforated-zinc tops can be used. The surface of the culture medium can be covered with damp squares of hessian, but opinion is divided as to the wisdom of this. Some workers maintain that this material helps to conserve moisture, keeping the surface of the compost dark and damp and encouraging maximum production of egg capsules, whereas others have found that besides twisting and wrinkling, the hessian tends to give an uneven distribution of water over the whole surface area.

Media for box production

The media should be kept at approximately pH 7, although slight acidity can be tolerated by some species.

Ordinary soil is the most easily obtained culture medium and it can be used alone if necessary. Sticky clays and heavy loams should, however, be avoided. The soil should preferably be light and sandy, and all lumps, stones and foreign bodies must be removed. It should be worked down finely by hand, spread out on shallow trays to a depth of about 2 cm, and allowed to dry in the air for some days before use (Guild, 1957). This helps to eliminate possible parasites and predators. It should then be remoistened until it contains about 30 per cent of water. The soil should never be cooked or baked.

Dung may be used alone or mixed with other substances. Any manure may be used, but horse or rabbit is recommended. This should be collected and dried out initially after which it is subsequently remoistened to give a water content of about 30 per cent.

Commercial peat moss is easily obtained, but it is best used only as an additive to the media mentioned above. It does, however, have a high acidity (about pH 3·5), and about 360 to 500 g of hydrated lime per 125 litres must be worked in in order to give a pH of approximately 7 or 7·5. Since it is dry on purchase it requires no preliminary drying out, for it contains no harmful organisms. The necessary moisture-content of approximately 30 per cent takes a little while to achieve, and the water should be added gradually in order to bring about an even distribution. It may be necessary to stand the wetted moss for some time in a closely packed heap.

Temperature
Normally no artificial heating is required except in cases where rapid breeding is desired. If necessary the cultures can be warmed by passing electrical heating wires through the bases of the boxes.

Culture

Breeding and rearing media
The recommended medium for breeding and rearing most species of earthworm is: 1 part of dung, 3 parts of soil, 5 parts of peat moss (treated with hydrated lime to give a pH of 7·0 or 7·5), and a small sprinkling of sharp sand. If dung is unobtainable, a very satisfactory food substitute can be made from a mixture of 25 parts (by volume) of biscuit meal, one part of Bemax, one part of potato powder and one part of powdered milk. The whole is mixed to a stiff paste with water and used as required.

Starting the cultures
When ready, the boxes are filled to approximately 6 cm from the top with a mixture of the soil and treated peat moss. The dung or other food is placed in a layer over this. The next step is to stock the boxes with mature worms. These may be purchased from a commercial worm farm or collected from the land surrounding the laboratory. If they are purchased, it should be possible to order preferred species but, of course, a random collection will most likely consist of a heterogeneous mixture of large and small species of varying ages. The length of mature earthworms varies from approximately 17 mm in the case of *Bimastus* and *Dendrobaena* to 300 mm for *Lumbricus terrestris*. A mature worm can be recognized by the presence of the clitellum, a glandular swelling on certain of the anterior segments of the body. The clitellum may be solely on the dorsal side, or it may completely surround the body. The position and number of segments contained by the clitellum is a character used in identifying the various species.

The selected worms should be placed on the surface of the compost, and any which have not burrowed down within fifteen minutes should be removed as unsuitable.

The size of box recommended (base 46×38 cm, height 20 cm) should easily take up to 200 large worms such as *Lumbricus terrestris*, or up to 400 small ones such as *L. rubellus* or *Allolobophora caliginosa*. As a rough guide, large species need approximately 145 cm^3 of compost per worm, small species about half this volume. Of course young specimens from a known stock can be reared in large numbers per unit; they must, however, be separated as their size increases with age.

Breeding
In nature, the periods of maximum activity and reproduction of worms are normally the spring and autumn. In the laboratory, on the other hand, the higher temperature, the maintenance of an optimum moisture content in the compost,

and the presence of abundant food supplies allow breeding to continue over a longer period, and it is also more rapid. Incubation of the eggs takes from 7 to 14 days in summer and up to a month in winter, whereas in nature it takes several weeks in summer and several months in winter. Smaller species, which take up to a year to mature in nature, will mature in 6 to 8 weeks in the laboratory, and the larger species will mature in the laboratory in 4 to 5 months. The brandling, *Eiseniella foetida*, which is normally a fast-growing worm, only takes from 4 to 6 weeks in the laboratory. *Allolobophora* spp. have periods of diapause, during which they stop feeding and breeding and curl up in holes in the soil; this condition may last for from several days to many weeks, but normal activity is later resumed. The diapause does not occur in *E. foetida*, nor in species of *Lumbricus* (Guild, 1957).

Earthworms, and the *Oligochaeta* in general, are hermaphrodite, and cocoon production may begin as soon as the clitellum is fully developed. Cross-fertilization is normal, and the eggs are fertilized in the cocoon, a ring-like structure secreted by the clitellar glands. The worm withdraws backwards through the ring and the eggs are released into the cocoon as it passes the opening of the oviducts. The spermatozoa obtained from the partner at mating, and subsequently stored, enter the cocoon as it passes the openings of the spermathecae. The cocoon closes as the worm withdraws its body. In most species only one or two worms will develop from the eggs in a single cocoon, although up to seven have been recorded for the brandling, *Eiseniella foetida*.

Maintenance of the cultures

Properly prepared cultures are not difficult to maintain, provided the moisture content is kept within the correct range. Laboratory cultures should be watered daily with the aid of a fire sprinkler, but waterlogging must be avoided. When watering it helps if the top 2 cm or so of compost is gently raked to allow easy penetration. Soil cultures tend to dry out more rapidly than those which are based on a high proportion of organic material.

It is also advisable to turn over at least the upper half of the culture periodically in order to facilitate a uniform distribution of moisture and an adequate exchange of gases.

Addition of fresh food

The dung or other food, such as corn meal or chicken mash, should be placed on the surface of the compost rather than mixed in with the culture. This will usually prevent the worms from working only at the bottom of the box, leaving the top layers comparatively untouched. It is always better to regulate the food-supply and to avoid adding large amounts of soggy food at any one time. For up to 300 worms in peat moss, no more than about 40 g of air-dried dung should be added at any one time. Overfeeding of wet food leads to fungal growths which spread out and spoil the cultures.

Examination of culture, and collecting cocoons

Cultures should be examined at least once a fortnight to see that they do not deteriorate, and also to gather cocoons. This gives an opportunity to check the water balance and to turn and aerate the culture. If this is not done, the compost will become foul owing to the packing down of the medium and the growth of undesirable fungi. The soil or other compost should be changed every six months.

For examination, the culture medium should be tipped out on to a tray to check the condition of the worms and to gather the cocoons for future breeding. The intervals between examinations should at no time exceed three weeks, as that would give some cocoons a chance to hatch in the culture, and the resultant young would not then be recovered easily. The cocoons are roughly spherical in shape, and they vary in size according to the species. In smaller species such as *Allolobophora caliginosa*, *A. chlorotica*, *Lumbricus rubellus*, and *Eiseniella foetida*, the cocoons vary in length from 2·5 mm to just over 3 mm, whereas those of *L. terrestris* are 6 to 7 mm long.

The cocoons vary in colour from brown to orange and yellow. Gathering the cocoons is rather a tedious process, but they can be easily seen and extracted quite cleanly. With cultures consisting mainly of soil, however, it may be necessary to wash the compost through a sieve (mesh 1·5 mm) with a jet of water, in order to get the cocoons clean. If the washing method is used it is advisable to remove all the worms first, and then soak the culture for 24 hours before using the jet and sieve. If washing is carried out, new compost will be needed after each examination (Evans & Guild, 1948).

After the cocoons have been removed from the compost and counted, they should be carefully placed on very damp filter paper or moist sand and enclosed in a glass or plastic container where their progress can be observed; or, if desired, they can be placed out immediately into fresh culture boxes. The cocoons are at first opaque but later become somewhat transparent, allowing the developing worms to be seen.

Killing and preservation

Earthworms can be killed by exposure to the fumes of chloroform, carbon tetrachloride, ethyl acetate or tetrachloroethylene. If dissection is not contemplated, then formalin is suitable as a preservative; Guild recommends a 4-per-cent solution initially, followed by transfer to a 5-per-cent solution after 2 to 3 days. For immediate dissection 60-per-cent ethyl alcohol as a preservative allows worms to remain pliable for several months. If they are wanted for dissection after a longer period of time they should be preserved in 70-per-cent alcohol.

REFERENCES

Barrett, T. J. (1948). *Earthworms. Their intensive propagation and use in soil building.* Roscoe, Calif.: Earthmaster.

Barrett, T. J. (1949). *Harnessing the Earthworm.* London: Faber.

Evans, A. C. & Guild, W. J. McL. (1947). Studies on the relationships between earthworms and soil fertility. No. 1. Biological studies in the field. *Ann. appl. Biol.,* **34,** 307–30.

Evans, A. C. & Guild, W. J. McL. (1948). Studies on the relationship between earthworms and soil fertility. No. 4. On the life cycles of some British Lumbricidae. *Ann. appl. Biol.,* **35,** 471–84.

Gerard, B. M. (1964). Lumbricidae (Annelida) with a key to common species. In *Synopsis of the British Fauna,* No. 6. The Linnean Society of London.

Guild, W. J. McL. (1957). Earthworms (Family Lumbricidae). In *UFAW Handbook on the Care and Management of Laboratory Animals,* 2nd ed. London: UFAW.

Addendum

New developments in euthanasia

Carbon dioxide

It has been found, in work on pigs (Mullenax & Dougherty, 1963) that animals given a mixture of carbon dioxide (68 per cent) and oxygen (32 per cent) tend to collapse with less convulsive activity, and, therefore, presumably, with less distress, than animals given the same proportions of carbon dioxide and air.

UFAW therefore decided to develop an apparatus in which cats and other small animals could be anaesthetised in a mixture of carbon dioxide 70 per cent and oxygen 30 per cent prior to killing in a carbon dioxide chamber. The apparatus, which consists of two separate chambers, is a modification of the cabinet for the destruction of chicks which is described on page 167. This method proved to be highly satisfactory and has been described in detail by McArthur (1976). The method and apparatus have been patented and will shortly be commercially available.

Captive-bolt pistol

A captive-bolt pistol has been developed which is suitable for the destruction of horses. Since it causes extensive brain damage, including destruction of the medulla, it kills rather than stuns, thus obviating the need for cutting the throat, except where exsanguination is necessary because the animal is to be used for food. The instrument has been described by Watts (1976).

Enquiries as to the commercial availability of both the carbon dioxide cabinet and the captive-bolt pistol should be made to: Accles & Shelvoke Ltd., Talford Street Works, Aston, Birmingham 6.

REFERENCES

MacArthur, Judy M. (1976). Carbon dioxide euthanasia of small animals. In *Humane Destruction of Unwanted Animals*. Proceedings of UFAW Symposium. Potters Bar: UFAW.

Mullenax, C. H. & Dougherty, R. W. (1963). Physiological responses of swine to high concentrations of inhaled carbon dioxide. *Am. J. vet. Res.*, **24,** 329–333.

Watts, R. Z. (1976). The development of a captive-bolt instrument for killing horses. In *Humane Destruction of Unwanted Animals*. Proceedings of UFAW Symposium. Potters Bar: UFAW.

Copies of the Symposium Proceedings *Humane Destruction of Unwanted Animals* can be obtained from UFAW, 8 Hamilton Close, Potters Bar, Herts., EN6 3QD, price 75p per copy, plus 11p postage and packing.

Index

Index